Lecture Notes in Computer Science 4993

Commenced Publication in 1973
Founding and Former Series Editors:
Gerhard Goos, Juris Hartmanis, and Jan van Leeuwen

Hang Li Ting Liu Wei-Ying Ma
Tetsuya Sakai Kam-Fai Wong
Guodong Zhou (Eds.)

Information Retrieval Technology

4th Asia Information Retrieval Symposium, AIRS 2008
Harbin, China, January 15-18, 2008
Revised Selected Papers

Volume Editors

Hang Li
Wei-Ying Ma
Microsoft Research Asia
Sigma Center 4F, Zhichun Road No. 49
Haidian District, Beijing, China 100080
E-mail:
{hangli, wyma}@microsoft.com

Kam-Fai Wong
The Chinese University of Hong Kong
Department of Systems Engineering
and Engineering Management
Shatin, N.T., Hong Kong
E-mail: kfwong@se.cuhk.edu.hk

Ting Liu
Harbin Institute of Technology
P. Box 321, 150001 Harbin P.R.China
E-mail: tliu@ir.hit.edu.cn

Guodong Zhou
Suzhou University, School of Computer
Science and Technology
1 ShiZi Street, 215006 Suzhou
E-mail: gdzhou@suda.edu.cn

Tetsuya Sakai
NewsWatch Inc.
Ginza Toshiba Building
5-2-1 Ginza, Chuo-ku Tokyo 104-0061,
Japan
E-mail: tetsuyasakai@acm.org

Library of Congress Control Number: 2008927190

CR Subject Classification (1998): H.3, H.4, F.2.2, E.1, E.2

LNCS Sublibrary: SL 3 – Information Systems and Application, incl. Internet/Web and HCI

ISSN 0302-9743
ISBN-10 3-540-68633-9 Springer Berlin Heidelberg New York
ISBN-13 978-3-540-68633-0 Springer Berlin Heidelberg New York

Springer is a part of Springer Science+Business Media

springer.com

Printed in Germany

Typesetting: Camera-ready by author, data conversion by Scientific Publishing Services, Chennai, India
Printed on acid-free paper SPIN: 12270675 06/3180 5 4 3 2 1 0

Preface

Asia Information Retrieval Symposium (AIRS) 2008 was the fourth AIRS conference in the series established in 2004. The first AIRS was held in Beijing, China, the second in Jeju, Korea, and the third in Singapore. The AIRS conferences trace their roots to the successful Information Retrieval with Asian Languages (IRAL) workshops, which started in 1996.

The AIRS series aims to bring together international researchers and developers to exchange new ideas and the latest results in information retrieval. The scope of the conference encompasses the theory and practice of all aspects of information retrieval in text, audio, image, video, and multimedia data.

We are pleased to report that AIRS 2006 received a large number of 144 submissions. Submissions came from all continents: Asia, Europe, North America, South America and Africa. We accepted 39 submissions as regular papers (27%) and 45 as short papers (31%). All submissions underwent double-blind reviewing. We are grateful to all the area Co-chairs who managed the review process of their respective area efficiently, as well as to all the Program Committee members and additional reviewers for their efforts to get reviews in on time despite the tight time schedule. We are pleased that the proceedings are published by Springer as part of their *Lecture Notes in Computer Science (LNCS)* series and that the papers are EI-indexed.

We thank Cheng Niu for managing the tutorials and Xueqi Cheng for coordinating demonstrations. The Publications Co-chairs, Guodong Zhou and Tetsuya Sakai, put in a lot of effort to compile the camera-ready papers and liaise with the Springer publisher. We thank Ting Liu and Youqi Cao, who chaired the local organization efforts. We also thank Muyun Yang for local management; Wanxiang Che for maintaining the AIRS 2008 website; Tetsuya Sakai and Dawei Song for publicity; and Le Sun and Munkew Leong for overseeing the financial aspects of the conference. We are particularly thankful to Jun Xu for managing the submission website and provide the necessary technical support.

We are grateful to Harbin Institute of Technology for hosting the conference and to Sheng Li for accepting to be the honorary Conference Chair. We thank Microsoft Research Asia for sponsoring the conference and Springer for publishing the conference proceedings.

Finally, we acknowledge, and are inspired by, the many authors who submitted papers and who continue to contribute in this Asia community of IR research and development.

Wei-Ying Ma
Kam-Fai Wong
Hang Li
Jian-Yun Nie

Organization

Steering Committee

Jun Adachi	National Institute of Informatics, Japan
Hsin-Hsi Chen	National Taiwan University, Taiwan
Lee-Feng Chien	Academia Sinica, Taiwan
Gary Geunbae Lee	POSTECH, Korea
Mun-Kew Leong	Institute for Infocomm Research, Singapore
Helen Meng	The Chinese University of Hong Kong
Sung Hyon Myaeng	Information and Communication University, Korea
Hwee Tou Ng	National University of Singapore, Singapore
Tetsuya Sakai	NewsWatch, Inc., Japan
Kam-Fai Wong	The Chinese University of Hong Kong, China
Ming Zhou	Microsoft Research Asia, China

Honorary Conference Chair

Sheng Li	Harbin Institute of Technology, China

General Co-chair

Wei-Ying Ma	Microsoft Research Asia
Kam-Fai Wong	The Chinese University of Hong Kong, China

Program Committee Co-chairs

Hang Li	Microsoft Research Asia
Jian-Yun Nie	University of Montreal, Canada

Organization Committee

Youqi Cao, Chinese Information Processing Society of China, China [Co-Chair]
Ting Liu, Harbin Institute of Technology, China [Co-Chair]

Tetsuya Sakai, NewsWatch, Inc., Japan [Publicity co-chair and ECIR liasion]
Dawei Song, Open Univesrity, UK [Publicity co-chair]

Le Sun, Institute of Software, Chinese Academy of Sciences, China [Finance Co-Chair]
Munkew Leong, Institute for Infocomm Research, Singapore [Finance Co-chair]

Muyun Yang, Harbin Institute of Technology, China [Local Organisation Chair]
Wanxiang Che, Harbin Institute of Technology, China [Webmaster]

Guodong Zhou, Suzhou University, China [Publication co-chair]
Tetsuya Sakai, NewsWatch, Inc., Japan [Publication co-chair]

Area Chairs

Peter Bruza	Queensland University of Technology, Australia
Hsin-His Chen	National Taiwan University, Taiwan
Jianfeng Gao	Microsoft Research, USA
Jimmy Huang	York University, Canada
Noriko Kando	National Institute of Information, Japan
Mounia Lalmas	Queen Mary, University of London, UK
Wai Lam	Chinese University of Hong Kong, China
Gary Lee	Pohang University of Science and Technology (POSTECH), Korea
Wenyin Liu	City University of Hong Kong, China
Shaoping Ma	Tsinghua University, China
Luo Si	Purdue University, USA
Christopher Yang	Chinese University of Hong Kong, China
Tie-Jun Zhao	Harbin Institute of Technology, China
Justin Zobel	NICTA, Australia

Tutorial Chair

Cheng Niu	Microsoft Research Asia, China

Demo Chair

Xueqi Cheng	Institute of Computing Technologies, China

Hosted by

Harbin Institute of Technology (HIT)
Chinese Information Processing Society of China (CIPSC)

Program Committee

Gianni Amati, Fondazione Ugo Bordini, Italy
Bill Andreopoulos, Biotechnology Centre (BIOTEC), Germany
Javed Aslam, Northeastern University, USA

Leif Azzopardi, University of Glasgow, UK
Jing Bai, University of Montreal, Canada
Holger Bast, Max-Planck-Institute for Informatics, Germany
Richard Cai, Micorsoft Research Asia, China
Yunbo Cao, Microsoft Research Asia, China
Ki Chan, Chinese University of Hong Kong, China
Kuiyu Chang, Nanyang Technological University, Singapore
Pu-Jen Cheng, National Taiwan University, Taiwan
Xueqi Cheng, Institute of Computing Technology, Chinese Academy of Sciences, China
Lee-Feng Chien, Academia Sinica, Taiwan
Charles Clarke, University of Waterloo, Canada
Arjen de Vries, CWI, Amsterdam, Netherlands
Pavel Dmitriev, Cornell University, USA
David Kirk Evans, National Institute of Informatics, Japan
Hui Fang, Ohio State University, USA
Cathal Gurrin, Dublin City University, Ireland
Donna Harman, NIST, USA
David Hawking, CSIRO ICT Centre, Australia
Bin He, IBM Research, USA
Djoerd Hiemstra, University of Twente, Netherlands
Eduard Hoenkamp, University of Maastricht, Netherlands
Xiansheng Hua, Microsoft Research Asia, China
Xuanjing Huang, Fudan University, China
Donghong Ji, Wuhan University, China
Jing Jiang, UIUC, USA
Gareth Jones, Dublin City University, Ireland
Jaap Kamps, University of Amsterdam, Netherlands
Tapas Kanungo, Yahoo!, USA
Gabriella Kazai, Microsoft Research, Cambridge, UK
Jun'ichi Kazama, Japan Advanced Institute of Science and Technology, Japan
Christopher Khoo, Nanyang Technological University, Singapore
Kevin Knight, USC/Information Sciences Institute, USA
Jeongwoo Ko, Oracle, USA
June-Jie Kuo, National Chung Hsing University, Taiwan
Kui-Lam Kwok, Queens College, City University of New York, USA
Andre Kushniruk, University of Victoria, Canada
Wai Lam, The Chinese University of Hong Kong, China
Jong-Hyeok Lee, Pohang University of Science & Technology (POSTECH), Korea
Mun-Kew Leong, Institute for Infocomm Research, Singapore
Nicholas Lester, Microsoft, USA
Gina-Anne Levow, University of Chicago, USA
Juanzi Li, Tsinghua University, China
Yuefeng Li, Queensland University of Technology, Australia

Ee-Peng Lim, Nanyang Technological University, Singapore
Chin-Yew Lin, Microsoft Research Asia, China, China
Chuan-Jie Lin, National Taiwan Ocean University, Taiwan
Shou-De Lin, National Taiwan University, Taiwan
Ting Liu, Harbin Institute of Technology, China
Jie Lu, IBM Research, USA
Lie Lu, Micorsoft Research Asia, China
Robert Luk, Hong Kong Polytechnic University, China
Jun Ma, Shandong University, China
Shaoping Ma, Tsinghua University, China
Andrew MacFarlane, City University, UK
Massimo Melucci, University of Padua, Italy
Hiroshi Nakagawa, University of Tokyo, Japan
Atsuyoshi Nakamura, Hokkaido University, Japan
Hwee Tou Ng, National University of Singapore, Singapore
Masayuki Okabe, Toyohashi Univ. of Technology, Japan
Bo Pang, Yahoo! Research, USA
Benjamin Piwowarski, Yahoo! Research Latin America, Chile
Vassilis Plachouras, Yahoo! Research, Spain
Ivana Podnar, University of Zagreb, Croatia
Hae-Chang Rim, Korea University, Korea
Thomas Roelleke, Queen Mary University London, UK
Stephen Robertson, Microsoft Research, UK
Ian Ruthven University of Strathclyde, UK
Tetsuya Sakai, NewsWatch, Inc., Japan
Mike Shepherd, Dalhousie Univ., Canada
Timothy Shih, Tamkang University, Taiwan
Fabrizio Silvestri, ISTI, National Research Council, Italy
Dawei Song, Open University, UK
Jian Su, Institute for Infocomm Research, Singapore
Aixin Sun, Nanyang Technological University, Singapore
Le Sun, Institute of Software, Chinese Academy of Sciences, China
Chunqiang Tang, IBM T.J. Watson Research Center, USA
Andrew Trotman, University of Otago, New Zealand
Richard Tzong-Han Tsai, Yuan Ze University, Taiwan
Yuen-Hsien Tseng, National Taiwan Normal University, Taiwan
Bin Wang, Institute of Computing Technology, Chinese Academy of Sciences, China
Fu Lee Philips Wang, City University of Hong Kong, China
Hsin-Min Wang, Academia Sinica, Taiwan
Lusheng Wang, City University of Hong Kong, China
Raymond Wong, City University of Hong Kong, China
Wensi Xi, Google, USA
Changsheng Xu, Institute for Infocomm Research, Singapore
Yue Xu, Queensland University of Technology, Australia

Seiji Yamada, National Institute for Informatics, Japan
Rong Yan, IBM T. J. Watson Research Center, USA
Jun Yang, CMU, USA
Tian-fang Yao, Shanghai Jiaotong University, China
Minoru Yoshida, University of Tokyo, Japan
Shipeng Yu, Siemens, USA
Min Zhang, Tsinghua University, China
Jun Zhao, Institute of Automation, Chinese Academy of Sciences, China
Ying Zhao, Tsinghua University, China

Additional Reviewers

Roi Blanco
Fabrizio Falchi
Frederik Forst
Jun Goto
Zhiwei Gu
Qiang Huang
Jing Liu
Hung-Yi Lo
Takashi Onoda
Yasufumi Takama
Hui Tan
Edwin Teng
Ming-Feng Tsai
Jinqiao Wang
Wei Xiong
Wai Keong Yong
Liang-Chih Yu
Yantao Zheng
Ye Zhou
Bin Zhu

Table of Contents

Session 1A: IR Models

Session 1B: Image Retrieval

Session 1C: Text Classification

Session 1D: Chinese Language Processing

Session 1E: Text Processing

Session 1F: Application of IR

Session 2A: Machine Learning

Session 2B: Taxonomy

Session 2C: IR Models

Session 2D: Information Extraction

Session 3A: Summarization

Session 3B: Multimedia

Session 3C: Web IR

Session 3D: Text Clustering

Poster Session

Improving Expertise Recommender Systems by Odds Ratio

Zhao Ru, Jun Guo, and Weiran Xu

Beijing University of Posts and Telecommunications,
100876 Beijing, China
rudjao@hotmail.com, guojun@bupt.edu.cn, xuweiran@pris.edu.cn

Abstract. Expertise recommenders that help in tracing expertise rather than documents start to apply some advanced information retrieval techniques. This paper introduces an odds ratio model to model expert entities for expert finding. This model applies odds ratio instead of raw probability to use language modeling techniques. A raw language model that uses prior probability for smoothing has a tendency to boost up "common" experts. In such a model the score of a candidate expert increases as its prior probability increases. Therefore, the system would trend to suggest people who have relatively large prior probabilities but not the real experts. While in the odds ratio model, such a tendency is avoided by applying an inverse ratio of the prior probability to accommodate "common" experts. The experiments on TREC test collections shows the odds ratio model improves the performance remarkably.

1 Introduction

Traditionally, recommender systems are an approach to solving problems of information overload. But in large organizations, they become one technology that assists to efficient locate people with needed expertise [4]. Such expertise recommender systems are gaining increasing importance as organizations continue to look for better ways to exploit their internal knowledge capital and facilitate collaboration among their employees for increased productivity [11]. As information technology advances, expertise recommendation approaches have moved from employing a database housing the knowledge of each employee to automated mining expertise representations in heterogeneous collections [5],[10].

The core of expertise recommender systems is expert finding that searches appropriate experts for recommendation. Increasing interest in expert finding led to the launch of an expert search task as a part of the enterprise track at TREC in 2005 [3]. It formulated the problem of expert finding as a retrieval task to rank people who are candidate experts given a particular topical query. Most systems that participated in this task applied advanced information retrieval (IR) approaches, such as language modeling techniques. Typically, a two-stage language model proposed by Cao et al. [2] showed excellent performance in 2005, and then became popular at TREC 2006. In this model, each document is represented as both a bag of topical words and a bag of expert entities. So for each

H. Li et al. (Eds.): AIRS 2008, LNCS 4993, pp. 1–9, 2008.

document, a language model of topical words which is called a relevance model is created, as well as a language model of expert entities called co-occurrence model. The probabilistic score of a candidate expert is then calculated from these two kinds of language models by assuming conditional independence between the candidate and the query. A general representation of this method is

$$P(e|q) = \sum_{d} P(d|q)P(e|d) \tag{1}$$

where e denotes a candidate expert, q a topical query, and d a document. A similar model was introduced by Balog et al. [1] as their Model 2.

In this paper, we analyze the language modeling approach to expert finding, taking the two-stage model as an example. As the relevance model $P(d|q)$ can be devised as a simple document retrieval system using language modeling techniques, the smoothing methods applied in this model have been studied in [12]. In this paper, our focus is on the co-occurrence model $P(e|d)$ which models expert entities. Our study reveals that the interpolation smoothing used in the co-occurrence model over-magnifies the affection of expert prior probability to give the model a tendency to prefer "common" experts. To avoid this shortcoming, we present an approach that uses odds ratio, a relative value of the smoothed probability to the prior probability, to replace the smoothed probability itself. We also present a more general model by modeling expertise on a set of documents in which it occurs.

2 Modeling Analysis

2.1 Language Model for Expert Finding and Its Smoothing

The problem of identifying candidates who are experts for a given topic can be stated as "what is the probability of a candidate e being an expert given the topical query q". That is, we rank candidates according to the probability $P(e|q)$. The candidates in the top are deemed the most probable expert for the given topic.

For expert ranking, language modeling techniques are applied to estimate the likelihood that one is an expert to a given topic. Such probability is computed based on the document collection which both the candidate and the topic are associated to, in a two-stage manner.

$$\begin{aligned} P(e|q) &= \sum_{d} P(e,d|q) \\ &= \sum_{d} P(d|q)P(e|d,q) \end{aligned} \tag{2}$$

where d denotes a document and $P(d|q)$ is the same as a traditional document retrieval.

In the following discussion, we focus on $P(e|d,q)$ which models the candidates. $P(e|d,q)$, as the co-occurrence model, captures the association between a candidate and a document. For simplification, people usually assume the query and the candidate are conditionally independent. Then we get equation (1). Therefore, we can treat each candidate identifier as a term and model it using language modeling techniques to get the probability $P(e|d)$. The maximum likelihood estimate of this probability is

$$P_{ml}(e|d) = \frac{tf_{e,d}}{|d|} \tag{3}$$

where $tf_{e,d}$ is the raw term frequency of candidate e in document d, $|d|$ is the total frequency of candidates in d. Usually, the empirical estimator $P_{ml}(e|d)$ is not accurate enough when dealing with sparse data like candidate identifiers. It needs smoothing.

Most smoothing methods make use of two distributions, a model used for "seen" words that occur in the document, and a model for "unseen" words that do not [12]. We use $P_s(e|d)$ for "seen" candidates and $P_u(e|d)$ for "unseen" candidates. The probability $P(e|q)$ can be written in terms of these models as follows

$$P(e|q) = \sum_{d:tf_{e,d}>0} P(d|q)P_s(e|d) \; + \sum_{d:tf_{e,d}=0} P(d|q)P_u(e|d) \tag{4}$$

The probability of an unseen candidate is typically taken as being proportional to the collection language model of the candidate, i.e. $P_u(e|d) = \lambda P(e|C)$. In interpolation smoothing, the probability of a seen candidate is represented as

$$P_s(e|d) = (1-\lambda)P_{ml}(e|d) + \lambda P(e|C) \tag{5}$$

where λ is constant parameter. In this case, we get

$$P(e|q) = \sum_{d:tf_{e,d}>0} (P(d|q) \times (1-\lambda)P_{ml}(e|d)) \; + \sum_{d} (P(d|q) \times (\lambda P(e|C))) \tag{6}$$

As the query is supposed to be created from the collection, we can just define $\sum_d P(d|q) = 1$. Then equation (6) becomes

$$P(e|q) = \sum_{d:tf_{e,d}>0} (1-\lambda)P(d|q)P_{ml}(e|d) \; + \; \lambda P(e|C) \tag{7}$$

It can be seen that the smoothing term is only proportional to $P(e|C)$. It seems to be similar with the language model used to retrieve documents. However, they have different effects. In the language model of topical words, the task is ranking documents; whilst in the language model of expert entities, the task is ranking experts. So the smoothing in the former is aimed to accommodate common words to reduce their affection, but in the latter there is an inverse effect that gives more weight to "common" experts.

How comes such a conclusion? Let us assume that a candidate whose prior probability is small has a big maximum likelihood probability in a document while another candidate whose prior probability is large has a small maximum likelihood probability in that document. Apparently, the first person would be more likely an expert of the document than the second one. But the second person would possibly get a larger score if we use (7) for ranking. That smoothing strategy over-magnifies the affection of $P(e|C)$ and the results would be given a bad tendency to the candidates with a relatively lager prior probability. The influence of the smoothing is shown in Fig. 1. It makes the results decline. Balog et al.'s Model 2 which only applies maximum likelihood estimate gets the best performance.

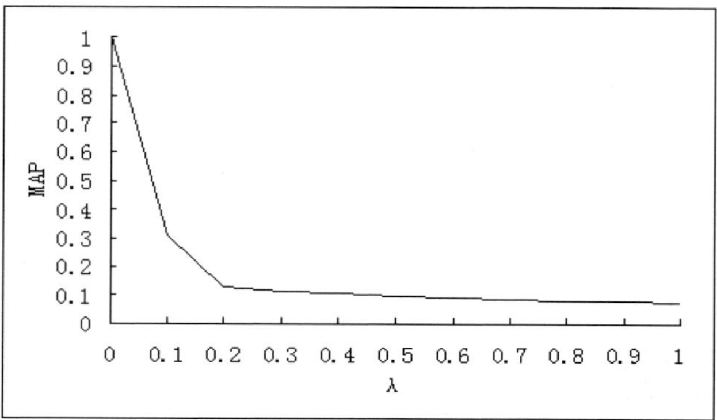

Fig. 1. The performance of the interpolation smoothing in expert finding systems. Note that the average precision results are normalized using the maximum likelihood estimate result ($\lambda = 0$). And that result would be reported as MLE in our experiments in next section.

2.2 Using Odds Ratio

To avoid the flaw that boosts up "common" experts, we try to reduce the affection of the prior probability. Furthermore, it is better to make the prior probability hold an inverse effect that accommodates "common" experts as it dose in document retrieval. So one of the most effective solutions is to apply a ratio of the seen probability and the collection background probability. That is what we called odds ratio in this paper, shown as

$$\begin{aligned} \frac{P(e|q)}{P(e|C)} &= \sum_{d:tf_{e,d}>0} P(d|q)(1-\lambda)\frac{P_{ml}(e|d)}{P(e|C)} + \lambda \\ &\propto \sum_{d:tf_{e,d}>0} P(d|q)\frac{P_{ml}(e|d)}{P(e|C)} \end{aligned} \qquad (8)$$

Note that λ is the const parameter which does not affect the results, so we use $\propto$ here to show the same ranking. The underlying meaning of the odds ratio is how much a document votes to a candidate depends on how this document is relatively important to the candidate but not just their association. So it would act as the "idf" in vector model to accommodate the experts who occur commonly in every documents.

From equation (8), we can see the score of a candidate expert is only associated with the documents in which it occurs. So it is better to just consider these documents and model the expert on them to get a more general model. Thus, the model is like

$$
\begin{aligned}
\frac{P(e|q)}{P(e|C)} &= \sum_{d:tf_{e,d}>0} P(d|q) \frac{P_s(e|d)}{P(e|C)} \\
&= \sum_{d:tf_{e,d}>0} P(d|q) \left((1-\lambda) \frac{P_{ml}(e|d)}{P(e|C)} + \lambda \right)
\end{aligned}
\tag{9}
$$

Note that equation (9) becomes (8) when the parameter λ is 0. This is our general model that applies odds ratio to improve language modeling techniques for expert finding. For convenience, in the experiments next section, we will not explicitly give the results of (8) but mention them as included in the results of (9).

3 Experimental Evaluation

We now present an experimental evaluation of the improved model for expert finding. First of all, we address the following research questions. We have reported the experimental results of the original model, we then use its best setting (i.e., maximum likelihood estimate) for comparison. The question is how the improved model compares with the original model. We also state an interpolation smoothing model which models a candidate on the documents it occurs. We compare this model with the improved model to show the effectiveness of the odds ratio in a range of settings. We then describe the experiments on TREC collections.

W3C corpus comprises 331,037 documents crawled from the public W3C (*.w3c.org) sites in June 2004 [3]. This heterogeneous repository contains several types of web pages, such as emails, personal homepages, wiki pages and so on. All of such pages were processed in our experiments. The evaluation was on Ent05 collection, which included W3C corpus and queries EX1-EX50, and Ent06 collection, which included W3C corpus and queries EX51-EX105. Queries EX1-EX50 contain only a "title" section; whilst queries EX51-EX105 contain "title", "description" and "narrative" sections.

We implemented our approach on version 4.2 of the Lemur Toolkit [9] which was used for the relevance model. First of all, the named entities of the candidates were extracted. We employed a list of 1,092 candidates, including full name and email. It is a specialized named entity recognition task, for which Rule-based name matching mechanisms described in [7] were applied. In practice, 795 of 1,092 candidates were found in the corpus and they appeared 8,587,453 times. When building index, we ignored case, stemmed with Krovetz's algorithm, and removed useless words by a list of 571 stop words. The same pre-processing was applied to the queries. We built another index for expert candidates, just treating each entity as a term. Language modeling approach was implemented on this index too.

We compared three expertise modeling approaches: maximum likelihood estimate (MLE), interpolation smoothing (IS) and odds ratio (OR). The OR model was formulized in (9). The MLE model would be stated as

$$P(e|q) = \sum_{d:tf_{e,d}>0} P(d|q)P_{ml}(e|d) \tag{10}$$

and the IS model is

$$P(e|q) = \sum_{d:tf_{e,d}>0} P(d|q)\left((1-\lambda)P_{ml}(e|d) + \lambda P(e|C)\right) \tag{11}$$

As all the models described above are general, they need to be detailed in the experiments. For convenience, we just applied the specific formula introduced by Cao et al. [2].

$$P(e|d) = (1-\lambda)\frac{tf_{e,d}}{|d|} + \lambda\frac{\sum_{d':tf_{e,d'}>0} \frac{tf_{e,d'}}{|d'|}}{df_e} \tag{12}$$

Here Dirichlet prior were used in smoothing of parameter λ:

$$\lambda = \frac{\kappa}{|d| + \kappa} \tag{13}$$

where κ is average length of term frequency of candidates in the collection. In our experiments, it is 10,801. Note that we use a flexible λ which is assumed to fit the general models.

The performances were tested on using top n documents retrieved by the relevance model as we found the performances would change according to returning different subsets of documents. We tried different value of n and got the result curves in Fig. 2. It reveals that the performance of the OR model is significantly better than the other two. In average, the OR model exceeds the MLE by over 12.0 percentage points. And if only the specific model of (8) is considered, it would exceed the MLE by about 10.6 percentage points in average. Meanwhile, the OR model exceeds the IS model by about 16.9 percentage points. The more significant improvement was supposed to be due to the IS model uses the prior probability for interpolation smoothing.

So is the OR model better than the IS model in any settings? To answer it, we relaxed the specification of λ in the two models and tested them by tuning λ, based on the condition of returning 200 documents as relevant. Such is the condition on which IS model almost gets its best results. Fig. 3 shows the results. It reveals that on all values of λ the OR model outperforms the IS model. Something interesting in Fig. 3 is that the OR model outperforms the IS model more significantly on Ent05 than Ent06. When looking into it, we found that such difference may be caused by the different ways how the judgements of these two collections created. In Ent05, the queries are the names of W3C groups and the judgements are

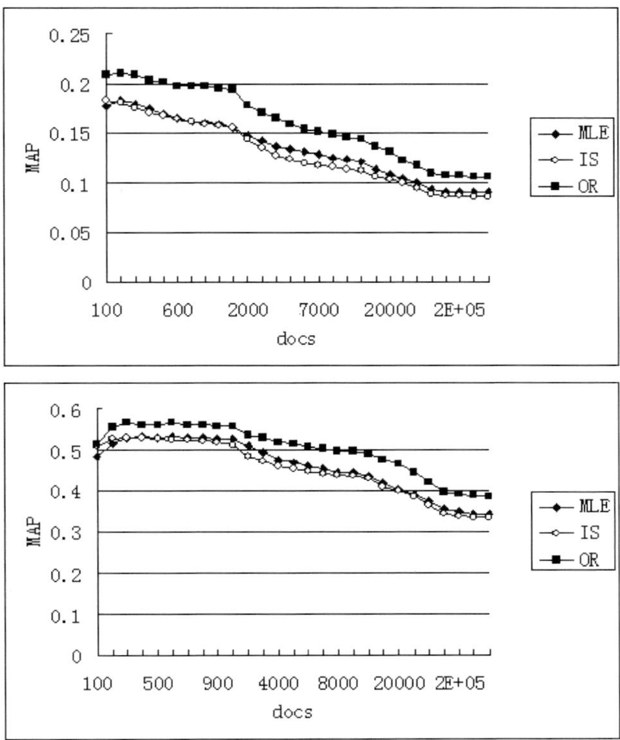

Fig. 2. The comparison of the three approaches on different subset of relevant documents. At the top is on Ent05 and at the bottom is on Ent06. Note that the size of the subset is not proportionally increasing.

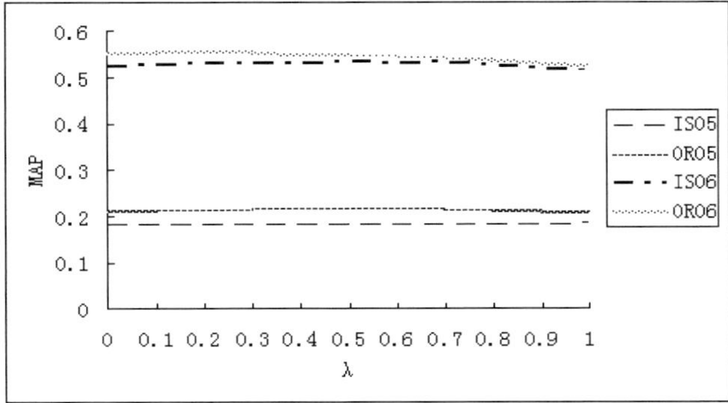

Fig. 3. The comparison of the IS model and the OR model on different value of λ

independent of the documents. But in Ent06, the judgements are contributed by the participants according the information from the documents. As a result, the judgements for Ent06 would show more preference to the experts who occur more times in the corpus, i.e., the "common" experts. So there seems less improvement for Ent06 because the IS model gets more benefit from the judgements.

Comparing our best results with the official results of the TREC Enterprise track, we found that using OR model improves the performances from the fifth best to the third best in TREC 2005 and 2006. As it is much harder to improve a top 5 system, we believe that is an obvious improvement. Also consider that it is a general model for expert finding, so its performance can be further improved by applying some elaborate mechanisms.

4 Conclusion

We analyzed the language modeling approach in expertise recommender systems. Taking the two-stage expert finding system as an example, our study was focused on the language model created for expert entities. In such a model, the smoothing is different from that used in the language model for retrieving documents. It wrongly uses expert prior probability which would be supposed to act as the "idf" in vector model to accommodate "common" experts. As a result, people who have relatively larger prior probability become more probable to be an expert. To avoid such a bias, we suggested an odds ratio model that applies odds ratio instead of raw probability to build the association between experts and documents. Additionally, we modeled one's expertise on a set of documents related to that person to get a general model. Our experiments on TREC collections reveals notable improvement using the odds ratio model. Compared with other systems that participated in TREC, the odds ratio model is promising.

Though our study showed that the two-stage expert finding system boosts up "common" experts, we believe the problem also exists in other kinds of language modeling approaches to expert finding. We will further study several other expert recommender systems to prove this assumption.

Acknowledgements

This work was supported by National Natural Science Foundation of China (Grant No.60475007, 60675001).

References

1. Balog, K., Azzopardi, L., de Rijke, M.: Formal models for expert finding in enterprise corpora. In: Proceedings of the 29th Annual International ACM SIGIR Conference on Research and Development in Information Retrieval, pp. 43–50 (2006)
2. Cao, Y., Li, H., Liu, J., Bao, S.: Research on expert search at enterprise track of TREC 2005. In: Proceedings of the Fourteenth Text REtrieval Conference (TREC 2005) (2005)

3. Craswell, N., de Vries, A.P., Soboroff, I.: Overview of the TREC 2005 enterprise track. In: Proceedings of the Fourteenth Text REtrieval Conference (TREC 2005) (2005)
4. McDonald, D.W.: Recommending expertise in an organizational setting. In: Conference on Human Factors in Computing Systems, pp. 63–64 (1999)
5. McDonald, D.W., Ackerman, M.S.: Expertise recommender: A flexible recommendation and architecture. In: Proceedings of the 2000 ACM Conference on Computer Supported Cooperative Work (CSCW 2000) (2000)
6. Ponte, J.M., Croft, W.B.: A language modeling approach to information retrieval. In: Proceedings of the 21st Annual International ACM SIGIR Conference on Research and Development in Information Retrieval, pp. 275–281 (1998)
7. Ru, Z., Chen, Y., Xu, W., Guo, J.: TREC 2005 enterprise track experiments at BUPT. In: Proceedings of the Fourteenth Text REtrieval Conference (TREC 2005) (2005)
8. Soboroff, I., de Vries, A.P., Craswell, N.: Overview of the TREC 2006 enterprise track. In: Proceedings of the Fifteenth Text REtrieval Conference (TREC 2006) (2006)
9. The Lemur toolkit for language modeling and information retrieval. `http://www.lemurproject.org/`
10. Yimam, D., Kobsa, A.: Centralization vs. decentralization issues in internet-based knowledge management systems: experiences from expert recommender systems. In: Workshop on Organizational and Technical Issues in the Tension Between Centralized and Decentralized Applications on the Internet (TWIST 2000) (2000)
11. Yimam, D., Kobsa, A.: DEMOIR: A hybrid architecture for expertise modeling and recommender systems. In: Proceedings of the IEEE 9th International Workshops on Enabling Technologies: Infrastructures for Collaborative Enterprises, pp. 67–74 (2000)
12. Zhai, C., Lafferty, J.: A study of smoothing methods for language models applied to ad hoc information retrieval. In: 24th ACM SIGIR Conference on Research and Development in Information Retrieval (SIGIR 2001), pp. 334–342 (2001)
13. Zhai, C., Lafferty, J.: The dual role of smoothing in the language modeling approach. In: Proceedings of the Workshop on Language Models for Information Retrieval (LMIR), pp. 31–36 (2001)

Exploring the Stability of IDF Term Weighting

Xin Fu* and Miao Chen*

University of North Carolina, Chapel Hill, NC 27599 USA
{xfu,mchen}@unc.edu

Abstract. TF·IDF has been widely used as a term weighting schemes in today's information retrieval systems. However, computation time and cost have become major concerns for its application. This study investigated the similarities and differences between IDF distributions based on the global collection and on different samples and tested the stability of the IDF measure across collections. A more efficient algorithm based on random samples generated a good approximation to the IDF computed over the entire collection, but with less computation overhead. This practice may be particularly informative and helpful for analysis on large database or dynamic environment like the Web.

Keywords: term weighting, term frequency, inverse document frequency, stability, feature oriented samples, random samples.

1 Introduction

Automatic information retrieval has long been modeled as the match between document collection and user's information needs. In any implementation based on this model, the representation of document collection and users' information need is a crucial consideration. Two main questions are involved in the representation: decision on what terms to include in the representations and determination of term weights [1].

TF·IDF is one of the most commonly used term weighting schemes in today's information retrieval systems. Two parts of the weighting were proposed by Gerard Salton [2] and Karen Spärck Jones [3] respectively. TF, the term frequency, is defined as the number of times a term in question occurs in a document. IDF, the inverse document frequency, is based on counting the number of documents in the collection being searched that are indexed by the term. The intuition was that a term that occurs in many documents is not a good discriminator and should be given lower weight than one that occurs in few documents [4]. The product of TF and IDF, known as TF·IDF, is used as an indicator of the importance of a term in representing a document.

The justification for and implementation of IDF has been an open research issue in the past three decades. One thread of research focuses on IDF calculation itself and proposes alternative IDF computation algorithms [5]. The other thread of research seeks theoretical justifications for IDF and attempts to understand why TF·IDF works so well although TF and IDF exist in different spaces [6].

* These authors contribute equally to this paper.

H. Li et al. (Eds.): AIRS 2008, LNCS 4993, pp. 10–21, 2008.

There have been a vast number of studies on distribution of word frequencies and other man made or naturally occurring phenomena. Studies found that these phenomena often follow a power-law probability density function and a Zipf or a Poisson mixture frequency rank distribution [7, 8]. However, there are different opinions on values of the parameters in the distribution function. It was also noted that the parameters might change across genre, author, topic, etc. [7]. Finally, many early experiments were conducted on abstracts rather than full text collections. The language patterns in the full text can be very different from the abstracts.

This study does not intend to test the term frequency distributions, or to derive the estimators for distribution parameters. Instead, it aims to investigate the similarities and differences between IDF distributions based on the global collection and on different samples and to test the stability of the IDF measure across collections. The study examines how IDF varies when it is computed over different samples of a document collection, including feature-oriented samples and random samples. As Oard and Marchionini [9] pointed out, estimates of IDF based on sampling earlier documents can produce useful IDF values for domains in which term usage patterns are relatively stable.

The motivation of this study comes from the observation that advance knowledge of IDF is either impossible for a real world collection or too expensive to obtain. The practical aim of the study is to develop a more efficient algorithm that requires less computational time/cost, but at the same time, generates a good approximation to the IDF computed over the entire collection. In a dynamic world where new information is added accretionally to a collection, it would be informative to understand how collection based weights will evolve when new information is added. With this understanding, we may make recommendations such as if the collection size increases by more than x percent, then the IDF weights should be updated. A recommendation like this will be particularly useful in a dynamic environment like the Web.

2 Methods

The experiments were conducted on a 1.16GB collection of full text journal articles published mostly between 1997 and 2006. The articles came from 160 journals and all of them were available in XML format. In order to allow for comparison of IDF computed from different components of articles (title, abstract, reference and table or figure caption), articles which missed any of the above four components were discarded. After the pruning, 15,132 articles were left and used for experiments. The rest of this section describes the pre-processing and the statistical methods.

2.1 Preprocessing

The pre-processing of the collection consists of four steps: extracting, tokenizing, removing stop words, and stemming. The first step was to extract from each document the information that would be used in analysis. First, information from the following XML tags was extracted: journal ID, journal title, article identifier, article title, abstract, titles of cited works and table/figure captions tags. If a cited work was a journal article, the title of the article was used. If a cited work was a book, the book title was used. Then, for the <body> section which contained the text of the articles, all information was extracted unless it was under headings which were non-topic in nature (e.g., those for structural or administrative information). Since there was variation across articles and across journals in the use of headings, we manually examined all the headings that appeared five times or more in the collection and identified non-topic headings for removal. We also sorted the headings alphabetically and identified variants of these headings and occasional misspellings.

The second step of pre-processing was to stripe off XML tags and tokenize the extracted parts into terms. We used space as the delimiter; therefore, only single words (instead of phrases) were considered. The tokenizer removed all the punctuation. In specific, this means that all hyphenated terms were broken into two. Next, we applied a basic stop word list called Libbow Stop Word List [10] which we expanded by adding all-digit terms, such as "100" and "1000", and octal control characters, such as "+". We also removed all the DNA sequences (strings of 8 characters or longer which were formed solely by letters a, c, g, and t) and information in the <inline-formula> tags, which contained noise.

Finally, we used the popular Porter stemming algorithm [11] to stem the words. All the stemmed words were stored into a table in the Oracle database. For each stemmed term, the article ID and source ('T' for title, 'A' for abstract, 'C' for caption, 'R' for reference, and 'B' for body) were stored. Body is the part of an article other than the title, abstract, table/figure and reference. Headings are regarded as part of the body.

In sum, the experiment collection consisted of 15,132 full text articles of 3,000 words long on average. Each article contained a title, an abstract, a body, some table/figures and a list of references. For tables and figures, we were only interested in their captions, so we discarded table/figure contents. For references, we were only interested in titles of cited works (either book titles or titles of journal articles) and did not consider other information, such as authors and journal names. The vocabulary size was 277,905, which was the number of unique stemmed terms in the collection.

2.2 Statistical Methods

Various ways have been proposed to calculate IDF. A basic formula was given by Robertson [4]. A later discussion between Spärck Jones [12] and Robertson resulted in the following formula of IDF:

$$idf(t_i) = \log_2(\frac{N}{n_i}) + 1 = \log_2(N) - \log_2(n_i) + 1 \quad (1)$$

where N is the total number of documents in the collection and n_i is the number of documents that contain at least one occurrence of the term t_i.

For a particular collection (fixed N), the only variable in the formula is n_i, the number of documents in which the term t_i appears. Let us call it document frequency (DF). IDF is then a monotone transformation to the inverse of the document frequency. When a different collection is used, the IDF will differ only by a constant, $\log_2(N/N')$.Therefore, we can approach the IDF comparison problem by comparing the document frequency distribution in the global collection and in each of the subsampling collections.

There were 14 collections involved in the analyses: the global collection, four feature oriented sub-sampling collection (title, abstract, caption and reference) and nine random sample collection (from 10% to 90% at 10% interval). Each collection can be represented as N feature-value pairs with the feature being a term in the vocabulary and value being the document frequency of the term. Formally, the document frequency representation of a collection is: DF_xx {($term_1$, df_1), ($term_2$, df_2), ..., ($term_{277905}$, df_{277905})}, with xx being G (global), T (title), A (abstract), C (caption), R (reference), 10 (10% random sample), ..., 90 (90% random sample). For example, the global collection document frequency is represented as: DF_G {(a, 10), (a0, 17), ..., (zzw, 1), (zzz, 2)}. When a term was missing in a sub-sampling collection, its document frequency was defined as 0.

With the above representation, the problem of comparing document frequency features of two collections (the global collection and a sub-sampling collection) was abstracted as comparing two data sets, DF_G and DF_sub, and asking if they follow the same distribution.

For each data set, we first summarized the data by plotting document frequencies against their ranks and showing a histogram of document frequency distribution. Some summary statistics were also reported to characterize the distribution.

To compare the two data sets, we first looked at their histograms to visually compare the shapes of the distributions. Then we generated scatter plots and computed correlation coefficients to estimate the strength of the linear relationship between the two data sets. As the data were by far not normally distributed, we used Spearman Rank Order Correlation Coefficient, instead of Pearson's Correlation Coefficient.

Finally, we computed the Kullback-Leibler distance between the two distributions as a numeric characterization of how close the distribution of the sub-sampling data set was from the distribution of the global data set. The Kullback-Leibler distance, also called the relative entropy [13], between the probability mass function based on a sub-sampling collection p(x) and that based on the global collection q(x) was defined as:

$$D(p \parallel q) = \sum_{x \in X} p(x) \log \frac{p(x)}{q(x)} = E_p \log \frac{p(X)}{q(X)} \qquad (2)$$

in which X was the vocabulary space, x was a term in the vocabulary space, p(x) and q(x) were the document-wise probability that the term occurred in the sub-sampling collection and the global collection, respectively. Specifically, p(x) was computed as $n_i/\sum n_i$, so all p(x) added up to 1. q(x) was computed in a similar way for the global collection. Note that in this definition, the convention (based on continuity arguments) that 0log(0/q)=0 was used. As Cover and Thomas [13] pointed out, the Kullback-Leibler distance is not a true distance between distributions since it is not symmetric and does not satisfy the triangle inequality; nonetheless, it is a useful and widely used measure of the distance between distributions.

Although there are many studies on the distribution of word frequency and frequency rank, it remains unclear which distribution and which parameters are the most appropriate for what type of collection. In this study, we used non-parametric methods without assuming any particular form of distribution.

3 Results

3.1 Feature Oriented Samples

We carried out the first set of experiments on feature based sub-sampling collections. We started with calculating document frequency, n_i in formula (1), for each stemmed term. The vocabulary size (i.e., the number of rows in the df_xx table) for the global collection and each of the special feature collections are summarized in Table 1.

Table 1. Vocabulary size for each collection

Data Set Code	Global	T	A	C	R
Collection Name	Global	Title	Abstract	Caption	Reference
Size	277,905	16,655	44,871	71,842	94,007

Figure 1 displays the scatter plots between the global DF and sample DFs. When a term was missing from a sub-sampling collection, its sample DF was defined as 0. The plots showed that the linear relationship between the global DF and any of the sample DFs was not very strong. The Spearman Correlation Coefficients were 0.5691 (global vs. reference) > 0.5373 (abstract) > 0.4669 (caption) > 0.3845 (title).

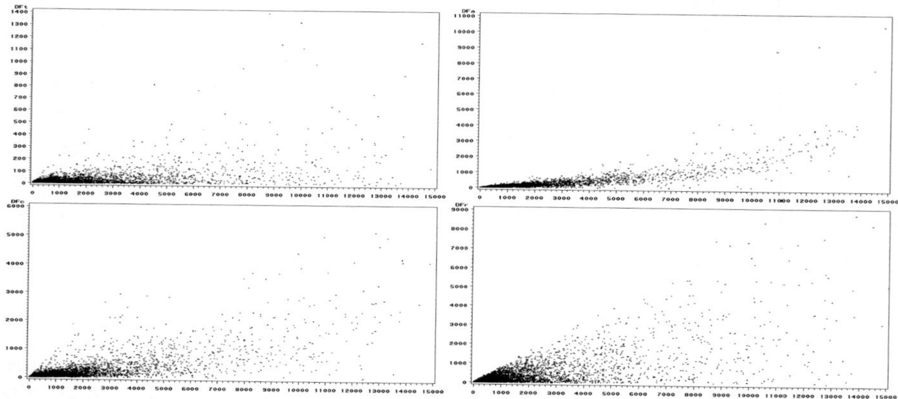

Fig. 1. Scatter plots between global DF and DF of Title (upper left), Abstract (upper right), Caption (lower left), Reference (lower right) collections

To help understand why the linear relationship was not very strong, we replotted the four pairs of DFs, but on a log-log scale. The plots are displayed in Figure 2. Since the logarithm function has a singularity at 0, these plots excluded all the terms that only appeared in the global collection, but not in a sub-sampling collection.

From these plots, we could easily see that the deviation from linear relationship happened more with terms that had low document frequencies in the sub-sampling collections. For example, in the upper left plot, y value of 0 (i.e., DF_T=1)

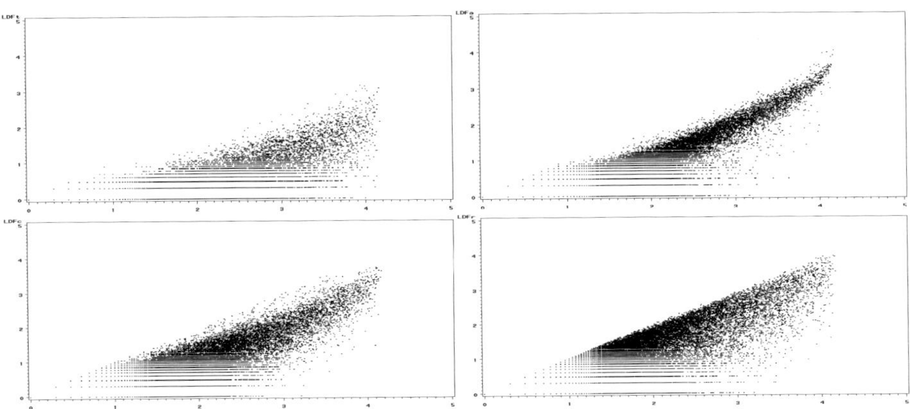

Fig. 2. Scatter plots between global DF and DF of Title (upper left), Abstract (upper right), Caption (lower left), Reference (lower right) collections on a log-log scale (base 10)

corresponded to a wide range of x values from 0 up to about 4 (i.e., DF_Global from 1 to 9,950). This means that a term that only appeared once in document titles appeared somewhere else in 9,950 documents.

Next, we computed the relative entropy for each of the feature-oriented samples compared to the global collection. Recall that the relative entropy was defined as:

$$D(p \parallel q) = \sum_{x \in X} p(x) \log \frac{p(x)}{q(x)} = E_p \log \frac{p(X)}{q(X)} \tag{2}$$

in which X was the vocabulary space and x was a term in the vocabulary space. For each term t_i, p(x) and q(x) were computed as $n_i/\sum n_i$ with n_i being the document frequency for term t_i in the sub-sampling collection and in the global collection, respectively. The results were: 0.2565 (abstract) < 0.3838 (caption) < 0.3913 (reference) < 0.7836 (title). This means that the abstract data set distributed most similarly to the global data set, followed by the caption data set and the reference data set. Note that this order was different from the order for Spearman Correlation Coefficients.

In summary, the above analyses compared the four sample distributions with the distribution of the global data set from different approaches. All five sets of data were heavily skewed. The scatter plots and the correlation coefficients showed that the linear relationships between the sample data sets and the global data set were not very strong. So, IDFs calculated based on those samples may not be very good estimates for those of the global collection. In particular, the title was probably too crude to be a good compression of the full text. Besides, qualitative analyses suggested that there were some systematic missing of terms in the title and reference collections. In other words, their language patterns were not the same as the rest of the collection. Furthermore, distribution distance measures led to different results than the correlation coefficients. The disagreement might be an indication that there were more complex relationships in the data.

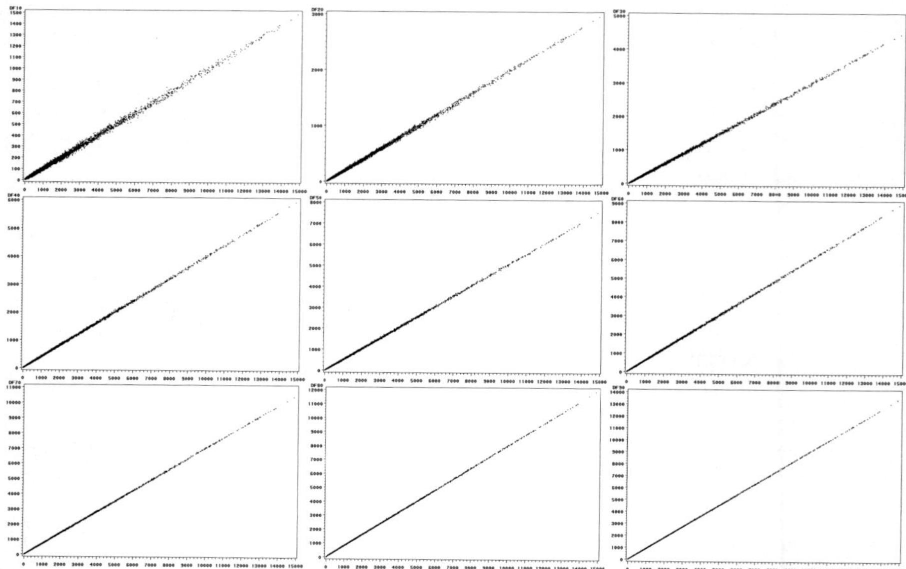

Fig. 3. Scatter plots between global DF on x-axis and DF of random 10% to 90% collections on y-axis (from left to right, then top to bottom)

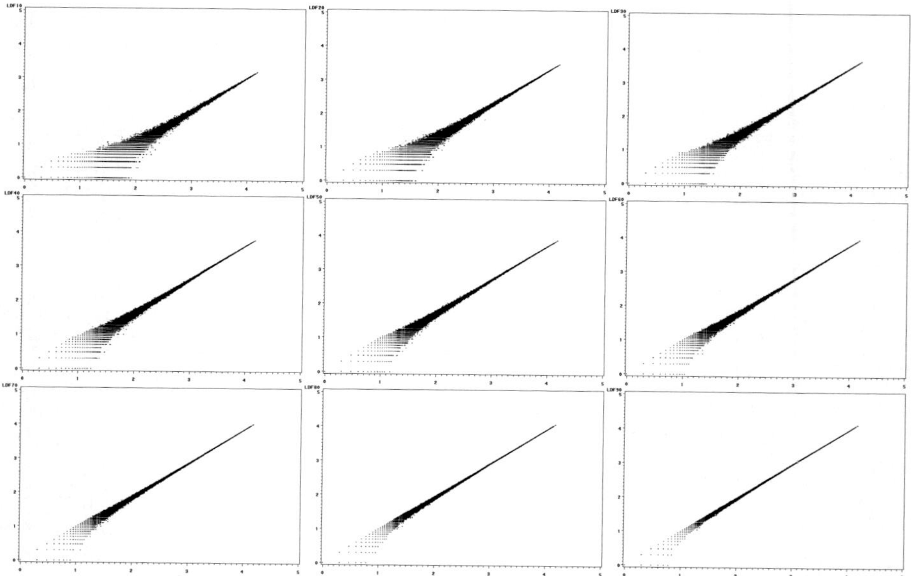

Fig. 4. Scatter plots between global DF and DF_10 to DF_90 (from left to right, then top to bottom) on a log-log scale (base 10)

3.2 Random Samples

In this set of experiments, we used random samples of the collection. For a 10% sample, we randomly picked 1,513 articles out of the 15,132 articles and built the data set with terms that came from this subset of articles. The vocabulary size for the global collection and each of the random sample collections are summarized below.

Table 2. Vocabulary size for the global collection and each of the random sample collections

Data Set	10%	20%	30%	40%	50%
# articles	1,513	3,026	4,540	6,053	7,566
Vocabulary	68,724	102,208	130,001	158,318	181,041
Data Set	60%	70%	80%	90%	Global
# articles	9,079	10,592	12,106	13,619	15,132
Vocabulary	202,236	222,170	241,675	260,225	277,905

In Figure 3, we display the scatter plots between the global DF and random sample DFs. When a term was missing from a random sample collection, its sample DF was defined as 0. The plots showed that the linear relationship between the global DF and any of the random sample DFs was stronger than those between the global DF and DFs of feature oriented samples. As the sample size increased, the Spearman Correlation Coefficient kept increasing from 0.5591 (10%) to 0.9299 (90%). It should be noted that although all these scatter plots look much closer to a straight line than the previous set, the smallest Spearman Correlation Coefficient between the random data set and the global data set [Corr(DF_10, DF_Global)=0.5591] was actually smaller than the largest one between the feature oriented data set and the global data set [Corr(DF_R, DF_Global)=0.5691]. We attributed this to the fact that Spearman Correlation Coefficient was rank based statistic while our data sets were seriously skewed and had a significant portion of "ties" in DF rank. Therefore, the scatter plots might be better descriptions of the relationship between data sets than simply looking at the correlation coefficients.

We also plotted the nine pairs of data sets on a log-log scale (base 10). The plots are displayed in Figure 4. Again, these plots excluded all the terms that only appeared in the global collection, but not in a sub-sampling collection.

The deviation from the linear relationship happened again with terms that had low document frequencies in the sub-sampling collections. However, unlike the title and reference collection in the first experiment, we did not notice any systematic missing of terms in the random samples. For example, the terms which were missing from the 10% collection while appearing in the largest numbers of documents in the global collection (number in parenthesis) were: array (64), darwin (62), bisulfit (59), subgraph (57), repercuss (56), Shannon (56), notochord (53), peyer (52), ced (51), puc (50). These numbers were also much smaller than those associated with the feature oriented samples.

Next, we computed the relative entropy for each of the random sample data set distribution compared to the global data set distribution. The values went down consistently as the sample size increased, from 0.0883 (10%) to 0.0025 (90%). Also, all of these relative entropies were much smaller than those we obtained in the previous section.

Overall, the results suggested that compared to distributions of feature oriented sample data sets, distributions of random sample data sets were much more similar to that of the global data set. IDFs calculated based on those samples should be better estimates for the global IDF. It is not exactly clear how large a sample will be "sufficient" to represent the whole; however, it is clear that terms that have low document frequencies in the sub-sampling collections are less stable then those that appear in a lot of the documents in the samples. In other words, we can safely use high-DF terms in a sample collection to predict the DF or IDF in the global collection, while we are much less confident in using low-DF terms.

Finally, back to the ultimate task of using sample data to predict global IDF, we ran a simple linear regression of IDF Global on IDF_10 to IDF_90. The regression equations are:

$$
\begin{aligned}
IDF\_G &= -2.4096 + 1.3661 \bullet IDF\_10 \\
IDF\_G &= -1.9296 + 1.2615 \bullet IDF\_20 \\
IDF\_G &= -1.5332 + 1.1957 \bullet IDF\_30 \\
IDF\_G &= -1.2108 + 1.1507 \bullet IDF\_40 \\
IDF\_G &= -0.9306 + 1.1134 \bullet IDF\_50 \\
IDF\_G &= -0.7199 + 1.0850 \bullet IDF\_60 \\
IDF\_G &= -0.5164 + 1.0595 \bullet IDF\_70 \\
IDF\_G &= -0.3171 + 1.0366 \bullet IDF\_80 \\
IDF\_G &= -0.1614 + 1.0179 \bullet IDF\_90
\end{aligned}
$$

R-Squares for the regressions ranged from .8076 (10%), .9211 (40%), .9555 (60%), to .9907 (90%). This means that on average IDFs calculated with 10% of the sample can predict 80 percent of the variation in IDFs calculated with the entire collection. We did not use the same model for feature based sampling data sets because we found that the error distributions were seriously non-normal for those data sets.

4 Conclusions and Discussions

This study explored the relationship between IDF distributions (via DF distributions) in sub-sampling collections and in the global collection with the intention to derive an optimal sampling method that used a minimal portion of the entire collection, but generated a satisfactory approximation to the IDF generated over the whole collection. We looked at two different sampling methods, feature based sampling and random sampling. Feature based sampling resulted in four sample data sets: title, abstract, caption, and reference. Random sampling resulted in nine sample data sets: from 10% to 90% at 10% intervals. Several strategies were used to compare the distribution of each of the sample data sets with the distribution of the global data set. Each data set was first summarized using two graphs: a plot of ranked document frequency against the rank and a histogram of the document frequency. The relationship between the two data sets in question was then characterized in three ways: a scatter plot, a Spearman Correlation Coefficient, and a Kullback-Leibler distance between their distributions. Finally, for random sampling data sets, we performed simple linear regression models to predict global IDFs from sample IDFs.

The results suggested that IDFs computed on random sample collections had stronger association with the global IDF than IDFs computed on feature based sample collections. On average, the IDF computed on the 10% random sample could explain about 80% of the variations in IDFs calculated with the global collection. We also noted from scatter plots that high DF terms in the samples (i.e., terms which appeared in a lot of articles in the sample collection) were much more reliable than low DF terms in predicting global DFs.

At the beginning of the study, we were interested in finding out an optimal sampling method which would begin with a collection as small as possible, but generate IDFs close enough to the global IDF that no significant difference can be found in their distributions. This then became a hypothesis testing problem: "can we disprove, to a certain required level of significance, the null hypothesis that two data sets were drawn from the same population distribution function?" It required establishing a difference measure and choosing a test statistic so that we could compare the difference to a critical value of the test statistic distribution.

To achieve this goal, we considered two goodness-of-fit tests: the Chi-square goodness-of-fit test and the Kolmogorov-Smirnov test. The Chi-square test is used to test if a sample of data came from a population with a specific distribution [14]. The Kolmogorov-Smirnov test has a similar purpose, but is based on the maximum distance between two cumulative distribution functions. The test statistics will tell us whether the sample DF follows the same distribution as the global DF.

The Chi-square test could potentially be carried out at three levels. At the first level, we could look at the raw data, i.e., the document frequency of each individual term, and directly compared the distribution of DF_G and DF_Sub. This was like measuring the association between two categorical variables, each having 277,905 categories. The frequency count in each cell was simply the document frequency of the term in that collection.

At the second level, we could look at the summary data of document frequency versus the number of terms with that document frequency. The method put the data into bins, each bin corresponding to a unique value of the document frequency. Then it counted the number of terms that fell into each bin. We can think of this as generating a histogram and forcing each bar to be of unit width. So, the feature space became unique document frequency values (not necessarily consecutive) and the value was the number of terms with that feature.

At the third level, we could form wider bins for either of the first two methods. Binning for the second method was easy to understand and was what statistical software would do by default (choosing an optimal bar width to plot the histogram), but grouping "bag of words" in the raw data, as at the first level, could only be arbitrary.

There were problems with each of the three levels. For the first level, there were two serious problems. Firstly, all the data sets contained significant proportion of low DF terms, which translated into cells with low frequency in the bivariate table. An appropriate Chi-square test would require that expected frequencies in each cell be at least 5 (possibly allowing a few exceptions if there are a large number of categories).

Obviously, our data do not meet the requirement. Secondly, it is argued that a Chi-square test should only be used when observations are independent, i.e., no category or response is dependent upon or influenced by another [15]. This assumption is seriously in jeopardy in our data. It is well known that some words turn to occur together, e.g., forming a phrase, while other words rarely co-occur. The test results would be very unreliable if we ignored this important linguistic phenomenon and assumed that term frequencies were independent. A direct consequence of working with dependent data was the difficulty in choosing the appropriate number of degrees of freedom, which was determined by the number of independent observations in the data.

The second and the third methods, both looking at distribution of data, instead of the raw data, introduced another problem. To give a simple example, assume there were two terms in the vocabulary. In the global collection, Term 1 appeared in 100 documents ($DF\_Global_1=100$) and Term 2 appeared in 10 documents ($DF\_Global_2=10$). In one sub collection (SubOne), $DF\_SubOne_1=9$ and $DF\_SubOne_2=1$. In another sub collection (SubTwo), $DF\_SubTwo_1=1$ and $DF\_SubTwo_2=9$. If we looked at the raw data, we would probably conclude that the SubOne collection had closer document frequency distribution with the global collection than the SubTwo collection. However, if we only looked at the summary data, the histograms of the two sub collections were identical. What was missing in the summary data was the mapping of terms between the global collection and the sub-sampling collection. Even if we found that the summaries of two data sets had same or similar distributions, we could not know for sure if the raw data had similar distributions.

Compared with the first and the second method, the third method should eliminate low count cells, but as in any other statistical methods, binning involves a loss of information. Plus, as we mentioned above, binning based on the raw data was very arbitrary.

Due to these problems, we felt that it was inappropriate to use Chi-square tests on the current data. The Kolmogorov-Smirnov test would be inappropriate either due to the dependency of terms. The Kolmogorov-Smirnov test is based on cumulative density functions. Since the probabilities of some terms are dependent on each other, the cumulative density functions generated from the data are unreliable.

To address the term dependence problem, we will consider in a future study using the principal component analysis (PCA) technique to project the original term space into an orthogonal space and use the principal components (linear combination of terms) to compute correlation coefficients and relative entropies, and perform goodness-of-fit tests. Analyses in that space should be more reliable than what has been done with the current data sets.

We are also aware of other possibilities for a future study. Instead of using simple random sampling to form the random sample collections, we can consider stratified sampling methods such that the 10% sample collection is formed by 10% of articles from Journal 1, 10% of articles from Journal 2, etc. This would account for the possibility that language patterns are different across journals.

This study only considered single terms. Taking phrases into consideration in a future study may better model the language use. In this study, we noted that terms that appeared in a lot of articles in the sample collection were more reliable than terms that appeared in a few articles in predicting global DFs. It would be interesting to follow up this with a term level analysis and see if there are certain terms that are more or less stable.

Acknowledgments. We thank Drs. Chuanshu Ji, Cathy Blake and Haipeng Shen for their guidance on this work and the anonymous reviewers for their feedback.

References

1. Salton, G., Buckley, C.: Term-weighting approaches in automatic text retrieval. Information Processing & Management 24(5), 513–523 (1988)
2. Salton, G.: Automatic information organization and retrieval. McGraw-Hill, New York (1968)
3. Spärck Jones, K.: A statistical interpretation of term specificity and its application in retrieval. Journal of Documentation 28, 11–21 (1972)
4. Robertson, S.: Understanding inverse document frequency: on theoretical arguments for IDF. Journal of Documentation 60, 503–520 (2004)
5. Wang, J., Rölleke, T.: Context-specific frequencies and discriminativeness for the retrieval of structured documents. In: Lalmas, M., MacFarlane, A., Rüger, S.M., Tombros, A., Tsikrika, T., Yavlinsky, A. (eds.) ECIR 2006. LNCS, vol. 3936, pp. 579–582. Springer, Heidelberg (2006)
6. Blake, C.: A Comparison of document, sentence and term event spaces. In: Coling & ACL joint conference, Sydney, Australia (2006)
7. Church, K.W., Gale, W.A.: Poisson mixtures. Natural Language Engineering 1(2), 163–190 (1995)
8. Newman, M.E.J.: Power laws, Pareto distributions and Zipf's law (2005), Available at: http://aps.arxiv.org/PS_cache/cond-mat/pdf/0412/0412004.pdf
9. Oard, D., Marchionini, G.: A conceptual framework for text filtering (1996), Available at: http://hcil.cs.umd.edu/trs/96-10/node10.html#SECTION00051000000000000000
10. McCallum, A.K.: Bow: A toolkit for statistical language modeling, text retrieval, classification and clustering (1996), Available at: http://www.cs.cmu.edu/_mccallum/bow
11. Porter, M.F.: An algorithm for suffix stripping. Program 14(3), 130–137 (1980)
12. Spärck Jones, K.: IDF term weighting and IR research lessons. Journal of Documentation 60, 521–523 (2004)
13. Cover, T.M., Thomas, J.A.: Elements of Information Theory. John Wiley, New York (1991)
14. Snedecor, G.W., Cochran, W.G.: Statistical Methods. Iowa State University Press, Ames (1989)
15. Conner-Linton, J.: Chi square tutorial (2003), Available at: http://www.georgetown.edu/faculty/ballc/webtools/web_chi_tut.html

Completely-Arbitrary Passage Retrieval in Language Modeling Approach

Seung-Hoon Na[1], In-Su Kang[2], Ye-Ha Lee[1], and Jong-Hyeok Lee[1]

[1] Department of Compueter Science, POSTECH, AITrc, Republic of Korea
{nsh1979,sion,jhlee}@postech.ac.kr
[2] Korea Institute of Science and Technology Information(KISTI), Republic of Korea
dbaisk@kisti.re.kr

Abstract. Passage retrieval has been expected to be an alternative method to resolve length-normalization problem, since passages have more uniform lengths and topics, than documents. An important issue in the passage retrieval is to determine the type of the passage. Among several different passage types, the arbitrary passage type which dynamically varies according to query has shown the best performance. However, the previous arbitrary passage type is not fully examined, since it still uses the fixed-length restriction such as *n* consequent words. This paper proposes a new type of passage, namely *completely-arbitrary passages* by eliminating all possible restrictions of passage on both lengths and starting positions, and by extremely relaxing the type of the original arbitrary passage. The main advantage using completely-arbitrary passages is that the proximity feature of query terms can be well-supported in the passage retrieval, while the non-completely arbitrary passage cannot clearly support. Experimental result extensively shows that the passage retrieval using the completely-arbitrary passage significantly improves the document retrieval, as well as the passage retrieval using previous non-completely arbitrary passages, on six standard TREC test collections, in the context of language modeling approaches.

Keywords: passage retrieval, complete-arbitrary passage, language modeling approach.

1 Introduction

Many retrieval functions are using the length-normalization to penalize the term frequencies of long-length documents, since documents are diversely distributed in lengths and topics [1-4]. The pivoted length normalization has been the most popular normalization technique which was applied to traditional retrieval models such as vector space model and probabilistic retrieval model [1]. The length-normalization is inherent in the recent language modeling approaches as well, within its smoothing methods. However, the length-normalization problem is still a non-trivial issue. For example, term frequencies in documents having redundant information will be high in proportional to lengths of the documents, while term frequencies in topically-diverse documents may not be high in proportional to the lengths.

H. Li et al. (Eds.): AIRS 2008, LNCS 4993, pp. 22–33, 2008.

The passage-based document retrieval (we call it passage retrieval) has been regarded as alternative method to resolve the length-normalization problem. The passage retrieval uses a query-relevant passage in document, instead of using the entire document, and sets the similarity between the document and the query by the similarity between the query-relevant passage and the query [5-8]. Different from documents, passages are more uniformly distributed in terms of length and topics. Thus, the passage retrieval has been the promising technique to appropriately handle the length-normalization problem.

The most critical issue in the passage retrieval is the definition of passage; what type of passage should we use for providing a good retrieval performance? For this issue, several types of passages have been proposed; a semantic passage using a paragraph or a session, a window passage which consists of subsequent words with fixed-length, and an arbitrary passage which consists of subsequent words with fixed length starting at arbitrary position in a document. Among them, the arbitrary passage was the more effective [7]. From extensive experiments on the arbitrary passage, Kaszkiel and Zobel showed that as passage is less restricted, the passage retrieval becomes more effective [7]. Their result is intuitive, since as the passage is less-restricted, the more query-specific parts in a document can be selected as the best passage. Generalizing the Kaszkiel and Zobel's result, a more query-specific passage provides better performance, while a more query-independent passage provides worse performance. At this point, Kaszkiel and Zobel's arbitrary passage could be more improved, since their proposed arbitrary passage still has the restriction of fixing the passage-length such as using 50, 100, …, 600. If we drop this restriction, then a more query-specific passage could become the best passage, and thus could result in a more retrieval effectiveness.

Regarding that, this paper proposes a new type of passage, namely *completely-arbitrary passages* by eliminating all possible restrictions of passage on both lengths and starting positions which still has remained in original arbitrary passage. Continuity of words comprising a passage is the only remaining restriction in the completely-arbitrary passage. Since there is no further restriction to the passage, we can expect that a more query-specific passage can be selected as the best passage, resulting in a better retrieval effectiveness.

As another advantage, *completely-arbitrary passages* can well-support the proximity between query terms which has been a useful feature for a better retrieval performance. Note that the length of passage can play a role for supporting the proximity (because the general scoring function of a passage is dependent on the length of the passage). Based on this passage length, original arbitrary passage cannot well-support the proximity feature, since original arbitrary passage has a lower bound on its length such as at least 50 words. On *completely-arbitrary passage*, extremely short passages including two or three terms only are possible, thus passages of query terms alone can be selected as the best passage. For example, suppose that a query consisting of "search engine", and the following two different documents (Tao's example [11]).

D_1: "………….. search engine ………………….."

D_2: "engine ………………….. search".

Normally, we can agree that D_1 is more relevant than D_2, since D_1 represents more exactly the concept of the given query. When two query terms in a D_2 are less-distant than 50 words, Kaszkiel's arbitrary passage cannot discriminate D_1 and D_2 at passage-retrieval level. On the other hand, completely-arbitrary passage can discriminate them, since it can construct the best passage of the length of 2.

Of course, we can directly model the proximity feature by using a function such as Tao and Zhai's method (simply, Tao's method) [11]. However, Tao's method heuristically models the proximity, with an unprincipled way by using a function which cannot be naturally derived from the original retrieval model [11]. In addition, Tao's method does not consider the importance of each query term, but simply representing a distance as the proximity-driven scoring function. At this naïve setting, when verbose queries including common terms are processed, the distance cannot be reasonably estimated, so that the effectiveness will not be improved. Compared to the direct method such as Tao's method, the passage retrieval has an advantage that the importance of query terms can be considered, resulting in robustness even for verbose queries.

When we use completely-arbitrary passages, an efficiency problem is raised since the number of completely-arbitrary passages is exponentially increasing according to the length of a document. Regarding this, we further propose a *cover-set ranking algorithm* by which the ranking problem of completely-arbitrary passages can be efficiently implemented. We will provide a simple proof of equivalence between cover-set ranking and finding the best passage among the set of completely-arbitrary passages.

Although the proposed passage retrieval can be utilized to any retrieval models, we apply it to the language modeling approach, since it is state of the art. Experimental result is encouraging, showing that the proposed passage retrieval using completely-arbitrary passage significantly improves the baseline, as well as that using the previous arbitrary passage, in the most of test collections.

Note that our work is the first successful passage retrieval in the context of language modeling approaches. To our best knowledge, Liu's work is a unique study for the passage retrieval in the language modeling approaches [8]. However, Liu's work did not show the effectiveness of passage retrieval on test collections, resulting in worse performances over the baseline in some test collections.

2 Complete Variable Passages and Cover-Set Ranking Algorithm

2.1 Passage Retrieval Based on Best-Passage Evidence

In this paper, we assume that the passage-level evidence is the score of the best passage, since it is well-motivated from the concept of passage retrieval and is empirically well-performed. Here, the best passage indicates the passage that maximizes its relevance to query. Let Score(Q, P) be the similarity function between passage P (or document P) and query Q. The passage-level evidence using the best passage is formulated as follows:

$$Score_{Passg}(Q,D) = \max_{P \in SP(D)} Score(Q,P) \tag{1}$$

In above Eq. (1), Q is a given query, $\text{Score}_{\text{passg}}(Q, D)$ indicates the passage-level evidence of document D. SP(D) is the pre-defined set of all possible passages. Let $\text{Score}_{\text{Doc}}(Q, D)$ be the final similarity score of document D. As for $\text{Score}_{\text{Doc}}(Q, D)$, we use the interpolation of the document-level evidence - Score(Q, D) - and the passage-level evidence - $\text{Score}_{\text{passg}}(Q, D)$ - as follows:

$$Score_{Doc}(Q, D) = (1-\alpha)Score(Q, D) + \alpha\ Score_{Passg}(Q, D) \tag{2}$$

where α is the interpolation parameter, controlling the impact of the passage-level evidence on the final similarity score. When α is 1, the final document score is completely substituted to the passage-level evidence only. We call it *pure passage retrieval.* Score(Q,D) is differently defined according to the retrieval model.

Of Eq. (1), note that SP(D) is dependent to types of passage. As for previous several passage types, we use notations of SP_{SEM}, SP_{WIN} and SP_{VAR} for discourse or semantic passage [9], for window passage [6], and for arbitrary passage [7], respectively. SP_{SEM}, SP_{WIN} and SP_{VAR} are defined as follows.

$SP_{SEM}(D)$: It consists of paragraphs, sessions, and automatically segmented semantic passages in D.

$SP_{WIN(N)}(D)$: It consists of N-subsequent words in D. It is further divided to an *overlapped window* and a *non-overlapped window*. In the overlapped window, overlapping among passages is allowed. For example, Callan used the overlapped window by overlapping $N/2$ sub-window in original N-length window to the previous and the next passage [6]. The non-overlapped window does not make any overlapping between passages. Note that the final passage in the end of document may have a shorter length than N for either a non-overlapped passage or an overlapped passage. An overlapped window includes a specialized type of passage – a *completely-overlapped window* which allows overlapping N-1 sub-window to previous and next passages, thus all passages have the same length of N, without having a different length of the passage at the end of document. This completely-overlapped passage is called fixed-length arbitrary passage in Kaszkiel's work [7]. From this point on, After now, we will refer to window passage as completely-overlapped passage.

$SP_{VAR(N1,\ldots Nk)}(D)$: It is defined as union of N_i-length window passages - $\cup$ $SP_{WIN(Ni)}(D)$. Kaszkiel called it variable length arbitrary passage [7]. His experimental results of $SP_{VAR(N1,\ldots Nk)}(D)$ showed the best performance over the other methods.

Now, completely-arbitrary passage, the proposed type of passage, is denoted as $SP_{COMPLETE}(D)$. Formally, $SP_{COMPLETE}(D)$ is defined as follows.

$SP_{COMPLETE}(D)$: It consists of all possible subsequent words in document D, and equivalently denoted as $SP_{VAR(1,\ldots|D|)}(D)$, where $|D|$ is the length of document. It is a special type of arbitrary passage.

Although completely-arbitrary passage is conceptually simple, the number of possible passages in D is $2^{|D|}$-1, which is an exponentially increasing function of the length of document. Therefore, it may be intractable to perform the passage retrieval based on completely-arbitrary passages, since we cannot pre-index all possible arbitrary passages, and cannot compare the scores of all possible passages at retrieval time.

To make the passage retrieval tractable, we propose a *cover-set ranking algorithm* to equivalently find the best passage by comparing cover sets only. Whenever the retrieval function produces the score of document inversely according to the length of the document, we can always use the cover-set ranking algorithm.

2.2 Cover Set Ranking

A cover is defined as follows:

Definition 1 (*Cover*). A cover is a special passage such that all terms at boundary of passage are query terms.

For example, let us consider the following document "*abbbcccbbddaaa*", where we regard *a*, *b*, *c*, and *d* as a term. When *Q* consists of *c* and *d*, *bbbcccbb*, *abbb*, and *ddaaa* are not covers, while *cccbbdd*, *cbbdd*, *c*, *d*, *cc*, *dd* are.

Let SC(*D*) be a set of all covers in *D*. Then, the following equation indicates that the best passage is a cover.

$$\max_{P \in SP_{COMPLETE}(D)} Score(Q,P) = \max_{P \in SC(D)} Score(Q,P) \tag{3}$$

Now, what is the class of retrieval functions which satisfies the above equation? For this question, we define the new class of retrieval function by the *length-normalized scoring function*.

Definition 2 (*Length-normalized scoring function*): Let c(*w*,*D*) be the frequency of term *w* in document *D*, and $|D|$ be the length of document *D*. Let us suppose that term frequencies for all query terms are the same for D_1 and D_2. In other words, $c(q,D_1) = c(q,D_2)$. If the scoring function produces Score(*Q*,*D*) which is inversely proportional to the length of document *D*, that is when $|D_1| < |D_2|$, $Score(Q,D_1) > Score(Q,D_2)$, then we call Score(*Q*,*D*) as a length-normalized scoring function.

We assume the following reasonable statement for the best passage.

Assumption 1 (*Minimum requirement of query term occurrence in the best passage*). The best passage should contain at least one query term.

Now, we can simply prove the following theorem.

Theorem 1. If Score(*Q*,*D*) is a length-normalized scoring function, then the best passage is cover.

Proof. Assume that the best passage is not a cover. Then, among two terms at the boundary of passage, at least one term is not a query term. According to Assumption 1, this best passage contains a query term. We can construct a new passage by decreasing the length of passages until all terms at the boundary become query terms. Then, the length of this new passage will be shorter than that of the best passage. But, since Score(*Q*,*D*) is assumed to be a length-normalized scoring function, this new passage has a more score than the best passage. However, this result contradicts our definition of the best passage, since the best passage should have the maximum score. Thus, the best passage should be a cover. □

Fortunately, most of the retrieval methods such as pivoted normalized vector space, model, probabilistic retrieval model, and language modeling approaches belong to the length-normalized scoring function. Thus, Theorem 1 is satisfied in the most cases.

From Theorem 1, the problem of finding the best passage among completely-arbitrary passages is equivalently converted to the ranking of covers, namely *cover-set ranking*. When *m* is the number of occurrences of query terms in documents i.e. *m*

$= \sum_{q \in Q} c(q,D)$, the number of covers is $2^m - 1 = {}_mC_1 + {}_mC_2 + \ldots + {}_mC_m$. This quantity is much smaller than the original number of completely-arbitrary passages - $2^{|D|} - 1$. For practical implementation of cover-set ranking, we should pre-index term position information for all documents. Note that our concept of cover is slightly different from Clarke's cover [10]. For a given subset T of query terms, Clarke's cover is defined as a *minimal passage* that contains all query terms in T, but does not contain a shorter passage contain T [10].

3 Passage-Based Language Models

Although the proposed passage retrieval is utilized to all retrieval models, we focus on the language modeling approach. The model-dependent part is the scoring function - Score(Q,D) for document D, and Score(Q,P) for passage P. In the language modeling approaches, the relevance of passage P is similarly defined as log-likelihood of query of passage P. Thus, the same smoothing techniques as document models are utilized to estimate passage models. From now on, we will briefly review two popular smoothing techniques [4]. In this section, we assume that a document also becomes a passage (a large passage). Jelinek-Mercer smoothing is the linear interpolation of a document model from MLE (Maximum Likelihood Estimation) and the collection model as follows:

$$P(w \mid \theta_D) = (1-\lambda)P(w \mid \hat{\theta}_D) + \lambda P(w \mid \theta_C) \tag{4}$$

where $\hat{\theta}_D$ indicates the document model from MLE, and θ_C is the collection model.

Dirichlet-prior smoothing enlarges the original document sample by adding a pseudo document of length μ, and then uses the enlarged sample to estimate the smoothed model. Here, the pseudo document has the term frequency, which is in proportional to the term probability of the collection model.

$$P(w \mid \theta_D) = \frac{c(w,D) + \mu P(w \mid \theta_C)}{|D| + \mu} \tag{5}$$

We can easily show that the Score(Q,D) using two smoothing methods is a length-normalized scoring function. The following presents the brief proof that Dirichlet-prior smoothing method is a length-normalized scoring function.

Proof. Dirichlet-prior smoothing induces the length-normalized scoring function. The retrieval scoring function of Dirichlet-prior smoothing is written by.

$$Score(Q,D) \propto \log P(Q \mid \theta_D) = \sum_q \log P(w \mid \theta_D) \tag{6}$$

If term frequencies of query term q are the same for document D_1 and D_2, that is c(q,D_1) is c(w,D_2), and $|D_1| < |D_2|$, then

$$P(w \mid \theta_{D_1}) = \left(\frac{c(w,D_1) + \mu P(w \mid \theta_C)}{|D_1| + \mu} \right) > \left(\frac{c(w,D_2) + \mu P(w \mid \theta_C)}{|D_2| + \mu} \right) = P(w \mid \theta_{D_2})$$

From Eq. (6), Score(Q,D_1) > Score(Q,D_2). Thus, Dirichlet-prior smoothing belongs to a length-normalized scoring function. Similarly, we can show that Jelinek-Mercer smoothing is also length-normalized one. □

4 Experimentation

4.1 Experimental Setting

For evaluation, we used six TREC test collections. Table 1 summarizes the basic information of each test collection. TREC4-AP is the sub-collection of Associated Press in disk 1 and disk 2 for TREC4. In columns, *#Doc*, *avglen*, *#Terms*, *#Q*, and *#R* are the number of documents, the average length of documents, the number of terms, and the number of topics, and the number of relevant documents, respectively.

Table 1. Collection summaries

Collection	*# Doc*	*avglen*	*# Term*	*# Q*	*# R*
TREC4-AP	158,240	156.61	268,232	49	3,055
TREC7	528,155	154.6	970,977	50	4,674
TREC8				50	4,728
WT2G	247,491	254.99	2,585,383	50	2,279
WT10G	1,692,096	165.16	13,018,003	50	2,617
WT10G-2				50	3,363

The standard method was applied to extract index terms; We first separated words based on space character, eliminated stopwords, and then applied Porter's stemming. For all experiments, we used Dirichlet-prior for document model due to its superiority over Jelinek-Mercer smoothing. For passage models, Jelinek-Mercer and Dirichlet-prior smoothings were applied, namely PJM and PDM, respectively. Similarly, for document model, we call JM for Jelinek-Mercer smoothing and DM for Dirichlet-prior smoothing. Interpolation methods using passage-level evidence are denoted by DM+PJM when DM and PJM are combined, and by DM+PDM when DM and PDM are combined.

As for retrieval evaluation, we use MAP (Mean Average Precision), Pr@5 (Precision at 5 documents), and Pr@10 (Precision at 10 documents).

4.2 Passage Retrieval Using DM+PDM, DM+PJM

First, we evaluated DM+PDM based on PDM using different μs (among 0.5, 1, 5, 10, 50 and 70). For queries, we used title field, except for TREC4-AP using description field. Table 2 shows retrieval performances of DM+PDM when the best interpolation parameter α is used, compared to DM [1]. DM indicates the baseline using document-level evidence based on Dirichlet-prior smoothing for estimating document model, in which the smoothing parameter μ is selected as the best performed one (among 22 different values between 100 and 30000). Bold-face numbers indicate the best performance in each test collection. To check whether or not the proposed method

[1] To find the best α, we examined 9 different values between 0.1 and 0.9.

(DM+PDM) significantly improves the baseline (DM), we performed the Wilcoxon sign ranked test to examine at 95% and 99% confidence levels. We attached ↑ and ¶ to the performance number of each cell in the table when the test passes at 95% and 99% confidence level, respectively. As shown in Table 2, DM+PDM clearly shows the better performance than DM on many parameters as well as the best parameter. We can see that the best performances for each test collection are obtained for extremely small μs which are less than 10. Although performances are different across μ, their sensitivity of μ is not much serious.

Table 2. Performances (MAP) of DM+PDM across different μs (smoothing parameter for passage language model), compared to DM (↑ and ¶ indicate a statistical significance respectively for 95% and 99% confidence level in the Wilcoxon signed rank test)

Collection	DM	μ=0.5	μ=1	μ=5	μ=10	μ=50	μ=70
TREC4-AP	0.2560	0.2632↑	0.2651¶	**0.2668↑**	0.2652	0.2595	0.2580
TREC7	0.1786	0.1910↑	0.1903¶	**0.1916¶**	0.1881↑	0.1848¶	0.1844¶
TREC8	0.2480	**0.2545**	0.2538↑	0.2536	0.2535	0.2528	0.2517
WT2G	0.3101	0.3480¶	**0.3485¶**	0.3451¶	0.3436¶	0.3381¶	0.3378¶
WT10G	0.1965	0.2101¶	0.2094¶	**0.2129¶**	0.2116	0.2098	0.2091
WT10G-2	0.1946	0.1983	**0.1989**	0.1966	0.1971	0.1953	0.1949

In the second experiment, we evaluated DM+PJM by using PJM across different smoothing parameters λ between 0.1 and 0.9. Table 3 shows the retrieval performance of DM+PJM when the best interpolation parameter α for each λ is used. As shown in Table 3, the best λ value is small, except for TREC4-AP. This exception is due to the characteristics of the description field of query topic used for TREC4-AP.

Table 3. Performances (MAP) of DM+PJM across different λs (smoothing parameter for passage language model), compared to DM

Collection	DM	λ=0.1	λ=0.2	λ=0.3	λ=0.5	λ=0.7	λ=0.9
TREC4-AP	0.2560	0.2622	0.2639↑	0.2654	0.2661↑	0.2685¶	**0.2706↑**
TREC7	0.1786	0.1924↑	0.1927¶	0.1927¶	**0.1931¶**	0.1929¶	0.1931¶
TREC8	0.2480	**0.2541↑**	0.2538↑	0.2536↑	0.2534↑	0.2537↑	0.2529↑
WT2G	0.3101	0.3480¶	**0.3488¶**	0.3485¶	0.3488¶	0.3484¶	0.3441¶
WT10G	0.1965	0.2110¶	0.2114¶	0.2117¶	**0.2121¶**	0.2115↑	0.2117¶
WT10G-2	0.1946	0.1972	0.1980	0.1988	0.2001	0.2016	**0.2017↑**

The performance improvement is similar to DM+PDM. At the best, there is no serious difference between DM+PDM and DM+PJM. In TREC4-AP, the retrieval performance is highly sensitive on smoothing parameter for DM+PDM and DM+PJM, since the description type of queries requires the smoothing which mainly plays a query modeling role.

4.3 Pure Passage Retrieval Versus Interpolated Passage Retrieval

In this experiment, we evaluated the pure passage retrieval (α is fixed to 1 in Eq. (2)) by comparing it to the previous result of the interpolation method (DM+PDM or

Table 4. Performance (MAP) of PDM (Pure passage retrieval) across different smoothing parameters (μs)

Collection	DM	μ=10	μ=50	M=100	μ=500	μ=1000	μ=3000
TREC4-AP	0.2560	0.1887	0.2024	0.2114	0.2382	0.2452	**0.2514**
TREC7	0.1786	0.1594	0.1688	0.1731	0.1757	**0.1761**	0.1704
TREC8	0.2480	0.2155	0.2322	0.2395	**0.2480**	0.2462	0.2333
WT2G	0.3101	0.2862	0.3125	0.3185	**0.3325**↑	0.3277¶	0.3155
WT10G	0.1965	0.1551	0.1825	0.2004	**0.2096**	0.1994	0.1963
WT10G-2	0.1946	0.1298	0.1493	0.1571	0.1767	0.1783	0.1744

DM+PJM). We used the same queries as section 4.2. Table 4 shows the retrieval performance of the pure passage retrieval (i.e. PDM) among different smoothing parameters.

We can see from Table 4 that pure passage retrieval based on PDM does not improve full document retrieval using DM. Although there is some improvement of pure passage retrieval on WT2G and WT10G, this result is not statistically significant. For this reason, we think, the interpolation method can efficiently divide the roles for making a good retrieval function into document models and passage models by differentiating smoothing parameters. However, the pure method should reflect all relevant features into a single smoothing parameter only. In the pure method, smoothing parameter of the passage model is moved towards the best parameter of document model, such that it also plays the roles of document model.

4.4 Comparison with Kaszkiel's Variable-Length Arbitrary Passages and Tao's Proximity Method

We compared different passage retrievals using completely-arbitrary passage and Kaszkiel's variable-length arbitrary passages [7]. Kaszkiel's arbitrary passages allow 12 different lengths of passages from 50 to 600. According to our notation, Kaszkiel's arbitrary passages correspond to $SP_{VAR(50,100,150,\ldots 550,600)}(D)$. Different from Kaszkiel's original experiment, we did not set the maximum length of passage by 600. Instead, we allowed the length of passage to the full length of the document. This slight modification is trivial, because it does not lose the original concept of arbitrary passage.

We formulate Kaszkiel's arbitrary passages to $SP_{VAR2(U)}(D)$, U-dependent set of passages that is equal to $SP_{VAR(U,2U,3U,4U,\ldots)}(D)$ in the section 2.1 where U means the increment unit of passage length. In our notation, Kaszkiel's arbitrary passages correspond to $SP_{VAR2(50)}(D)$ using U of 50, and our completely-arbitrary passage ($SP_{COMPLETE}(D)$) corresponds to $SP_{VAR2(50)}(D)$ using U of 1. Note that parameter U in $SP_{VAR2(U)}(D)$ can control the degree of completeness of arbitrary passages. In other words, as U is changed from 50 to 1, $SP_{VAR2(U)}(D)$ becomes more close to completely-arbitrary passages from simple arbitrary passage.

Table 5 and Table 6 show results of passage retrievals (DM+PJM) for different Us from 1 and 50 when using keyword queries (title field), and verbose queries (description field), respectively. For smoothing parameter of PJM, we used 0.1 for keyword queries, and 0.99 for verbose queries. Exceptionally, we used 0.9 for keyword queries in WT10G-2, and 0.9 for TREC4-AP, in order to obtain a more improved performance.

Table 5. For keyword queries, performance (MAP) of DM+PJM using variable-length arbitrary passages $P_{VAR2(U)}(D)$ by changing U. Our completely-arbitrary passage and Kaszkiel's arbitrary passages corresponds to the cases when U is 1 and 50, respectively.

Collection	DM	U = 1	U = 3	U = 5	U = 10	U = 20	U = 30	U = 50
TREC7	0.1786	**0.1924**↑	0.1899	0.1874	0.1855	0.1842	0.1835	0.1821
TREC8	0.2480	**0.2541**↑	0.2525	0.2517	0.2516	0.2511	0.2506	0.2507
WT2G	0.3101	**0.3480**¶	0.3449¶	0.3387¶	0.3342↑	0.3313↑	0.3268¶	0.3251¶
WT10G	0.1965	**0.2110**¶	0.2082¶	0.2069↑	0.2042	0.2006	0.1993	0.1951
WT10G-2	0.1946	**0.2017**↑	0.2012	0.2003	0.1980	0.1978	0.1974	0.1973

Table 6. For verbose queries, performance (MAP) of DM+PJM using variable-length arbitrary passages $P_{VAR2(U)}(D)$. Our completely-arbitrary passage and Kaszkiel's arbitrary passages corresponds to the cases when U is 1 and 50, respectively.

Collection	DM	U = 1	U = 3	U = 5	U = 10	U = 20	U = 30	U = 50
TREC4-AP	0.2560	**0.2706**↑	0.2680↑	0.2672¶	0.2647↑	0.2591	0.2581	0.2573
TREC7	0.1791	**0.1960**↑	0.1930	0.1895	0.1877	0.1868	0.1855	0.1851
TREC8	0.2294	0.2404	0.2409	**0.2411**	0.2401	0.2387	0.2386	0.2374
WT2G	0.2854	**0.3282**¶	0.3257	0.3236	0.3170	0.3130	0.3100	0.3062
WT10G	0.1955	**0.2462**¶	0.2436¶	0.2443¶	0.2397¶	0.2340↑	0.2202↑	0.2263↑
WT10G-2	0.1866	0.2008¶	0.2005	**0.2033**	0.1999	0.1992	0.1978	0.1967

As shown in tables, Kaszkiel's arbitrary passages (when U is 50), statistically significant improvement is found in only WT2G test collection. Overall, as U is smaller, the performances become better, and show more frequently statistically significant improvements.

As mentioned in introduction, the superiority of the completely-arbitrary passage to Kaszkiel's arbitrary passage can be explained by effects of proximity of query terms. We know that this proximity can be well-reflected in the completely-arbitrary passage, since it allows any possible lengths of passages. On the other hand, Kaszkiel's arbitrary passage has the restriction of lengths (i.e. such as at least 50 words), so it cannot fully support effects of proximity. Overall, as U becomes larger, the passage retrieval will more weakly reflect effects of the proximity. From the viewpoint of proximity, the passages with fixed-lengths such as Kaszkiel's one has limitation to obtain a better retrieval performance.

For further comparison, we re-implemented Tao's work [11], which is a recent work using the proximity. Tao's approach can be described by modifying Eq. (2) as follows [11].

$$Score_{Doc}(Q,D) = (1-\alpha)Score(Q,D) + \alpha\,\pi(Q,D)$$

where $\pi(Q,D)$ is the proximity-based additional factor. Tao proposed $\log(\beta + \exp(\delta(Q,D)))$ for $\pi(Q,D)$ where $\delta(Q,D)$ indicates proximity of query terms in document D. We used Tao's *MinDist* for $\delta(Q,D)$ due to its better performance. We performed several different runs using various parameters α from 0.1 to 0.9, and β from 0.1 to 1.5, and selected the best run for each test collection. Table 7 shows Tao's proximity-based result, comparing with two passage retrievals using Kaszkiel's arbitrary passage (U = 50) and our completely-arbitrary passage (U = 1), respectively.

Table 7. For Keyword queries, comparison with Tao's Proximity, Kaszkiel's Arbitrary Passage, and completely-arbitrary passage (MAP, Pr@10). In TREC4-AP, verbose queries are used. For Tao's proximity method and Kaszkiel's method (arbitrary passage), statistical significant symbols (↑ for 95% and ¶ for 99%) are attached by regarding DM as the baseline. Note that, for our method (completely-arbitrary passage), statistical significant symbols are attached by regarding Tao's method (↑ for 95% and ¶ for 99%) and Kaszkiel's method († for 95% and ‡ for 99%) as the baseline, respectively.

Collection	DM (baseline)	Tao's Proximity [11]		Kaszkiel's Arbitrary Passage (U = 50) [7]		Completely-Arbitrary Passage (U = 1)	
	MAP	MAP	Pr@5	MAP	Pr@5	MAP	Pr@5
TREC4-AP	0.2560	0.2624↑	0.4408	0.2573	0.4367	**0.2706 †**	**0.4408**
TREC7	0.1786	0.1871	0.4480	0.1821	0.4640	**0.1924 ‡**	**0.4680**
TREC8	0.2480	0.2506↑	0.4840	0.2507	0.4800	**0.2541 †**	**0.4880**
WT2G	0.3101	0.3165	0.5200	0.3251¶	0.5200	**0.3480¶‡**	**0.5520**
WT10G	0.1965	0.2013¶	0.2960	0.1951	0.3080	**0.2110↑‡**	**0.3000**
WT10G-2	0.1946	0.1965	0.3520	0.1973	0.3480	**0.2017**	**0.3880↑†**

For Tao's proximity method and Kaszkiel's method (arbitrary passage), statistical significant symbols (↑ for 95% and ¶ for 99%) are attached by regarding DM as the baseline. We can see from Table 7 that Tao's proximity-based method does show significant improvement over the baseline (DM) only in TREC4-AP, TREC8, and WT10G, showing less-significant effectiveness in other test collections. In terms of effectiveness, Tao's proximity method is better than Kaszkiel's method. Remarkably, the proposed approach showed the significant improvement for all test collections. From the viewpoint of improvement over DM, the proposed approach is clearly better than these two methods.

5 Conclusion

This paper proposed a completely-arbitrary passage as a new type of passage, presented a cover-set ranking to efficiently perform the passage retrieval based on this new type, and formulated the passage retrieval in the context of language modeling approaches. Experimental results showed that the proposed passage retrieval consistently shows the improvement in most standard test collections. From comparison of the pure passage retrieval and the interpolation method, we confirmed that the interpolation method is better than the pure version. The smoothing role in passage language models tends to be similar to the role of document language models, differentiating the best smoothing parameters for keyword queries and verbose queries. In addition, we showed that our passage retrieval using completely-arbitrary passage is better than those using Kaszkiel's arbitrary passage [7], as well as Tao's method [11].

The strategy which this work adopts is to use the best passage only. However, the best passage cannot fully cover all contents in documents, since query-relevant contents may separately appear by multiple-passages in documents, not by single best passage. It is especially critical to long-length queries, which consists of several number of query terms, since this type of query causes more possibility of such separation. In this regard, one challenging research issue would be to develop a new passage retrieval method using multiple-passages for reliable retrieval performance, and to examine its effectiveness.

Acknowledgements

This work was supported by the Korea Science and Engineering Foundation (KOSEF) through the Advanced Information Technology Research Center (AITrc), also in part by the BK 21 Project and MIC & IITA through IT Leading R&D Support Project in 2007.

References

1. Singhal, A., Buckley, C., Mitra, M.: Pivoted document length normalization. In: SIGIR 1996: Proceedings of the 19th annual international ACM SIGIR conference on Research and development in information retrieval, pp. 21–29 (1996)
2. Robertson, S.E., Walker, S.: Some simple effective approximations to the 2-poisson model for probabilistic weighted retrieval. In: SIGIR 1994: Proceedings of the 17th annual international ACM SIGIR conference on Research and development in information retrieval, pp. 232–241 (1994)
3. Ponte, J.M., Croft, W.B.: A language modeling approach to information retrieval. In: SIGIR 1998: Proceedings of the 21st annual international ACM SIGIR conference on Research and development in information retrieval, pp. 275–281 (1998)
4. Zhai, C., Lafferty, J.: A study of smoothing methods for language models applied to ad hoc information retrieval. In: SIGIR 2001: Proceedings of the 24th annual international ACM SIGIR conference on Research and development in information retrieval, pp. 334–342 (2001)
5. Salton, G., Allan, J., Buckley, C.: Approaches to passage retrieval in full text information systems. In: SIGIR 1993: Proceedings of the 16th annual international ACM SIGIR conference on Research and development in information retrieval, pp. 49–58 (1993)
6. Callan, J.: Passage-level evidence in document retrieval. In: SIGIR 1994: Proceedings of the 17th annual international ACM SIGIR conference on Research and development in information retrieval, pp. 302–310. Springer-Verlag New York, Inc., New York (1994)
7. Kaszkiel, M., Zobel, J.: Effective ranking with arbitrary passages. Journal of the American Society for Information Science and Technology (JASIST) 52(4), 344–364 (2001)
8. Liu, X., Croft, W.B.: Passage retrieval based on language models. In: CIKM 2002: Proceedings of the eleventh international conference on Information and knowledge management, pp. 375–382 (2002)
9. Hearst, M.A., Plaunt, C.: Subtopic structuring for full-length document access. In: SIGIR 1993: Proceedings of the 16th annual international ACM SIGIR conference on Research and development in information retrieval, pp. 59–68 (1993)
10. Clarke, C.L.A., Cormack, G.V., Tudhope, E.A.: Relevance ranking for one to three term queries. Inf. Process. Manage. 36(2), 291–311 (2000)
11. Tao, T., Zhai, C.: An exploration of proximity measures in information retrieval. In: SIGIR 2007: Proceedings of the 30th annual international ACM SIGIR conference on Research and development in information retrieval, pp. 295–302 (2007)

Semantic Discriminative Projections for Image Retrieval

He-Ping Song, Qun-Sheng Yang, and Yin-Wei Zhan

Faculty of Computer, GuangDong University of Technology
510006 Guangzhou, P.R. China
hepingsong@gmail.com, jsjqsy@sina.com, ywzhan@gdut.edu.cn

Abstract. Subspace learning has attracted much attention in image retrieval. In this paper, we present a subspace learning approach called Semantic Discriminative Projections (SDP), which learns the semantic subspace through integrating the descriptive information and discriminative information. We first construct one graph to characterize the similarity of contented-based features, another to describe the semantic dissimilarity. Then we formulate constrained optimization problem with a penalized difference form. Therefore, we can avoid the singularity problem and get the optimal dimensionality while learning a semantic subspace. Furthermore, SDP may be conducted in the original space or in the reproducing kernel Hilbert space into which images are mapped. This gives rise to kernel SDP. We investigate extensive experiments to verify the effectiveness of our approach. Experimental results show that our approach achieves better retrieval performance than state-of-art methods.

1 Introduction

With the development of digital imaging technology and the popularity of World Wide Web, Gigabytes of images are generated every day. It is a challenge that manage effectively images visual content. Content Based Image Retrieval (CBIR) is receiving research interest for this purpose [1,2,3,4]. However, there are still many open issues to be solved. Firstly, the visual content such as color, shape, texture, is extracted from an image as feature vectors. The dimensionality of feature space is usually very high. It ranges from tens to hundreds of thousands in most cases. Traditional machine learning approaches fail to learn in such a high-dimensional feature space. This is the well-known curse of dimensionality. Secondly, the low-level image features used in CBIR are often visual characterized, but it doesn't exist the directly connection with high-level semantic concepts, i.e. so-called semantic gap.

To alleviate the open issues, more and more attention has been drawn on the dimensionality reduction techniques. ISOMAP [5], Locally Linear Embedding (LLE) [6] and Laplacian eigenmaps [7] usher in manifold learning, these algorithms discover the intrinsic structure and preserve the local or global property of training data. Tenenbaum et al. [5] uses geodesic distance instead of Euclidean

H. Li et al. (Eds.): AIRS 2008, LNCS 4993, pp. 34–43, 2008.

distance to estimate distance between points. Multidimensional Scaling [8] is applied to discover the embedding space. Saul et al. [6] assumes there are smooth local patches that could be approximately linear, meanwhile a point in the training space could be approximated by linear combination of its neighbors. The projected space minimizes the reconstruction error using neighborhood correlation. Laplacian eigenmaps preserves locality information and makes neighbors close to each other in the projected space. These algorithms are unsupervised and limited to a nonlinear map. It is hard to map entire data space to low-dimensional space.

Locality Preserving Projections (LPP) [9], and Local Discriminant Embedding (LDE) [10] are proposed to extend the nonlinear learning approaches. These algorithms are all motivated by Laplacian eigenmaps. He et al. [9] uses a neighborhood graph to characterize locality preserving property that nearest neighbors in the original space should be nearest neighbors in the projected space. LDE constructs two neighborhood graphs, one prevents neighbors from different category and another preserves the locality through the affinity matrix using neighborhood information. LPP and LDE are effectively used to map data in entire image space. But only one neighborhood graph is used to discover the intrinsic structure, and LLE doesn't utilize the label information. LDE only keeps the neighborhood images from different classes away. LPP and LDE need to compute the inverse matrix, suffering from the singularity problem.

Bridge low-level visual feature to the high-level semantic is a great challenge in CBIR. We use Laplacian to learn the images semantic subspace in order to achieve more discriminative image representation for CBIR. In our work, both visual similarity and semantic dissimilarity are applied to construct neighborhood graph since they not only contain the descriptive information of the unlabeled images but also the discriminative information of the labeled images utilized in learning. We introduce a penalty γ to formulate a constrained optimization problem in the difference form, so that the optimal projection can be found by eigenvalue decomposition. Information of conjunctive graphs is represented by a affinity matrix, and it is much more computationally efficient in time and storage than LPP and LDE. On the other hand,the learnt subspace can preserve both local geometry and relevance information. Previous works often neglect the singularity problem and the optimal dimensionality, but we will determine the optimal dimensionality and avoid the singularity problem simultaneously.

This paper is organized as follows. In section 2, we introduce the SDP approach, kernel trick is used to the nonlinear learning approach in section 3, followed by the experiment results are discussed in section 4, and lastly we conclude our paper in section 5.

2 Laplacian Based Subspace Learning

In this section, we introduce our learning approach for image retrieval which respects the local descriptive and discriminative information of the original image space.

Suppose n training images, $\{\mathrm{x}_i\}_{i=1}^n \in \mathbb{R}^D$. we can construct two graph. G^S denotes the semantic similarity via semantic label and G^V denotes the visual resemblance by exploring the neighborhood of each image in the geometric space. W^S and W^V denotes the affinity matrix of G^S and G^V respectively. W^S and W^V are computed as follows:

$$W_{ij}^S = \begin{cases} 1 & \mathrm{x}_i\ ,\mathrm{x}_j \text{ share the same class label,} \\ 0 & \text{otherwise.} \end{cases} \tag{1}$$

$$W_{ij}^V = \begin{cases} 1 & \mathrm{x}_i \in k\text{-NN of } \mathrm{x}_j \text{ or } \mathrm{x}_j \in k\text{-NN of } \mathrm{x}_i, \\ 0 & \text{otherwise.} \end{cases} \tag{2}$$

Where $k-\mathrm{NN}(\cdot)$ denotes the k-nearest-neighbors.
We integrate two kinds of information:

$$\begin{aligned} W^- &= \overline{W^S} \\ W^+ &= W^S \wedge W^V \end{aligned} \tag{3}$$

Where "$\wedge$" denotes the Meet of two zero-one matrices.
We utilize the penalized difference form to formulate following constrained optimization problem.

$$P = \arg \max_{P^T P = I} \sum_{i,j} \|P^T \mathrm{x}_i - P^T \mathrm{x}_j\|^2 (W_{ij}^- - \gamma W_{ij}^+) \tag{4}$$

where γ is a penalized coefficient, the constraint $P^T P = I$ avoids trivial solution, and I is the $d \times d$ identity matrix, d is the reduced dimensionality.

The above formulation exhibits the implication that local neighbors with semantic dissimilarity should separate each other and different semantic classes are far away from each other; the images with similar semantic and visual content will be clustered together, preserving the intrinsic structure.
We rewrite (4) in the form of trace, and get the following formulation:

$$\begin{aligned} \mathcal{J} &= \sum_{i,j} \|P^T \mathrm{x}_i - P^T \mathrm{x}_j\|^2 (W_{ij}^- - \gamma W_{ij}^+) \\ &= \sum_{i,j} tr\{(P^T \mathrm{x}_i - P^T \mathrm{x}_j)(P^T \mathrm{x}_i - P^T \mathrm{x}_j)^T\}(W_{ij}^- - \gamma W_{ij}^+) \\ &= \sum_{i,j} tr\{P^T (\mathrm{x}_i - \mathrm{x}_j)(\mathrm{x}_i - \mathrm{x}_j)^T P\}(W_{ij}^- - \gamma W_{ij}^+) \\ &= tr\{P^T \sum_{i,j} (\mathrm{x}_i - \mathrm{x}_j)(W_{ij}^- - \gamma W_{ij}^+)(\mathrm{x}_i - \mathrm{x}_j)^T P\} \\ &= 2tr\{P^T [(XD^- X^T - XW^- X^T) - \gamma(XD^+ X^T - XW^+ X^T)] P\} \\ &= 2tr\{P^T X (L^- - \gamma L^+) X^T P\} \end{aligned} \tag{5}$$

Where $L^- = D^- - W^-$, and D^- is a diagonal matrix with $D_{ii}^- = \sum_j W_{ij}^-$. Analogously, $L^+ = D^+ - W^+$ with $D_{ii}^+ = \sum_j W_{ij}^+$. Thus the optimization problem can be formulated as:

$$
\begin{aligned}
P = &\quad \arg \max_{P^T P = I} tr\{P^T X(L^- - \gamma L^+)X^T P\} \\
\Rightarrow P = &\qquad\qquad \mathrm{eig}(X(L^- - \gamma L^+)X^T)
\end{aligned}
\tag{6}
$$

We take no dimensionality reduction as the baseline, therefore (5) could be zero. We could get a positive scalar γ when the dimensionality is reduced [11]. Then we have:

$$
\mathcal{J} = tr(X(L^- - \gamma L^+)X^T) = 0 \quad \Longrightarrow \quad \gamma = \frac{trL^-}{trL^+} \tag{7}
$$

It is obviously that $X(L^- - \gamma L^+)X^T$ is a $D \times D$, sparse, symmetric and positive semidefinite matrix. According to the result of Rayleigh quotient, the optimization problem can be calculated by eigenvalue decomposition.

Denote $P \in \mathbb{R}^{D \times d}$ by $P = [p_1, p_2, \cdots, p_d]$, where $p_i(i = 1, \cdots, d)$ is the d largest eigenvectors corresponding to the d largest eigenvalues of $X(L^- - \gamma L^+)X^T$. $\sum \lambda_i$ is the optimal value of the above optimization problem, where $\lambda_i(i = 1, \cdots, d)$ are the d largest eigenvalues. d is the number of positive eigenvalues and $\mathcal{J}$ reaches the maximum.

We can see that the singularity problem in LPP, LDE does not exist in our approach, meanwhile we find the optimal dimensionality.

To get returns for the query in image retrieval, we project any test image $\overline{\mathrm{x}} \in \mathbb{R}^D$ to $\mathbb{R}^d$ via $\overline{\mathrm{y}} = P^T \overline{\mathrm{x}}$ and will find the nearest neighbors of Euclidean distances. Those images corresponding to the nearest neighbors will be the top-ranking returns.

3 Kernel SDP

As the kernel trick [12] successfully applied to Kernel LPP [13], Kernel LDE [10], we generalize SDP to kernel SDP, in which kernel transformation is applied to handle nonlinear data.

Denote $\Phi : \mathbb{R}^D \rightarrow \mathcal{F}$ is a nonlinear mapping, so the image feature vectors in the high-dimensionality feature space is denoted as $\Phi(\mathrm{x}_i), (i = 1, \cdots, n)$. The inner product in $\mathcal{F}$ can be computed by the kernel function. we specify the RBF kernel $k(\mathrm{x}_i, \mathrm{x}_j) = \Phi(\mathrm{x}_i)^T \Phi(\mathrm{x}_j) = exp(-\|\mathrm{x}_i - \mathrm{x}_j\|^2/t)$ in our work. we find the optimal projection $V, (\mathrm{v}_i, i = 1, \cdots, d)$ in $\mathcal{F}$, the v_i is spanned by $\Phi(\mathrm{x}_i), i = 1, \cdots, n$, and assume v_i is the linear combination of $\Phi(\mathrm{x}_i)$ in the projected space $\mathcal{F}$:

$$
\mathrm{v}_i = \sum_{j=1}^{n} \alpha_{ij} \Phi(\mathrm{x}_j) = \Phi(\mathrm{X}) \alpha_i \tag{8}
$$

we have:

$$
(\mathrm{V}^T \Phi(\mathrm{X}))_{ij} = \mathrm{v}_i^T \Phi(\mathrm{x}_j) = (AK)_{ij} \tag{9}
$$

where $A = [\alpha_i, \cdots, \alpha_n]^T$ denotes the linear coefficient in vector, $K_{ij} = k(\mathrm{x}_i, \mathrm{x}_j)$ is kernel matrix.

Replacing X with $\Phi(X)$, we rewrite (5), and consider the kernel-based optimization problem:

$$\begin{aligned} A = & \arg\max_{A^T A = I} tr\{AK(L^- - \gamma L^+)KA\} \\ \Rightarrow A = & \quad \text{eig}(K(L^- - \gamma L^+)K) \end{aligned} \tag{10}$$

where the constraint $A^T A = I$ avoids trivial solution, I is the $d \times d$ identity matrix.

Analogously according to the result of Rayleigh quotient, the optimization problem can be calculated by eigenvalue decomposition. We select the d largest eigenvalues of $K(L^- - \gamma L^+)K$, where d is the number of positive eigenvalues. Our approach doesn't suffer from the singularity problem, and get the optimal dimensionality.

To get returns for the query in image retrieval, we map any test image $\overline{\mathrm{x}}$ to by $\overline{\mathrm{y}} = V^T \overline{\mathrm{x}}$ with the ith dimensionality computed by $\overline{\mathrm{y}}_i = \mathrm{v}_i^T \overline{\mathrm{x}} = \sum_{i=1}^{n} \alpha_{ij} k(\mathrm{x}_j, \overline{\mathrm{x}})$, and find the nearest neighbors of Euclidean distances. Those images corresponding to the nearest neighbors will be the top-ranking returns.

4 Experimental Results

In this section, we experimentally evaluate the performance of SDP on COREL dataset and compare with LPP, LDE in order to demonstrate effectiveness of our approach for image retrieval. The COREL dataset is widely used in many CBIR systems [18]. In our experiments, the dataset consists of 2500 color images, including 25 categories, each category contains 100 samples. Those images in the same category share the same semantic concept, but they have their individual varieties. Images from the same category are considered relevant, and otherwise irrelevant.

In our experiments, we only consider these queries which don't exist in the training images. Five-fold cross validation is used to evaluate the retrieval performance. We pick one set as the testing images, and leave the other four sets as the training images. Table 1 shows the features of which the dimensionality is 145.

Precision-Recall curve (PRC) is widely used as a performance evaluation metrics for image retrieval [19]. In many cases, PRC is overlapped in high recall,

Table 1. Image features used in experiment

Name	Description &Dimension
Normalized Color Histogram [14]	HSV Space 64 dimension
Normalized LBP Histogram [15]	HSV Space 54 dimension
Normalized Color Moments [16]	HSV Space 9 dimension
Normalized Wavelet Moments [17]	RGB Space 18 dimension

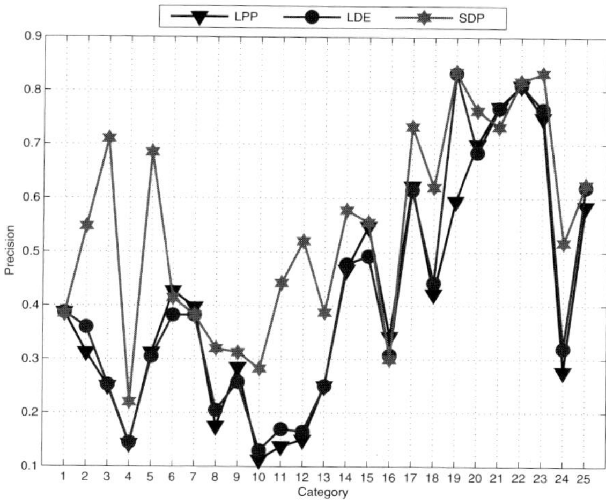

Fig. 1. The plots of precision versus category for SDP, LDE, and LPP

moreover, it is unreasonable to calculate the mean average precision among different categories. We alternatively adopt the precision-category. Given a specified number N, we define the precision as following:

$$precision = \frac{\text{The number of relevant images in top } N \text{ returns}}{N}$$

Users are usually interested in the first screen results just like Google^{TM} Image. We have N=20 in our experiments.

4.1 Comparisons with LPP, LDE

Model Selection is very important in many subspace learning algorithms. In our experiments, it is not very sensitive to tuning parameter k. we set $k = 10$. We adopt Gaussian heat kernel to compute the W_{ij}, $W_{ij} = exp(-\|x_i - x_j\|^2/c)$, where c is a positive scalar. the aforementioned W_{ij} is superior to 0/1 and is not sensitive to c.

In this experiment, we compare SDP with LPP, LDE. LPP and LDE have the limitation of singularity problem due to compute inverse matrix. Both LPP and LDE adopt PCA to overcome the limitation, retaining the 98% principal components [13]. The optimal dimensionality of SDP is 64, as shown in Fig. 1. SDP achieves much better retrieval performance than other approaches. As a matter of fact, we gain much more discriminating image representation by SDP. Next, we give the experiment result of Kernel SDP in Fig. 2, the optimal dimensionality of Kernel SDP is 145, Except for the category 15, Kernel LDE performs marginally better than Kernel SDP. We can conclude that Kernel SDP outperforms other approaches.

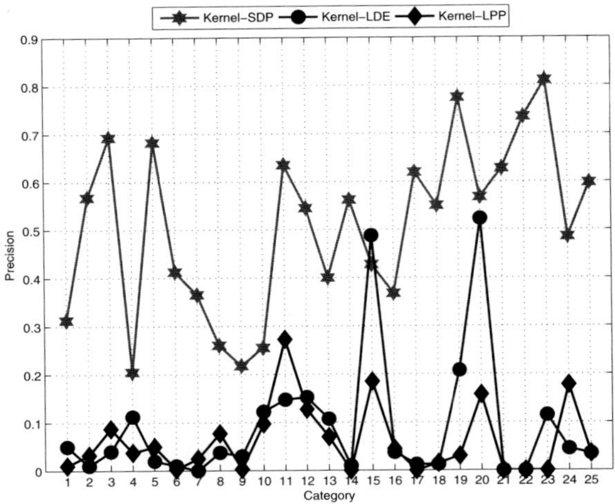

Fig. 2. The plots of precision versus category for Kernel SDP, Kernel LDE, and Kernel LPP

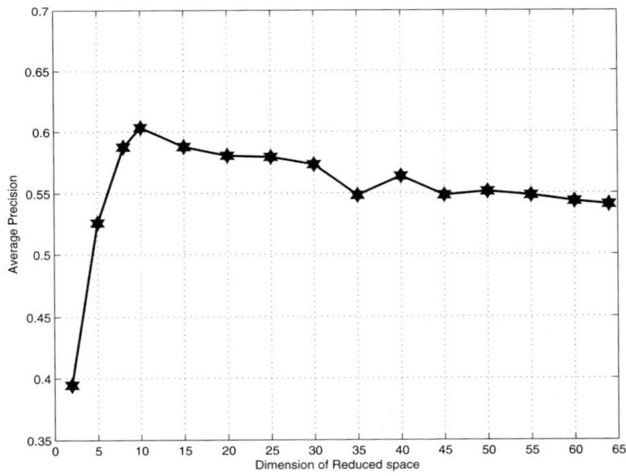

Fig. 3. Average Precision in reduced space with different dimension for SDP

4.2 Reduced Dimensionality

Even though our approach can determine the optimal dimensionality, the dimensionality of reduced space is important tradeoff between retrieval precision and the computational complexity. In this experiment, we investigate the relation between dimension and precision. The precision of SDP reaches its peak at 10 dimensions in Fig. 3. The precision of Kernel SDP converges rapidly from 2 to 10 dimensions, and then achieves reliable results from 10 to 64 dimensions. As shown in Fig. 4, The precision reaches its peak at 30 dimensions, converges

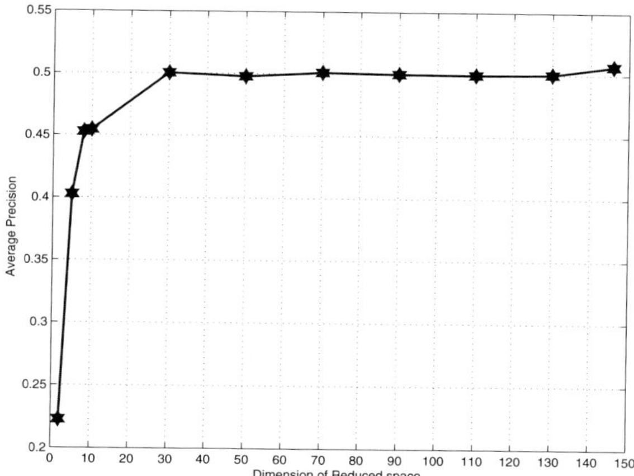

Fig. 4. Average Precision in reduced space with different dimension for Kernel SDP

rapidly from 2 to 30 dimensions, and then achieves smooth results from 30 to 145 dimensions. As shown in Fig. 3 and Fig. 4, we can gain the lower dimensionality while the precision is stable, even relatively higher.

5 Conclusions and Future Work

We have introduced a subspace learning approach SDP for image retrieval. The image structure is approximated by the adjacent graph integrating descriptive and discriminative information of the original image space. SDP focuses on the improvement of the discriminative performance of image representation. As previous work neglect the singularity problem and optimal dimensionality, SDP avoid the singularity problem and determine the optimal dimensionality simultaneously. Moreover, we extend our approach and present kernel SDP. Experimental results have revealed the effectiveness of our approach.

Owing to the effectiveness of SDP, further work will be investigated as following:

1. Feature selection is an open issue in CBIR. In this work, we only adopt global features, which only describe the respective overall statistics for a whole image. A great of previous work has applied local features to CBIR [20,21,22]. We augment local features to improve retrieval performance.
2. Representing an image as a matrix intrinsically, and extending the subspace learning algorithm with tensor representation [23,24].
3. Utilizing the user's interaction, A possible extension of our work is to incorporate the feedback information to update the affinity matrix, might achieve higher precision [25,26].

Acknowledgments

This work was supported in part by the National Natural Science Foundation of China (Grant No. 60572078) and by Guangdong Natural Science Foundation (Grant No. 05006349).

References

1. Rui, Y., Huang, T.S., Chang, S.F.: Image retrieval: current techniques, promising directions and open issues. Journal of Visual Communication and Image Representation 10(4), 39–62 (1999)
2. Smeulders, A.W.M., Worring, M., Santini, S., Gupta, A., Jain, R.: Content-based image retrieval at the end of the early years. IEEE Transactions on Pattern Analysis and Machine Intelligence 22(12), 1349–1380 (2000)
3. Datta, R., Joshi, D., Li, J., Wang, J.Z.: Image Retrieval: Ideas, Influences, and Trends of the New Age. ACM Computing Surveys 39(20), 1–77 (2007)
4. Liu, Y., Zhang, D.S., Lu, G.J., Ma, W.Y.: Asurvey of content-based image retrieval with high-level semantics. Pattern Recognition 40(1), 262–282 (2007)
5. Tenenbaum, J.B., de Silva, V., Langford, J.C.: A Global Geometric Framework for Nonlinear Dimensionality Reduction. Science 290(5500), 2319–2323 (2000)
6. Saul, L.K., Roweis, S.T.: Nonlinear dimensionality reduction by locally linear embedding. Science 290(5500), 2323–2326 (2000)
7. Belkin, M., Niyogi, P.: Laplacian eigenmaps and spectral techniques for embedding and clustering. Advances in Neural Information Processing Systems 14, 585–591 (2002)
8. Cox, T., Cox, M.: Multidimensional Scalling (1994)
9. He, X.F.: Laplacian Eigenmap for Image Retrieval. Master's thesis, Computer Science Department, The University of Chicago (2002)
10. Chen, H.T., Chang, H.W., Liu, T.L.: Local discriminant embedding and its variants. In: IEEE Conference on Computer Vision and Pattern Recognition, vol. 2, pp. 846–853 (2005)
11. Nie, F.P., Xiang, S.M., Song, Y.Q., Zhang, C.S.: Optimal Dimensionality Discriminant Analysis and Its Application to Image Recognition. In: IEEE Conference on Computer Vision and Pattern Recognition, Minneapolis, MN, USA, June17-22, pp. 1–8 (2007)
12. Scholkopf, B., Smola, A.J.: Learning with Kernels: Support Vector Machines, Regularization, Optimization, and Beyond. MIT Press, Cambridge (2001)
13. He, X.F.: Locality Preserving Projections. PhD thesis, Computer Science Department, the University of Chicago (2005)
14. Swain, M.J., Ballard, D.H.: Color indexing. International Journal of Computer Vision 7(1), 11–32 (1991)
15. Ojala, T., Pietikainen, M., Maenpaa, T.: Multiresolution gray-scale and rotation invariant texture classification with local binary patterns. IEEE Transactions on Pattern Analysis and Machine Intelligence 24(7), 971–987 (2002)
16. Stricker, M., Orengo, M.: Similarity of color images. In: Proceedings SPIE Storage and Retrieval for Image and Video Databases, San Jose, CA, USA, vol. 2420, pp. 381–392 (1995)
17. Smith, J.R., Chang, S.F.: Transform features for texture classification and discrimination inlarge image databases. In: IEEE International Conference Image Processing, vol. 3, pp. 407–411 (1994)

18. Muller, H., Marchand-Maillet, S., Pun, T.: The truth about corel-evaluation in image retrieval. In: Proceedings of the International Conference on Image and Video Retrieval, pp. 38–49. Springer, Heidelberg (2002)
19. Smith, J.R., Chang, S.F.: Tools and techniques for color image retrieval. In: Storage & Retrieval for Image and Video Databases IV, vol. 2670, pp. 426–437 (1996)
20. Tuytelaars, T., Van Gool, L., et al.: Content-based image retrieval based on local affinely invariant regions. In: Int. Conf. on Visual Information Systems, pp. 493–500. Springer, Heidelberg (1999)
21. Jing, F., Li, M., Zhang, H.J., Zhang, B.: An effective region-based image retrieval framework. In: Proceedings of the tenth ACM international conference on Multimedia, pp. 456–465. ACM Press, New York (2002)
22. Mikolajczyk, K., Schmid, C.: A performance evaluation of local descriptors. IEEE Transactions on Pattern Analysis and Machine Intelligence 27(10), 1615–1630 (2005)
23. Xia, J., Yeung, D.Y., Dai, G.: Local Discriminant Embedding with Tensor representation. In: IEEE International Conference on Image Processing, pp. 929–932 (2006)
24. He, X.F., Cai, D., Niyogi, P.: Tensor subspace analysis. Advances in Neural Information Processing Systems 18 (2005)
25. He, X.F.: Incremental semi-supervised subspace learning for image retrieval. In: Proceedings of the 12th annual ACM international conference on Multimedia, pp. 2–8. ACM Press, New York (2004)
26. Lin, Y.Y., Liu, T.L., Chen, H.T.: Semantic manifold learning for image retrieval. In: Proceedings of the 13th annual ACM international conference on Multimedia, pp. 249–258. ACM Press, New York (2005)

Comparing Dissimilarity Measures for Content-Based Image Retrieval

Haiming Liu, Dawei Song, Stefan Rüger, Rui Hu, and Victoria Uren

Knowledge Media Institute
The Open University, Walton Hall
Milton Keynes, MK7 6AA, UK
{h.liu,d.song,s.rueger,r.hu,v.s.uren}@open.ac.uk
http://kmi.open.ac.uk/mmis

Abstract. Dissimilarity measurement plays a crucial role in content-based image retrieval, where data objects and queries are represented as vectors in high-dimensional content feature spaces. Given the large number of dissimilarity measures that exist in many fields, a crucial research question arises: Is there a dependency, if yes, what is the dependency, of a dissimilarity measure's retrieval performance, on different feature spaces? In this paper, we summarize fourteen core dissimilarity measures and classify them into three categories. A systematic performance comparison is carried out to test the effectiveness of these dissimilarity measures with six different feature spaces and some of their combinations on the Corel image collection. From our experimental results, we have drawn a number of observations and insights on dissimilarity measurement in content-based image retrieval, which will lay a foundation for developing more effective image search technologies.

Keywords: dissimilarity measure, feature space, content-based image retrieval.

1 Introduction

Content-based image retrieval is normally performed by computing the dissimilarity between the data objects and queries based on their multidimensional representations in content feature spaces, for example, colour, texture and structure. There have been a large number of dissimilarity measures from computational geometry, statistics and information theory, which can be used in image search. However, only a limited number of them have been widely used in content-based image search. Moreover, the performance of a dissimilarity measure may largely depend on different feature spaces. Although there have been some attempts in theoretically summarizing existing dissimilarity measures [6], and some evaluation to find which dissimilarity measure for shape based image search [13], there is still lack of a systematic investigation into the applicability and performance of different dissimilarity measures in image retrieval field and the investigation into various dissimilarity measures on different feature spaces for large-scale image retrieval.

H. Li et al. (Eds.): AIRS 2008, LNCS 4993, pp. 44–50, 2008.

In this paper, we systematically investigate 14 typical dissimilarity measures from different fields. Firstly, we classify them into three categories based on their theoretical origins. Secondly, we experimentally evaluate these measures in content-based image retrieval, based on six different typical feature spaces from colour, texture and structure category and some of their combinations, on the standard Corel image collection. Our systematic empirical evaluation provides initial evidence and insights on which dissimilarity measure works better on which feature spaces.

2 Classification of Dissimilarity Measures

Based on McGill and others' studies on dissimilarity measures [6,4,12], we choose 14 typical measures that have been used in information retrieval.

2.1 Geometric Measures

Geometric measures treat objects as vectors in a multi-dimensional space and compute the distance between two objects based along pairwise comparisons on dimensions.

Minkowski Family Distances (d_p)

$$d_p(A, B) = (\sum_{i=1}^{n} |a_i - b_i|^p)^{\frac{1}{p}} \tag{1}$$

Here $A = (a_1, a_2, ..., a_n)$ and $B = (b_1, b_2, ..., b_n)$ are the query vector and test object vector respectively. The Minkowski distance is a general form of the **Euclidean** (p=2), **City Block** (p=1) and **Chebyshev** ($p = \infty$) distances. Recent research has also suggested the use of **fractional** dissimilarity (i.e., $0 < p < 1$) [3], which is not a metric because it violates the triangle inequality. Howarth and Rüger [3] have found that the retrieval performance would be increases in many circumstances when p=0.5.

Cosine Function Based Dissimilarity (d_{cos}). The cosine function computes the angle between the two vectors, irrespective of vector lengths [13]:

$$s_{cos}(A, B) = \cos\theta = \frac{A \cdot B}{|A| \cdot |B|}$$

$$d_{cos}(A, B) = 1 - \cos\theta = 1 - \frac{A \cdot B}{|A| \cdot |B|} \tag{2}$$

Canberra Metric (d_{can}) [4]

$$d_{can}(A, B) = \sum_{i=1}^{n} \frac{|a_i - b_i|}{|a_i| + |b_i|} \tag{3}$$

Squared Chord (d_{sc}) [4]

$$d_{sc}(A,B) = \sum_{i=1}^{n} (\sqrt{a_i} - \sqrt{b_i})^2 \tag{4}$$

Obviously, this measure is not applicable for feature spaces with negative values.

Partial-Histogram Intersection (d_{p-hi}): This measure is able to handle partial matches when the sizes of the two object vectors are different [13]. When A and B are non-negative and have the same size, in terms of the City Block metric ($|x| = \sum_i |x_i|$), it is equivalent to the City Block measure. [12, 9]

$$d_{p-hi}(A,B) = 1 - \frac{\sum_{i=1}^{n}(\min(a_i, b_i))}{\min(|A|, |B|)} \tag{5}$$

2.2 Information Theoretic Measures

Information-theoretic measures are various conceptual derivatives from the Shannon's entropy theory and treat objects as probabilistic distributions. Therefore, again, they are not applicable to features with negative values.

Kullback-Leibler (K-L) Divergence (d_{kld}). From the information theory point of view, the K-L divergence measures how one probabilistic distribution diverges from the other. However, it is non-symmetric. [7]

$$d_{kld}(A,B) = \sum_{i=1}^{n} a_i \log \frac{a_i}{b_i} \tag{6}$$

Jeffrey Divergence (d_{jd})

$$d_{jd}(A,B) = \sum_{i=1}^{n} (a_i \log \frac{a_i}{m_i} + b_i \log \frac{b_i}{m_i}), \tag{7}$$

where $m_i = \frac{a_i + b_i}{2}$, Jeffrey divergence, in contrast to the Kullback-Leibler divergence, is numerically stable and symmetric. [10]

2.3 Statistic Measures

Unlike geometric measures, statistical measures compare two objects in a distributed manner rather than simple pair wise distance.

χ^2 Statistics (d_{χ^2})

$$d_{\chi^2}(A,B) = \sum_{i=1}^{n} \frac{(a_i - m_i)^2}{m_i}, \tag{8}$$

where $m_i = \frac{a_i+b_i}{2}$. It measures the difference of query vector (observed distribution) from the mean of both vectors (expected distribution). [13]

Pearson's Correlation Coefficient (d_{pcc}). A distance measurement derived from Pearson correlation coefficient [5] is defined as

$$d_{pcc}(A, B) = 1 - |p|, \tag{9}$$

where

$$p = \frac{n\sum_{i=1}^{n} a_i b_i - (\sum_{i=1}^{n} a_i)(\sum_{i=1}^{n} b_i)}{\sqrt{[n\sum_{i=1}^{n} a_i^2 - (\sum_{i=1}^{n} a_i)^2][n\sum_{i=1}^{n} b_i^2 - (\sum_{i=1}^{n} b_i)^2]}}$$

Note the larger $|p|$ is the more correlated the vectors A and B. [1]

Kolmogorov-Smirnov (d_{ks}). Kolmogorov-Smirnov distance is a measure of dissimilarity between two probability distributions [2]. Like K-L divergence and Jeffrey divergence, it is defined only for one-dimensional histograms [12]:

$$d_{ks}(A, B) = \max_{1 \le i \le n} |F_A(i) - F_B(i)| \tag{10}$$

$F_A(i)$ and $F_B(i)$ are the simple probability distribution function (PDF) of the object vectors, which are interpreted as probability vectors of one-dimensional histogram.

Cramer/von Mises Type (CvM) (d_{cvm}). A statistics of the Cramer/von Mises Type(CvM) is also defined based on probability distribution function (PDF) [11]:

$$d_{cvm}(A, B) = \sum_{i=1}^{n} (F_A(i) - F_B(i))^2 \tag{11}$$

3 Empirical Performance Study

Our experiment aims to address the performance of 14 dissimilarity measures on different feature spaces. We use mean average precision as the performance indicator.

3.1 Experimental Setup

Data Set. In this experiment, we use a subset of the Corel collection, developed by [8]. There are 63 categories and 6192 images in the collection, which is randomly split into 25% training data, and 75% test data. We take the training set as queries to retrieve similar images from the test set.

Features. Six typical image feature spaces are applied in the experiment.

- Colour feature spaces: RGB is three-dimensional joint colour histogram, which contains a different proportion of red, green and blue; MargRGB-H does a one-dimensional histogram for each component individually; MargRGB-M only records the first several central moments; HSV is similar to RGB, which are hue, saturation and value of colour-space.
- Texture feature spaces: Gabor, is a texture feature generated using Gabor wavelets; Tamura is a texture feature generated by statistical processing points of view.
- Structure feature space: Konvolution (Konv), discriminates between low level structures in an image, which designed to recognize horizontal, vertical and diagonal edges.

Approach. Here, we use the vector space model approach for image retrieval. The difference from [8] is that we aim to test various dissimilarity measures instead of using traditional cosine based or city block measures.

3.2 Single Feature Spaces

We investigate the performance of the 14 dissimilarity measures on 6 single image feature spaces as described above.

3.3 Combined Feature Spaces

In a further experiment, we picked up three typical features from colour, texture and structure, respectively. This experiment we use the same set up on the three and their combined feature spaces, HSV and Gabor, HSV and Konv, Gabor and Konv, and HSV, Gabor and Konv.

3.4 Results

Table 1 and Table 2 show the experimental results, from which the following observations can be made. Firstly, most of the dissimilarity measures from the geometric category have better performance than other two categories; Secondly, the performance of most of the dissimilarity measures in the color feature spaces outperform the other feature space; Finally, after identifying the top five performing dissimilarity measures on every feature space, we find Canberra metric, Squared Chord from the geometric measures category, Jeffrey Divergence from the information-theoretic measures category, and χ^2 from the statistical measures category have better performance than Euclidean and City Block dissimilarity measures, which have been most widely used in image retrieval field. Significance tests, using the paired Student's t-test (parametric test), the sign test and the paired Wilcoxon signed-rank test (non-parametric test), have shown that the improvements over the city-block measure are statistically significant (p-value less than 0.05). Therefore we would recommend them for image retrieval applications.

Table 1. Mean Average Precision on Single Feature Spaces

	HSV	margRGB-H	margRGB-M	gabor	tamura	konv
Geometric Measures						
Fractional(p=0.5)	0.1506	0.1269	0.0871	0.1490	0.1286	0.0731
City Block(p=1)	0.1682	0.1207	0.0912	0.1350	0.0949	0.0951
Euclidean(p=2)	0.1289	0.1128	0.0917	0.1161	0.0678	0.0761
Chebyshev(p=∞)	0.1094	0.1013	0.0886	0.0615	0.0358	0.0555
Cosine Similarity	0.1345	0.1204	0.0778	0.1057	0.0671	0.0716
Canberra Metric	0.1568	0.1333	0.0824	0.1496	0.1267	0.0709
Squared Chord	0.1876	0.1294	0.0967	0.1259	0.0880	0.0984
Partial-Histogram	0.1682	0.1207	0.0566	0.0320	0.0209	0.0301
Information-Theoretic Measures						
Kullback-Leibler Divergence	0.1779	0.1113	0.0893	0.1019	0.0528	0.0948
Jeffrey Divergence	0.1555	0.1185	0.0902	0.1353	0.0960	0.0950
Statistic Measures						
χ^2 Statistics	0.1810	0.1282	0.0832	0.1303	0.0897	0.0984
Pearson's Correlation	0.1307	0.1182	0.0818	0.1035	0.0692	0.0763
Kolmogorov-Smirnov	0.0967	0.1041	0.0750	0.0575	0.0426	0.0598
Cramer/von Mises Type	0.0842	0.1077	0.0724	0.0529	0.0406	0.0516

Table 2. Mean Average Precision on Combined Feature Spaces

	HSV	Gabor	Konv	HSV +Gabor	HSV +Konv	Gabor +Konv	HSV+Gabor +Konv
Geometric Measures							
Fractional(p=0.5)	0.1506	0.1490	0.0731	0.0693	0.0733	0.0686	0.0686
City Block(p=1)	0.1682	0.1350	0.0951	0.1350	0.0964	0.1396	0.1397
Euclidean(p=2)	0.1289	0.1161	0.0761	0.1163	0.0782	0.1198	0.1199
Chebyshev(p=∞)	0.1094	0.0615	0.0555	0.0623	0.0576	0.0721	0.0727
Cosine Similarity	0.1345	0.1057	0.0716	0.1542	0.1435	0.1164	0.1617
Canberra Metric	0.1568	0.1496	0.0709	0.1573	0.0765	0.1617	0.1627
Squared Chord	0.1876	0.1259	0.0984	0.1261	0.1116	0.1304	0.1306
Partial-Histogram	0.1682	0.0320	0.0301	0.0301	0.0320	0.0209	0.0205
Information-Theoretic Measures							
Kullback-Leibler Divergence	0.1779	0.1019	0.0948	0.0411	0.0306	0.0414	0.0414
Jeffrey Divergence	0.1555	0.1353	0.0950	0.1283	0.1085	0.1329	0.1330
Statistic Measures							
χ^2 Statistics	0.1810	0.1303	0.0984	0.1304	0.1062	0.1351	0.1352
Pearson's Correlation	0.1307	0.1035	0.0763	0.0529	0.1083	0.0316	0.0528
Kolmogorov-Smirnov	0.0967	0.0575	0.0598	0.1099	0.1155	0.0438	0.1163
Cramer/von Mises Type	0.0842	0.0529	0.0516	0.1291	0.1420	0.0529	0.1422

4 Conclusion and Future Work

We have reviewed fourteen dissimilarity measures, and divided them into three categories: geometry, information theory and statistics, in terms of their theoretical characteristic and functionality. In addition, these dissimilarity measures have been empirically compared on six typical content based image feature spaces, and their combinations on the standard Corel image collection.

Interesting conclusions are drawn from the experimental results, based on which we recommend Canberra metric, Squared Chord, Jeffrey Divergence, and χ^2 for future use in the Content based Image Retrieval.

This work will be a foundation for developing more effective content-based image information retrieval systems. In the future, we are going to test how the dissimilarity measures work on multi-image queries, and what their performances are on different data collections.

Acknowledgments

This work was funded in part the European Union Sixth Framework Programme (FP6) through the integrated project Pharos (IST-2006-045035). In addition, we would like to thank Peter Howarth for helping construct the feature spaces and setting up the experiments.

References

1. Chen, C.-C., Chu, H.-T.: Similarity measurement between images. In: Proceedings of the 29th Annual International Computer Software and Applications Conference (COMPSAC 2005), IEEE, Los Alamitos (2005)
2. Geman, D., Geman, S., Graffigne, C., Dong, P.: Boundary Detection by Constrained Optimization. IEEE Transactions on Pattern Analysis and Machine Intelligence 12(7), 609–628 (1990)
3. Howarth, P., Rüger, S.: Fractional distance measures for content-based image retrieval. In: Losada, D.E., Fernández-Luna, J.M. (eds.) ECIR 2005. LNCS, vol. 3408, Springer, Heidelberg (2005)
4. Kokare, M., Chatterji, B., Biswas, P.: Comparison of similarity metrics for texture image retrieval. In: Proceeding of IEEE Conf. on Convergent Technologies for Asia-Pacific Region, vol. 2, pp. 571–575 (2003)
5. Luke, B.T.: Pearson's correlation coefficient. Online (1995)
6. Noreault, T., McGill, M., Koll, M.B.: A performance evaluation of similarity measures, document term weighting schemes and representations in a Boolean environment. In: Proceeding of the 3rd annual ACM Conference on Research and development in inforamtion retreval, SIGIR 1980, Kent, UK, pp. 57–76. ACM, Butterworth Co. (1980)
7. Ojala, T., Pietikainen, M., Harwood, D.: Comparative study of texture measures with classification based on feature distributions. Pattern Recognition 29(1), 51–59 (1996)
8. Pickering, M.J., Rüger, S.: Evaluation of key frame-based retrieval techniques for video. Computer Vision and Image Understanding 92(2-3), 217–235 (2003)
9. Puzicha, J.: Distribution-Based Image Similarity, ch. 7, pp. 143–164. Kluwer Academic Publishers, Dordrecht (2001)
10. Puzicha, J., Hofmann, T., Buhmann, J.M.: Non-parametric similarity measures for unsupervised texture segmentation and image retrieval. In: Proceedings of the IEEE International Conference on Computer Vision and Pattern Recognition, San Juan (1997)
11. Puzicha, J., Rubner, Y., Tomasi, C., Buhmann, J.M.: Empirical evaluation of dissimilarity measures for color and texture. In: Proceeding of the international conference on computer vision, vol. 2, pp. 1165–1172 (September 1999)
12. Rubner, Y., Tomasi, C., Guibas, L.J.: The earth mover's distance as a metric for image retrieval. International Journal of Computer Vision 40(2), 99–121 (2004)
13. Zhang, D., Lu, G.: Evaluation of similarity measurement for image retrieval. In: Procedding of IEEE International Conference on Neural Networks Signal, Nanjing, December 2003, pp. 928–931. IEEE, Los Alamitos (2003)

A Semantic Content-Based Retrieval Method for Histopathology Images

Juan C. Caicedo, Fabio A. Gonzalez, and Eduardo Romero

Bioingenium Research Group
Universidad Nacional de Colombia
{jccaicedoru,fagonzalezo,edromero}@unal.edu.co
http://www.bioingenium.unal.edu.co

Abstract. This paper proposes a model for content-based retrieval of histopathology images. The most remarkable characteristic of the proposed model is that it is able to extract high-level features that reflect the semantic content of the images. This is accomplished by a *semantic mapper* that maps conventional low-level features to high-level features using state-of-the-art machine-learning techniques. The semantic mapper is trained using images labeled by a pathologist. The system was tested on a collection of 1502 histopathology images and the performance assessed using standard measures. The results show an improvement from a 67% of average precision for the first result, using low-level features, to 80% of precision using high-level features.

Keywords: content-based image retrieval, medical imaging, image databases.

1 Introduction

Medical images have been supporting clinical decisions in health care centres during the last decades, for instance the Geneve University Hospital reported a production rate of 12.000 daily images during 2.002 [10]. Traditional medical image database systems store images as a complementary data of textual information, providing the most basic and common operations on images: transfer and visualization. Usually, these systems are restricted to query a database only through keywords, but this kind of queries limits information access, since it does not exploit the intrinsic nature of medical images.

A recent approach to medical image database management is the retrieval of information by content, named Content-Based Image Retrieval (CBIR)[10] and several systems such as ASSERT [12], IRMA [8] or FIRE [4] work following this approach. These systems allow evaluation of new clinical cases so that when similar cases are required, the system is able to retrieve comparable information for supporting diagnoses in the decision making process. One drawback of current CBIR systems is that they are based on basic image features that capture low-level characteristics such as color, textures or shape. This approach fails to capture the high-level patterns corresponding to the semantic content of the image, this may produce poor results depending on the type of images the system deals with.

H. Li et al. (Eds.): AIRS 2008, LNCS 4993, pp. 51–60, 2008.

On the other hand, it is well known that the diagnosis process in medicine is mainly based on semantic or semiotic knowledge, difficult issues to deal with when image knowledge contents has to be organized for retrieval tasks. To extract image semantics is a great challenge because of the semantic gap [9], that is to say, the existing distance between conceptual interpretation at a high level and the low-level feature extraction. One strategy to bridge the semantic gap in image retrieval is the automatic image annotation, investing efforts to assign labels to images as accurately as possible to support keyword-based image search.

The problem of extracting semantic features from images may be approached from two different perspectives: an analytic approach and an inductive approach. The analytic approach requires to understand, with the help of an expert, what a given pattern is; then a model to decide whether the pattern is present or not is built, based on this knowledge. On the other hand, the inductive approach, or machine-learning approach, requires to collect enough image samples where the pattern is present or absent, and to train a model able to discriminate both situations. The inductive approach has many advantages: it just relays on the expert for labeling the samples; the model may be easily retrained when new data is available; and there is not need for dealing directly with the complexity of the patterns. In this work, the inductive approach is followed.

This paper presents the design, implementation and evaluation of a new method for the semantic analysis of a basal-cell-carcinoma database. The whole system is modeled as to map a set of low-level features into high-level semantic properties for a collection of basal-cell-carcinoma images, which were previously annotated by an expert pathologist. The semantic mapper is able to recognize which concepts are present in an image, quantifying a degree of certainty about those decisions using a binary SVM per concept. Although classifiction methods have been previously used for automatic annotation of image concepts [5], our approach builds a new semantic feature space instead of assigning keywords to the image. An image similarity measure is calculated in the semantic feature space. This similarity measure provides a finer mechanism for ranking similar images than keyword-matching-based retrieval.

The reminder of this paper is organized as follows. In Section 2, the problem of content-based retrieval in histopathology is introduced. In Section 3, the model for feature extraction is presented. Methods for compare images are in Section 4, Section 5 presents results of the experimental evaluation and some concluding remarks are presented in Section 6.

2 The Problem of Accessing Histopathology Images by Content

Medical practice constantly requires access to reference information for the decision making process in diagnostics, teaching and research. Previous works have designed CBIR systems for medical image databases providing services such as query-by-example, query-by-regions or automatic image annotations among others [14,11]. This kind of tools helps physicians to take informed decisions and to make case-based reasoning.

The aim of this work is to include domain-specific knowledge to improve the performance of a medical CBIR system. Specifically, the proposed system deals with histopathology images, so some details of this kind of images need to be studied and understood. A basic concept in histology is that there exist four basic types of tissue: epithelial, connective, muscle, and nerve[7]. With very few exceptions, all organs contain a different proportion of these four basic tissues. In general, histological techniques highlight these tissues with few colours since dyes are designed to specifically arise a particular tissue feature. In terms of image processing, histological images are distinguished by having more or less homogeneous textures or repeated patterns, which may be used to characterise the image. Main information in histological images lyes on repeated patterns of textures, with particular edges and slight color differences.

Histopathology images used in this work were acquired to diagnose a special skin cancer called basal-cell carcinoma. Slides were obtained from biopsy samples which were fixed in paraphin, cut to a 5 mm thickness, deposited onto the glass slides and finally colored with Hematoxilin-Eosin. The whole collection is close to 6.000 images associated with clinical cases. A subset of the collection consisting of 1.502 images were annotated and organized in semantic groups by a pathologist. The groups are representative of the semantic categories that are relevant in the scenario of a content-based image retrieval system, according to the expert. Each group is composed of a number of image samples of a histopathology concept. A particular image may belong to many groups simultaneously, because each image may contain more than one concept.

3 Feature Extraction

This section is devoted to describe how low-level features are transformed into semantic characteristics. The whole process starts by a conventional low-level feature extraction phase that reduces image dimensionality: histograms of predefined edge, texture and color characteristics. Dimensionality is further reduced using statistical descriptors of the histogram up to a fourth order along with its entropy. The resulting feature vector, herein called meta-features, grossly describes the underlying probability distribution associated with each different histogram. Once images are expressed as meta-features, a semantic mapper transforms them into semantic features. This mapper is devised for capturing the pathologist knowledge and is composed of 19 SVM classifiers, each specialized upon different concepts previously defined by an expert pathologist. Figure 1 illustrates the feature extraction process.

3.1 Low-Level Feature Extraction

A very convenient approach to face feature extraction consists in using a statistical frame: images are modeled as random variables so that histograms stand for the probability distribution of any of the selected features i.e. edges, textures and

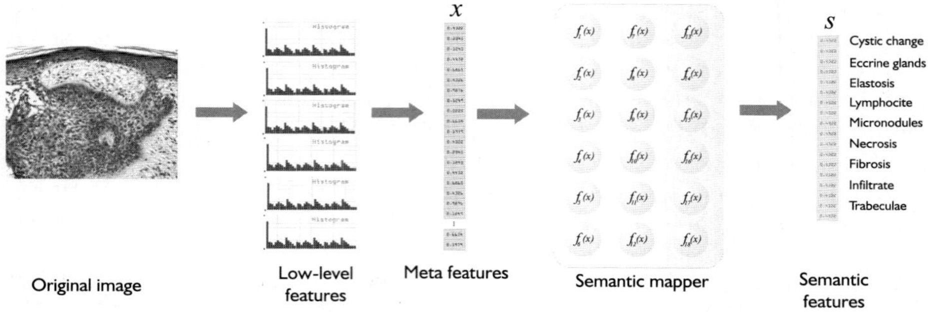

Fig. 1. Feature extraction model. Histogram features are extracted from the original image. Then histograms are processed to obtain meta features, which are the input for the semantic mapper to produce semantic features.

colors. Histogram features have been traditionally used in content-based image retrieval to calculate similarity measures and to rank images [3]. The following histogram features were used:

- *Gray histogram*: Luminance intensities in a 256 scale.
- *Color histogram*: In the RGB color model with a partition space of $8 \times 8 \times 8$
- *Local binary partition*: A local texture analysis to determine neighbor dominant intensities
- *Tamura texture histogram*: Composed of contrast, directionality and coarseness
- *Sobel histogram*: Edge detection
- *Invariant feature histograms*: Local invariant transformations such as rotation and translation

A set of meta-features are calculated from the information of each histogram h as follows (k is a index for histogram bins):

- *Mean*:

$$\sum_k kh(k)$$

- *Deviation*:

$$\sum_k (k - \mu)h(k)$$

- *Skewness*: the third central moment.

$$\frac{\mu^3}{\sigma^3}$$

- *Kurtosis*: the fourth central moment.

$$\frac{\mu^4}{\sigma^4} - 3$$

– *Entropy*:

$$-\sum_{k} h(k)ln[h(k)]$$

All meta-features are calculated on each of the six histogram features, which amounts to a total of 30 meta-features per image.

3.2 Semantic Mapper

Overall, every pathologist follows a standard training addressed to strength out both diagnosis precision and velocity. An efficient management of these two complementary issues is based on four classic steps: look, see, recognize and understand [2]. A pathologist that evaluates an image is basically looking for patterns and his/her decisions are usually based on the presence or absence of a specific pattern. These patterns are associated with concepts, that is, pathologists give meaning to the image or in other words, they "understand" the image contents. Patterns may correspond to simple low-level features, but in most cases they are a complex combination of them. Features are usually made up of many of this kind of patterns and are called high-level or semantic features. The main goal of this work is to design a model to capture the semantic interpretation of histopathology images, in order to achive a better CBIR effectivity.

The core of the proposed semantic model is the semantic mapper. Since groups are non disjoint, the semantic mapper is not a single classifier but a model of many learning algorithms identifying the different concepts present in an image. This mapper is composed of 18 Support Vector Machine (SVM) classifiers [13], each specialized on deciding whether or not one image contains one of the eighteen possible concepts, and a extra classifier to detect noise[1]. When the image representation is processed through this semantic mapper, meta-features are individually processed by each of the 19 classifiers. In this model, each classifier outputs a score value indicating whether the image contains the concept or not. With the output of each classifier, the semantic model builds a semantic feature vector containing the membership degree of one image to every semantic group.

3.3 Semantic Mapper Training

The dataset used to train each classifier is composed of 1.502 images, organized in 19 different groups (corresponding to the 19 different categories defined by the pathologist). The training dataset is composed of meta-features with their corresponding labels and each group has a specific dataset. Each dataset is entailed with exactly the same attributes except for the class label which can only have two possible values: positive if the example belongs to this group and negative otherwise. In most of the groups there is a considerable amount of imbalance between negative and positive classes, this is solved by resampling the class with less elements. Each dataset is split, using a stratified sampling approach, into two subsets: 20% is used for testing and 80% for training and validation. A 10-fold

[1] Noisy images are those that do not contain any important histopathology concept.

cross validation scheme on the training set is used to evaluate the classification rate. The test dataset is set aside and used at the end for calculating the final error rate of the classifier.

A SVM classifier is provided with different parameters (herein called hyper-parameters), such as the regularization parameter λ along with the kernel and its particular parameterization. In order to select the best configuration for each SVM classifier, generalizing the nature of the associated concept, the following domains for the hyperparameters were studied:

- *Polynomial Kernel.* (inhomogeneous) with exponent d set to 5 and 20
$$k(x, x') = (x \cdot x' + 1)^d$$
- *Radial Basis Function Kernel.* with four different values for γ: 0.001, 0.1, 0.5 and 1
$$k(x, x') = exp(-\gamma \left\| x - x' \right\|^2)$$
- *Lambda.* The regularization parameter was evaluated in the interval $[1, 50]$ using increment steps of 0.1

The purpose of the training evaluation is to identify the best model for each histopathology concept. In total, there were 5 different models to evaluate per concept: one with a polynomial kernel and four with a RBF kernel for each of the parameters shown above. Each model has a different complexity value which is found as the minimum value drawn from the error-complexity plot. When an optimal model value is obtained, the minimum error rate between models allows to select the best model among the 5 possible ones.

4 Metrics for Image Content

Similarity evaluation of image contents is achieved using metrics. For image retrieval, metrics are designed to detect differences between the available features. This work uses metrics for two type of contents: low-level features and semantic features as follows.

4.1 Low-Level Feature Metrics

Since low-level features are histograms, they require metrics evaluating differences between probability distributions. Evaluated metrics were Relative Bin Deviation D_{rbd}, Jensen-Shannon Divergence D_{JSD} and Euclidean Distance L_2. For each feature, we experimentally found the most appropriate metric capturing the feature topology in the image collection, obtaining a set of feature-metric pairs able to rank histopathology images. Many features can be evaluated in an individual metric using a linear combination approach of the feature-metric pairs. If x and y are images; F_k is a function to extract a low-level feature k; and

M_k is a metric able to compare the feature k, then, a metric to evaluate many low level features is:

$$d(x, y) = \sum_k w_k M_k \left(F_k(x), F_k(y)\right)$$

where w_k is a factor that controls the relative importance of each feature-metric pair. The best values for all w_k were found by exhaustive search.

4.2 Semantic Metric

Semantic features are codified in a vector per image, in which each position represents a value of membership degree to the corresponding group. These values are produced by each component of the semantic mapper and are scaled to fit the $[0, 1]$ range. Each image may belong to many groups at the same time, providing information about the content and interpretation of the overall scene. To compare images in a semantic way, the Tanimoto coefficient was selected, which is a generalization of the Jaccard coefficient [1]. In this problem, Tanimoto coefficient can interpret, how many positions in the semantic vectors are showing coincidences, emphasizing the similarity between concepts shared by both images. Given two semantic vectors A and B, each with 19 positions, the Tanimoto coefficient assigns a similarity score to the associated images as follows:

$$T(A, B) = 1 - \frac{A \cdot B}{\|A\|^2 + \|B\|^2 - A \cdot B}$$

5 Experimentation and Results

MEler et al [6] presents a framework to evaluate CBIR systems in order to report comparable results from different research centers in a standardized way. The most representatives of those performance measures are precision and recall. Since precision can be measured for different values of recall, the average precision of the n-th result is reported to compare experiments, named $P(n)$. Also a precision vs recall graph may be drawn, which provides information about the behavior of the system in many points. Also, the rank of relevant results is used for measuring performance; in this work, the rank of the first relevant result (Rank1) and the average, normalized rank (NormRank) were used.

Each experiment was configured to select 30 random queries in the collection, through a query-by-example approach. Results associated to each query were evaluated as relevant or irrelevant against the ground truth, and performance measures were averaged to obtain the final result of the experimentation.

Table 1 shows the performance of the CBIR system. In one case, only low-level features were used. In the other case, semantic features were used. For all the measures, the semantic features outperform the low-level features. This is corroborated by the precision vs. recall graph (Fig. 2).

The low-level-feature system performance serves a bottom line to assess the real contribution of incorporating domain knowledge to the system. The results

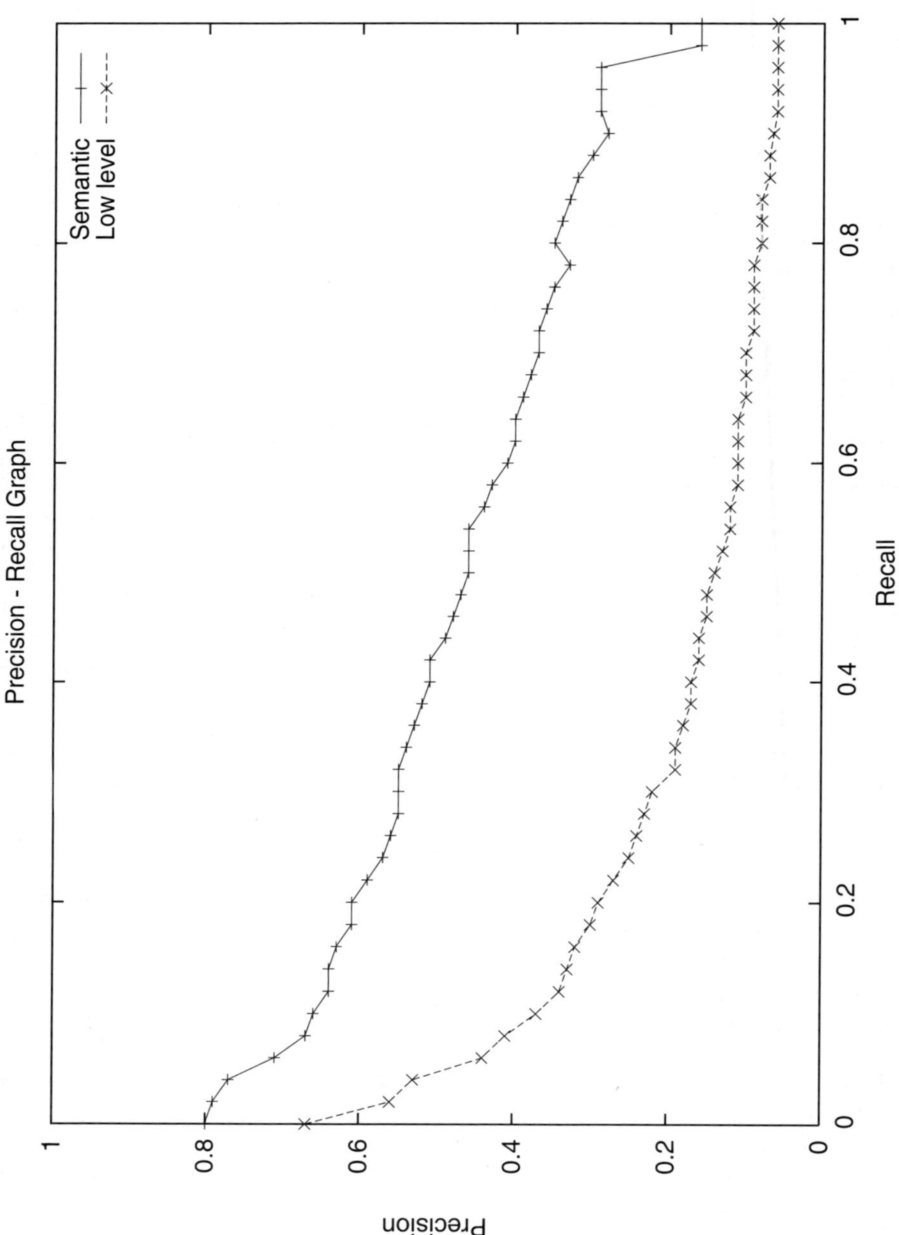

Fig. 2. Precision vs Recall graph comparing the system performance using two types of features: low-level features and semantic features

Table 1. Performance measures for image retrieval

Model	Rank1	NormRank	P(1)	P(20)	P(50)	P(100)
Low level features	8.22	0.28	0.67	0.30	0.21	0.16
Semantic features	1.96	0.07	0.80	0.59	0.51	0.45

show that the proposed model to represent this domain knowledge in a semantic feature space, greatly improve the performance of the system.

6 Conclusions and Future Work

The paper presented a novel approach to represent histopathology knowledge, which is naturally included into a CBIR system. The approach allows to bridge the semantic gap preserving in the same context both, the low-level features and the semantic features. This was accomplished by a semantic mapper based on SVM classifiers. This mapper allows the building of a new semantic feature space in which a metric is used to calculate the similarity between images.

The experimental results show that this approach can effectively model the histopathology knowledge, that is to say, images are automatically interpreted and compared as pathologists does. This strategy provides a semantic analysis of image contents, allowing a highly improved operation of the CBIR system, in terms of precision.

The future work includes exploring richer semantic representations, using other low-level features, performing a more extensive evaluation with a larger bank of images and additional pathologists to test the system.

References

1. Bow, S.T. (ed.): Pattern Recognition and Image Preprocessing. Marcel Dekker. Inc., New York (2002)
2. Bussolati, G.: Dissecting the pathologists brain: mental processes that lead to pathological diagnoses. Virchows Arch. 448(6), 739–743 (2006)
3. Deselaers, T.: Features for Image Retrieval. PhD thesis, RWTH Aachen University. Aachen, Germany (2003)
4. Deselaers, T., Weyand, T., Keysers, D., Macherey, W., Ney, H.: Fire in imageclef 2005: Combining content-based image retrieval with textual information retrieval. Image Cross Language Evaluation Forum (2005)
5. Feng, H., Chua, T.-S.: A bootstrapping approach to annotationg large image collection. In: ACM SIGMM International Workshop on Multimedia Information Retrieval, pp. 55–62 (2003)
6. Müller, H., Müller, W., Marchand-Maillet, S., McG Squire, D., Pun, T.: A framework for benchmarking in visual information retrieval. International Journal on Multimedia Tools and Applications 21, 55–73 (2003)
7. Junqueira, L.C., Carneiro, J.: Basic Histology, 10th edn. MacGraw Hill (2003)

8. Lehmann, T., Güld, M., Thies, C., Fischer, B., Spitzer, K., Keysersa, D., Neya, H., Kohnen, M., Schubert, H., Weinb, B.: The irma project: A state of the art report on content-based image retrieval in medical applications. In: In Korea-Germany Workshop on Advanced Medical Image, pp. 161–171 (2003)
9. Liu, Y., Zhang, D., Lu, G., Ma, W.-Y.: A survey of content-based image retrieval with high-level semantics. Pattern Recognition 40, 262–282 (2007)
10. Müller, H., Michoux, N., Bandon, D., Geissbuhler, A.: A review of content based image retrieval systems in medical applications clinical bene ts and future directions. International Journal of Medical Informatics 73, 1–23 (2004)
11. Petrakis, E., Faloutsos, C.: Similarity searching in medical image databases. IEEE Transactions on Knowledge and Data Engineering 9 (1997)
12. Shyu, C.-R., Brodley, C., Kak, A., Kosaka, A., Aisen, A., Broderick, L.: Assert: A physician-in-the-loop content-based retrieval system for hrct image databases. Computer Vision and Image Understanding 75, 111–132 (1999)
13. Smola, A.J., Schölkopf, B.: Learning with kernels: Support Vector Machines, Regularization, Optimization, and Beyond. The MIT Press, Cambridge (2002)
14. Wang, J.Z.: Region-based retrieval of biomedical images. In: International Multimedia Conference Proceedings of the eighth ACM international conference on Multimedia (2000)

Integrating Background Knowledge into RBF Networks for Text Classification

Eric P. Jiang

University of San Diego, 5998 Alcala Park
San Diego, California 92110, United States of America
jiang@sandiego.edu

Abstract. Text classification is a problem applied to natural language texts that assigns a document into one or more predefined categories, based on its content. In this paper, we present an automatic text classification model that is based on the Radial Basis Function (RBF) networks. It utilizes valuable discriminative information in training data and incorporates background knowledge in model learning. This approach can be particularly advantageous for applications where labeled training data are in short supply. The proposed model has been applied for classifying spam email, and the experiments on some benchmark spam testing corpus have shown that the model is effective in learning to classify documents based on content and represents a competitive alternative to the well-known text classifiers such as naïve Bayes and SVM.

Keywords: Radial basis function networks, text classification, clustering, information retrieval.

1 Introduction

Automatic text classification is a problem applied to natural language texts that assigns a document into one or more predefined categories, based on its content. This is typically accomplished by machine learning algorithms and involves models built on the top of category-labeled training data. With the growth of the Internet and advances of computer technology, more textual documents have been digitized and stored electronically, and digital libraries and encyclopedias have become the most valuable information resources. Recently, the US Library of Congress has started a project named World Digital Library, which aims to digitize and place on the Web significant primary materials from national libraries and other institutions worldwide. It is quite clear that text classification plays an increasingly important role in this digital phenomenon, and it has been widely applied in many areas that include Web page indexing, document filtering and management, information security, business and marketing intelligence mining, customer survey and customer service automation.

Over the years, a number of machine learning algorithms have been used in text classification problems [8]. Among them, naïve Bayes [6], nearest neighbor [1], decision tree with boosting [7], Support Vector Machines (SVM) [4] are the most cited. As supervised learning algorithms, they all require some labeled training sets and in general, the quantity and quality of the training data have an important impact

H. Li et al. (Eds.): AIRS 2008, LNCS 4993, pp. 61–70, 2008.

on their classification effectiveness. In this paper, a text classification model, based on the radial basis function (RBF) networks, is proposed. It uses a labeled training set and also integrates additional unlabeled background knowledge to aid in its classification task. In Section 2, an overview of RBF networks is provided. In Section 3, the proposed model is described, and its application on one benchmark email corpus and performance comparisons with two popular text classifiers are presented in Section 4. Some concluding remarks are provided in Section 5.

2 Radial Basis Function Networks

As a popular feed-forward neural network paradigm, the radial basis function (RBF) networks have been applied in many fields in science and engineering which include telecommunication, signal and image processing, time-series modeling, control systems and computer visualization. The architecture of a typical RBF network is presented in Fig. 1.

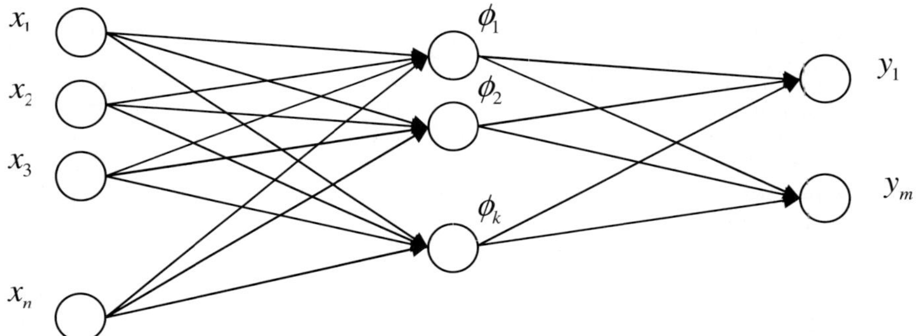

Fig. 1. The RBF network architecture

The network consists of three layers: an input layer, a hidden layer of nonlinear processing neurons and an output layer. The input to the network is a vector $x \in \Re^{n \times 1}$, and the output of the network, $y \in \Re^{m \times 1}$, is calculated according to

$$y_i = \sum_{j=1}^{k} w_{ij}\phi_j(x, c_j) = \sum_{j=1}^{k} w_{ij}\phi_j(\| x - c_j \|_2), \quad i=1, 2, \ldots m \tag{1}$$

where c_j, j = 1, 2, …, k are the RBF centers in the input space, w_{ij} are the weights for the output layer, and ϕ_j is a basis function in the jth hidden neuron that maps a distance in the Euclidean norm to a real value. For each neuron in the hidden layer, the distance between its associated center and the input to the network is computed. The activation of the neuron is a nonlinear basis function of the distance; the closer the distance, the stronger the activation. The most commonly used basis function is the Gaussian

$$\phi(x) = e^{\frac{-x^2}{2\sigma^2}} \tag{2}$$

where σ is the width parameter that controls the smoothness properties of the basis function. Finally, the output of the network is computed as a weighted sum of the hidden layer activations, as shown in Equation (1).

3 Text Classification Model

The radial basis function (RBF) networks can be applied for text classification, and for a given document collection, they are used to build models that characterize principal correlations of terms and document in the collection. In this section, an RBF network based text classifier is described in terms of its structure and major components.

The classifier is constructed by feature selection, document vector formulation and network training. First, it preprocesses input data by a rigorous feature selection procedure. In this paper, a feature or a term is referred to as a word, a number, or simply a symbol in a document. For a given document collection, the dimensionality of feature space is generally very large. The procedure aims to reduce the input feature dimensionality and to remove irrelevant features from network training and deploying. It consists of two feature selection steps: unsupervised and supervised. In the first unsupervised selection step, features such as common words and words with very low and very high frequencies are eliminated from training data. Terms with a common stem are also substituted by the stem. In the second step, the selection is conducted on those retained features by their frequency distributions between document classes in a training data set. This supervised selection intends to further identify the features that distribute most differently between document classes and uses the well-known Information Gain [8] as the selection criterion.

After feature selection, each document is encoded as a numerical vector of values of the selected features. More precisely, each vector component represents a combination of the local and global weights of a retained term and is computed by the log(tf)-idf weighting scheme [5].

RBF networks are typically trained by some nonlinear comprehensive algorithms that involve the entire networks. It can be very computationally intensive for large training data. Alternatively, it can be trained by a much less expensive two-stage training process [3]. More specifically, the first training stage determines the RBF centers and widths through some unsupervised algorithms. In our model, a variant of *k*-means algorithm is used in this stage. It is used to construct a representation of the density distribution in input training data space and is accomplished by clustering each of the document classes in a collection. It is noted that in many cases, documents in a class, though may vary in content, can likely be grouped by topics into a number of clusters where some underlying semantic term and document correlations are present. Since this is done after feature selection, the resulting cluster centroids are effectively the encoded content vectors representing the most important document features and subsequently, they summarize topics within the document classes. With the RBF network parameters being determined and fixed, the second training stage selects the weights for the network output layer. This is essentially a linear modeling problem and is solved by a logistic algorithm in our model.

There are many text classification problems where unlabeled data are readily available and labeled data may be limited in quantity due to labeling cost or difficulty. For instance, labeling newsgroup articles by interest of a news reader can be quite tedious and time-consuming. As another example, categorization of Web pages into subclasses for a search engine is very desirable. However, given the exponential growth rate of the Web, only a tiny percentage of Web content can realistically be hand-labeled and classified. All of these problems require solutions that can learn accurate text classification through limited labeled training data and additional related unlabeled data (background knowledge).

The two-stage training process described above particularly facilitates an integration of additional background unlabeled data. The data used in the first training stage, for determining the network basis functions of the hidden layer, are not required to be labeled and in fact, it is done by a clustering algorithm. However, some empirical analysis on our model in this regard has indicated that far better classification performance is generally achieved if the stage is carried out on a combined training data set that includes both labeled and unlabeled documents. Of course, some labeled data are needed in the second training stage. Overall, the available background unlabeled documents can be used to compensate for insufficient labeled training data and also to further improve the quality of classification decisions of the proposed model.

The model training process can be summarized as follows:

Model Training Process

1. Select data features on all labeled and unlabeled training data
2. Construct training document content vectors that combine feature local and global weights
3. Cluster labeled content vectors in each document class and background unlabeled content vectors by the *k*-means clustering algorithm, and then determine the RBF parameters
4. Determine the network output layer weights by the logistic regression algorithm on labeled content vectors

It should be pointed out that the proposed model that applies both feature selection and the two-stage training is effective in significantly reducing the computational workload for network training and hence, it provides a practical text classifier for applications with large training data.

3 Experiments

The proposed RBF network based text classification model has been applied to email spam filtering, a special and important two-category text classification problem. In this section, the experiments of the model on the benchmark spam testing corpus PU1 and its comparison with the popular SVM and naïve Bayes approaches are described.

4.1 Experiment Settings

The corpus PU1 [2] is made up of 1099 real email messages, with 618 legitimate and 481 spam. The messages in the corpus have been preprocessed with all attachments, HTML tags and header fields except subject being removed, and with all retained words being encoded into numbers for privacy protection. The experiments are performed using the stratified 10-fold cross validation. In other words, the corpus is partitioned into ten equally-sized subsets. Each experiment takes one subset for testing and the remaining nine subsets for training, and the process repeats ten times with each subset takes a turn for testing. The performance is then evaluated by averaging over the ten experiments. Various feature sets are used in the experiments ranging from 50 to 650 with an increment of 100.

The performance of a text classifier can be evaluated by precision and recall. These measurements, however, do not take an unbalanced misclassification cost into consideration. Spam filtering is a cost sensitive learning process in the sense that misclassifying a legitimate message to spam is typically a more severe error than misclassifying a spam message to legitimate. In our experiments, a unified cost sensitive weighted accuracy [2] is used as the performance criterion and it can be defined as

$$WAcc(\lambda) = \frac{\lambda n_{L->L} + n_{S->S}}{\lambda(n_{L->L} + n_{L->S}) + (n_{S->S} + n_{S->L})} \tag{3}$$

where $n_{L->L}$, $n_{L->S}$, $n_{S->S}$ and $n_{S->L}$ denotes the classified message count of correct legitimate, incorrect legitimate, correct spam, and incorrect spam, respectively, and λ is a cost parameter. The *WAcc* formula assumes that the misclassification error on legitimate is λ times more costly than the error on spam. In our experiments, the popular $\lambda = 9$ is used.

4.2 Classification Performance with and without Background Knowledge

As discussed in Section 3, the proposed classification model is capable of incorporating both labeled and unlabeled data (background knowledge) in its learning effectively, and this can be particularly advantageous for the applications where labeled training data are in short supply.

The first part of our experiments was to compare classification performance of the model with and without using background knowledge. The experiments were conducted with the training set being further divided into a labeled and an unlabeled subset. Those known email class labels for the data assigned to the unlabeled subset are ignored in the experiments. When the model is trained with background knowledge, both subsets are used in the first stage of training and only the labeled subset is used in the second stage of training. When the model is trained without background knowledge, only the labeled training subset is used in both stages of training.

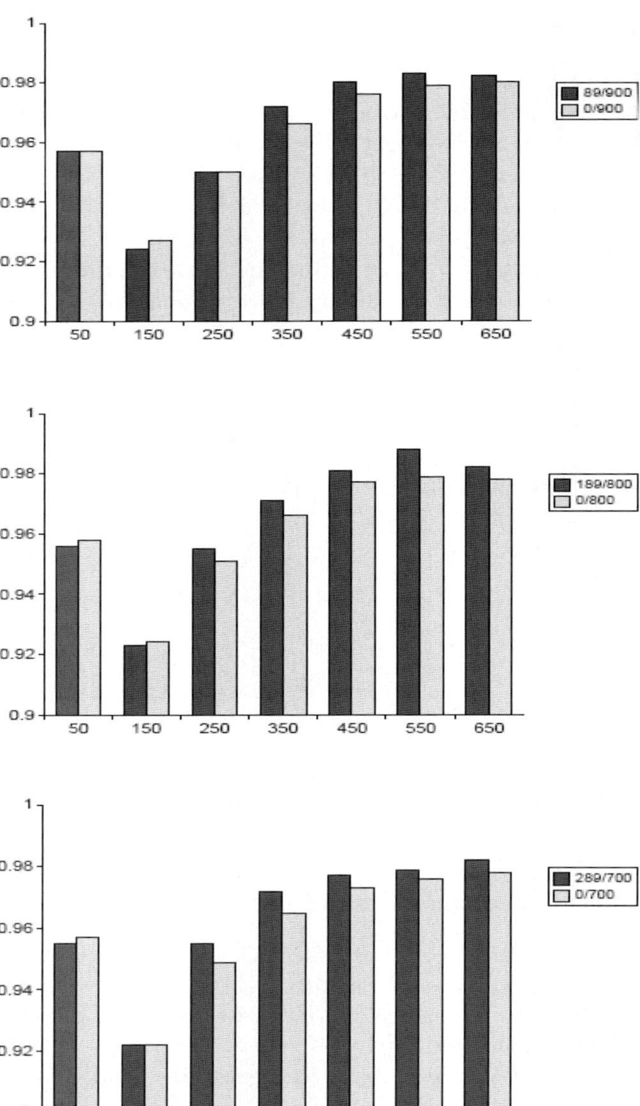

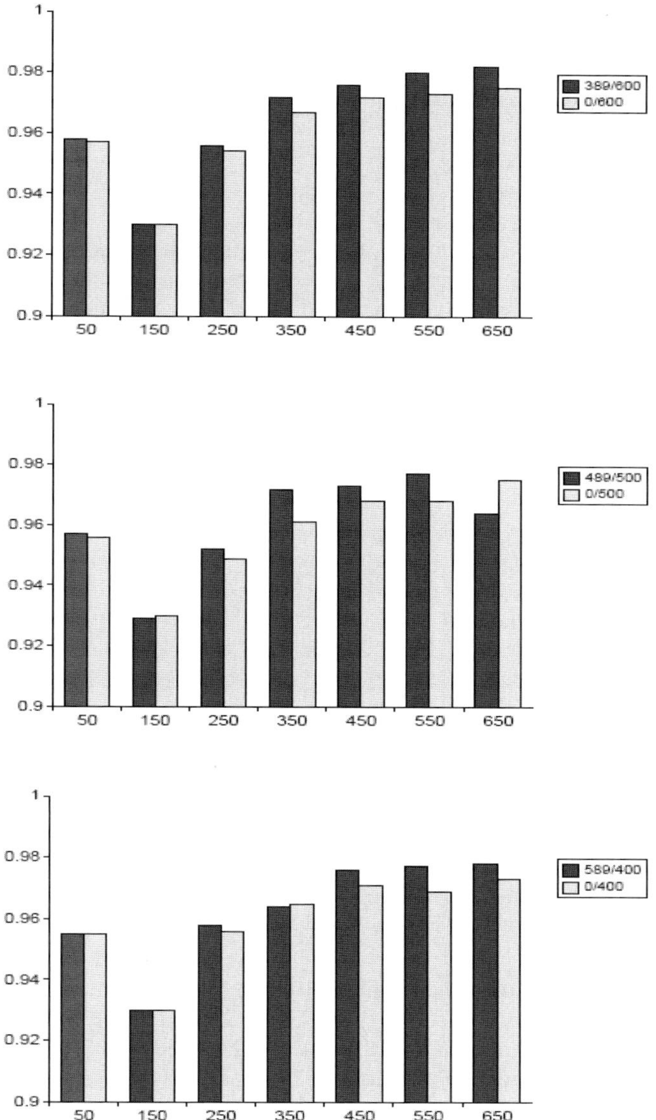

Fig. 2. Summarized classification accuracy values with and without background knowledge

Fig. 2 summarizes the average weighted accuracy results (in y-axis) obtained by the model over various feature set sizes (in x-axis) and several combined unlabeled/labeled training set sizes (as n1/n2 in legend). The labeled training size n2 varies from 900 to 400 with a decrement of 100 whereas the unlabeled training size n1 varies from 89 to 589 with a corresponding increment of 100, and the total combined size of n1 and n2 is 989. It can be observed from Fig. 2 that, as the feature size increases, a trend on accuracy starts with a decent initial value, dips at the size of

150 and then lifts up gradually. For small feature sets, the training with additional background knowledge might not be very beneficial as expected. However, for relatively larger feature sets, background knowledge can help improve the quality of classification decisions, and that includes the cases where labeled training sets are relatively small.

4.3 Comparison with Other Text Classifiers

The second part of our experiments was to compare the proposed model with two well-known classification algorithms: naïve Bayes and SVM [9]. The former is a standard implementation of naïve Bayes and the latter is an implementation of SVM using sequential minimization optimization and a polynomial kernel. Note that the input data to all three classifiers are the same set of document vectors after feature selection and term weighting.

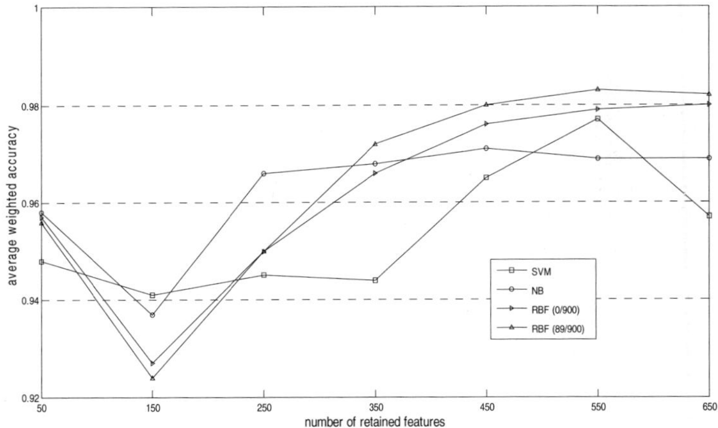

Fig. 3. Average weighted accuracy values with a labeled training set of size 900

The experiments also used the stratified 10-fold cross validation procedure. With a labeled training set of size 900, Fig. 3 compares SVM, naïve Bayes (NB) with two versions of the RBF based model (RBF): one is trained only by a labeled set (marked as 0/900 in the aforementioned notation) and the other is trained by both the labeled set and an additional unlabeled background set of size 89 (marked as 89/900). This setting represents the model training using a relatively small background set and a relatively large labeled set. Clearly, Fig. 3 shows that background knowledge is useful for the RBF model to outperform other algorithms as the feature size gets large enough.

We also looked at different situations for model training where background data sets are relatively large, compared to sets of labeled training data. Fig. 4 shows the performance comparison of these models on a small labeled training set of size 400. Note that the model RBF (598/400) uses an additional background set of size 598 in

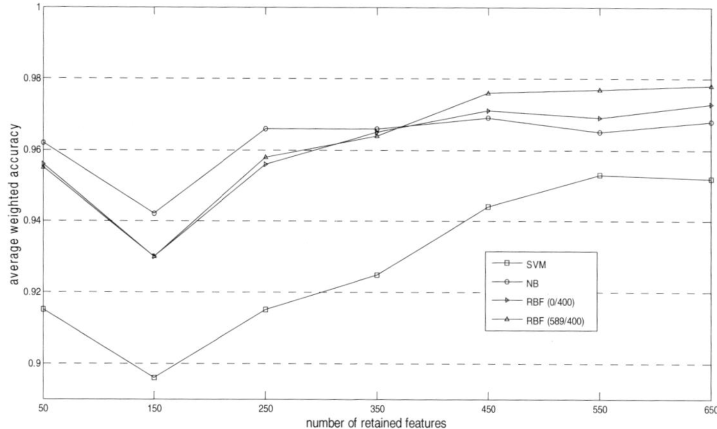

Fig. 4. Average weighted accuracy values with a labeled training set of size 400

the training. Overall, Fig. 4 demonstrates the similar classification outcomes as in Fig. 3, except that SVM seems suffering more noticeably from the reduced labeled training data. Both Fig.3 and Fig.4 have indicated that, with some appropriate feature selection, the training with background knowledge can indeed be beneficial for the RBF model in learning the associations between a class and its constituent documents and in supporting its classification task. It is also interesting to observe that in both cases, naïve Bayes performs very well and its weighted accuracy values stabilize after the feature size reaches 250. Likely, its performance is enhanced by our rigorous feature selection process.

5 Conclusions

The radial basis function (RBF) networks have been successfully used in many applications in science and engineering. In this paper, an RBF-based text classification model is proposed. It is capable of integrating background knowledge into its learning and utilizing valuable discriminative information in training data. The experiments of the model on email spam classification and a comparison with some popular text classifiers have shown that it is very effective in learning to classify documents based on content and represents a competitive text classification approach. As future work, we plan to conduct experiments of the model with general text classification corpora and improve the accuracy and efficiency of the model by further exploring feature-document underlying associations. We also plan to perform some empirical work in comparing the model with other more closely related semi-supervised classification algorithms.

References

1. Aha, W., Albert, M.: Instance-based Learning Algorithms. Machine Learning 6, 37–66 (1991)
2. Androutsopoulos, I., Paliouras, G., Michelakis, E.: Learning to filter unsolicited commercial e-mail, Technical Report 2004/2, NCSR Demokritos (2004)

3. Bishop, C.: Neural Networks for Pattern Recognition. Oxford University Press, Oxford (1995)
4. Christianini, B., Shawe-Taylor, J.: An introduction to Support Vector Machines and other kernel-based learning methods. Cambridge University Press, Cambridge (2000)
5. Jiang, E., Berry, M.: Solving Total Least-Squares Problems in Information Retrieval. Linear Algebra and its Applications 316, 137–156 (2000)
6. Sahami, M., Dumais, S., Heckerman, D., Horvitz, E.: A Bayesian Approach to Filtering Junk E-mail. In: Proceedings of AAAI workshop, pp. 55–62 (1998)
7. Schapier, R., Singer, Y.: BoosTexter: A Boosting-based System for Text Categorization. Machine Learning, 135–168 (2000)
8. Sebastiani, F.: Machine Learning in Automated Text Categorization. ACM Computing Surveys 1, 1–47 (2002)
9. Witten, I., Frank, E.: Data mining, 2nd edn. Morgan Kaufmann, San Francisco (2005)

An Extended Document Frequency Metric for Feature Selection in Text Categorization

Yan Xu, Bin Wang, JinTao Li, and Hongfang Jing

Institute of Computing Technology,Chinese Academy of Sciences
No.6 Kexueyuan South Road, Zhongguancun,Haidian District, Beijing,China
{xuyan,wangbin,jtli,jinghongfang}@ict.ac.cn

Abstract. Feature selection plays an important role in text categorization. Many sophisticated feature selection methods such as Information Gain (IG), Mutual Information (MI) and $\chi 2$ statistic measure (CHI) have been proposed. However, when compared to these above methods, a very simple technique called Document Frequency thresholding (DF) has shown to be one of the best methods either on Chinese or English text data. A problem is that DF method is usually considered as an empirical approach and it does not consider Term Frequency (TF) factor. In this paper, we put forward an extended DF method called TFDF which combines the Term Frequency (TF) factor. Experimental results on Reuters-21578 and OHSUMED corpora show that TFDF performs much better than the original DF method.

Keywords: Rough Set, Text Categorization, Feature Selection, Document Frequency.

1 Introduction

Text categorization is the process of grouping texts into one or more predefined categories based on their content. Due to the increased availability of documents in digital form and the rapid growth of online information, text categorization has become one of the key techniques for handling and organizing text data.

A major difficulty of text categorization is the high dimensionality of the original feature space. Consequently, feature selection-reducing the original feature space,is seriously projected and carefully investigated.

In recent years, a growing number of statistical classification methods and machine learning techniques have been applied in this field. Many feature selection methods such as document frequency thresholding, information gain measure, mutual information measure, $\chi 2$ statistic measure, and term strength measure have been widely used.

DF thresholding, almost the simplest method with the lowest cost in computation, has shown to behave comparably well when compared to more sophisticated statistical measures [13], it can be reliably used instead of IG or CHI while the computation of these measures are more expensive. Especially, experiments show that DF has better performance in Chinese text categorization [1][11] than IG, MI and CHI. In one

H. Li et al. (Eds.): AIRS 2008, LNCS 4993, pp. 71–82, 2008.

word, DF, though very simple, is one of the best feature selection methods either for Chinese or English text categorization.

Due to its simplicity and effectiveness, DF is adopted in more and more experiments[7][4][2][6]. However, this method is only based on an empirical assumption that rare terms are noninformative for category prediction. In addition, like most feature selection methods, DF does not consider the Term Frequency (TF) factor, which is considered to be a very important factor for feature selection[12].

Rough Set theory, which is a very useful tool to describe vague and uncertain information, is used in this paper to give a theoretical interpretation of DF method. In Rough Set theory, knowledge is considered as an ability to partition objects. We then quantify the ability of classify objects, and call the amount of this ability as knowledge quantity. We use the knowledge quantity of the terms to rank them, and then put forward an extended DF method which considers the term frequency factor. Experiments show the improved method has notable improvement in the performances than the original DF.

2 Document Frequency Thresholding and Rough Set Theory Introduction

2.1 Document Frequency Thresholding

A term's document frequency is the number of documents in which the term occurs in the whole collection. DF thresholding is computing the document frequency for each unique term in the training corpus and then removing the terms whose document frequency are less than some predetermined threshold. That is to say, only the terms that occur many times are retained. DF thresholding is the simplest technique for vocabulary reduction. It can easily scale to very large corpora with a computational complexity approximately linear in the number of training documents.

At the same time, DF is based on a basic assumption that rare terms are noninformative for category prediction. So it is usually considered an empirical approach to improve efficiency. Obviously, the above assumption contradicts a principle of information retrieval (IR), where the terms with less document frequency are the most informative ones [9].

2.2 Basic Concepts of Rough Set Theory

Rough set theory, introduced by Zdzislaw Pawlak in 1982 [5][8], is a mathematical tool to deal with vagueness and uncertainty. At present it is widely applied in many fields, such as machine learning, knowledge acquisition, decision analysis, knowledge discovery from databases, expert systems, pattern recognition, etc. In this section, we introduce some basic concepts of rough set theory which used in this paper.

Given two sets U and A, where $U = \{x_1, ..., x_n\}$ is a nonempty finite set of objects called the universe, and $A = \{a_1, ..., a_k\}$ is a nonempty finite set of attributes, the attributes in A is further classified into two disjoint subsets, condition attribute set C and decision attribute set D, $A=C\cup D$ and $C\cap D = \Phi$. Each attribute $a\in A$, V is the domain of values of A, V_a is the set of values of a, defining an information function $f_a, : U\rightarrow V_a$, we call 4-tuple <U,A,V, f > as an information system. a(x) denotes the value of attribute a for object x .

Any subset $B \subseteq A$ determines a binary relation Ind(B) on U, called indiscemibility relation:

$$Ind(B)=\{ (x,y) \in U\times U \mid \forall a\in B , a(x) = a(y) \}$$

The family of all equivalence classes of Ind(B), namely the partition determined by B, will be denoted by U/B . If (x , y) $\in$ Ind(B), we will call that x and y are B-indiscernible .Equivalence classes of the relation Ind(B) are referred to as B - elementary sets.

3 A Rough Set Interpretation of Document Frequency Thresholding

Given a 4-tuple <U,A,V, f > information system for text categorization, where $U=\{D_1, ..., D_n\}$ is a set of documents, $A = \{ t_1,..., t_k \}$ is a set of features (terms), V is the domain of values of t_i ($1\leq i\leq k$), $V=\{0,1\}$, An information function f, $U\rightarrow V$, can be defined as:

$$f(D_i) = \{ \begin{matrix} 0, & t \text{ dosen't occurs in } D_i \\ 1, & t \text{ occurs in } D_i \end{matrix}$$

An example of such an information table is given in Table 1. Rows of Table 1, labeled with D_1, D_2, …D_6, are documents, the features are T_1, T_2, T_3 and T_4.

3.1 The Ability to Discern Objects

The important concept in rough set theory is indiscernibility relation. For example, in Table 1, (D_1, D_2) is T_1-indiscernible, (D_1, D_3) is not T_1-indiscernible.

Table 1. An information table: terms divide the set of documents into two equivalence classes

	T_1	T_2	T_3	T_4
D_1	0	0	1	1
D_2	0	1	0	1
D_3	1	0	1	1
D_4	0	0	1	1
D_5	0	0	0	1
D_6	0	1	0	1

In Table 1, T_1 only occurs in D_3, so T_1 divides $\{D_1, D_2, \ldots D_6\}$ into two equivalence classes $\{D_1, D_2, D_4, D_5, D_6\}$ and $\{D_3\}$. That is to say, T_1 can discern D_3 from D_1, D_2, D_4, D_5, D_6. Similarly, T_2 can discern D_2, D_4 from D_1, D_3, D_5, D_6. T_3 can discern D_1, D_3, D_4 from D_2, D_5, D_6. T_4 can not discern each document from the other, because T_4 divides $\{D_1, D_2, \ldots D_6\}$ into only one equivalence class. Now we quantify the ability of discerning objects for a feature T_i or a set of features P, we call the amount of the ability of discerning objects as *knowledge quantity*.

3.2 Knowledge Quantity

This section will be discussed on information table (Let decision feature set $D = \Phi$).

Definition 1. The object domain set U is divided into m equivalence classes by the set P (some features in information table), the number of elements in each equivalence class is: n_1, n_2,..., n_m, let $W_{U,P}$ denotes the knowledge quantity of P, $W_{U,P}=W(n_1, n_2, \ldots, n_m)$, and it satisfies the following conditions:

1) $W(1,1) = 1$
2) if $m = 1$ then $W(n_1)=W(n) = 0$
3) $W(n_1, \ldots, n_i, \ldots, n_j, \ldots, n_m) = W(n_1, \ldots, n_j, \ldots, n_i, \ldots, n_m)$
4) $W(n_1,n_2, \ldots, n_m) = W(n_1,n_2+\ldots+n_m) + W(n_2,\ldots, n_m)$
5) $W(n_1,n_2+n_3) = W(n_1,n_2) + W(n_1,n_3)$

Conclusion 1. If the domain U is divided into m equivalence classes by some feature set P, and the element number of each equivalence class is $n_1,n_2, \ldots n_m$, then the knowledge quantity of P is: $W(n_1,n_2,\ldots,n_m)= \sum_{1\le i<j\le m} n_i \times n_j$.

3.3 Interpretation of Document Frequency Thresholding

In Table 1, T_1 only occurs in D_3, T_1 divides $\{D_1, D_2, \ldots D_6\}$ into two equivalence classes $\{D_1, D_2, D_4, D_5, D_6\}$ and $\{D_3\}$, the number of each equivalence classes is $n_1=5$, $n_2=1$. According to Conclusion 1, the ability of discern $\{D_1, D_2, \ldots, D_6\}$ for T_1 (the knowledge quantity of T_1) is: $W_{U,T_1} = \sum_{1\le i<j\le 2} n_i \times n_j = 5\times1=5$

Let U denote a set of all documents in the corpus, n denotes the number of documents in U, m demotes the number of documents in which term t occurs, the knowledge quantity of t is defined to be:

$$W_{U,t} = m(n-m) \tag{1}$$

$\because$ m=DF

$\therefore$ When $m\le n/2$, DF $\propto W_{U,t}$

After stop words removal, stemming, and converting to lower case, almost all term's DF value is less than n/2.

We compute the knowledge quantity for each unique term by (1) in the training corpus and remove those terms whose knowledge quantity are less than some predetermined threshold, this is our Rough set-based feature selection method which do not consider term frequency information(RS method). Feature selected by DF is the same as selected by this RS method. This is an interpretation of DF method.

4 An Extended DF Method Based on Rough Set

DF method does not consider the term frequency factor, however, a term with high frequency in a document should be more informative than the term that occurs only once. So, we think that terms divide the set of documents into not only two equivalence classes, but more than two equivalence classes, a term occurs in a document at least twice should be different from once occurs in the document, so there are 3 equivalence classes in our method.

Given a 4-tuple <U,A,V, f > information system, U={D_1, ..., D_n} is a set of documents, A = { t_1,..., t_k } is a set of features (terms), V is the domain of values of t_i (1≤i≤k), V={0,1,2}, defining an information function f, U→V:

$$f(D_i) = \begin{cases} 0, & t \text{ dosen't occurs in } D_i \\ 1, & t \text{ once occurs in } D_i;\ 2,\ t \text{ occurs in } D_i \text{ at least twice} \end{cases}$$

An example of such an information system is given in Table 2.

Table 2. An information table: terms divide the set of documents into three equivalence classes

	T_1	T_2
D_1	1	1
D_2	0	0
D_3	0	0
D_4	0	0
D_5	1	2
D_6	0	0

In Table 2, term T_1 occurs once both in D_1 and D_5, T_2 occurs once in D_1 but occurs more than once in D_5, the document frequency of term T_1 and T_2 is the same, but

$$W_{U,T_1} = \sum_{1\le i<j\le 2} n_i \times n_j = 2\times4=8$$

$$W_{U,T_2} = \sum_{1\le i<j\le 3} n_i \times n_j = (1\times1+1\times4+1\times4)=9,$$

$$W_{U,T_2} > W_{U,T_1}$$

Let n denotes the number of documents in the corpus, term t divides the documents into 3 equivalence classes, and the number of elements in each equivalence class is: n_1, n_2, n_3. n_1 denotes the number of documents which t does not occurs, n_2 denotes the number of documents which t occurs only once, n_3 denotes the number of documents which t occurs at least twice. The knowledge quantity of t is defined as:

$$W_{U,t} = \sum_{1 \le i < j \le 3} n_i \times n_j \tag{2}$$

In order to emphasize the importance of multiple occurences of a term, equation (2) can be changed to:

$$\text{TFDF(t)} = (n_1 \times n_2 + c(n_1 \times n_3 + n_2 \times n_3)) \tag{3}$$

Here c is a constant parameter(c≥1). As the value of c increases, we give more weight for multiple occurrences of a term.

Given a training corpus, we compute the TFDF(t) by (3) for all terms and rank them, then remove those terms which are in an inferior position from the feature space, this is our feature selection method based on rough set theory, we call it term frequency-based document frequency(TFDF).

5 Experimental Results

Our objective is to compare the original DF with the TFDF method. A number of statistical classification and machine learning techniques have been applied to text categorization, we use two different classifiers, k-nearest-neighbor classifier (kNN) and Naïve Bayes classifier. We use kNN, which is one of the top-performing classifiers, evaluations [14] have shown that it outperforms nearly all the other systems, and we selected Naïve Bayes because it is also one of the most efficient and effective inductive learning algorithms for classification [16]. According to [15], micro-averaging precision was widely used in Cross-Method comparisons, here we adopt it to evaluate the performance of different feature selection methods.

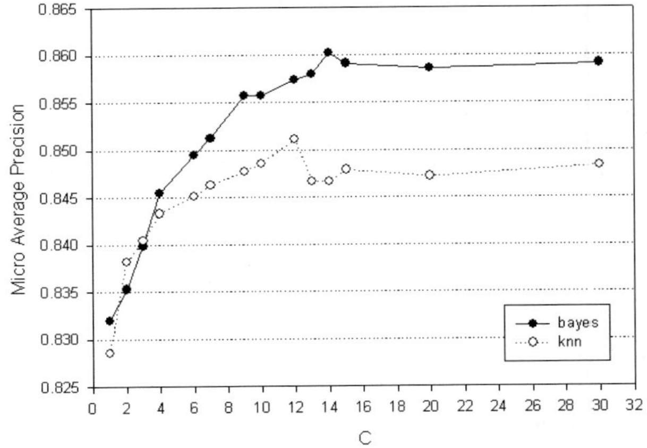

Fig. 1. Average precision of KNN and Naïve Bayes vary with the parameter c in Reuters (using TFDF method)

5.1 Data Collections

Two corpora are used in our experiments: Reuters-21578 collection[17] and the OHSUMED collection[19].

The Reuters-21578 collection is the original Reuters-22173 with 595 documents which are exact duplicates removed, and has become a new benchmark lately in text categorization evaluations. There are 21578 documents in the full collection, less than half of the documents have human assigned topic labels. In our experiment, we only consider those documents that had just one topic, and the topics that have at least 5 documents. The training set has 5273 documents, the testing set has 1767 documents. The vocabulary number is 13961 words after stop words removal, stemming, and converting to lower case.

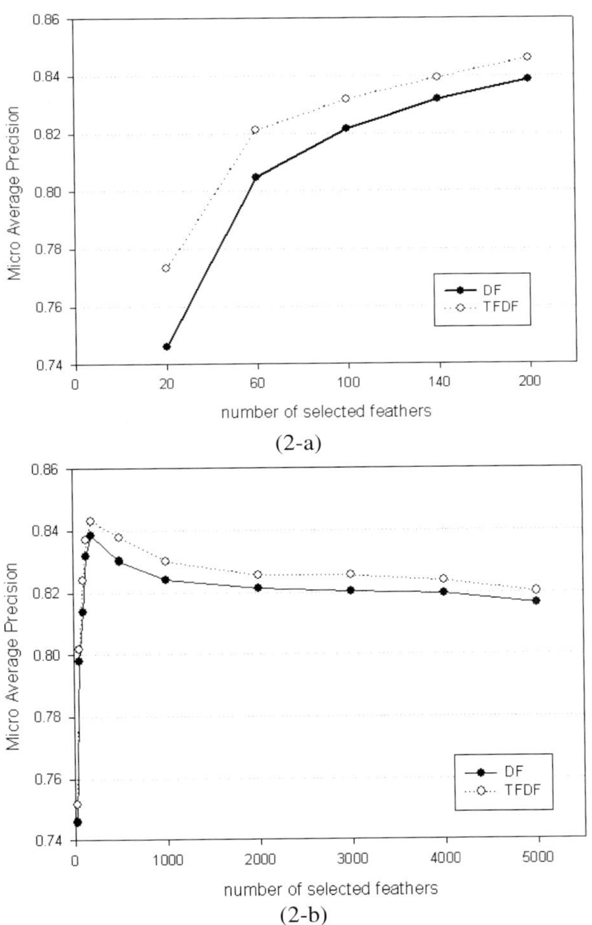

Fig. 2. Average precision of KNN vs. DF and TFDF number of selected features in Reuters

OHSUMED is a bibliographical document collection. The documents were manually indexed using subject categories in the National Library of Medicine. There are about 1800 categories defined in MeSH, and 14321 categories present in the OHSUMED document collection. We used a subset of this document collection. 7445 documents as a training set and the 3729 documents as the test set in this study. There are 11465 unique terms in the training set and 10 categories in this document collection.

5.2 Experimental Setting

Before evaluating the feature selection methods, we use the same selected feature number in both DF method and the TFDF method for the experiment. Weka[18] is used as our experimental platform.

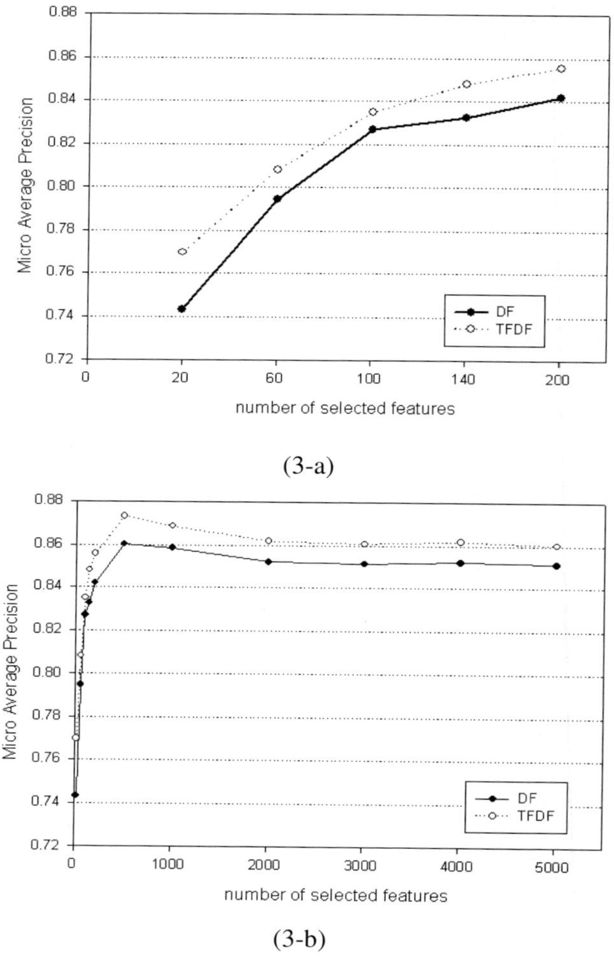

Fig. 3. Average precision of Naïve Bayes vs. DF and RS number of selected features in Reuters

5.3 Results

Figure 1 shows that the Average precision of KNN and Naïve Bayes varying with the parameter c in Reuters (in equation (3), using TFDF method) at a fixed number of selected features, here, the fixed number of selected features is 200. We can notice that when c≤12, as c increases, the Average precision increases accordingly.

Figure 2 and Figure 3 exhibit the performance curves of kNN and Naïve Bayes on Reauters-21578 after feature selection DF and TFDF(c=10). We can note from figure 2 and figure 3 that TFDF outperform DF methods, specially, on extremely aggressive reduction, it is notable that TFDF prevalently outperform DF((2-a),(3-a)).

Figure 4 and Figure 5 exhibit the performance curves of kNN and Naïve Bayes on OHSUMED after feature selection DF and TFDF(c=10). We can also note from figure 2 and figure 3 that TFDF outperform DF methods, specially, on extremely aggressive reduction, it is notable that TFDF prevalently outperform DF((4-a),(5-a)).

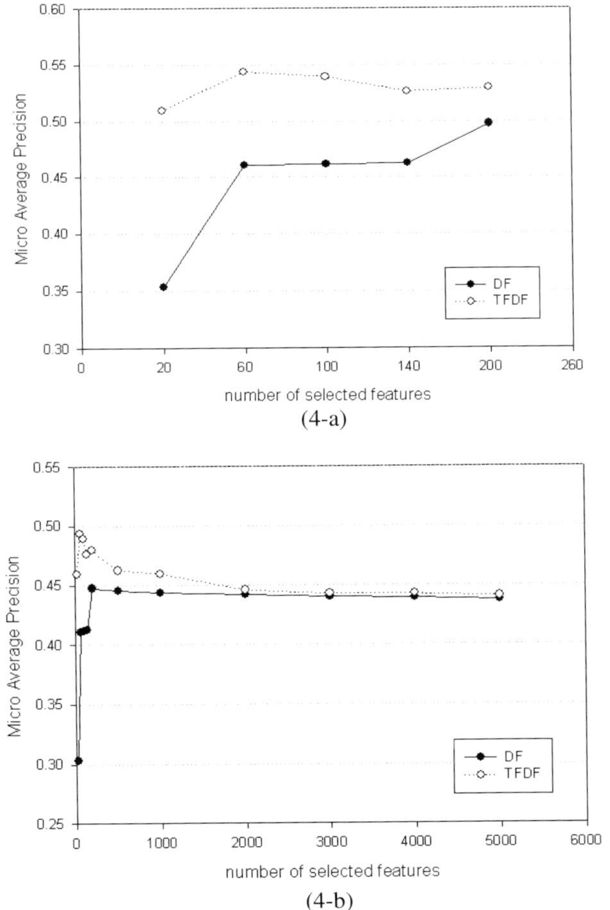

Fig. 4. Average precision of KNN vs. DF and TFDF number of selected features on OHSUMED

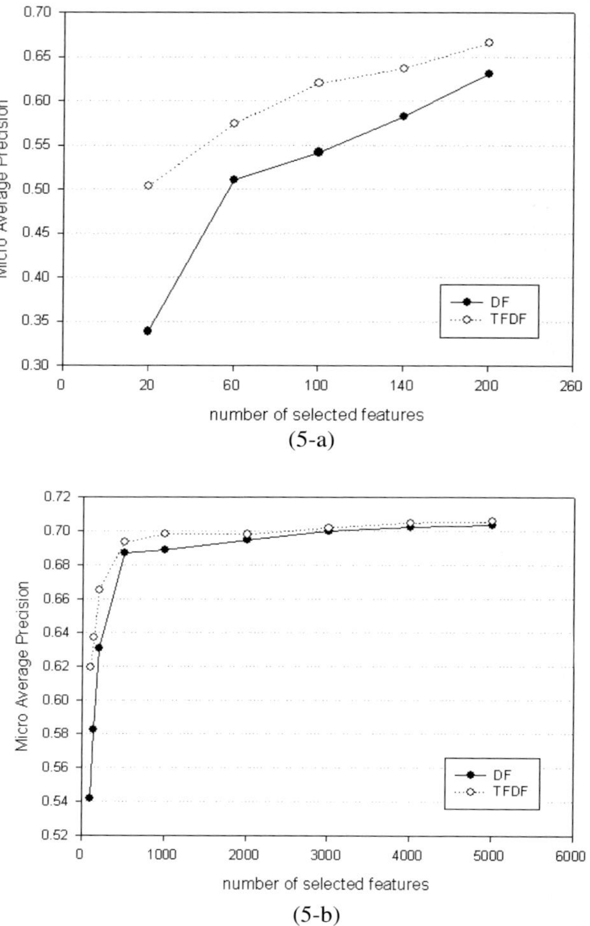

Fig. 5. Average precision of Naïve Bayes vs. DF and TFDF number of selected features on OHSUMED

6 Conclusion

Feature selection plays an important role in text categorization. DF thresholding, almost the simplest method with the lowest cost in computation, has shown to behave well when compared to more sophisticated statistical measures, However, DF method is usually considered as an empirical approach and does not have a good theoretic interpretation, and it does not consider Term Frequency (TF) factor, in this paper: we put forward an extended DF method called TFDF which combines the Term Frequency (TF) factor. Experiments on Reuters-21578 collection and OHSUMED collection show that TFDF perform much better than the original DF method, specially, on extremely aggressive reduction, it is notable that TFDF prevalently outperform DF. The experiments also show that Term Frequency factor is important for feature selection.

Many other feature selection methods such as information gain measure, mutual information measure, $\chi 2$ statistic measure, and term strength measure have been widely used in text categorization, but none of them consider Term Frequency (TF) factor, In the future research we will investigate to use TF in these feature selection methods.

Acknowledgments. This work is supported by the National Natural Science Fundamental Research Project of China (60473002, 60603094), the National 973 Project of China (2004CB318109)and the National Natural Science Fundamental Research Project of Beijing (4051004).

References

1. Liu-ling, D., He-yan, H., Zhao-xiong, C.: A comparative Study on Feature Selection in Chinese Text Categorization. Journal of Chinese Information Processing 18(1), 26–32 (2005)
2. Dumais, S.T., Platt, J., Heckerman, D., Sahami, M.: Inductive learning algorithms and representations for text categorization. In: Proceedings of CIKM 1998, pp. 148–155 (1998)
3. Sebastiani, F.: Machine learning in automated text categorization. ACM Computing Surveys 34(1), 1–47 (2002)
4. Itner, D.J., Lewis, D.D.: Text categorization of low quality images. In: Proceedings of SDAIR 1995, pp. 301–315 (1995)
5. Komorowski, J., Pawlak, Z., Polkowski, L., Skowron, A.: Rough sets: A tutorial. In: A New Trend in Decision-Making, pp. 3–98. Springer, Singapore (1999)
6. Li, Y.H., Jain, A.K.: Classification of text documents. Comput. J. 41(8), 537–546 (1998)
7. Maron, M.: Automatic indexing: an experimental inquiry. J. Assoc. Comput. Mach. 8(3), 404–417 (1961)
8. Pawlak, Z.: Rough Sets. International Journal of Computer and Information Science 11(5), 341–356 (1982)
9. Salton, G., Buckley, C.: Term-weighting approaches in automatic text retrieval. Inform. Process. Man 24(5), 513–523 (1988)
10. Salton, G., Wong, A., Yang, C.: A vector space model for automatic indexing. Commun. ACM 18(11), 613–620 (1975)
11. Songwei, S., Shicong, F., Xiaoming, L.: A Comparative Study on Several Typical Feature Selection Methods for Chinese Web Page Categorization. Journal of the Computer Engineering and Application 39(22), 146–148 (2003)
12. Yang, S.M., Wu, X.-B., Deng, Z.-H., Zhang, M., Yang, D.-Q.: Modification of Feature Selection Methods Using Relative Term Frequency. In: Proceedings of ICMLC 2002, pp. 1432–1436 (2002)
13. Yang, Y., Pedersen, J.O.: Comparative Study on Feature Selection in Text Categorization. In: Proceedings of ICML 1997, pp. 412–420 (1997)

14. Yang, Y., Liu, X.: A re-examination of text categorization methods. In: SIGIR 1999, pp. 42–49 (1999)
15. Yang, Y.: An evaluation of statistical approaches to text categorization. Journal of Information Retrieval 1(1/2), 67–88 (1999)
16. Zhang, H.: The optimality of naive Bayes. In: The 17th International FLAIRS conference, Miami Beach, May 17-19 (2004)
17. Reuters 21578, http://www.daviddlewis.com/resources/testcollections/reuters21578/
18. Weka, http://www.cs.waikato.ac.nz/ml/weka/
19. OHSUMED, http://www.cs.umn.edu/%CB%9Chan/data/tmdata.tar.gz

Smoothing LDA Model for Text Categorization

Wenbo Li[1,2], Le Sun[1], Yuanyong Feng[1,2], and Dakun Zhang[1,2]

[1] Institute of Software, Chinese Academy of Sciences, No. 4, Zhong Guan Cun South 4th Street, Hai Dian, 100190, Beijing, China
[2] Graduate University of Chinese Academy of Sciences, No. 19, Yu Quan Street, Shi Jin Shan, 100049, Beijing, China
{liwenbo02,sunle,yuanyong02,dakun04}@iscas.cn

Abstract. Latent Dirichlet Allocation (LDA) is a document level language model. In general, LDA employ the symmetry Dirichlet distribution as prior of the topic-words' distributions to implement model smoothing. In this paper, we propose a data-driven smoothing strategy in which probability mass is allocated from smoothing-data to latent variables by the intrinsic inference procedure of LDA. In such a way, the arbitrariness of choosing latent variables' priors for the multi-level graphical model is overcome. Following this data-driven strategy, two concrete methods, Laplacian smoothing and Jelinek-Mercer smoothing, are employed to LDA model. Evaluations on different text categorization collections show data-driven smoothing can significantly improve the performance in balanced and unbalanced corpora.

Keywords: Text Categorization, Latent Dirichlet Allocation, Smoothing, Graphical Model.

1 Introduction

Text representation is one of the main difficulties in current text categorization research [1]. The document is usually represented as a vector by weighted index terms. A typical method, the Bag of Words (BOW), identifies all the words occurring in the corpus as index terms. For its simplicity and usability, BOW has been widely adopted in text categorization. However, BOW results in the high dimensions of feature space and information loss of the original texts. In order to overcome these drawbacks, many text representation methods have been proposed in two main directions:

In one direction, diverse language units, such as phrase [2], word senses and syntactic relations [3], are selected as index terms of documents rather than words. Extensive experiments have shown that these complex unites do not yield significantly better effectiveness. Likely reasons for the discouraging result, as stated in [4], is that although these higher level units have superior semantic qualities for human, they have inferior statistical qualities against words for classification algorithm to tackle.

In the other direction, topic-models, such as, Distributional Words Clustering (DWC) [5], Latent Semantic Indexing (LSI) [6] and LDA [7] [8] etc. are proposed to

H. Li et al. (Eds.): AIRS 2008, LNCS 4993, pp. 83–94, 2008.

extract some kinds of latent information structures of documents beyond words. Among these models, LDA has attracted much attention for its dimension reducing power, comprehensibility and computability. Whereas LDA has been extensively applied in machine learning, information retrieval [9] and some NLP tasks, the potential of LDA in text categorization isn't systematically explored. In addition, the size of corpora and/or categories in former related studies [8] [10] are relative small.

In this paper, we will study how to effectively smooth LDA model for improving text categorization under the generative framework. Our works concentrate on followings: we analyze the irrationality of directly smoothing the latent variable distributions from inside of multi-level graphical models. Accordingly, we propose the data-driven smoothing strategy for LDA model and its two instances. Experiment evaluations show this data-driven strategy can significantly improve the performance on both balanced and unbalanced text categorization corpora.

This paper is arranged as follows. Section 2 shows a brief of LDA model. Our proposed data-driven smoothing strategy and two concrete methods, L_LDA model and JM_LDA model, are presented in section 3. Experiment evaluations and analyses are given in section 4. Section 5 is the related work, and section 6 gives a conclusion.

2 The LDA Model

LDA [8] is a document-level generative probabilistic language model (in Figure 1), in which each document is modeled as a random mixture over latent topics. Each topic is, in turn, modeled as a random mixture over words. The generative process for each document **w** can be described as following steps:

1: Select a latent topics mixture vector $\boldsymbol{\theta}$ from the Dirichlet distribution $Dir(\boldsymbol{\alpha})$.
2: For each word w_n
  2.1: Select a latent topic z_n from the multinomial distribution *Multinomial* $(\boldsymbol{\alpha})$.
  2.2: Select the word w_n from the multinomial distribution *Multinomial* $(\boldsymbol{\beta}, z_n)$.

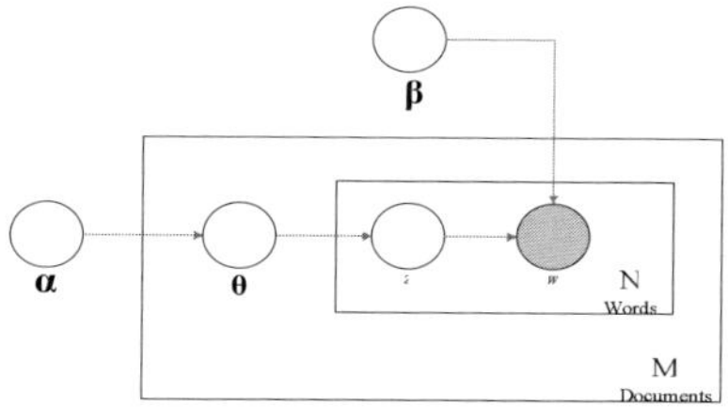

Fig. 1. LDA model, which has 3 tiers: *corpus* ~ $(\boldsymbol{\alpha}, \boldsymbol{\beta})$, *doc* ~ $\boldsymbol{\theta}$ and *word* ~ (z, w)

LDA is a directed graphical model and the joint distribution is given by formula (1):

$$p(\boldsymbol{\theta},\mathbf{z},\mathbf{w}|\boldsymbol{\alpha},\boldsymbol{\beta}) = p(\boldsymbol{\theta}|\boldsymbol{\alpha})\left(\prod_{n=1}^{N} p(z_n|\boldsymbol{\theta})p(w_n|z_n,\boldsymbol{\beta}) \right) \tag{1}$$

For probabilistic language models, the critical task is to calculate the probability of the marginal distribution of a document. This can be obtained for LDA by:

$$p(\mathbf{w}|\boldsymbol{\alpha},\boldsymbol{\beta}) = \int p(\boldsymbol{\theta}|\boldsymbol{\alpha})\left(\prod_{n=1}^{N}\sum_{z_n} p(z_n|\boldsymbol{\theta})p(w_n|z_n,\boldsymbol{\beta}) \right) d\boldsymbol{\theta} \tag{2}$$

As the formula (**2**) shows: Given a LDA model instance defined by parameters (**α, β**), the whole generative procedure has two sub-procedures: one is integrating over latent topics mixture vector **θ** and the other is summing over latent topic z_n. In such a way, LDA model can synthesizes the topic structures implicated in the document and the distributional information of words over these latent topics together. This is the key advantage compared with other language models like N-gram model which only consider the distribution of words.

2.1 Applying LDA Model in Text Categorization

Under the generative probability framework of classification can be formulated as (**3**), in which $p(\mathbf{w}|c_i)$ is the generative probability of the document **w** belong to class c_i.

$$c = \arg\max_i p(c_i|\mathbf{w}) = \arg\max_i p(\mathbf{w}|c_i) * p(c_i) \tag{3}$$

Furthermore, the choice of the generative probability model of the document **w** is the key point. When LDA model is selected for $p(\mathbf{w}|c_i)$ as: $p(\mathbf{w}|c_i)= p_{lda}(\mathbf{w}|\boldsymbol{\alpha}^{(i)},\boldsymbol{\beta}^{(i)})$, we obtain a new generative classifying model in formula (4):

$$c = \arg\max_i p_{lda}(\mathbf{w}|\boldsymbol{\alpha}^{(i)},\boldsymbol{\beta}^{(i)}) * p(c_i) \tag{4}$$

3 Smoothing LDA Model by Data-Driven Strategy

Before using the LDA model, parameters (**α, β**) must be determined. As the other language models (like n-gram, hmm, etc.), A smoothing procedure must be added to the parameter **β** to overcome the zero probability caused by OOV(out of vocabulary). Let v_i denote the vocabulary of the documents in category c_i ,thus the OOV of category c_i, is formulated as (5):

$$oov_i = \bigcup_{j \neq i} v_j - v_i \tag{5}$$

For multi-class text categorization problem, the increase of the class number will increase the oov_i number in every class. More uncertainty will be introduced. As a result, the deviation of the training of the LDA smoothing method will also increase.

In order to obtain a tractable posterior in the LDA model setting, the Dirichlet prior $Dir(\eta)$ is placed on the multinomial parameter **β**. Then the variational Bayesian inference method [11] [12] can be used to execute on the LDA model and derive the estimation equation of parameter **β** in formula (6):

$$\beta_{ij} = \eta + \sum_{d=1}^{M}\sum_{n=1}^{N(d)} \varphi_{dni} w_{dn}^{*j} \tag{6}$$

It is obvious that $\boldsymbol{\beta}_{ij}$ can avoid zero probability for any word index j because of the positive η . Although this additional Dirichlet prior can play some role of model smoothing, it is arbitrary to operate the latent variable distributions directly in the multi-level graphical model. In the iterative procedure of training the LDA model, **β** is coupled with other latent variables such as $\boldsymbol{\theta}$, $\mathbf{z}$ and $\boldsymbol{\alpha}$. In principle, the smoothing of $\boldsymbol{\beta}$ should take account of these latent variables. But in the formula (5), only a fixed const is added in the estimation for the reason of simplicity of calculation.

3.1 Data-Driven Strategy

This can't be overcome directly by the Dirchlet smoothing method from inside the LDA model. We construct virtual instances which cover all words in global vocabulary and add them to the training set. The LDA model would be smoothed and avoid the zero probability problem during predicting phase. We call this strategy as data-driven smoothing. The most virtue of this data-driven smoothing is that it can take use of the inference mechanism rooted in LDA model to integrate with the inference of $\boldsymbol{\theta}$ and $\mathbf{z}$. In such a natural way, appropriate probability mass are allocated toward OOV. So, this strategy can avoid that the OOV's probabilities are calculated isolated in the traditional LDA model.

In this paper, two concrete smoothing methods following the data-driven strategy are introduced. We name them as Laplace LDA model (L_LDA) and Jelinek-Mercer LDA model (JM_LDA) which are discussed in detail in following.

3.2 Laplacian Smoothing LDA (L_LDA) Model

We first choose the Laplacian smoothing method, in formula (7), for the every class of documents in the corpus:

$$\hat{p}(w_t|c_i) = (n_i(w_t) + a_t) \,/\, (n(c_i) + A),\ A = \sum\nolimits_{t=1}^{N} a_t \tag{7}$$

The smoothing procedure is not embeded in the LDA model. In implement, a virtual instance is constructed and added to the class c_i. All words in the global vocabulary are selected as features in this instance with $|w_t|=a_t$. In such a way, proper probability is set to OOV of each class. Iit is guaranteed that the probability mass of OOV will not decrease to a too low level in the iteration of the training of the model.

3.3 Jelinek-Mercer Smoothing LDA (JM_LDA) Model

From another way, we can use the L_LDA as a background model under the Jelinek-Mercer Smoothing framework which has been used widely in information retrieval. We formulize this in (8):

$$p_{JM\_LDA}(w_t \mid c_i) = \lambda p_{MLE}(w_t \mid c_i) + (1\text{-}\lambda)\, p_{L\_LDA}(w_t \mid c_i),\ 0<\lambda<1 \tag{8}$$

For the Jelinek-Mercer Smoothing on 1-gram language model, there is an assumption that the occurrence of the words in a document is position-free and then we obtain the multinomial generative model [13] for a specific class where w_{t1} is statistically conditional independent on the occurrence of any other w_{t2} given the document's type c_i. We can formulate this idea in formula (9) which will be used in the phrase of predicating:

$$p(\mathbf{w}|c_i) = \prod\nolimits_{t=1}^{|V|} p(w_t|c_i)^{n(t)} = \prod\nolimits_{t=1}^{|V|} p_{JM\_LDA}(w_t|c_i)^{n(t)} \tag{9}$$

We can use the MLE to estimate $p_{JM\_LDA}(w_t|c_i)$ by formula (10):

$$\hat{p}_{MLE}(w_t|c_i)= n_i(w_t) / n(c_i) \tag{10}$$

This item must be attached some other smoothing method. Here, L_LDA is introduced to do this job in formula (11):

$$\hat{p}_{LDA}(w_t|c_i) = \int p(\boldsymbol{\theta}|\boldsymbol{\alpha}^{(i)})(\sum_{z_n} p(z_n|\boldsymbol{\theta})p(w_n|z_n,\boldsymbol{\beta}^{(i)}))\, d\boldsymbol{\theta} \tag{11}$$

As we know that MLE parameter estimating method only uses the explicit word frequency information. In contrast, the LDA parameter estimating method uses both latent topics distribution information $\boldsymbol{\alpha}^{(i)}$ and word distribution information $\boldsymbol{\beta}^{(i)}$ of the latent topics related to the corresponding category c_i to estimate the probability $p(w_t|c_i)$, as shown in formula (11).

The integral in formula (11) take into account the marginalization to synthetically consider all the possible combination given the latent topics distribution $\boldsymbol{\theta}$. Existed research[8] indicates that LDA can effectively dig out the internal structures (latent topics) of the document. Hence this method can more accurately estimate the weight of the specific word by using the internal structural information. This is the advantage compared with the MLE estimate method which only uses the frequency information of words.

The integration (11) is normally calculated by variational method [11] [12]. One pass of variational inference for LDA requires $O((N+1)k)$ operations. Empirically, the number of iterations required for a single document is on the order of the number of words in the document. This yields a total number of operations roughly on the order of $O(N^2k)$. Thus, the parameters estimation for Jelinek-Mercer Smoothing has the order of about $O(k)$ because that the "document" in (11) have only 1 word, so the N=1. It is obviously that this is a quickly calculating procedure.

Consequently, applying LDA to parameter estimating process but not to predicating phase will increase the predicating speed as in formula (9). The order at predication phrase of JM_LDA Model is $O(N)$, which is equal to the order of classical NaïveBayes multinomial model whereas the order at predication phrase of L_LDA model is $O(N^2k)$. This is very important for the real application environment.

4 Experience Research

4.1 Experiment Setting

In our experiments, the following test collections are used:

20newsgroup: It is collected by Ken Lang for text categorization research. All documents distributed across 20 categories evenly **(balanced)**. We remove all headers, common stop words by the tools "***rainbow***"[1] of CMU. The last vocabulary size is 47802. We randomly select 50% of documents per class for training and the remaining 50% for testing.

Fudan: It is published on http://www.nlp.org.cn. This is a Chinese text categorization corpus with 20 categories and nearly 20,000 documents totally. This corpus classes

[1] Rainbow -d ~/model -h -O 3 --index ~/20_newsgroups/*.

are **unbalanced**. We remove common stop words and the last vocabulary size is 79093. We also randomly select 50% of documents per class for training and the remaining 50% for testing.

These are summarized in the table 1:

Table 1. The basic situation of two data sets: 20Newsgroups & fudan

	20Newsgroups	**fudan**
Category Size	20	20
Corpus Size	about 20,000	about 20,000
Vocabulary Size	47802	79093
Train/Test	50% vs. 50%	50% vs. 50%
Distributation	even(balanced)	unbalanced

In this section, we evaluate the traditional LDA model and two LDA smoothing methods: L_LDA and JM_LDA. We firstly concentrate on the **synthesis** classification performance measured by $\boldsymbol{micro\_F_1}$ and $\boldsymbol{Macro\_F_1}$ across all classes on each corpus. Then we dive into the **detailed** classification performance of every class on the **unbalanced** classification to enucleate the significant improvement obtained on **fudan** corpus.

In all our experiments, the *LDA-C* (implemented by D. Blei) with default settings is chosen as the basic software. On the other side as you have seen, the JM_LDA model has two super parameters: topic number and mix coefficient λ. As in other models, we can use a two-dimensional grid search to find the optimum value for λ. So, after you fix the topic number, you can only search the optimum mix coefficient λ.

4.2 Experiment & Analysis

Synthesis Classification Performance

In this part of experiment, the $\boldsymbol{micro\_F_1}$ and the $\boldsymbol{Macro\_F_1}$ measures are used to evaluate the synthesis classification performance on **20newsgroup** and **fudan** corpus. We perform the comparison among three methods: traditional LDA model and our L_LDA model and JM_LDA model as figure 2. The *X*-axis is the number of latent topics used in the three models and the *Y*-axis is the synthesis performance measure ($\boldsymbol{micro\_F_1}$ or $\boldsymbol{Macro\_F_1}$).

For the LDA model, the number of latent topics is an important factor which defines the granularity of the model. So, we evaluate models on different topic number. From the figure 2.1 and 2.2, we can get the following things:

(1) On the whole, the L_LDA model and JM_LDA model have a very near performance where the L_LDA has a small superiority of about 1% than JM_LDA. At the same time, both L_LDA and JM_LDA have an evident higher performance of about 5%~6% than LDA model. This occurs across all value of topic number.

(2) When different latent topic number is selected, the behaviors of the three models are different. With the increase of topic number, the LDA vibrates in a narrow interval and the JM_LDA remains much stably. Among the three models, only the L_LDA model increases monotonously along with this model expanding.

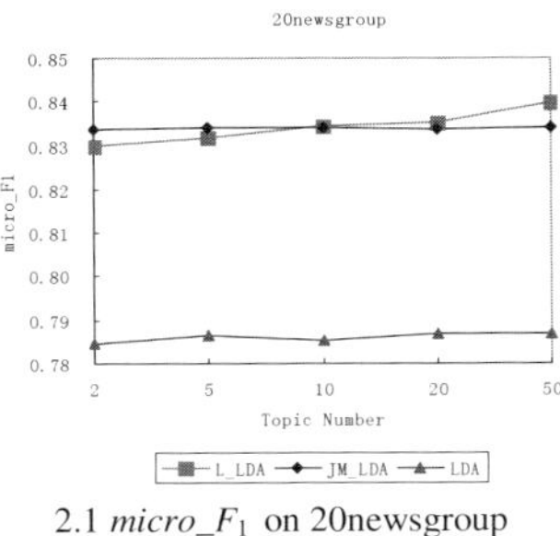

2.1 *micro_F*$_1$ on 20newsgroup

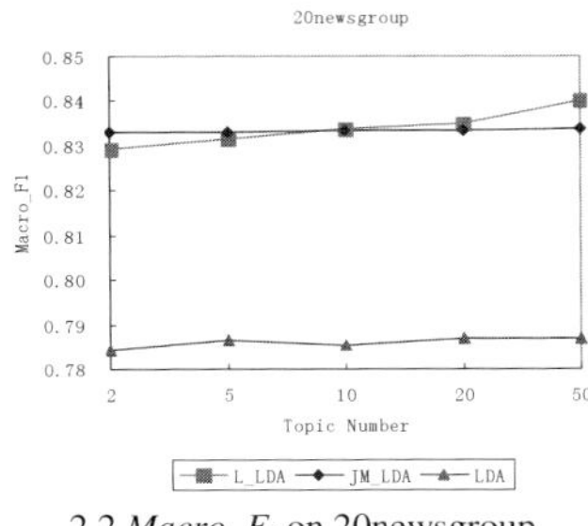

2.2 *Macro_F*$_1$ on 20newsgroup

Fig. 2. Comparison between LDA model and two Smoothed LDA models, L_LDA and JM_LDA, on 20newsgroup corpus

(3) Comparing the figure 2.1 and 2.2, we can find that the performances measured by ***micro_F*$_1$** and ***Macro_F*$_1$** have the very consistent results, namely: the absolute performance value of every model, the relative performance difference among three models and the performance tendencies exhibited on the sequence of topic numbers.

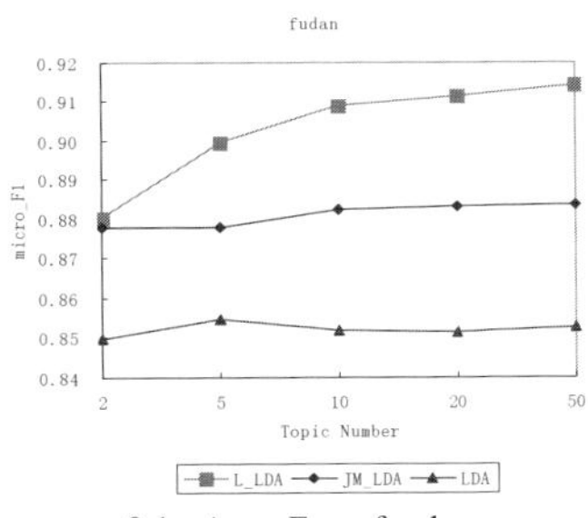

3.1 *micro_F*$_1$ on fundan

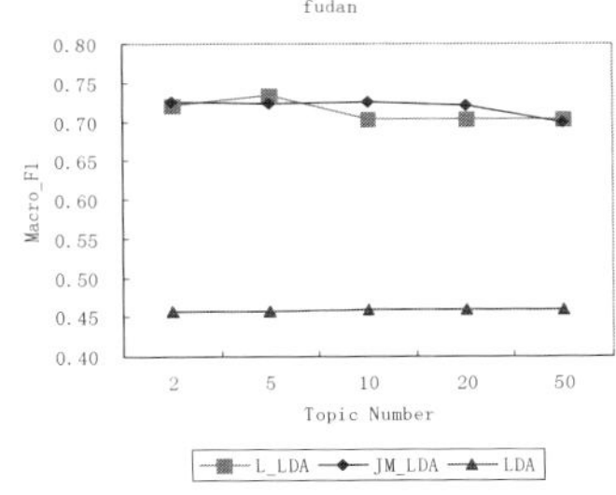

3.2 *Macro_F*$_1$ on fundan

Fig. 3. Comparison between LDA model vs. two smoothed LDA models, L_LDA and JM_LDA, on fudan corpus

In conclusion, just as we have discussed in the former section 3, the basic LDA model is harmed by the over-fitting of its naïve smoothing method, but our L_LDA model and JM_LDA model can alleviate this difficulty obviously.

Subsequently, we will compare the performances of the three models on the **fudan** corpus as shown in the figure 3.

In this part, the experiment setting is identical with the former 20newsgroup's. From the figure 3.1 and 3.2, we can obtain the following things:

On the whole, both L_LDA and JM_LDA have evident highly performance (***micro_F*$_1$** and ***Macro_F*$_1$**) than LDA model across all value of topic number. Nevertheless, except this point of similarity, figure 3.1 and 3.2 show several differences:

(1) The most significant one is the different improvement extent between the measures of ***micro_F*$_1$** and ***Macro_F*$_1$**. The ***micro_F*$_1$** improves about 3%~6%, yet the

***Macro_F*$_1$** obtains an approximate 30% improvement for both L_LDA and JM_LDA. This result shows that the data-driven smoothing strategy can advance the Macro performance on the unbalanced corpus.

(2) For the L_LDA model, the tendencies across the sequence of topic numbers are different between the measures of ***micro_F*$_1$** and ***Macro_F*$_1$**. The ***micro_F*$_1$** increases monotonously along with the topic number but the ***Macro_F*$_1$** is insensitive to the change of topic number.

(3) Comparing between the L_LDA model and JM_LDA model, the ***Macro_F*$_1$** measures of them are very near but the improvement extent of L_LDA's ***micro_F*$_1$** is twice of JM_LDA's. This shows that L_LDA has some superiority than JM_LDA.

Summarizing on the experiments with respect to the two corpora, 20newsgroup and fudan, we can draw the following main conclusions:

(1) Both L_LDA and JM_LDA can significantly improve the synthesis performance measured by ***micro_F*$_1$** and ***Macro_F*$_1$**. This result corroborates the effectiveness of our data-driven smoothing strategy on LDA model.

(2) On the balanced corpus, our smoothing methods show highly identical between ***micro_F*$_1$** and ***Macro_F*$_1$**. On the unbalanced corpus, however, the ***Macro_F*$_1$** improve much more than the ***micro_F*$_1$**. So, the data-driven smoothing strategy exhibits the superiority of realizing the classes' equitableness which is very important in the unbalanced classification task.

In order to explorer the mechanism of this superiority, we construct the following experiments (section 4.2.2) to dig in the data-driven smoothing strategy at a more detailed granularity where the performance of every class is examined.

Detailed Analysis Classification Performance on Unbalanced Corpus

In this part of experiment, we will explorer the classification performance of our data-driven smoothing strategy on unbalanced corpus (**fudan** corpus). ***F*$_1$** measure is used for every class and ***Macro_F*$_1$** measure is used for groups of classes. We also compare the three methods: traditional LDA model and our L_LDA model & JM_LDA model. In the following experiments, the topic number of three models is fixed on a specific value: 5. Under this topic number, we lay out several aspects of the three models to demonstrate the superiority of data-driven smoothing for unbalanced classification task.

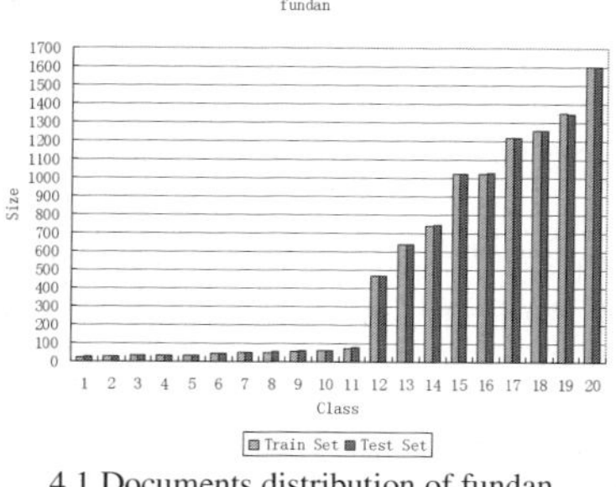

4.1 Documents distribution of fundan

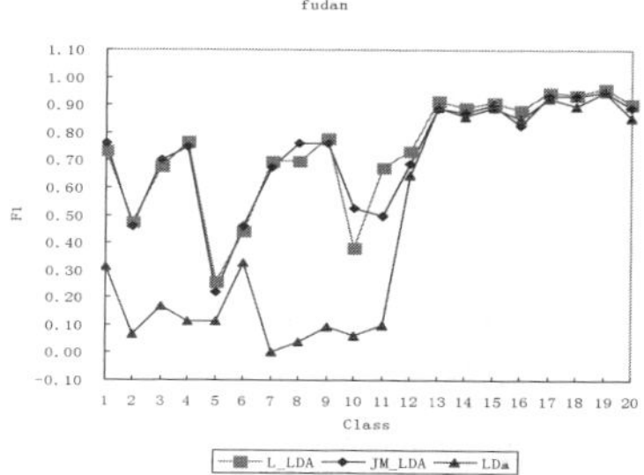

4.2 F_1 on fundan across all classes(topics=5)

Fig. 4. 4.1 The distribution of documents (training set & testing set) across all categories on fudan corpus. 4.2 A comparison between models (LDA, L_LDA and JM_LDA) on fudan corpus.

First of all, we show the unbalanced distribution of documents quantities across all categories on **fudan** corpus in figure 4.1 and corresponding performance comparison between models (LDA, L_LDA and JM_LDA) as 4.2:

Figure 4.1 shows that the classes of corpus are evidently divided into 2 parts: there are 11 small classes (1-11), where every one has about tens of documents in training & testing set; the other 9 classes are big classes.

Table 2. A comparison between small classes (1-11) and big classes (12-20) by $Macro\_F_1$. Note: the number of topics is fixed at 5

	Small: 1-11	**Big: 12-20**	**Whole Corpus**
L_LDA	0.599	0.898	0.734
JM_LDA	0.598	0.876	0.723
LDA	**0.127**	0.862	**0. 457**

From table 2, we can firstly get that on the whole corpus, the $\boldsymbol{Macro\_F_1}$ obtains an improvement over LDA about 26% for JM_LDA and about 28% for L_LDA. Other details are discussed as following:

(1) The performances of L_LDA model and JM_LDA model measured by $\boldsymbol{Macro\_F_1}$ are very near on both small classes and big classes. So, these two models can get the closer performance on the whole corpus.

(2) As we only focus on the results of the big classes, we find that three models only have little difference: the $\boldsymbol{Macro\_F_1}$ of L_LDA is higher than JM_LDA by 2.2%. In turn, the $\boldsymbol{Macro\_F_1}$ of JM_LDA is higher than LDA by 1.4%. These improvements are insufficient for the large increase of about 26%~28% on the whole corpus. The key reason is small classes.

(3) On small classes, the LDA model has a very poor performance which can only attain 12.7% measured by $\boldsymbol{Macro\_F_1}$. After smoothed by L_LDA and JM_LDA, this performance advances to about 60% which result in an improvement of 47%. It is just this great contribution urges the whole corpus' $\boldsymbol{Macro\_F_1}$ goes up about 26%~28%.

Next step, we will look into the detail views across all classes by different aspects to explore the effect of data-driven strategy. Firstly, let's examine the F_1 on every class in figure 4.2: the *X*-axis is the index of classes in fudan corpus and the *Y*-axis is the measure $\boldsymbol{F_1}$.

We can see that the $\boldsymbol{F_1}$ of L_LDA model is always higher than LDA model across all classes where the $\boldsymbol{F_1}$ of most small classes increase greatly and L_LDA model can obtains some tiny increase on every big class at the same time. This property of L_LDA model shows that the main principle of the whole improvement is digging inside the model to rectify the deviation on small classes without depressing the big classes. The similar principle is also the same with JM_LDA.

We know that the measure $\boldsymbol{F_1}$ is composed by ***Recall*** and ***Precision***. We will look into these two measures to find the factors that prompt the synthesis performance. The ***Recall*** result is shown in figure 5.1 and the ***Precision*** result is shown in figure 5.2.

The most noticeable thing is that the figure 5.1 is almost the same as the figure 4.2. So, we can determine that the promotion of ***Recall*** on the unbalanced corpus greatly improve performance of $\boldsymbol{F_1}$. So, L_LDA and JM_LDA can find documents of small size classes significantly. This is the key function for unbalanced classification.

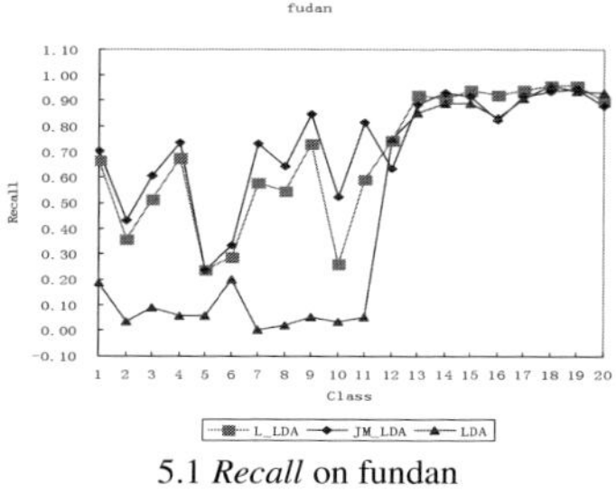

5.1 *Recall* on fundan

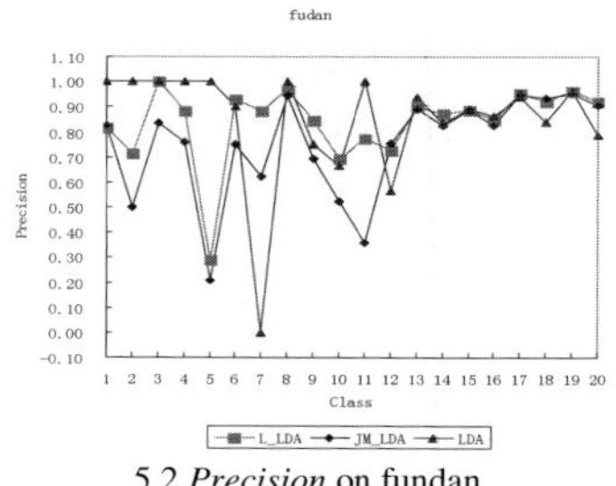

5.2 *Precision* on fundan

Fig. 5. A comparison between LDA model vs. two smoothed LDA models, L_LDA and JM_LDA, under the measure ***Recall*** & ***Precision*** of every class on fudan corpus

On the contrary, the figure 5.2 is much different with figure 4.2 and figure 5.1. The precisions of L_LDA and JM_LDA of small classes vibrate severely and deviate from the LDA's to much extent. As whole, the ***Precision*** of L_LDA and JM_LDA are lower than LDA model's. But in fact, the relative lower ***Precision*** of L_LDA and JM_LDA are more effective than LDA model's because that the ***Precision*** of LDA is based on the very low ***Recall*** which makes the high ***Precision*** meaningless.

For the unbalanced corpus, the OOV amounts of the small classes are much more than those of the huge classes. Thus over-fitting phenomenon will happen more significantly when applying LDA to small classes. As a result, the applicability of the learning algorithms for small classes will decrease severely. For the smoothing procedure of the corpus, probability distribution has been performed for all word across all classes. This procedure has different influences between huge classes and small classes. Huge classes have bigger vocabularies, and words appear more frequently. So the smoothing procedure has little influence on big classes. By contrast small classes have relative much smaller vocabulary, and words appear less frequently because fewer documents involved. Thereby the smoothing procedure has bigger influence on small classes. According to the experiments, the default smoothing approach of LDA has severe over-fitting on small classes. Our data-driven strategy smoothing methods can alleviate this defect to much extent.

5 Related Work

LDA model has been examined in some researches of text categorization. Blei et al. [8] conduct binary classification experiments on the Reuters-21578 dataset. They compared two type text presentations, classical BOW and low-dimensional representation derived from LDA model. Their performances are very near (with a difference of about 1%~2%). This shows the potential of the LDA in using as dimension reducing method for text categorization.

Wei Li et al. [10] conduct a 5-way classification experiment on the ***comp*** subset of the **20newsgroup** dataset to compare LDA with PAM (Pachinko Allocation Model) which tries to capture correlations between topics by the DAG structure. The PAM gets an improvement of about 3% over LDA by **Accuracy** measure. Our experiments show data-driven smoothing LDA exceeds traditional LDA about 4% measured by ***micro_F*$_1$** (in single-labeled setting, ***micro_F*$_1$** is equivalent to the **Accuracy** [14]).

In above experiments, the quality of both classes and/or documents is too small to exhibit the advantages or disadvantages of LDA. So, we study LDA on corpora with more classes and documents. Moreover, their works have not dig into unbalanced classification as our works. For unbalanced classification problem, there are two basic strategies [15]: classifier modification and re-sampling. Our work is different from them and similar with the idea of the feature processing as [16].

Our JM_LDA model is inspired by the work of Wei, X and Croft, W.B. [9]. However there is evident differences: they estimate one LAD model $p_{lda}(d_i)$ for each document for inferencing the probability of query $p_{lda}(q|d_i)$. Our work use the LDA model at class level $p_{lda}(c_i)$ and make inference by the class-conditional probability $p_{lda}(w_t|c_i)$.

6 Conclusion and Future Work

In this paper, we propose the data-driven smoothing strategy for LDA model to overcome the arbitrariness originated from operating latent variable distributions directly in multi-level graphical model. Laplacian smoothing and Jelinek-Mercer smoothing are introduced to LDA model following this data-driven smoothing strategy. Laplacian smoothing alleviates the over-fitting of LDA model evidently; and moreover, Jelinek-Mercer smoothing decreases the time complexity of Laplacian at predicting phase without damaging the performance on the whole. The evaluation shows that the data-driven smoothing strategy can significantly improve the performance on both balanced and unbalanced text categorization corpora.

Future research will be carried out on following aspects. Firstly, since the current work shows that the smoothing is very important for the application of the LDA model to text categorization, further researches of different smoothing methods should be done. Secondly, topic model is a rich area where many instances have been proposed [17] [18] [10]. Testing data-driven smoothing strategy on these models will be concerned.

Acknowledgments. This work is partially supported by National Natural Science Foundation of China with the contract No. 60773027 and No. 60736044 and by the 863 project of China with the contract No. 2006AA010108.

References

1. Sebastiani, F.: Text categorization. In: Text Mining and its Applications, pp. 109–129. WIT Press, Southampton (2005)
2. Koster, C.H., Seutter, M.: Taming wild phrases. In: Sebastiani, F. (ed.) ECIR 2003. LNCS, vol. 2633, pp. 161–176. Springer, Heidelberg (2003)
3. Moschitti, A., Basili, R.: Complex linguistic features for text classification: a comprehensive study. In: McDonald, S., Tait, J.I. (eds.) ECIR 2004. LNCS, vol. 2997, pp. 181–196. Springer, Heidelberg (2004)

4. Lewis, D.D.: An evaluation of phrasal and clustered representations on a text categorization task. In: 15th annual international ACM SIGIR conference on Research and development in information retrieval, pp. 37–50. ACM Press, New York (1992)
5. Baker, L.D., McCallum, A.K.: Distributional clustering of words for text classification. In: 21st annual international ACM SIGIR conference on Research and development in information retrieval, pp. 96–103. ACM Press, New York (1998)
6. Deerwester, S., Dumais, S.T., Furnas, G.W., Landauer, T.K., Harshman, R.: Indexing by latent semantic indexing. J. Amer. Soc. Inform. Sci. 41, 391–407 (1990)
7. Blei, D.: Probabilistic Models of Text and Images. PhD thesis, U.C. Berkeley (2004)
8. Blei, D., Ng, A., Jordan, M.I.: Latent Dirichlet allocation. J. Journal of Machine Learning Research 3, 993–1022 (2003)
9. Wei, X., Croft, W.B.: LDA-based Document Models for Ad-hoc Retrieval. In: 29th annual international ACM SIGIR conference on Research and development in information retrieval, pp. 178–185. ACM Press, New York (2006)
10. Wei, L., McCallum, A.: Pachinko Allocation: DAG-Structured Mixture Models of Topic Correlations. In: 23rd International Conference on Machine Learning, pp. 577–584. ACM Press, New York (2006)
11. Jordan, M.I., Ghahramani, Z., Jaakkola, T.S., Saul, L.K.: An introduction to variational methods for graphical models. In: Learning in Graphical Models, pp. 105–161. MIT Press, Cambridge, USA (1999)
12. Wainwright, M.J., Jordan, M.I.: Graphical models, exponential families, and variational inference. Technical report, 649, University of California, Berkeley (2003)
13. McCallum, A., Nigam, K.: A comparison of event models for naive Bayes text classification. Technical report, WS-98-05, AAAI-98 Text Categorization Workshop (1998)
14. Yang, Y., Liu, X.: A re-examination of text categorization methods. In: 22nd annual international ACM SIGIR conference on Research and development in information retrieval, pp. 42–49. ACM Press, New York (1999)
15. Japkowicz, N., Stephen, S.: The Class Imbalance Problem: A Systematic Study. J. Intelligent Data Analysis Journal 6, 429–449 (2002)
16. Zhuang, L., Dai, H., Hang, X.: A Novel Field Learning Algorithm for Dual Imbalance Text Classification. In: Wang, L., Jin, Y. (eds.) FSKD 2005. LNCS (LNAI), vol. 3614, pp. 39–48. Springer, Heidelberg (2005)
17. Blei, D.: Probabilistic Models of Text and Images. PhD thesis, U.C. Berkeley, Division of Computer Science (2004)
18. Blei, D., Lafferty, J.: Correlated topic models. J. Advances in Neural Information Processing Systems 18, 147–154 (2006)

Fusion of Multiple Features for Chinese Named Entity Recognition Based on CRF Model*

Yuejie Zhang[1], Zhiting Xu[1], and Tao Zhang[2]

[1] Department of Computer Science & Engineering,
Shanghai Key Laboratory of Intelligent Information Processing,
Fudan University, Shanghai 200433, P.R. China
[2] School of Information Management & Engineering,
Shanghai University of Finance & Economics, Shanghai 200433, P.R. China
{yjzhang,zhiting}@fudan.edu.cn, taozhang@mail.shufe.edu.cn

Abstract. This paper presents the ability of Conditional Random Field (CRF) combining with multiple features to perform robust and accurate Chinese Named Entity Recognition. We describe the multiple feature templates including local feature templates and global feature templates used to extract multiple features with the help of human knowledge. Besides, we show that human knowledge can reasonably smooth the model and thus the need of training data for CRF might be reduced. From the experimental results on People's Daily corpus, we can conclude that our model is an effective pattern to combine statistical model and human knowledge. And the experiments on another data set also confirm the above conclusion, which shows that our features have consistence on different testing data.

Keywords: Named Entity Recognition; Conditional Random Field; multiple features.

1 Introduction

With the development of Natural Language Processing (NLP) technology, the need for automatic Named Entity Recognition (NER) is highlighted in order to enhance the performance of information extraction systems [1]. The task of NER is to tag each Named Entity (NE) in documents with a set of certain NE types. It's impossible to collect all the existing NEs in a dictionary. One reason is because the amount of NEs is tremendous, and another reason is due to the variability of NE set, new NEs appear while old NEs are eliminated. As a result, NER plays a vital role in NLP.

In the 7th Message Understanding Conference (MUC-7), NER mainly aims at seven classes of NE including Person Name (PN), Location Name (LN), Organization Name (ON), Time, Date, Money and Percentage [2]. Chinese PN can be further divided into two parts, that is, Beginning of PN (PN-B) and Inner of PN (PN-I). Similarly, LN can be divided into Beginning of LN (LN-B) and Inner of LN (LN-I). Since

* This paper is supported by National Natural Science Foundation of China (No. 60773124, No. 70501018).

H. Li et al. (Eds.): AIRS 2008, LNCS 4993, pp. 95–106, 2008.

ONs usually consist of several words and the last word usually has obvious characteristic (e.g. "国防部/Department of Defense"), ONs are divided into three parts, that is, Beginning of ON (ON-B), Inner of ON (ON-I) and Ending of ON (ON-E). Since time and digital expressions are relatively easy to be recognized, our work focuses on subcategories related to PN, LN and ON.

In comparison with NER for Indo-European languages, there are more difficulties for Chinese NER, such as no space between words and no capitalization information that can play very important roles in recognizing NEs. Further, the use of character forming NEs is wide, which causes the problem of data sparseness, and the training usually calls for abundance training data to overcome these problems.

This paper presents a fusion pattern of multiple features for Chinese NER based on Conditional Random Field (CRF) model. First, lists of human knowledge are constructed, and details of such lists will be described in section 4. Then, a series of feature templates are constructed based on the lists of human knowledge to extract both local features and global features. After that, we use CRF++ Toolkit[1] to train our model. Using the trained model, the most possible label of each character can be calculated, and we combine characters based on their labels to get NEs.

After trained on the NEs labeled corpus from People's Daily and tested on corpus from People's Daily as Open-Test-I and corpus from PKU in SIGHAN 2005 as Open-Test-II, the model achieves consistent results, from which we can conclude that the performance of our model is better than pure rule-based or statistics-based models, or other models that only use one kind of feature. Moreover, the comparison of raise speed of performance between the model without human knowledge and the model with human knowledge suggests that human knowledge is a reasonable way to smooth model and the training data might be reduced if human knowledge can be combined in the model.

2 Related Work

Recently, the researches on English NER have achieved impressive result. The best English NER system in MUC-7 achieved 95% precision and 92% recall [3]. However, research on Chinese NER is still at its early stage. At present, the performance (Precision, Recall) of the best Chinese NER system is (66%, 92%), (89%, 88%) and (89%, 88%) for PN, LN and ON respectively [4].

Nowadays, the method of NER is taking a transition from simple rule-based models to statistics-based models, and is becoming a hybrid model combining statistical model and human knowledge. The pure statistical models that use single features are fit for large-scale text processing, but it may suffer from several problems such as data sparseness and overfitting, which may reduce the accuracy. Thus, linguistic features are usually combined with pure statistics-based models to achieve better results.

Among various statistical models, Maximum Entropy (ME) model has many merits such as flexibility of incorporating arbitrary features and it performs well in many problems of NLP and has been proven to perform well on NER [5]. However, ME models can only take account of single state, say, the probability of transferring

[1] http://crfpp.sourceforge.net/

between different states cannot be considered. Thus, Maximum Entropy Markov Model (MEMM) was proposed to combine the probability of transferring between states [3]. Nevertheless, MEMMs tend to get an optimal solution based on the optimization of each pair of state, which may cause the problem of label bias. As a result, Conditional Random Field (CRF) was proposed to overcome such problem [6]. CRF is widely used in many NLP tasks such as Segmentation, Shallow Parsing and Word Alignment, and it has been proven to be one of the best classifiers in NLP [7].

Besides the choice of classifier, another crucial aspect that greatly affects the performance of the model is the choice of features. A good classifier cannot work without good features, and a less powerful classifier may also perform well with a set of deliberate chosen features [8].

In fact, owing to the nature of Chinese, the use of character to construct NEs, especially ONs, is too wide for limited training data to estimate. Consider, for example, two words of person title "部长/minister" and "外交官/diplomats", the former may appear frequently in the training data while the latter may rarely appear, which may lead to the weight of two features differ a lot, but as a common sense, the probability of a person following those two words should be equal. As a result, we think that the introduction of a feature to indicate whether a word is in a person title list can improve the performance of the model. The details about such list will be discussed in the section 4.

3 Conditional Random Fields

Conditional Random Fields (CRFs) are a class of undirected graphical models with exponent distribution [6]. A common special case of CRFs is linear chain, which has similar graph structure with linear HMM except that HMM is directed graph while CRF is undirected graph. A linear chain CRF has form described as follows:

$$P_{\Lambda}(\vec{y} \mid \vec{x}) = \frac{1}{Z_{\vec{x}}} \exp\left(\sum_{t=1}^{T} \sum_{k} \lambda_k f_k(y_{t-1}, y_t, \vec{x}, t)\right) \tag{1}$$

where $f_k(y_{t-1}, y_t \vec{x}, t)$ is a function which is usually an indicator function, λ_k is the learned weight of feature f_k, and $Z_{\vec{x}}$ is the normalization which ensures that the answer is a probability. The feature function actually consists of two kinds of feature, including the feature of single state and the feature of transferring between states. How to get these features will be discussed in section 5.

The parameter λ_k can be estimated in many ways and one common way is ME, which is the dual form of Conditional Maximum Likelihood. Also, there are many ways to estimate parameters of the object function (e.g. GIS, IIS, L-BFGS) [9]. CRF++ Toolkit uses L-BFGS for optimization, which converges much faster but requires more memory.

4 Human Knowledge

In this section, we introduce human knowledge we used based on which features are extracted. All these word lists are constructed from the training data.

(1) Word List – Includes 108,778 items. This list includes most frequently used words. It is actually a Chinese dictionary. This list aims to enhance accuracy of segmentation of words in dictionary.
(2) Chinese Surname List – Includes 887 items. Chinese surnames such as "王/wang" and "张/zhang" are included. Characters in this list are more likely to be the PN-B.
(3) Chinese Name Characters List – Includes 2,578 items. This list includes common used characters of Chinese given name.
(4) Person Title List – Includes 401 items. Person titles such as "部长/Minister" and "总统/President" are listed. PNs tend to appear after such words.
(5) Transliteration Characters List – Includes 799 items. Since the characters used to construct transliteration name differ a lot from those characters used to construct Chinese name, a list for transliteration name characters is needed.
(6) Location Name List – Includes 2,059 items. Since many LNs do not have obvious common feature, we introduced a list of frequent LNs to overcome sparseness.
(7) Location Suffix List – Includes 89 items. This list includes suffix of LNs. For example, "峰/mountain" and "江/river". Most LNs outside Location Name List have a suffix in this list. Therefore, we can use this list to help detect LNs that are out of Location Name List.
(8) Organization Suffix List – Includes 188 items. Suffixes of ONs such as "国务院/State Department" and "国防部/Defense Department" are included in this list.
(9) General Word List – Includes 305 items. This list represents those characters that maybe out of words, such as "的" and "说/say".

5 Feature Representation

NER can be viewed as a classification problem that should predict the category of the candidate word using various features in the context. The features used in the model can be divided into two classes: Local Feature (LF) and Global Feature (GF). The former is constructed based on neighboring tokens and the token itself, and the latter is extracted from other occurrences of the same token in the whole document.

To take advantage of the ability of sequence labeling of CRFs, we give each character a label; that is, the B (Begin), I (Inner) and E (Ending) plus its POS tag. For example, the PN "胡锦涛/Jintao Hu" will be tagged as "胡/PN-B", "锦/PN-I" and "涛/PN-I". Since the length of ON could be very long and suffix characteristics of ON are obvious (in our Organization Suffix List), the characters in the suffix of ON are tagged as ON-E.

5.1 Local Feature

There are two types of contextual information to be considered when extracting local features, namely, internal lexical information and external contextual information. For example, in the character sequence "记者陈贻宁报道", "陈贻宁" is a candidate word, "记者" and "报道" provide the external contextual information, and "陈", "贻", "宁" provide the internal lexical information.

Candidate features are extracted from the training corpus with feature templates. A feature template is a pattern to extract features. It may correspond to several features, while a pair of context and candidate word may correspond to several features. Feature templates are described as follows:

(1#) C_{-4} – The forth previous character.
(2#) C_{-3} – The third previous character.
(3#) C_{-2} – The second previous character.
(4#) C_{-1} – The previous character.
(5#) C_0 – The current character.
(6#) C_0 – Whether current word is in List (2).
(7#) C_0 – Whether current word is in List (3).
(8#) C_0 – Whether current word is in List (5).
(9#) C_0 – Whether current character is in List (6).
(10#) C_0 – Whether current character is in List (7).
(11#) C_0 – Whether current character is in List (8).
(12#) C_0 – Whether current character is in List (9).
(13#) C_1 – The next character.
(14#) C_2 – The second next character.
(15#) C_3 – The third next character.
(16#) C_4 – The forth next character.
(17#) $C_{-1,0}$ – Whether this word is in List (1).
(18#) $C_{-1,0}$ – Whether previous character is in List (2) and current character in List (3).
(19#) $C_{-1,0}$ – Whether previous character is in List (2) and current character in List (5).
(20#) $C_{-1,0}$ – Whether this word is in List (4).
(21#) $C_{-1,0}$ – Whether this word is in List (6).
(22#) $C_{-1,0}$ – Whether this word is in List (7).
(23#) $C_{-1,0}$ – Whether this word is in List (8).
(24#) $C_{-2,-1,0}$ – Whether the second previous character is in List (2) and both previous character and current character in List (3).
(25#) $C_{-2,-1,0}$ – Whether second previous character is in List (2) and both previous character and current character in List (5).
(26#) $C_{-2,-1,0}$ – Whether this word is in List (4).
(27#) $C_{-2,-1,0}$ – Whether this word is in List (6).
(28#) $C_{-2,-1,0}$ – Whether this word is in List (7).
(29#) $C_{-2,-1,0}$ – Whether this word is in List (8).
(30#) $C_{-3,-2,-1,0}$ – Whether this word is in List (4).
(31#) $C_{-3,-2,-1,0}$ – Whether this word is in List (6).
(32#) $C_{-3,-2,-1,0}$ – Whether this word is in List (7).
(33#) $C_{-3,-2,-1,0}$ – Whether this word is in List (8).
(34#) $C_{-4,-3,-2,-1,0}$ – Whether this word is in List (4).

(35#) $C_{-4,-3,-2,-1,0}$ – Whether this word is in List (6).
(36#) $C_{-4,-3,-2,-1,0}$ – Whether this word is in List (7).
(37#) $C_{-4,-3,-2,-1,0}$ – Whether this word is in List (8).

By using these feature templates, several thousands of features can be extracted from the training corpus and thus form a candidate feature base.

5.2 Global Feature

Within a same document, a certain NE often occurs repeatedly and it usually appears in the abbreviated forms in the latter text. Furthermore, some candidate words appear in an ambiguous context, which enhances the difficulty of recognition. Consider the following three sentences:

Sentence (1) – "来自*北京大学*、*清华大学*等200所院校现场接受考生的咨询。"
Sentence (2) – "许多内地优秀学生舍弃*北京大学*与*清华大学*而转投香港高校。"
Sentence (3) – "在如火如荼的高考招生季节，人们开始替*北大*、*清华*担忧。"

The above three sentences come from the same news document, ordered by their order in the document. NEs in Sentence (1) are easy to be recognized, and NEs in (2) are also not difficult if the result of (1) can be used. NEs in (3) seems to be difficult at the first glance since both external feature and internal feature provide limited useful information, which makes two candidate words can be either PN, LN or ON. But if the related context information in the same document (such as in (1) and (2)) can be used, this problem can be solved.

From the above examples, it can be concluded that the information from global features may play a crucial role. So we need global features to take full use of global information. Two global feature templates are constructed as follows:

(1#) Other Occurrences with Same Form – Mainly aims to provide information of the recognized NEs with the same form in the previous part of the document. Construct a Dynamic Word List (DWL) and recognized NEs are stored. This feature is set to 1 when candidate word appears in DWL. Think of previous example, "北京大学" and "清华大学" are stored in DWL after they are recognized in sentence (1), which can help the recognition of candidate words in sentence (2). This template actually extracts features of indicators of whether candidate word (C_0, $C_{-1,0}$, $C_{-2,-1,0}$, $C_{-3,-2,-1,0}$, $C_{-4,-3,-2,-1,0}$) is in DWL.

(2#) Other Occurrences for Prefix and Suffix – Mainly aims to provide the important affix information of the recognized NEs. In the previous example, the suffix word of "清华大学" supplies the helpful context information to the candidate word "清华" in sentence (3). The features extracted by this template are actually indicators of whether candidate word (C_0, $C_{-1,0}$, $C_{-2,-1,0}$, $C_{-3,-2,-1,0}$, $C_{-4,-3,-2,-1,0}$) is the prefix or suffx of the words in DWL.

6 Reliability Evaluation

The label of each character is evaluated during the testing. After that, the continuous characters with the same POS label are combined. For example, continues three words "江", "泽", "民" with the label PN-B, PN-I and PN-I, PN-B and PN-Is are combined to form a PN "江泽民". During the step of tagging, the weight of each feature has already been estimated in the step of training. Then each conditional probability can be computed as follows:

$$p(y \mid x) = \frac{1}{Z(x)} \prod_{j=1}^{k} \lambda_j^{f_j(x,y)} \tag{2}$$

The value of $p(y_i \mid x)$ is the probability of NE class y_i for the candidate word under the current contexts. And the y with the largest probability is the label of current character.

7 Experiments

7.1 Data Sets and Evaluation Metrics

The training corpus contains data from People's Daily of the first half year of 1998, and is tagged with POS according to Chinese Text POS Tag Set provided by PeiKing University (PKU) of China. The size of the corpus is 2,248,595 bytes (including POS tags). It contains 13,751 sentences (about 350,000 Chinese characters). This corpus includes 4,907 PNs, 10,120 LNs and 2,963 ONs.

One testing corpus Open-Test-I is randomly selected from the raw texts of People's Daily of the latter half year of 1998, with the size of 1,319,496 bytes, containing 2,672 PNs, 4,928 LNs and 1,480 ONs. In addition, another wide coverage testing corpus Open-Test-II is built, which mainly comes from corpus provided by PKU in SIGHAN 2005. It is a balanced test set covering several different domains, such as politics, economy, entertainment, sports, and so on. The size of this corpus is 121,124 bytes, including 270 PNs, 508 LNs and 118 ONs.

Three parameters are used in the experimental evaluation. Precision (P) is the percent of the correctly recognized NEs in all the recognized NEs. Recall (R) is the percent of the correctly recognized NEs in all the NEs included in the testing corpus. F-measure (F) is a weighted combination of precision and recall, $F = \frac{(\beta^2+1) \times P \times R}{(\beta \times P) + R}$, where β is the relative weight of precision and recall, and is usually set as 1. Noted especially in the evaluation, PNs and LNs in nested form are also included, and only recognized PNs and LNs with correct boundary and class tag are considered as the correct recognitions.

7.2 Experimental Results Based on Open-Test-I

The experimental testing on Open-Test-I are based on the following three patterns.

(1) ***CRF-based*** – Based on the basic CRF-based model only estimated with the training data. The only used features are characters within the window of length

Table 1. Baseline performance

NE Class	*P* (%)	*R* (%)	$F_{\beta=1}$ (%)
PN	84.31	55.72	67.10
LN	70.79	74.49	72.59
ON	85.11	50.47	63.37
Total	75.44	65.06	69.87

4 around current candidate character (that is, C_n (n=-4,-3,-2,-1,0,1,2,3,4)). We get the baseline performance, as shown in Table 1.

(2) ***CRF-based+LF*** – Based on the above baseline experiment, we integrated with using local features described in section 5.1. As shown in Table 2, the experiment results indicate that almost all the precision, recall and F value of PN, LN and ON recognition have obtained improvements to some extent.

Table 2. Results of Local Features integrated into the CRF-based pattern

NE Class	*P* (%)	*R* (%)	$F_{\beta=1}$ (%)
PN	87.99	82.92	85.38
LN	82.68	87.70	85.12
ON	90.73	63.82	74.94
Total	85.17	82.41	83.77

(3) ***CRF-based+LF+GF*** – In this part, we want to see the effects of using global feature described in section 5.2. The results are shown in Table 3.

Table 3. Results of Local Features and Global Features integrated into the CRF-based pattern

NE Class	*P* (%)	*R* (%)	$F_{\beta=1}$ (%)
PN	91.82	84.87	88.21
LN	89.16	88.24	88.70
ON	91.46	69.44	78.95
Total	89.36	86.38	87.84

Furthermore, to show that human knowledge can smooth the model, we divided training data into 10 pieces, and we compare the speed of raise in the performance of Base model and Base+LF+GF separately. The results are shown in Fig. 1.

From Fig. 1 above, we can observe that the performance of the base model rises rapidly when the size of training data increases; on the contrary, the performance grows slowly with the increase of training data. And thus we can conclude that the introduction of human knowledge can help smooth the model and we can get a high performance only using a little amount of training data.

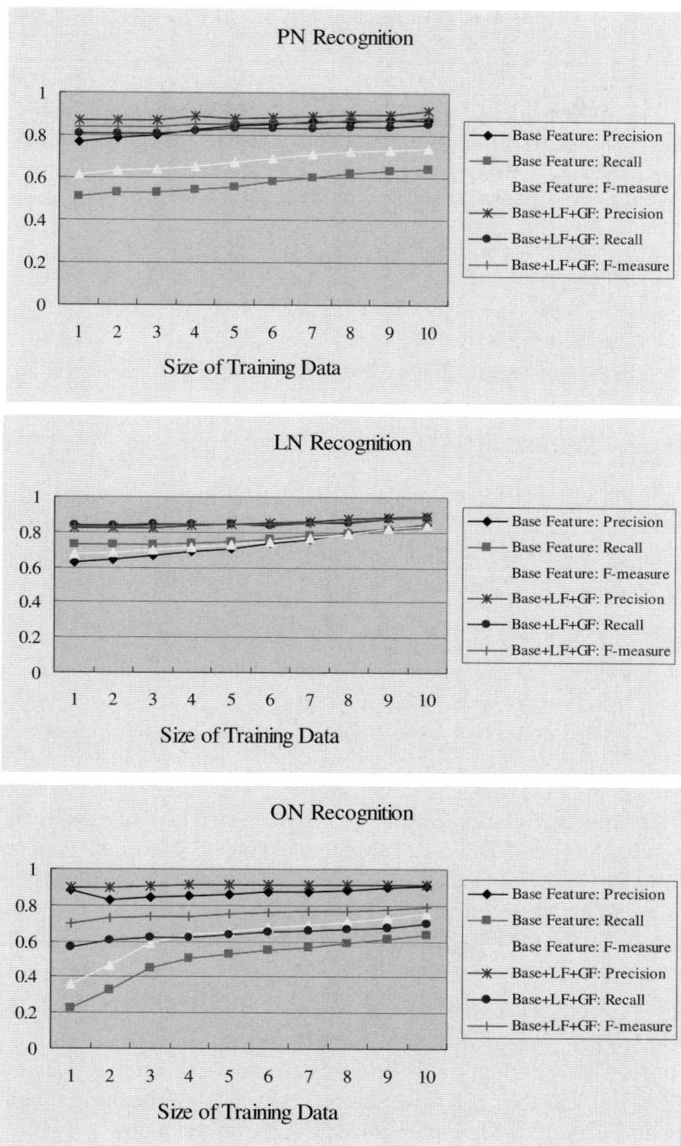

Fig. 1. Performance of PN, LN and ON Recognition using different size of training data

7.3 Other Experimental Results Based on Open-Test-II

In order to validate whether the established Hybrid Model based on different testing set could achieve better performance consistent with the tests on Open-Test-I, some experiments on Open-Test-II are also completed. The experimental results are shown in Table 4.

Table 4. Experimental results on Open-Test-II

NE Class	*Pattern*	*P (%)*	*R (%)*	$F_{\beta=1}$ (%)
	CRF-based	89.58	63.70	74.46
PN	***CRF-based+LF***	91.26	69.63	78.99
	CRF-based+LF+GF	91.74	82.22	86.72
	CRF-based	96.47	96.85	96.67
LN	***CRF-based+LF***	96.47	96.85	96.67
	CRF-based+LF+GF	96.47	96.85	96.67
	CRF-based	70.28	47.25	58.33
ON	***CRF-based+LF***	71.19	61.76	66.14
	CRF-based+LF+GF	74.57	63.77	68.75

7.4 Analysis and Discussion

Comparing the baseline performance of *CRF-based* pattern (as shown in Table 1) with the performance of the current best Chinese NER system (as described in Section 2), we find that *CRF model is exactly an effective statistical model for Chinese NER, and can get much better performance as well as or than other machine learning models.*

Comparing with Table 2 and Table 1, it can be observed that *human knowledge can not only reduce the search space and improve the processing efficiency, but also increase the recognition performance significantly.*

Comparing Table 3 and Table 2 with Table 1, it can be seen that *information from a sentence is sometimes insufficient to classify a NE correctly and global context from the whole document is available, so high performance can be achieved through combining local features and global features in the proposed hybrid model.*

Comparing Table 4 with Table 1, Table 2 and Table 3, *the same conclusions as experiments on Open-Test-I can be confirmed, which shows the consistence of the established model on different testing data.*

Through analysis for the recognized NEs with failure and error, it can be found that there are some typical errors shown as follows.

(1) Conflicts between NEs and existing words – For example, "记者/ 王/ 大方/" and "李/ 洪亮/". This type of mistakes mainly caused by over weighting of Vocabulary words. The introduction of Word List may make each Vocabulary word has a high weight, and thus if a given name is actually a Vocabulary word, it is very likely that this name will be recognized as a common word.

(2) No separation between consecutive NEs – For example, "记者/ 李学军董芳忠/". This type of error might be avoided by taking into account the features of NE length. However, it is also a fact that the recognition result of this sample sentence is not lexically wrong. The string "李学军董芳忠' can be viewed as a transliterated PN or a legal Chinese PN.

(3) Relatively low precision and recall in Japanese and Russian PN – The main cause is that the characters used in Japanese and Russian PNs are dispersed and different from other transliteration PNs. They are trained with transliteration PN

as a category together. That makes the internal lexical information relatively feeble and leads to a lower reliability value. Low reliability values will reduce recall, and might also reduce the precision by confusing PNs, LNs and ONs. This type of error maybe avoided by training Japanese and Russian PNs separately.

(4) Global features may reduce precision – If a NE was falsely recognized at first time and was included in DWL, this word may trigger the chain-false in the latter classification.

(5) Confusing between PN and LN – Some words such as "华盛顿" may be PN as well as LN, but they are more likely to be tagged as a LN since they are in Location Name List.

(6) Fuzzy bound of ON – Current means used to choose candidate word may cause the problem that candidate is too long though it has lots of characteristics of ON. For example, the candidate "美国可能会" has the prefix "美国" and the suffix "会", which are characteristics of ON. But we cannot limit the length of ON candidate since the structure of ON is complex. Splitting ON into several parts and training them respectively might solve this problem.

(7) Lack of available features – In some context, either global or local feature is limited. For example, "博卡萨/ 上台后" almost provides no specific information that can be used to decide category of "博卡萨". Actually, this is the most stubborn problem in NER.

(8) Sparse training of ONs – Since ONs are constructed with several words and our training is based on the character level, that is, the beginning of an ON might have little impact on the ending of this ON during the training if the length of this ON is too long. Besides, the number of ON samples is small in the training data compared with the number of their possible structures, and thus it is hard to catch their characteristics. This type of error is stubborn owning to the nature of the ONs. And Dynamic Conditional Random Fields (D-CRF), which allows a state with several labels, might be able to solve this problem [10]. However, D-CRF is difficult to implement and train because of its complex structure.

8 Conclusions

This paper introduces some research work on Chinese NER based on CRF model. An approach which has a CRF framework with the fusion of multiple features and applies human knowledge as the optimization strategy is presented. The experimental results have shown that the approach has achieved some satisfactory performance. Besides showing the advantage of CRF approach, this model also shows that it's an effective pattern to combine human knowledge in statistical model. In addition, the comparison of effects of training data size on the performance between base model and model with human knowledge shows that human knowledge can not only enhance performance but also smooth the model to overcome the problem of data sparseness.

Up to now, the features used in our model not only include word information, but also include part-of-speech information. The difficulty of the latter is that part-of-speech tagging also has dependency on the correctness of NER. Also the feature templates need extension, such as considering word length, word frequency in the same text, different window size, and so on. In addition, human knowledge used in feature representation is also a kind of feature. Theoretically, they could be used as the form of feature, and all words in the same list would be assigned the same weight. That will further reduce the manual interference and enhance the cohesion of the model. All the aspects above will be our research focuses in the future.

References

1. Volk, M., Clematide, S.: Learn - Filter - Apply – Forget Mixed Approaches to Named Entity Recognition. In: Proceeding of the 6th International Workshop on Applications of Natural Language for Information Systems (2001)
2. Berger, A.L., Pietra, S.A.D., Pietra, V.J.D.: Della Pietra, and Vincent J. Della Pietra: A Maximum Entropy Approach to Natural Language Processing. Computational Linguistics 22(1) (1996)
3. Bender, O., Och, F.J., Ney, H.: Maximum Entropy Models for Named Entity Recognition. In: Proceeding of CoNLL 2003 (2003)
4. Zhang, H.-P., Liu, Q., Zhang, H., Cheng, X.-Q.: Automatic Recognition of Chinese Unknown Words Based on Roles Tagging. In: Proceeding of the 19th International Conference on Computational Linguistics (2002)
5. Wu, Y., Zhao, J., Xu, B., Yu, H.: Chinese Named Entity Recognition Based on Multiple Features. In: Proceedings of Human Language Technology Conference and Conference on Empirical Methods in Natural Language Processing (HLT/EMNLP) (2005)
6. Lafferty, J., McCallum, A., Pereira, F.: A McCallum, and F. Pereira: Conditional Random Fields: Probabilistic Models for Segmenting and Labeling Sequence Data. In: Proceedings of the 18th International Conf. on Machine Learning (ICML) (2001)
7. Sha, F., Pereira, F.: Shallow Parsing with Conditional Random Fields. In: Proceedings of HLT-NAACL 2003 (2003)
8. Smith, N., Vail, D., Lafferty, J.: Computationally Efficient M-Estimation of Log-Linear Structure Models. In: Proceedings of the Annual Meeting of the Association for Computational Linguistics (2007)
9. Darroch, J.N., Ratcliff, D.: Generalized Iterative Scaling for Log-Linear Models. The Annals of Mathematical Statistics 43(5) (1972)
10. Sutton, C.A., Rohanimanesh, K., McCallum, A.: Dynamic Conditional Random Fields: Factorized Probabilistic Models for Labeling and Segmenting Sequence Data. In: International Conference on Machine Learning (ICML) (2004)

Semi-joint Labeling for Chinese Named Entity Recognition

Chia-Wei Wu[1], Richard Tzong-Han Tsai[2,*], and Wen-Lian Hsu[1,3]

[1] Institute of Information Science, Academia Sinica, Taipei, Taiwan
[2] Department of Computer Science and Engineering, Yuan Ze University, Chung-Li, Taiwan
[3] Department of Computer Science, National Tsing-Hua University, Hsinchu, Taiwan
cwwu@iis.sinica.edu.tw, thtsai@saturn.yzu.edu.tw,
hsu@iis.sinica.edu.tw

Abstract. Named entity recognition (NER) is an essential component of text mining applications. In Chinese sentences, words do not have delimiters; thus, incorporating word segmentation information into an NER model can improve its performance. Based on the framework of dynamic conditional random fields, we propose a novel labeling format, called *semi-joint labeling* which partially integrates word segmentation information and named entity tags for NER. The model enhances the interaction of segmentation tags and NER achieved by traditional approaches. Moreover, it allows us to consider interactions between multiple chains in a linear-chain model. We use data from the SIGHAN 2006 NER bakeoff to evaluate the proposed model. The experimental results demonstrate that our approach outperforms state-of-the-art systems.

Keywords: Named entity recognition, Chinese Word Segmentation.

1 Introduction

Named entity recognition (NER) is widely used in text mining applications. English NER achieves a high performance, but Chinese NER needs to be improved substantially. A named entity (NE) is a phrase that contains predefined names, such as person names, location names, and organization names. Named Entity Recognition (NER) is the process used to extract named entities in many applications, such as question answering systems, relation extraction, and social network analysis. Several conferences have been held to evaluate NER systems, for example, CONLL2002, CONLL2003, ACE (automatic context understanding), and SIGHAN 2006 NER Bakeoff. In many works, the NER task is formulated as a sequence labeling problem. Such problems have been discussed extensively in the past decade and several practical machine learning models have proposed, for example, the maximum entropy (ME) model[1], the hidden Markov model (HMM) [8], memory-based learning[5], support vector machines (SVM)[6] and conditional random fields (CRFs)[13].

Chinese NER is particularly difficult because of the word segmentation problem. Unlike English, Chinese sentences do not have spaces to separate words. Therefore,

[*] Corresponding author.

H. Li et al. (Eds.): AIRS 2008, LNCS 4993, pp. 107–116, 2008.

word segmentation information is important in many Chinese natural language applications. Depending on the way such information is incorporated, NER approaches can be classified as either character-based or word-based. In character-based approaches, segmentation information is used as features, whereas word-based methods use the output of the word segmentation tagger as the basic tagging unit. However, irrespective of the method used, the interactions between NER and word segmentation tags can not be considered jointly and dynamically.

One solution for handling multiple related sub-tasks like word segmentation and named entity recognition is to use joint learning methods, for example, jointly tagging parts-of-speech and noun phrase chucking using dynamic CRFs [13], incorporating features into different semantic levels using a log-linear joint model [3], and using a re-ranking model to jointly consider parsing and semantic role labeling [12]. These joint learning methods yield richer interactions between sub-tasks, which they consider dynamically.

In this paper, based on the concept of joint learning, we propose a novel Chinese NER tagging presentation, called the *semi-joint labeling* which partially integrates segmentation labels and named entity labels. The format can represent the interactions between the named entity and word segmentation states. It also facilitates dynamic consideration of NER and word segmentation states in a linear chain to alleviate the problem of potentially higher computation costs incurred by multiple layer tagging. Because it uses semi-joint tagging, the proposed system outperforms state-of-the-art systems.

The remainder of this paper is organized as follows. In Section 2, we introduce Chinese NER and word segmentation. In Section 3, we describe the proposed method. Then, in Section 4, we discuss the features of our system. Section 5 details the experiment results, and Section 6 contains our conclusions.

2 Chinese Word Segmentation and Named Entity Recognition

In this section, we introduce the Chinese word segmentation and named entity recognition task, and consider existing approaches that incorporate word segmentation information in NER models. In Table 1, the first row shows a series of Chinese characters with word segmentation and named entity labels. We list two segmentation tagging formats, BI and IE, in the next two rows. In the BI format, B denotes that *a character is at beginning of a word* and I denotes that *a character is in a word.* In the EI format, E denotes that *a character is at the end of a word* and I denotes *the inside character of a word.* In the named entity tagging format, a label is defined as a named entity type extended with a boundary tag. For example, B-LOC denotes that a character is at the beginning of a location name, while O denotes that the character is not part of a named entity. Word segmentation can provide valuable information for NER. For example, the boundaries between a word and a named entity can not cross or overlap. Previous works, such as Guo et. al.[4] and Chen et. al.[2], have shown that word segmentation information can improve NER performance.

There are two ways to incorporate word segmentation information into an NER model, namely, character-based approaches and word-based approaches. Unlike

Table 1. Examples of word-based and character-based tagging representation and their corresponding English phrases with NER tags

Character-based	**Characters**	俄	罗	斯	总	统	普	京	说
	BI-format Word Segmentation	B	I	I	B	I	B	I	B
	IE-format Word Segmentation	I	I	E	I	E	I	E	E
	Named entity labels	B-LOC	I-LOC	I-LOC	O	O	B-PER	I-PER	O
Word-based	**Words**	俄罗斯		总统		普京		说	
	Named entity labels	B-LOC		O		B-PER		O	
English	**Words**	Russian		president		Putin		says	
	Named entity labels	LOC		O		PER		O	

English NER, Chinese character-based NER uses characters as the basic tokens in the tagging process. Chen et. al.[2] and Wu et. al.[14] use a character-based approach in their NER models. The advantage of this approach is that it avoids the propagation of potential errors by the segmentation tagger. However, this approach does not consider the word segmentation information. One common approach employs a cascaded training and testing method that uses the output of the segmentation tagger as a feature in the NER model. For example, Guo et. al.[4] use word segmentation information as a feature in a character-based model.

Word-based NER uses words as the basic tokens. A number of systems, like those of Ji and Grishman [7] and Sun et. al. [11] use the word-based approach. In Figure 1, the first row of the word-based section shows an example of a phrase with word-based NER tags. Comparison with the first row of the English section shows that the NER tags of Chinese word-based and English word-based segmentation are the same. However, since word-based segmentation needs the output of a word segmentation tagger as the basic tagging token, propagated errors will be passed on to the NER model.

No matter whether the cascaded approach uses word segmentation information in character-based tagging or uses word-based tagging directly, the interactions between word segmentation and NER can be represented in is limited and can not be considered dynamically. Next, we introduce dynamic CRFs and the semi-joint labeling format used to represent more complex interactions.

3 Methods

3.1 Dynamic Conditional Random Fields

Dynamic conditional random fields (DCRF) [13] are generalizations of linear-chain conditional random fields (CRF) in which each time slot contains a set of state variables and edges. The form of a dynamic CRF can be written as follows:

$$P(y \mid x) = \frac{1}{Z(x)} \prod_{t=1} \prod_{c \in C} \exp\left(\sum_k \lambda_k f_k \left(y_{t,c}, x, t \right) \right) \tag{1}$$

where $\boldsymbol{y}$ is a label sequence over observation sequence $\boldsymbol{x}$; c denotes a clique in a graph; λ_k and f_k are, respectively, the weights and feature function associated with the clique index k; t denotes a time slot; and Z is a normalization constant. By using different definitions of c, DCRFs can represent various interactions within a time slot. For example, if we define c as a combination of labels in multiple tagging layers, then $\boldsymbol{y}_{t,c}$ denotes a joint label of multiple layers in time slot t. Therefore, we can identify rich interactions between word segmentation information and named entity recognition.

We use DCRFs to present a graphical model that considers the interactions of named entities and word segmentation tags in a multiple chain structure. In Figure 2a, the two chains correspond to the state sequences of named entities and word segmentation tags. Using DCRFs, we can represent this structure by the following equation:

$$P(y \mid x) = \frac{1}{Z(x)} \prod_{t=1}^{T-1} \Omega\left(y_{t,n}, y_{t+1,n}, x, t \right) \prod_{t=1}^{T-1} \omega\left(y_{t,n}, y_{t,s}, x, t \right) \tag{2}$$

where $\boldsymbol{y}_n$ denotes the named entity label sequence and $\boldsymbol{y}_s$ denotes the segmentation label sequence; Ω denotes the function of the interactions between $\boldsymbol{y}_{n,t}$ and $\boldsymbol{y}_{n,t+1}$, the labels of named entities in the adjacent time slot; and ω denotes the function of the interactions between the named entities and word segmentation labels in the same time slot. Based on the feature f_k and the parameter λ_k, Ω and ω can also be presented as:

$$\begin{aligned} \Omega\left(y_{t,n}, x, t \right) &= \exp\left(\sum_k \lambda_k f_k \left(y_{t,n}, y_{t+1,n}, x, t \right) \right) \\ \omega\left(y_{t,n}, y_{t,s}, x, t \right) &= \exp\left(\sum_k \lambda_k f_k \left(y_{t,s,n}, x, t \right) \right), \end{aligned} \tag{3}$$

Figure 1(b) is a three-chain structure in which the chains corresponding to the tagging sequence from the top to the bottom represent named entity segmentation, BI-format word segmentation, and EI-format segmentation, respectively. Using DCRFs, we can represent this structure by the following equation,

$$P(y \mid x) = \frac{1}{Z(x)} \prod_{t=1}^{T-1} \Omega\left(y_{t,n}, y_{t+1,n}, x, t \right) \prod_{t=1}^{T-1} \omega\left(y_{t,n}, y_{t,s}, x, t \right) \prod_{t=1}^{T-1} \Psi\left(y_{t,n}, y_{t,s}, x, t \right) \tag{4}$$

where Ψ denotes the interactions between the state sequence of named entity labels and the EI-format word segmentation labels.

3.2 Semi-joint Labels in Linear Conditional Random Fields

If DCRFs are used to represent complicated structures, such as multiple layers of tags, the number of cross-products of states between the layers will cause the inference space increases. For example, the cross-product space of the segmentation labels and named entity labels is twice as large as the original named entity label space. We propose a semi-joint model to reduce the inference spaces.

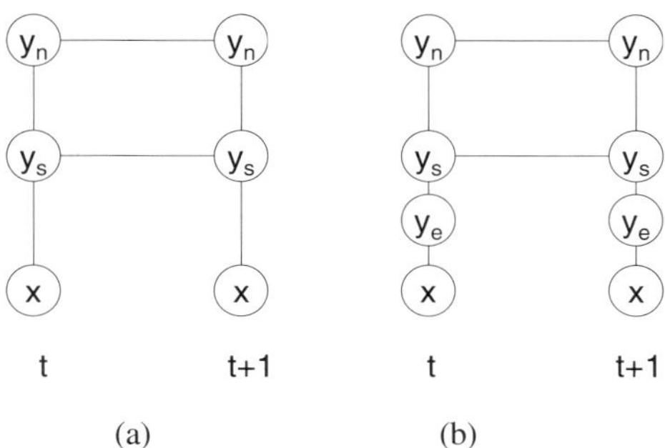

Fig. 1. The two tagging structures

A semi-joint label set partially integrates the original labels in different layers. Semi-joint labeled CRFs are linear-chain CRFs transformed from multiple chain CRFs by applying the semi-joint label set. Next, we define a semi-joint label set.

Let a semi-joint label set $\boldsymbol{q}$ be { $\boldsymbol{q}_1, \boldsymbol{q}_2, \ldots, \boldsymbol{q}_m$ } where $\boldsymbol{q}_k$ is a vector of the label set selected from the Cartesian product of the original label set. The selection rule can be decided manually or systematically For example, Table 1 shows a Chinese phrase with word segmentation tags and named entity tags that are integrated by semi-joint labeling tags. The second column shows the phrase's corresponding English translation. Each character has segmentation tags in two formats, a named entity tag and two semi-joint labeling tags, as shown in the last two columns Note that semi-joint labeling only integrates a segmentation tag with a named entity tag if the named entity is "O". Other named entity tags will be reserved. The number of distinct tags in the semi-joint labeling format is only one more than in the original named entity format. Even in semi-joint labeling II, which integrates two kinds of segmentation format, there are only three more distinct tags than in the original format.

In this example, we also find that integrating the word segmentation tag with the named entity tag "O" can provide boundary information, which can not be derived from the original tag "O". For example, the named entity tag sequence of the first three characters is (B-LOC, I-LOC, I-LOC); hence, the next tag can not be I-O it would be against the constraint that the word boundary can not cross or overlap the named entity boundary. This constraint helps us rule out impossible inference paths and thereby improve the precision of named entity diction boundaries.

Next, we replace $\boldsymbol{y}$ with the semi-joint labels q_{kj} in linear-CRFs, as shown in Equation 5.

$$P(\boldsymbol{y} \mid \boldsymbol{x}) = \frac{1}{Z(\boldsymbol{x})} \exp\left(\sum_k \lambda_k f_k \left(q_t, q_{t+1}, \boldsymbol{x}, t \right) \right) \quad (5)$$

By combining the BI and IE formats, we can identify more significant interactions, such as constraints, when considering the transition of labels. For example, in the last row of Table 2, the tag before B-PER can only be I-E-O and B-E-O; otherwise, it would be against the cross-overlap constraint on words and named entities.

Table 1. An example of a Chinese phrase with different tagging representation

Token	Meaning	Segmentation Tag		NER	Semi-joint Label	Semi-joint Label II
		BI format	IE format			
俄	Russian	B	I	B-LOC	B-LOC	B-LOC
罗		I	I	I-LOC	I-LOC	I-LOC
斯		I	E	I-LOC	I-LOC	I-LOC
总	president	B	I	O	B-O	B-I-O
统		I	E	O	I-O	I-E-O
普	Putin	B	I	B-PER	B-PER	B-PER
京		I	E	I-PER	I-PER	I-PER
说	says	B	E	O	B-O	B-E-O

4 Features

4.1 Basic Features

The basic features of our NER model are:

State features

- C_n (n = -2, -1, 0, 1, 2)
- C_nC_{n+1} (n = -2, -1, 0, 1)

Transition features

- C_n (n = -1, 0, 1) ,

where C denotes a character and n denotes its position. For example, C_0 denotes the current character and C_nC_{n+1} denotes its bi-gram feature, which is a combination of the previous character and the current character. A state feature is a feature that only corresponds to the current label, whereas transition features relate the previous and current labels.

4.2 Knowledge Features

Knowledge features are semantic hints that help an NER model identify more named entities. Several Chinese NER models use knowledge features; for example, Youzheng Wu [14] collects named entity lists to identify more named entities and thereby resolves the data sparseness problem in Chinese NE.

To compare our system with other approaches, we observe the closed task rules, which do not allow the use of external resources. Therefore, we only generate knowledge lists from the training corpus. For example, we compile the surname list from the tagged person names in the corpus. The knowledge feature types are listed in the Table 3. Since the features are generated automatically, we filter out those that occur less than twice [2], [14]. The table also shows the number of each distinct feature that we obtain from the training corpus.

Next, we consider how we represent knowledge in feature functions. If a character is included in a list of knowledge features, the feature's value is set at 1; otherwise, it is set at 0.

Table 2. The list of knowledge feature types

Feature type	Description	#
Person surname	The first character of a person name	678
Person name	The characters of a person name, except the first character.	1374
Previous characters of a person name	The previous single character of a person name and the previous two characters of the name	1847
The characters after a person name	The first character after a person name and the first bigram characters after the name	2467
Location name	The characters of a location name	778
Organization name	The characters of an organization name	823
Suffix characters of an organization name	The last two characters of an organization name	417

5 Experiment

In this section, we describe the experimental data, introduce the parameters used in the CRF model, and detail the experiment results.

5.1 Data Source

To evaluate our methods, we use the City University of Hong Kong (CityU) Chinese corpus from SIGHAN 2006 [10], the Special Interest Group for Chinese Language Processing of the Association for Computational Linguistics. We choose the CityU corpus because it provides both segmentation tags and named entity tags. The corpus contains 1.6M words and 76K named entities in the training part, and 220K words and 23K named entities in the test part. It also contains three named entity types: person names, organization names, and location names.

5.2 Settings

We use CRF++[1] to implement our CRF models. The parameters we use in CRF++ are f, the cut-off threshold, which is set to 1; and c, the C value that prevents over fitting, which is set to 3. The maximum number of training iterations is 1000, and the training environment is Windows Server 2003 with an AMD 2.39GHz CPU and a 10 Gigabyte RAM.

---

[1] Information about CRF++ can be found at http://chasen.org/~taku/software/CRF++/

5.3 Results and Discussion

Table 4 shows the results achieved by the three NER models. Each row shows the performance of an NER model for three types of NE with specific tagging formats, as well as the model's overall performance. The models are evaluated on the full test set (220K words and 23K NEs) of the CityU corpus. BIO uses the traditional format, i.e., a named entity type extended with a boundary, while the Semi-Joint labeling and Semi-Joint II labeling formats use the methods proposed in Section 3.2. Basic and knowledge features are included in all three models. The only difference is that the models using semi-joint formats do not include word segmentation features. By contrast, in the model that uses the BIO format, the output of a segmentation tagger includes word segmentation features. The results show that the Semi-joint format outperforms the BIO format for all three NE types with an overall F-score of approximately 1.41%. Meanwhile, the Semi-joint II format outperforms the Semi-joint format with an overall F-score of approximately 0.24%.

Table 3. The results of the BIO, semi-joint, and semi-joint II formats

		precision	recall	F-score
Baseline System	**PER**	91.42	85.35	88.28
	ORG	90.31	77.19	83.24
	LOC	92.09	91.85	91.97
	Overall	**91.49**	**86.29**	**88.82**
Semi-Joint Labeling A	**PER**	93.50	88.25	90.80
	ORG	90.43	78.14	83.89
	LOC	92.35	92.74	92.55
	Overall	**92.27**	**87.83**	**89.99**
Semi-Joint Labeling B	**PER**	93.51	89.32	91.37
	ORG	90.05	77.81	83.48
	LOC	92.43	93.32	92.87
	Overall	**92.23**	**88.31**	**90.23**

Since the proposed semi-joint labeling method integrates word segmentation with an NER model, and word segmentation can help detect the boundaries of named entities, it is worth discussing changes in the error rates due to named entity boundary detection. We define a boundary error as a named entity is identified and their lengths are different with the correct ones. Each row in Table 5 shows the reduced boundary error rate achieved by using semi-joint labeling. The error rate is computed by dividing the number of named entities with boundary errors in the semi-joint labeling method by those in the baseline system. We observe that semi-joint labeling reduces boundary errors, especially the semi-joint labeling II, which integrates two word segmentation formats.

Next we consider different types of boundary error. Suppose the boundary of a named entity in a sentence is $<i, j>$ where i is start position and j is end position. We define boundary detection error type I as $i_{\text{guessed entity}} = i_{\text{correct entity}}$ and $j_{\text{guessed entity}} \neq j_{\text{correct entity}}$, and boundary error type II as $i_{\text{guessed entity}} \neq i_{\text{correct entity}}$ and $j_{\text{guessed entity}} = j_{\text{correct entity}}$. Semi-joint B in boundary error type II is more significant than semi-joint A. We infer that, with the IE-format word segmentation information, the beginning character of a named entity can be identified more easily by the "E" label, which refers to the end of a word.

Table 4. Reduced boundary error rates achieved by the two semi-joint methods

	Reduced Boundary Error Rates
Semi-Joint A	3.90 %
Semi-Joint B	8.62 %

Table 5. Reduced error rates of boundary errors achieved by the two semi-joint methods

	Reduced Error Rates of Boundary Error Type I	**Reduced Error Rates of Boundary Error Type II**
Semi-Joint A	8.33 %	2.33 %
Semi-Joint B	15.87 %	18.14 %

We list the performance of the top five teams at the SIGAHN NER bakeoff for the CityU corpus. The performance of the proposed model is better than the top one in 1.2% F-scores. The major difference between our results and those of NII is that the latter approach uses word segmentation information as features, while we partially join word segmentation tags with named entity tags.

Table 6. The performance of the top five teams for the CityU corpus at the SIGHAN 2006 NER bakeoff

Team Name	**Precision**	**Recall**	**F-score**
Our Results	92.23	88.31	90.23
NII	91.43	86.76	89.03
Yahoo!	92.66	84.75	88.53
Chinese Academy of Sciences	92.76	81.81	86.94
Alias-i, Inc.	86.90	84.17	85.51

6 Conclusion

We propose a semi-joint tagging format that partially combines word segmentation and named entity recognition labels. The format allows us to consider the interactions between multiple labeling layers in a linear-chain CRF model. To evaluate our model, we use the CityU corpus of SIGHAN 2006 NER bakeoff. The model based on semi-joint labeling outperforms the model that uses word segmentation tags as features, with an overall F-score of approximately 1.41%. Because of the novel labeling format, the proposed model outperforms the top one system by about 1.2% in terms of the F-score.

In our future work, we will explore other possible interactions between word segmentation information and NER. We also plan to apply our method to other applications that would be improved by incorporating word segmentation information.

Acknowledgements

We are grateful for the support of thematic program of Academia Sinica under Grant AS 95ASIA02 and the Research Center for Humanities and Social Sciences, Academia Sinica.

Reference

1. Borthwick, A.: A Maximum Entropy Approach to Named Entity Recognition. New York University, New York (1999)
2. Chen, W., Zhang, Y., Isahara, H.: Chinese Named Entity Recognition with Conditional Random Fields. In: Proceedings of the Fifth SIGHAN Workshop on Chinese Language Processing, pp. 118–121 (2006)
3. Duh, K.: A Joint Model for Semantic Role Labeling. In: Proceedings of the 9th Conference on Computational Natural Language Learning, pp. 173–176 (2005)
4. Guo, H., Jiang, J., Hu, G., Zhang, T.: Chinese Named Entity Recognition Based on Multilevel Linguistic Features. In: International Joint Conference on Natural Language Processing, pp. 90–99 (2004)
5. Hendrickx, I., Bosch, A.v.d.: Memory-based One-step Named-entity Recognition: Effects of Seed List Features, Classifier Stacking, and Unannotated Data. In: Proceedings of the seventh conference on Natural language learning at HLT-NAACL, pp. 176–179 (2003)
6. Isozaki, H., Kazawa, H.: Efficient support vector classifiers for named entity recognition. In: Proceedings of the 19th international conference on Computational linguistics, pp. 1–7 (2002)
7. Ji, H., Grishman, R.: Improving Name Tagging by Reference Resolution and Relation Detection. In: Proceedings of the 43rd Annual Meeting of the ACL, pp. 411–418 (2005)
8. Klein, D., Smarr, J., Nguyen, H., Manning, C.D.: Named Entity Recognition with Character-Level Models. In: Conference on Computational Natural Language Learning, pp. 180–183 (2003)
9. Lafferty, J., McCallum, A., Pereira, F.: Conditional random fields: Probabilistic models for segmenting and labeling sequence data. In: International Conference on Machine Learning, pp. 282–289 (2001)
10. Levow, G.-A.: The Third International Chinese Language Processing Bakeoff: Word Segmentation and Named Entity Recognition. In: Proceedings of the Fifth SIGHAN Workshop on Chinese Language Processing, pp. 108–117 (2006)
11. Sun, J., Gao, J., Zhang, L., Zhou, M., Huang, C.: Chinese named entity identification using class-based language model. In: Proceedings of the 19th international conference on Computational linguistics, pp. 1–7 (2002)
12. Sutton, C., McCallum, A.: Composition of Conditional Random Fields for Transfer Learning. In: Proceedings of Human Language Technologies / Empirical Methods in Natural Language Processing, pp. 748–754 (2005)
13. Sutton, C., Rohanimanesh, K., McCallum, A.: Dynamic Conditional Random Fields: Factorized Probabilistic Models for Labeling and Segmenting Sequence Data. In: Proceedings of the Twenty-First International Conference on Machine Learning, pp. 99–107 (2004)
14. Wu, Y., Zhao, J., Xu, B.: Chinese Named Entity Recognition Combining Statistical Model wih Human Knowledge. In: Dignum, F.P.M. (ed.) ACL 2003. LNCS (LNAI), vol. 2922, Springer, Heidelberg (2004)

On the Construction of a Large Scale Chinese Web Test Collection

Hongfei Yan, Chong Chen, Bo Peng, and Xiaoming Li

School of Electronics Engineering and Computer Science,
Peking University, Beijing 100871, P.R. China
{yhf,cc,pb,lxm}@net.pku.edu.cn

Abstract. The lack of a large scale Chinese test collection is an obstacle to the Chinese information retrieval development. In order to address this issue, we built such a collection composed of millions of Chinese web pages, known as the Chinese Web Test collection with 100 gigabyte (CWT100g) in data volume, which is the largest Chinese web test collection as of this writing, and has been used by several dozen research groups besides being adopted in the evaluation of the SEWM-2004 Chinese Web Track[1] and the HTRDPE-2004[2]. We present the total solution for constructing a large scale test collection like the CWT100g. Further, we found that: 1) the distribution of the number of pages within sites obeys a Zipf-like law instead of a power law proposed by Adamic and Huberman [3, 4]; 2) and an appropriate filtering method on host alias will economize resources for about 25% while crawling pages. The Zipf-like law and the method of filtering host alias proposed in the paper will facilitate both to model the Web and to perfect a search engine. Finally, we report on the results of the SEWM-2004 Chinese Web Track.

Keywords: Test Collection, Documents, Zipf-like law.

1 Introduction

Test collections are considered as the standard dataset for evaluating the effectiveness of information retrieval (IR) systems. A large scale test collection is the basis of accelerating the research of IR. The good quality of an IR system performed in small scale test collections by no means represent the same thing would happen in a large scale test data. An IR system can be taken as practicable one only when it is capable to give a satisfying result under the evaluation of a large scale test collection, which is the key to validate and improve IR techniques and systems.

Because the Web enjoys a rapid growth on information volume and a great diversity of subjects, a popular way to construct a large scale test collection is to make the Web the superset of the collection – that is the reason why it is known as the web test collection. The English web test collections provided by TREC[5] orient the English IR, and the Japanese web test collection provided by NII[6] can serve for the Japanese IR. But a coequal scale Chinese web test collection could not be available by researchers before our work. To promote technologies of the Chinese IR, we built the CWT100g, and in this paper we introduce the methodology on building such a large

H. Li et al. (Eds.): AIRS 2008, LNCS 4993, pp. 117–128, 2008.

scale web test collection. Although the dataset itself orients the Chinese IR, the method is not limited to the Chinese language; and we hope it contribute to build test collections of other languages.

In this paper, after introducing terms, we first present the goals and design principles of the CWT100g in section 3, then introduce how to construct its three parts, the documents, the queries and the relevance judgments in section 4 and 5. Finally, we cover related works in section 6 and conclude in section 7.

2 Terms

Ever since the Cranfield experiment at the end of 1950, it has been widely accepted that an IR test collection should include three components - the document set, the query set and the relevance judgment set [7].

A document set is a collection of documents whose contents are used for text analyses by IR systems, i.e. they are the direct objects processed by IR systems. This set is the epitome of its superset.

A query set is a collection of questions asking the IR system for results. The queries of a test collection are created by assessors who also assess the relevance judgment set. Due to the difficulty in getting relevance judgments, the number of queries is usually between a few score to several hundred queries.

A relevance judgment set is a collection of standard answers for queries. It is used to compare IR systems' returning results under given queries. The more an IR system is close to the relevance judgment set, the higher quality it gets. To keep the authority of the relevance judgment set, they are usually obtained manually or semi-automatically in terms of the document set. It is clear that if an IR system can get a satisfying quality in the evaluation of a large scale document set, the system will be convictive in practice.

Herein, we sometimes use the terms "*documents*", "*queries*" and "*relevance judgments*" to denote the above three sets respectively.

Host name is an identification of a certain web site. Usually, a site can be identified by one or several host names and sometimes by its IP address. In this paper, we use the most popular host name as the identification of each web site. For example, both "www.pku.edu.cn" and "gopher.pku.edu.cn" point to the web server of Peking University. While the number of URLs including the former string is far larger than that of the latter, "www.pku.edu.cn" is taken as the popular host name of this physical web site and the "gopher.pku.edu.cn" is filtered in the process of the site selection.

3 Design Principles of the CWT100g

3.1 Three Considerations to the Documents

Usually a document set plays a more important role than the other components in a collection. To build a web test collection with a high quality, a document set should be good enough to model the Web. To qualify the epitome of the Web, the documents should be broad in subject domains and large enough in size. Kennedy and Huang et al [8, 9] proposed three aspects on the consideration of the representative of the documents – 1) should the documents be sampled statically or dynamically? 2) to what extent do the documents represent its superset? 3) and what is a suitable size of documents satisfying the needs of both general and special IR goals?

For the first consideration, we prefer a static rather than a dynamic sampling method in getting a large scale documents from the Web. A static method means web pages are crawled during a specific period meanwhile a dynamic method means pages are crawled at a random time and added to the documents continuously. It seems that a dynamic method is better than a static one. But its inconvenient is obvious for evaluations of IR systems. Because when the scale of documents is large, manpower consumed of finding relevance judgments are difficult in acquiring even if the documents do not change, let alone the documents vary as the time goes. Our selection is a trade-off consideration for the representative of documents to its superset as well as the convenience of evaluations of IR systems.

The data size of documents is a couple of orders of magnitude smaller than the Web. So the static sampling is not absolute, we are able to rebuild documents as occasion requires. Generally, constructing a large collection adopts a static manner.

For the second consideration, the scale of the documents should be large enough, and subjects of the documents should be diverse and keep the balance. Although it is hardly to measure whether the document set is representative to its superset objectively, accommodating the large, diverse and balance sampling subjects is definitely a sure warrant of the representative of the documents. But the size of the documents is in contradiction with manpower exhausted in confirming the relevance judgments. The more documents are sampled, the higher the representative would get. Although this circumstance brings a higher representative and balance, too many documents will return a mass of results for a query and exhaust much human labor in finding out relevance judgments when doing evaluations. Fortunately, by using the pooling method [10], the range of constructing relevance judgments can be concentrated, and human labor can be cut down.

For the third consideration, the data size of documents should be kept in an easy-to-use scale in practice. For example, in 2004, 100 gigabyte disk capacity is of the mainstream. A dataset with this order of magnitude not only can be delivered conveniently, but also is capable to meet the demand of the link analysis of a web IR system. The link analysis proved to improve the precision of a web IR system.

3.2 The Practice of the CWT100g

The CWT100g is the achievement of practicing to model the Chinese Web. It includes the documents, the queries, and the relevance judgments.

The documents come from 17,683 sites out of the 1,000,614 web sites that are identified by different host names. Those names are found within the Chinese scope by Tianwang Search Engine [11] before Feb. 1, 2004. After selecting and processing to both sites and web pages, we sampled 5,712,710 pages, about 90 gigabyte disk spaces. Each page in the collection has a "text/html" or “text/plain" MIME type received from the HTTP server response message. If the crawling capacity of Tianwang, 0.2 billion unique URL pages accumulated on Feb.1, 2004, is equally as large as the corpora of the Chinese Web, we presume that the nearly 6 million pages is a moderate size to represent the Chinese web pages.

The queries were numbered TD1-TD50 for a Topic Distillation (TD) task and NP1-NP285 for a navigational task. The topic distillation task involves finding relevant homepages, and the navigational task is to find out a Home/Named Page (NPHP) [12, 13]. They are sampled from user query logs of Tianwang between Apr. 2002 and June 2004, and then compiled by assessors.

The relevance judgments, 1,178 for the TD queries and 285 for the NPHP queries, are constructed by taking advantage of both Tianwang techniques on the web search and the improved pooling method.

After the brief illustration, we will focus on the construction of the documents, and then show how to deal with queries and judgments. The documents are the basis of a test collection and require a reasonable constructing strategy, while the queries would keep changing, and the relevance judgments are up to the concrete queries in the evaluation phase.

4 Constructing the Documents

The first important thing in constructing the CWT100g is to ensure the document collection representative to the Chinese Web, and thus provide the IR researchers a highly simulating web test-bed. It comes down to how to sample web sites and crawl pages within them.

The characteristics of the Chinese Web help determine the sampling strategy. In this section, we show our study on the distribution of pages within sites before choosing the candidate sites to form the CWT100g, and then discuss the choice strategies to sample sites and pages.

4.1 The Characteristics of the Chinese Web

Several famous works have been known in detecting the characteristics of the Web in [3, 4, 14, 15].

As to the study of the Chinese Web, Yan and Li [16] found that in the early 2002 there were about 50 million pages and 50 thousand web sites with unique IP addresses, the average size of a page was 13KB, and the Chinese Web was highly connected; Meng et al [17] validated that in 2002 Tianwang could cover 37% of the total Chinese web pages, and 50% of the high ranked pages. To keep an overall coverage and a fresh update, Tianwang adopts an incremental crawling strategy and provides a service of registering sites. Until Feb. 2004, we had collected 0.2 billion unique URLs among 1,000,614 different sites.

Based on the above works, we got a report on the relationship between the site number and the page number within various sites. All the data were fetched from the Tianwang data on Feb. 1, 2004. Figure 1 shows the distribution of pages within sites in a linear scale plot.

Y-axis denotes the site number and x-axis denotes the number of pages within a site. Figure 1 shows that sites with a large number of pages are small amount and those with a few pages are large amount. For example, we found only 1.9% sites held upward of 500 pages, whereas most sites held pages less than 500. The distribution is so extreme that if the full range was shown on the axes, the curve would be a perfect L shape.

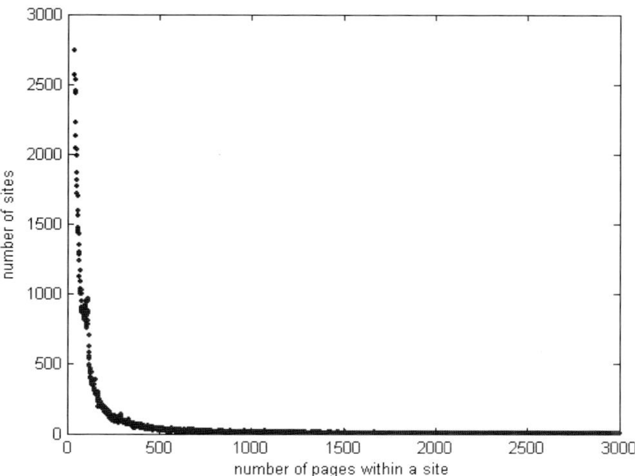

Fig. 1. Linear distribution of pages within sites

To verify whether the distribution accords with a power law as being claimed by Adamic and Huberman [3, 4], we plotted Figure 2 according to the data provided by Tianwang.

If we directly show a log-log scale plot for the same data of Figure 2, the tail end of the distribution is 'messy' - there are only a few sites with a large amount of pages. Because there are so few data points in that range, simply fitting a straight line to the data gives a slope that is too shallow. To get a proper fit, we need to bin the data into exponentially wider bins (They will appear evenly spaced on a log scale) as Adamic [18] does. A linear relationship now extends over 6 decades ($1\text{-}10^6$) pages vs. the earlier 4 decades: ($1\text{-}10^4$) sites. The slope is 0.68, which makes it a Zipf-like [19] distribution.

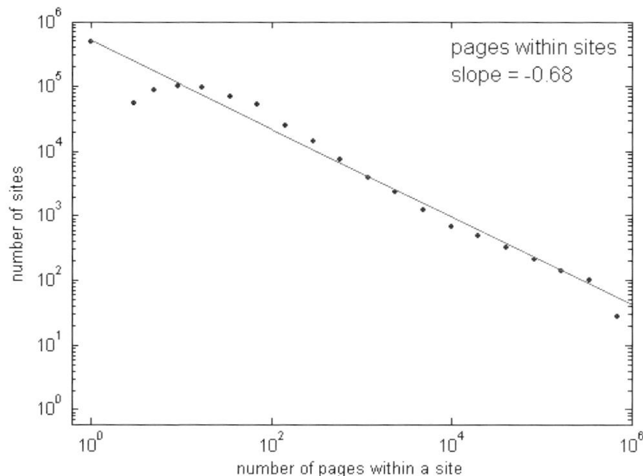

Fig. 2. Binned distribution of pages within sites

The answer to the characteristic of the Web will facilitate processes of the site selection and the web crawler, which are illustrated as follows.

4.2 The Process of the Site Selection

To validate how to satisfy the representative of the documents, we present the process of the site selection shown as Figure 3.

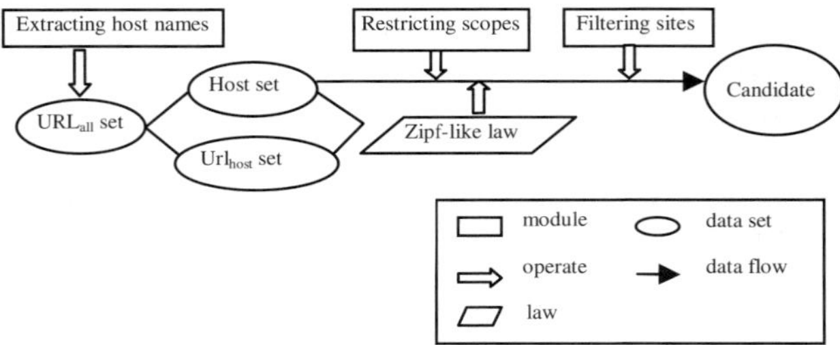

Fig. 3. The process of the site selection

The *URL_{all} set*, the superset (the Chinese Web) of the CWT100g's documents, is the collection of 0.2 billion unique URLs crawled by Tianwang. The host names were extracted from the URL set and formed the initial *host set*, and the *URL_{host} sets* is acquired according to host names. The *Zipf-like law*, the distribution of pages within sites mentioned above, facilitates to eliminate the disturbance of sites with extremely fewer or larger pages. Besides, the module of the *filtering sites* and the module of the *restricting scopes* operate together on the initial *host set* to get rid of the redundant, unstable and spamming sites and limit the remnant *candidate* sites within the Chinese scope in order to build the model of the Chinese Web.

4.2.1 The Module of Restricting Scopes

Since the superset provided by Tianwang is within the Chinese scope. The module of restricting scopes is simplified to remove the host names with weird suffixes. According to the supplier of domain name registrations, http://www.net.cn, and the characteristics of host names with Hong Kong, Macao and Tai Wan, the sites of the CWT100g are restricted in host names ending with "cn", "com", "net", "org", "info", "biz", "tv", "cc", "hk", "tw".

4.2.2 The Module of Filtering Sites

The module of *filtering sites* in Figure 3 takes charge of 1) eliminating the errors brought by the *URL_{all} set*; 2) filtering out host names with non-80 ports, 3) removing names that are IP addresses in standard dot notation, 4) detecting DNS validation of names, 5) distinguishing the importance of names, 6) and finding out the sites built by templates in batches and automatically. Step 1), 2), and 3) are verified offline by parsing strings of names, step 4) and 5) are finished online by site detecting programs, and the step 6) is done semi-automatically.

Table 1. Filtering sites for the CWT100g

Step	D_i	E_i (comments)
0	1,000,614	1736 (host name error)
1	998,878	8506 (non-80 port)
2	990,372	19,135 (IP address)
3	971,237	97,898 (DNS validation)
4	873,339	314,799 (no service & host alias)
5	558,540	119,542 (template sites)
6	438,998	

Supposing $\boldsymbol{D_0}$ denotes the initial URL_{all} set. After each step, it will produce a new URL set named $\boldsymbol{D_i}$, $\boldsymbol{i=0,1,\ldots 7}$ ($\boldsymbol{D_7}$ will appear in section 4.2.3 and 4.3), and the following steps is based on the previous ones.

To make a summary, filtering sites include two phrases. Firstly, it is necessary to find out different host names standing for the same site, and keep the popular name as the unique identifier. Secondly, it is required to eliminate non-representative sites to improve the precision of sampling sites. A typical representative site provides stable services and maintains a number of meaningful pages. The remaining sites in every step are calculated as:

$$D_i = D_{i-1} - E_{i-1} \tag{1}$$

where $1 \le i \le 6$, $\boldsymbol{D_i}$ is the set of every step, and $\boldsymbol{E_i}$ is the set found dissatisfied with the rule of the step. The sizes of $\boldsymbol{D_i}$ and $\boldsymbol{E_i}$ in six steps are listed in Table 1.

4.2.3 Determining Sampling Sites

To support algorithms based on linkage analyses, it needs to preserve sites with enough links. Each site has more than 10 out links.

For determining sample sites for the CWT100g, the first step is to estimate the number of sites needed by it. Supposing the size of a single web page (a document) is 15KB[16], the total 100 gigabyte will contain 7 million pages. According to the above deduction of the preprocessing host names, 0.2 billion pages are distributed in about 500,000 validate sites, 7 million pages will be distributed in about 17,500 (7m/0.2b=X/500000 => X =(7/200)*500000) sites. Considering some sites maybe fail to be crawled, we sample about 20,000 sites to ensure enough sites to be crawled, known as $\boldsymbol{D_7}$. The ratios of sampling is about 3.5% (7/200) for pages and 4% (20000/500000) for sites that are big enough to guarantee the representative of the CWT100g.

Up to now, results are nearly out of noise data. Again we study the distribution of pages within sites. Its fitting coefficient is 0.64 which is nearly the same result in figure 2. The result repeatedly shows that the distribution of pages within sites of the Chinese Web accords with a Zipf-like law. Meanwhile, it illustrates that the existence of noise sites does not affect the distribution.

4.3 The Strategy of Document Selection

After determining sampling sites $\boldsymbol{D}_7$, the next step is to crawl pages in terms of them. Not all pages are crawled from the selected sites. Because web pages with "plain/html" and "plain/text" types are widely used and easily indexed by IR systems, we only preserve pages whose MIME types are one of them. If the pages are written by non-Chinese languages but belong to the Chinese scope, the pages are also kept. Files with DOC and PDF formats are excluded from the documents. Due to difficult in judging the validity of a dynamic page, URLs with question marks are gotten rid of the documents.

There are two main issues in the crawling process, the storage format for raw web pages and the strategy of crawling pages. The former requires: 1) less disk spaces, 2) fault tolerances, and 3) convenient usages. The latter requires: 1) crawling pages in parallel so that all pages crawled within a short interval in time, such as 10 to 15 days, to stand for a web snapshot of a certain period; 2) crawling all static pages on sites as completely as possible. To limit redundant pages from a large site, the crawling process will be broken off when the size of the log file is larger than 2 gigabyte.

The storage format for raw web pages: the CWT100g adopts Tianwang storage format [20] that satisfies with above three requirements of storing web pages.

The strategy of crawling pages: we use the modified version of TSE [21] as the crawling system whose strategy is similar to the breadth first crawling method. Because sites in $\boldsymbol{D}_7$ are all independent each other, it is easy to parallel the process of crawling pages. The detail is described as follows.

1) Separating $\boldsymbol{D}_7$ into groups, each group with 100 sites, we get 205 groups. Each group is tokenized by an unique filename to stand for a subtask, such as, xaa, xab,…,xba, xbb,…,xha, xhb, xhw, the number of sites in the last group may be less than 100. Each group is a subtask at a fine granularity. The task is defined as integrating subtasks with the same middle letter into a big group. Each task is at a coarse granularity. We use a task as a basic unit to allocate the workload. Because the distribution of pages within sites in the Chinese scope obeys a Zipf-like law, the sites sampled later will have larger amount of pages than those before and need more crawling resources to crawl pages. Through assigning different number of tasks among crawling machines, we make the workload balanceable.

2) In the machine executing one or more tasks, we start a TSE module for each subtask. Each subtask is only taking charge of crawling pages within their scope. Each TSE module has 10 threads working together, and pages crawled are stored into ten files corresponding to ten threads. The TSE stops when the subtask is over. For those large sites, the TSE stops when the size of its log file is larger than 2 gigabyte.

3) Finally, we integrate all pages crawled into the CWT100g's documents. To avoid overriding files with same names, each file name is renamed to append a suffix of the subtask name.

4.4 The Properties of Documents and Experiments

Based on the initial host set D_0 provided by Tianwang, there are 5,712,710 pages among 17,683 sites in the CWT100g's documents.

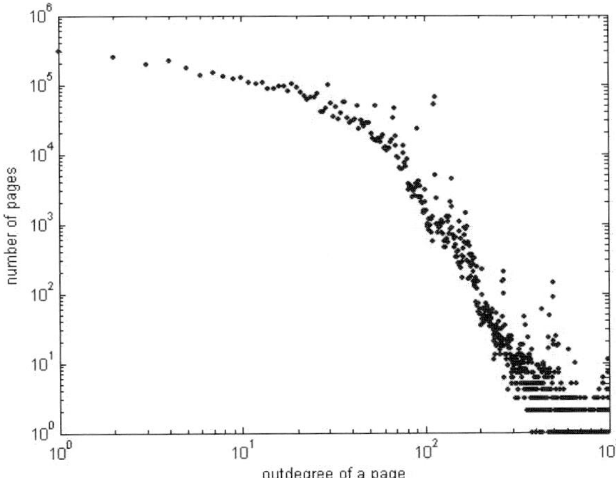

Fig. 4. Log-log distribution of out-degree among pages

In the documents, there are 4,814,747 pages with a non-zero out-degree, and the total number of links is 127,519,080. Each page averagely has 26 links. The distribution of out-degree among pages is shown in Figure 4. It is a power-law distribution.

There are 4,252,696 (74%) pages encoded with GB2312, 295,274 pages with GBK, 12,230 pages with BIG5, and 254,547 pages with other types, such as ISO-8859-1, ISO-2022-CN, and US-ASCII.

The CWT100g is now freely available to social circles [1], and has been designated as the web test collection of SEWM-2004 Chinese Web Track [1] and HTRDPE-2004 [2] for evaluating all participant IR systems.

5 Constructing the Queries and the Relevance Judgments

After completing the first part of the CWT100g, the documents, we will illustrate how to obtain the other two parts, the queries and the relevance judgments.

5.1 The Construction of the Queries

Queries should also be representative to characterize the behavior of netizens. The CWT100g's queries are sampled from user query logs of Tianwang between Apr. 2002 and June. 2004. The queries coming from the real web users represent the real state of the Web. To avoid queries includes the barren ones whose relevance judgments are difficult to obtain or otherwise, all sampled queries are checked by assessors.

A query is also known as a topic in the TREC. Referring to the information on the TREC-2003 Web Track [12, 13], we constructed 50 queries for a topic distillation task and 285 queries for a navigational task.

5.2 The Construction of the Relevance Judgments

Relevance judgments require completeness and a consistency. Completeness means that all relevance documents are judged relevantly. Consistency means that all persons will subjectively consider the relevance of some documents with the same judgment.

The CWT100g's relevance judgments include two parts. One is for the topic distillation task; the other is for the navigational task. The latter is easily constructed by recording corresponding pages while creating the queries. The former is not easy to create. Here we use the improved pooling method to get the final set.

The traditional judgment pools are created as follows. 1) Participants submit their retrieval runs. 2) Assessors choose a number of runs and merge them as a pool for the specific queries. 3) For each selected run, the top X documents are added to the pool and judged. This method is not effective when the participants are not sufficient because the relevant information is incomplete in the pool and the evaluation will be less authoritative. To address the issue, we extended the *pooling* method to the *pooling plus* method that yields effective judgments with the help of search engines, whose ranking results can be treated as the retrieval runs of virtual participants. By introducing the results of different search engines with same queries, many virtual participants are simulated.

The pools of the relevance judgments of the topic distillation queries consists of the results of 8 groups (http://www.cwirf.org/2004WebTrack/result.htm) and those of five search engines including Google, Yisou, Baidu, Sogou, Zhongsou. Across the 75 TD queries, 1,178 pages were judged relevant to 52 queries (average of 22.65 pages per query). 23 (75-52) queries were discarded due to the queries with too many or too few relevance judgments.

6 Related Works

Ever since the Cranfield experiment at the end of 1950, it has been widely accepted that an IR test collection should include three components - the document set, the query set and the relevance judgment set [7].

Web test collections developed include: VLC2, WT2g, WT10g, .GOV, .GOV2[5], NTCIR-3[6]. Form the point of the view of the trend, the data size of web test collections is getting bigger and more complete.

In the areas of constructing test collections, Kennedy and Huang et al [8, 9] proposed three aspects on the consideration of the representative of documents. Though their viewpoints aim at linguistic corpus, it is also helpful for constructing web test collections. The WT10g [22] has 1.7 million web pages, taking VLC2 as its superset, and the relevance judgments coming from TREC451-500(from search logs) . However, the superset, VLC2 crawled in 1997 by Internet Archive, is very stale, and the WT10g only contain English web pages and has a small size of the documents. Although the data size of the .GOV2 is larger than 400 gigabyte, it is limited to the English pages and ".gov" domain. It is neither a model of the Web nor a suitable test collection for other language IR systems. The pooling method [10] makes the range of constructing relevance judgments concentrate, and cut down human labor, but it will fail whenever there are few participants.

To better model the current Web, it requires to crawl the latest web pages. The CWT100g takes the Chinese Web as its superset, and aims to sample 100 gigabyte web pages to model the Chinese Web. If we follow the way of the TREC to build the web test collection by sampling pages archived several years ago or crawling pages from a special domain name, the data content would be neither popular nor wide enough to represent the Web. Thus the CWT100g adopts the strategy of sampling sites and crawling all their pages. For those extreme large sites, the crawling process stops when the size of log file is larger than 2 gigabyte.

7 Conclusion

The CWT100g has been functioned as a test-bed and benchmark of the Chinese IR and proved valid and useful. Its documents, queries and relevance judgments, and the pooling plus method are applied to evaluate eight different systems.

The documents of the CWT100g, sampled from the overall China-wide web pages, contribute a model of the Chinese Web. The total 90 gigabyte, 5.71 million documents, provides a reusable test collection for the Chinese IR. The approaches on building the documents enable the collection to be easily rebuilt and grown in size when new trends come. All the toolkits we used are freely available.

Further, we discovered the distribution of pages within sites of the Chinese Web obeys a Zipf-like law instead of a power law ever been considered and put forward a method of filtering host alias to save about 25% cost while crawling pages.

Acknowledgements

We would like to thank Tao Meng for providing the updated site lists he crawled in the Chinese scope. This work described therein is supported in part by by NSFC grant 60435020, NSFC Grant 60603056 and 863 Grant 2006AA01Z196.

References

1. CWT100g, Chinese Web test collection (2004)
2. HTRDPE. HTRDP Chinese Information Processing and Intelligent Human-Machine Interface Technology Evaluation (2004)
3. Huberman, B.A., Adamic, L.A.: Growth dynamics of the World-Wide Web. Nature 401, 131 (1999)
4. Adamic, L.A., Huberman, B.A.: Zipf's law and the Internet. Glottometrics 3, 143–150 (2002)
5. CSIRO, TREC Web Tracks Homepage (2004)
6. NTCIR, NTCIR (NII-NACSIS Test Collection for IR Systems) Project (2004)
7. Cleverdon, C.W.: The significance of the Cranfield tests on index languages. In: Proceedings of the 14th annual international ACM SIGIR conference on Research and development in information retrieval, Chicago, Illinois, United States (1991)

8. Kennedy, G.: An Introduction to Corpus Linguistics, vol. 280. Longman, London (1998)
9. Huang, C., Li, J.: Linguistic corpse: Business publisher (2002)
10. Jones, K.S., Rijsbergen, C.v.: Report on the need for and provision of an 'deal' information retrieval test collection. British Library Research and Development Report 5266, Computer Laboratory, University of Cambridge (1975)
11. Tianwang, Tianwang Search Engine (2004)
12. Craswell, N., et al.: Overview of the TREC-2003 Web Track. In: Proceedings of TREC 2003, Gaithersburg, Maryland USA (2003)
13. Hawking, D., Craswell, N.: Very Large Scale Retrieval and Web Search (Preprint version) (2004)
14. Lawrence, S., Giles, C.L.: Accessibility of information on the web. Nature 400(6740), 107–109 (1999)
15. Broder, A., et al.: Graph structure in the web: experiments and models. In: Proceedings of the 9th World-Wide Web Conference, Amsterdam (2000)
16. Yan, H.F., Li, X.: On the Structure of Chinese Web 2002. Journal of Computer Research and Development, 2002 39(8), 958–967 (2002)
17. Meng, T., Yan, H.F., Li, X.: An Evaluation Model on Information Coverage of Search Engines. ACTA Electronica Sinaca 31(8), 1168–1172 (2003)
18. Adamic, L.A.: Zipf, power-laws, and pareto - a ranking tutorial, Tech. Rep., Xerox Palo Alto Research Center (2000)
19. Breslau, L., et al.: Web Caching and Zipf-like Distributions: Evidence and Implications. Proc. IEEE Infocom 99, 126–134 (1999)
20. Yan, H.F., et al.: A New Data Storage and Service Model of China Web InfoMall. In: the 4th International Web Archiving Workshop (IWAW 2004) of 8th European Conference on Research and Advanced Technologies for Digital Libraries (ECDL 2008), Bath, UK (2004)
21. TSE, Homepage of Tiny Search Engine (2004)
22. Bailey, P., Craswell, N., Hawking, D.: Engineering a multi-purpose test collection for Web retrieval experiments. Information Processing & Management 39(6), 853–871 (2003)

Topic Tracking Based on Keywords Dependency Profile

Wei Zheng, Yu Zhang, Yu Hong, Jili Fan, and Ting Liu

School of Computer Science and Technology, Harbin Institute of Technology, 150001 Harbin, China
{zw,yzhang,hy,jlfan,tliu}@ir.hit.edu.cn

Abstract. Topic tracking is an important task of Topic Detection and Tracking (TDT). Its purpose is to detect stories, from a stream of news, related to known topics. Each topic is "known" by its association with several sample stories that discuss it. In this paper, we propose a new method to build the keywords dependency profile (KDP) of each story and track topic basing on similarity between the profiles of topic and story. In this method, keywords of a story are selected by document summarization technology. The KDP is built by keywords co-occurrence frequency in the same sentences of the story. We demonstrate this profile can describe the core events in a story accurately. Experiments on the mandarin resource of TDT4 and TDT5 show topic tracking system basing on KDP improves the performance by 13.25% on training dataset and 7.49% on testing dataset comparing to baseline.

Keywords: Topic Detection and Tracking, topic tracking, word co-occurrence, keywords dependency profile.

1 Introduction

Topic detection and tracking (TDT) automatically organizing a stream of news as it arrives over time [1]. Topic tracking is a primary task of TDT, whose purpose is to associate incoming stories with known topics. A topic is a seminal event with all directly related events [2]. It is "known" by association with on-topic and off-topic sample stories. A story is "on-topic" if its events relate to the topic's seminal event.

Currently, most studies break each story into a "bag of words" and judge if the story is related to a topic by their overlapping words [1]. But it assumes that words are independent. It ignores that words in the same event have strong dependency relations important for describing the event. It causes the semantic confusion and words in different events can form wrong semantics. For example, the seminal event of topic 41002 in TDT4 is: "*Kim Dae-jung received Nobel Peace Prize*" and two related events are "*South Korean President Kim Dae-jung won the Nobel Peace Prize for efforts toward reconciliation with North Korea*" and "*Chen Shui-bian sent telegrams of congratulations to Kim Dae-jung and forecasted cross-strait relations between Taiwan and mainland China*". The "bag of words" can form wrong semantic: "*efforts of mainland China toward reconciliation in cross-strait relations*". So, stories about this are judged on-topic by mistake. "Bag of words" can't represent events accurately.

An event can be captured by a few semantic elements, such as what, who, where and when [3], called keywords. Their relations are important for describing events,

H. Li et al. (Eds.): AIRS 2008, LNCS 4993, pp. 129–140, 2008.

which can make the tracking easy. For example, a story is judged on-topic where "Kim Dae-jung" and "Nobel Peace Prize" depend on each other. But the story where "Nobel Peace Prize" and "United Nations" have strong relations is judged off-topic.

This paper proposes a tracking algorithm based on keywords dependency profile with three main characteristics. First, document summarization technology is utilized to select keywords, which finds indicative content of story [4]. Second, dependency relations of keywords are evaluated by their co-occurrence frequency in the same sentences. The relations are considered strong if they often co-occur in the same sentences. Third, profile of each story is built basing on keywords, the dependency relations and their contexts in the story. The topic model built by profiles of sample stories can describe the seminal event and topic tracking based on it is more effective.

This paper is organized as follows: Section 2 introduces the related works. Section 3 describes the method to build the profile basing on co-occurrence of keywords. Section 4 describes how to use profile to complete topic tracking task. Section 5 and 6 discuss the experiment and results. Finally, we draw conclusions in Section 7.

2 Related Works

Most previous approaches in topic tracking are based on statistical strategy [5]. Some methods in pattern classification are utilized. CMU applies k-NN algorithm to topic tracking [6]. It selects k most similar stories of current story and relates current story to the topic which most of the k stories are related to. Binary classification [7] is also applied. But sparsity of annotated stories is a main factor limiting their performance.

Another trend of research is to use Natural Language Processing technology. Named Entity Recognition (NER) [8] and time information extraction [9] are utilized [8]. POLYU [10] uses NER to build profiles of stories. It calculates information gains of NE to evaluate their importance, and selects words in important sentences as features. The profile is built by clustering features to different groups. But it ignores the relations among features which are important for describing the events.

Some research finds the relations among words to improve performance. N-gram language model [11] is applied to topic tracking which considers words' relations. Class-based models [12] use dependency relations to divide words to different groups. UMass [13] uses co-occurrence among NE in the same story to find their relations. But it doesn't apply that to sentence which is a more integrated semantic unit.

3 Keywords Dependency Profile

As mentioned above, keywords dependency relations can describe the core events of a story and co-occurrence of these keywords in the same sentences can reflect their dependency relations. For example, there are two stories:

story1 :金大中昨天荣获本年度诺贝尔和平奖…… 诺贝尔和平奖委员会的奖状说"金大中…… 金大中荣获21 世纪首个诺贝尔和平奖……

story2:美国助理国务卿阿米蒂奇上星期在华盛顿同韩国总统金大中.. 朝鲜领袖金正日曾滥用金大中的'阳光政策' 金大中和克林顿所提倡的同朝鲜建立的

Story1 introduces the reasons why Kim Dae-jung received Nobel Peace Prize. *Story2* discusses activities between America and South Korea. In *story1*, "Kim Dae-jung(金大中)" and "Nobel Peace Prize(诺贝尔和平奖)" (as the Chinese words with underlines in *story1*) mostly co-occur in the same sentences, so this story is related to topic 41002. In *story2*, "Kim Dae-jung" appears frequently and co-occurs with "America (美国)", "South Korea (韩国)" and "North Korea (朝鲜)", but seldom with "Nobel Peace Prize". So *story2* is off-topic.

On the whole, keywords and their dependency relations can describe core events in the story and seminal event in the topic accurately. In the rest part of section 3, the method to build the keywords dependency profile (KDP) of a story is introduced.

3.1 Selecting Keywords

MEAD, an important method in document summarization, is utilized to select keywords in a story. The reason is that keywords are content which can describe events completely [3] and the goal of MEAD exactly is to find the indicative sentences of the story. It assumes that sentences similar to the centroid are more indicative, where centroid consists of $N_{centroid}$ words appearing most frequently in the story. The score of each sentence is calculated as follows [4]:

$$SCORE(s_i) = \eta_c C_i + \eta_p P_i + \eta_f F_i \tag{1}$$

Where $SCORE(s_i)$ is the score of sentence s_i. C_i is the similarity between centroid vector and s_i. P_i is calculated by C_1 multiplying $(n-i+1)/n$ and n is the total number of sentences in the story. F_i is the similarity between s_i and the first sentence.η_c,η_p, andη_f are parameters which are set to 1, 2 and 1 [4].

The sentence with the highest score is chosen. After word segmentation and removing stop words, the remaining words in that sentence are chosen as candidate keywords: $K=\{key_1, key_2, ..., key_m\}$. The set K chosen in *story1* is {"South Korea", "President", "Kim Dae-jung", "received", "this year", "Nobel" and "Peace Prize"}.

3.2 Building Keywords Dependency Profile

Building the profile of a story by dependency relations of keywords includes three steps. Firstly, the network of dependency relations among candidate keywords in a story is built. Figure 1 is the network of *story1*. $N=\{n_1, n_2, ..., n_m\}$ denotes the nodes set. $n_l (1 \le l \le m)$ is a node representing key_l in candidate keywords set K and $tf(n_k)$ is its frequency. $E=\{e_{1,2}, e_{1,3}, \ldots, e_{m-1,m}\}$ denotes edges set. $e_{k,l}$ is the edge between n_k and n_l. $tf(e_{k,l})$ is the frequency that key_k and key_l co-occur in the same sentences.

Secondly, the weights of nodes and edges are calculated. KDP determines weight of each node by two factors:

- The more important are nodes which a node adjacent to, the higher is the weight of this node. The idea is that the weight of a keyword is high if it strongly depends on other "important" keywords, and the initial "important" of these keywords are evaluated by their *tf*idf*. For example, the weight of "received" (n_4 in figure 1) should be high since it co-occurs with important words "Kim Dae-jung", "Nobel" and "Peace Prize", although it only appears three times.

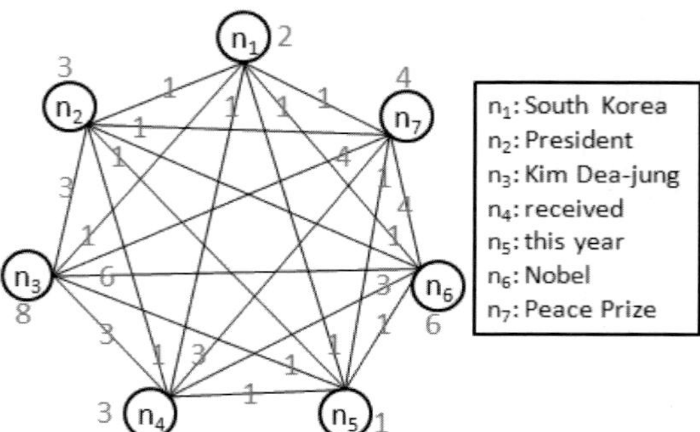

Fig. 1. The network of dependency relations in *story1*. The numbers recorded besides nodes and edges are the values of their *tf*.

- The more edges connecting to a node, the higher is the weight of this node. The idea is that a keyword is important if it mostly co-occurs with other keywords.

The equation to calculate the weight of each node is:

$$w(n_k) = \sum_{l=1, l\neq k}^{m} \log(tf(n_l) \times idf(n_l) \times tf(e_{k,l}) + 1) \tag{2}$$

Where $w(n_k)$ is the weight of node n_k and m is the total number of nodes. n_l adjacent to n_k. $tf(n_l)$ and $idf(n_l)$, term frequency and inverse document frequency of n_l, are utilized to represent its initial importance. $e_{k,l}$ is the edge connecting to n_k and n_l. $tf(e_{k,l})$ is the co-occurrence frequency of keywords key_k and key_l in the same sentences. After calculation, the nodes in figure 1 with the highest weights are n_3, n_4, n_6 and n_7.

The rule to determine the weight of each edge is: the higher are weights of the nodes that an edge connects to, the higher is the weight of this edge. An edge is important if it reflects dependency relation of two important keywords. For example, the weight of edge $e_{3,7}$, “Kim Dae-jung”-“ Peace Prize ”, is high since the weights of nodes it connects to are the highest. The equation to calculate the weight of edge is:

$$w(e_{k,l}) = w(n_k) \times \frac{tf(e_{k,l})}{tf(n_k)} + w(n_l) \times \frac{tf(e_{k,l})}{tf(n_l)} \tag{3}$$

Where $w(e_{k,l})$ is the weight of edge $e_{k,l}$. $w(n_k)$ and $w(n_l)$ are weights of nodes which $e_{k,l}$ connects to and are used to evaluate their importance. $tf(e_{k,l})$, $tf(n_k)$ and $tf(n_l)$ are their frequency. $tf(e_{k,l})$ divided by $tf(n_k)$ shows the importance of $e_{k,l}$ comparing to other edges connecting to n_k.

After calculation, the edges with highest weights in figure 1 are $e_{3,7}$, $e_{6,7}$, $e_{4,7}$, $e_{3,6}$, $e_{3,4}$ and $e_{4,6}$. These can describe core events accurately. Especially, although the important keyword “Nobel Peace Prize” is mistakenly broken into “Nobel” and “Peace Prize” by word segmentation system, they still depend strongly on each other ($e_{6,7}$). Then weights of nodes and edges are normalized by dividing the total weights.

At last, KDP of the story is built by appending contexts to nodes and edges. The context of node n_l is sentences containing the keyword key_l. The context of edge $e_{k,l}$ is sentences containing both key_k and key_l. Each context is a vector including Chinese words in it, and the word's weight is its value of *tf*idf*. The profile is denoted by $KDP=\{N, E, C, W\}$. Where N is the nodes set and E is edges set. $C=\{\vec{c}(n_1), \vec{c}(n_2), \ldots, \vec{c}(n_m); \vec{c}(e_{1,2}), \vec{c}(e_{1,3}) \ldots, \vec{c}(e_{m-1,m})\}$ contains their context vectors. $W=\{w(n_1), w(n_2), \ldots, w(n_m); w(e_{1,2}), w(e_{1,3}) \ldots, w(e_{m-1,m})\}$ contains the weight of each node and edge.

4 Keywords Dependency Profile for Topic Tracking

In this section, KDP is utilized in topic tracking which includes three steps:

- Firstly, the topic model is built by profiles of all sample stories in the topic. The topic model, namely topic's profile, contains keywords and their dependency relations useful for describing seminal event of the topic.
- Secondly, the profile of each incoming story is built according to the keywords in the topic model.
- At last, the topic model and the profile of each incoming story are compared to judge if the story is on-topic.

4.1 Building Topic Model

The topic model is built by the KDP of sample stories in the topic and denoted by $KDP_T=\{N_T, E_T, C_T, W_T\}$. The nodes set N_T is formed by selecting nodes in KDP of each on-topic sample story. The edges set E_T is formed in the same way. The contexts set C_T is formed by adding all words in the context of the same node or edge in on-topic sample stories. W_T, weights of all nodes and edges, is calculated next.

A node's weight is determined by the rule: the higher is the weight of a node in on-topic sample stories and lower in off-topic ones, the higher is the node's weight in topic model. The idea is that nodes abundant in on-topic stories are important for describing seminal event, and those in off-topic stories are unimportant according to definition of "on-topic". The weight of each node is calculated as follows:

$$w_{KDP_T}(n_k) = \frac{\sum_{KDP_i \in S_{on}} w_{KDP_i}(n_k)}{|S_{on}|} - \frac{\sum_{KDP_j \in S_{off}} w_{KDP_j}(n_k)}{|S_{off}|} \tag{4}$$

Where n_k is a node in KDP_T and $w_{KDP_T}(n_k)$ is its weight. $w_{KDP_i}(n_k)$ is its weight in KDP_i. S_{on} contains all KDP of on-topic sample stories and $|S_{on}|$ is the number of stories in it. S_{off} contains KDP of off-topic sample stories. The nodes with weights less than zero are deleted. The weights of words in context of node are calculated similarly:

$$\vec{c}_{KDP_T}(n_k) = \frac{\sum_{KDP_i \in S_{on}} \vec{c}_{KDP_i}(n_k)}{|S_{on}|} - \frac{\sum_{KDP_j \in S_{off}} \vec{c}_{KDPj}(n_k)}{|S_{off}|} \tag{5}$$

Where n_k is a node in topic model. $\vec{c}_{KDP_T}(n_k)$, $\vec{c}_{KDP_i}(n_k)$ *and* $\vec{c}_{KDPj}(n_k)$ are n_k's contexts vectors in topic model, on-topic and off-topic sample stories.

The weights of all edges and words in their context are calculated in the same way. At last the whole topic model is built up.

4.2 Building Profile of Incoming Story

All keywords in topic model are extracted and then these keywords are utilized to build KDP of each incoming story by the method introduced in section 3.2.

The nodes and edges in topic model are useful to describe the seminal event of the topic. The aim of building profile of each incoming story is to evaluate if its events are related to seminal event, but don't directly evaluate if its core events are the seminal event. So whether keywords in topic model appear in the story and whether their dependency relations are the same as those in the topic model is evaluated. Therefore, keywords in topic model are utilized to build the profile of each incoming story, instead of keywords selected from the story as introduced in section 3.1.

4.3 Measuring Similarity

Topic model is denoted by $KDP_T=\{N_T, E_T, C_T, W_T\}$ and the profile of an incoming story by $KDP_S=\{N_S, E_S, C_S, W_S\}$. The similarity between the incoming story and the topic is measured by similarity between the story and edges set of topic model, as well as similarity between the story and notes set.

$$Sim(T,S) = \alpha Sim'(E_T,S) + (1-\alpha)Sim'(N_T,S) \tag{6}$$

Where *Sim(T, S)* is the similarity between topic *T* and an incoming story *S*. E_T is the edges set in topic model and $Sim'(E_T, S)$ is its similarity with the story. N_T is the nodes set in topic model. α is a parameter representing the contribution of $Sim'(E_T, S)$ to *Sim(T, S)*. Its value is set to 0.9 according to experiments in section 6. The story will be judged on-topic if the similarity is higher than the threshold.

$$Sim'(E_T,S) = \frac{\sum_{e_{k,l}\in E_T} w_{KDP_T}(e_{k,l}) * w_{KDP_S}(e_{k,l}) * \cos ine(\vec{c}_T(e_{k,l}), \vec{c}_S(e_{k,l})) * \cos ine(\vec{c}_S(e_{k,l}), \vec{S})}{\sqrt{\sum_{e_{k,l}\in E_T} w_{KDP_T}(e_{k,l})^2} * \sqrt{\sum_{e_{g,h}\in E_S} w_{KDP_S}(e_{g,h})^2}} \tag{7}$$

Where $e_{k,l}$ is an edge. $w_{KDP_T}(e_{k,l})$ and $w_{KDP_S}(e_{k,l})$ are its weights in KDP_T and KDP_S. $\vec{c}_T(e_{k,l})$ and $\vec{c}_S(e_{k,l})$ are the contexts vectors of $e_{k,l}$ in KDP_T and KDP_S. $\cos ine(\vec{c}_T(e_{k,l}), \vec{c}_S(e_{k,l}))$, the cosine of the two vectors, is utilized to evaluate their similarity in VSM. $\vec{S}$ is the vector space of the whole story where the weight of each word is its value of *tf*idf*. $\cos ine(\vec{c}_S(e_{k,l}), \vec{S})$ is the similarity between $\vec{c}_S(e_{k,l})$ and $\vec{S}$.

In the numerator of equation (7), the third multiplier in numerator represents the similarity between the same edge in topic and story by its different contexts. The fourth multiplier is the similarity between the edge's context in story and the story vector, which can be viewed as to what degree its context in *S* can represent the story. So the multipliers in numerator can calculate the similarity between the story and an edge of the topic. Therefore equation (7) calculates the similarity between the story and the edges set of the topic. *Sim'*(N_T,S) is calculated in the same way.

5 Experiment

5.1 Dataset

Mandarin resource of TDT4 corpus is used as the training dataset and TDT5 as testing dataset. Both the TDT4 and TDT5 contain English, Mandarin and Arabic stories. TDT4 contains more than 20,000 mandarin stories collected from seven sources: newswire, radio and television sources. Radio and television sources were manually transcribed at closed-caption quality. TDT5 contains about 50,000 mandarin stories.

TDT4 contains 53 annotated topics for mandarin stories and TDT5 contains 50, each of which was defined by at most 4 on-topic sample stories and 2 off-topic ones.

5.2 Evaluation Metrics

Each incoming story will be judged if it is related to given topics. The detection cost [2] is one way to evaluate the performance.

$$C_{Det} = C_{Miss} \times P_{Miss} \times P_{t\arg et} + C_{FA} \times P_{FA} \times P_{non-t\arg et} \tag{8}$$

C_{Miss} and C_{FA}, costs of a miss and false alarm, are 1.0 and 0.1 in TDT5. P_{targe} and $P_{non\text{-}targe}$, priori target probabilities, are 0.02 and 0.98. P_{Miss} and P_{FA} are the conditional probabilities of miss and false alarms. Where P_{Miss} = *#(Missed Detections) / #Targets*, P_{Fa} = *#(False Alarms) / # Non_Targets*. Then C_{Det} is normalized to $(C_{Det})_{Norm}$.

$$(C_{Det})_{Norm} = C_{Det} / \min(C_{Miss} \times P_{t\arg et}, C_{FA} \times P_{non-t\arg et}) \tag{9}$$

The decision error tradeoff (DET) curve is another method to measure the performance. It's a visualization of the trade-off between the P_{Miss} and P_{FA}. Each point on it corresponds to a P_{Miss} and P_{FA} with a certain threshold. The closer is a curve to the lower-left corner of the graph, the better is the system's performance. The minimum value of $(C_{Det})_{Norm}$ found on the curve, called $MIN((C_{Det})_{Norm})$, is the optimal value that a system could reach at the best possible threshold.

5.3 Experiment Setup

The baseline system is tracking system of Carnegie Mellon University [14]. In year 2004, this system won the first place in TDT5 evaluation. It used an improved Rocchio algorithm to build and adjust the topic model. A story is judged on-topic if its similarity with the topic model is higher than a fixed threshold.$\gamma$is a parameter in improved Rocchio algorithm which needs to be trained.

The parameters needed to be trained in KDP algorithm include $N_{centroid}$ in section 3.1 and α in equation (6). To verify the effectiveness of KDP algorithm, several systems are experimented on in section 6. Their abbreviations are:

- *KDP*: Topic tracking system based on keywords dependency profile.

- *KDP-NoCacuWeight*: System using KDP algorithm but using *tf*idf* of each node and *tf* of each edge as their weight, instead of the equation calculating weights in section 3.2. It is utilized to prove the idea of evaluating weights of nodes and edges by the co-occurrence information is effective.
- *KDP-NoOffTrain*: System using KDP algorithm but building topic model just with on-topic sample stories, instead of with all sample stories in section 4.1. It is utilized to prove the importance of off-topic stories in building topic model.
- *KDP-NoTestProfile*: System using KDP algorithm but building profile of each incoming story by keywords selected from itself instead of from topic model introduced in section 4.2. It is utilized to prove our method to build profile of incoming story is reasonable.
- *KDP-NoContext*: System using KDP algorithm but building the profile only with the dependency network of keywords and don't append context to it. This is to prove the importance of the context in building topic model.

6 Result and Discussion

In this section, results of experiments introduced in section 5.3 are represented.

Figure 2 shows the $MIN((C_{Det})_{Norm})$ in systems with different $N_{centroid}$ and α. The lower is the $MIN((C_{Det})_{Norm})$, the better is the performance. The minimum value is 0.376363 when α equals 0.9 and $N_{centroid}$ equals the number of all the words, so 0.9 and *MAX* are set as their values. The systems perform better with increase of $N_{centroid}$ because keywords selected by MEAD are more indicative when more words are added into the centroid. Also, the systems perform better with the increase of α at first. This shows keywords dependency relations play a more important role than independent keywords in measuring similarity. The performance reaches best as α equals 0.9 but not 1 because sparsity of the same edges between topic and story has a negative influence. But it can be mitigated by combination of edges and nodes.

Figure 3 shows the training result of γ in baseline. The system performs best when γ 0.8 and threshold equals 0.3. So 0.8 is set as γ's value.

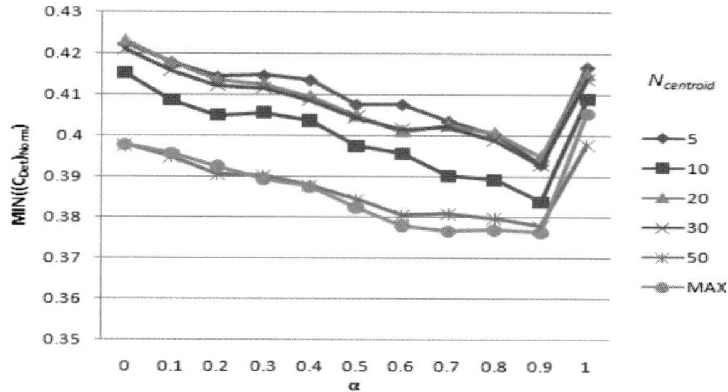

Fig. 2. MIN((CDet)Norm) with different Ncentroid and α on TDT4

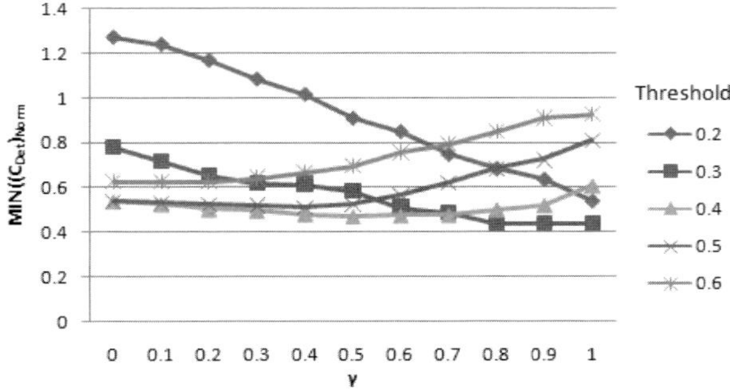

Fig. 3. MIN((CDet)Norm) with different γand threshold in baseline system

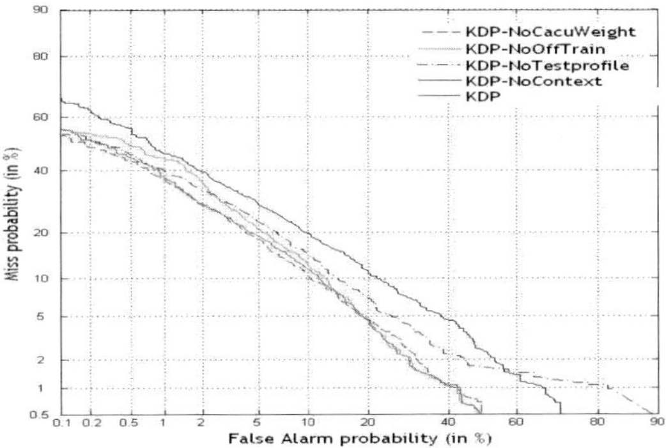

Fig. 4. Detection Error Tradeoff on TDT4 of systems removing different parts in KDP

Next, several parts of the algorithm are inspected by systems introduced in section 5.3. Figure 4 shows their DET curves and table 1 shows their best results. The results show the KDP's performance decreased after removing any part of the algorithm.

Table 1. Performance of systems on TDT4

	Threshold	P_{Miss}	P_{FA}	$MIN((C_{Det})_{Norm})$
KDP-NoCacuWeight	0.07220	0.28758	0.01879	0.37966
KDP-NoOffTrain	0.06158	0.29102	0.02726	0.42460
KDP-NoTestProfile	0.02487	0.33262	0.01884	0.42492
KDP-NoContext	0.1951	0.36095	0.02453	0.48115
KDP	0.02926	0.27566	0.02055	0.37636

The performance of *KDP-NoCacuWeight* system decreases by 0.88% comparing to *KDP*. This explains it reasonable to evaluate weights in KDP by the co-occurrence information. The reason is that if a keyword strongly depends on important keywords, it is also important to describe the core events. Also, their relations are important.

The performance of *KDP-NoOffTrain* decreases by 12.82% comparing to *KDP*. This explains the importance of using off-topic sample stories in building topic model. On-topic sample stories can be used to extract valuable keywords and their relations. Off-topic sample stories can be used to reduce the weight of noises and their relations.

The performance of *KDP-NoTestProfile* decreases by 12.90%. Building profile of incoming story by topic's keywords is reasonable because it judges on-topic story by if seminal event is involved in story but not if core event in the story is seminal event.

The performance of *KDP-NoContext* decreases by 27.84%. This proves contexts are important for building profile. A dependency relation of two keywords in different stories may express different meanings. So contexts of the same dependency relation in different stories must participate in the similarity measuring between two profiles.

Table 2 shows the importance of each part in KDP in another form. It is completed by adding each part to the algorithm one by one and evaluating the improvement after adding each part. *Context* denotes the part to append context. *TestProfile* is the part to build profiles of incoming stories. *OffTrain* is the part to build topic model. *CacuWeight* is the part to calculate the weights.δis the percentage of improvement comparing to baseline. The results show each part improves the system's performance.

Table 2. Performance and improvement of difference systems

	MIN($(C_{Det})_{Norm}$)	δ
Baseline	0.433821	--
Context	0.427833	1.38%
Context+TestProfile	0.420187	3.14%
Context+TestProfile+OffTrain	0.379662	12.48%
Context+TestProfile+OffTrain+CacuWeight	0.376363	13.25%

Figure 5 and 6 show the performances of the baseline and KDP algorithm on the training and testing dataset. Almost all values of $(C_{Det})_{Norm}$ in KDP are less than that in the baseline. KDP outperforms baseline by 13.25% on TDT4 and 7.49% on TDT5.

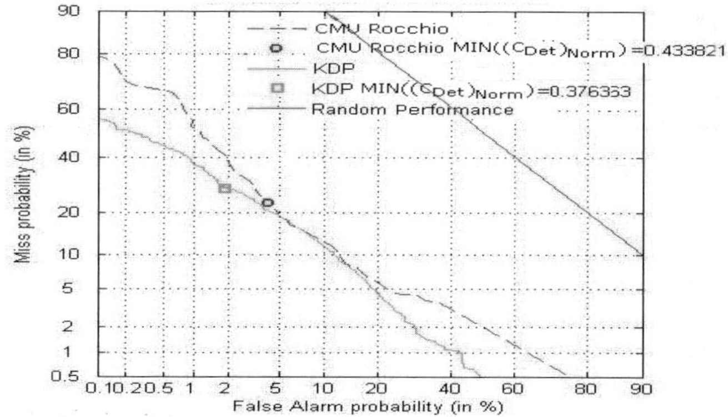

Fig. 5. DET curves of KDP and baseline on mandarin resource of TDT4

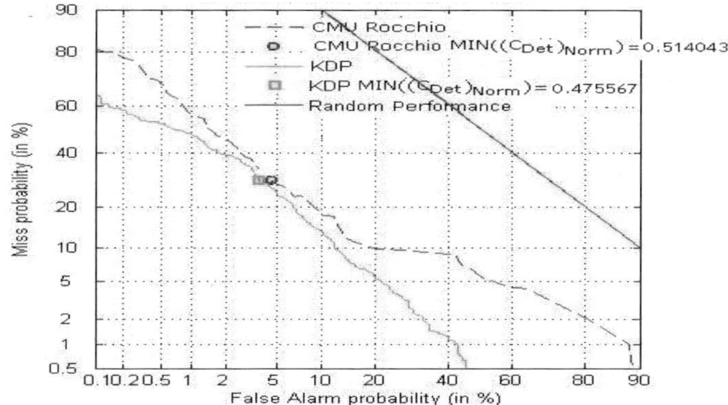

Fig. 6. DET curves of KDP and baseline on mandarin resource of TDT5

At the same time, the speed of our system is faster than baseline. Time spending in processing one story is about 0.1 second in our system and 0.3 second in baseline. It shows that our system can get better performance without adjust topic model. But the baseline system has to adjust topic model continuously which leads to its low speed.

7 Conclusions

This paper proposes a method to build the keywords dependency profile of topic by keywords co-occurrence frequency in topic tracking. It was found that the relations can describe events accurately. But building the profile with just keywords and their relations is not good enough since the same relation in different stories may express different meanings. This can be solved by appending contexts to them. Also, we find it preferable to reevaluate the importance of keywords by their dependency relations in the profile. Our algorithm outperforms baseline almost at all threshold. This proves the keywords dependency profile is a good representation of the topic and story.

But the topic model using for describing the seminal event is sensitive to the quality of sample stories. In the future work, we will use incoming stories to reinforce the topic's profile and reduce the profile's sensitivity to sample stories.

Acknowledgments. This research was supported by National Natural Science Foundation of China (60736044, 60503072, 60435020).

References

1. Allan, J.: Topic Detection and Tracking: Event-based Information Organization. Kluwer Academic Publishers, Massachusetts (2002)
2. The, Topic Detection and Tracking (TDT2004) Task Definition and Evaluation Plan (2004), `http://www.nist.gov/speech/tests/tdt/tdt2004/TDT04.Eval.Plan v1.2.pdf`

3. Allan, J., Papka, R., Lavrenko., V.: On-line New Event Detection and Tracking. In: Proceedings of the 21st Annual International ACM SIGIR Conference on Research and Development in Information Retrieval, pp. 37–45. ACM Press, Melbourne (1998)
4. Radev, R., Jing, H., Budzikowska, M.: Centroid-based Summarization of Multiple Documents: Sentence Extraction, Utility-based Evaluation, and User Studies. In: ANLP/NAACL 2000 Workshop on Automatic Summarization, Association for Computational Linguistics, Seattle, pp. 21–29 (2000)
5. Masand, B., Linoff, G., Waltz, D.: Classifying News Stories Using Memory Based Reasoning. In: Proceedings of the 15th Annual International ACM SIGIR Conference on Research and Development in Information Retrieval, pp. 59–65. ACM Press, Copenhagen
6. Carbonell, J., Yang, Y., Lafferty, J., Brown, R.D., Pierce, T., Liu, X.: CMU Report on TDT-2: Segmentation, Detection and Tracking. In: Proceedings of the DARPA Broadcast News Workshop, pp. 117–120. Morgan Kauffman Publishers, San Francisco (1999)
7. Kupiec, J., Pedersen, J., Chen, F.: A Trainable Document Summarizer. In: Proceedings of the 18th Annual International ACM SIGIR Conference on Research and Development in Information Retrieval, pp. 68–73. ACM Press, Seattle (1995)
8. Yang, Y., Zhang, J., Carbonell, J., Jin, C.: Topic-conditioned Novelty Detection. In: Proceedings of the 8th ACM SIGKDD International Conference on Knowledge Discovery and Data Mining, pp. 668–693. ACM Press, Edmonton (2002)
9. Li, B., Li, W., Lu, Q.: Topic Tracking with Time Granularity Reasoning. ACM Transactions on Asian Language Information Processing (TALIP) 5, 388–412 (2006)
10. Li, B., Li, W., Lu, Q., Wu, M.: Profile-based Event Tracking. In: Proceedings of the 28th Annual International ACM SIGIR Conference on Research and Development in Information Retrieval, pp. 631–632. ACM Press, Salvador (2005)
11. Larkey, L., Feng, F., Connell, M., Lavrenko, V.: Language-specific Models in Multilingual Topic Tracking. In: Proceedings of the 27th Annual International ACM SIGIR Conference on Research and Development in Information Retrieval, pp. 402–409. ACM Press, Sheffield (2004)
12. Makkonen, J.: Investigations on Event Evolution in TDT. In: Proceedings of the 2003 Conference of the North American Chapter of the Association for Computational Linguistics on Human Language Technology, Association for Computational Linguistics. Edmonton, pp. 43–48 (2003)
13. Shah, C., Croft, W.B., Jensen, D.: Representing Documents with Named Entities for Story Link Detection (SLD). In: Proceedings of the 15th ACM International Conference on Information and Knowledge Management, pp. 868–869. ACM Press, Virginia (2006)
14. CMU TEAM-A in TDT 2004 Topic Tracking, http://www.nist.gov/speech/tests/tdt/tdt2004/papers/CMU_A_tracking_TDT2004.ppt

A Dynamic Programming Model for Text Segmentation Based on Min-Max Similarity

Na Ye[1], Jingbo Zhu[1], Yan Zheng[1],
Matthew Y. Ma[2], Huizhen Wang[1], and Bin Zhang[3]

[1] Institute of Computer Software and Theory,
Northeastern University, Shenyang 110004, China
[2] IPVALUE Management Inc. 991 Rt. 22 West, Bridgewater, NJ 08807, USA
[3] Institute of Computer Applications, Northeastern University, Shenyang 110004, China
yn.yena@gmail.com,
{zhujingbo,wanghuizhen,zhangbin}@mail.neu.edu.cn,
mattma@ieee.org

Abstract. Text segmentation has a wide range of applications such as information retrieval, question answering and text summarization. In recent years, the use of semantics has been proven to be effective in improving the performance of text segmentation. Particularly, in finding the subtopic boundaries, there have been efforts in focusing on either maximizing the lexical similarity within a segment or minimizing the similarity between adjacent segments. However, no optimal solutions have been attempted to simultaneously achieve maximum within-segment similarity and minimum between-segment similarity. In this paper, a domain independent model based on min-max similarity (MMS) is proposed in order to fill the void. Dynamic programming is adopted to achieve global optimization of the segmentation criterion function. Comparative experimental results on real corpus have shown that MMS model outperforms previous segmentation approaches.

Keywords: text segmentation, within-segment similarity, between-segment similarity, segment lengths, similarity weighting, dynamic programming.

1 Introduction

A natural language discourse is usually composed of multiple subtopics, which in turn may convey only one main topic. In traditional text processing tasks such as information retrieval (IR), question answering (QA) and text summarization, if the subtopic structure of a text can be identified and consequently its semantic segments can be used in the basic processing unit, the performance of the system will be greatly improved [1][2]. In addition, the segment-based IR will provide users with answers of higher accuracy and less redundancy results. The core technology involved in the identification of subtopic structure and therefore semantic segments of a text is called text segmentation, which is the focus of this paper.

In text segmentation, it becomes critical how to design a good criterion to evaluate the subtopic coherence of a document. According to the definition of text segmentation

H. Li et al. (Eds.): AIRS 2008, LNCS 4993, pp. 141–152, 2008.

task, in an appropriate segmentation, sentences within a segment convey the same subtopic while sentences among different segments belonging to different subtopics. Therefore, in order to achieve good separation over all segments, both high within-segment similarity and low between-segment similarity should be achieved. However, in previous literature [2-10], no solutions have been given to simultaneously optimize both within-segment and between-segment similarities.

Another important issue in text segmentation is the strategy for finding the best segmentation. Some algorithms use local optimization approaches, such as sliding window [2-3]and divisive clustering [6-7], to detect subtopic changes. Some models use more global strategy by representing lexical distribution on a *dotplot* [4]. However this is still not a complete globalization strategy. A truly global optimization searching strategy is dynamic programming [5] [8] [10]

In this paper we present a global optimization model, MMS, for text segmentation. This model adopts a segmentation criterion function attempting to optimize both within-segment lexical similarity and between-segment lexical similarity. Segmentation with maximum within-segment and minimum between-segment similarity is selected as the optimal one. In MMS model, additional text structure factors, such as segment lengths and lexical similarity weighting strategy based on sentence distance, are also incorporated as part of lexical similarity weighting strategy. To achieve global optimization, we implemented our MMS model using the dynamic programming searching strategy, with which the number of segments can be determined automatically. Experimental results show that our MMS model outperforms other popular approaches in terms of P_k [11]and *WindowDiff* [12] measure.

The remainder of this paper is organized as follows. Literature research is briefly reviewed in Section 2. In Section 3, the proposed MMS model and a complete text segmentation algorithm are described in detail. In Section 4, experimental results are given to compare our approach with other popular systems. At last, we draw conclusion and address future work in Section 5.

2 Related Work

Existing text segmentation algorithms can be classified into two categories with respect to the segmentation criteria being employed. One is to make use of the property that lexical similarity within a segment is high. Lexical densities within segments are measured to find lexically homogeneous text fragments [6-10]. The second approach assumes that lexical similarity between different segments is low, and subtopic boundaries correspond to locations where adjacent text fragments have the lowest lexical similarity [2-5].

In contrast to previous work, the focus of our work is not only lexical relations within a segment or between different segments, but appropriate combination of the two factors. Fundamental structural factors of written texts, such as segment length and sentence distance are also taken into account in the design of the segmentation criterion. In analogy, our work is similar to [13], which measured homogeneity of a segment not only by the similarity of its words, but also by their relation to words in other segments of the text. However, their method is designed for spoken lecture segmentation and can not address the problems of written texts very well. Zhu [14] used Multiple Discriminant Analysis (MDA) criterion function to find the best segmentation by means of the largest word similarity within a segment and the smallest word similarity between segments. However, the algorithm applied a full search to find the optimal segmentation, which is an NP problem with high computational complexity. In comparison, MMS model adopts dynamic programming strategy, which greatly reduces the time cost. Fragkou[10] also used dynamic programming in optimizing the segmentation cost function. But their method only considers within-segment similarity and needs prior information about segment length.

3 The Segmentation Algorithm

3.1 Problem Definition

Let's assume a text consists of K sentences, denoted by $S = \{1, 2, ..., K\}$, and has a vocabulary of T distinct words $V = \{w_1, w_2, ..., w_T\}$. Each sentence can be represented as a point in a T-dimensional data space. Assume that the topic boundaries occur at the ends of sentences and there are N segments in the text, then the task of text segmentation is to partition the sentences into N groups $G = \{1, 2, ..., p_1\}$, $\{p_1 + 1, p_1 + 2, ..., p_2\}, ..., \{p_{N-1} + 1, p_{N-1} + 2, ..., K\}$. Each group $G_i = \{p_{i-1} + 1, p_{i-1} + 2, ..., p_i\}$ is a segment that reflects an individual subtopic. A shorter representation of the segmentation can be given as $G = \{p_0, p_1, ..., p_N\}$, where $p_0, p_1, ..., p_N$ are segment boundaries with $p_0 = 0$ and $p_N = K$. Text segmentation aims at finding the best segmentation G^* among all possible segmentations.

In this paper we design a criterion function J to evaluate segmentations of a text. Thus the process of finding the best segmentation can be viewed as the process of finding the segmentation with the highest evaluation score as follows:

$$G^* = \arg\max_{G} J(S, G) . \tag{1}$$

In the following section we will introduce our motivation in designing the criterion.

3.2 Motivation

It will reasonably hold true that in an appropriately segmented text, sentences within a single segment are topically related and sentences that belong to adjacent segments are topically unrelated conveying different subtopics. In much of previous work[4] [6-10], the lexical similarity is a natural candidate in measuring the topical relation of sentences. If two sentences describe the same topic, words used in them tend to be related to one another. Thus, within a segment, vocabulary tends to be cohesive and repetitive, leading to significant within-segment lexical similarity; whereas between adjacent segments, the vocabulary tends to be distinct, leading to dismal between-segment similarity. We believe that the above lexical similarity property must exist for a good segmentation strategy.

3.3 Segmentation Criterion Function

Following the lexical similarity property stated above, we propose our MMS model to comprise of the segmentation evaluating criterion function as follows:

$$J = F(Sim_{Within}, Sim_{Between}) \ . \tag{2}$$

where Sim_{Within} refers to the within-segment similarity, $Sim_{Between}$ refers to the between-segment similarity. F is a function whose value increases as Sim_{Within} increases, and decreases as $Sim_{Between}$ increases. The best segmentation can be achieved by maximizing the value of F, which is expressed as:

$$F(Sim_{Within}, Sim_{Between}) = \alpha \cdot Sim_{Within} - (1-\alpha) \cdot Sim_{Between} \ . \tag{3}$$

where α and $1-\alpha$ are the relative weight of within-segment lexical similarity and between-segment lexical similarity, respectively.

Within-segment lexical similarity is:

$$Sim_{Within} = \sum_{i=1}^{N} \frac{\sum_{m=p_{i-1}+1}^{p_i} \sum_{n=p_{i-1}+1}^{p_i} D_{m,n}}{(p_i - p_{i-1})^2} \ . \tag{4}$$

where m and n are the mth and n^{th} sentence in the text. $D_{m,n}$ is the lexical similarity between sentence m and sentence n. The value of $D_{m,n}$ equals to one if there exist one or more words in common between sentence m and n, and zero otherwise. Sim_{Within} represents the global density of word repetition within segments.

Similarly, between-segment lexical similarity is defined as:

$$Sim_{Between} = \sum_{i=1}^{N} \frac{\sum_{m=p_i+1}^{p_{i+1}} \sum_{n=p_{i-1}+1}^{p_i} D_{m,n}}{(p_{i+1} - p_i)(p_i - p_{i-1})} \ . \tag{5}$$

$Sim_{Between}$ represents the global lexical similarity between adjacent segments.

Combining Eq. 3 to Eq. 5, the segmentation evaluation function is computed:

$$J = \alpha \cdot \sum_{i=1}^{N} \frac{\sum_{m=p_{i-1}+1}^{p_i} \sum_{n=p_{i-1}+1}^{p_i} D_{m,n}}{(p_i - p_{i-1})^2} - (1-\alpha) \cdot \sum_{i=1}^{N} \frac{\sum_{m=p_i+1}^{p_{i+1}} \sum_{n=p_{i-1}+1}^{p_i} D_{m,n}}{(p_{i+1} - p_i)(p_i - p_{i-1})} \ . \tag{6}$$

3.4 Text Structure Weighting Factors

In addition to segment lexical similarity, there are other text structure factors that are weighted into the proposed text segmentation algorithm.

- *Segment Length Factor*

In text segmentation, text pieces that are too short do not adequately describe an independent subtopic. For example, if there is a sentence in a text, and is not closely related with its adjacent text, and then it is likely a parenthesis or a connecting link between its preceding and its successive segments to enhance coherence. To address this phenomenon, we should avoid introducing too many segments. Restriction on segment number is added into the segmentation criterion function. It penalizes segmentation choices with too many segments by assigning a small evaluation function score to it.

For example, we have a segmentation $G=\{G_1, G_2, \ldots, G_N\}$, each segment G_i has length L_i, and the length of the whole text is L. Then the length factor can be defined as $\sum_{i=1}^{N}(\frac{L_i}{L})^2$, where $L=\sum_{i=1}^{N}L_i$. This factor achieves low value when too many segments are produced.

- *Distance-based Similarity Weighting*

If we randomly select two sentences from a discourse, the probability of them belonging to the same segment varies greatly with the distance between them. Two sentences far apart are unlikely to belong to the same segment, whereas two adjacent sentences are much more likely. Therefore, we add a distance-based weighting factor to the density function, thus the lexical similarity of two sentences fluctuate with the distance between them.

Having incorporated the above factors, the overall segmentation evaluating function for our proposed MMS model becomes:

$$J=\alpha\cdot\sum_{i=1}^{N}\frac{\sum_{m=p_{i-1}+1}^{p_i}\sum_{n=p_{i-1}+1}^{p_i}W_{m,n}D_{m,n}}{(p_i-p_{i-1})^2} \\ -(1-\alpha)\cdot\sum_{i=1}^{N}\frac{\sum_{m=p_i+1}^{p_{i+1}}\sum_{n=p_{i-1}+1}^{p_i}W_{m,n}D_{m,n}}{(p_{i+1}-p_i)(p_i-p_{i-1})}+\beta\cdot\sum_{i=1}^{N}(\frac{L_i}{L})^2\ . \tag{7}$$

where $W_{m,n}$ is the weighting factor, and based on the distance between the sentence m and sentence n. The values of m and n represent the positions of each corresponding sentence. An exemplary definition of $W_{m.n}$ is as follows:

$$W_{m,n}=\begin{cases}1 & if\ |m-n|\le 2\\ \dfrac{1}{\sqrt{|m-n|-1}} & else\end{cases} \tag{8}$$

We see that in our MMS model, rich information such as the within-segment similarity and between-segment similarity, segment length and the distance between sentences, are considered to discover topical coherence.

Eq. 7 represents the final form of the evaluation function in our MMS model, in which N stands for the desired number of segments.

3.5 Text Segmentation Algorithm

To optimize the segmentation evaluating function (Eq. 7) globally, we provide an implementation using the dynamic programming searching strategy to find the best segmentation. Since there is a between-segment similarity item in the function, two-dimension dynamic programming is applied. The complete text segmentation algorithm is shown in Figure 1, followed by detailed explanation.

Input: The $K \times K$ sentence similarity matrix D; the parameter α, β

```
Initialization
For t = 1, 2, ..., K
  For s = 1, 2, ..., t
    Sum = 0;
    For k = s, ..., t
      For w = s, ..., t
        Sum = Sum + W_{w,k}·D_{w,k};
      End
    End
    S_{s,t}= Sum;
  End
End
Maximization
For t = 1, 2, ..., K
  For s = 0, 1, ..., t-1
```

$C_{t,s} = 0;$

$C_{t,0} = \alpha \cdot \frac{S_{1,t}}{t^2};$

```
    For w = 0, 1, ..., s-1
```

If $C_{s,w} + \alpha \cdot \frac{S_{s+1,t}}{(t-s)^2} - (1-\alpha) \cdot \frac{S_{w+1,t} - S_{w+1,s} - S_{s+1,t}}{(t-s)(s-w)} + \beta \cdot (\frac{t-s}{K})^2 \geq C_{t,s}$

$C_{t,s} = C_{s,w} + \alpha \cdot \frac{S_{s+1,t}}{(t-s)^2} - (1-\alpha) \cdot \frac{S_{w+1,t} - S_{w+1,s} - S_{s+1,t}}{(t-s)(s-w)} + \beta \cdot (\frac{t-s}{K})^2;$

$Z_{t,s} = w;$

```
      EndIf
    End
  End
End
Backtracking
e = -∞ ; k = 0; N=1;
For t = 0, 1, ..., T-1
```

If $C_{T,t} \geq e$

```
    k = t;
  EndIf
```

```
End
```
$S_{K-1} = T$;

$S_K = k$;

While $Z_{S_{N-1},S_N} > 0$

$\quad N = N + 1$;

$\quad S_N = Z_{S_{N-2},S_{N-1}}$;

```
End
```
$N = N+1$;

$\hat{t} = 0$;

For $k = 1, 2, \ldots, N$

$\hat{t}_k = S_{N-k}$;

```
End
```
Output: The optimal segmentation vector $\hat{t} = (\hat{t}_1, \hat{t}_2, \ldots, \hat{t}_N)$

Fig. 1. MMS text segmentation algorithm using dynamic programming scheme

In the above procedure, maximization and backtracking are the basic parts. During maximization, we recursively compute $C_{t,s}$, which is the optimal (maximum) value of the segmentation criterion function of the subtext starting from sentence 1 and ending at sentence t (with the second to the last boundary of s). That is, $C_{t,s}$ is the maximum (with respect to s and w) value of $C_{s,w} + \alpha \cdot \frac{S_{s+1,t}}{(t-s)^2} - (1-\alpha) \cdot \frac{S_{w+1,t} - S_{w+1,s} - S_{s+1,t}}{(t-s)(s-w)} + \beta \cdot (\frac{t-s}{K})^2$, where w is the best boundary before t and s. This sum is the optimal evaluation score to segment sentences 1 to s (with the second to last boundary of w) plus the score of creating a segment including sentences $s+1$ until t. $Z_{t,s}$ is the segment boundary preceding s (with the next boundary of t) in the optimal segmentation. Upon completion of the maximization part of the algorithm we have computed the maximum segmentation criterion function value for sentences 1 until K. The backtracking part produces the optimal segmentation $\hat{t} = (\hat{t}_0, \hat{t}_1, \ldots, \hat{t}_N)$ in reverse order. In this process, the optimal number of segments N is computed automatically. The time complexity of the algorithm is $O(K^3)$ (K is the number of sentences in the text).

4 Evaluation

The evaluation has been conducted systematically under a strict guideline in order to compare our approach with other state of the art algorithms on a fair basis. The key requirements are: 1) Evaluation should be conducted using a sizable testing data in order to generate meaningful results; 2) The testing data should be publicly available; 3) In order to compare with other people's work, we attempt to use their own implementations or published results as these are likely optimized for taking maximum advantages of their merits.

4.1 Experiment Settings

In our experiments, we use two suites of corpus including English one and Chinese one to evaluate the proposed model. The English testing corpus is the publicly available book *Mars* written by Percival Lowell in 1895. There are 11 chapters in all and the body of the book (6 chapters) is extracted for testing. Each chapter is composed of 2-4 sections. Chinese testing corpus is the scientific exposition *Exploring Nine Planets*. There are 10 chapters in the corpus and each chapter consists of 2-6 sections. The boundaries of paragraphs in the sections are taken as the subtopic boundaries for reference. Sections with few paragraphs (less than 3) are excluded.

In fact, there is another testing corpus developed by Choi[1], which is widely used for the evaluation of text segmentation algorithms. This is a synthetic corpus in which each article is a concatenation of ten text segments. A segment is the first *n* sentences of a randomly selected document from the Brown corpus. The motivation of constructing corpus in this way is to avoid the difficulty of judging subtopic boundaries by human beings. However this strategy has introduced obvious limitations to the corpus. Namely the subtopic similarity in the article is artificially enhanced, making the boundaries more distinct. This is quite different from real corpus and cannot represent the performance on real corpus by reducing the difficulty of segmentation. Therefore we used the real corpus instead of the synthetic one in our experiments.

To evaluate text segmentation algorithms, using precision and recall is inadequate because inaccurately identified segment boundaries are penalized equally regardless of their distance from the correct segment boundaries. In 1997, Beeferman[11] proposed the P_k metric to overcome the shortcoming. P_k is the probability that a randomly chosen pair of words with a distance of k words apart is incorrectly segmented[2]. P_k metric is defined as:

$$P_k = \sum_{1 \le i \le j \le K} D_k(i, j)(\delta_{ref}(i, j) \oplus \delta_{hyp}(i, j)) . \quad (9)$$

where $\delta_{ref}(i, j)$ is an indicator function whose value is one if sentences i and j belong to the same segment and zero otherwise. Similarly, $\delta_{hyp}(i, j)$ is one if the two sentences are hypothesized as belonging to the same segment and zero otherwise. The $\oplus$ operator is the XOR operator. The function D_k is the distance probability distribution that uniformly concentrates all its mass on the sentences which have a distance of k. The value of k is usually selected as half the average segment length. Low P_k value indicates high segmentation accuracy.

This error metric was recently criticized by Pevzner [12] to have several biased flaws such as penalizing missed boundaries more than erroneous additional boundaries and a new metric called *WindowDiff* was proposed:

$$WindowDiff(ref, hyp) = \frac{1}{K-k} \sum_{i=1}^{K-k} (| b(ref_i, ref_{i+k}) - b(hyp_i, hyp_{i+k}) | > 0) . \quad (10)$$

[1] www.lingware.co.uk/homepage/freddy.choi/index.htm

[2] We use the implementation of P_k in Choi's software package. (www.lingware.co.uk/homepage/freddy.choi/index.htm).

where *ref* is the correct segmentation for reference, *hyp* is the segmentation produced by the model, *K* is the number of sentences in the text, *k* is the size of the sliding window and *b(i, j)* is the number of boundaries between sentence *i* and sentence *j*. Low *WindowDiff* value indicates high segmentation accuracy. We will make comparison under both metrics (P_k and *WindowDiff*) on the testing corpus.

In experiments, punctuation marks and stopwords are removed. The Porter[15] stemming algorithm is applied to the remaining words to obtain word stems for English experiments.

4.2 Experimental Results

In MMS model, there are two parameters α and β that affect the quality of segmentation over certain ranges. To obtain the best parameter we randomly selected 50% corpus for training. The algorithm is run on the training corpus with α and β varies (the variation is 0.1 each time). Appropriate combination of α and β value is selected as the one which yields the minimum *WindowDiff* value. The rest of the corpus is used as testing corpus.

We evaluate MMS model in comparison to the C99 [6] method and Dotplotting[4] method including minimization algorithm (D_Min) and maximization algorithm (D_Max). The experimental results of the two methods come from Choi's software package, and it is an exact implementation of the published method[3]. Table 1 and Table 2 summarize the experimental results of all the algorithms on English corpus and Chinese corpus, respectively.

In the above tables, C_mS_n refers to the nth section of the mth chapter in the corpus. From experimental results we can see that our MMS model performs better with more than 6% reduction on average error rate (*WindowDiff*) for min and max Dotplotting methods and up to 6.4% for C99 on English corpus. Similar results are obtained on

Table 1. Comparative Results on English Testing Corpus

Method	P_k				*WindowDiff*			
	MMS	C99	D_Min	D_Max	MMS	C99	D_Min	D_Max
C_1S_2	**0.3023**	0.5763	0.3709	0.3636	**0.4318**	0.5836	0.4562	0.4586
C_2S_1	**0.3664**	0.3990	0.4385	0.4604	0.4578	**0.4488**	0.5197	0.5401
C_3S_2	0.4609	0.4771	**0.4500**	0.4785	**0.4609**	0.4897	0.5349	0.5634
C_4S_2	**0.3095**	0.4691	0.4525	0.3580	0.4330	0.4947	**0.4259**	0.4938
C_5S_1	**0.4105**	0.4958	0.4562	0.4547	**0.4407**	0.5196	0.5351	0.6129
C_5S_2	**0.3293**	0.4127	0.4515	0.3985	**0.3822**	0.4468	0.5312	0.5029
C_6	0.4240	0.4356	**0.3619**	0.4555	**0.4240**	0.4857	0.4722	0.4918
Average	**0.3718**	0.4665	0.4259	0.4242	**0.4329**	0.4956	0.4965	0.5234

[3] For the Dotplotting method, Choi[6] developed a package that includes both the implementation of the original Dotplotting method and his own interpretation. In this paper we cite the experimental results of the original Dotplotting method as published in [4]. The implementation comes from the publicly available software package. (www.lingware.co.uk/homepage/freddy.choi/index.htm)

Table 2. Comparative Results on Chinese Testing Corpus

Method	P_k				*WindowDiff*			
	MMS	C99	D_Min	D_Max	MMS	C99	D_Min	D_Max
C_2S_1	**0.3922**	0.4122	0.3946	0.3673	**0.4079**	0.4696	0.4492	0.4468
C_2S_3	**0.1947**	0.3344	**0.1447**	0.6006	**0.2047**	0.3427	0.2129	0.6272
C_4S_4	**0.2900**	0.4342	0.4806	0.5269	**0.3583**	0.4831	0.5320	0.5564
C_4S_5	**0.2820**	0.3658	0.4899	0.3416	**0.3030**	0.4391	0.5705	0.4439
C_5S_2	**0.3934**	0.4812	0.4158	0.4019	**0.4341**	0.5064	0.4994	0.5101
C_6S_2	**0.2279**	0.5785	0.5247	0.4487	**0.3145**	0.6537	0.5353	0.4611
C_6S_4	0.4157	**0.4002**	0.5305	0.5610	0.4824	**0.4188**	0.6054	0.6287
C_9S_1	**0.2248**	0.3120	0.3736	0.4136	**0.2960**	0.3992	0.4568	0.4816
$C_{10}S_2$	**0.4224**	0.5136	0.4699	0.4589	**0.4308**	0.5668	0.5372	0.5106
$C_{10}S_3$	0.3872	**0.2673**	0.4895	0.4697	0.4343	**0.2772**	0.5566	0.5665
$C_{10}S_5$	0.4235	**0.3848**	0.5359	0.5523	0.4345	**0.3867**	0.5404	0.6114
Average	**0.3322**	0.4077	0.4409	0.4675	**0.3728**	0.4494	0.4996	0.5313

Chinese corpus. MMS model achieves more than 12% average error rate reduction from Dotplotting methods and up to 7.7% for C99. The tables also indicate that *WindowDiff* metric penalizes errors more heavily than P_k metric. However the overall rank of the algorithms remains approximately the same on both metrics.

On both corpora MMS achieves best performance on most chapters (5 out of 7 for English corpus and 8 out of 11 for Chinese corpus). This is because more weighting factors affecting the segmentation choices are considered in our model. As previously mentioned, either within-segment or between-segment lexical similarity is examined in Dotplotting and C99 while our MMS model examines both factors. In addition, using text structure factors such as segment lengths and sentence distances also leads to an improvement.

The dynamic programming searching strategy adopted in our model is a global optimization algorithm. Compared to the divisive clustering algorithm of C99 and *dotplot* algorithm of Dotplotting, our strategy is more accurate and effective due to the global perspective of dynamic programming. With this strategy, the number of segments can be determined automatically when the best segmentation is achieved. In contrast, the number of segments has to be given in advance for Dotplotting method because it cannot decide when to stop inserting boundaries. The same problem exists in C99 method. Although the author proposed an algorithm to determine the number of segments automatically, this algorithm has some negative influence on the performance of the method[6].

MMS model remedies two problems of Dotplotting. Ye [16] reported analysis of two problems in Dotplotting's segmentation evaluating function:

$$f_D = \sum_{i=2}^{N} \frac{V_{p_{i-1},p_i} \cdot V_{p_i,K}}{(p_i - p_{i-1})(K - p_i)} . \tag{11}$$

where K is the end of the whole text, and $V_{x,y}$ is a vector containing the word counts associated with word positions x through y in the article.

First, the above function is asymmetric. If the text is scanned from the end to the start, a "backward" function will be got in a form different from Eq. 11. This problem leads to the apparent illogical phenomenon that forward scan may result in different segmentation with backward scan. In MMS model, the segmentation criterion function (Eq. 7) is symmetrized.

Secondly, while determining the next boundary, the evaluating strategy of Dotplotting does not adequately take the previously located boundaries into account. For each candidate boundary p_i being examined, only the segment boundary before it (p_{i-1}) is taken into consideration, and may work less effectively because it ignores the restriction of the segment boundary after it (p_{i+1}). In our MMS model, the restrictions of adjacent segment boundaries on both sides are considered. From the within-segment and between-segment similarity function (Eq. 4 and Eq. 5) we can see the segmentation evaluating function value after locating a boundary at p_i is determined by p_{i-1} and p_{i+1}. In this way the restriction from the previously located boundaries is strengthened. The optimization process of dynamic programming also helps to select boundaries globally.

5 Conclusion and Future Work

In this paper we presented a dynamic programming model for text segmentation. This model attempts to simultaneously maximizing within-segment and minimizing between-segment similarity. An analytical form of the segmentation evaluation function is given and a complete text segmentation algorithm using two-dimension dynamic programming searching scheme is described. In addition, other text structure factors, such as segment length and sentence distance, are also incorporated in the model to capture subtopic changes.

Experimental results on the public available real corpora are provided and compared with popular systems. The MMS model is shown to be promising and effective in text segmentation that it outperforms all other systems in most testing data sets. In comparing with the best comparable system (C99), the MMS model has achieved a reduction of more than 6% in average error rate (*WindowDiff* metric).

In the future we plan to optimize our algorithm by incorporating more features of the subtopic distribution and text structure. It is demonstrated in [17] that semantic information trained from background corpus can help improve text segmentation performance. We will also consider introducing semantic knowledge in the model. Besides, the length factor in our model is in a simple form and more adequate segment length factor needs to be investigated. Applying the algorithm to other text segmentation tasks such as news stream and conversation segmentation is also an important research topic.

Acknowledgments. This work was supported in part by the National Natural Science Foundation of China under Grant No.60473140, the National 863 High-tech Project No. 2006AA01Z154 ; the Program for New Century Excellent Talents in University No. NCET-05-0287; and the National 985 Project No.985-2-DB-C03 .

References

1. Boguraev, B.K., Neff, M.S.: Discourse segmentation in aid of document summarization. In: 33rd Hawaii International Conference on System Sciences, Hawaii (2000)
2. Hearst, M.A.: Multi-paragraph segmentation of expository text. In: 32nd Annual Meeting of the Association for Computational Linguistics, pp. 9--16, New Mexico (1994)

3. Brants, T., Chen, F., Tsochantarides, I.: Topic-based document segmentation with probabilistic latent semantic analysis. In: 11th International Conference on Information and Knowledge Management, pp.211--218, Virginia (2002)
4. Reynar, J.C.: An automatic method of finding topic boundaries. In: 32nd Annual Meeting of the Association for Computational Linguistics, pp. 331--333, New Mexico (1994)
5. Heinonen, O.: Optimal Multi-Paragraph Text Segmentation by Dynamic Programming. In: 17th International Conference on Computational Linguistics, pp. 1484--1486, Montreal (1998)
6. Choi, F.Y.Y.: Advances in domain independent linear text segmentation. In: 1st Meeting of the North American Chapter of the Association for Computational Linguistics, pp. 26-33, Washington (2000)
7. Choi, F.Y.Y., Wiemer-Hastings, P., Moore, J.: Latent Semantic Analysis for Text Segmentation. In: 6th Conference on Empirical Methods in Natural Language Processing, pp. 109--117, Pennsylvania (2001)
8. Utiyama, M., Isahara, H. : A Statistical Model for Domain-Independent Text Segmentation. In: 9th Conference of the European Chapter of the Association for Computational Linguistics, pp. 491--498, Bergen (2001)
9. Ji, X., Zha, H.: Domain-independent Text Segmentation Using Anisotropic Diffusion and Dynamic Programming. In: 26th Annual International ACM SIGIR Conference on Research and Development in Information Retrieval, pp. 322--329, Toronto (2003)
10. Fragkou, P., Petridis, V., Kehagias, A.: A Dynamic Programming Algorithm for Linear Text Segmentation. Journal of Intelligent Information Systems, Vol. 23, pp. 179--197 (2004)
11. Beeferman, D., Berger, A., Lafferty, J.: Text Segmentation Using Exponential Models. In : 2nd Conference on Empirical Methods in Natural Language Processing, pp. 35--46, Rhode Island (1997)
12. Pevzner, L., Hearst, M.: A Critique and Improvement of an Evaluation Metric for Text Segmentation, Computational Linguistics, Vol. 28, pp.19--36 (2002)
13. Malioutov, I., Barzilay, R.: Minimum Cut Model for Spoken Lecture Segmentation. In: 21st International Conference on Computational Linguistics, pp. 25--32, Sidney (2006)
14. Zhu, J.B., Ye, N., Chang, X.Z., Chen, W.L., Tsou, B.K.: Using Multiple Discriminant Analysis Approach for Linear Text Segmentation. In: 2nd International Joint Conference on Natural Language Processing, pp. 292--301, Jeju (2005)
15. Porter, M.F.: An Algorithm for Suffix Stripping. Program, Vol. 14, pp. 130--137 (1980)
16. Ye, N., Zhu, J.B., Luo, H.T., Wang H.Z., Zhang, B.: Improvement of the dotplotting method for linear text segmentation. In: 2005 IEEE International Conference on Natural Language Processing and Knowledge Engineering, pp. 636--641, Wuhan (2005)
17. Bestgen, Y.: Improving Text Segmentation Using Latent Semantic Analysis: A Reanalysis of Choi, Wiemer-Hastings, and Moore (2001). Computational Linguistics, Volume 32, pp. 5--12 (2006)

Pronoun Resolution with Markov Logic Networks

Ki Chan and Wai Lam

Department of Systems Engineering and Engineering Management,
The Chinese University of Hong Kong
{kchan,wlam}@se.cuhk.edu.hk

Abstract. Pronoun resolution refers to the problem of determining the coreference linkages between the antecedents and the pronouns. We propose to employ a combined model of statistical learning and first-order logic, the Markov logic network (MLN). Our proposed model can more effectively characterize the pronoun coreference resolution process that requires conducting inference upon a variety of conditions. The influence of different types of constraints are also investigated.

1 Introduction

Understanding natural language has always been a challenging task. The variations in writing and the different means of conveying information pose huge difficulties in automatic understanding of text. To support language understanding tasks, different relations conveyed in text have to be identified and extracted. Among these relations, noun phrase coreference has been gaining increasing attention. Noun phrase coreference is the process of identifying the entities where different mentions belongs to. Coreference is a form of coherence in language representation. In representing ideas in language, a variety of forms is adopted in presenting the same idea. Different noun phrases may refer to the same entity. The resolving of the noun phrases is a crucial step for a broad range of language understanding processes such as relation extraction.

Human can understand noun phrase coreference via an inference process based on background knowledge of the noun phrases, agreement, such as gender and quantity, between phrases, the synonymity between phrases. Moreover, in maintaining the consistency of concepts, certain structures are usually being adopted for readers to follow the coherence within text.

Among the noun phrase coreference, pronoun resolution is a particularly important issue. A variety of pronouns may be used within a sentence, and may refer to different entities. In the following example, there exists three pronouns, namely, *who*, *his*, *it*. The pronouns *who* and *his* corefer with *John*, while *it* corefers with *the incident*.

John, *who* witnessed the incident, informed *his* friend about *it*.

Moreover, coreference between pronouns and entities are usually not restricted to the same sentence, exemplified as follows:

H. Li et al. (Eds.): AIRS 2008, LNCS 4993, pp. 153–164, 2008.

Mary met Susan yesterday. *She* was on *her* way home.

In this sentence, the pronoun *she* may refer to either *Mary* or *Susan*. Hence, the determination of which entity the pronoun corefers is an important issue.

Pronoun resolution is different from coreference resolution on proper nouns where surface features, such as string comparison, are not as significant. Despite the fact that pronouns are lack of rich semantic information, they are crucial in maintaining the coherence of knowledge representation in text. Hence, research from linguistic society has been keen on studying the characteristics of pronoun resolution, so as to discover the implicit relationship associated with the pronouns and their coreferred mentions. Based on those investigations, regularities of pronouns in language are studied and heuristic approaches are adopted in pronoun interpretation and on identifying pronoun coreferences.

However, there are no absolute rules on the way the pronouns corefer as there are infinite possibilities in language representation. Therefore, we propose to employ a combined model of statistical learning and first-order logic, the Markov logic network (MLN) [1]. Our proposed model can more effectively characterize the pronoun coreference resolution process that requires conducting inference upon a variety of conditions. The influence of different types of constraints are also investigated. With first-order logic, domain knowledge, such as, linguistic features or constraints as heuristic rules can be incorporated into coreference resolution, with the benefits of handling uncertainties.

We present how the problem of pronoun resolution can be formulated in MLN. An investigation on the adoption of pronoun resolution constraints will be presented. In next section, some related works regarding coreference resolution and pronoun resolution are included. In Section 3, background information on MLN will be introduced and a description on the formulation of pronoun resolution in MLN will be described in Section 4. Experiments and results will be presented in Section 5.

2 Related Work

For long, pronominal reference has been studied in the linguistic and cognitive society. A variety of views on the corresponding regularities are proposed [2]. Research on investigating the relations in pronominal reference, such as the clausal relationship and the structure, is still being studied [3].

While the works in the linguistic and cognitive society have been focused on the formal modeling of coreference relations, in the area of computational linguistic, research on performing automatic coreference resolution is being studied. The research in coreference resolution has been mainly focused on two directions, namely, linguistic and machines learning.

The linguistic approaches focus on adopting syntactic and semantic constraints on coreference resolution. The Hobb's algorithm [4] tackled pronoun resolution by searching through a syntactic parse tree of a sentence under some syntactic constraints. The centering theory [5] adopted the idea of coherence in texts and its idea is to trace the entities in focus.

In recent years, machine learning approaches are more widely adopted for coreference resolution, such as the Naive Bayes [6] and decision tree [7] approaches. Wellner and McCallum [8] tackled the coreference problems by using conditional models and graph partition approach. Besides pairwise resolution of mentions, coreference resolution is also considered as clustering mentions [9].

Moreover, much research has been carried on the investigation of features for the coreference resolution. A wide range of features has been experimented. Luo et al. [10] used syntactic features based on binding theory for improving pronoun resolution. Ng [11] investigated features with semantic knowledge. Ponzetto et al. [12] explored the use of semantic role labeling, and features with knowledge mined from WordNet and Wikipedia using a Maximum Entropy model.

Regarding pronoun resolution, both syntax-based and knowledge-based approaches are used. In particular, some works focused on resolving antecedents for third person pronouns. Lappin et al. [13] adopted a syntax-based approach which relies on syntactic information and determines the salience value of the candidate antecedents. In addition to syntactic information on texts, Bergsma et al. [14] proposed an approach based on syntactic paths, which analyze the dependency path information between potentially coreferent entities. Knowledge poor approaches with limited and shallow knowledge are also reported [15].

Moreover, world-knowledge is employed in retrieving the semantics-related information for pronoun resolution. Kehler et al. [16] investigated in the utility of corpus-based semantics for pronoun resolution and argued that the improvement is not significant. However, Yang et al. [17] investigated the use of semantic compatibility information obtained from web, and significantly improved the resolution of neutral pronouns, such as "it".

3 Background

3.1 Pronoun Resolution

From the linguistic point of view, the distribution and location of different mentions within texts are governed by certain restrictions. In other words, through identifying whether mentions satisfy the constraints or not, the referential linkage can be deduced. Noun phrase coreference resolution involves resolving coreference relations mainly between proper noun phrases, nominal noun phrases, and pronouns. This paper focuses on the task of pronoun coreference.

Pronoun resolution is usually defined as identifying or matching the corresponding antecedent of the pronouns. Since pronouns are substituents for nouns, noun phrases or pronouns, which can help maintain the coherence of the representation of ideas in language or text, pronoun resolution is crucial to the understanding of language. However, pronoun resolution is not a trivial task. The pronoun itself contains little semantic information, which hinders the relation resolving between the pronouns and their antecedents. This poses differences between the pronoun resolution problem and the noun phrase coreference resolution problem, since matching features, such as phrase matches, commonly used in noun phrase coreference problem, are not applicable in pronoun resolution.

Nevertheless, clues indicating the behaviors of different types of pronouns exist. These clues serve as the constraints or conditions for making the resolution decision. A knowledge base can be constructed with these constraints and hence corresponds to a logic network for reasoning. Hence, pronoun resolution can be well described in first-order logic. Also, the use of Markov logic network can support the handling of uncertainties in pronoun resolution.

3.2 First-Order Logic

For reasoning in First-Order Logic, sentences in the knowledge base are formed by atoms and terms. It enables the flexibility of incorporating domain knowledge. It consists of primitives, including constant symbols, function symbols, and predicate symbols. A term is a constant or a variable or a function of n-terms, where an atom is a predicate of n-terms. Constants are considered as objects, variables are ranges of objects and predicates are the mapping of objects to truth values. For example, $P(x)$ is an atom, where P is the predicate, and x is a variable. Sentences are constructed from atoms with connectives and quantifiers.

3.3 Markov Logic Network

Markov Logic Network is proposed by Richardson et al. [1]. It is referred to as a first-order knowledge base with a weight attached to each formula. It combines the representation power of wide variety of knowledge in first-order logic with the advantage of probabilistic model in handling uncertainties.

The probability distribution of a Markov network is:

$$P(X = x) = \frac{1}{Z} exp(\sum_{i}^{F} w_i n_i(x)) = \frac{1}{Z} \prod \phi_i(x_{\{i\}})^{n_i(x)} \quad (1)$$

where w_i is the weight for each formula, i. $n_i(x)$ is the number of true groundings of a formula in first-order logic in the possible world x, and $x_{\{i\}}$ is the truth value of the atoms appeared in the formula, and $\phi_i(x_{\{i\}}) = e^{w_i}$. Z is the normalizer.

In Equation 1, F represents the number of formulae in the corresponding network. A node corresponds to each grounding of the predicates specified in the formulae, and there is an edge between two nodes if their corresponding ground predicates appear together in a formula. As an example, for the following two formulae:

$\forall x \quad drives(x) \Rightarrow has\_car(x) \wedge adult(y)$

$\forall x, y \quad colleagues(x, y) \Rightarrow (drives(x) \Leftrightarrow drives(y))$

In clausal form:

$\neg drives(x) \vee has\_car(x)$

$\neg drives(x) \vee adult(y)$

$colleagues(x, y) \vee drives(x) \vee \neg drives(y)$

$colleagues(x, y) \vee \neg drives(x) \vee drives(y)$

Figure 1 shows a ground network for the above formulae and a finite set of constants, $C = \{\text{Alan}, \text{Ken}\}$. When the formula with weights are given, the

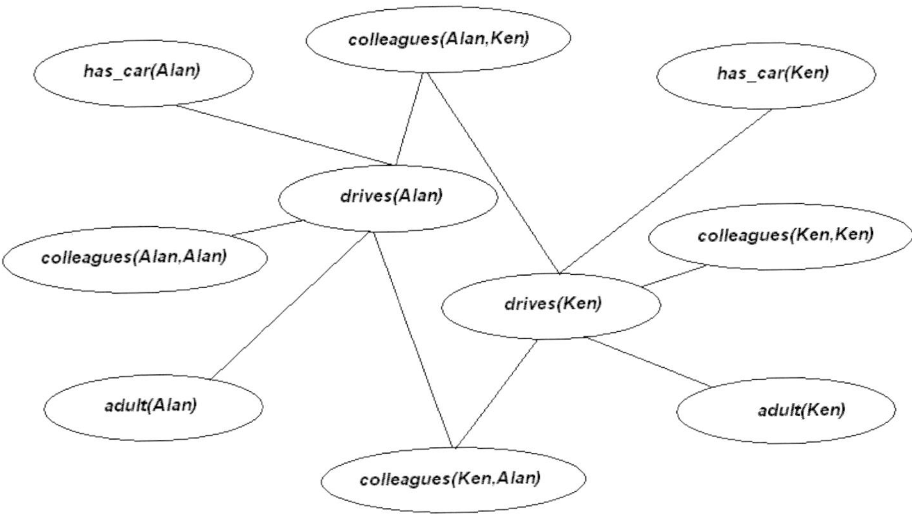

Fig. 1. An example of a ground Markov network

network can be used to infer the probability of the nodes, for example, the probability of whether Ken has a car, given that his colleague, Alan, drives.

4 Problem Formulation

Coreference can be viewed as a relation among entities within texts. The relations are believed to have special characteristics and can be described by constraints and conditions. Those constraints or conditions for pronoun resolution can be formulated into a set of formulae in first-order logic. Pronoun resolution decisions are made on the validation or violation of these formulae.

4.1 Coreference Relation

Coreference linkage can be regarded as the generative process of finding a replacement for a mention in succeeding texts. Suppose we would like to find a coreference for a mention, a corresponding and agreeing pronoun is identified and used in succeeding texts. As an example, when establishing the coreference with a person, a pronoun compatible with the person and the context is selected based on syntactic and semantic constraints. We design a first-order predicate $coref(x, y)$ to represent the coreference linkage between two mentions, x and y.

Pronouns are used when we would like to refer to the same concept by different kinds of mentions. From this process of coreference, some constraints on maintaining the consistency of concept can be deduced. When a mention x corefers with a mention y, the two mentions must have followed some syntactic and semantic constraints. Hence, in solving pronoun resolution, the following formulation is used:

$$\text{syntactic constraints on } x \;\&\; y \Rightarrow x \text{ corefer } y \tag{2}$$
$$\text{semantic constraints on } x \;\&\; y \Rightarrow x \text{ corefer } y \tag{3}$$

where the variable x represents the antecedent candidate, which can be pronouns, nouns, or noun phrases. The variable y refers to the pronouns to be resolved.

This forms the basic idea of our formulation, where predicates for syntactic constraints or semantic constraints will be introduced in Section 4.2.

It is noted that coreference linkages in pronoun resolution are not symmetric:

$$\forall x, y \quad coref(x, y) \neq coref(y, x) \tag{4}$$

In pronoun resolution, the process is to find the corresponding antecedents of each pronoun. Hence, there are two types of constants, pronouns (consequents), and antecedents. The antecedents include all the pronominal, nominal and proper nouns or noun phrases. As a result, some pronouns may be also be the antecedents of some other pronouns. Hence we design an equality predicate, $same(x, y)$, for indicating the pronoun y and the antecedent candidate x, though of different types, are actually the same mention.

Moreover, the dependency between predicates can also be represented in the Markov logic network. For example, the transitivity between coreference links in pronoun resolution can be formulated as:

$$\forall x_1, x_2, y_1, y_2 \quad coref(x_1, y_1) \wedge coref(x_2, y_2) \wedge same(x_2, y_1) \Rightarrow coref(x_1, y_2) \tag{5}$$

The above formulae form the basic relations among mentions, which ensures the validity of the pronoun coreference relations.

Besides the relations related on deciding the existence of the coreference linkage, relations regarding the negative existence of the coreference linkage are also important. We refer these relations as the **negative constraints**. As an example, the negative forms of Equations 2 and 3 are:

$$\neg\text{syntactic constraints on } x \;\&\; y \Rightarrow x \;\neg\text{corefer } y \tag{6}$$
$$\neg\text{semantic constraints on } x \;\&\; y \Rightarrow x \;\neg\text{corefer } y \tag{7}$$

4.2 Syntactic Constraints

In constructing the relations and constraints regarding pronoun resolution, two types of predicates, namely, grammatical and positional predicates, are defined.

Grammatical Predicates

The behavior and the relations between the antecedents of pronouns are highly affected by different types of pronouns. Hence, pronoun types are represented as:

- $personal\_pronoun(y)$ for personal pronouns which can be further classified into three types:

  - *subjective_pronoun*(y) for personal pronouns used as the subject. e.g. *I, we, you, he, she, they, it*, etc.
  - *objective_pronoun*(y) for personal pronouns used as the object. e.g. *me, us, you, him, her, them, it*, etc.
  - *reflexive_pronoun*(y) for personal pronouns which replaces the objective pronoun when referring to the same entity as the subject.
- *possessive_pronoun*(y) for pronouns used when it is the possessor of another noun. e.g. *my, our, mine, ours, hers, his, yours, theirs, its*, etc.
- *relative_pronoun*(y) for pronouns used when referring back to the noun or noun phrase previously mentioned.
- *noun_phrase*(y) for indicating the candidate is a noun phrase.

Positional Predicates

Positional information regarding the pronouns and their antecedent candidate are defined as:

- **Paragraph distance**: the predicate $same\_paragraph(x, y)$ represents that the two nouns are in the same paragraph.
- **Sentence distance**: the predicate $same\_sentence(x, y)$ represents that the two nouns are in the same sentence.
- **Relative position**: the predicate $precedes(x, y)$ represents that the noun x precedes noun y.

With the above predicates, constraints for pronoun resolution can be constructed.

cCommand Constraints

Unlike noun phrase coreference, matching features are not the most influential factor for pronoun resolution. Instead, theories concerning pronouns provide clues in governing the pronoun behavior. The binding theory [2] provides some principles on pronoun interpretation, and defines the relations between two nouns.

- A non-reflexive pronoun should be free within its local domain.
- A reflexive pronoun should be bound within its local domain.

For example: The cat did *it itself*.

The pronoun *it* cannot be coreferent with *the cat*, while *itself* certainly means *the cat*. Through the binding theory, the two coreference linkages can be deduced.

For defining the binding theory, a noun n_1 is said to bind another noun n_2 if and only if (1) n_1 c-commands n_2 (2) n_1 and n_2 are coindexed. C-command represents the relation between two nodes in a parse tree. n_1 is said to c-command n_2 if and only if the first branching node that dominates n_1 also dominates n_2. The c-command prevents coreference between a c-commanding noun phrase with a c-commanded noun phrase, except when it is a reflexive pronoun. The $cCommands(x, y)$ predicate represents the relation that x c-commands y.

With the above definitions, the constraints can be described by the following formulae using the $cCommands(x, y)$ predicate and grammatical predicates:

– Non-reflexive pronouns

$$\forall x, y \quad pronoun(y) \wedge \neg cCommands(x, y) \wedge coindexed(x, y) \Rightarrow coref(x, y) \tag{8}$$

– Reflexive pronouns

$$\forall x, y \quad reflexive\_pronoun(y) \wedge cCommands(x, y) \wedge coindexed(x, y) \Rightarrow coref(x, y) \tag{9}$$

Moreover, the negative form of the above formulae are included for indicating the non-existence of a coreference linkage. The **negative constraints** for syntactic constraints are:

$$pronoun(y) \wedge cCommands(x, y) \wedge coindexed(x, y) \Rightarrow \neg coref(x, y) \tag{10}$$

$$reflexive\_pronoun(y) \wedge \neg cCommands(x, y) \wedge coindexed(x, y) \Rightarrow \neg coref(x, y) \tag{11}$$

For pronoun resolution, it is apparent that these syntactic constraints are the most crucial factors governing the coreference linkages.

Filtering Constraints
In MLN, formulae with finite weight can be regarded as constraints in capturing the possibilities of those conditions, while formulae with infinite weight are hard constraints. They can be regarded as filtering constraints in ensuring the violation of these constraints will cause the query to have zero probability.

The addition of formulae with infinite weight serves as a filtering process, which is usually performed as a separate step in other pronoun resolution algorithms. Hence for handling pronoun resolution, we have to limit the reference candidate for the pronouns as only the antecedents of the pronouns. The following formula is added to filter out the non-antecedent mentions using the positional predicates.

$$\forall x, y \quad \neg precedes(x, y) \Rightarrow \neg coref(x, y) \tag{12}$$

4.3 Semantic Constraints

Besides syntactic constraints, two nouns have to be semantically compatible for them to refer to the same entity. Despite that pronouns are lack of semantic information, two kinds of information, Gender and number, can be obtained from their definitions. And these information about the mentions provides important clues to whether two mentions agreed semantically.

In our pronoun resolution formulation, three types of gender are used: masculine, feminine, and neutral, and two types of number information are used: singular, plural. The two types of information are specified using the predicates, $gender(x, a)$ and $number(x, b)$, respectively. Variable x refers to the pronouns or antecedents, where variable a refers to the three gender types, and variable b

refers to the two number types. $gender(x, a)$ indicate that x is of gender type a, and $number(x, a)$ indicate that x is of number type b.

The recognition of gender and number types for pronouns is relatively straightforward. Pronouns can be classified into gender-specific or gender-neutral. For gender-specific pronouns, they can be further classified into three gender types: masculine(*he/him*), feminine(*she/her*) and neuter(*it*). Gender-neutral pronouns refer to those pronouns which did not distinguish the gender(*you, they*). And all pronouns are well defined to be either singular or plural.

However, the recognition of gender and number types for other noun phrases involves a lot of background knowledge and uncertainties. We employ the noun gender and number data developed by Bergsma, et al. [14]. The corpus is generated from a large amount of online news articles by using web-based gender-indicating patterns. It contains the numbers of times a noun phrase is connected to a masculine, feminine, neutral and plural pronoun. With this corpus, we obtain the gender and number information by matching the noun phrases..

The semantic constraints are hence defined as:

$$gender(x, a) \wedge gender(y, c) \wedge a = c \Rightarrow coref(x, y) \tag{13}$$

$$number(x, b) \wedge number(y, d) \wedge b = d \Rightarrow coref(x, y) \tag{14}$$

and the negative constraints are:

$$gender(x, a) \wedge gender(y, c) \wedge a \neq c \Rightarrow \neg coref(x, y) \tag{15}$$

$$number(x, b) \wedge number(y, d) \wedge b \neq d \Rightarrow \neg coref(x, y) \tag{16}$$

5 Experiments

5.1 Experimental Setup

We have conducted experiments to evaluate our proposed model. The coreference chains obtained are evaluated. We used the noun coreference ACE 2004 data corpus for our experiments. The dataset is split into training and testing datasets randomly. We used 159 articles and 44 articles from the broadcast news (BNEWS) source as training and testing datasets respectively. We consider only the true mentions annotated in the ACE corpus. For c-command predicate generation, the Charniak parser [18] is used for generating the parse tree.

As pronouns can corefer with pronouns and hence a coreference chain will be formed. We would like to evaluate the coreference chains formed in addition to the individual coreference links between pronouns and their antecedents. Hence, the results are evaluated using recall and precision following the standard model-theoretic metric [19] adopted in the MUC task. This evaluation algorithm focuses on assessing the common links between the true coreference chain and the coreference chain generated by the system output.

The Alchemy system [20], which provides algorithms in statistical relational learning on the Markov Logic Networks, was used in our experiments. Discriminative learning are adopted for weight learning during training.

5.2 Results

The results are depicted in Table 1. Since MLN has the benefits of enabling the incorporation of domain knowledge such as formulae describing relations between coreference links, we carried out experiments using different combination of constraints. Three sets of results are given in Table 1. A baseline experiment is conducted assuming coreference linkage existed if the pronoun and its antecedent are in the same sentence. Next, as mentioned before, negative constraints for deducing the non-existence of coreference linkage can be crucial to pronoun resolution. As a result, we excluded the negative constraints to investigate their influence. Lastly, an experiment with the complete formulation including both the syntactic and semantic constraints is conducted.

Table 1. Performance of Pronoun Resolution on the BNEWS dataset

	Recall	Precision	F-measure
Baseline	33.9%	42.5%	37.7%
Without negative constraints	46.1%	45.0%	45.5%
With the negative constraints	56.5%	47.2%	51.4%

The performance results in the last row demonstrate that the resolution performance can be greatly improved in recall and precision with the semantic and syntactic constraints, including their negative constraints. The second row shows the experimental results on excluding the use of negative constraints. The decrease in performance with respect to the second row demonstrates that the negative constraints are also critical in the pronoun resolution formulation with MLN, as inferencing on the decision of not having a coreference link is equally important to the decision of having a coreference link. From analyzing the system generated coreference links, it is observed that as BNEWS contains transcripts for spoken dialogues, extra consideration should be carried on the coreference linkage within conversation by different persons.

6 Conclusions and Future Work

In this paper, we have investigated the application of Markov Logic Network on the resolution of pronoun coreference. The experiments show the characteristics of the formulation of pronoun resolution with Markov logic network in modeling dependencies between entities. It provides an encouraging direction on coreference and cohesion resolutions. Linguistic experts have long been studying the relation between entities in language. Coherence of a text must be maintained for language understanding. And for a text to be coherent, cohesion must be maintained. Textual cohesion refers to the focus of entities or concepts in texts for readers. Strong relations existed between textual cohesion and coreference. Cohesion strategies can be followed using parallelism and dependency. These cohesion cues are crucial for coreference resolution, especially on nominal and pronominal coreference. Heuristic approaches have been mostly employed for incorporating these cohesion cues [21]. These cues can be well represented in first-order logic representations, and will be beneficial by the probabilistic characteristics in Markov Logic Network.

Also, detailed analysis on the strategies governing the referential linkages between noun phrases are proposed by many linguistic experts. The binding theory proposed by Chomsky [2] is one of the well know model. The binding theory provides well defined syntactic constraints to coreference resolution. The c-command concept is currently implemented in our system as a binary predicate. But as determining the binding among noun phrases is not a straightforward task, the uncertainties can be handled through the use of MLN. Moreover, we believe by further investigating the features or relations best suited to the logic network, the performance could be further improved.

Our future directions include further investigating the incorporation of cohesion cues in Markov logic network for coreference resolutions, and further expanding the coreference resolution on not only pronoun resolution, but also nominal and proper noun coreference resolutions.

Acknowledgement

The work described in this paper is substantially supported by grants from the Research Grant Council of the Hong Kong Special Administrative Region, China (Project Nos: CUHK 4179/03E, CUHK4193/04E, and CUHK4128/07) and the Direct Grant of the Faculty of Engineering, CUHK (Project Codes: 2050363 and 2050391). This work is also affiliated with the Microsoft-CUHK Joint Laboratory for Human-centric Computing and Interface Technologies.

References

1. Richardson, M., Domingos, P.: Markov logic networks. Machine Learning 62, 107–136 (2006)
2. Chomsky, N.: The Minimalist Program. MIT Press, Cambridge (1995)
3. Harris, C., Bates, E.A.: Clausal backgrounding and pronominal reference: A functionalist approach to c-command. Language and Congitive Processes 17(3), 237–269 (2002)
4. Hobbs, J.: Resolving pronoun references. Readings in natural language processing, 339–352 (1986)
5. Grosz, B.J., Weinstein, S., Joshi, A.K.: Centering: a framework for modeling the local coherence of discourse. Comput. Linguist. 21(2), 203–225 (1995)
6. Ge, N., Hale, J., Charniak, E.: A statistical approach to anaphora resolution. In: Proceedings of the Sixth Workshop on Very Large Corpora (1998)
7. Soon, W.M., Ng, H.T., Lim, D.C.Y.: A machine learning approach to coreference resolution of noun phrases. Comput. Linguist. 27(4), 521–544 (2001)

8. Wellner, B., McCallum, A.: Towards conditional models of identity uncertainty with application to proper noun coreference. In: IJCAI Workshop on Information Integration and the Web (2003)
9. Cardie, C., Wagstaff, K.: Noun phrase coreference as clustering. In: Proceedings of the EMNLP and VLC (1999)
10. Luo, X., Zitouni, I.: Multi-lingual coreference resolution with syntactic features. In: Proceedings of Human Language Technology Coreference and Coreference on Empirical Methods in Natural Language Processing(HLT/EMNLP), Vancouver, october 2005, pp. 660–667 (2005)
11. Ng, V.: Shallow semantics for coreference resolution. In: Proceedings of the Twentieth International Joint Conference on Artificial Intelligence (IJCAI) (2007)
12. Ponzetto, S.P., Strube, M.: Exploiting semantic role labeling, wordnet and wikipedia for coreference resolution. In: Proceedings of NAACL (2006)
13. Lappin, S., Leass, H.: An algorithm for pronominal anaphora resolution. Computational Linguistics 20(4), 535–561 (1994)
14. Bergsma, S.: Boostrapping path-based pronoun resolution. In: Proceedings of the 21st International Conference on Computational Linguistics and 44th Annual Meeting of the ACL, pp. 33–40 (2006)
15. Mitkov, R.: Robust pronoun resolution with limited knowledge. In: Proceedings of the 17th International Conference on Computational Linguistics, pp. 869–875 (1998)
16. Kehler, A., Appeit, D., Taylor, L., Simma, A.: The (non)utility of predicate-arguement frequencies for pronoun interpretation. In: Proceedings of 2004 North American chapter of the Association for Computational Linguistics annual meeting (2004)
17. Yang, X., Su, J., Tan, C.L.: Improving pronoun resolution using statistic-based semantic compatibility information. In: Proceedings of the 43rc Annual Meeting of the ACL, pp. 33–40 (2005)
18. Charniak, E.: A maximum-entropy-inspired parser. In: Proceedings of NAACL (2000)
19. Vilain, M., Burger, J., Connolly, D., Hirschman, L.: A model-theoretic coreference scoring scheme. In: Proceedings of MUC-6, pp. 176–183 (1995)
20. Kok, S., Domingos, P.: The alchemy system for statistical relational AI, Technical report, Department of Computer Science and Engineering, University of Washington, Seattle, WA (2005) http://www.cs.washington.edu/ai/alchemy/
21. Harabagiu, S., Maiorano, S.: Knowledge-lean coreference resolution and its relation to textual cohesion and coherence. In: Proceedings of the ACL 1999 Workshop on the Relation of Discourse Dialogue Structure and Reference, pp. 29–38 (1999)

Job Information Retrieval Based on Document Similarity

Jingfan Wang[1], Yunqing Xia[2], Thomas Fang Zheng[2], and Xiaojun Wu[2]

[1] Department of Computer Science and Technology, Tsinghua University, Beijing 100084, China
wangjf@cst.cs.tsinghua.edu.cn
[2] Center for Speech and Language Technologies, RIIT, Tsinghua University, Beijing 100084, China
{yqxia,fzheng}@tsinghua.edu.cn
wuxj@cst.cs.tsinghua.edu.cn

Abstract. Job information retrieval (IR) exhibits unique characteristics compared to common IR task. First, searching precision on job posting full text is low because job descriptions cannot be properly used in common IR methods. Second, job names semantically similar to the one mentioned in the searching query cannot be detected by common IR methods. In this paper, job descriptions are handled under a two-step job IR framework to find job postings semantically similar to seeds job posting retrieved by the common IR methods. Preliminary experiments prove that this method is effective.

1 Introduction

Similar to common information retrieval (IR), job information retrieval aims to help job seekers to find job postings on the Web promptly. The task is made unique due to the following two characteristics. Firstly, job names are usually used as search queries directly in job IR. However, they can be expressed by numerous alternatives in natural language. For example, *manager* can be worded as "经理","主管", "总监" and "主任" in Chinese. As an extreme case, the job name "美工(*art designer*)" holds nine semantically similar job names. According to our study on query log, people with different background hold different preferences in selecting job names. This brings job IR a challenging issue to find job postings with conceptually similar job names but not necessarily with literally same job name in the query.

Secondly, job posting usually comprises of two fields, i.e. title and description. The title field is pretty short (1 to 6 words) and presents the most important points for the job while the description is a bit longer (20 to 100 words) and provides detailed requirements of the job. The most interesting point is that, the job name is usually contained in the title only and is scarcely mentioned in the description. To summarize, title and description depict the same job but share very few common words.

Problems arising from the two characteristics of job IR are two-fold. First, the title field provides too short text for the vector space model (VSM) to locate similar job postings. Second, as it shares little common word with job name, job description

H. Li et al. (Eds.): AIRS 2008, LNCS 4993, pp. 165–175, 2008.

provides very little contribution in finding the relevant job postings directly. This is also proved by our experiments (see Section 4), which shows that searching in job posting full text (title and description) yields very little performance gain over searching merely in the title. Discarding the job descriptions is certainly not a good idea, then how could we make use of job description properly?

In this paper we propose to make use of document similarity to locate relevant job postings. The basic assumption is that job description usually provides sufficient and unambiguous information, referred to as semantic clues behind the job name. We argue that the semantic clues can be used to find similarity job postings. In our job IR system, a two-step framework is designed to retrieve this goal. In the first step, queries are used to locate literally relevant job names. In the second step, the job posting full text is used to find relevant job postings. To re-rank all relevant job postings, a combined ranking model is proposed, which considers query-document relevance score and document-document similarity score in one formula.

The rest of the paper is organized as follows. The unique job IR task is described in Section 2. Then the two-step method for finding the similar job postings is presented in Section 3. In Section 4, experiments and discussions are presented. We summarize the related works in Section 5 and conclude this paper in Section 6.

2 Job Information Retrieval, a Unique IR Task

Job information retrieval system aims to facilitate job seekers to find job postings in a large scale online job posting collection. Basically, the job seekers type in job names as the queries directly.

The job posting is a piece of natural language text that contains two fields, title and job description. Two typical example job postings are given as follows.

Job posting example 1:

<title>软件工程师 (Software Engineer)</title>

<descrtion>熟练掌握Java,j2EE；熟悉Eclipse插件开发优先；富有激情；良好的团队合作精神,中 英文流利 (Strong in programming with Java, J2EE; Priority to those who are familiar with Eclipse plug-in development; Self-motivated; Excellent teamwork spirit and communication skills; Fluent in English and Chinese)</description>

Job posting example 2:

<title>程序员(Programmer)</title>

<description>须有Java开发及Eclipse插件开发相关经验，有RFT使用经验者优先；计算机或相关专业研究生(Experienced in Java programming ; Experience in Eclipse plug-in development, Priority to those familiar with RFT; Master of Computer Science or related) </description>

As shown in the two examples, job names mostly appear only in the title field. Since most users use job name as query keywords directly, only job postings containing the job name within title field can be successfully retrieved by the traditional VSM. Another finding is that, the two job postings are semantically similar. Users who are interested in the one may also be interested in the other. Unfortunately, they can not be retrieved with one query using VSM because their job name strings are literally different.

Text in the description field is a bit longer, and semantic clues can be found such as professional experience, technical skills and education background. The semantic clues cannot be properly used in the VSM based query-document relevance measuring scheme, but helpful in finding semantically relevant job postings.

Our observations on job postings provide two assumptions: 1) similar job names hold semantically similar job descriptions; 2) semantically similar job descriptions in turn determine similar items. Enlightened by the two assumptions, we designed a two-step framework for job IR. The traditional VSM is applied on the title field in the first step, and similarity between job postings over full text is calculated to find the semantically similar job postings in the second step.

3 Finding Relevant Job Postings

3.1 The Two-Step Framework and the Combined Ranking Model

The key ideas of the two-step framework are summarized as follows.

(1) Each job posting is considered as a piece of semi-structured data comprising of two fields, i.e. title and description, which are treated differently in two steps.
(2) Query-document relevance score and document-document similarity score are combined to find semantically similar job postings.

Objective of job IR is achieved in two steps. In the first step, the standard VSM is applied on the title to retrieve relevant job postings according to query-document relevance score. Then job postings with relevance scores bigger than the threshold are selected as seeds for searching result expansion. In the second step, we calculate document similarity on full text to find the semantically similar job postings to the seed ones.

To re-rank the relevant job postings, a combined ranking model is proposed as follows, considering query-document relevance and document-document similarity in one formula as follow:

$$rel^*(q,d) = \max_{d_i}\{rel(q,d), rel(q,d_i) \times sim(d,d_i)\} \tag{1}$$

where $rel^*(q,d)$ denotes final relevance score between document d and query q, $rel(q,d)$ the general relevance score between d and q calculated in the first step, and $sim(d,d_i)$ ($\in [0,1]$) the similarity score between document d and d_i calculated in the second step.

3.2 The First Step Job Information Retrieval Based on VSM

In the first step, we retrieve job postings using the VSM. Two query-document relevance measures are implemented, i.e. cosine and inner product. As shown in [1][2], vector-length normalization causes a drop for cosine similarity when it is applied to very short string. So the inner product might be a good choice in our case. We calculate the classical *tf-idf* value as term weight.

As a result, relevant job postings are retrieved as well as relevance scores. We setup a threshold to get the seed job postings for further process in the second step.

3.3 Expanding Relevant Jobs Using Similarity between Job Postings

In this step, we use full text of each job posting to construct a *tf-idf* weighted feature vector, and attempt to find the job postings that are semantically similar to the seed ones by document similarity within the VSM.

Features and Similarity Measures

We choose two kinds of features, i.e. word and character bi-gram, which are proved by [3] to be the best feature types for Chinese text classification. We apply stop word list and finally obtain 25,000 word features and 140,000 character bi-gram features.

Two similarity measures are implemented in this paper, i.e. cosine and the extended Jaccard, which are found to be the best measures in the document cluster [5] and commonly used in Chinese text processing.

Feature Selection

A major characteristic of VSM is the high dimensionality rendering spare data problem. This problem is usually addressed by some automatic feature selection schemes. Yang and Pedersen [4] prove that feature selection technology can improve performance of text classification. In our work, two feature selection schemes are implemented, i.e. *DF* (document frequency) and χ^2 statistics (Chi-square) [4].

DF is the number of documents where a feature occurs. Terms with low *DF* score will be eliminated in this feature selection scheme.

χ^2 statistics originally estimates how one feature is independent from one class in text classification. For our case, we apply a clustering algorithm to generate the class labels required in χ^2 calculation. The χ^2 score for the term t and the class label c is defined as follows.

$$\chi^2(t,c) = \frac{N(AD - BC)^2}{(A+C)(B+D)(A+B)(C+D)} \tag{2}$$

where A denotes the number of documents with class label c and containing feature t, B is the number of documents with class label c while not containing feature t, C is the number of documents without class label c and containing feature t, D is the number of documents without class label c and not containing feature t, and N is the total number of documents. Finally goodness score for each feature is defined as the maximum cluster-specific χ^2 score as follows.

$$\chi^2_{\max}(t) = \max_c\{\chi^2(t,c)\} \tag{3}$$

We compute *DF* and χ2 score for each unique feature and remove a certain proportion of features.

Feature Re-weighting for Ad-hoc Retrieval

The document similarity measure discussed above is independent from query thus can be calculated off-line. However, query contributes more or less to feature weighting. Carpineto [6] shows that features actually play different roles in automatic query expansion for ad-hoc retrieval. We thus propose to make use of query to re-weight features.

In our case, we use the top *N* job postings obtained in the first step to select useful features. The usefulness score is calculated by the Rocchio's formula [7] as follows.

$$w_i^* = \alpha * w_{iq} + \frac{\beta}{|R|}\sum_{d_j \in R} w_{ij} \tag{4}$$

where *R* denotes the pseudo-feedback job posting sets retrieved in the first step, w_{iq} denotes the weight of term t_i in the original query, and w_{ij} the weight of term t_i in document d_j, α and β are two constants.

The top *K* features with high score are deemed useful and their weights are doubled in our work.

3.4 Some Critical Issues

In the two-step framework, two critical issues are worth noting. We in fact combine the IR model and the similarity measure of two piece of document into one model. In the first step, no extra calculation is involved compared to VSM. In the second step, several similarity measures for relevant job expansion are implemented, most of which are independent from the query thus can be calculated off-line. The exception is the re-weighting scheme, where the similarity scores can be updated for the selected features, rather than be re-calculated between every pairs of documents on-line. Therefore, computational complexity of our method can be appropriately controlled.

The second issue is retrieval quality. In the two-step framework, quality of the first retrieval is crucial. We set an appropriate threshold to get enough number of the relevant job postings as accurate as possible in the first retrieval. Meanwhile, the combined ranking model (see Section 3.1) is helpful to discard the false job postings.

4 Experiments

4.1 Setup

Data

Our job posting collection contains around 55,000 Chinese job postings downloaded from job-hunting websites including *ChinaHR* (www.chinahr.com) and *51Job* (www.51job.com). Title and description filed of each job posting can be detected by an HTML document parser. The query set contains 100 random queries, which in real applications are actually job names.

Evaluation Criteria

We use precision at top ranked *N* feedbacks, i.e. *p@N*, as evaluation criteria in our experiments. That is, for each of the 100 queries, we compute searching precision as percentage of job postings correctly retrieved in *top ranked N feedbacks*. To be practical, we set *N* as 1, 5, 10, 20, 30 and 40 our evaluation. Around 5000 job postings are judged manually whether they are relevant to the 100 queries.

4.2 Experiment 1: The First VSM Retrieval

In this experiment, we evaluate job IR methods on the title field vs. the full text using VSM. We use words as features and two query-document relevance measures, i.e. cosine and inner product. Experimental results are shown in Fig. 1.

Fig. 1 shows that searching on the full text obtains very little performance gain over that on title only. Two conclusions can be drawn. First, search intension can be reflected by the title rather than the description. Second, the description filed contributes very little in matching to the query using VSM though it is longer. This stimulates the idea to make use of the description in other manners.

Note that we use the VSM based on "*title + inner product*" as our baseline in the following experiments since it achieves relatively better performance at most points in Fig. 1.

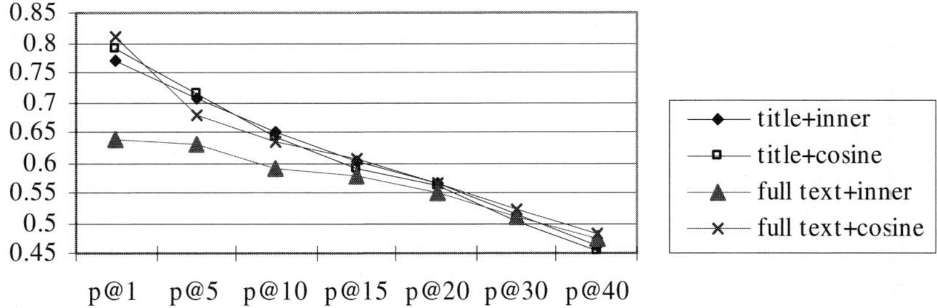

Fig. 1. Searching precision on title vs. full text using two similarity measures

4.3 Experiment 2: Relevant Job Expansion

In this experiment we attempt to expand the relevant job postings starting from the seed job postings using document similarity.

We first evaluate our method on different features types, i.e. words and character bi-grams, with cosine as similarity measure. Experimental results are shown in Fig. 2.

It is shown that 1) using similar job posting as expansion for seed job postings can improve searching quality; 2) word outperforms character bi-gram as feature type for document similarity measuring.

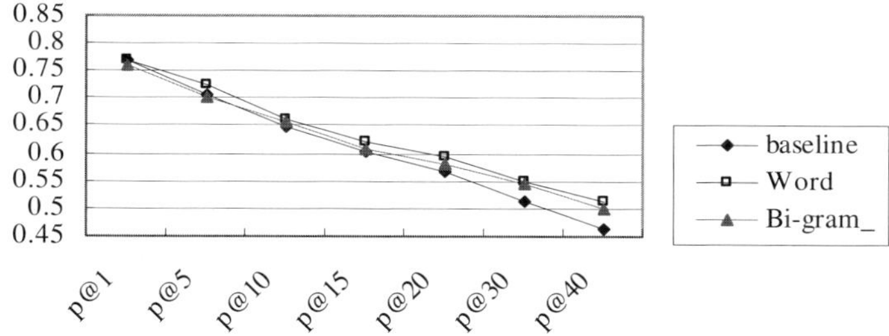

Fig. 2. Searching precision by expanding seed job postings using two feature types

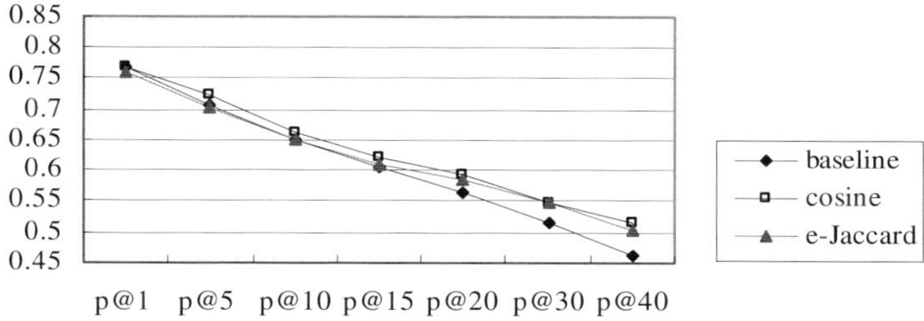

Fig. 3. Searching precision by expanding seed job postings using cosine vs. the extended Jaccard as document similarity measure

Using word as feature type, we then compare two document measures, i.e. cosine and the extended Jaccard. Experimental results are presented in Fig. 3. It is shown that cosine outperforms the extended Jaccard at all points.

4.4 Experiment 3: Feature Selection

In the following experiment, two feature selection schemes on word features are compared, i.e. *DF* and χ^2 statistics (CHI). For χ^2 statistics, we select k-1 repeated-bisection clustering method by the CLUTO package [8] to generate class labels. The experimental results are presented in Fig. 4.

Fig. 4 shows that both *DF* and χ^2 statistics can remove more than 70% terms and improves searching quality. χ^2 statistics on word improves most over baseline by 0.06 and over all-words by 0.011 at p@40.

It should be pointed out that the motivation to incorporate the clustering technique in our method is to separate the data set into a finite set of "natural" structure, namely clusters or subsets within job postings holding internal homogeneity and external separation, rather than accurate characterization or class label predefined as classification, so that the χ^2 statistics based supervised feature selection methods can make use of the labels to estimate goodness score of each feature. We have tried several clustering algorithms in CLUTO to obtain these labels. It is disclosed in our experiments that goodness of the clusters does not bias the feature selection much.

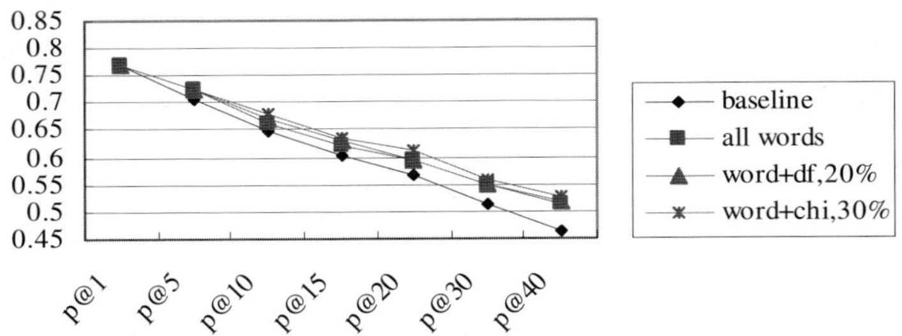

Fig. 4. Searching precision by expanding baseline searching results using two feature selection schemes. The percentages represent the proportions of features that remain after feature selection that yield best searching quality with certain setup.

4.5 Experiment 4: Feature Re-weighting

In this experiment we investigate on the feature re-weighting scheme. We apply Rocchio's formula to select features with high usefulness score and double their weights if they are determined as useful. Experimental results are presented in Fig. 5.

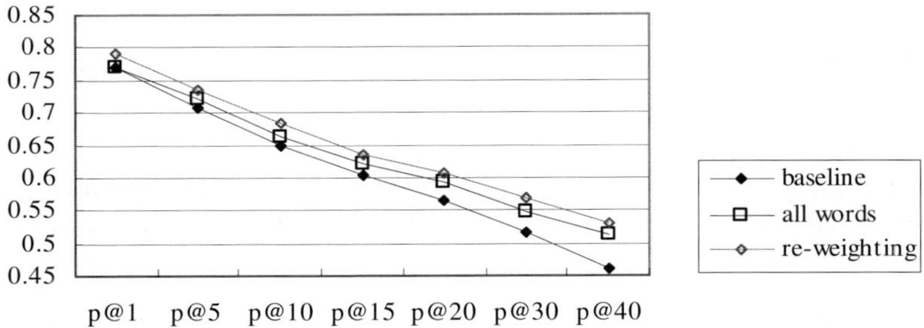

Fig. 5. Searching precision by expanding baseline searching results using Rocchio's formula

It is shown in Fig.5 that feature re-weighting scheme improves by around 0.02 at every point over the method using all words as features. Compared to the baseline method, feature re-weighting scheme improves most by 0.06 at p@40.

4.6 Experiment 5: The All-in-One System

In this experiment, we evaluate our all-in-one job IR system which uses word as feature type, cosine as similarity measure, integrating χ^2 statistics feature selection scheme and Rocchio's formula based feature re-weighting scheme.

To compare our method against the traditional query expansion method, the method based on traditional pseudo-relevance feedback is implemented. The method is briefed as follows. First, the initial search is performed to obtain the top K relevant documents, referred as pseudo-relevance feedback. Second, a number of terms are selected and reweighed from the feedback documents using certain scoring criteria to expand the initial query. Third, the expanded query is used to perform new search to get relevant documents. Corresponding to the first step within our method, we implement two methods to perform the initial search, i.e. using cosine as similarity measure on full text and inner product on job posting title. We do not perform query expansion on merely title field because the title field is too brief to yield valid extra query terms.

In this experiment, we implement several term-scoring functions [6][11] such as Rocchio, RSV, CHI, KLD, etc., in which Rocchio is found best in our job IR case. The experimental results are presented in Fig. 6.

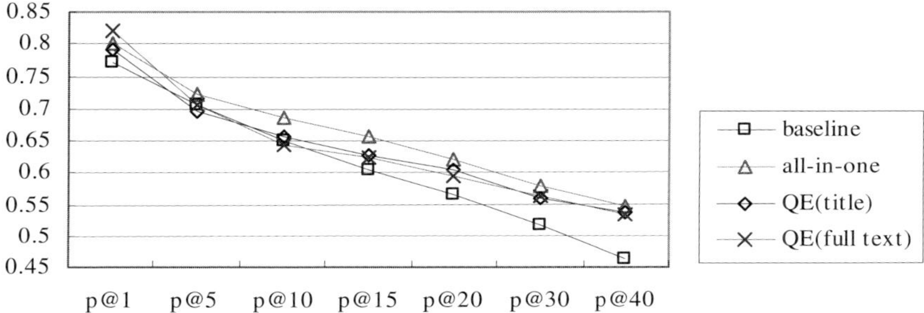

Fig. 6. Searching precision in baseline system vs. all-in-one system. QE(title) represents the query expansion method inner product as similarity measure on job posting title for initial search and QE(full text) the one using cosine on full text.

It is shown in Fig. 6 that our all-in-one scheme outperforms both traditional query expansion schemes at most points, in which p@40 is improved most by around 0.08325. This provides sufficient proof for the claim that our method for job IR is effective.

The second finding is that both query expansion methods outperform the baseline, in which the QE(title) outperforms the QE(full text). This accords to our results in the Experiment 1 where the "title+inner product" method outperforms "full text+cosine" at most points.

5 Related Works

The two-step framework we present in this paper is enlightened by the query expansion techniques [6][11], which have been used in the IR community for ages. The pseudo-feedback query expansion techniques also make use of the top documents to improve search performance, however, in a different way, that is, to use these documents to re-construct a new query first, while we apply document similarity to the pseudo-feedback documents to find the semantically similar job postings.

Feature and similarity measures within the VSM are explored in both IR and text categorization/cluster field. Term weighting in query-document relevance measuring is studies by [2][9]. Li et al. compare two feature types [3], i.e. word and character bi-gram in Chinese text categorization. Yang and Pedersen evaluate five feature selection methods [4], i.e. DF, IG, CHI, TS and MI, to reduce dimensionality of features in document categorization. In this work, we select DF and CHI as our feature selection method because it is an unsupervised method and CHI yields best performance in Yang's experiments. Strehl et al. compare four similarity measures on web-page clustering [5], we use the cosine and e-Jaccard in our work which lead to best performance in their work. Besides, Liu et al. make use of clustering results as class labels so that the supervised feature selection methods can be applied in unsupervised way [10].

6 Conclusions and Feature Works

This paper presents a two-step framework for job IR, which in fact combine the IR model and the similarity measure schemes of two semi-structure documents together. In this work, we investigate on the most popular IR model, i.e.VSM, in job IR. Several document similarity measures commonly used in NLP fields are implemented including cosine and extended Jaccard. We also investigate on several feature selection and term re-weighting schemes in this work. The experiment results show that our all-in-one system outperforms all other methods in performing the task of job IR. Several other conclusions can be drawn as follows. Firstly, word is a better feature type than character bi-gram. Secondly, cosine is a better document similarity measure than the extended Jaccard here. Thirdly, feature selection schemes are helpful to improve accuracy of document similarity, in which χ^2 statistics outperforms *DF*. Fourthly, feature re-weighting method is helpful for document similarity measuring. Finally, the traditional query expansion techniques are inferior to our method in the special job IR task.

Several future works are described as follows. Firstly, we will investigate on other IR models for the job IR task, such as the probabilistic models and language models. Secondly, we will investigate on information extraction techniques for the job IR task because the job postings are semi-structured and some job related information such as company information, responsibility, requirements, etc. can be easily recognized. We will try to use information of this kind to improve accuracy in job posting similarity measuring.

Acknowledgement

Research work in this paper is partially supported by NSFC (No. 60703051) and Tsinghua University under the Basic Research Foundation (No. JC2007049).

References

1. Yuwono, B., Lee, D.L.: WISE: A World Wide Web Resource Database System. In: Proc. of ICDE 1996, New Orleans, Louisiana (1996)
2. Salton, G., Buckley, C.: Term-weighting Approaches in Automatic Text Retrieval. Information Processing & Management 24(5), 513–523 (1988)

3. Li, J., Sun, M., Zhang, X.: A Comparison and Semi-quantitative Analysis of Words and Character-bigrams Features in Chinese Text Categorization. In: Proc. Of ACL 2006, Sydney, Australia (2006)
4. Yang, Y., Pedersen, J.O.: A Comparative Study on Feature Selection in Text Categorization. In: Proc. of ICML 1997, pp. 412–420 (1997)
5. Strehl, A., Ghosh, J., Mooney, R.: Impact of Similarity Measures on Web-page Clustering. In: Proc. of AAAI-2000 Workshop for Web Search, Austin (2000)
6. Carpineto, C., de Mori, R., Romano, G., Bigian, B.: Information Theoretic Approach to Automatic Query Expansion. ACM trans. 19(1), 1–27 (2001)
7. Rocchio, J.J.: Relevance Feedback in Information Retrieval. In: Salton, G. (ed.) The SMART Retrieval System, pp. 313–323. Prentice-Hall, Inc, Englewood Cliffs, N.J (1971)
8. Karypis, G.: CLUTO: A Clustering Toolkit. Dept. of Computer Science, University of Minnesota (May 2002)
9. Zobel, J., Moffat, A.: Exploring the Similarity Space. In: Proc. Of SIGIR 1998, Melbourne, Australia (1998)
10. Liu, T., Wu, G., Chen, Z.: An Effective Unsupervised Feature Selection Method for Text Clustering. Journal of Computer Research and Development 42(3), 381–386 (2005)
11. Lee, J.H.: Combining the evidence of different relevance feedback methods for information retrieval. Information Processing and Management 34(6), 681–691

Discrimination of Ventricular Arrhythmias Using NEWFM

Zhen-Xing Zhang, Sang-Hong Lee, and Joon S. Lim

Division of Software, Kyungwon University, Sungnam 461-701, Korea
jjh@ku.kyungwon.ac.kr, shleedosa@kyungwon.ac.kr,
jslim@kyungwon.ac.kr

Abstract. The ventricular arrhythmias including ventricular tachycardia (VT) and ventricular fibrillation (VF) are life-threatening heart diseases. This paper presents a novel method for detecting normal sinus rhythm (NSR), VF, and VT from the MIT/BIH Malignant Ventricular Arrhythmia Database using the neural network with weighted fuzzy membership functions (NEWFM). This paper separates pre-processing into 2 steps. In the first step, ECG beasts are transformed by using Filtering Function [1]. In the second step, transformed ECG beasts produce 240 numbers of probability density curves and 100 points in each probability density curve using the probability density function (PDF) processing. By using three statistical methods, 19 features can be generated from these 100 points of probability density curve, which are the input data of NEWFM. The 15 generalized features from 19 PDF features are selected by non-overlap area measurement method [4]. The BSWFMs of the 15 features trained by NEWFM are shown visually. Since each BSWFM combines multiple weighted fuzzy membership functions into one using bounded sum, the 15 small-sized BSWFMs can realize NSR, VF, and VT detection in mobile environment. The accuracy rates of NSR, VF, and VT is 98.75 %, 76.25 %, and 63.75 %, respectively.

Keywords: fuzzy neural networks, NSR, VT, VF, filtering transform, PDF, weighted fuzzy membership function, feature selection.

1 Introduction

Classifying cardiac arrhythmias using the electrocardiogram (ECG) is in great need of an adaptive decision support tool that can handle noise, baseline drift, and artifacts. Fuzzy neural networks (FNN) can be effectively used for this type of tool as a major pattern classification and predictive rule generation tool for cardiac pattern analysis [2] [6] [9] [10] [11] [13]. Since the ECG signal includes noise, baseline drift, and abnormal behavior, the Filtering Transforms (FT) as filtering process is needed. The filter function, filtfilt function, and butter function are used in the FT filtering process.

Ventricular tachycardia (VT) is a potentially lethal disruption of normal heartbeat (arrhythmia) that may cause the heart to become unable to pump adequate blood through the body. If the VT appears a period of time, VT will induce a Ventricular Fibrillation (VF), the most dangerous type of heart arrhythmia. The VF is represented

H. Li et al. (Eds.): AIRS 2008, LNCS 4993, pp. 176–183, 2008.

by fast rhythm, abnormal and ineffective contractions of the ventricles and it finishes in a systole .The VF within a few minutes or a few days will lead to cardiac sudden death. The survival probability for a human who has a VF attack outside the hospital ranges between 2-25% [14].

This paper presents a set of FT and probability density function (PDF) processing result as common input features for automatic NSR, VF, and VT detection using neural network with weighted membership functions (NEWFM) and the non-overlap area distribution measurement method [5]. The method extracts minimum number of input features each of which constructs an interpretable fuzzy membership function.

Methods of feature extraction of ECG are categorized into four functional groups including direct, transformation, and characteristic parameter estimation methods. FT belongs to the transformation method that filtering process is a reasonable defibrillator method. Chen *et al.* [7] used the PDF of the Blanking Variability and sequential probability ratio test (SPRT) method for detecting arrhythmia classification. Chowdhury *et al.* [3] used Fuzzifier transformation and Fuzzy Rule Base method for detecting arrhythmias classification. This study has quondam accuracy rates of NSR, VF, and VT is 94.3 %, 78.0 %, and 82 %, respectively. But this result doesn't count the classification CT decision (the classification CT implies that no decision can be reached for the interval.)[3], so the recalculated accuracy rates counted of NSR, VF, and VT are 82.5 %, 58 %, and 62.5 %, respectively.

In this paper, the 15 generalized features from 19 PDF features are selected by non-overlap area measurement method [4]. A set of 15 extracted coefficient features are presented to predict NSR, VF, and VT classification using the FT, PDF, and the neural network weighted fuzzy membership functions (NEWFM) [6][5]. The 15 features are interpretably formed in weighted fuzzy membership functions preserving the disjunctive fuzzy information and characteristics, locally related to the time signal, of ECG patterns. Although reducing the number of features is advantageous to computation and complexity, it becomes one of main causes of increasing dependency on data sets or patients.

2 Pre-process of ECG Signals

2.1 Filtering Transformation

NEWFM uses the same filtering transformation in Amann et al [1]. The filtering transformation process contained a filtering function file. The filtering function works in four steps as follows:

A. Remove the average value of the signal from the signal.
B. Apply a moving averaging filter in order to remove high frequency noise.
C. Carry out the drift suppression. It removes slow signal changes, which are not produced by the heart and originate from external sources.
D. Make a butterworth filter with a remove frequency of 30 Hz eliminates frequencies higher than 30 Hz, which seem to be of no relevance in our simulations. By applying this filtering process, the behavior of the signal acquisition by a defibrillator is reasonably simulated.

2.2 Probability Density Function (PDF) Transformation

This PDF process is based on sampling the amplitude distribution of the same baseline cardiac rhythm signal. The distinct NSR, VT, and VF rhythm signal probability density curve has been shown in Fig.1. There are eight ranges in Y coordinate of the curve like [0,0.5],[0.5,1],[1,2],[2,3],[3,4],[4,5],[5,6],[6, +∞]. The number of Y coordinates in the every interval of the eight ranges are counted and then the every interval's average of the eight ranges are counted as all Y coordinates divided by the number of Y coordinates. The number of Y coordinates in the [0.5, +∞], [1, +∞] interval and maximum value of the curve are counted. The appearing 19 data are selected after some tests and are used as the NEWFM input data.

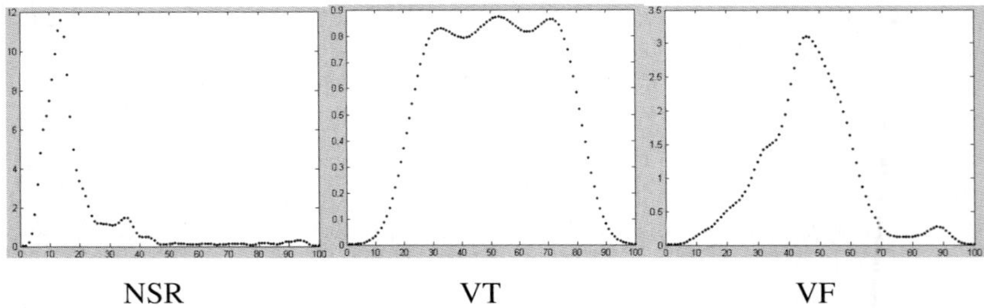

Fig. 1. NSR, VT, and VF probability density curves

3 Neural Network with Weighted Fuzzy Membership Function (NEWFM)

3.1 The Structure of NEWFM

Neural network with weighted fuzzy membership function (NEWFM) is a supervised classification neuro-fuzzy system using bounded sum of weighted fuzzy membership functions (BSWFM in Fig. 3) [5][6]. The structure of NEWFM, illustrated in Fig. 2, comprises three layers namely input, hyperbox, and class layer. The input layer contains n input nodes for an n featured input pattern. The hyperbox layer consists of m hyperbox nodes. Each hyperbox node B_l to be connected to a class node contains n BSWFMs for n input nodes. The output layer is composed of p class nodes. Each class node is connected to one or more hyperbox nodes. An h-th input pattern can be recorded as $I_h=\{A_h=(a_1, a_2, \ldots , a_n), class\}$, where class is the result of classification and A_h is n features of an input pattern.

The connection weight between a hyperbox node B_l and a class node C_i is represented by w_{li}, which is initially set to 0. From the first input pattern I_h, the w_{li} is set to 1 by the winner hyperbox node B_l and class i in I_h. C_i should have one or more than one connections to hyperbox nodes, whereas B_l is restricted to have one connection to a corresponding class node. The B_l can be learned only when B_l is a winner for an input I_h with class i and $w_{li} = 1$.

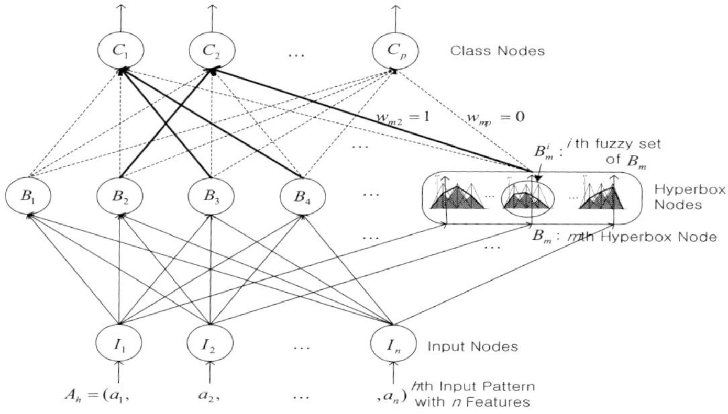

Fig. 2. Structure of NEWFM

3.2 Learning Scheme

A hyperbox node B_l consists of n fuzzy sets. The ith fuzzy set of B_l, represented by B_l^i, has three *weighted fuzzy membership functions* (WFM, grey triangles ω_{l1}^i, ω_{l2}^i, and ω_{l3}^i in Fig. 3) which randomly constructed before learning. Each $\omega_{l\,j}^i$ is originated from the original membership function $\mu_{l\,j}^i$ with its weight $W_{l\,j}^i$ in the Fig. 3. The *bounded sum of three weighted fuzzy membership functions* (BSWFM, bold line in Fig. 3) of B_l^i combines the fuzzy characteristics of the three WFMs. The BSWFM value of B_l^i, denoted as $BS_l^i(.)$, and is calculated by formulas (1) where a_i is an ith feature value of an input pattern A_h for B_l^i.

$$BS_l^i(a_i) = \sum_{j=1}^{3} \omega_{l\,j}^i(a_i), \tag{1}$$

The winner hyperbox node B_l is selected by the *Output* (B_l) operator. Only the B_l, that has the maximum value of *Output* (B_l) for an input I_h wtih class i and w_{li} = 1, among the hyperbox nodes can be learned. For the hth input A_h= (a_1, a_2… a_n) with n features to the hyperbox B_l, output of the B_l is obtained by formulas (2)

$$Output\ (B_l) = \frac{1}{n}\sum_{i=1}^{n} BS_l^i(a_i). \tag{2}$$

Then, the selected winner hyperbox node B_l is learned by the *Adjust* (B_l) operation. This operation adjusts all B_l^i s according to the input a_i, where i=1, 2… n. The membership function weight $W_{l\,j}^i$ (where $0 \le W_{l\,j}^i \le 1$ and j=1, 2, 3) represents the strength of $\omega_{l\,j}^i$. Then a WFM $\omega_{l\,j}^i$ can be formed by ($v_{l\,j-1}^i$, $W_{l\,j}^i$, $v_{l\,j+1}^i$). As a result of *Adjust* (B_l) operation, the vertices $v^i_{l\,j}$ and weights $W^{\,i}_{l\;j}$ in Fig. 4 are adjusted by the following expressions:

$$
\begin{aligned}
v^i_{l\,j} &= v^i_{l\,j} + s\times\alpha\times E^i_{l\,j}\times\omega^i_{l\,j}(a_i)\\
&= v^i_{l\,j} + s\times\alpha\times E^i_{l\,j}\times\mu^i_{l\,j}(a_i)\times W^{\,i}_{l\;j}, where
\end{aligned}
$$

$$
\begin{cases}
s=-1, E^i_{l\,j}=\min(\left|v^i_{l\,j}-a_i\right|,\left|v^i_{l\,j-1}-a_i\right|),\ \ if\ v^i_{l\,j-1}\le a_i < v^i_{l\,j}\\
s=1, E^i_{l\,j}=\min(\left|v^i_{l\,j}-a_i\right|,\left|v^i_{l\,j+1}-a_i\right|),\ \ if\ v^i_{l\,j}\le a_i < v^i_{l\,j+1}\\
E^i_{l\,j}=0,\ \ otherwise
\end{cases}
\tag{3}
$$

$$
W^{\,i}_{l\;j} = W^{\,i}_{l\;j} + \beta\times(\mu^{\,i}_{l\;j}(a_i) - W^{\,i}_{l\;j})
$$

where the α and β are the learning rates for $v^i_{l\,j}$ and $W^{\,i}_{l\;j}$ respectively in the range from 0 to 1 and j=1,2,3.

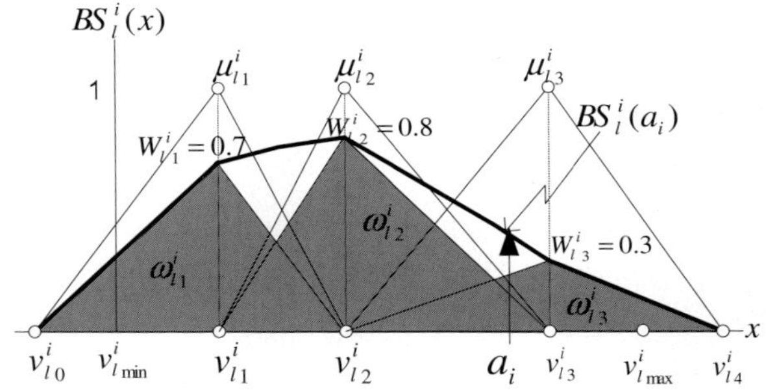

Fig. 3. An Example of Bounded Sum of Weighted Fuzzy Membership Functions (BSWFM, Bold Line) of B^i_l and $BS^i_l(a_i)$

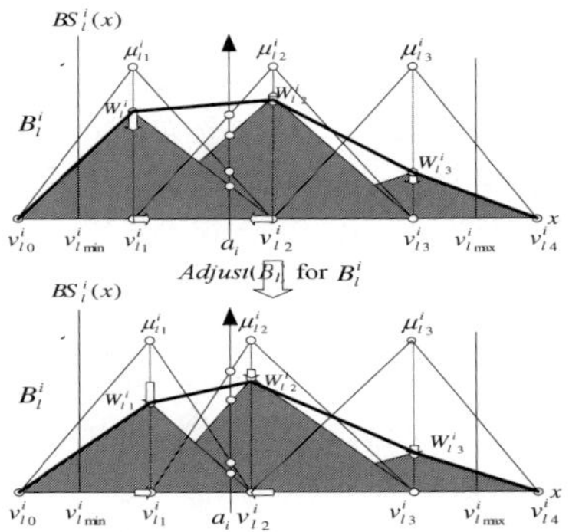

Fig. 4. An Example of Before and After Adjust (B_l) Operation for B^i_l

Fig. 4 shows BSWFMs before and after *Adjust* (B_l) operation for B_l^i with an input a_i. The weights and the centers of membership functions are adjusted by the *Adjust* (B_l) operation, e.g., W_{l1}^{i}, W_{l2}^{i}, and W_{l3}^{i} are moved down, v_{l1}^{i} and v_{l2}^{i} are moved toward a_i, and v_{l3}^{i} remains in the same location.

The Adjust (B_l) operations are executed by a set of training data. If the classification rate for a set of test data is not reached to a goal rate, the learning scheme with Adjust(B_l) operation is repeated from the beginning by randomly reconstructing all WFMs in B_l s and making all connection weights to 0 ($w_{li} = 0$) until the goal rate is reached.

4 Experimental Results

In this section, the NSR, VF, and VT data sets, which were used in Chowdhury [3], are used to evaluate the accuracy of the proposed NEWFM. The 15 generalized features among 19 generalized features are selected by non-overlap area measurement method [4].

The analyzed data set is taken from MIT/BIH Malignant Ventricular Arrhythmia Database [8]. This data base consists of 22 thirty-five-minute records and 80 four-second samples of each NSR, VT, and VF episodes in the 22 records are selected at random. Fig. 5 shows BSWFMs of the generalized 6 features among the generalized 15 features. The solid lines, broken lines and dotted lines represent NSR, VF, and VT characteristics of ECG visually, which enables the features to interpret explicitly.

Chowdhury has quondam accuracy rates of NSR, VF, and VT is 94.3 %, 78.0 %, and 82 %, respectively. But this result doesn't count the classification CT decision (the classification CT implies that no decision can be reached for the interval.) [3], so the recalculated accuracy rates of NSR, VF, and VT is 82.5 %, 58 %, and 62.5 %, respectively. Table 1 shows the recalculated accuracy rates. The average accuracy rate is 67.5%.

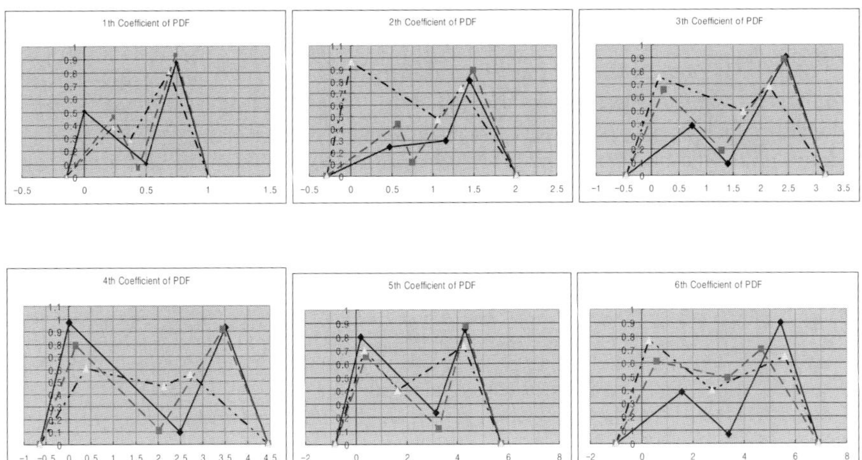

Fig. 5. Trained BSWFMs of the 1- 6th Features for NSR, VF, and VT Classification

Table 1. Chowdhury's results of evaluating the NSR, VT, and VF detection algorithm using the MIT/BIH Malignant Ventricular Arrhythmia ECG DATABASE

True Events& Tot.NO.of Samples	NSR	VT	VF	CT	Accuracy Rate
NSR(80)	66	4	0	10	82.5
VT(80)	1	50	10	19	62.5
VF(80)	0	13	46	21	58

Table 2. Comparisons of performance results for NEWFM with CHOWDHURY'S

True Events& Tot.NO.of Samples	NSR	VT	VF	Accuracy Rate
NSR(80)	79	1	0	98.75
VT(80)	22	51	7	63.75
VF(80)	7	12	61	76.25

Table 2 shows the performance results of NEWFM using the generalized 15 features. In this paper, there is no CT decision. NEWFM classifies NSR, VF, and VT on all data sets. The accuracy rates of NSR, VF, and VT are 98.75 %, 76.25 %, and 63.75 %, respectively and the average accuracy rate is 79.16%.

5 Concluding Remarks

The BSWFMs of the 15 features trained by NEWFM are shown visually, which makes the features interpret explicitly. Since each BSWFM combines multiple weighted fuzzy membership functions into one using bounded sum, the 15 small-sized BSWFMs can realize real-time NSR, VT, and VF detection in mobile environment. These algorithms are pivotal component in Automated External Defibrillators (AED). To improve the accuracy rates of NSR, VF, and VT, some kinds of mathematics' method instead of PDF will be needed to study in real application of AED. On the other hand, some good results are achieved on economy index and stock forecasting which using NEWFM.

References

1. Amann, A., Tratnig, R., Unterkofler, K.: Detecting Ventricular Fibrillation by Time-Delay Methods. IEEE Trans. on Biomedical Engineering 54(1), 174–177 (2007)
2. Engin, M.: ECG beat classification using neuro-fuzzy network. Pattern Recognition Letters 25, 1715–1722 (2004)
3. Chowdhury, E., Ludeman, L.C.: Discrimination of cardiac arrhythmias using a fuzzy rule-based method. Computers in Cardiology, 549–552, September 25-28 (1994)

4. Lim, J.S., Gupta, S.: Feature Selection Using Weighted Neuro-Fuzzy Membership Functions. In: The 2004 International Conference on Artificial Intelligence(IC-AI 2004), Las Vegas, Nevada, USA, June 21-24, vol. 1, pp. 261–266 (2004)
5. Lim, J.S., Ryu, T.-W., Kim, H.-J., Gupta, S.: Feature Selection for Specific Antibody Deficiency Syndrome by Neural Network with Weighted Fuzzy Membership Functions. In: Wang, L., Jin, Y. (eds.) FSKD 2005. LNCS (LNAI), vol. 3614, pp. 811–820. Springer, Heidelberg (2005)
6. Lim, J.S., Wang, D., Kim, Y.-S., Gupta, S.: A neuro-fuzzy approach for diagnosis of antibody deficiency syndrome. Neurocomputing 69(7-9), 969–974 (2006)
7. Clarkson, S.-W.C., Clarkson, P.M., Fan, Q.: A robust sequential detection algorithm for cardiac arrhythmia classification. IEEE Transactions on Biomedical Engineering 43(11), 1120–1124 (1996)
8. MIT-BIH Malignant Ventricular Arrhythmia database directory, Document BMEC TR010, Mass. Inst. Technol, Cambridge, July (1992)
9. Minami, K., Nakajima, H., Toyoshima, T.: Real-Time Discrimination of Ventricular Tachyarrhythmia with Fourier-Transform Neural Network. IEEE Trans. on Biomedical Engineering 46(2), 176–185 (1999)
10. Linh, T.H., Osowski, S., Stodolski, M.: On- Line Heart Beat Recognition Using Her8e Polynomials and Neuro-Fuzzy Networks. IEEE Trans. on Instrumentation and Measurement 52(4), 1224–1231 (2003)
11. Osowski, S., Linh, T.H.: ECG beat recognition using fuzzy hybrid neural network. IEEE Trans. on Biomedical Engineering 48(4), 1265–1271 (2001)
12. Ramirez-Rodriguez, C., Hernandaz-Silveria, M.: Multi-Thread Implementation of a Fuzzy Neural Network for Automatic ECG Arrhythmia Detection. Proceedings in Computers in Cardiology 2001, 297–300 (September 2001)
13. Silipo, R., Marchesi, C.: Artificial Neural Networks for Automatic ECG Analysis. IEEE Trans. on Signal Processing 46(5), 1417–1425 (1998)
14. Fernandez, A.R., Folgueras., J., Colorado., O.: Validation of a set of algorithms for ventricular fibrillation detection: experimental results. In: Proceedings of the 25th Annual International Conference of the IEEE, September 17-21 2003, vol. 3, pp. 2885–2888 (2003)

Efficient Feature Selection in the Presence of Outliers and Noises*

Shuang-Hong Yang and Bao-Gang Hu

National Lab of Pattern Recognition(NLPR) & Sino-French IT Lab(LIAMA)
Institute of Automation, Chinese Academy of Sciences
{shyang,hubg}@nlpr.ia.ac.cn

Abstract. Although regarded as one of the most successful algorithm to identify predictive features, *Relief* is quite vulnerable to outliers and noisy features. The recently proposed *I-Relief* algorithm addresses such deficiencies by using an iterative optimization scheme. Effective as it is, I-Relief is rather time-consuming. This paper presents an efficient alternative that significantly enhances the ability of Relief to handle outliers and strongly redundant noisy features. Our method can achieve comparable performance as I-Relief and has a close-form solution, hence requires much less running time. Results on benchmark information retrieval tasks confirm the effectiveness and efficiency of the proposed method.

1 Introduction

Feature subset selection is a process of identifying a small subset of highly predictive features out of a large set of candidate features which might be strongly irrelevant and redundant [2,3]. It plays a fundamental role in data mining, information retrieval, and more generally machine learning tasks for a variety of reasons [3]. In the literature, many feature selection methods approach the task as a search problem [3,4], where each state in the search space is a possible feature subset. Feature weighting simplify this problem by assigning to each feature a real valued number to indicate its usefulness, making possible to select a subset of features efficiently by searching in a continuous space rather than a discrete state space.

Among the existing feature weighting methods, *Relief* [5,7,9] is considered one of the most successful ones due to its effectiveness, simplicity and efficiency. Suppose we are given a set of input vectors $\{\mathbf{x}_n\}_{n=1}^N$ along with corresponding targets $\{y_n\}_{n=1}^N$, where $\mathbf{x}_n \in \mathbf{X} \subset \mathcal{R}^D$ is a training instance (e.g., the *vector space model* of a document) and $y_n \in \mathbf{Y}=\{0,1,\ldots,C\text{-}1\}$ is its label (e.g., the category of the document), N, D, C denote the training set size, the input space dimensionality and the total number of categories respectively. The d-th feature of $\mathbf{x}$ is denoted as $x^{(d)}$, d=1,2,...,D. Relief ranks the features according to the weights w_d's obtained from a convex optimization problem [9]:

* This work is supported in part by NSFC (#60073007, #60121302).

H. Li et al. (Eds.): AIRS 2008, LNCS 4993, pp. 184–191, 2008.

$$\begin{aligned} \mathbf{w} = \arg\max \ & \textstyle\sum_{n=1}^{N} \mathbf{w}^T \mathbf{m}_n \\ s.t. : & \ ||\mathbf{w}|| = 1, w_d \geq 0, d = 1, 2, ..., D \end{aligned} \tag{1}$$

where $\mathbf{w}=(w_1, w_2, ..., w_D)^T$, $\mathbf{m}_n = |\mathbf{x}_n - M(\mathbf{x}_n)| - |\mathbf{x}_n - H(\mathbf{x}_n)|$ is called the margin for the pattern $\mathbf{x}_n$, $H(\mathbf{x}_n)$ and $M(\mathbf{x}_n)$ denote the nearest-hit (the nearest neighbor from the same class) and nearest-miss (the nearest neighbor form different class) of $\mathbf{x}_n$ respectively.

However, a crucial drawback [9] of the standard Relief algorithm is that it lacks mechanisms to tackling outliers and redundant features, which heavily degrade its performance in practice.

- The success of Relief hinges largely on its attempting to discriminate between neighboring patterns (nearest-miss and nearest-hit). However, the nearest neighbors are defined in the original feature space. When there are a large number of redundant and/or noisy features present in the data, it is less likely that the nearest neighbors in the original feature space will be the same as those in the target feature space. As a consequence, the performance of Relief can be degraded drastically;
- The objective function of the Relief algorithm, Eq.(1), is to maximize the average margin of the training samples. This formulation makes it rather vulnerable to outliers, because the margins of outlying patterns usually take very negative values (thus can heavily affect the performance of Relief).

The recently proposed Iterative-Relief algorithm (I-Relief, [9]) addresses these two problems by introducing three latent variables for each pattern and employing the Expectation-Maximization (EM) principal to optimize the objective function. Powerful as it is, this algorithm surfers two drawbacks: (*i*) It is very time-consuming since there is no close-form solution. Therefore, iterative optimization scheme must be employed. In particular, within each iteration, the I-Relief algorithm involves at least $O(N^2D)$ times of computation, which is only tractable for very small data set; (*ii*) I-Relief requires storing and manipulating three $N \times N$-sized matrix at each iteration, which is infeasible for large data set.

In this paper, we propose efficient alternative approaches to address the deficiencies of Relief in tackling outliers and noises. In particular, in order to handle outliers, we borrow the concept of margin-based loss function [1,6] from the supervise learning literature, and integrate a loss function into the objective function of Relief, i.e.: instead of maximizing the average margin, this method minimizes the empirical sum of a specific loss function. Since the resulted problem has a close-form solution, this method is much more efficient (in fact, it is of the same complexity as the standard Relief). In the meanwhile, when appropriate loss functions are chosen, this method can achieve comparable performance as I-Relief. In addition, to tackling noisy features, we propose a novel algorithm, named Exact-Relief, which is based on a new perspective of Relief as a greedy nonparametric Bayes error minimization feature selection approach. We finally conduct empirical evaluations on various benchmark information retrieval tasks. The results confirm the advantages of our proposed algorithms.

2 The Proposed Algorithms

2.1 Against Outliers: Ramp-Relief

Relief maximizes the empirical average margin on the training set (see Eq.(1)). An alternative (and equivalent) way to view this is to minimize the empirical sum of a margin-based loss function:

$$\begin{aligned} &\min \ \textstyle\sum_{n=1}^{N} l(\mathbf{w}^T \mathbf{m}_n) \\ &s.t.: ||\mathbf{w}|| = 1, w_d \geq 0, d = 1, 2, ..., D \end{aligned} \tag{2}$$

where $l(\cdot)$ is a margin-based loss function [1,6]. In this viewpoint, the standard Relief is a special case of the above formulation, i.e., it uses a simple linear loss function $l(z)$=-z.

To minimize empirical sum of a specific margin-based loss function has been extensively studied in supervised learning literature both theoretically and empirically. This methodology offers various advantages. We refer the interested readers to [6,1] and the references therein for more detailed discussions.

The new perspective of Relief allows us to extend Relief from using linear loss function to other more extensively studied loss functions. For computational simplicity, we solve an approximate problem in this paper, i.e.:

$$\begin{aligned} &\min \ \textstyle\sum_{n=1}^{N} \mathbf{w}^T l(\mathbf{m}_n) \\ &s.t.: ||\mathbf{w}|| = 1, w_d \geq 0, d = 1, 2, ..., D \end{aligned} \tag{3}$$

and a variation of the Ramp loss function used in ψ-learning [8] is employed:

$$\begin{aligned} r(z) &= \max(z_2, \min(z_0 - z, z_1)) \\ &= \begin{cases} z_1, z < z_0 - z_1 \\ z_2, z > z_0 - z_2 \\ z_0 - z, else \end{cases} \end{aligned} \tag{4}$$

where z_0, z_1 and z_2 are three constants. By using the Lagrangian technique, a quite simple close-form solution to problem Eq.(4) can be easily derived, i.e.:

$$\mathbf{w} = (\boldsymbol{\gamma})^+ / ||(\boldsymbol{\gamma})^+|| \tag{5}$$

where $\boldsymbol{\gamma} = \sum_{n=1}^{N} -r(|\mathbf{x}_n - H(\mathbf{x}_n)| - |\mathbf{x}_n - M(\mathbf{x}_n)|)$, and $(\cdot)^+$ denotes the positive part.

We term this algorithm as **Ramp-Relief** (**R-Relief**). We will show that the R-Relief algorithm is able to deal with outliers as well as I-Relief but is much more efficient and simpler to compute.

2.2 Against Noisy Features: Exact-Relief

Recently, we found that Relief greedily attempts to minimize the nonparametric Bayes error estimated by k-nearest-neighbor (kNN) methods with feature

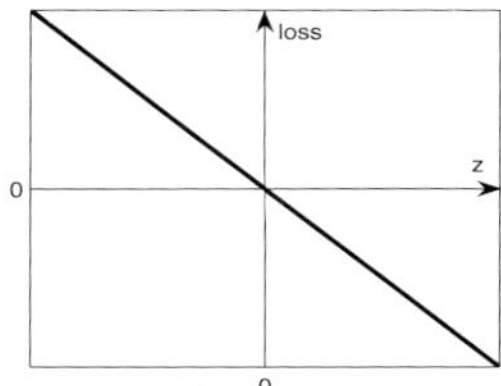

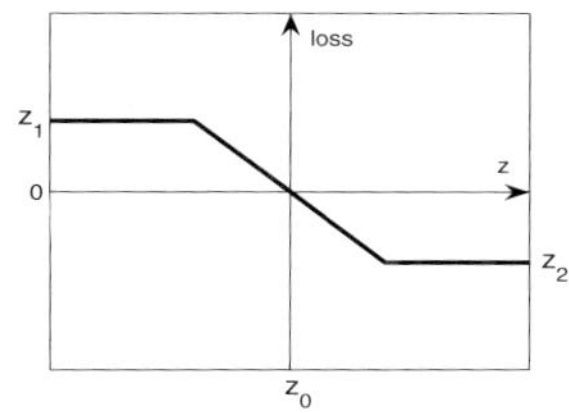

Fig. 1. Linear loss function (left) and ramp loss function (right)

weighting as the search strategy [10]. One of the assumptions made by Relief is that the nearest neighbor of a pattern $\mathbf{x}$ locates close to $\mathbf{x}$ in any single dimensional space. For instance, suppose $\mathbf{x}_a$ is the nearest neighbor of $\mathbf{x}$: $||\mathbf{x}_a - \mathbf{x}|| \leq ||\mathbf{x}_n - \mathbf{x}||$ for $n = 1, 2, \ldots, N$, Relief implicitly assumes that $x_a^{(d)}$ is also, approximately, the nearest neighbor of $x^{(d)}$, that is $x_a^{(d)} \approx x_b^{(d)}$, where $|x_b^{(d)} - x^{(d)}|| \leq ||x_n^{(d)} - x^{(d)}||$ (b is dependent on d). Therefore, Relief approximates $x_b^{(d)}$ with $x_a^{(d)}$ for all $d = 1, 2, \ldots, D$. Although this approximation can reduce the computation complexity significantly, it also pays prices. In particular, if the feature set is strongly redundant such that a large proportion of features are irrelevant, noisy, or useless. In that case, $x_a^{(d)}$ is highly unlikely to locate close to $x^{(d)}$, which can heavily degrade the performance of the solutions. Therefore, it may be preferable to eliminate this assumption. For this purpose, we propose an algorithm refereed as '**Exact-Relief**' (**E-Relief**), which resemble the standard Relief algorithm except using a different margin definition: $\mathbf{m}_n = (m_n^{(d)})_{D\times 1}$, $m_n^{(d)} = |x_n^{(d)} - M_n^{(d)}| - |x_n^{(d)} - H_n^{(d)}|$, where $M_n^{(d)}$ and $H_n^{(d)}$ denote the nearest-miss and nearest-hit of $\mathbf{x}_n$ in the d-th dimension.

2.3 Against Both Outliers and Noisy Features

In practice, it is quite possible that both outliers and noisy features are present in the data. For instance, in spam filtering, junk mails usually contain a large amount of noisy characters in order to cheat the filter. On the other hand, legitimate mails may only have very few words but contain many hyperlinks. Such mails not only contain many noisy features but can also be easily detected as outliers. To handle both factors, an obvious strategy is to combine the R-Relief and E-Relief algorithm, i.e.:

$$\begin{aligned} &\max \sum_{d=1}^{D} w_d \sum_{n=1}^{N} r(|x_n^{(d)} - M_n^{(d)}| - |x_n^{(d)} - H_n^{(d)}|) \\ &s.t. : \mathbf{w} \geq \mathbf{0}, ||\mathbf{w}|| = 1 \end{aligned} \tag{6}$$

We term this algorithm as **ER-Relief**. It can be easily seen that ER-Relief and E-Relief are of the same complexity, i.e., $O(N^2 D)$, which is much more efficient compared to I-Relief, whose worst case complexity is $O(N^3 D)$.

Table 1. Characteristics of data sets

Data Set	#Train	#Test	#Feature	#Class
Spam	1000	3601	57	2
LRS	380	151	93	48
Vowel	530	460	11	11
Trec11	114	300	6429	9
Trec12	113	200	5799	8
Trec23	84	120	5832	6
Trec31	227	700	10127	7
Trec41	178	700	7454	10
Trec45	190	500	8261	10

3 Experiments

In this section, we conduct extensive experiments to evaluate the effectiveness and efficiency of the proposed methods in comparison with state-of-art algorithms in Relief family.

3.1 Experiments on UCI Data Sets

To demonstrate the performance of the proposed algorithms in different information retrieval tasks, we first perform experiments on three benchmark UCI data sets, namely, the spam filtering data set (`Spam`), the low-resolution satellite image recognition data set (`LRS`) and the speaker-independent speech recognition data set (`Vowel`). To conduct comparison in a controlled manner, fifty irrelevant features (known as 'probes') are added to each pattern, each of which is an independently Gaussian distributed random variable, i.e., $\mathcal{N}(0{,}20)$. The efficiency of a feature selection algorithm can be directly measured by its running time. To evaluate the effectiveness, two distinct metrics are used. One is the classification accuracy estimated by kNN classifier, where k is determined by five-fold cross validation. The other metric is the Receiver Operating Characteristic (ROC) curve [9], which is used to indicate the abilities of different feature selection algorithms in identifying relevant features and at the same time ruling out useless ones. To eliminate statistical deviations, all the experiments are repeated for 20 runs. In each run, the data set is randomly partitioned into training and testing data, and only the training data are used to learn the feature selector. Three groups of experiments have been done:

1. **Against outliers.** Relief, I-Relief and R-Relief are compared. A randomly selected subset of 10% training samples are mislabelled. The testing data is kept intact. No probe is added. The testing errors is shown in the top line of Fig.2. We can see R-Relief improves the performance of Relief significantly. It performs comparably with I-Relief when outliers are present.
2. **Against noisy features.** E-Relief is compared with Relief and I-Relief. 50 probes are added to each example, but no mislabelling is conducted. The

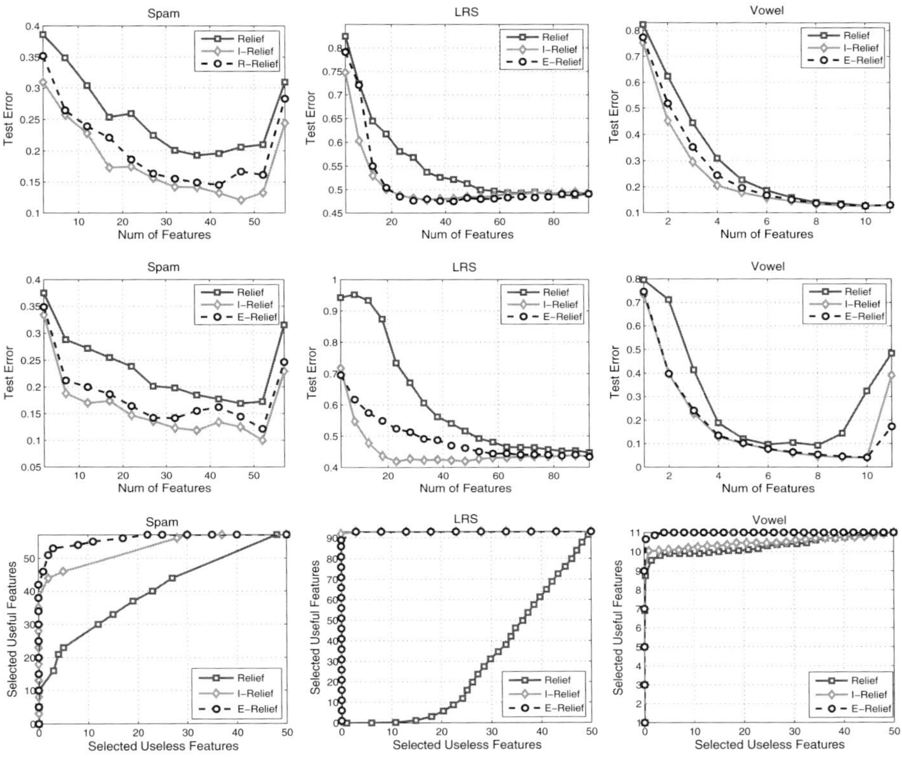

Fig. 2. Comparison of R-Relief/E-Relief I-Relief and Relief on UCI data set

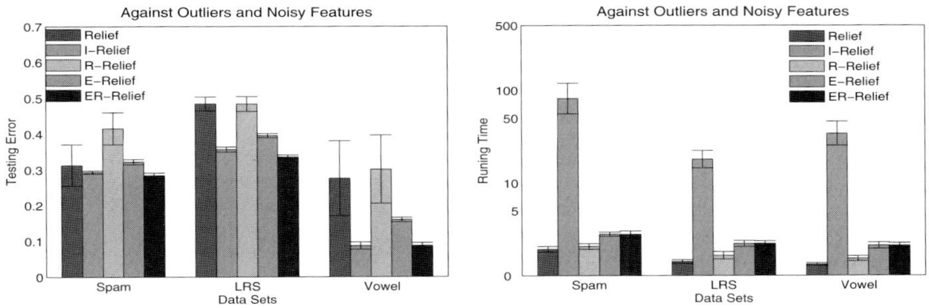

Fig. 3. Average testing errors and running times (sec.) as well as standard deviations on UCI data sets when both outliers and noisy features are involved

testing errors are shown in Fig.2 (middle line). The ROC curves are also plotted in Fig.2(bottom line). We can see that E-Relief performs comparably with I-Relief (much better than Relief) when noisy features are involved.

3. **Against outliers and noisy features.** E-Relief, R-Relief and their combination, ER-Relief, along with I-Relief and Relief are compared. 50 probes are added to each pattern (both training and testing), and 10% of training samples are mislabeled. The testing errors and running times of each algorithm, with average values and standard deviations, are shown by two bar plot, in Fig.3. We can see that in the presence of both outliers and noisy features, the performance of Relief is degraded badly. R-Relief and E-Relief do not necessarily improve the performance. However, their combination, ER-Relief, improves the performance drastically. In most cases, ER-Relief performs comparably with I-Relief. In some cases, it performs the best. With respect to computational efficiency, we can see that E-Relief, R-Relief and ER-Relief do not introduce a large increase of computational expense compared to Relief, while I-Relief is far more time-consuming.

3.2 Document Clustering and Categorization

We then apply the algorithms to document clustering and categorization tasks. For this purpose, six benchmark text data sets from **Trec** (the Text REtrieval Contest, http://trec.nist.gov) collection that are frequently used in information retrieval research are selected. The information of each data set is also summarized in Table.1.

The Relief, I-Relief and ER-Relief algorithms are compared, with no probe or mislabelling. For text clustering, C-mean algorithm is employed to get the clustering result after dimensionality reduction. For simplicity, the number of cluster, C, is set to be the true number of classes. For document categorization, the nearest-neighbor classifier is applied for final classification. Each experiment is repeated for 20 runs, each of which is based on a random splitting of the data set. The $Macro_{ave}F_1$ and $Micro_{ave}F_1$ are used to assess the classification results, and ARI (Adjusted Rand Index) and NMI (Normalized Mutual Information) are used to evaluate the clustering results. Table.2 presents the best average result of each algorithm.

Again, we observe that (i) ER-Relief performs much better than Relief, and that (ii) ER-Relief has achieved comparable performances comparably to I-Relief in most cases, although its computation complexity and operating time are much less than I-Relief. Note that the results about the running time are not given due to space limitation.

In information retrieval, *huge amount of data* and *extremely high dimensionality* are two core challenges (and are also becoming increasingly challenging). Therefore, the efficiency of ER-Relief as well as its effective ability to identify a small subset of predictive features (out of a huge amount of redundant ones) may make it a rather appealing and encouraging tool for both challenges, i.e., it is efficient with respect to data set size, and, it is able to effectively reduce the dimensionality. This confirms our attempting in applying ER-Relief to information retrieval tasks and encourages us to investigate its performance in extensive IR applications in the future.

Table 2. Comparison of feature weighting algorithms: Relief (RLF), I-Relief (IRLF) and ER-Relief (ERRL), in text categorization and clustering tasks. Best results are highlighted in bold.

	Categorization						**Clustering**					
	$Macro_{avg}F_1$			$Micro_{avg}F_1$			ARI			NMI		
	RLF	IRLF	ERRL	RLF	IRLF	ERRL	RLF	IRLF	ERRL	RLF	IRLF	ERRL
Trec11	**0.50**	0.43	0.45	**0.61**	0.54	0.57	**0.17**	0.13	0.11	**0.25**	0.16	0.16
Trec12	0.58	0.58	**0.64**	0.59	0.58	**0.60**	0.04	0.06	**0.07**	0.10	0.13	**0.15**
Trec23	0.49	**0.53**	0.42	0.62	**0.66**	0.59	0.04	**0.07**	0.05	0.09	**0.12**	0.10
Trec31	0.66	0.66	**0.71**	0.82	0.80	**0.86**	0.07	0.08	**0.09**	0.13	0.12	**0.15**
Trec41	0.65	**0.68**	0.64	0.74	0.77	**0.78**	0.16	**0.17**	**0.17**	0.20	**0.33**	0.29
Trec45	**0.63**	0.54	0.61	0.68	0.61	**0.71**	0.06	0.05	**0.07**	0.19	0.16	**0.21**

4 Conclusion

Fast growing internet data poses a big challenge for information retrieval. Feature selection, for the purpose of defying curse of dimensionality among others, plays a fundamental role in practice. Relief is an appealing feature selection algorithm. However, it lacks mechanisms to handle outliers and noisy features. In this paper, we have established two algorithms to address these two factor respectively. Compared with the recently proposed I-Relief, our algorithms are able to achieve comparable performance, while operating much more efficiently, which is proved by extensive experiments on various benchmark information retrieval tasks.

References

1. Bartlett, P., Jordan, M.I., McAuliffe, J.D.: Convexity, Classification and Risk Bounds. J. of American Stat. Assoc. 101(473), 138–156 (2006)
2. Dash, M., Liu, H.: Feature Selection for Classification. Intelligent Data Analysis (IDA) 1, 1131–1156 (1997)
3. Guyon, I., Elissee, A.: An Introduction to Variable and Feature Selection. JMLR 3, 1157–1182 (2003)
4. Hall, M.A., Holmes, G.: Benchmarking Attribute Selection Techniques for Discrete Class Data Mining. IEEE Trans. KDE 15(3), 1437–1447 (2003)
5. Kira, K., Rendell, L.A.: A Practical Approach to Feature Selection. In: Proc. of Ninth ICML, pp. 249–256 (1992)
6. Lin, Y.: A Note on Margin-based Loss Functions in Classification. Statistics and Probability Letters 68, 73–82 (2004)
7. Robnik-Šikonja, M., Kononenko, I.: Theoretical and Empirical Analysis of ReliefF and RRlief. J. Machine Learning 53(1-2), 23–69 (2003)
8. Shen, X., Tseng, G., Zhang, X., Wang, W.: On ψ-learning. J. of American Stat. Assoc., 724–734 (1998)
9. Sun, Y.J.: Iterative Relief for Feature Weighting: Algorithms, Theories, and Applications. IEEE Trans. PAMI 29(6), 1035–1051 (2007)
10. Yang, S.H., Hu, B.G.: Feature Selection by Nonparametric Bayes Error Minimization In: Proc. of the 12th PAKDD (2008)

Domain Adaptation for Conditional Random Fields

Qi Zhang, Xipeng Qiu, Xuanjing Huang, and Lide Wu

Department of Computer Science and Engineering, Fudan University
{qi_zhang,xpqiu,xjhuang,ldwu}@fudan.edu.cn

Abstract. Conditional Random Fields (CRFs) have received a great amount of attentions in many fields and achieved good results. However, a case frequently encountered in practice is that the test data's domain is different with the training data's. It would affect negatively the performance of CRFs. This paper presents a novel technique for maximum a posteriori (MAP) adaptation of Conditional Random Fields model. The background model, which is trained on data from a domain, could be well adapted to a new domain with a small number of labeled domain specific data. Experimental results on tasks of chunking and capitalizing show that this technique can significantly improve performance on out-of-domain data. In chunking task, the relative improvement given by the adaptation technique is 56.9%. With two in-domain sentences, it also can achieve 30.2% relative improvement.

1 Introduction

Conditional Random Fields (CRFs) are undirected graphical models that were developed for labeling relational data [1]. A CRF has a single exponential model for the joint probability of the entire sequence of labels given the observation sequence. Therefore, the weights of different features at different states can be traded off against each other. CRFs modeling technique has received a great amount of attentions in many fields, such as part-of-speech tagging [1], shallow parsing [2], named entity recognition [3,4], bioinfomatics [5], Chinese word segmentation [6,7], and Information Extraction [8]. It achieves good results in them.

Similar to most of the classification algorithms, CRFs also have the assumption that training and test data are drawn from the same underlying distributions. However, a case frequently encountered in practice is that the test data is drawn from a distribution that is related but not identical with the training data's. For example, one may wish to use a POS tagger trained with WSJ corpus to label email or bioinformatics research papers. This typically affects negatively the performance of a given model. From the experimental results we can know that the performance of the chunker trained with WSJ corpus can achieve 96.2% in different part of WSJ corpus. While performance of the same chunker in BROWN corpus is only 88.4%.

In order to achieve better results in a specific domain, labeled in-domain data is needed. Although large scale in-domain labeled corpus is hard to get, a small number of in-domain labeled data(*adaptation data*) and a large number of domain related labeled data(*background data*) is easier to get. For example Penn Treebanks [9] can be used as background training data for POS tagging, chunking, parsing and so on. This

H. Li et al. (Eds.): AIRS 2008, LNCS 4993, pp. 192–202, 2008.

kind of adaptation technique is used in many fields, such as language modelling [10], capitalization [11], automatic speech recognition [12], parsing [13,14] and so on.

Directly combining background and adaptation data together is a way to use the in-domain data. But if the scale of adaptation data is much smaller than the background data, the adaptation data's impact would be low. It can be seen from the experimental results. Another disadvantage of this method is that this technique need to retrain the whole model. It would waste a lot of time. In order to take advantage of the in-domain labeled data, a maximum a-posteriori (MAP) adaptation technique for Conditional Random Fields models is developed, following the similar idea with adaptation of Maximum Entropy [11]. The adaptation procedure proves to be quite effective in further improving the classification result on different domains. We evaluate the performance of this adaptation technique in chunking, capitalizing. The relative chunking's performance improvement of the adapted model over the background model is 56.9%. In capitalization task, the adapted model achieves 29.6% relative improvement.

The remainder of this paper is organized as follows: Section 2 describes the related works. The CRFs modeling technique is briefly reviewed in Section 3. Section 4 describes the MAP adaptation technique used for CRFs. The experimental results are presented in Section 5. Conclusions are presented in the last section.

2 Related Works

Leggetter and Woodland [12] introduced a method of speaker adaptation for continuous density Hidden Markov Models (HMMs). Adaptation statistics are gathered from the available adaptation data and used to calculate a linear regression-based transformation for the mean vectors.

Several recent papers also presented their works on modifying learning approaches-boosting [15], naive Bayes [16], and SVMs [17] - to use domain knowledge in text classification. Those methods all modify the base learning algorithm with manually converted knowledge about words.

Chelba and Acero [11] presented a technique for maximum a posteriori (MAP) adaptation of maximum entropy (MaxEnt) and maximum entropy Markov models (MEMM). The technique was applied to the problem of recovering the correct capitalization of uniformly cased text. Our work has similarities to Chelba and Acero's.

Daume and Marcu [18] presented a framework for domain adaptation problem. They treat the in-domain data as drawn from a mixture of "truly in-domain" distribution and a "general domain" distribution. Similarly, the out-of-domain are also drawn from a "truly out-of-domain" distribution and a "general domain" distribution. Then they apply EM method to estimate parameters. However, this framework used in CRF is computationally expensive.

3 Conditional Random Fields

Conditional Random Fields (CRFs) are undirected graphical models trained to maximize a conditional probability [1]). CRFs avoid a fundamental limitation of maximum entropy Markov models (MEMMs), which can be biased towards states with few successor states.

Let $X = x_1...x_n$ and $Y = y_1...y_n$ represent the generic input sequence and label sequence. The cliques of the graph are now restricted to include just pairs of states (y_{i-1}, y_i) that are neighbors in the sequence. Linear-chain CRFs thus define the conditional probability of a state sequence given an input sequence to be

$$P_\Lambda(Y|X) = \frac{1}{Z_x} \exp\left(\sum_{i=1}^{n}\sum_{k=1}^{m} \lambda_k f_k(y_{i-1}, y_i, x, i)\right)$$

where Z_x is a normalization factor over all state sequences, $f_k(y_{i-1}, y_i, x, i)$ is an arbitrary feature function over its arguments, and λ_k (ranging from $-\infty$ to ∞) is a learned weight for each feature function. A feature function is either a state feature $s(y_i, x, i)$ or a transition feature $t(y_{i-1}, y_i, x, i)$.

Then, the CRF's global feature vector for input sequence X and label sequence Y is given by

$$F(Y, X) = \sum_i f(y_{i-1}, y_i, x, i)$$

where i ranges over input positions. Using the global feature vector, $P_\Lambda(Y|X) = \frac{1}{Z_X} \exp(\Lambda \cdot F(Y, X))$. The most probable path $\hat{Y}$ for input sequence X is then given by

$$\hat{Y} = \arg \max_{Y \in Y(x)} P(Y|X) = \arg\max_{Y} \lambda \cdot F(Y, X)$$

which can be found by Viterbi algorithm.

3.1 Parameter Estimation

CRFs can be trained by the standard maximum likelihood estimation, i.e., maximizing the log-likelihood $\mathcal{L}_\Lambda$ of a given training set $T = \{< X_j, Y_j >\}_{j=1}^{N}$.

$$\hat{\Lambda} = \arg \max_{\Lambda \in \mathbb{R}^k} \mathcal{L}_\Lambda,$$

where

$$\begin{aligned}\mathcal{L}_\Lambda &= \sum_j \log(P(Y_j|X_j)) \\ &= \sum_j \left[\Lambda \cdot F(Y_j, X_j) - \log(Z_{X_j})\right].\end{aligned}$$

To perform the optimization, we seek the zero of the gradient

$$\begin{aligned}\frac{\partial \mathcal{L}_\Lambda}{\partial \lambda_k} &= \sum_j \left(F_k(Y_j, X_j) - E_{P(Y|X)}[F_k(Y, X_j)]\right) \\ &= O_k - E_k = 0,\end{aligned}$$

where $O_k = \sum_j F_k(Y_j, X_j)$ is the count of feature k observed in the training data T, and $E_k = E_{P(Y|X)}[F_k(Y, X_j)]$ is the expectation of feature k over the model distribution $P(Y|X)$ and T. The expectation can be efficiency calculated using a variant of the forward-backword algorithm.

$$E_{P(Y|X)}[F_k(Y, X)] = \sum_i \frac{\alpha_i (f_i * M_i) \beta_i^T}{Z_X}$$

$$Z_X = \alpha_n \cdot 1^T$$

where α_i and β_i are the forward and backward state-cost vectors defined by

$$\alpha_i = \begin{cases} \alpha_{i-1} M_i & 0 < i \leq n \\ 1 & i = 0 \end{cases}$$

$$\beta_{i-1}^T = \begin{cases} M_{i+1} \beta_{i+1}^T & 0 \leq i < n \\ 1 & i = n \end{cases}$$

To avoid over fitting, we also use Gaussian weight prior [19]:

$$\mathcal{L}_\lambda{}' = \sum_j \log(P(Y_j|X_j) - \frac{\|\lambda\|^2}{2\sigma^2} + const$$

with gradient

$$\nabla \mathcal{L}_\lambda{}' = O_k - E_k - \frac{\lambda}{\sigma^2}$$

The optimal solutions can be obtained by using traditional iterative scaling algorithms (e.g., IIS or GIS [20]) or quasi-Newton methods(e.g., L-BFGS [21]).

4 MAP Adaptation of Conditional Random Fields

The overview of adaptation stages is shown in Figure 1. A simple way to accomplish this is to use MAP adaptation using a prior distribution on the model parameters [11]. A Gaussian prior for the model parameters Λ has been previously used to smooth CRFs models. The prior has 0 mean and diagonal covariance: $\Lambda \sim \mathcal{N}(0, diag(\sigma_i^2))$. In the adaptation part, the prior distribution is centered at the parameter Λ^0 estimated from the background data:$\Lambda \sim \mathcal{N}(\Lambda^0, diag(\sigma_i^2))$. For the features generated only from the adaptation, the prior distribution is still centered at 0. In our experiments the variances were tied to $\sigma_i = \sigma$ whose value was determined by line search on development data drawn from the background data or adaptation data.

Different from the Chelba and Acero's method [11], we use both σ_a and σ_m here. In their method, σ is used not only to balance the background and adaptation data, but also to represent the variance of the adaptation data. However they are different in most of circumstance. In order to overcome this problem we use two σ in adaptation step.

The log-likelihood $\mathcal{L}_\Lambda$ of the given adaptation data set becomes:

$$\mathcal{L}_\lambda' = \sum_j \log(P(y_j|x_j) - \sum_{i=1}^{F_{background}} \frac{\|\lambda_i - \lambda_i^0\|^2}{2\sigma_m^2} - \sum_{i=1}^{F_{adaptation}} \frac{\|\lambda_i\|^2}{2\sigma_a^2}$$

Therefore the gradient becomes:

$$\nabla\mathcal{L}_\lambda' = O_k - E_k - \sum_{i=1}^{F_{background}} \frac{\lambda_i - \lambda_i^0}{\sigma_m^2} - \sum_{i=1}^{F_{adaptation}} \frac{\lambda_i}{\sigma_a^2},$$

where $F_{background}$ is the features generated from the background data, $F_{adaptation}$ is the features generated only from the adaptation data, σ_a represents the variance of the adaptation data, and σ_m is used to balance the background and adaptation model. A small variance σ_m will keep the weight λ_m close to the background model, while a large variance σ_m will make the model sensitive to adaptation data. With $\mathcal{L}_\lambda'$ and $\nabla\mathcal{L}_\lambda'$, λ can be iteratively calculated through L-BFGS.

Algorithm MAP Adaptation of CRFs
$\mathcal{F}_{background}$ = Feature set generated from background data
$\mathcal{F}_{adaptation}$ = Feature set generated from adaptation data
λ_i = f_i's corresponding weight

Generate $\mathcal{F}_{background}$ from background data

Estimate λ_i^0 for $\mathcal{F}_{background}$

Generate $\mathcal{F}_{adaptation}$ from adaptation data

Let $\mathcal{F} = \mathcal{F}_{background} \bigcup \mathcal{F}_{adaptation}$

Let $\lambda_i = \lambda_i^0$ if $f_i \in \mathcal{F}_{background}$
$\lambda_i = 0$, otherwise

Estimate λ_i with equation $\mathcal{L}_\lambda'$ and $\nabla\mathcal{L}_\lambda'$

Fig. 1. Algorithm of MAP Adaptation of CRFs

Table 1. Feature templates used by Chunker

type	template
Base features	w_{-2}, w_{-1}, w_0,w_1, w_2
	p_{-2}, p_{-1}, p_0,p_1, p_2
Bi-gram features	$w_{-2}w_{-1}$, $w_{-1}w_0$,
	w_0w_1,w_1w_2
	$p_{-2}p_{-1}$, $p_{-1}p_0$,
	p_0p_1,p_1p_2
	$p_{-2}w_{-1}$, $p_{-1}w_0$,
	p_0w_1,p_1w_2
	$w_{-2}p_{-1}$, $w_{-1}p_0$,
	w_0p_1,w_1p_2
Tri-gram features	$w_{-2}w_{-1}w_0$, $w_{-1}w_0w_1$,
	$w_0w_1w_2$
	$p_{-2}p_{-1}p_0$, $p_{-1}p_0p_1$,
	$p_0p_1p_2$
	$p_{-2}p_{-1}w_0$, $p_{-1}w_0p_1$,
	$p_0w_1p_2$
	$w_{-2}w_{-1}p_0$, $w_{-1}p_0w_1$,
	$w_0p_1w_2$

w_0 is the word at current position, w_1 is the word instant after w_0, w_{-1} is the word instant before it, p_* represents word's POS tags.

5 Experiments

To evaluate the MAP adaptation of CRFs, we did several experiments on chunking and capitalizing. Penn Treebanks III [9] is used to train chunker. Capitalizer's training data comes from Tipster corpus [22]. We will introduce the detail steps and features used in the following parts.

5.1 Experiments on Chunking

The goal of chunking is to group sequences of words together and classify them by syntactic labels. Various NLP tasks can be seen as a chunking task, such as English base noun phrase identification (base NP chunking), English base phrase identification (chunking), and so on. Because chunking technique is used in many different fields, we choose chunking task to evaluate the adaptation methods.

The background data used for chunker is generated from WSJ data(wsj_0200.mrg - wsj_2172.mrg). The in-domain test data is from wsj_0000.mrg to wsj_0199.mrg. The others are used to tune parameters. Bracketed representation is converted into IOB2 representation [23,24].

For adaptation experiments we use BROWN data in Penn Treebanks III. As Brown Corpus dataset contains eight types of articles, we extract one article from each type(C* _01.mrg), which are used as adaptation data. The second articles from each type(C*_02. mrg) are used as development data. The others are used for evaluations.

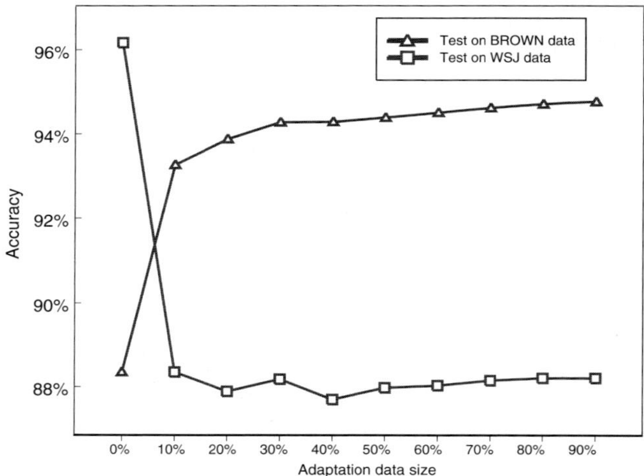

Fig. 2. The impact of the adaptation data's size

Table 2. Chunking Results on in-domain(WSJ) and out-of-domain data(BROWN)

Background data	Adaptation data	Evaluation data	Accuracy
WSJ	NONE	B-tst	88.4%
WSJ	B-ada	B-tst	**95.0%**
WSJ	NONE	wsj-tst	**96.2%**
WSJ	B-ada	wsj-tst	88.4%

where "wsj-tst" represents the test part of WSJ, "B-tst" represents the test part of BROWN.

The templates used in chunking experiments are shown is Table 1.

Results of both in-domain and out-of-domain are shown in Table 2. The σ^2 used in background model is selected by in-domain development data. σ_a^2, and σ_m^2 are selected by development data extracted from BROWN data. From the result we observe that the performance of background model in in-domain data is significantly better than in out-of-domain data. Adaptation improves the performance on Brown data by 56.9% relative.

Figure 2 shows the result of the impact of the adaptation data's size. X axis represents the percentage of the adaptation data in BROWN corpus(B-ada). Y axis represents the accuracy. Two lines represent the results of test data set on BROWN (B-tst) and WSJ (wsj-tst) corpus. The result in 0% is got by the background model. The result in 10% is got by the model adapted by 10 percents B-ada data. We observe from the result that the larger adaptation data are used the higher accuracy in this domain could be get. When the size of the adaptation data is very small, this technique can also achieve good result. We use two sentences extracted from B-ada data to adapt the background model. The adapted model also achieves 30.2% relative improvement.

Then we evaluate the impact of σ_m^2 to the performance. The result is shown in Figure 3. X axis represents σ_m^2. Y axis represents the accuracy. As expected low values of

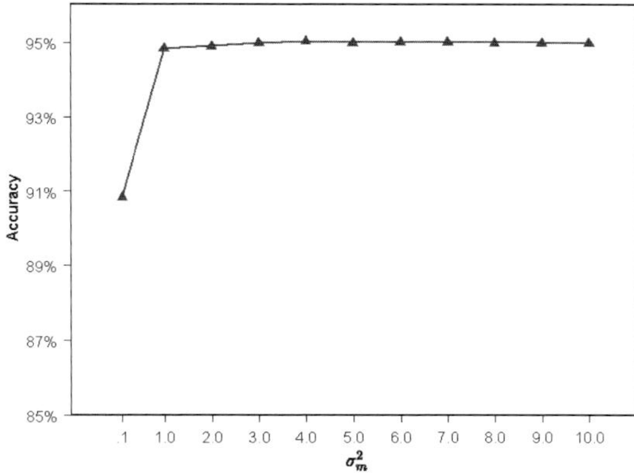

Fig. 3. The impact of the σ_m^2

σ_m^2 result little adaptation. When the σ_m^2 is between 1 and 10, the accuracy does have significant changes. Therefore this parameter can be easily set in the real system.

5.2 Experiments on Capitalizing

Capitalization can be converted to sequence tagging problem. Each low case words receive a tag which represents its capitalization form. It's also domain dependent. For example, in bioinformatics domain "gene" is almost low case form. It represents a concept. While in some domains, "gene" is usually capitalized, which represents a human name. Therefor we did some experiments to show the impact of model adaptation technique on this task.

The TIPSTER copra are used to generate both background and adaptation data for the capitalizer. The background data is WSJ data from 1987 - files from WSJ7_001 to WSJ7_127 in TIPSTER Phrase I. The in-domain test data is WSJ_0402 and WSJ_0403, which belong to WSJ 1990 in TIPSTER Phrase II. WSJ_0404 and WSJ_0405 are in-domain development data. The out-of-domain adaptation data is the combination of AP880212 and AP880213, which belong Associated Press 1988 in TIPSTER Phrase II. Files AP880214 and AP880215 are out-of-domain test data.

We use the same tag set with the set used in [11]. Each word in a sentence is labeled with one of the tags:

- `LOC` lowercase
- `CAP` capitalized
- `MXC` mixed case; no further guess is made as to the capitalization of such words.
- `AUC` all upper case
- `PNC` punctuation;

The feature templates we used are shown in table 3.

Table 4 shows results of in-domain and out-of-domain data. The σ^2 in background model we use in this experiment is 5, which is selected by development data. The σ_a^2

Table 3. Feature templates used by Capitalizer

type	template
Base features	w_{-1}, w_0,w_1
Bi-gram features	$w_{-1}w_0$, w_0w_1

used in adaptation part is set to 5. The σ_m^2 is 10. We can get the same trend with chunking's results. The adapted model gives 29.6% relative improvement. The size of adaptation data is less than 1% of the background WSJ data's size.

Table 4. Capitalizing Results on in-domain and out-of-domain data

Background data	Adaptation data	Evaluation data	Accuracy
WSJ	NONE	AP-tst	94.6%
WSJ	AP-ada	AP-tst	**96.2%**
WSJ	NONE	wsj-tst	**96.8%**
WSJ	AP-ada	wsj-tst	96.4%
WSJ+AP-ada	NONE	AP-tst	94.7%
WSJ+AP-ada	NONE	wsj-tst	96.8%

where "wsj-tst" represents the test part of WSJ,"AP-ada" represents the adaptation data, "AP-tst" represents the test part of Associated Press.

Then we combine adaptation data(AP-ada) with the background data(WSJ) and train a capitalizer with it. The accuracy of capitalizing wsj-tst is 96.8%. In AP-tst data, the accuracy is 94.7%. Comparing with the results got by background model, the capitalizer trained by combined data couldn't significantly improve the performance.

6 Conclusions

In this paper we present a novel technique for maximum a posteriori (MAP) adaptation of Conditional Random Fields Model. Through experimental results,we observe that this technique can effectively adapt a background model to a new domain with a small amount of domain specific labeled data. We did several experiments in three different fields: chunking and capitalizing. The relative chunking's performance improvement of the adapted model over the background model is 56.9%. With two in-domain sentences, it also can achieve 30.2% relative improvement. The relative improvement of capitalizing experiment is 29.6%. The experimental results prove that the MAP adaptation of Conditional Random Fields Model technique can benefit the performances in different tasks.

Acknowledgements

This work was partially supported by Chinese NSF 60435020 and 60673038. The authors would like to thank Bingqing Wang and Lin Zhao for their carefully proofreading. Thanks the anonymous reviewers for their valuable comments and suggestions.

References

1. Lafferty, J., McCallum, A., Pereira, F.: Conditional random fields: Probabilistic models for segmenting and labeling sequence data. In: Proc. 18th International Conf. on Machine Learning (2001)
2. Sha, F., Pereira, F.: Shallow parsing with conditional random fields. In: Proceedings of Human Language Technology-NAACL 2003 (2003)
3. Carreras, X., Márquez, L., Padró, L.: Learning a Perceptron-Based Named Entity Chunker via Online Recognition Feedback. In: Association with HLT-NAACL 2003 (2003)
4. Okanohara, D., Miyao, Y., Tsuruoka, Y., Tsujii, J.: Improving the scalability of semi-markov conditional random fields for named entity recognition. In: Proceedings of COLING-ACL 2006 (2006)
5. Settles, B.: Biomedical named entity recognition using conditional random fields and rich feature sets. In: COLING 2004 International Joint workshop on Natural Language Processing in Biomedicine and its Applications (NLPBA/BioNLP 2004) (2004)
6. Peng, F., Feng, F., McCallum, A.: Chinese Segmentation and New Word Detection using Conditional Random Fields. In: Proceedings of COLING 2004 (2004)
7. Feng, Y., Sun, L., Lv, Y.: Chinese word segmentation and named entity recognition based on conditional random fields models. In: Proceedings of the Fifth SIGHAN Workshop on Chinese Language Processing (2006)
8. Peng, F., McCallum, A.: Accurate information extraction from research papers using conditional random fields. In: HLT-NAACL 2004: Main Proceedings (2004)
9. Marcus, M.P., Santorini, B., Marcinkiewicz, M.A.: Building a large annotated corpus of english: The penn treebank. Computational Linguistics 19, 313–330 (1993)
10. Clarkson, P., Robinson, A.J.: Language model adaptation using mixtures and an exponentially decaying cache. In: Proc. ICASSP 1997 (1997)
11. Chelba, C., Acero, A.: Adaptation of maximum entropy capitalizer: Little data can help a lot. In: Proceedings of EMNLP 2004 (2004)
12. Leggetter, C., Woodland, P.: Maximum likelihood linear regression for speaker adaptation of continuous density hidden markov models. Journal of Computer Speech and Language (1995)
13. McClosky, D., Charniak, E., Johnson, M.: Reranking and self-training for parser adaptation. In: Proceedings of COLING-ACL 2006 (2006)
14. Lease, M., Charniak, E.: Parsing biomedical literature. In: Dale, R., Wong, K.-F., Su, J., Kwong, O.Y. (eds.) IJCNLP 2005. LNCS (LNAI), vol. 3651, Springer, Heidelberg (2005)
15. Schapire, R.E., Rochery, M., Rahim, M.G., Gupta, N.: Incorporating prior knowledge into boosting. In: Proceedings of the ICML 2002 (2002)
16. Liu, B., Li, X., Lee, W.S., Yu, P.S.: Text classification by labeling words. In: Proceedings of the Eighteenth National Conference on Artificial Intelligence (2005)
17. Wu, X., Srihari, R.: Incorporating prior knowledge with weighted margin support vector machines. In: Proceedings of the tenth ACM SIGKDD (2004)
18. Daumé III, H., Marcu, D.: Domain Adaptation for Statistical Classifiers. Journal of Artificial Intelligence Research (2006)
19. Chen, S.F., Rosenfeld, R.: A gaussian prior for smoothing. maximum entropy models. Technical Report CMU-CS-99-108 (1999)
20. Della-Pietra, S., Della-Pietra, V., Lafferty, J.: Inducing features of random fields. IEEE Transactions on PAMI (1997)

21. Liu, D.C., Nocedal, J.: On the limited memory BFGS method for large scale optimization. Mathematical Programming
22. Harman, D., Liberman, M.: Tipster complete. In: Linguistic Data Consortium catalog number LDC93T3A and ISBN: 1-58563-020-9 (1993), http://www.ldc.upenn.edu/Catalog/CatalogEntry.jsp?catalogId=LDC93T3A
23. Ramshaw, L., Marcus, M.: Text chunking using transformation-based learning. In: Proceedings of the Third Workshop on Very Large Corpora, Somerset, New Jersey, pp. 82–94 (1995)
24. Sang, E.F.T.K., Veenstra, J.: Representing Text Chunks

Graph Mutual Reinforcement Based Bootstrapping

Qi Zhang, Yaqian Zhou, Xuanjing Huang, and Lide Wu

Department of Computer Science and Engineering, Fudan University
{qi_zhang,xpqiu,xjhuang,ldwu}@fudan.edu.cn

Abstract. In this paper, we present a new bootstrapping method based on Graph Mutual Reinforcement (GMR-Bootstrapping) to learn semantic lexicons. The novelties of this work include 1) We integrate Graph Mutual Reinforcement method with the Bootstrapping structure to sort the candidate words and patterns; 2) Pattern's uncertainty is defined and used to enhance GMR-Bootstrapping to learn multiple categories simultaneously. Experimental results on MUC4 corpus show that GMR-Bootstrapping outperforms the state-of-the-art algorithms. We also use it to extract names of automobile manufactures and models from Chinese corpus. It achieves good results too.

1 Introduction

Learning semantic lexicons is the task of automatically acquiring words with semantic classes (e.g. "gun" is a WEAPON). It has been proved to be useful for many natural language processing tasks, including question answering [1,2], information extraction [3] and so on. In recent years, several algorithms have been developed to automatically build semantic lexicons using supervised or semi-supervised methods [4,5,6,7]. As unsupervised method dispenses with the manually labeled training data, more and more methods have been proposed [8,9,10,11,12,13]. There exist some semantic dictionaries (e.g., WordNet [14], HowNet [15]). However most of them don't contain resources from specialized domain.

In this paper we propose a weakly supervised learning method, GMR-Bootstrapping, to learn semantic lexicons. It begins with unlabeled corpus and a few of seed words. Then it automatically generates a lists of words with the same category with seed words. From analyzing procedure of the similar bootstrapping algorithm, Basilisk [10], we found that some of the good patterns are given very low score at the beginning stage by Basilisk. Because Basilisk's scoring functions give low scores to all the patterns with large amount of extractions at the beginning stage. However some of them are good ones, whose extractions almost belong to the same category. Therefore we incorporate Graph Mutual Reinforcement to weight candidate words and extraction patterns, in order to partially overcome this problem. Evaluations on MUC4 corpus [16] show that incorporating Graph Mutual Reinforcement to weight the candidate words and extraction patterns enables substantial performance gains in extracting BUILDING, EVENT, HUMAN, LOCATION, TIME and WEAPON lexicons.

Another novelty of GMR-Bootstrapping is that pattern's uncertainty is added into scoring functions to learn multiple categories simultaneously. Normally, if a pattern's extractions belong to several different categories, the pattern's correctness should be

H. Li et al. (Eds.): AIRS 2008, LNCS 4993, pp. 203–212, 2008.

low. In order to use this information, we integrate a scoring functions to measure pattern's uncertainty. The experimental results show that adding patterns' uncertainty into scoring functions improves the performance too.

We also evaluate GMR-Bootstrapping method on extracting automobile manufacture names and automobile models Chinese corpus (details in Section 4). From the experimental results we also observe that the quality of lexicons of extracted by GMR-Bootstrapping is better than quality of lexicons extracted by Basilisk.

The reminder of the paper is organized as follows: Section 2 discussed the related works. In section 3, we introduce our bootstrapping structure and scoring functions. In section 4, experiments are given to show the improvements. Section 5 concludes the paper.

2 Related Work

Several weakly supervised classifier algorithms have been proposed to learning semantic lexicons with a small set of labeled data and a large number of unlabeled data, such as Co-training and Bootstrapping. Co-training [17] alternately learns using two orthogonal views of data in order to utilize unlabeled data. This enables bootstrapping from a small set of labeled training data via a large set of unlabeled data.

Meta-bootstrapping [18] is a Bootstrapping algorithm that uses a two layer bootstrapping structure to learn a dictionary of extraction patterns and a domain specific semantic lexicon. Snowball [19], a system for extracting relations from large collections of plain-text documents, uses standard bootstrapping structure and introduces novel techniques for evaluating the quality of the patterns and tuples generated at each step of the extraction process. The KnowItAll [12] utilizes a set of domain-independent extraction patterns to generate candidate facts. Then the candidate facts are evaluated by point wise mutual information (PMI) statistics. Hassan et al.(2006) presented an unsupervised method, which does not require seeds or examples. Instead, it depends on redundancy in large data sets and graph based mutual reinforcement to acquire extraction patterns.

The algorithm most closely related to our method is Basilisk [10], which is also a bootstrapping algorithm. While meta-bootstrapping trusts individual extraction patterns to make unilateral decisions, Basilisk gathers collective evidence from a large set of extraction patterns. We also use the same idea and structure. While there are some differences between GMR-Bootstrapping and Basilisk. Firstly, our method incorporates Graph Mutual Reinforcement to weight candidate words and extraction patterns. Another difference is that we enhance the GMR-Bootstrapping with pattern's uncertainty to learn multiple categories simultaneously.

3 GMR-Bootstrapping

GMR-Bootstrapping is a weakly supervised learning method. It is used to generate semantic lexicons. The input to GMR-Bootstrapping are few manually selected *seed words* for each semantic category and an unlabeled text corpus. Figure 1 shows the structure of GMR-Bootstrapping process. In this section, we describe details of GMR-Bootstrapping algorithm.

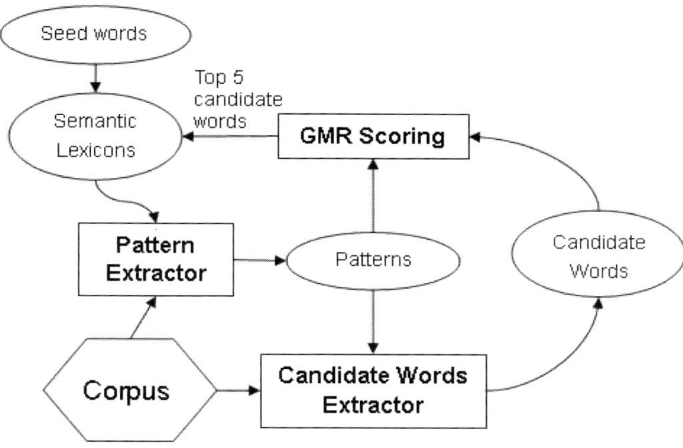

Fig. 1. GMR-Bootstrapping Algorithm

3.1 Structure and Algorithm of GMR-Bootstrapping

As shown in Figure 1, the GMR-Bootstrapping begins by extracting a number of the extraction patterns that can match the seed words. After that candidates for the lexicon are extracted with these patterns. Then a bipartite graph (Figure 2) is built, which represents the matching relation between patterns and candidate words. Finally GMR Scoring is applied to iteratively assign correctness weights of patterns and candidate words. The five best candidate words are added to the lexicon.Then process starts over again.

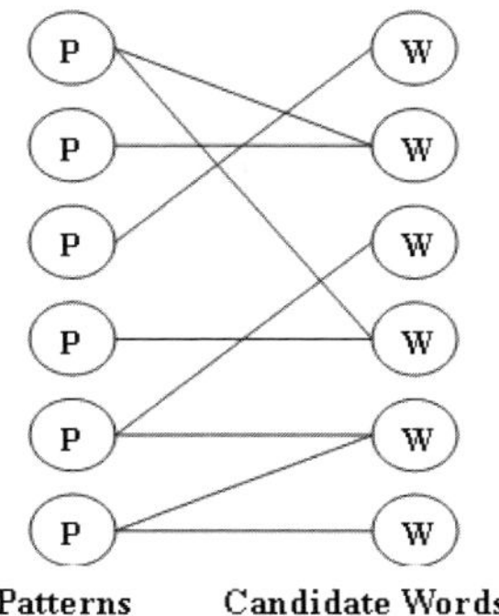

Fig. 2. A bipartite graph representing patterns and candidate words

3.2 Pattern

In order to find new lexicon entries, extraction patterns are used to provide contextual evidence that a word belongs to which semantic classes. There are two commonly used patterns, ***Syntactic Pattern*** and ***Context Pattern***. Both of them are used in our experiments.

Syntactic Pattern. Syntactic Pattern is used by many other Bootstrapping methods [20,18,10]. We follow Thelen and Riloff (2002)'s methods, which used the AutoSlog system [21], to represent extraction patterns. AutoSlog's extraction patterns represent linguistic expressions that extract a noun phrase's head noun in one of three syntactic roles: subject, direct object, or prepositional phrase object. For example, three patterns that would extract weapon are: [subject]was fired, carry [direct object], wounded by [pp object].

Context Pattern. Different from the Syntactic Pattern, Context Pattern uses words only. Syntactic roles are not included in it. In our implementations, if X_0 belongs to lexicons, a number of patterns will be generated based on the following templates in MUC4 corpus:

$$X_{-3}X_{-2}X_{-1}[X_0],\quad X_{-2}X_{-1}[X_0],$$
$$[X_0]X_1X_2\,,\quad [X_0]X_1X_2X_3.$$

X_{-1} represents the word immediately before current word,while X_1 represents the word immediately after current word.

Different from English, Chinese words are not delimited by spaces. There is no clue to tell where the word boundaries are. Automobile manufacture names and automobile models names can not be segmented well by most of the state-of-the-art Chinese word segmenters. Therefore, patterns we used in Chinese corpus contain characters before and after the target words. Chinese word segmenter could be omitted through this kind of patterns. The following templates are used in Chinese corpus:

$$X_{-3}X_{-2}X_{-1}[ext]X_1X_2X_3,$$
$$X_{-2}X_{-1}[ext]X_1X_2,$$
$$X_{-2}X_{-1}[ext]X_1X_2X_3,$$
$$X_{-3}X_{-2}X_{-1}[ext]X_1X_2.$$

where X_* represents the Chinese character. If a article fragment matches both the part before and after the $[ext]$, the characters between them are extracted as candidate word.

3.3 GMR Scoring

GMR Scoring is used to iteratively assign scores of patterns and candidate words. We assume that patterns that match many words from the same category tend to be important. Similarly, words matched by many patterns that belong to same category tend to be correct. A bipartite graph is build to represent the relation between pattern and candidate word. Each pattern or candidate word is represented by a node in the graph. Edges represent matching between patterns and candidates. Then the problem is transferred to hubs (patterns) and authorities (words) problem which can be solved using the Hypertext Induced Topic Selection (HITS) algorithm [22].

Each pattern *p* in *P* is associated with a weight $sp(p)$ denoting its correctness. Each candidate words *w* in *W* has a weight $sw(w)$ which express the correctness of the word. The weights are calculated iteratively through equation 1 to equation 5 as follows:

$$F^{(i+1)}(p) = \sum_{u \in W(p)} sw^{(i)}(u) \tag{1}$$

$$sp^{(i+1)}(p) = \frac{F^{(i+1)}(p) \cdot \log F^{(i+1)}(p)}{|W(p)| \cdot SP^{(i)}} \tag{2}$$

$$sw^{(i+1)}(w) = \frac{\sum_{p \in P(w)} \log \left(F^{(i+1)}(p) + 1\right)}{|P(w)| \cdot SW^{(i)}} \tag{3}$$

where $sw(u)$ is initialized to 1 if $u \in Semantic Lexicons$ and to 0 if $u \notin SemanticLexicons$, $W(p)$ is the set of words matched by p, $P(w)$ is the set of patterns matching w, $sp^{(i)}(p)$ is the correctness weight of pattern p in the iteration i, and $sw^{(i)}(w)$ is the correctness weight of word w in the iteration i, $SW^{(i)}$ and $SP^{(i)}$ are the normalization factors defined as:

$$SW^{(i)} = \sum_{p=1}^{|P|} \sum_{u=1}^{W(p)} sw^{(i)}(u) \tag{4}$$

$$SP^{(i)} = \sum_{w=1}^{|W|} \sum_{v=1}^{P(w)} sp^{(i)}(v) \tag{5}$$

Equation 2 is similar with $RlogF$, which has been used to score patterns [10], except that we changed the F_i in $RlogF$ to Equation 1. $RlogF$ is

$$RlogF(pattern_i) = \frac{F_i}{N_i} * log(F_i)$$

where F_i is the number of category members extracted by $pattern_i$ and N_i is the total number of nouns extracted by $pattern_i$. From the definition of $RlogF$, we observe that the patterns which contain a large amount of extractions would be given low scores by $RlogF$ at the beginning stage. Although some of them are good ones. Equation 3 is changed from $AvgLog$, which has been used to score candidate words [10]. Through those changes the scoring functions of pattern and candidate words are connected and can be iteratively calculated. The scores of good patterns with amount extractions will increase through iterations. The problem of $RlogF$ could be partially overcame.

3.4 Learning Multiple Semantic Categories

From and Thelen and Riloff (2002)'s analysis and results of Basilisk-MACT+ [10], we observe that learning multiple semantic categories can improve all the categories' results. We also extend GMR-Bootstrapping to learn multiple semantic categories simultaneously (GMR-M-Bootstrapping). Normally, if extractions of a pattern belong to several different categories, the pattern's correctness should be low.

We use L_p to represent labels of pattern p's extractions, $H(L_p)$ is the entropy of L_p, which is calculated only in the extractions which have been labeled to a semantic category. For example: pattern p, whose extractions are $w_1, w_2, ...w_n$. We can find its extractions' labels through semantic lexicons at this stage.$L_p = l_1, l_2, ..., l_n$, where

$$l_i = \begin{cases} Label_j\text{,if } w_i \in Lexicon_j; \\ NULL\text{,otherwise.} \end{cases}$$

Then the entropy of L_p is calculated through Equation 6.

$$H(L_p) = -\sum_{k=1}^{|\chi|} p(Label_k) \cdot log(p(Label_k)) \quad (6)$$

where $p(Label_k) = \frac{C_k}{\sum_{k=1}^{|\chi|} C_k}$, C_k denotes the number of times $Lable_k$ occurs in L_p. Through we could define the patterns' uncertainty, $(1 - \frac{H(L_p)}{\log_2 |\chi|})$, which varies from 0 when L_p is uniform to 1 when L_p contains one types of labels [23].

Therefore equation 2 and 3 are changed into:

$$sp^{(i+1)}(p) = \frac{F^{(i+1)}(p) \cdot \log F^{(i+1)}(p) \cdot (1 - \frac{H(L_p)}{\log_2 |\chi|})}{|W(p)| \cdot SP^{(i)}} \quad (7)$$

$$sw^{(i+1)}(w) = \frac{\sum_{p \in P(w)} \log (F^{(i+1)}(p) + 1) \cdot (1 - \frac{H(L_p)}{\log_2 (|\chi|)})}{|P(w)| \cdot SW^{(i)}} \quad (8)$$

which are modified by multiplying patterns' uncertainty. The experiments and results using E.q 7 and E.q 8 are shown in Section 4.

4 Experiments

To compare the performance of GMR-Bootstrapping with other weakly supervised methods, we design several experiments to evaluate with two corpora. One is the MUC-4 corpus [16], which contains 1700 texts (includes both test and training parts) in terrorism domain. All the words in the corpus are divided into nine semantic categories [10]: BUILDING, EVENT, HUMAN, LOCATION, ORGANIZATION, TIME, VEHICLE, WEAPON and OTHER. A few of semantic lexicon learners have previously been evaluated on the this corpus [20,4,18,10], and of these Basilisk achieved the best results. We reimplemented the Basilisk algorithm to compare it with GMR-Bootstrapping. Another one, CRCV (Chinese Review Corpus about Vehicle), which was collected by ourselves, contains about 500,000 articles in around 500 automobile domain forums. All the articles are reviews about vehicle. GMR-Bootstrapping and Basilisk are used to learn MANUFACTURE and MODEL categories in this corpus.

4.1 Results in MUC4 Corpus

Figure 3 shows results of Repeated Basilisk (R-Basilisk) with different patterns, Context Pattern (*CP*) and Syntactic Pattern (*SP*). For each category, 10 most frequent nouns that belong to the category are extracted as seed words. It is the same as Thelen and Riloff (2002)'s way. We ran the algorithm with different patterns for 200 iterations , so 1000 words are extracted. The X axis shows the number of words were extracted. The Y axis shows the number of correct ones. From the results we know that the performance of SP is better than CP's in all categories except the time category. The results reflect that syntactic roles are useful to learn semantic lexicons. Experimental results also show that the R-Basilisk's performance is similar with the Basilisks' results reported by Thelen and Riloff (2002) in all categories.

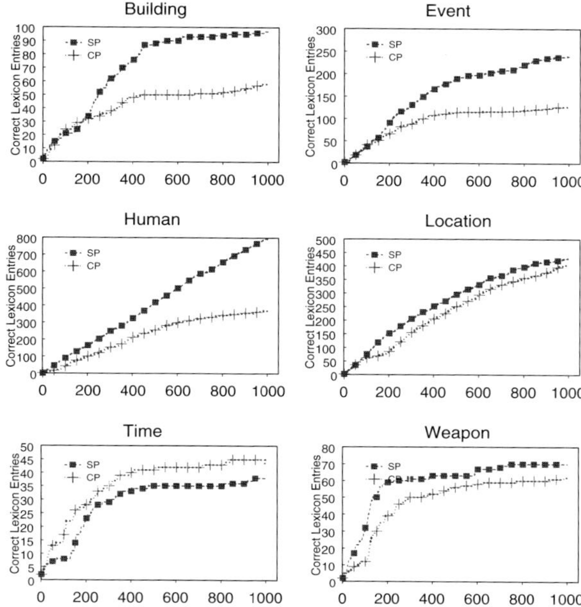

Fig. 3. Repeated Basilisk (R-Basilisk) Results with different Patterns

Table 1. The impact of number of Seed words

#Seed Words	GMR-Bootstrapping	R-Basilisk	#Seed Words	GMR-Bootstrapping	R-Basilisk
building			**location**		
2	78	58	2	452	397
5	114	98	5	452	435
10	109	94	10	463	436
50	75	58	50	426	452
event			**time**		
2	250	70	2	39	14
5	270	158	5	50	25
10	258	238	10	46	39
50	240	222	50	16	17
human			**weapon**		
2	691	456	2	72	39
5	786	702	5	80	68
10	821	795	10	86	72
50	793	758	50	45	36

The next experiment we did is to evaluate the impact of the number of seed words on both our method and Basilisk. Table 1 shows the result. The left-hand column represents the number of seed words given and each cell represents the number of correct words

Table 2. GMR-M-Bootstrapping vs Basilisk-MACT+

Category	GMR	GMR-M	RI	B-MACT+
building	109	**115**	5.50%	109
event	258	**273**	5.81%	266
human	821	**852**	3.78%	829
location	463	**510**	10.15%	509
time	46	**48**	4.35%	45
weapon	86	**92**	6.98%	88

Table 3. GMR-Bootstrapping vs Basilisk in Chinese CRCV Corpus

Category	Total Words	GMR	GMR-M	R-Basilisk
manufacture	100	47	**61**	44
	200	67	**82**	61
	500	93	**112**	87
	944	107	**120**	92
model	100	63	**71**	50
	200	115	**121**	66
	500	196	**216**	131
	1000	253	**299**	210

learned by different algorithms. Same as the previous experiments, the seed words are also selected by their frequency in MUC-4 Corpus. From the result we can know that performance of GMR-Bootstrapping is better than Basilisk in all the conditions. The improvements are more significant when the number of seed words is small. GMR-Bootstrapping's result is good even when only 2 seed words are given. We can also find that the sum of the number of seed words and correct extractions doesn't have significant increase in building, time and weapon categories when the number of seed words is more than 5. It shows that both GMR-Bootstrapping and Basilisk have a bottleneck. Even a lot of seed words are given, low frequency words are hard to extract.

Then we evaluated GMR-M-Bootstrapping to learn multiple semantic categories simultaneously. The results are shown in Table 2, where the column GMR represents the result of GMR-Bootstrapping, GMR-M represents GMR-M-Bootstrapping's results, RI represents the relative improvement of GMR-M-Bootstrapping over GMR-Boot strapping, Basilisk-MACT+'s results [10], which is the previous best result in MUC-4 corpus, are shown in column B-MACT+. Experiments show that GMR-M-Boot strapping's results are better than GMR-Bootstrapping's in all the categories. Those results indicate that our method can improve the results and including pattern's uncertainty into scoring functions can benefit the final results. We know that GMR-M-Bootstrapping's results are better than the previous best results in all the categories and GMR-Bootstrapping' results are similar with the Basilisk MACT+'s.

4.2 Results in CRCV Corpus

Finally we compare the results of GMR-Bootstrapping with R-Basilisk in CRCV Corpus (details in the beginning of the Section 4). Both of them ran 200 iterations to learn MANUFACTURE and MODEL categories. 10 seed words are given for each category. From the results, which are shown in Table 3, we can know that GMR-M-Bootstrapping's performance is better than GMR-Bootstrapping's and GMR-Boot strapping's performance is better than R-Basilisk's in both of the categories. The trend is same as the results in MUC-4 corpus.

5 Conclusion and Feature Work

In this work, we present a novel bootstrapping method, GMR-Bootstrapping, to learn semantic lexicons. Through changing the candidate words and patterns scoring functions, we incorporate Graph Mutual Reinforcement to weight the correctness of candidate word and extraction patterns. The motivation for our approach is provided from graph theory and graph link analysis. We also enhance the GMR-Bootstrapping to learn multiple categories simultaneously by adding patterns' uncertainty into scoring functions. Another contribution of this work is that we present the number of seed words' impact on Basilisk and our method. Experimental results show that GMR-Bootstrapping's results are better than previous best algorithm's results in MUC4 corpus. Experiments in Chinese corpus also show that GMR-Bootstrapping outperform the state-of-the-art technique Basilisk.

In our future work, we plan to focus on Chinese corpus, try to find more generalized feature and pattern.

Acknowledgements

This work was partially supported by Chinese NSF 60673038 and 60503070. The authors would like to thank Ellen Riloff for sharing AutoSlog and labeled data with us. Thanks Chaofeng Sha and Xian Qian for their valuable comments and suggestions. Thanks the anonymous reviewers for their valuable comments and suggestions.

References

1. Hirschman, L., Light, M., Breck, E., Burger, J.D.: Deep read: A reading comprehension system, University of Maryland, pp. 325–348 (1999)
2. Moldovan, D., Harabagiu, S., Pasca, M., Mihalcea, R., Goodrum, R., Girju, R., Rus, V.: Lasso: A tool for surfing the answer net. In: Proceedings of the Eighth Text REtrieval Conference (TREC-8) (1999)
3. Riloff, E., Schmelzenbach, M.: An empirical approach to conceptual case frame acquisition. In: Proceedings of the Sixth Workshop on Very Large Corpora (1998)
4. Roark, B., Charniak, E.: Noun-phrase co-occurence statistics for semi-automatic semantic lexicon construction. In: Proceedings of ACL 1998 (1998)

5. Skounakis, M., Craven, M., Ray, S.: Hierarchical hidden markov models for information extraction. In: Proceedings of the 18th International Joint Conference on Artificial Intelligence (2003)
6. Florian, R., Hassan, H., Ittycheriah, A., Jing, H., Kambhatla, N., Luo, X., Nicolov, N., Roukos, S.: A statistical model for multilingual entity detection and tracking. In: HLT-NAACL 2004: Main Proceedings, pp. 1–8 (2004)
7. Kambhatla, N.: Combining lexical, syntactic, and semantic features with maximum entropy models for information extraction. In: Proceedings of ACL 2004, pp. 178–181 (2004)
8. Collins, M., Singer, Y.: Unsupervised models for named entity classification. In: Proceedings of the Joint SIGDAT Conference on EMNLP/VLC (1999)
9. Riloff, E., Wiebe, J., Wilson, T.: Learning subjective nouns using extraction pattern bootstrapping. In: Proceedings of the Seventh Conference on Natural Language Learning (2003)
10. Thelen, M., Riloff, E.: A bootstrapping method for learning semantic lexicons using extraction pattern contexts. In: Proceedings of EMNLP 2002, Philadelphia (July 2002)
11. Widdows, D., Dorow, B.: A graph model for unsupervised lexical acquisition. In: Proceedings of COLING 2002 (2002)
12. Etzioni, O., Cafarella, M., Downey, D., Popescu, A.: Unsupervised named-entity extraction from the web: An experimental study. Artificial Intelligence 165(1), 91–134 (2005)
13. Hassan, H., Hassan, A., Emam, O.: Unsupervised information extraction approach using graph mutual reinforcement. In: Proceedings of the EMNLP 2006, pp. 501–508 (2006)
14. Miller, G.: Wordnet: An on-line lexical database. International Journal of Lexicography (1990)
15. Dong, Z., Dong, Q.: HowNet (1999), `http://www.HowNet.com`
16. MUC-4 Proceedings: Muc-4 proceedings. In: proceedings of the Fourth Message Understanding Conference (MUC-4) (1992)
17. Blum, A., Mitchell, T.: Combining labeled and unlabeled data with co-training. In: Proceedings of the 1998 Conference on Computational Learning Theory (July 1998)
18. Riloff, E., Jones, R.: Learning dictionaries for information extraction by multi-level bootstrapping. In: Proceedings of the 16th National Conference on Artificial Intelligence (1999)
19. Agichtein, E., Gravano, L.: Snowball: Extracting relations from large plain-text collections. In: Proceedings of the 5th ACM International Conference on Digital Libraries (July 2000)
20. Riloff, E., Shepherd, J.: A corpus-based ap- proach for building semantic lexicons. In: Proceedings of EMNLP 1997, pp. 117–124 (1997)
21. Riloff, E.: Automatically generating extraction patterns from untagged text. pattern bootstrapping. In: Proceedings of the Thirteenth National Conference on Artificial Intelligence (1996)
22. Kleinberg, J.: Authoritative sources in a hyperlinked environment. In: Proceedings of the 9th ACM-SIAM Symposium on Discrete Algorithms (1998)
23. Cover, T.M., Thomas, J.A.: Elements of Information Theory. John Wiley & Sons, Inc., N.Y (1991)

Combining WordNet and ConceptNet for Automatic Query Expansion: A Learning Approach

Ming-Hung Hsu, Ming-Feng Tsai, and Hsin-Hsi Chen

Department of Computer Science and Information Engineering
National Taiwan University
Taipei, Taiwan
{mhhsu,mftsai}@nlg.csie.ntu.edu.tw
hhchen@csie.ntu.edu.tw

Abstract. We present a novel approach that transforms the weighting task to a typical coarse-grained classification problem, aiming to assign appropriate weights for candidate expansion terms, which are selected from WordNet and ConceptNet by performing spreading activation. This transformation benefits us to automatically combine various features. The experimental results show that our approach successfully combines WordNet and ConceptNet and improves retrieval performance. We also investigated the relationship between query difficulty and effectiveness of our approach. The results show that query expansion utilizing the two resources obtains the largest improving effect upon queries of "medium" difficulty.

Keywords: Query Difficulty, Query Expansion, WordNet, ConceptNet.

1 Introduction

Query expansion (QE) has been a well-known and popular technique to improve performance of typical Information Retrieval (IR) systems. The effectiveness of QE comes from that users' queries (especially short queries) usually cannot describe their information needs clearly, and on the other hand, sometimes the vocabulary in a query is inconsistent with that in relevant documents. Typical sources of terms for automatic QE include 1) query logs; 2) statistical thesauri constructed from corpus; 3) top-ranked documents of initial retrieval; and 4) general-purpose knowledgebase such as WordNet [13] and ConceptNet [10]. Hsu et al [7] studied the characteristics of WordNet and ConceptNet and show that these two resources are intrinsically complementary for QE. Intuitively commonsense knowledge in ConceptNet is useful for IR. However, it is still challenging to introduce "Concept" for automatic QE. This paper explores the combination of these two resources automatically to improve retrieval performance.

In the past researches, candidate terms for expansion are usually ranked and selected according to their co-occurrence correlations with the regarded query. The threshold determining whether to select a candidate term or not must be carefully tuned. The weight of expansion terms are decided by the correlation [11] and/or by a constant parameter [21, 24]. However, Peat and Willett [15] have investigated that there are limitations with QE

H. Li et al. (Eds.): AIRS 2008, LNCS 4993, pp. 213–224, 2008.

methods based on co-occurrence statistics. Their study indicates that some good expansion terms may not frequently co-occur with the regarded query terms, and some terms which have high co-occurrence correlations with the query may not be suitable for expansion with high weights. Therefore, typical co-occurrence-based correlation estimations may not be a desirable manner to select and to weight expansion terms. While the weights of expansion terms may cause different effects on retrieval performance, investigating a better weighting method for QE will support the IR system more strongly.

Given a retrieval environment, weight of an expansion term represents its importance to the regarded query. In this paper, we investigate if there are some characteristics, which are dependent only on the regarded query, would determine the appropriate weights of expansion terms for this query. For example, intuitively a "simple" query (e.g., average precision=0.9) would need "slightly" (i.e., lightly weighted) helpful expansion terms, but a "difficult" (e.g., average precision=0.1) query would need "greatly" (i.e. heavily weighted) helpful ones. Such characteristics, which are independent of expansion terms, are rarely studied in previous works involving QE. However, it's strongly connected with the investigations about predicting query difficulty, which has been paid much attention recently [1, 20, 22].

Aiming to combine WordNet and ConceptNet for automatic QE by assigning appropriate weights for expansion terms, we transform the weighting task to a typical coarse-grained classification problem. This transformation benefits us to automatically combine various features by employing the state-of-art classification algorithm. Some of the features have been shown significantly correlated with query difficulty, and the others have been shown useful for QE in previous works.

1.1 Related Work

There are rich investigations regarding QE, e.g. [2, 4, 8]. Rocchio [17] proposed the well-known pseudo-relevance feedback approach and then Salton addressed an improved version [18]. In Rocchio's approach, candidate expansion terms are ranked according to their TF-IDF values averaged over the top-*N* initially retrieved documents, and the expansion terms are weighted by a constant parameter. Voorhees [23] expanded queries by utilizing lexical semantic relations defined in WordNet. Her experimental results showed that automatically fitting a general-purpose knowledge base to a specific environment of ad hoc retrieval is certainly a challenge. Liu et al. [11] addressed the first approach that effectively adopted WordNet for automatic QE. They performed phrase recognition and sophisticated word sense disambiguation on queries, and then selected highly-correlated terms of the same sense with query terms. The weight of an expansion term was the probability that it has the same sense as the regarded query term. Xu and Croft [24] compared the performances of utilizing local and global document analysis for QE. Their experiments concluded that local analysis is more effective than global analysis. In their approach, candidate expansion terms are ranked by their co-occurrence correlations with the query and weighted by a constant according to their rank. Sun [21] mined dependency relations in passages to rank candidate terms with a weighting scheme similar to the work of Xu and Croft.

As ConceptNet has been adopted in many interactive applications [9], Hsu et al [7] investigated the intrinsic comparison of WordNet and ConceptNet for QE by using quantitative measurements. Their experimental results showed that WordNet and

ConceptNet are intrinsically complementary. However, as far as our investigations, there have been no approaches successfully utilizing ConceptNet for automatic QE.

On the other hand, the goal of past Robust Retrieval Track in TREC [22] is to identify "difficult" queries and then IR systems can handle those queries specifically to improve system robustness. There have been some linguistic or statistical features shown to be significantly correlated to query difficulty; for example, document frequencies (DF) and polysemy values of query terms [14], retrieval scores of top-ranked documents [5] and the consistency between top-ranked documents retrieved by using the whole query and sub-queries respectively [20]. We simplified and adopted some of these features in our learning model.

1.2 Our Approach

As discussed above, previous researches on QE focus on how to rank candidate expansion terms, and the weight of an expansion term is often decided by its correlation or similarity with the regarded query, or by a constant parameter. In order to introduce ConceptNet for automatic QE, we explore a novel approach to assign appropriate weights to expansion terms. We transform the task of weighting to a typical coarse-grained classification problem. The basic idea is to classify candidate expansion terms to be as one of the discrete weight classes, e.g., greatly important (heavy weight), slightly important (light weight), or not important (no weight, i.e., not selected for expansion). The classified instances, i.e., candidate terms for expansion, are selected from WordNet and ConceptNet, via *spreading activation* [19] on the two knowledge bases. We use the state-of-art classification learning algorithm, support vector machine (SVM), to automatically combine features of various types extracted from several sources.

This paper is arranged as follows. Section 2 shows the framework of our approach. In Section 3, we give a brief introduction to ConceptNet and describe the way we perform spreading activation to select candidate expansion terms. Section 4 describes the various features we utilized. Section 5 shows and discusses the experimental results. Section 6 concludes the remarks.

2 The Framework

Figure 1 shows the procedures of our approach. In the pre-processing stage, the original query is POS-tagged and an n-gram in the query is identified as a phrase if it occurs in WordNet or ConceptNet. Words and phrases with their POS-tags are treated as concepts in the following procedures. After pre-processing, we perform spreading activation in ConceptNet and WordNet, and concepts in the queries are the activation origins. Words with the top N activation scores form the set of candidate expansion terms. For each candidate, we automatically label its suitable weight versus the original query and extract its features from global analysis, local analysis, and knowledge base analysis (Section 4). Candidates along with their labels and features are then sent to SVM [3] for classification training and testing, i.e., weight prediction.

Candidate terms with their predicted weights are combined with the original query to form the expanded query, and it is sent to an IR system for further evaluation of retrieved result.

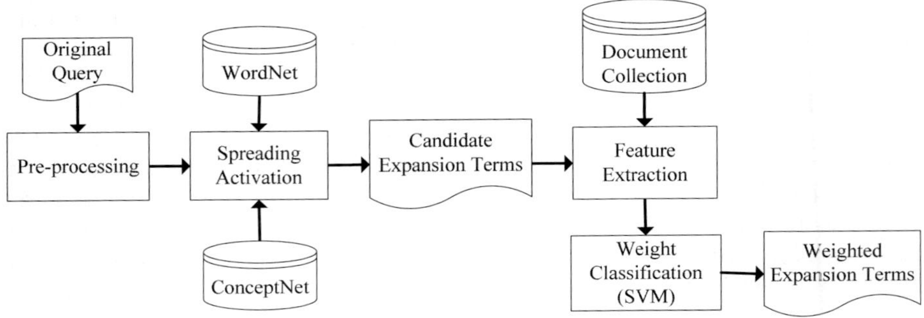

Fig. 1. The framework of our approach

3 Selecting Candidate Terms

3.1 A Brief Introduction to ConceptNet

ConceptNet, presently the largest commonsense knowledge base, is developed by MIT Media Laboratory. It is a relational semantic network automatically generated from the Open Mind Common Sense corpus[1]. Nodes in ConceptNet are compound concepts formed as natural language fragments (e.g. "buy food" and "grocery store"). Aiming to cover pieces of commonsense knowledge to describe the real world, ConceptNet covers 20 kinds of relations categorized as causal, spatial, etc. ConceptNet has been adopted in many interactive applications [9]. Hsu et al [6] utilized it to expand image annotations and got slight improvement. As commonsense knowledge is deeply context-sensitive, the suitability of ConceptNet for QE is still uncertain. Note that ConceptNet bears word sense ambiguities without differentiation.

3.2 Spreading Activation

We perform spreading activation in a semantic network to select candidate terms. The node of *activation origin* represents the concept of the given query. The activation origin is the first to be activated, with an initial activation score (e.g., 1.0). Next, nodes one link away from the activation origin are activated, then two links away, and so on. Equation (1) determines the activation score of node *j* by three factors: (i) a constant $C_{dd} \leqq 1$, which is called *distance discount* that causes a node closer to the activation origin to get a higher activation score; (ii) the activation score of node *i*; (iii) *W*(*i*,*j*), the weight of the link from *i* to *j*. Different relations in the semantic network are of different weights. *Neighbor*(*j*) represents the nodes connected to node *j*.

$$Activation\_score(j) = C_{dd} \times \sum_{i \in Neighbor(j)} Activation\_score(i) \times W(i,j) \cdot \quad (1)$$

Since most traditional IR systems are of a bag-of-words model, we select the top *N* words with the higher activation scores as the candidate expansion terms. Note that for

[1] http://openmind.media.mit.edu/

a word w, its activation score is the sum of scores of the nodes (i.e., synsets in WordNet) that contain w, in our implementation.

3.3 Spreading Activation in ConceptNet

In addition to commonsense concepts and relations, ConceptNet also provides a set of tools for reasoning over text. One of these tools, *Get_context (concepts)*, performs spreading activation on all kinds of relations in ConceptNet, to find contextual neighborhood relevant to the concepts as parameters. Different relations are set to different weights in default setting. For example, the weight of 'IsA' relation is 0.9 and the weight of 'DefinedAs' is 1.0, etc. We adopt this tool directly for selecting candidate expansion terms. In our experiments, each word in a compound concept has the same activation score as that of the compound concept. More details about the reasoning tools in ConceptNet please refer to [10].

3.4 Spreading Activation in WordNet

Since each node in WordNet is a synset that contains synonyms of a certain sense, spreading activation in WordNet is naturally performed on the unit of synset. Because ConceptNet covers most relations in WordNet, we determine the weights of relations in WordNet, shown in Table 1, by referring to the settings in ConceptNet. The 'Definition' relation is abstractly utilized but not defined in WordNet. For an activated node i, we implement spreading activation through the 'Definition' relation by activating the first sense of distinct concepts (word with POS-tag) in the definition of i.

Table 1. Weights of relations in WordNet for spreading activation

Relation Type	Causes	Definition	Holonyms	Hypernyms	Hyponyms	Meronyms	Pertainyms
Weight	0.8	1.0	0.8	0.8	0.9	0.9	0.7

For each concept in a query, all its different senses (synsets) of the corresponding POS-tag are the activation origins. Suppose $S = \{S_1, S_2, ..., S_n\}$ is the set of all the n synsets (senses) of query term t. According to the frequency information provided by WordNet, Equation (2) determines the initial activation score of synset S_i, *Initial_score*(S_i), where *Freq*(S_i) represents the frequency of S_i in WordNet.

$$Initial\_score(S_i) = \frac{Freq(S_i)^c}{\sum_{S_k \in S} Freq(S_k)^c} . \quad (2)$$

Equation (2) indicates that the initial activation score of activation origin S_i is positively correlated to its frequency. We do not perform word sense disambiguation (WSD) for query terms. The constant c is set as 0.5 in our experiments.

4 Labeling and Feature Extraction

4.1 Labeling the Weight Class

In this study, the weighting task is transformed to a coarse-grained classification problem. As a preliminary study of such transformation for weighting, we introduce only three weight classes here. That is, we classify each candidate expansion term with corresponding query as one of the three weight classes: *no weight* (not important), *light weight* (slightly important), and *heavy weight* (greatly important). In our implementation, we set the light weight to 0.1 because this small value was shown to be a good constant weight for typical pseudo relevance feedback approaches [2]. The heavy weight is set to 0.5, which causes an expansion term more effective than lightly weighted ones but less effective than original query terms.

Given the IR system, Okapi [16] we adopted, and the relevant assessment of the regarded query, labeling of the suitable weight class for each candidate term is automatically carried out, by examining which of the three weights assigned to the candidate would lead to the best improvement in traditional IR evaluation metrics. For example, for the query "Hubble Telescope Achievement", the performance without expansion is 0.1424 in average precision (AP). One of the candidate terms is "orbit", which improves the performance to 0.1472 with weight 0.5 and 0.1569 with weight 0.1; hence "orbit" would be labeled as the light weight class.

4.2 Extraction of Various Features

We categorize the adopted features as two types: query-analysis and candidate-dependent. Features of query-analysis type are dependent on 1) the query terms, 2) the IR system, and 3) the initial retrieval result without expansion, but independent of the candidate expansion term. Most of the adopted query-analysis features are shown to be significantly correlated to query difficulty in previous works. Features of the two types are described in detail in the following sub-sections.

Suppose $Q= \{q_1, q_2 \dots q_m\}$ is the regarded query of m terms and $E= \{e_1, e_2 \dots e_n\}$ is the set of n candidate expansion terms for Q. Features of query-analysis type depict: 1) some properties of q_i; and 2) some relations between q_i and q_j. The other feature type, candidate-dependent, covers features that describe: 1) properties of the candidate e_k; and 2) relations between e_k and Q. The co-occurrence correlation $COR(t_1, t_2)$ between two terms t_1 and t_2 is calculated by Equation (3), the typical t-test estimation:

$$COR(t_1,t_2) = \frac{p(t_1,t_2) - p(t_1)\cdot p(t_2)}{\sqrt{\frac{p(t_1,t_2)\cdot(1-p(t_1,t_2))}{N}}}, \tag{3}$$

where $p(t_1, t_2)= n(t_1, t_2)/N$, and $n(t_1, t_2)$ is the number of documents that contain both t_1 and t_2. N is the number of documents in the corpus.

Some features (QDIST in Table 2 and CDIST in Table 3) require calculating the average distance (measured in character) of two terms t_1 and t_2 in a document. We implement it by looking for the positions of t_1 occurrences in the document, summing up the distances between t_1 and t_2 in those t_1 positions, and then averaging the sum of

Table 2. Description of query-analysis features

Feature	Description
QCOR	the co-occurrence correlations between q_i and q_j, i.e., $COR(q_i, q_j)$ for $q_i, q_j \in Q$
QDIST	the average distance between q_i and q_j in top 20 documents of initial retrieval result, for $q_i, q_j \in Q$
QIDF	the logarithm of inverse document frequency (IDF) of q_i for $q_i \in Q$
QMPS	the maximum possible retrieval score of returned documents for Q
QMXS	the retrieval score of document with rank 1 in the initial result of Q
QNS	the number of senses of q_i in WordNet, for $q_i \in Q$
QNT	the number of query terms, i.e., $\|Q\|$
QPS	the probability of q_i being its first sense for $q_i \in Q$

Table 3. Description of candidate-dependent features

Feature	Description
CACR	the average of co-occurrence correlations between e_k and q_i for $q_i \in Q$
CCOR	the co-occurrence correlation between e_k and q_i, i.e., $COR(e_k, q_i)$ for $q_i \in Q$
CCRK	rank of e_k in the spreading activation of Q in ConceptNet
CCSA	score of e_k in the spreading activation of Q in ConceptNet
CDIST	average distance between e_k and q_i in top 20 documents of initial retrieval result of Q, for $q_i \in Q$
CIDF	logarithm of IDF of the candidate term e_k
CMPS	the maximum possible retrieval scores for Q expanded by e_k with the two weights
CMXS	the retrieval scores of documents ranked 1 in retrieved results for Q expanded by e_k with the two weights (0.1 and 0.5)
CNS	the number of senses of e_k in WordNet
CPS	the probability of e_k being its first sense
CPTD	the probability of documents that contain e_k in the initially retrieved top 20 documents of Q
CPTT	the probability of e_k occurrences in all terms in the initially retrieved top 20 documents of Q
CWRK	rank of e_k in the spreading activation of Q in WordNet
CWSA	score of e_k in the spreading activation of Q in WordNet

distances over the number of t_1 occurrences. For example, suppose the positions of t_1 occurrences are (5, 66, 173) and the positions of t_2 occurrences are (35, 102), then the average distance between t_1 and t_2 is $(|35-5|+|35-66|+|102-173|)/3=44$.

Query-Analysis Features. Table 2 shows the query-analysis features and their descriptions. As described in previous sections, some of the query-analysis features, e.g., QIDF, QNS and QMXS, have been shown to be correlated to difficulty of the regarded query Q. QMPS and QMXS utilize the retrieval scores provided by the IR system. QMPS is special, because it's supported by Okapi only. The extraction of QPS relies on the information of sense frequency provided by WordNet. Let the frequency of the first sense of q_i in WordNet be f_{i1} and the sum of frequencies of all senses of q_i be f_i. QPS is the value of (f_{i1} / f_i) for $q_i \in Q$.

Candidate-Dependent Features. Table 3 shows the candidate-dependent features. The extraction of CPS is similar to that of QPS of query-analysis type. CCRK, CCSA, CWRK and CWSA are extracted from the result of spreading activation of Q in ConceptNet and in WordNet. The extraction of CMPS and CMXS is similar to that of QMPS and QMXS, except that the regarded query is expanded by e_k with assigned weight. CPTD and CPTT calculate the probabilities that e_k occurs in top-ranked documents of initial retrieval result of Q.

5 Experiments

5.1 Experimental Environment

We evaluate our approach with the topics of TREC-6, TREC-7 and TREC-8, i.e., total 150 topics. Only the title part of each topic is used to simulate short queries in web search. The IR system is Okapi-BM25. The words with top N activation scores in WordNet or ConceptNet, except those in the query, are selected as the set of candidate terms for each query. We use the state-of-art classification algorithm, SVM [3], for our coarse-grained weight classification. The default kernel function, radial basis function, is adopted. When TREC-6 topics are used for testing, TREC-7 and TREC-8 topics are for training, and vice versa. Aiming to improve retrieval performance, in the training phase, instances of heavy weight would cause higher misclassified penalty than those of light weight, and light weight ones would cause higher penalty than no weight ones. We adjust the penalty ratio between heavy and light weight classes (and between light and no weight ones) through cross-validation.

5.2 Experiment Results and Discussions

Number of Candidate Terms vs. Retrieval Performance. The number of candidate terms (the *N* in Subsection 5.1) is an important parameter in our approach. It reflects both the amount of training instances in the training stage and the number of candidate expansion terms for the testing (i.e. QE) stage. Table 4 shows the "upper bound" performance (in average precision) of our model, which happens when the accuracy of weight classification is 100%, when adopting various number of candidate terms. BL is the result of the baseline performance, i.e. word-based document retrieval using Okapi-BM25 without expansion. U50 shows the upper bound performance obtained with N=50, and U100 represents N=100, etc. Table 4 shows that when N increases, the upper bound also rises. When N is equal to or larger than 100, the upper bound results (the bold) are higher than the best known result of the same topic set, 0.3199 in the

Table 4. Upper bound performance of our approach with various number of candidate terms

Topic set	BL	U50	U100	U200	U300	U400
TREC-6	0.2209	0.3383	0.3613	0.3952	0.4065	0.4011
TREC-7	0.1637	0.2875	0.3093	0.3263	0.3370	0.3401
TREC-8	0.2196	0.3123	0.3318	0.3516	0.3590	0.3664
All	0.2014	0.3127	**0.3341**	**0.3577**	**0.3675**	**0.3692**

Table 5. Retrieval performance of our approach with various number of candidate terms

Topic set	BL	N50	N100	N200	N300	N400
TREC-6	0.2209	0.2076	**0.2398**	0.2273	0.2350	0.2345
TREC-7	0.1637	**0.1865**	**0.1891**	**0.1955**	**0.1937**	**0.1941**
TREC-8	0.2196	0.2277	0.2362	**0.2473**	**0.2413**	0.2386
All	0.2014	0.2073	**0.2217**	**0.2234**	**0.2233**	**0.2224**

previous works [12]. It shows the potential of our model and the importance of appropriate weighting method. However, taking U100 for example, 0.3341 is not the best performance that can be achieved by using the same set of candidate terms. We observed some topics in this result perform worse than in the baseline "BL", indicating that the weights of individual candidate terms are not optimal for their combination. This problem would be a direction of future work.

Table 5 shows the performances of our weight classification model versus the values of *N*. The bold results are significant (with confidence over 99% by two-tailed *t*-test) improvements over the baseline. Note that we didn't perform WSD in queries, indicating that our approach is certainly feasible and effective. On the other hand, dissimilar to the upper bound results, the performance of our model doesn't rise when *N* increases larger than 100. The reason is that a big value of *N* would introduce noises in the training stage, decreasing the quality of training data. Therefore, we set the value of *N* to 100 in the later experiments.

Query Difficulty vs. Suitable Weights of Candidates. We plan to observe the correlation between query difficulty and suitable weights of candidate expansion terms. The difficulty of a query is estimated according to its baseline performance in average precision, i.e., the lower the average precision, the higher the difficulty. Therefore, all the 150 topics are sorted according to their baseline performance and divided equally into three sets of difficulty classes: High, Medium and Low. Each set of difficulty class contains 50 topics. For each topic, E is the set of candidate expansion terms, $W(E)$ represents the sum of the suitable (i.e., labeled) weights of all terms in E, and E_u represents the subset of E that contains only useful (i.e., with positive suitable weight) candidates.

Table 6 shows the mean average precision (MAP), the average $W(E)$ and the average $|E_u|$ for each difficulty class. It shows the "Medium" and the "High" difficulty classes have almost the same average $W(E)$. However, the two classes are significantly distinct

Table 6. Statistics of query difficulty vs. weights of candidates

Topic set	MAP	Average $W(E)$	Average $\lvert E_u \rvert$
HIGH	0.0285	13.958	49.5
MEDIUM	0.1432	13.946	**59.94**
LOW	0.4327	**9.592**	53.28

from each other in average $|E_u|$, with the confidence of over 98% by two-tailed t-test. This result indicates that for the "High" difficulty topics, useful candidates selected from the two resources are much fewer than those for the "Medium" difficulty topics. The difference of average $|E_u|$ between the "Low" and the "Medium" classes is not significant. The average $W(E)$ of the "Low" class is significantly distinct from both of the "Medium" and the "High" classes. This result exhibits that the difficulty of the query is indeed correlated to the suitable weights of its candidate expansion terms, especially those selected from WordNet and ConceptNet. Overall, Table 6 brings out an interesting summary: the "High" and the "Medium" classes are distinct from each other in their average $|E_u|$, while the "Low" and the "Medium" classes are distinct in their average $W(E)$.

Figure 2 shows in detail the correlation between $W(E)$ and query difficulty, in terms of average precision, for all the 150 topics. The "Trend "curve shown in the figure is the 3rd-order polynomial regression of these points. The curve shows the abstract trend that $W(E)$ decreases with the increase of AP. However, $W(E)$ varies widely between queries which have similar AP values.

Influence of Query Analysis Features. As Table 6 and Figure 2 indicate the correlation between query difficulty and *sum* of suitable weights of candidates, Table 7 shows the performance of our model with or without ("N100-No-QA") query-analysis features. Surprisingly, our approach achieves slightly higher MAP when not adopting query-analysis features. By observing the weight prediction result for each topic, we found that when adopting query-analysis features, our model tends to predict the candidate terms as higher-weighted for most of the topics. It may indicate a phenomenon of "over fitting" or that the query-analysis features are too rough to predict query difficulty. This problem requires more investigations.

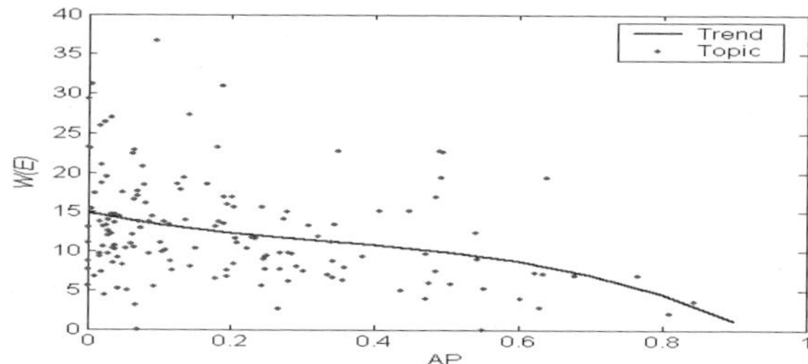

Fig. 2. Query difficulty vs. sum of suitable weights of candidate expansion terms

Table 7. Performance with or without query analysis features

Topic set	MAP	N100	N100-No-QA	U100
HIGH	0.0285	**0.0400**	**0.0394**	**0.1643**
MEDIUM	0.1432	**0.1725**	**0.1760**	**0.3098**
LOW	0.4327	0.4526	**0.4640**	**0.5282**
All	0.2014	**0.2217**	**0.2264**	**0.3341**

6 Conclusions

In this paper, we transform the weighting task to a coarse-grained multi-class classification. While the goal, i.e., predicting suitable weight classes for candidate expansion terms, is difficult and the idea of our approach is simple, we achieved significant improvement over the baseline, without performing WSD in queries and with a small amount of training queries and features. The experimental results show that we successfully combined WordNet and ConceptNet, indicating that commonsense knowledge in ConceptNet is feasible for automatic QE.

Acknowledgments. Research of this paper was partially supported by Excellent Research Projects of National Taiwan University, under the contract 96R0062-AE00-02.

References

1. Carmel, D., Yom-Tov, E., Darlow, A., Pelleg, D.: What makes a query difficult? In: Proceedings of the 29th Annual International ACM SIGIR Conference, pp. 390–397 (2006)
2. Carpineto, C., et al.: An information-theoretic approach to automatic query expansion. ACM Transactions on Information Systems 19(1), 1–27 (2001)
3. Chang, C.-C., Lin, C.-J.: LIBSVM: a library for support vector machines (2001)
4. Cui, H., et al.: Query expansion by mining user logs. IEEE Transaction on Knowledge and Data Engineering 15(4), 829–839 (2003)
5. Grivolla, J., Jourlin, P., De Mori, R.: Automatic classification of queries by expected retrieval performance. In: ACM SIGIR 2005 Workshop on Predicting Query Difficulty – Methods and Applications (2005)
6. Hsu, M.-H., Chen, H.-H.: Information retrieval with commonsense knowledge. In: Proceedings of the 29th Annual International ACM SIGIR Conference, pp. 651–652 (2006)
7. Hsu, M.-H., Tsai, M.-F., Chen, H.-H.: Query expansion with ConceptNet and WordNet: an intrinsic comparison. In: Ng, H.T., Leong, M.-K., Kan, M.-Y., Ji, D. (eds.) AIRS 2006. LNCS, vol. 4182, pp. 1–13. Springer, Heidelberg (2006)
8. Kwok, K.L., et al.: TREC 2004 Robust Track experiments using PIRCS. In: Proceedings of the 13th Text Retrieval Conference (TREC 2004) (2005)
9. Lieberman, H., Liu, H., Singh, P., Barry, B.: Beating Common Sense into Interactive Applications. AI Magazine 25(4), 63–76 (2004)
10. Liu, H., Singh, P.: ConceptNet: a practical commonsense reasoning toolkit. BT Technology Journal 22(4), 211–226 (2004)

11. Liu, S., Liu, F., Yu, C.T., Meng, W.: An effective approach to document retrieval via utilizing WordNet and recognizing phrases. In: SIGIR 2004, pp. 266–272 (2004)
12. Liu, S., et al.: Word sense disambiguation in queries. In: Proceedings of the 14th ACM International Conference on Information and Knowledge Management, pp. 525–532 (2005)
13. Miller, G.A.: WordNet: an on-line lexical database. International Journal of Lexicography (1990)
14. Mothe, J., Tanguy, L.: Linguistic features to predict query difficulty – a case study on previous TREC campaigns. In: ACM SIGIR 2005 Workshop on Predicting Query Difficulty – Methods and Applications (2005)
15. Peat, J.H., Willett, P.: The limitations of term co-occurrence data for query expansion in document retrieval systems. Journal of American Society for Information Science 42(5), 378–383 (1991)
16. Robertson, S.E., et al.: Okapi at TREC-7: Automatic Ad Hoc, Filtering, VLC and Interactive. In: Proceedings of the 7th Text Retrieval Conference, pp. 253–264 (1998)
17. Rocchio, J.: Relevance feedback in information retrieval. In: Salton, G. (ed.) The Smart Retrieval System: Experiments in Automatic Document Processing, pp. 313–323. Prentice-Hall, Englewood Cliffs
18. Salton, G., Buckley, C.: Improving retrieval performance by relevance feedback. Journal of American Society for Information Science 41(4), 288–297 (1990)
19. Salton, G., Buckley, C.: On the use of spreading activation methods in automatic information retrieval. In: Proceedings of the 11th ACM-SIGIR Conference, pp. 147–160 (1988)
20. Steve, C.-T., Zhou, Y., Croft, W.B.: Predicting query performance. In: Proceedings of the 29th Annual International ACM SIGIR Conference, pp. 299–306 (2002)
21. Sun, R., et al.: Mining dependency relations for query expansion in passage retrieval. In: Proceedings of the 29th Annual International ACM SIGIR Conference, pp. 382–389 (2006)
22. Voorhees, E.M.: Overview of the TREC 2005 Robust Retrieval Track. In: Proceedings of the 14th Text Retrieval Conference (TREC 2005) (2006)
23. Voorhees, E.M.: Query expansion using lexical-semantic relations. In: Proceedings of the 17th Annual International ACM SIGIR Conference, pp. 61–69 (1994)
24. Xu, J., Croft, W.B.: Query expansion using local and global document analysis. In: Proceedings of the 19th Annual International ACM SIGIR Conference, pp. 4–11 (1996)

Improving Hierarchical Taxonomy Integration with Semantic Feature Expansion on Category-Specific Terms

Cheng-Zen Yang, Ing-Xiang Chen, Cheng-Tse Hung, and Ping-Jung Wu

Department of Computer Science and Engineering
Yuan Ze University, Taiwan, R.O.C.
{czyang,sean,chris,pjwu}@syslab.cse.yzu.edu.tw

Abstract. In recent years, the taxonomy integration problem has obtained much attention in many research studies. Many sorts of implicit information embedded in the source taxonomy are explored to improve the integration performance. However, the semantic information embedded in the source taxonomy has not been discussed in the past research. In this paper, an enhanced integration approach called SFE (Semantic Feature Expansion) is proposed to exploit the semantic information of the category-specific terms. From our experiments on two hierarchical Web taxonomies, the results are positive to show that the integration performance can be further improved with the SFE scheme.

Keywords: hierarchical taxonomy integration, semantic feature expansion, category-specific terms, hierarchical thesauri information.

1 Introduction

In recent years, the taxonomy integration problem has obtained much attention in many research studies (e.g. [1,2,3,4,5,6,7]). A taxonomy, or catalog, usually contains a set of objects divided into several categories according to some classified characteristics. In the taxonomy integration problem, the objects in a taxonomy, the *source* taxonomy $\mathcal{S}$, are integrated into another taxonomy, the *destination* taxonomy $\mathcal{D}$. As shown in past research, this problem is more than a traditional document classification problem because the implicit information in the source taxonomy can greatly help integrate source documents into the destination taxonomy. For example, a Naive Bayes classification approach can be enhanced with the source implicit importation to achieve accuracy improvements [1], and SVM (Support Vector Machines) approaches have similar improvements [6].

The implicit source information studied in previous enhanced approaches generally includes following features: (1) co-occurrence relationship of source objects [1,6], (2) latent source-destination mappings [2,4], (3) inter-category centroid information [3], and (4) parent-children relationship in the source hierarchy [5,7]. To the best of our survey, however, the semantic information embedded in the source taxonomy has not been discussed. Since different applications have shown that the semantic information can benefit the task performance [8,9], such information should be able to achieve similar improvements for taxonomy integration.

H. Li et al. (Eds.): AIRS 2008, LNCS 4993, pp. 225–236, 2008.

In this paper, we propose an enhanced integration approach by exploiting the implicit semantic information in the source taxonomy with a semantic feature expansion (SFE) mechanism. The basic idea behind SFE is that some semantically descriptive terms can be found to represent a source category, and these representative terms can be further viewed as the additional common category labels for all documents in the category. Augmented with these additional semantic category labels, the source documents can be more precisely integrated into the correct destination category.

To study the effectiveness of SFE, we implemented it based on a hierarchical taxonomy integration approach (ECI) proposed in [7] with the Maximum Entropy (ME) model classifiers. We have conducted experiments with real-world Web catalogs from Yahoo! and Google, and measured the integration performance with precision, recall, and F_1 measures. The results show that the SFE mechanism can further consistently improve the integration performance of the ECI approach.

The rest of the paper is organized as follows. Section 2 describes the problem definition and Section 3 reviews previous related research. Section 4 elaborates the proposed semantic feature expansion approach and the hierarchical integration process. Section 5 presents the experimental results, and discusses the factors that influence the experiments. Section 6 concludes the paper and discusses some future directions of our work.

2 Problem Statement

Following the definitions in [7], we assume that two *homogeneous* hierarchical taxonomies, the source taxonomy $\mathcal{S}$ and the destination taxonomy $\mathcal{D}$, participates in the integration process. The taxonomies are said to be *homogeneous* if topics of two taxonomies are similar. The taxonomies under consideration are additionally required to be overlapped with a significant number of common documents. In our experimental data sets, 20.6% of the total documents (436/2117) in the `Autos` directory of Yahoo! also appear in the corresponding Google directory.

The source taxonomy $\mathcal{S}$ has a set of m categories, or directories, S_1, S_2, ..., S_m. These categories may have subcategories, such as $S_{1,1}$ and $S_{2,1}$. Similarly, the destination catalog $\mathcal{D}$ has a set of n categories. The integration process is to directly decide the destination category in $\mathcal{D}$ for each document d_x in $\mathcal{S}$. In this study, we allow that d_x can be integrated into multiple destination categories because a document commonly appears in several different directories in a real-world taxonomy.

Fig. 1 depicts a typical scenario of the integration process on two hierarchical taxonomies. For illustration, we assume that the source category $S_{1,1}$ has significantly some overlapped documents with the destination categories $D_{1,1}$ and $D_{2,2}$. This means that the documents appear in $S_{1,1}$ should have similar descriptive information as the documents in $D_{1,1}$ and $D_{2,2}$. Therefore, a non-overlapped document d_x^1 in category $S_{1,1}$ should be intensively integrated into both two destination categories $D_{1,1}$ and $D_{2,2}$.

3 Related Research

In previous studies, different sorts of implicit information embedded in the source taxonomy are explored to help the integration process. As described in Section 1, these

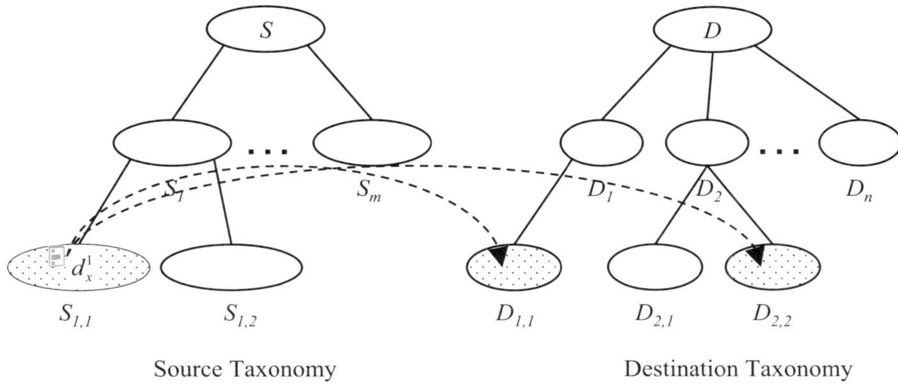

Fig. 1. A typical integration scenario for two hierarchical taxonomies

implicit source features can be mainly categorized into four kinds. (1) co-occurrence relationship of source objects, (2) latent source-destination mappings, (3) inter-category centroid information, and (4) parent-children relationship in the source hierarchy. The co-occurrence relationships of source objects are first studied to enhance a Naive Bayes classifier based on an intuition that if two documents are in the same source category, they are more likely to be in the same destination category [1]. The enhanced Naive Bayes classifier (ENB) is shown to have more than 14% accuracy improvement on average. The work in [6] also has the similar concept in its iterative pseudo relevance feedback approach. As reported in [6], the enhanced SVM classifiers consistently achieve improvements.

Latent source-destination mappings are explored in [2,4]. The cross-training (CT) approach [2] extracts the mappings from the first semi-supervised classification phase using the source documents as the training sets. Then the destination documents are augmented with the latent mappings for the second semi-supervised classification phase to complete the integration. The co-bootstrapping (CB) approach [4] exploits the predicted source-destination mappings to repeatedly refine the classifiers. The experimental results show that both CT and CB outperform ENB [2,4].

In [3], a *cluster shrinkage* (CS) approach is proposed in which the feature weights of all objects in a documents category are shrunk toward the category centroid. Therefore, the cluster-binding relationships among all documents of a category are strengthened. The experimental results show that the CS-enhanced Transductive SVMs have significant improvements to the original T-SVMs and consistently outperform ENB.

In [5,7], the parent-children information embedded in hierarchical taxonomies is intentionally extracted. Based on the hierarchical characteristics, Wu et al. extend the CS and CB approach to improve the integration performance. In [7], an enhanced approach called ECI is proposed to further extract the hierarchical relationships as a conceptual thesaurus. Their results show that the implicit hierarchical information can be effectively used to boost the accuracy performance.

The semantic information embedded in the source taxonomy has not been discussed in past studies. This observation motivates us to study the embedded taxonomical semantic information and its effectiveness.

4 Hierarchical Taxonomy Integration with Semantic Feature Expansion

In our work, the proposed semantic feature expansion (SFE) approach is currently studied with a hierarchical taxonomy integration approach (ECI) to further study the hierarchical integration problem. The Maximum Entropy (ME) model is used because of its prominent performance in many tasks, such as natural language processing [10] and flattened taxonomy integration [5]. In the following, the ME model and the ECI approach are introduced first. Then, SFE is presented in detail.

4.1 The Maximum Entropy Model

Here we briefly describe the ME model according to [10] where more details can be found. In ME, the *entropy* $H(p)$ for a conditional distribution $p(y|x)$ is used to measure the uniformity of $p(y|x)$, where y is an instance of all outcomes Y in a random process and x denotes a contextual environment of the contextual space X, or the history space. To express the relationship between x and y, we can have an indicator function $f(x, y)$ (usually known as *feature* function) defined as

$$f(x,y) = \begin{cases} 1 & \text{if } (x,y) \text{ has the defined relationship} \\ 0 & \text{else} \end{cases} \tag{1}$$

The entropy $H(p)$ is defined by

$$H(p) = -\sum_{x \in X} p(y|x) \log p(y|x) \tag{2}$$

The Maximum Entropy Principle is to find a probability model $p^* \in C$ such that

$$p^* = \underset{p \in C}{\text{argmax}} H(p) \tag{3}$$

where C is a set of allowed conditional probabilities. However, there are two constraints:

$$E_{\tilde{p}}\{f\} = E_p\{f\} \tag{4}$$

and

$$\sum_{y \in Y} p(y|x) = 1 \tag{5}$$

where $E_{\tilde{p}}\{f\}$ is the expected value of f with the empirical distribution $\tilde{p}(x, y)$ as defined in Equation 6 and $E_p\{f\}$ is the observed expectation of f with the observed distribution $\tilde{p}(x)$ from the training data as defined in Equation 7.

$$E_{\tilde{p}}\{f\} \equiv \sum_{x,y} \tilde{p}(x,y) f(x,y) \tag{6}$$

$$E_p\{f\} \equiv \sum_{x,y} \tilde{p}(x)p(y|x)f(x,y) \tag{7}$$

As indicated in [10], the conditional probability $p(y|x)$ can be computed by

$$p(y|x) = \frac{1}{z(x)} \exp\left(\sum_i \lambda_i f_i(x,y)\right) \tag{8}$$

where λ_i is the Lagrange multiplier for feature f_i, and $z(x)$ is defined as

$$z(x) = \sum_y \exp\left(\sum_i \lambda_i f_i(x,y)\right) \tag{9}$$

With the *improved iterative scaling* (IIS) algorithm [10,11], the λ_i values can be estimated. Then the classifiers are built according to the ME model and the training data.

4.2 The ECI Integration Schemes

In ECI, the conceptual relationships (category labels) is first extracted from the hierarchical taxonomy structure as a thesaurus [7]. Then the features of each document are extended with the thesaurus by adding the weighted label features. A weighting formula is designed to control the impact of the semantic concepts of each hierarchical level. Equation 10 calculates the ECI feature weight $f^e_{x,d}$ of each term x in document d, where L_i is the relevant label weight assigned as $1/2^i$ with an i-level depth, $f_{x,d}$ is the original weight, and λ is used to control the magnitude relation. The weight $f_{x,d}$ is assigned by $TF_x/\sum TF_i$, where TF_x is the term frequency of x, and i denotes the number of the stemmed terms in each document. The label weight L_i of each thesaurus is exponentially decreased and accumulated based on the increased levels.

$$f^e_{x,d} = \lambda \times \frac{L_x}{\sum_{i=0}^{n} L_i} + (1-\lambda) \times f_{x,d} \tag{10}$$

Table 1 shows the label weights of different levels, where L_0 is the document level, L_1 is one level upper, and so on to L_n for n levels upper. To build the enhanced classifiers for destination categories, the same enhancement on hierarchical label information is also applied to the destination taxonomy to strengthen the discriminative power of the classifiers.

4.3 Semantic Feature Expansion

To further improve the integration performance, the semantic information of inter-taxonomy documents is explored in the proposed approach to perform semantic feature expansion (SFE). The main idea is to augment the feature space of each document with representative topic words. As noted in [12], the hypernyms of documents can be considered as the candidates of the representative topic words for the documents. Hereby,

Table 1. The weights assigned for different labels

Hierarchical Level	Label Weight
Document Level (L_0)	$1/2^0$
One Level Upper (L_1)	$1/2^1$
Two Levels Upper (L_2)	$1/2^2$
$\vdots$	$\vdots$
n Levels Upper (L_n)	$1/2^n$

SFE adopts the similar approach as in [12] to first select important term features from the documents and then decide the representative topic terms from hypernyms.

According to previous studies [12,13,14], although the χ^2-test (chi-square) method is very effective in feature selection for text classification, it cannot differentiate negatively related terms from positively related ones. For a term t and a category c, their χ^2 measure is defined as:

$$\chi^2(t,c) = \frac{N \times (N_T^+ \times N_T^- - N_F^+ \times N_F^-)^2}{(N_T^+ + N_F^-)(N_F^+ + N_T^-)(N_T^+ + N_F^+)(N_F^- + N_T^-)} \tag{11}$$

where N is the total number of the documents, N_T^+ (N_F^+) is the number of the documents of category c (other categories) containing the term t, and N_T^- (N_F^-) is the number of the documents of category c (other categories) not containing the term t.

Therefore, the *correlation coefficient* (CC) method is suggested to filter out the negatively related terms [12,13]. Since N is the same for each term, we can omit it and get the following equation to calculate the CC value for each term:

$$CC(t,c) = \frac{(N_T^+ \times N_T^- - N_F^+ \times N_F^-)}{\sqrt{(N_T^+ + N_F^-)(N_F^+ + N_T^-)(N_T^+ + N_F^+)(N_F^- + N_T^-)}} \tag{12}$$

Since the categories in a taxonomy are in a hierarchical relationship, SFE only considers the categories of the same parent in the CC method.

Then five terms with the highest CC values are selected to perform semantic feature expansion. As indicated by [12,13], the terms selected with CC are highly representative for a category. However, the category-specific terms of a source category may not be topic-genetic to the corresponding destination category. Therefore, SFE uses them as the basis to find more topic-indicative terms for each category.

Some lexical dictionaries, such as InfoMap [15] and WordNet [16], can be used to extract the hypernyms of the category-specific terms to get the topic indicative features of a category. For example, if a category has the following five category-specific terms: *output*, *signal*, *circuit*, *input*, and *frequency*, SFE gets the following hypernyms from InfoMap: *signal*, *signaling*, *sign*, *communication*, *abstraction*, *relation*, etc. These hypernyms are more topic-generic than the category specific terms. Then SFE calculates the weight HW_x of each extracted hypernym x by

$$HW_x = \frac{HF_x}{\sum\limits_{i=1}^{n} HF_i} \tag{13}$$

Table 2. The experimental categories and the numbers of the documents

	Yahoo!	|Y-G|	|Y Class|	|Y Test|	Google	|G-Y|	|G Class|	|G Test|
Autos	/automotive/	1681	24	436	/autos/	1096	12	426
Movies	/movies_film/	7255	27	1344	/movies/	5188	26	1422
Outdoors	/outdoors/	1579	19	210	/outdoors/	2396	16	208
Photo	/photography/	1304	23	218	/photography/	615	9	235
Software	/software/	1876	15	691	/software/	5829	27	641
Total		13695	108	2918		15124	90	2932

where HF_x is the term frequency of x, and i denotes the number of the hypernyms in each category.

For each document d_k, its SFE feature vector $\mathbf{sf}_k$ is changed by extending Equation 10 as follows:

$$\mathbf{sf}_k = \lambda \times \mathbf{l}_k + (1 - \lambda) \left[\alpha \times \mathbf{h}_k + (1 - \alpha) \times \mathbf{f}_k \right] \tag{14}$$

where $\mathbf{l}_k$ denotes the feature vector of the hierarchical thesaurus information computed from the left term of Equation 10, $\mathbf{h}_k$ denotes the feature vector of the topic-generic terms of the category computed from Equation 13), and $\mathbf{f}_k$ denotes the original feature vector of the document derived from the right term of Equation 10.

5 Experimental Analysis

We have conducted experiments with real-world catalogs from Yahoo! and Google to study the performance of the SFE scheme with an Maximum Entropy classification tool from Edinburgh University (ver. 20041229) [17]. Two versions were implemented. The baseline is ME with ECI (ECI-ME), and the other is ME with ECI and SFE (SFE-ME). We measured three scores with different λ and α settings: precision, recall, and F_1 measures. The experimental results show that SFE-ME can effectively improve the integration performance. In precision and recall, SFE-ME outperforms ECI-ME in more than 70% of all cases. SFE-ME can also achieve the best recall and precision performance. In F_1 measures, SFE-ME outperforms ECI-ME in nearly all the cases. The experiments are detailed in the following. Due to the paper length limitation, this paper only reports part of our results of integrating Google taxonomies into Yahoo! taxonomies.

5.1 Data Sets

In the experiments, five directories from Yahoo! and Google were extracted to form two experimental taxonomies (Y and G). Table 2 shows these directories and the number of the extracted documents after ignoring the documents that could not be retrieved. As in previous studies [1,2,7], the documents appearing in only one category were used as the training data (|Y-G| and |G-Y|), and the common documents were used as the testing data (|Y Test| and |G Test|). Since some documents may appear in more than one category in a taxonomy, |Y Test| is slightly different with |G Test|. For simplicity

Table 3. The macro-averaged recall (MaR) measures of ECI-ME and SFE-ME from Google to Yahoo!

	ECI-ME					SFE-ME ($\alpha = 0.4$)				
	λ_d=0.10	λ_d=0.30	λ_d=0.50	λ_d=0.70	λ_d=0.90	λ_d=0.10	λ_d=0.30	λ_d=0.50	λ_d=0.70	λ_d=0.90
λ_s=0.10	0.8023	0.7491	0.7320	0.7334	0.7175	0.8935	0.8618	0.8500	0.8489	0.8678
λ_s=0.20	0.7636	0.7342	0.7274	0.7331	0.7192	0.7867	0.7845	0.7769	0.7742	0.7950
λ_s=0.30	0.7481	0.7336	0.7315	0.7333	0.7210	0.7347	0.7501	0.7511	0.7476	0.7539
λ_s=0.40	0.7422	0.7329	0.7283	0.7313	0.7197	0.7185	0.7367	0.7403	0.7374	0.7398
λ_s=0.50	0.7362	0.7299	0.7272	0.7301	0.7204	0.7085	0.7310	0.7340	0.7337	0.7346
λ_s=0.60	0.7317	0.7261	0.7262	0.7292	0.7207	0.7081	0.7284	0.7338	0.7338	0.7338
λ_s=0.70	0.7262	0.7242	0.7233	0.7263	0.7191	0.6941	0.7227	0.7333	0.7338	0.7338
λ_s=0.80	0.7231	0.7205	0.7232	0.7253	0.7235	0.6922	0.7208	0.7277	0.7304	0.7338
λ_s=0.90	0.7192	0.7205	0.7191	0.7262	0.7243	0.6922	0.7146	0.7224	0.7275	0.7304
λ_s=1.00	0.7186	0.7200	0.7181	0.7216	0.7211	0.7020	0.7138	0.7214	0.7223	0.7243

consideration, the level of each hierarchy was controlled to be at most three in the experiments. If the number of the documents of a certain subcategory is less than 10, the subcategory would be merged upward to its parent category.

Before the integration, we used the stopword list in [18] to remove the stopwords, and the Porter algorithm [19] for stemming. In the integration process, we allow that each source document d_x can be integrated into multiple destination categories (one-to-many) as what we can find in real-world taxonomies. Different λ values from 0.1 to 1.0 were applied to the source taxonomy (λ_s) and the destination taxonomy (λ_d). To both taxonomies, the same α value ranging from 0.1 to 1.0 was applied for semantic feature expansion. The lexical dictionary used in the experiments was InfoMap [15] to get hypernyms. As reported in [12], we believe that WordNet will result in similar hypernym performance.

In the experiments, we measured the integration performance of ECI-ME and SFE-ME in six scores: macro-averaged recall (MaR), micro-averaged recall (MiR), macro-averaged precision (MaP), micro-averaged precision (MiP), macro-averaged F_1 measure (MaF) and micro-averaged F_1 measure (MiF). The standard F_1 measure is defined as the harmonic mean of recall and precision: $F_1 = 2rp/(r + p)$, where recall is computed as $r = \frac{correctly\ integrated\ documents}{all\ test\ documents}$ and precision is computed as $p = \frac{correctly\ integrated\ documents}{all\ predicted\ positive\ documents}$. The micro-averaged scores were measured by computing the scores globally over all categories in five directories. The macro-averaged scores were measured by first computing the scores for each individual category, and then averaging these scores. The recall measures are used to reflect the traditional performance measurements on integration accuracy. The precision measures show the degrees of false integration. The standard F_1 measures show the compromised scores between recall and precision.

5.2 Experimental Results and Discussion

Although we have measured the integration performance with different λ values, this paper only lists part of the results in which λ_d is 0.1, 0.3, 0.5, 0.7, and 0.9, respectively. Considering α, we have also measured the integration performance with different values ranging from 0.1 to 1.0. When α is between 0.2 to 0.5, SFE-ME is superior to ECI-ME. Here we only report the $\alpha = 0.4$ case to save paper space.

Table 4. The micro-averaged recall (MiR) measures of ECI-ME and SFE-ME from Google to Yahoo!

	ECI-ME					SFE-ME ($\alpha = 0.4$)				
	λ_d=0.10	λ_d=0.30	λ_d=0.50	λ_d=0.70	λ_d=0.90	λ_d=0.10	λ_d=0.30	λ_d=0.50	λ_d=0.70	λ_d=0.90
λ_s=0.10	0.8561	0.7999	0.7873	0.7945	0.7718	0.9301	0.9096	0.8972	0.8969	0.9109
λ_s=0.20	0.8174	0.7807	0.7770	0.7934	0.7732	0.8369	0.8400	0.8325	0.8133	0.8284
λ_s=0.30	0.7989	0.7797	0.7787	0.7907	0.7746	0.7838	0.8030	0.8058	0.7921	0.7962
λ_s=0.40	0.7921	0.7797	0.7777	0.7873	0.7742	0.7698	0.7831	0.7917	0.7835	0.7849
λ_s=0.50	0.7866	0.7787	0.7773	0.7862	0.7746	0.7640	0.7804	0.7821	0.7814	0.7825
λ_s=0.60	0.7801	0.7773	0.7770	0.7859	0.7742	0.7650	0.7790	0.7818	0.7818	0.7818
λ_s=0.70	0.7780	0.7766	0.7760	0.7831	0.7739	0.7585	0.7756	0.7814	0.7818	0.7818
λ_s=0.80	0.7763	0.7739	0.7760	0.7828	0.7763	0.7575	0.7746	0.7780	0.7790	0.7818
λ_s=0.90	0.7736	0.7739	0.7732	0.7831	0.7766	0.7575	0.7715	0.7760	0.7777	0.7790
λ_s=1.00	0.7729	0.7736	0.7725	0.7801	0.7749	0.7619	0.7715	0.7753	0.7753	0.7766

Table 5. The macro-averaged precision (MaP) measures of ECI-ME and SFE-ME from Google to Yahoo!

	ECI-ME					SFE-ME ($\alpha = 0.4$)				
	λ_d=0.10	λ_d=0.30	λ_d=0.50	λ_d=0.70	λ_d=0.90	λ_d=0.10	λ_d=0.30	λ_d=0.50	λ_d=0.70	λ_d=0.90
λ_s=0.10	0.1936	0.3273	0.3356	0.3426	0.3425	0.2122	0.2980	0.3139	0.3158	0.3557
λ_s=0.20	0.3491	0.3482	0.3475	0.3459	0.3559	0.3664	0.3696	0.3572	0.3477	0.3510
λ_s=0.30	0.3890	0.3537	0.3486	0.3460	0.3547	0.4707	0.3960	0.3793	0.3523	0.3486
λ_s=0.40	0.4090	0.3613	0.3497	0.3482	0.3543	0.5794	0.4137	0.3797	0.3723	0.3531
λ_s=0.50	0.4253	0.3657	0.3515	0.3521	0.3560	0.6279	0.4649	0.3971	0.3778	0.3552
λ_s=0.60	0.4373	0.3734	0.3565	0.3588	0.3603	0.6613	0.4918	0.4192	0.3556	0.3624
λ_s=0.70	0.4455	0.3811	0.3611	0.3681	0.3655	0.6600	0.5592	0.4397	0.3663	0.3916
λ_s=0.80	0.4532	0.3876	0.3686	0.3735	0.3559	0.6607	0.6403	0.4872	0.3876	0.3333
λ_s=0.90	0.4548	0.3904	0.3747	0.3853	0.3607	0.6636	0.6543	0.5738	0.4321	0.3548
λ_s=1.00	0.4565	0.4125	0.3862	0.4070	0.3625	0.6662	0.6575	0.5955	0.5043	0.4304

Table 6. The micro-averaged precision (MiP) measures of ECI-ME and SFE-ME from Google to Yahoo!

	ECI-ME					SFE-ME ($\alpha = 0.4$)				
	λ_d=0.10	λ_d=0.30	λ_d=0.50	λ_d=0.70	λ_d=0.90	λ_d=0.10	λ_d=0.30	λ_d=0.50	λ_d=0.70	λ_d=0.90
λ_s=0.10	0.1156	0.2504	0.2715	0.2740	0.2817	0.1205	0.2835	0.3099	0.2782	0.3687
λ_s=0.20	0.2253	0.3018	0.2947	0.2822	0.3080	0.1569	0.3661	0.3570	0.3504	0.3629
λ_s=0.30	0.2741	0.3170	0.2984	0.2858	0.3107	0.2737	0.3777	0.3946	0.3515	0.3642
λ_s=0.40	0.3136	0.3329	0.3002	0.2897	0.3115	0.4721	0.3834	0.3776	0.3866	0.3688
λ_s=0.50	0.3494	0.3390	0.3033	0.2965	0.3135	0.5581	0.4556	0.3862	0.3879	0.3666
λ_s=0.60	0.3763	0.3475	0.3101	0.3101	0.3199	0.6061	0.4663	0.4032	0.3147	0.3700
λ_s=0.70	0.3906	0.3583	0.3192	0.3336	0.3317	0.6041	0.4924	0.3952	0.3180	0.4151
λ_s=0.80	0.3966	0.3759	0.3334	0.3485	0.3414	0.6016	0.5824	0.4335	0.3229	0.3452
λ_s=0.90	0.3987	0.3826	0.3402	0.3734	0.3540	0.6078	0.5871	0.4726	0.3509	0.3800
λ_s=1.00	0.3992	0.4332	0.3772	0.4198	0.3568	0.5999	0.5894	0.4937	0.3879	0.3974

Table 3 and Table 4 show the macro-averaged and micro-averaged recall results of ECI-ME and SFE-ME in different λ settings. Table 5 and Table 6 show the macro-averaged and micro-averaged precision results of ECI-ME and SFE-ME in different λ settings. Table 7 and Table 8 show the macro-averaged and micro-averaged F_1 measure results of ECI-ME and SFE-ME in different λ settings.

From Table 3 and Table 4, we can notice that SFE-ME is superior to ECI-ME in more than 75% of all MaR scores and in more than 60% of all MiR scores. Among

Table 7. The macro-averaged F_1 (MaF) measures of ECI-ME and SFE-ME from Google to Yahoo!

	ECI-ME					SFE-ME ($\alpha = 0.4$)				
	λ_d=0.10	λ_d=0.30	λ_d=0.50	λ_d=0.70	λ_d=0.90	λ_d=0.10	λ_d=0.30	λ_d=0.50	λ_d=0.70	λ_d=0.90
λ_s=0.10	0.3119	0.4556	0.4602	0.4670	0.4637	0.3430	0.4428	0.4585	0.4603	0.5046
λ_s=0.20	0.4792	0.4724	0.4703	0.4700	0.4761	0.5000	0.5025	0.4894	0.4799	0.4870
λ_s=0.30	0.5118	0.4773	0.4722	0.4702	0.4755	0.5738	0.5184	0.5041	0.4789	0.4767
λ_s=0.40	0.5274	0.4840	0.4725	0.4718	0.4748	0.6415	0.5299	0.5020	0.4948	0.4780
λ_s=0.50	0.5391	0.4872	0.4739	0.4751	0.4765	0.6658	0.5684	0.5154	0.4983	0.4789
λ_s=0.60	0.5475	0.4932	0.4782	0.4810	0.4804	0.6839	0.5871	0.5336	0.4790	0.4852
λ_s=0.70	0.5522	0.4994	0.4817	0.4886	0.4847	0.6766	0.6305	0.5497	0.4887	0.5106
λ_s=0.80	0.5572	0.5040	0.4884	0.4931	0.4771	0.6761	0.6782	0.5836	0.5065	0.4584
λ_s=0.90	0.5572	0.5064	0.4927	0.5035	0.4816	0.6776	0.6831	0.6396	0.5422	0.4776
λ_s=1.00	0.5583	0.5245	0.5022	0.5205	0.4825	0.6836	0.6845	0.6525	0.5939	0.5399

Table 8. The micro-averaged F_1 (MiF) measures of ECI-ME and SFE-ME from Google to Yahoo!

	ECI-ME					SFE-ME ($\alpha = 0.4$)				
	λ_d=0.10	λ_d=0.30	λ_d=0.50	λ_d=0.70	λ_d=0.90	λ_d=0.10	λ_d=0.30	λ_d=0.50	λ_d=0.70	λ_d=0.90
λ_s=0.10	0.2037	0.3814	0.4037	0.4075	0.4128	0.2133	0.4322	0.4607	0.4246	0.5249
λ_s=0.20	0.3533	0.4353	0.4273	0.4163	0.4406	0.2642	0.5099	0.4997	0.4897	0.5047
λ_s=0.30	0.4082	0.4508	0.4314	0.4199	0.4435	0.4058	0.5138	0.5298	0.4869	0.4998
λ_s=0.40	0.4493	0.4666	0.4332	0.4236	0.4443	0.5852	0.5148	0.5113	0.5177	0.5018
λ_s=0.50	0.4838	0.4723	0.4363	0.4306	0.4463	0.6450	0.5753	0.5170	0.5184	0.4993
λ_s=0.60	0.5077	0.4803	0.4433	0.4447	0.4527	0.6764	0.5834	0.5320	0.4487	0.5023
λ_s=0.70	0.5201	0.4904	0.4523	0.4679	0.4644	0.6725	0.6024	0.5249	0.4521	0.5422
λ_s=0.80	0.5250	0.5060	0.4664	0.4823	0.4742	0.6706	0.6649	0.5568	0.4565	0.4789
λ_s=0.90	0.5262	0.5121	0.4725	0.5057	0.4863	0.6744	0.6668	0.5874	0.4835	0.5108
λ_s=1.00	0.5264	0.5554	0.5069	0.5458	0.4887	0.6713	0.6682	0.6032	0.5171	0.5257

these cases, SFE-ME can achieve the best MaR of 0.8915 and the best MiR of 0.9301 when $\lambda_s = 0.1$ and $\lambda_d = 0.1$. When $\lambda_d = 0.1$ and $\lambda_s \geq 0.3$, ECI-ME outperforms SFE-ME. It appears that the imbalanced weighting between λ_d and λ_s seriously impairs the SFE improvement.

From Table 5 and Table 6, we can notice that SFE-ME is superior to ECI-ME in more than 80% of all MaP and MiP scores. Among these cases, SFE-ME can achieve the best MaR of 0.6662 $\lambda_s = 1.0$ and $\lambda_d = 0.1$ and the best MiR of 0.6078 when $\lambda_s = 0.9$ and $\lambda_d = 0.1$. It appears that if the portion of the mis-integrated documents is reduced, SFE has lower recall scores but better precision performance.

For many applications, a compromised performance may be required with a high F_1 score. From Table 7 and Table 8, we can notice that SFE-ME is superior to ECI-ME in 88% of all MaF and MiF scores. In our experiments with $\alpha = 0.4$, SFE-ME achieves the best MaF (0.6839) and the best MiF (0.6764) when $\lambda_s = 0.6$ and $\lambda_d = 0.1$. It reveals that the SFE scheme can mostly get more balanced improvements in both recall and precision considerations.

We have also measured these six scores for the $\lambda_s = 0.0$, $\lambda_d = 0.0$, and $\alpha = 0.0$ case which means that the integration is performed by only ME without ECI and SFE enhancements. In this configuration, ME can achieve the highest MaR (0.9578) and MiR (0.9616) but with very low MaP (0.0111) and MiP (0.0111). Its MaF and MiF are

0.022 and 0.0219, respectively. Although ME can get the best recall performance, it allows many documents of other categories to be mis-integrated.

The experimental results are positive to show that SFE-ME can get more improved integration performance with the SFE scheme. Compared with ECI-ME, SFE-ME shows that the semantic information of the hypernyms of the category-specific terms can be used to facilitate the integration process between two hierarchical taxonomies.

6 Conclusions

In recent years, the taxonomy integration problem has been progressively studied for integrating two homogeneous hierarchical taxonomies. Many sorts of implicit information embedded in the source taxonomy are explored to improve the integration performance. However, the semantic information embedded in the source taxonomy has not been discussed in the past research.

In this paper, an enhanced integration approach (SFE) is proposed to exploit the semantic information of the hypernyms of the category-specific terms. Augmented with these additional semantic category features, the source documents can be more precisely integrated into the correct destination category in the experiments. The experimental results show that SFE-ME can achieve the best macro-averaged F_1 score and the best micro-averaged F_1 score. The results also show that the SFE scheme can get precision and recall enhancements in a significant portion of call cases.

There are still some issues left for further discussion. For example, we do not yet clearly know whether SFE can be applied to other classification schemes, such as SVM and NB. In addition, an open issue is whethere there is other implicit information embedded in the source taxonomy with more powerful discriminative capability. We believe that the integration performance can be further improved with appropriate assistance of more effective auxiliary information and advanced classifiers.

Acknowledgement

This work was supported in part by National Science Council of R.O.C. under grant NSC 96-2422-H-006-002 and NSC 96-2221-E-155-067. The authors would also like to express their sincere thanks to anonymous reviewers for their precious comments.

References

1. Agrawal, R., Srikant, R.: On Integrating Catalogs. In: Proceedings of the 10th International Conference on World Wide Web, pp. 603–612 (2001)
2. Sarawagi, S., Chakrabarti, S., Godbole, S.: Cross-training: Learning Probabilistic Mappings between Topics. In: Proceedings of the 9th ACM SIGKDD International Conference on Knowledge Discovery and Data Mining, pp. 177–186 (2003)
3. Zhang, D., Lee, W.S.: Web Taxonomy Integration using Support Vector Machines. In: Proceedings of the 13th International Conference on World Wide Web, pp. 472–481 (2004)
4. Zhang, D., Lee, W.S.: Web Taxonomy Integration Through Co-Bootstrapping. In: Proceedings of the 27th annual international ACM SIGIR Conference on Research and Development in Information Retrieval, pp. 410–417 (2004)

5. Wu, C.W., Tsai, T.H., Hsu, W.L.: Learning to Integrate Web Taxonomies with Fine-Grained Relations: A Case Study Using Maximum Entropy Model. In: Lee, G.G., Yamada, A., Meng, H., Myaeng, S.-H. (eds.) AIRS 2005. LNCS, vol. 3689, pp. 190–205. Springer, Heidelberg (2005)
6. Chen, I.X., Ho, J.C., Yang, C.Z.: An Iterative Approach for Web Catalog Integration with Support Vector Machines. In: Lee, G.G., Yamada, A., Meng, H., Myaeng, S.-H. (eds.) AIRS 2005. LNCS, vol. 3689, pp. 703–708. Springer, Heidelberg (2005)
7. Ho, J.C., Chen, I.X., Yang, C.Z.: Learning to Integrate Web Catalogs with Conceptual Relationships in Hierarchical Thesaurus. In: Ng, H.T., Leong, M.-K., Kan, M.-Y., Ji, D. (eds.) AIRS 2006. LNCS, vol. 4182, pp. 217–229. Springer, Heidelberg (2006)
8. Krikos, V., Stamou, S., Kokosis, P., Ntoulas, A., Christodoulakis, D.: DirectoryRank: Ordering Pages in Web Directories. In: Proceedings of 7th ACM International Workshop on Web Information and Data Management (WIDM 2005), pp. 17–22 (2005)
9. Hsu, M.H., Tsai, M.F., Chen, H.H.: Query Expansion with ConceptNet and WordNet: An Intrinsic Comparison. In: Ng, H.T., Leong, M.-K., Kan, M.-Y., Ji, D. (eds.) AIRS 2006. LNCS, vol. 4182, pp. 1–13. Springer, Heidelberg (2006)
10. Berger, A.L., Pietra, V.J.D., Pietra, S.A.D.: A Maximum Entropy Approach to Natural Language Processing. In: Computational Linguistics, pp. 39–71 (1996)
11. Darroch, J.N., Ratcliff, D.: Generalized Iterative Scaling for Log-linear Models. Annals of Mathematical Statistics (43), 1470–1480 (1972)
12. Tseng, Y.H., Lin, C.J., Chen, H.H., Lin, Y.I.: Toward Generic Title Generation for Clustered Documents. In: Ng, H.T., Leong, M.-K., Kan, M.-Y., Ji, D. (eds.) AIRS 2006. LNCS, vol. 4182, pp. 145–157. Springer, Heidelberg (2006)
13. Ng, H.T., Goh, W.B., Low, K.L.: Feature selection, Perception Learning, and a Usability Case Study for Text Categorization. In: Proceedings of the 20th Annual International ACM SIGIR Conference on Research and Development in Information Retrieval, pp. 67–73 (1997)
14. Yang, Y., Pedersen, J.O.: A Comparative Study on Feature Selection in Text Categorization. In: Proceedings of the 14th International Conference on Machine Learning (ICML 1997), pp. 412–420 (1997)
15. Information Mapping Project: Computational Semantics Laboratory, Stanford University, http://infomap.stanford.edu/
16. WordNet: A lexical database for the English language: Cognitive Science Laboratory, Princeton University, http://wordnet.princeton.edu/
17. Zhang, L.: Maximum Entropy Modeling Toolkit for Python and C++, http://homepages.inf.ed.ac.uk/s0450736/maxent.html
18. Frakes, W., Baeza-Yates, R.: Information Retrieval: Data Structures and Algorithms, 1st edn. Prentice Hall, PTR, Englewood Cliffs (1992)
19. The Porter Stemming Algorithm, http://tartarus.org/~martin/PorterStemmer

HOM: An Approach to Calculating Semantic Similarity Utilizing Relations between Ontologies

Zhizhong Liu, Huaimin Wang, and Bin Zhou

College of computer science, National University of defense Technology, Changsha China
liuzane@msn.com, whm_w@163.net, bin.zhou.cn@gmail.com

Abstract. In the Internet environment, ontology heterogeneity is inevitable due to many coexistent ontologies. Ontology alignment is a popular approach to resolve ontology heterogeneity. Ontology alignment establishes the relation between entities by computing their semantic similarities using local or/and non-local contexts of entities. Besides local and non-local context of entities, the relations between two ontologies are helpful for computing their semantic similarity in many situations. The aim of this article is to improve the performance of ontology alignment by using these relations in similarity computing. A hierarchical Ontology Model (HOM) which describes these relations formally is proposed followed by HOM-Matching, an algorithm based on HOM. It makes use of the relations between ontologies to compute semantic similarity. Two groups of experiments are conducted for algorithm validation and parameters optimization.

1 Introduction

Ontologies—an explicit specification of a conceptualization [1]—facilitate knowledge sharing and semantic interoperability between different systems. However, in the Semantic Web environment, data come from different ontologies, and information processing across ontologies is impossible without knowing the semantic mappings between them. Ontology alignment or ontology mapping[2-4]establishes relationships between entities in different ontologies according to semantic similarities between them. Most available approaches used local or/and non-local contexts of entities within ontology to compute semantic similarity. Besides those information, the relations between ontologies are helpful for computing the semantic similarity between different ontology entities.

For example, suppose two ontologies O_1, O_2, O_1 is the Economy ontology, O_2 is Transportation ontology, and both of them extend Mid-Level-Ontology. Mid-Level-Ontology provides much heuristic information to align O_1, O_2.

In this article, we focus on aligning ontologies utilizing the relations between them. The goal of our approach is not to provide a complete solution to automated ontology alignment but rather to augment existing methods by determining additional possible points of semantic similarity between ontology entitiese. To begin with, we propose a hierarchical ontology model (HOM) to express the relation between ontologies formally. Based on HOM, we present an algorithm HOM-Matching, which utilizes

H. Li et al. (Eds.): AIRS 2008, LNCS 4993, pp. 237–245, 2008.

these relations in computing similarity between ontology entities. Given a pair of entities, similarity computation consists of four steps: 1) identifying their tracks; 2) obtaining LUBC (Least Upper Bound Concept) of entity pair; 3) identifying different partition of the revised tracks of concepts; 4) computing external structure similarity.

The rest of this paper is organized as follows. Section 2 reviews the related works. The hierarchical ontology model is expounded in section 3. The detail of HOM-Matching is presented in section 4. And section 5 gives some experiments and analysis of their results. Finally conclusion was made in section 6.

2 Related works

In different context, there are different methods to compute similarity during ontology alignment. J. Euzenat etc.[2] classified them as six categories, namely terminological, internal structure comparison, external structure comparison, extensional comparison and semantic comparison.

The semi-automated approaches of ontology alignment that do exist today such as PROMPT[5] and Chimaera[6] analyze only local context in ontology structure, i.e. given two similar classes, the algorithms consider classes and slots that are directly related to the classes in question. External structure comparison compares the relations of the entities with other entities. Rose Dieng [7] describes ontology as conceptual graph, and matches conceptual graphs by comparing super-classes and subclasses. Steffen Staab[8] computes the dissimilarity between two taxonomies by comparing for each class the labels of their superclasses and subclasses. Anchor-Prompt[9] uses a bounded path comparison algorithm with the originality that anchor points can be provided by the users as a partial alignment. Anchor-PROMPT takes as input a set of anchors—pairs of related terms defined by the user or automatically identified by lexical matching. Anchor-PROMPT treated ontology as a graph which represents classes as nodes and slots as links. The algorithm analyzes the paths in the sub-graph limited by the anchors, and determines which classes frequently appear in similar positions on similar paths. These classes are likely to represent semantically similar concepts.

To align two representations of anatomy at the lexical and structural level, Songmao Zhang[10] presented a novel method. They considered alignment consists of the following four steps: 1) acquiring terms, 2) identifying anchors (i.e., shared concepts) lexically, 3) acquiring explicit and implicit semantic relations, and 4) identifying anchors structurally. Their method aligned the representations using the shared concepts.

Bernstein and colleagues created the SimPack framework[11] that uses a set of similarity measures to calculate the similarity of different concepts. SemMF[12] describes three kinds of concept matching techniques implemented: the string matcher, the numeric matcher, and the taxonomic matcher exploiting a concept hierarchy. The taxonomic matcher computes the similarity between two concepts c1 and c2 based on the distance dc(c1, c2) between them, which represents the path over the closest common parent (ccp). And it presented two alternative calculators to compute the distance. Mark Hefke and colleagues[13] present the conceptual basis and a prototypical implementation of a software framework for syntactical and semantic similarities between ontology instances. They calculated the taxonomic similarity of two instances by looking at the relative taxonomic position of the concepts of the regarded instances.

HOM-Matching, which we present here, complements these approaches by analyzing the relations between aligned ontologies, and by providing additional suggestions for possible matching terms.

3 Hierarchical Ontology Model (HOM)

Category theory, which is independent of special model and language, is a useful tools to express ontology[14-16]. In this section, we illustrate hierarchical ontology model (HOM), which formally describes the relations between ontologies sharing same imported ontologies, with category theory firstly. First of all, we give the definition of ontology formally.

Definition 1 Ontology (O). Ontology is 4-tuple $O := (C, R_C, H^C, A)$, consisting of two disjoint sets C and R_C, whose elements are called concepts and relation identifiers, respectively, a concept hierarchy H^C : H^C is a directed, transitive relation, $H^C \subseteq C \times C$,which is also called concept taxonomy. $H^C(C_1, C_2)$ means that C_1 is a super-concept of C_2, and a set of axioms A, whose elements characterize the relations. It is obvious that $H^C \subseteq R_C$. Therefore we denote ontology as 3-tuple $O := (C, R_C, A)$ unless focusing on hierarchy

Definition 2 Ontology Model (OM). Ontology Model is a tuple $OM := (O, R_o)$, where:
O is the set of ontologies used in the model;
$R_o \subseteq O \times O$ is binary relations on O.
We take hierarchical relation R_H into account. $R_H \subseteq R_O$, which is defined as follows.

Definition 3 Hierarchical Relation (R_H). Given two ontologies $O_1 = (C_1, R_{C_1}, A_1)$, $O_2 = (C_2, R_{C_2}, A_2)$ and any formula F, $(O_1, O_2) \in R_H$, if and only if: [1]

(1) $C_1 \subseteq C_2$

(2) $R_{C_1} \subseteq R_{C_2}$

(3) $A_1 \mapsto F \Rightarrow A_2 \mapsto F$

Formula (3) means that any formula inferred from Axioms A_1 can also be inferred from Axiom A_2.

Definition 4 Hierarchical Ontology Model (HOM). $HOM := (O, R_H)$ is a hierarchical ontology model, if R_H are the relation defined as definition 3.

Definition 5 Common Ancestor (CA). Given two ontologies $O_1 = (C_1, R_{C_1}, A_1)$, $O_2 = (C_2, R_{C_2}, A_2)$, $CA = \{O \mid R_H(O, O_1) \wedge R_H(O, O_2)\}$. That's to say, CA is the set of share imported ontologies of O_1, O_2.

---

[1] " $A \mapsto F$ " denotes formula F can be inferred from axioms A;" $\Rightarrow$ " is logical implication.

The structural topological dissimilarity δ^s [17] on a domain presented by Valtchev follows the graph distance. And Mädche and Zacharias[13] described the upward cotopy distance for computing the similarity between entities. These measures can be applied as such in the context of ontology alignment since the ontologies are supposed to share the same taxonomy H. In this paper, we revise the similarity used in SemMFP[16]P, where the similarity is calculated based on the path.

The concept similarity is defined as: $sim(C_1, C_2) = 1 - dc(C_1, C_2)$ within a single ontology. Similarity between two concepts from different ontologies is defined as: $sim(C_1, C_2) = sim(C_1, LUBC) * sim(C_2, LUBC)$ with the help of their LUBC. Therefore the similarity between two given concepts in different hierarchy is calculated as: $Sim(C_1, C_2) = (1 - d_{c1}(C_1, LUBC)) * (1 - d_{c2}(C_2, LUBC))$, where semantic distances $d_c(C, LUBC)$ are calculated with:

$$d_c(C, LUBC) = \frac{\lambda_1 * P_1 + \lambda_2 * P_2}{L(N)}$$

where $L(N)$ represents length of the track of concept and λ_1 is the weight for revised track in LCA ; λ_2 is the weight for revised track out of LCA; P_1 is the length of revised track in LCA; P_2 is the length of the revised track out of LCA. To ensure $1 > Sim(C_1, C_2) > 0$, the parameters' scope are set as $1 > \lambda_1, \lambda_2 > 0$. This formula implies that the semantic distance between concept and LUBC increases and semantic similarity decreases with the length of revised track of concepts.

4.2 Algorithm

In this section, we proposed the related algorithm for computing the similarity using LUBC. Figure 2 shows the detail of the algorithm.

Algorithm 1. Calculating semantic similarity between two concepts in different ontologies

```
ComputeSimilarity(Concept C1,Ontology O1,Concept C2,Ontology O2) : Similarity Sim
Get the track of C1 in ontology O1
Get the track of C2 in ontology O3
Get the Common Ancestor Concepts (CAC) of C1,C2
LUBC=thing
    Foreach concept LUBC' in CAC do
        If LUBC is parent LUBC' then
            LUBC=LUBC'
        endif
    endfor
Get the Latest Common Ancestor LCA
D1=Distance(C1,LUBC,O1,LCA)
D2=Distance(C2,LUBC,O2,LCA)
S=(1-D1)*(1-D2)
Return S
```

Fig. 2. Algorithm of Similarity Calculation

```
Algorithm 2. Computing distance between two entities within ontology
Distance( Concept C, Concept LUBC, Ontology O, OntologyLCA): Distance D
Get RevisedTrack of C with LUBC
    Foreach concept RC in RevisedTrack do
        If RC in LCA then
            D = D + λ1 / L(N)
        Else
            D = D + λ2 / L(N)
        Endif
    Endfor
Return D
```

Fig. 3. Algorithm of Semantic Distance Calculation

The algorithm computes the Tracks of the concepts respectively. Then LUBC and LCA of concept pair are calculated. Based on these, the distances between concepts and LUBC are accumulated along the revised track of the concept respectively. Finally, we get the similarity across ontologies based on these distances.

5 Evaluation

Our approach is to facilitate ontology alignment and improve its performance. In this section, we conduct two groups of experiments to validate our algorithm's efficiency and optimize the parameters of our algorithms.

5.1 Measure and Evaluation Scenarios

Standard information retrieval metrics are adopted to evaluate the performance of alignment:

$$\text{Precision } p = \frac{\# \textit{correct found mappings}}{\# \textit{found mappings}} \qquad \text{Recall } r = \frac{\# \textit{correct found mappings}}{\# \textit{existing mappings}}$$

$$\text{F-Measure } f = \frac{2pr}{p+r} \psi$$

Precision is the ratio of correct found mappings to found mappings; Recall is the ratio of correct found mappings to existing mappings; F-Measure is harmonic means of precision and recall. We consider the F-Measure as most relevant for our evaluation since it balances well precision and recall. If the focus were laid more onto precision or recall, as may be necessary for specific use cases, slight changes would be necessary.

When evaluating the performance of ontology alignment, we aggregate different local semantic similarities, including terminological similarity with and without lexicon; internal structure similarity and external structure similarity, to get global similarity. And the terminological similarity (Sim_T) is computed by longest-common-substring; further with the help of Java WordNet library (JWNL)[20], the terminological similarity

(Sim_{TM}) is computed; the internal structure comparison compares (Sim_{IS}) the properties of classes, include properties' name, rang and domain. The external structure similarity (Sim_{ES}) is calculated by our algorithm. The global similarity is the weighted sum of these local similarities:

$$Sim(C_1,C_2) = \omega_1 * Sim_T(C_1,C_2) + \omega_2 * Sim_{TW}(C_1,C_2) + \omega_3 Sim_{IS}(C_1,C_2) + \omega_4 Sim_{ES}(C_1,C_2)$$

where $\omega_1, \omega_2, \omega_3, \omega_4$ are the weights and $\omega_1 + \omega_2 + \omega_3 + \omega_4 = 1$.

In first group experiments, where different weights assignments for different local similarities are tested according to the performance of ontology alignment, we align two pair ontologies, BibTeX/MIT and ATO-Mission/ ATO-Task. ATO-Mission/ ATO-Task sharing same imported ontologies come from SUO ontology library. BibTeX/MIT, provided for the alignment contest I3Con, is independent ontologies. And the weights for other local similarities are equal except for external structure similarity. In these experiments, the parameters are set as follows: $0.8 = \lambda_2 > \lambda_1 = 0.6$, and $\omega_1 = \omega_2 = \omega_3 = (1 - \omega_4)/3$.

In our algorithm, the parameters λ_1, λ_2 impact the performance of ontology alignment. To get the optimal setting of λ_1, λ_2, we conduct the second group experiments, where we align ATO-Mission/ ATO-Task only and assign weights for different similarities as $\omega_1 = \omega_2 = \omega_3 = 0.2; \omega_4 = 0.4$, which is the optimal setting getting in first group experiments. These experiments show the relation between λ_1, λ_2 and the performance of ontology alignment.

5.2 Results and Discussion

In our experiments, we prune the available mappings by n-percent method[13]. In this method, the mapping whose global similarity is over the cut-off value are preserved and the cut-off value is defined by taking the highest similarity value of all and subtracting a fixed percentage from it. In our experiments, the fixed percentage is set to 50%.

The experimental results are shown as figure 4. Figure 4 (a) show the relation between the weights for local similarities and F-Measure of ontology alignment while aligning different pairs of ontologies. Figure4 (b) shows the relation between λ_1, λ_2 and F-Measure of ontology alignment while aligning the ontology sharing imported ontologies.

From those experimental results, we get following conclusions:

1) While aligning ontologies not sharing same imported ontologies, HOM-Matching almost does not affect the performance of ontology alignment, shown as fig.4 (a).
2) Assigned appropriate weight (about 0.4), the F-Measures of ontology alignment are improved observably while aligned ontologies share same imported ontologies(Fig.4(a)).
3) While aligning ontologies shared imported ontologies, assigning proper weights to different local similarities ($\omega_1 = \omega_2 = \omega_3 = 0.2$ and $\omega_4 = 0.4$), the weights for different parts of revised track of concept λ_1, λ_2 impact the performance of ontology alignment(Fig.5(b)). The figure shows that the optimum setting is about $0.8 = \lambda_2 > \lambda_1 = 0.6$.

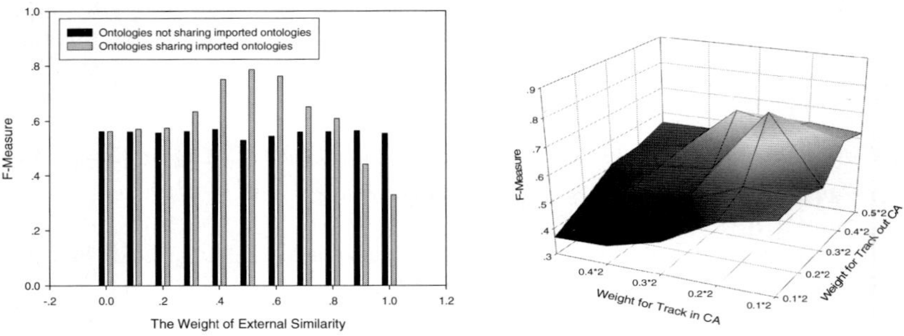

Fig. 4. Relation between parameters of algorithm and F-Measure of ontology alignment

In our approach, if $\lambda_1 = \lambda_2$, and $O_1 = O_2$, the similarity measure equals the taxonomic matcher of SemMF. That is to say, our approach adapts this matcher to computing semantic similarity between entities from different ontologies.

6 Conclusion and Future Works

In open and distributed computing environment, autonomic agents and services may employ different ontologies. High-quality alignment between ontologies is a necessary precondition to establish semantic interoperability between these agents or services. Ontologies describing the same or relevant domain share same imported ontologies in most situations in Semantic web. We propose a hierarchical ontology model (HOM) to make the relation between ontologies explicit. Basing on HOM, HOM-Matching utilizes the relations between ontologies in computing semantic similarity between entities. The experimental results show that HOM-Matching improves the performance of ontology alignment when the parameters are set rationally and aligned ontologies share imported ontologies.

In future, we will present and compare some other calculators that computing the semantic distance between entities in different context. Further, we will use the obtained alignment to discover and match web service semantically in multiple-ontology environment.

Acknowledgements. Research reported in this paper has been partially financed by National Basic Research Program of China under Grant, No.2005CB321800, National Natural Science Foundation of China under Grant, No.90412011 and National High-Tech Research and Development Plan of China under Grant, Nos. 2003AA115210, 2003AA115410.

References

Gruber, T.: Ontolingua: A translation approach to portable ontology specifications. Knowledge Acquisition 5(2), 199–220 (1993)

Bouquet, J.E.P., Franconi, E., Serafini, L., Stamou, G., Tessaris, S.: The state of art of ontology alignment. Deliverable D2.2.3. Knowledge web (2004)

Kalfoglou, Y., Schorlemmer, M.: Ontology mapping: the state of the art. The Knowledge Engineering Review 18(1), 1–31 (2003)

Ehrig, M., Sure, Y.: Ontology mapping - an integrated approach. In: Bussler, C.J., Davies, J., Fensel, D., Studer, R. (eds.) ESWS 2004. LNCS, vol. 3053, pp. 76–91. Springer, Heidelberg (2004)

N.F., Musen Noy, M.A.: PROMPT: Algorithm and Tool for Automated Ontology Merging and Alignmenteditors. In: Proceedings of the Seventeenth National Conference on Artificial Intelligence (AAAI-2000), Austin, TX (2000)

Fikes, R., Mcguinness, D.L., Rice, J., Wilder, S.: An environment for merging and testing large ontologieseditors. In: Proceeding of KR 2000, pp. 483–493 (2000)

Dieng, R., Hug, S.: Comparison of personal ontologies represented through conceptual graphs. In: Proc. of 13th ECAI 1998, Brighton, UK, pp. 341–345 (1998)

Staab, S., Mädche, A.: Measuring similarity between ontologies. In: Gómez-Pérez, A., Benjamins, V.R. (eds.) EKAW 2002. LNCS (LNAI), vol. 2473, pp. 251–263. Springer, Heidelberg (2002)

Noy, N., Musen, M.: Anchor-PROMPT: Using non-local context for semantic matchingeditors. In: Proc. IJCAI 2001 workshop on ontology and information sharing, Seattle, pp. 63–70 (2001)

Zhang, S.M., Bodenreider, O.: Aligning Representations of Anatomy using Lexical and Structural Methods. In: 2003 editors Proceedings of AMIA Annual Symposium, USA, pp. 753–757 (2003)

Bernstein, A., Kaufmann, E., Bürki, C., Klein, M.: Object Similarity in Ontologies: A Foundation for Business Intelligence Systems and High-Performance Retrieval. In: Proc. of 25th Int. Conf. on Information Systems, pp. 741–756 (2004)

Oldakowski, R., Bizar, C.: SemMF: A Framework for Calculating Semantic Similarity of Objects Represented as RDF Graphseditors. In: 4th Int. Semantic Web Conference (2005)

Hefke, V.Z.M., Abecker, A., Wang, Q.: An Extendable Java Framework for Instance Similarity in Ontologies. In: Yannis Manolopoulos, J.F., Constantopoulos, P., Cordeiro, J. (eds.) Proceedings of the Eighth International Conference on Enterprise Information Systems: Databases and Information Systems Integration, Paphos, Cyprus, pp. 263–269 (2006)

Krötzsch, P.H.M., Ehrig, M.: York Sure Category. Theory in Ontology Research: Concrete Gain from an Abstract Approach. AIFB, Universität Karlsruhe (2005)

Kent, R.: A KIF formalization of the IFF category theory ontology. In: Proc. IJCAI 2001 Workshop on the IEEE Standard Upper Ontology, Seattle Washington, USA (2001), http://citeseer.ist.psu.edu/kent01kif.html

Zimmermann, M.K.A., Euzenat, J., Hitzler, P.: Formalizing Ontology Alignment and its Operations with Category Theory. In: Fellbaum, B.B.a.C. (ed.) Proceedings of the Fourth International Conference on Formal Ontology in Information Systems (FOIS 2006). Frontiers in Artificial Intelligence and Applications, vol. 150, pp. 277–288. IOS Press, Amsterdam (2006)

Valtchev, P., Euzenat, J.: Dissimilarity Measure for Collections of Objects and Values. In: Liu, X., Cohen, P.R., R. Berthold, M. (eds.) IDA 1997. LNCS, vol. 1280, pp. 259–272. Springer, Heidelberg (1997)

JWNL: Java WordNet Library (2004), http://sourceforge.net/projects/jwordnet

A Progressive Algorithm for Cross-Language Information Retrieval Based on Dictionary Translation

Song An Yuan and Song Nian Yu

School of Computer Engineering and Science, Shanghai University,
149 Yan Chang Road, Shanghai, 200072, P.R. China
yuansongan@sina.com.cn, snyu@staff.shu.edu.cn

Abstract. Query translation is the mainstream in cross-language information retrieval, but ambiguity must be resolved by methods based on dictionary translation. In this paper, we propose a progressive algorithm for disambiguation which is derived from another algorithm we propose called the max-sum model. The new algorithm take a strategy called weighted-average probability distribution to redistribute the probabilities. Moreover, the new algorithm can be computed in a more direct way by solving an equation system. All the resource our method requires is a bilingual dictionary and a monolingual corpus. Experiments show it outperforms four other methods.

Keywords: Cross-language information retrieval, Co-occurrence measures, Max-sum model, Weighted-average distribution.

1 Introduction

The common way to overcome the language barrier in cross-language information retrieval (CLIR) is to translate either the query or the documents. Since the price of document translation is too high, query translation becomes the mainstream though the former can get a better translation. The problem with query translation is how to overcome translation ambiguity, and many methods are proposed for disambiguation.

All the methods use a corpus for training, and according to the language of the corpus, they are divided into two categories: one is to use parallel bilingual corpus; the other is to use monolingual corpus. In the former, there are approaches such as statistical translation models [1, 2] and relevance language models [3, 4]. However, all these methods are very time-consuming and it's hard to acquire large parallel bilingual corpus, especially of minor languages. On the other hand, monolingual corpus is easy to get and methods based on it always use term (or word) co-occurrence statistics to resolve the translation ambiguity [5, 6, 7, 8]. In general, monolingual corpus is more widely used.

Translation usually involves two aspects: term and grammar. In the case of query translation, the queries are short and often given in irregular grammar, sometimes even without any sentence structure, hence correct translation of a term is more important, while the grammar can be ignored somehow. In such situation, only a dictionary is needed, so we call it dictionary translation. However, this increases the difficulty of translation because each term maybe has several entries in the dictionary

H. Li et al. (Eds.): AIRS 2008, LNCS 4993, pp. 246–255, 2008.

and combinations of every translation candidate are large. To select the correct translation combination out of n^k candidate combinations (n is the average number of entries each term has and k is the number of terms a query has) is a very difficult task, and it's very probable to be wrong if only the best translation combination is returned by a method ([5, 6] use such strategy). Therefore, it's far more accepted to compute the probability distribution of translations for each term than make a simple binary decision among the translations.

In this paper, we propose a progressive algorithm for computing the translation probability. Its original form is from another unsatisfactory progressive algorithm we propose called max-sum model. By analyzing the model's drawbacks, we replace its strategy on redistribution of translation probability by a more reasonable one. What's more interesting is that the new progressive algorithm can be computed in a more direct way by solving an equation system. We inspect the performance of the algorithm on SougouT2.0. Experiments show it outperforms four other methods.

The remainder of this paper is organized as follows. Section 2 describes the max-sum model in detail and its drawbacks. Section 3 proposes the new algorithm and how to compute it directly. Section 4 presents the experiment setup and its results. In Section 4, we draw our conclusions and have an outlook of further work.

2 Max-Sum Model

To choose the best translation for a query by using a monolingual corpus, a simple way is to choose the most frequent translation combination. For example, a query has three source terms, s_1, s_2 and s_3, and each has 1, 2 and 3 translations respectively. Then the number of all the possible translation combinations is 6. For every combination, we count the number of documents where all the translation terms of this combination appear and the highest one is regarded as the best translation combination. But this method is infeasible, because the time it takes is huge and it can't be done beforehand. Moreover, it needs a very large corpus to avoid data-sparseness when the query is long. To overcome this problem, we do it in an approximate way. We only compute the co-occurrences of each pair of terms, and sum them up to approximate the frequency of a translation combination. This approximate method is more understandable in a graph as shown in Fig.1.

In Fig.1, $t_{i,j}$ stands for the translations of the source term s_i, and edges are only between translation terms of different source terms. We can define the graph in a rigid way. Let T be the collection of all the translation terms, S= $\{S_1, S_2,\ldots, S_k\}$ be a partition of T, and A be the adjacency matrix of the graph, then the graph $G(T,S)$ can be described as a graph satisfies:

$$a_{i,j} = \begin{cases} 1 & i \in S_l, j \in S_m, l \neq m \\ 0 & otherwise \end{cases} . \quad (1)$$

For convenience, we define $C(i)$ as the set of all the terms which have an edge with i, E be the set of all the edges, and $S(i)$ as the set S_j which contains term i. Each edge $e_{i,j}$ has a weight value $w_{i,j}$, namely the co-occurrences of the two terms. There are

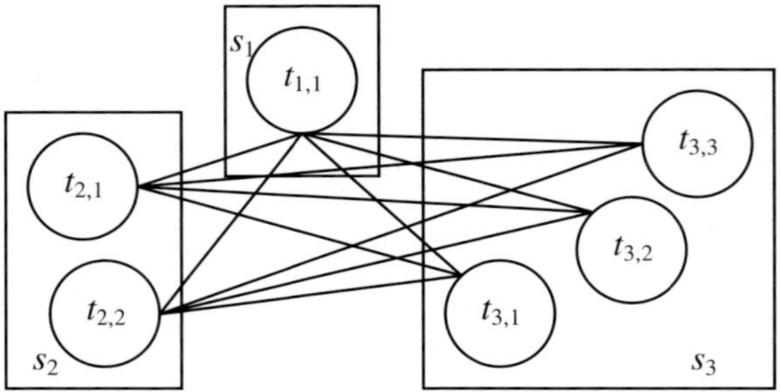

Fig. 1. Co-occurrences of each pair of terms

several kinds of measures, and here we use the measure called dice coefficient [9] which is defined in (2). The *freq(i,j)* is the frequency *i* and *j* co-occur in same document.

$$DC(i,j) = \frac{2 \cdot freq(i,j)}{freq(i) + freq(j)} \tag{2}$$

In *G*(*T*,*S*), each translation combination constitutes a *k*-complete subgraph (as shown in Fig.1, *k* is the number of source terms). By summing up all the edges' weights, a score approximate to the frequency of the combination can be obtained. Hence the *k*-complete subgraph *G*(*T*,*S*) with the highest score is regarded as the best translation. As mentioned above, we prefer to compute the probability distribution of a source term rather than make a simple decision. Some modification should be done to the highest-score *k*-complete subgraph method to suit the probability form. We specify a probability to each translation term *i*, notated by p_i, and all the p_i of some S_j sum to 1. Instead of computing the highest-score *k*-complete subgraph, we compute the maximum value of (3), which is also used in a similar form in [8].

$$\sum_{e_{i,j} \in E} p_i \cdot w_{i,j} \cdot p_j \ . \tag{3}$$

In fact, the computation of the highest-score *k*-complete subgraph is a special case of (3) when the probabilities are 0-1 distributions. But very strangely, though the formula generalized to the probability form, the maximum doesn't exceed the highest score of *k*-complete subgraph according to the following theorem.

Theorem 1. In *G*(*T*,*S*), the maximum of formula (3) equals to the highest score of *k*-complete subgraph.

Proof. Suppose there exists some probability distribution which makes the maximum of (3) larger than the highest score of *k*-complete. There must be some S_i where the probability is not 0-1 distribution. Rewrite (3) as following:

$$(3) = \sum_{j \in S_i} p_j \cdot \sum_{l \in C(j)} p_l \cdot w_{j,l} \le \max_{j \in S_i} \sum_{l \in C(j)} p_l \cdot w_{j,l} (\because \sum_{j \in S_i} p_j = 1) .$$

That is to say, we can redistribute the probability to make the inequality above equal, by letting p_j of the max j in the right side of inequality above be 1 and others be 0. Thus, the probability of S_i becomes a 0-1 distribution, and the value of (3) is no less than the original. Because the number of S_i is finite, we can redistribute the probability of all S_i to 0-1 distribution in finite steps and the value of (3) will not decrease in each step. So the original value of (3) can't be larger than some k-complete subgraph, and the supposition can't be true. □

From the proof of Theorem 1, a progressive algorithm for computing the maximum of (3) is derived.

Algorithm 1

---

1. For every p_i, $i \in T$, specify an initial value
2. While

$$\exists S_i, \sum_{j \in S_i} p_j \cdot \sum_{l \in C(j)} p_l \cdot w_{j,l} < \max_{j \in S_i} \sum_{l \in C(j)} p_l \cdot w_{j,l} \tag{4}$$

Do for every $j \in S_i$, let

$$p_j = \begin{cases} 1 & \max \sum_{l \in C(j)} p_l \cdot w_{j,l} \\ 0 & others \end{cases} \tag{5}$$

The value of (3) increases after every loop of step 2 and terminate on the maximum at last. From Theorem 1, we know that the highest-score k-complete subgraph is the result of Algorithm 1, so the answer is always the simple 0-1 distribution though we compute it in a probability form, which makes the method perform poor in the experiment (see Section 4). The main reason is that the relationship between the value of (3) and the correctness of the translation is not linear, even though we think that the value of (3) somehow reflects the correctness of the translation. For instance, one term has three translations in dictionary, a, b, c respectively, and the weights 0.127, 0.131, 0.028 respectively; according to the strategy of Algorithm 1, the probability distribution should be 0, 1, 0 respectively, while the correct answer is a though its weight is not the highest. Hence the strategy of computing the maximum is not so wise that other condition should be established to compute a reasonable result of the translation probability.

3 Weighted-Average Probability Distribution

The main idea of Algorithm 1 is to readjust the probability distribution of each S_i until there is none to be readjusted, but in the course of readjusting, it always adds all the probability to the term which have the highest weights (seen in (5) in Algorithm 1), which eventually makes the result 0-1 distribution. This strategy is not so fair because if the weights of two terms are close, the result is still the 0-1 distribution rather than specifying close probabilities to them. A fairer probability distribution seems more reasonable in practice. Hence, we use the following formula in place of (5) in Algorithm 1.

$$p_j = \frac{\sum_{l \in C(j)} p_l \cdot w_{j,l}}{\sum_{m \in S(j)} \sum_{l \in C(m)} p_l \cdot w_{m,l}} . \tag{6}$$

(6) redistributes the probability according to the weight each term has, so we call it the weighted-average probability distribution. In addition, the iterating condition of (4) should be modified as:

$$\exists j, f_j(T) \neq 0 . \tag{7}$$

If we let:

$$f_j(T) = p_j - \frac{\sum_{l \in C(j)} p_l \cdot w_{j,l}}{\sum_{m \in S(j)} \sum_{l \in C(m)} p_l \cdot w_{m,l}} . \tag{8}$$

Thus, we acquire a new algorithm, and besides, the new algorithm can be computed in a more direct way. Since (7) can be rewritten as "if any of the following equalities is not satisfied:

$$\begin{array}{l} f_1(T) = 0 \\ f_2(T) = 0 \\ \quad \ldots \\ f_n(T) = 0 \quad (n = | T |) \end{array} \text{."} \tag{9}$$

Actually, all these equalities constitute an equation system, and the new algorithm terminates on the solutions to these equations. Hence, we direct solve the equations of (9) to obtain the result.

3.1 Solving the Equations

Since the equations in (9) are nonlinear, we use the Newton Method [10]. In brief, the Newton Method is divided into two steps. First, specify the initial value:

$$p_i = \frac{1}{| S(i) |} . \tag{10}$$

Then iteratively calculate (11):

$$P^{(k+1)} = P^{(k)} - \delta^{(k)} . \tag{11}$$

Until (12) is met:

$$\max_{j \in (T)} | p_j^{(k+1)} - p_j^{(k)} | < \varepsilon . \tag{12}$$

Where $P^{(k)}$ is:

$$P^{(k)} = (p_1^{(k)}, p_2^{(k)}, \ldots, p_n^{(k)})^T \ . \tag{13}$$

$\delta^{(k)}$ is the solution to (14).

$$\begin{bmatrix} \frac{\partial f_1(T)}{\partial p_1} & \frac{\partial f_1(T)}{\partial p_2} & \cdots & \frac{\partial f_1(T)}{\partial p_n} \\ \frac{\partial f_2(T)}{\partial p_1} & \frac{\partial f_2(T)}{\partial p_2} & \cdots & \frac{\partial f_2(T)}{\partial p_n} \\ \vdots & \vdots & & \vdots \\ \frac{\partial f_n(T)}{\partial p_1} & \frac{\partial f_n(T)}{\partial p_2} & \cdots & \frac{\partial f_n(T)}{\partial p_n} \end{bmatrix} \bullet \begin{bmatrix} \delta_1^{(k)} \\ \delta_2^{(k)} \\ \vdots \\ \vdots \\ \delta_n^{(k)} \end{bmatrix} = \begin{bmatrix} f_1^{(k)}(T) \\ f_2^{(k)}(T) \\ \vdots \\ \vdots \\ f_n^{(k)}(T) \end{bmatrix} \ . \tag{14}$$

Since (14) is a linear equation system, there are many ways to solve it. Here we use the common method called Gauss elimination [11]. The calculation of the Jacobian matrix in (14) seems very time-consuming, but the partial derivative can be simplified according to (8).

$$\frac{\partial f_i(T)}{\partial p_j} = \begin{cases} 1 & i = j \\ 0 & j \in S(j) \\ \frac{(A_i \bullet B_i) d_{i,j} - (A_i \bullet C_i) w_{i,j}}{(A_i \bullet C_i)^2} & j \notin S(j) \end{cases} \ . \tag{15}$$

Where A_i is the vector form of $C(i)$; B_i, C_i and $d_{i,j}$ are defined as follows:

$$B_i = (w_{i,t_1}, w_{i,t_2}, \ldots, w_{i,t_n}) \qquad t \in C(i) \ . \tag{16}$$

$$d_{i,j} = \sum_{k \in S(i)} w_{k,j} \ . \tag{17}$$

$$C_i = (d_{i,t_1}, d_{i,t_2}, \ldots, d_{i,t_n}) \qquad t \in C(i) \ . \tag{18}$$

All the three vectors and d only need calculated once, and because the two vector products of (15) are same in every row of the Jacobian matrix, the number of vector product calculation in each iteration is only $2n$. To further simplify the computation, the divisor of (15) can be moved to the right side of equations (14). Therefore, the price for computing the Jacobian matrix in (14) is roughly $2n^2$ times of product, which is not so high as it seems. Besides, the Gauss elimination needs $(n^2+3n-1)n/3$ times of product or division [11], so the overall computational overhead in each iteration is roughly $n^3/3+3n^2-n/3$ times of product or division and $2n$ times of vector product.

4 Experiment

In this section, experiments are setup to verify the effect of the method in Section 3. Various methods are compared under the measure of precision curve by retrieving a given documents collection. There are three kinds of experiment data we need in the experiments: queries, corpus for training and retrieving, and on-line dictionary.

The queries we use are in English, come from a manual English translation of the 70 topics (TD216-TD285) from the Chinese Web Information Retrieval Forum (http://www.cwirf.org/). Each topic is divided into two fields. One is the "title", and the other is the "description". The "title" is short and concise, while the "description" is long and detailed. Both fields are used as queries. Since most words in a "title" are highly relevant to each other, while the "description" usually includes many irrelevant or only slightly relevant words, we expect translation disambiguation is a more challenging problem for the latter.

The online bilingual dictionary we use is Lexiconer (http://www.lexiconer.com/), and the translation is done in a word-by-word way. Processing of phrases is not considered in the experiments.

The corpus is in Chinese, comes from a simplified version of SogouT2.0. Its size is roughly 1.1 GB. We use it in two aspects: one is to compute the dice coefficient (2) between pairs of terms; the other is to retrieve the documents relevant to the queries.

4.1 Methods Compared

Five methods for translation disambiguation are compared. In addition, the original Chinese queries are used as a reference to all the methods. Apart from the max-sum model method of Section 2 and the weighted-average method of Section 3, the other three methods are:

Include All [12] makes no difference between any translations of a term. In fact, it includes all the translations of a term. Hence, it must include the correct translation combination, but also include much more wrong ones.

Simple Weighted is a simplified version of the weighted-average method. Instead of (6), it computes the probability distribution in a more direct way like (19).

$$p_j = \frac{\sum_{l \in C(j)} w_{j,l}}{\sum_{m \in S(j)} \sum_{l \in C(m)} w_{m,l}} . \tag{19}$$

Maximum Coherence Model [8] computes the maximum of a formula similar to (3), but it treats it as a quadratic programming problem. Its implementation is quite complex and the price for computation is also high.

The results of these methods are not all in probability form, but they are all treated in probability form in retrieving. All the probabilities of a query are viewed as a vector, and we use standard vector space model [13] for calculating the relevance of retrieved documents.

4.2 Results and Remarks

Precision curves are plotted using the precisions of top 5-100 ranked relevant documents. The results of short (using the "title" field of the topic) and long (using the "description" field) queries are presented in Fig.2 and Fig.3 respectively. Tab.1 lists the average precision of each method and the relative improvement our method achieves over others.

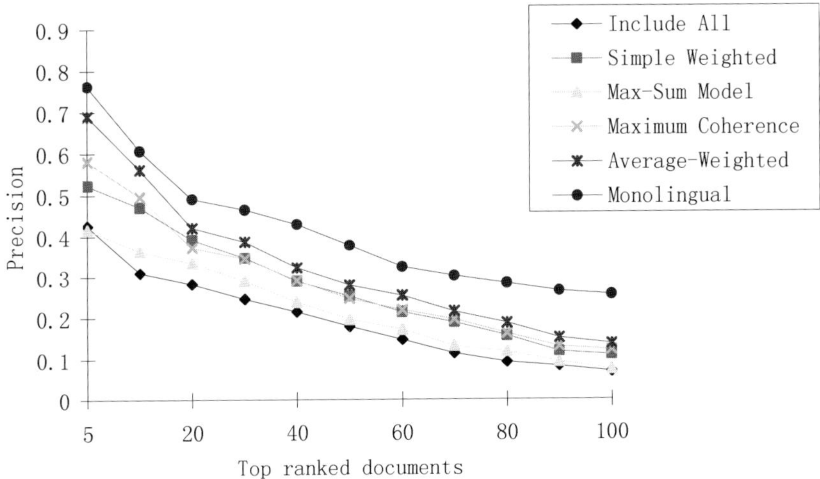

Fig. 2. Precision curve for short queries

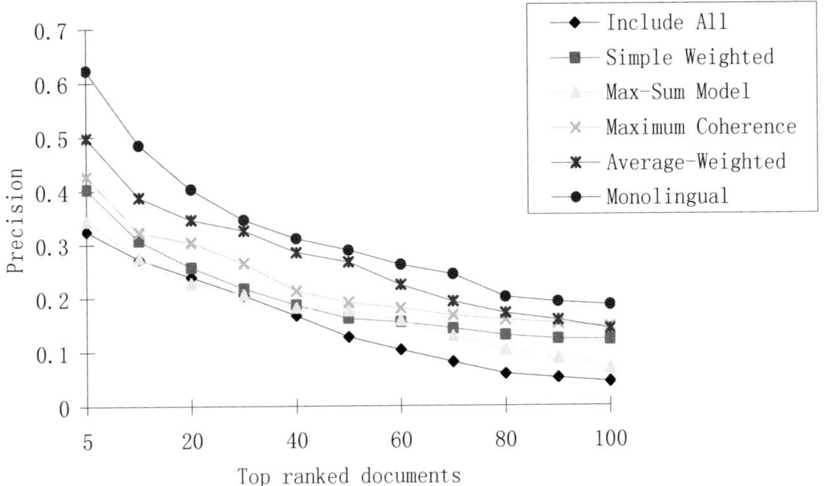

Fig. 3. Precision curve for long queries

Table 1. Average precision and the relative improvements of our algorithm over others

	Short Queries		Long Queries	
	Avg Prec.	Rel. impr.	Avg Prec.	Rel. impr.
Include All	0.196	+67.3%	0.153	+78.4%
Simple Weighted	0.279	+17.6%	0.201	+35.8%
Max-Sum Model	0.220	+49.1%	0.178	+53.4%
Max Coherence	0.288	+13.9%	0.230	+18.7%
Average-Weighted	0.328	--	0.273	--
Monolingual	0.415	--	0.322	--

In Fig.2 and Fig.3, the curve of monolingual (direct using Chinese queries to retrieve) represents the theoretic upper boundary of precision, so it always stays above other curves. Except for this, we can see our average-weighted algorithm performs better than other four and the average improvement over other methods is about 15%-60% from Tab.1. The max-sum model performs unexpectedly poor in the experiments, and this may be due to the 0-1 distribution of its results and the unreasonable strategy to redistribute the probabilities. Moreover, a better retrieval achieved by short queries proves that the long queries tend to include more irrelevant or only slightly relevant words and they are usually more difficult to disambiguate for algorithms based on co-occurrence statistics of terms.

5 Conclusion

In this paper, we first introduce a progressive algorithm called the max-sum model. Its main idea is to compute the maximum value of a formula, but its result is always a 0-1 distribution which proves incompetent in the experiments. Then we modify the algorithm by taking a strategy called weighted-average probability distribution, which seems fairer because it takes terms having close weights into account. The new algorithm outperforms four other methods in the experiments. Moreover, it can be computed in a more direct way by solving an equation system using Newton method.

The running efficiency of our algorithm is not so satisfactory, and we plan to improve the algorithm in the following fields in the future: 1) accelerate the convergence rate of Newton method; 2) optimization of time and space in Gauss elimination and the Jacobian matrix; 3) design a parallel version of the algorithm for longer query.

References

1. Federico, M., Bertoldi, N.: Statistical cross-language information retrieval using N-best query translations. In: SIGIR 2002: Proceedings of the 25th annual international ACM SIGIR conference on Research and development in information retrieval, pp. 167–174. ACM Press, New York (2002)
2. Xu, J., Weischedel, R.: TREC-9 cross-lingual retrieval at BBN. In: Proceedings of the TREC-9 Conference (2001)

3. Lavrenko, V., Choquette, M., Croft, W.B.: Cross-lingual relevance models. In: SIGIR 2002: Proceedings of the 25th annual international ACM SIGIR conference on Research and development in information retrieval, pp. 175–182. ACM Press, New York (2002)
4. Nie, J.-Y., Simard, M.: Using statistical translation models for bilingual ir. In: Peters, C., Braschler, M., Gonzalo, J., Kluck, M. (eds.) CLEF 2001. LNCS, vol. 2406, pp. 137–150. Springer, Heidelberg (2002)
5. Gao, J., Nie, J.-Y., Xun, E., Zhang, J., Zhou, M., Huang, C.: Improving query translation for cross-language information retrieval using statistical models. In: SIGIR 2001: Proceedings of the 24th annual international ACM SIGIR conference on Research and development in information retrieval, pp. 96–104. ACM Press, New York (2001)
6. Kraaij, W., Pohlmann, R., Hiemstra, D.: Twenty-one at TREC-8: using language technology for information retrieval. In: Voorhees, E.M., Harman, D.K. (eds.) The Eigth Text REtrieval Conference (TREC-8). National Institute of Standards and Technology, NIST, vol. 8, pp. 285–300. NIST Special Publication 500-246 (2000)
7. Maeda, A., Sadat, F., Yoshikawa, M., Uemura, S.: Query term disambiguation for web cross-language information retrieval using a search engine. In: IRAL 2000: Proceedings of the fifth international workshop on Information retrieval with Asian languages, pp. 25–32. ACM Press, New York (2000)
8. Liu, Y., Jin, R., Chai, J.Y.: A maximum coherence model for dictionary-based cross-language information retrieval. In: Proceedings of the 28th annual international ACM SIGIR conference on Research and development in information retrieval, Salvador, Brazil, August 15–19, 2005, pp. 536–543 (2005)
9. Adriani, M.: Dictionary-based CLIR for the CLEF multilingual track. In: Working Notes of the Workshop in Cross-Language Evaluation Forum (CLEF), September 2000, Lisbon (2000)
10. Mathews, J.H., Fink, K.D.: Numerical Methods Using MATLAB, 3rd edn. Prentice-Hall, Englewood Cliffs (2002)
11. Xu, S.: Computer Algorithms for Common Use, 2nd edn. Tsinghua University Press (1995)
12. Daelemans, W., Sima'an, K., Veenstra, J., Zavrel, J. (eds.): Different Approaches to Cross Language Information Retrieval, Language and Computers: Studies in Practical Linguistics, vol. 37. Rodopi, Amsterdam (2001)
13. Li, X., Yan, H., Wang, J.: Search Engine: Principle, Technology and Systems. Science Press (2004)

Semi-Supervised Graph-Ranking for Text Retrieval

Maoqiang Xie[1], Jinli Liu[1], Nan Zheng[1], Dong Li[2], Yalou Huang[1], and Yang Wang[1]

[1] College of Software, Nankai University, Tianjin, China
xiemq@nankai.edu.cn, yode2006@mail.nankai.edu.cn,
zhengnan@mail.nankai.edu.cn, huangyl@nankai.edu.cn,
wangyang022@mail.nankai.edu.cn

[2] College of Information Technology and Science, Nankai University, Tianjin, China
nodoubt@mail.nankai.edu.cn

Abstract. Much work has been done on supervised ranking for information retrieval, where the goal is to rank all searched documents in a known repository with many labeled query-document pairs. Unfortunately, the labeled pairs are lack because human labeling is often expensive, difficult and time consuming. To address this issue, we employ graph to represent pairwise relationships among the labeled and unlabeled documents, in order that the ranking score can be propagated to their neighbors. Our main contribution in this paper is to propose a semi-supervised ranking method based on graph-ranking and different weighting schemas. Experimental results show that our method called *SSG-Rank* on 20-newsgroups dataset outperforms supervised ranking (Ranking SVM and PRank) and unsupervised graph ranking significantly.

Keywords: Semi-Supervised Ranking, Graph Ranking, Text Retrieval.

1 Introduction

Ranking is the key problem for information retrieval, which sorts candidate documents searched from a large amount of corpus based on their relevance to the query submitted by user. Recently, many researches have been developed on it. In general, all of these researches can be categorized into two classes. One is unsupervised ranking which makes use of the link relations among the web pages to construct graph, such as PageRank [3] or HITS [8]. Based on the link relation graph, the influence of web pages can be propagated through the path of graph and the final ranking scores can be obtained from the stable graph. The other one is supervised ranking (also called "learning to rank"), which trains ranking models by learning from the labeled query-document pairs, such as Ranking SVM [6] or Prank [5]. Recently, learning to rank has attracted much attention. For example, Cao and Xu et al. has proposed cost-sensitive Ranking SVM that gives different penalties for the errors occurring at different positions [4] [13]. Xu and Li directly optimize ranking performance measure by Boosting method [14]. Furthermore, in SIGIR 2007, "learning to rank" workshop is held and supervised ranking has gained increasing attention.

H. Li et al. (Eds.): AIRS 2008, LNCS 4993, pp. 256–263, 2008.

From the above work, it can be found that unsupervised ranking has been used for the web page ranking successfully, but cannot be used for text retrieval directly, as link relations of documents cannot be obtained. Moreover, 'learning to rank' needs a lot of labeled query-document pairs which require expensive human labeling and much time. At the same time, large numbers of unlabeled data are far easier to obtain. Therefore, semi-supervised ranking for text retrieval should be deeply studied.

To address the above issues, a novel semi-supervised graph ranking (SSG-Rank) method is proposed in this paper, which re-weights the affinity matrix by using pairs of labeled documents, so that similarities between documents in the same class are enhanced, and vice versa. With the constrained graph, the ranking scores are propagated more precisely and efficiently.

The rest of paper is organized as follows: Section 2 introduces some basic notions on graph, and then a semi-supervised graph ranking method called SSG-Rank is proposed based on the graph Laplacian regularization. Section 3 presents the experimental results and the analysis of ranking for text retrieval, and this paper is concluded in Section 4.

2 Semi-Supervised Graph Ranking

An important problem for text retrieval is the insufficiency of labeled instances, since the labeling of documents are often expensive, difficult and time consuming, meanwhile unlabeled data are relatively easy to collect[18]. In addition, the generation of documents is much faster than the manual labeling on them. Hence it is important for the text retrieval to create an appropriate ranking model, in order that the large amount of unlabeled documents may help supervised ranking. Major research on semi-supervised learning is focused on classification and regression. For example, Blum and Mitchell exploit co-training method that splits the features into two sets, such as web link and content, and then two classifiers trained by using labeled data with these two different feature sets predict the unlabeled data, and teach the other one with predicted labels until most of predictions are agreed by these two independent classifiers[2]. Joachims applies SVM to semi-supervised classification by adding the unlabeled data constraint to the loss function[7]. Zhou, Z. employs co-training to semi-supervised regression[17]. Zhu adopts graph to semi-supervised learning[18]. Zhou, D. et al. and Agarwal propose the unsupervised graph ranking algorithms respectively[16][1]. And Wan et al. employ unsupervised graph ranking for document search[11]. Unfortunately, few semi-supervised ranking methods for text retrieval are proposed. To deal with above problems, we propose a novel semi-supervised graph-ranking algorithm by re-weighting the affinity graph on the basis of labeled document pairs.

2.1 Construction of Text Ranking Matrix

The key to semi-supervised learning is priori consistency (also referred to as cluster assumption): (1) two points are likely to have the same label if they are so near in a dense region; (2) points in the same dense connective region (formally called a manifold or a cluster) are likely to have the same label. With these two assumptions, semi-supervised learning methods can spread the label information to their neighbors through the global structure, so that the model trained by labeled data can be used to find credible predictions of unlabeled data, and then the training dataset is expanded by inserting predicted unlabeled data with high confidence. Motivated by it, the labeled pairs can be used as constraint when the affinity matrix is constructed on the basis of data features, which will make the ranking information spread more precisely and efficiently.

Let G = (V, E, w) denote a weighted graph with vertex set $V \subset \mathbb{R}^d$, edge set $E \subseteq V \times V$, and corresponding edge weights $w: E \mapsto \mathbb{R}^+$. The goal of graph ranking is to learn a ranking function $f: V \mapsto \mathbb{R}$, The graph G represents the similarity between documents by assigning weight to each edge. The degree of similarity often uses RBF kernel distance (such as exp($-|x_i - x_j|^k/\sigma$) , k = 1 or 2) or BM25(x_i, x_j) [10]. Note that relevant documents are not always near; therefore, the pairs between labeled documents should be used to revise the similarity graph. More concretely, the pairs from the same class are to be ranked with highest score (r_1 level), followed by pairs in different classes (r_m level). This weighting schema can also be extended to the task with hierarchical classes, and the pair from the same father class can be ranked as "r_h" ($1 < h < m$). Suppose that R = {r_1, r_2, …, r_h ,…, r_m} is the ranking level set, and r_i is preferred to r_j if $i < j$. Consequently, the weighting schema can be represented as $\tau(r(x_i, x_j))$ that will return high score if the ranking level is high, and vise versa, where function $r(\cdot,\cdot)$ denotes the ranking level of document pairs. After being re-weighted by $\tau(\cdot)$, the similarity graph G is revised under the constraint of known pairwise relationships and will be more suitable for the graph ranking.

2.2 Semi-Supervised Graph Ranking

A good ranking function can be considered as a ranking model that can minimize a suitable combination of the penalty of the model's complexity and the empirical ranking error. And the optimization function is:

$$Q(F) = \sum\sum w_{ij} \left\| \frac{F_i}{\sqrt{d_i}} - \frac{F_j}{\sqrt{d_j}} \right\|^2 + \mu \sum_{i=1}^{n} \| F_i - y_i \|^2 \tag{1}$$

where F can be considered as a vectorial function that assigns a ranking value vector Fi to each point xi, and di is given by equation (2).

$$d_i = \sum_{j:\{v_i, v_j\} \in E} w_{ij} \tag{2}$$

The first term of equation (1) is the penalty of model complexity, ensuring that a good ranking model should make little difference between adjacent documents, which is also referred to as smoothness constraint. The second term is the empirical risk used to punish the ranking error on labeled data as well as the great change from the initial ranking score. The positive regularization parameter μ is the trade-off between these two terms.

In order to minimize the $Q(F)$, $Q(F)$ should be differentiated about F as follows:

$$\left.\frac{\partial Q}{\partial F}\right|_{F=F^*} = F^* - SF^* + \mu(F^* - Y) = 0 \tag{3}$$

and the optimum F can be obtained.

$$F^* = (I - \alpha S)^{-1} Y \tag{4}$$

where $S = D^{-1/2}$ W $D^{-1/2}$.

In practice, F^* is often solved approximately by the iterative propagation, and the algorithm is illustrated as follows:

Algorithm: ***SSG-Rank*** (Semi-Supervised Graph Ranking)

Input: L={< x_i , y_i> | $\mathbf{x}_i \in \mathbb{R}^d$, $y_i \in \mathbb{R}$ } Labeled dataset

U={x_j | $\mathbf{x}_j \in \mathbb{R}^d$ } Unlabeled dataset

1. Construct the similarity matrix W, where w_{ij} is the similarity of vertex x_i and x_j, and w_{ii} = 0. In text retrieval, exp(-|x_i - x_j |/σ) or BM25(x_i, x_j) is often used as the similarity;
2. Calculate the τ(r(x_i, x_j)) by generating the pair from L, and re-weight the matrix W;
3. The matrix W is normalized to $S= D^{-1/2}WD^{-1/2}$, where D = diag{d_i | i = 1 … n};
4. Iterate $F_{t+1} = \alpha SF_t + (1-\alpha)Y$ until convergence;
5. Let F^* be the limit of the sequence {F_t}. Rank each query x_i according to F_i^*.

Output: The vectorial / discrete ranking function

Fig. 1. Algorithm SSG-Rank

When the iteration converges, the ranking score between query x_i and document x_j should be high if both of them belong to the same cluster or quite relevant, and vise versa.

3 Experiments and Discussions

In our experiments, we made use of 20-newsgroups dataset to test the performance of *SSG-Rank*, which consists of 20 hierarchical classes about 1000 documents per class. We selected 6 classes, including "comp.graphics", "comp.os.ms-windows.misc", "rec.motorcycles","rec.sport.baseball", "sci.space", and "talk.politics.mideast"(Fig. 2). Since Ranking SVM cannot handle large scale training dataset effectively and the count of document pairs is (n-1)/2 times of document, 360 documents are sampled from these 6 classes. There are 3 rank levels in our experiments, comprising r_1(query and document are in the same leaf class), r_2 (query and document are in the same father class, such as "comp" and "rec"), and r_3 (the other pairs), where $\tau(r_1) = 2$, $\tau(r_2) = 1$, and $\tau(r_3) = 0$. To test the average performance, 12 datasets are sampled randomly. The

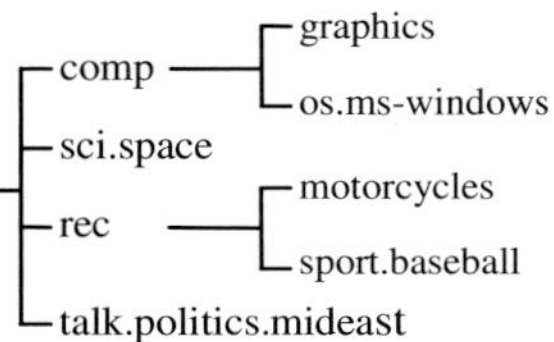

Fig. 2. Hierarchical classes of 20-newsgroups

Table 1. Datasets used in the experiments

Dataset Name	Size	Dataset Name	Size
L0	77	U0	82
U1	82	L1	83
U2	120	L2	125
U3	162	L3	159
U4	201	L4	197
U5	239	L5	239
U6	283	L6	278

name and the size of them are listed in Table 1, where L_i represents labeled document set and U_j represents unlabeled document set.

3.1 SSG-Rank v.s. Supervised Ranking Methods

The first experiment is performed for comparing with SSG-Rank and selected baselines, such as supervised Ranking SVM and PRank. Additionally, BM25 representing the state-of-the-arts IR method was also chosen as a baseline. Experiments on SSG-Rank and these baselines were conducted respectively on L_0U_i (i=1…6) and U_0L_j (j=1…6), and the Figure 3 shows the MAP and NDCG measures of results ranked by different methods. It is clear that SSG-Rank outperforms supervised ranking models and BM25 particularly in terms of MAP and NDCG. Additionally, even the least NDCG accuracy is above 90%.

We investigate and find the possible reason lying in the intrinsic manifold in the unlabeled data together with the constraint of labeled data. As the relationships between labeled data are credible, the ranking scores propagated from neighbors via the graph will be more credible. Furthermore, since SSG-Rank ran under the setting of transductive learning, the data to be tested also are involved in the spread of ranking score, while the supervised ranking methods never make any use of unlabeled data

3.2 Properties of SSG-Rank

The second experiment is conducted to test the properties of SSG-Rank by using different datasets and settings. Figure 4 (a) illustrates the ranking performance in terms of NDCG on the initial fixed labeled dataset and different unlabeled datasets with incremental size. It shows that NDCGs increase until the count of unlabeled data up to 162, and then decrease gradually, which indicates that SSG-Rank has an effective range with regard to the ratio of unlabeled data. Figure 4 (b) shows the ranking performance (MAP) on the same above datasets.

Figure 5 presents the improvements of NDCGs and MAP when the unlabeled dataset with fixed size and the labeled data added incrementally are given. It can be observed that the performance is improved significantly at first, and then become stable with the incremental labeled data, which implies that too many labeled data are not needed when the affinity graph is re-weighted well.

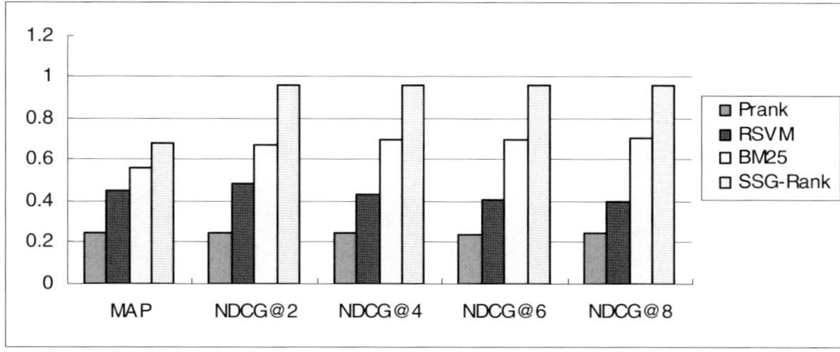

Fig. 3. Average ranking performance comparisons among Ranking SVM, PRank, BM25 and *SSG-Rank* on 12 datasets (L_0U_i (i=1…6) and U_0L_j (j=1…6)). Ranking SVM and PRank adopt the training set listed in table 1 with features used in [9], and the affinity matrix applied in SSG-Rank is calculated by using BM25.

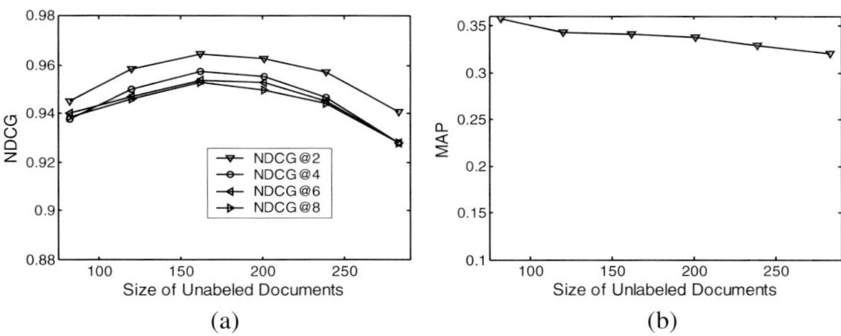

Fig. 4. NDCG(a) and MAP(b) performance on labeled data with fixed size and unlabeled data added incrementally

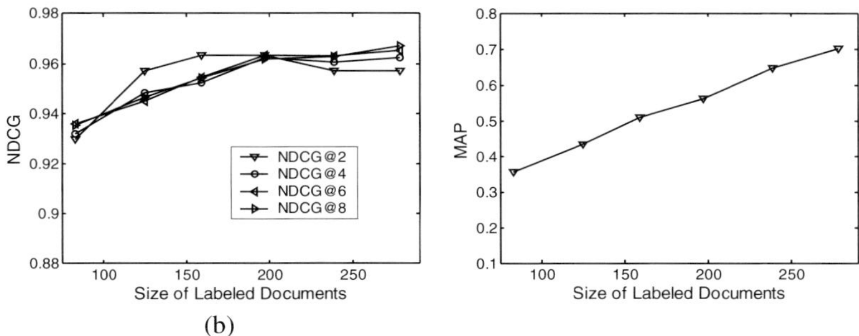

Fig. 5. NDCG(a) and MAP(b) performance on labeled data added incrementally and unlabeled data with fixed size

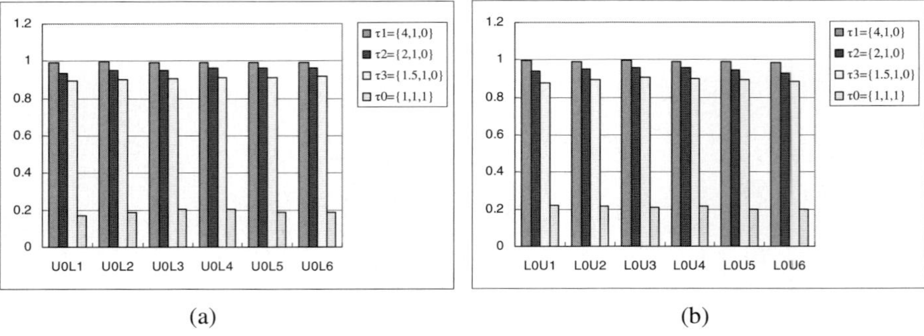

Fig. 6. NDCG performance on L0Ui (i=1…6) and U0Lj (j=1…6) datasets under the different weighting schemas

Finally, experiment is executed with different weighting schemas, and the results are illustrated in Figure 6. From the 12 results, it is evident that τ1 schema (τ(r1) = 4, τ(r2) = 1, and τ(r3) = 0) reveals the best performance for the reason that the ranking score propagated via definitely relevant pairs will be enhanced, and more scores can be revised by these paths. In addition, unsupervised graph ranking is the special case of SSG-Rank with τ(r1) = 1, τ(r2) = 1, and τ(r3) = 1. Figure 6 indicates that semi-supervised graph ranking benefiting from re-weighting of affinity graph on the basis of labeled pairs is much more effective than unsupervised graph ranking.

4 Conclusion

In this paper, we proposed a novel semi-supervised ranking algorithm referred to as *SSG-Rank* in text retrieval. In contrast to existing algorithms, on the basis of cluster assumptions and graph notions, *SSG-Rank* can modify the intrinsic manifold by using labeled pairs constraint. This is done by re-weighting the affinity matrix and propagating the ranking scores to their neighbors via the graph. The experimental results show that *SSG-Rank* executed on 20-newsgroups dataset outperforms supervised ranking (Ranking SVM and PRank) and unsupervised graph ranking significantly. Finally, we also analyze different weighting schemas for improving the ranking performance of top documents.

Future work includes theoretical analysis on the generalization error and other properties of the *SSG-Rank* algorithm, and further empirical evaluations of the algorithm including comparisons with other algorithms should be studied.

Acknowledgments. This work is supported by National Science Foundation of China under the grant 60673009 and the key project of the Tianjin Science and Technology Research Program under the grant 05YFGZGX24000. We would like to thank Rui Kuang for helpful discussion.

References

1. Agarwal, S.: Ranking on Graph Data. In: The proceedings of International Conference of Machine Learning 2006, pp. 25–32 (2006)
2. Blum, A., Mitchell, T.: Combining labeled and unlabeled data with co-training. In: Proceedings of Annual Conference on Computational Learning Theory, pp. 92–100 (1998)
3. Brin, S., Page, L.: The Anatomy of a Large Scale Hypertextual Web Search Engine. In: Proceedings of 7th International World Wide Web Conference, pp. 107–117 (1998)
4. Cao, Y., Xu, J., Liu, T.Y., Li, H., Huang, Y.L., Hon, H.W., Adapting Ranking, S.V.M.: to Document Retrieval. In: Proceedings of ACM SIGIR, vol. 29, pp. 186–193 (2006)
5. Crammer, K., Singer, Y.: PRanking with ranking. Advances in Neural Information Processing Systems, Canada (2002)
6. Herbrich, R., Graepel, T., Obermayer, K.: Large Margin Rank Boundaries for Ordinal Regression, Advances in Large Margin Classifiers, pp. 115–132. MIT Press, Cambridge (2000)
7. Joachims, T.: Transductive inference for text classification using support vector machine. In: Proceedings of 16th International Conference of Machine Learning, pp. 200–209 (1999)
8. Kleinberg, J.: Authoritative sources in a hyperlinked environment. In: Proceedings of the 9th ACM-SIAM Symposium on Discrete Algorithms, New Orleans, pp. 668–677 (1997)
9. Liu, T., Xu, J., Qin, T., Xiong, W., Li, H.: LETOR: Benchmark Dataset for Research on Learning to Rank for Information Retrieval. In: SIGIR 2007 Workshop on Learning to Rank for Information Retrieval (2007)
10. Robertson, S., Hull, D.: The TREC-9 filtering track final report. In: TREC, pp. 25–40 (2000)
11. Wan, X., Yang, J., Xiao, J.: Document Similarity Search Based on Manifold- Ranking of TextTiles. In: The 3rd Asia Information Retrieval Symposium, Singapore, pp. 14–25 (2006)
12. Wang, F., Zhang, C.: Label Propagation Through Linear Neighborhoods. In: Proceedings of 23rd International Conference of Machine Learning, pp. 985–992 (2006)
13. Xu, J., Cao, Y., Li, H., Huang, Y.: Cost-Sensitive Learning of SVM for Ranking. In: Fürnkranz, J., Scheffer, T., Spiliopoulou, M. (eds.) ECML 2006. LNCS (LNAI), vol. 4212, pp. 833–840. Springer, Heidelberg (2006)
14. Xu, J., Li, H.: AdaRank: A Boosting Algorithm for Information Retrieval. In: The proceedings of SIGIR 2007, pp. 391–398 (2007)
15. Zhou, D.Y., Bousquet, O., Lal, T.N., Weston, J., Schölkopf, B.: Learning with Local and Global Consistency. In: Advances in Neural Information Processing Systems 16, pp. 321–328 (2004)
16. Zhou, D.Y., Weston, J., Gretton, A., et al.: Ranking on Data Manifolds. In: Advances in Neural Information Processing System 16 (2003)
17. Zhou, Z.H., Li, M.: Semi-supervised regression with co-training. In: Proceedings of International Joint Conference on Artificial Intelligence 2005 (2005)
18. Zhu, X.J.: Semi-Supervised Learning Literature Survey, Computer Sciences Technical Report 1530, University of Wisconsin-Madison (2005)

Learnable Focused Crawling Based on Ontology

Hai-Tao Zheng, Bo-Yeong Kang, and Hong-Gee Kim*

Biomedical Knowledge Engineering Laboratory, Dentistry College, Seoul National University, 28 Yeongeon-dong, Jongro-gu, Seoul, Korea
hgkim@snu.ac.kr

Abstract. Focused crawling is proposed to selectively seek out pages that are relevant to a predefined set of topics. Since an ontology is a well-formed knowledge representation, ontology-based focused crawling approaches have come into research. However, since these approaches apply manually predefined concept weights to calculate the relevance scores of web pages, it is difficult to acquire the optimal concept weights to maintain a stable harvest rate during the crawling process. To address this issue, we propose a learnable focused crawling approach based on ontology. An ANN (Artificial Neural Network) is constructed by using a domain-specific ontology and applied to the classification of web pages. Experiments have been performed, and the results show that our approach outperforms the breadth-first search crawling approach, the simple keyword-based crawling approach, and the focused crawling approach using only the domain-specific ontology.

1 Introduction

Due to the tremendous size of information on the Web, it is increasingly difficult to search for useful information. For this reason, it is important to develop document discovery mechanisms based on intelligent techniques such as focused crawling [4,6,7]. In a classical sense, crawling is one of the basic techniques for building data storages. Focused crawling goes a step further than the classical approach. It was proposed to selectively seek out pages relevant to a predefined set of topics called crawling topics.

One of the main issues of focused crawling is how to effectively traverse off-topic areas and maintain a high harvest rate during the crawling process. To address this issue, an effective strategy is to apply background knowledge of crawling topics to focused crawling. Since an ontology [1] is defined as a well-organized knowledge scheme that represents high-level background knowledge with concepts and relations, ontology-based focused crawling approaches have come into research [6,8,11]. Such approaches use not only keywords, but also an ontology to scale the relevance between web pages and crawling topics.

In the ontology-based focused crawling approaches, because concept weights are heuristically predefined before being applied to calculate the relevance scores

* Corresponding author.

[1] http://www.w3.org/TR/owl-guide/

H. Li et al. (Eds.): AIRS 2008, LNCS 4993, pp. 264–275, 2008.

of web pages, it is difficult to acquire the optimal concept weights to maintain a stable harvest rate during the crawling process. To address this issue, a learnable focused crawling approach based on ontology is proposed in this paper. Since the ANN is particularly useful for solving problems that cannot be expressed as a series of steps, such as recognizing patterns, classifying into groups, series prediction and data mining, it is suitable to be applied to focused crawling. In our approach, an ANN (Artificial Neural Network) is constructed by using a domain-specific ontology and applied to the classification of web pages.

The outline of this paper is as follows: Section 2 reviews the related work to provide an overview of focused crawling. Section 3 describes the framework of our learnable focused crawling approach and elaborates the mechanism of relevance computation. Section 4 provides an empirical evaluation that makes a comparison with the breadth-first search crawling approach, the simple keyword-based crawling approach, and the focused crawling approach using only the domain-specific ontology. Section 5 discusses the conclusion and future work.

2 Related Work

Focused crawlers aim to search only the subset of the web related to a specific category. There are lots of methods that have been proposed for focused crawling. A generic architecture of focused crawling is proposed by S. Chakrabarti [4]. This architecture includes a classifier that evaluates the relevance of a web page with respect to the focused topics and a distiller that identifies web nodes with high access points to many relevant pages within links. M. Diligenti [5] proposes a focused crawling algorithm that constructs a model named context graph for the contexts in which topically relevant pages occur on the web. This context graph model can capture typical link hierarchies in which valuable pages occur, and model content on documents that frequently occur with relevant pages. In [7], focused crawling using a context graph is improved by constructing a relevancy context graph, which can estimate the distance and the relevancy degree between the retrieved document and the given topic. J. Rennie [10] proposes a machine learning oriented approach for focused crawling: the Q-learning algorithm is used to guide the crawler to pass through the off-topic document to highly relevant documents. A domain search engine, MedicoPort [3], uses a topical crawler, Lokman crawler, to build and update a collection of documents relevant to the health and medical field.

However, most of the above approaches do not consider applying background knowledge such as ontology to focused crawling. Since an ontology is a description of the concepts and relationships that can well represent the background knowledge, a good way to improve the harvest rate of focused crawling is to use ontology for crawling. Several focused crawling approaches based on ontology have been proposed [6,8,11].

The most prevalent focused crawling approach based on ontology is called ontology-focused crawling approach [6]. Figure 1 depicts the ontology-focused crawling framework. In this approach, the crawler exploits the web's hyperlink

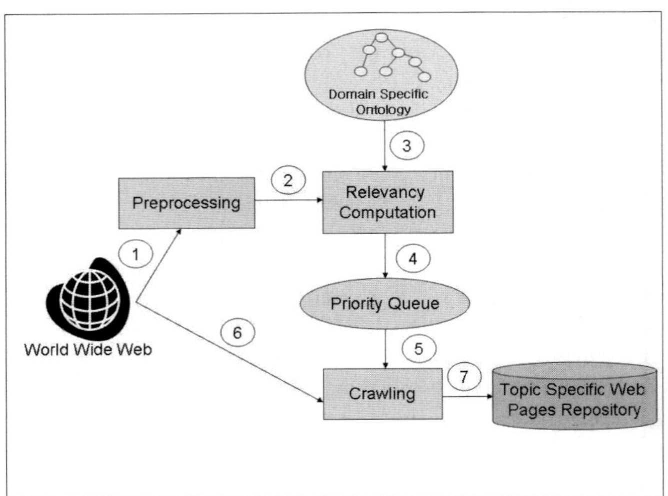

Fig. 1. Ontology-focused crawling framework

structure to retrieve new pages by traversing links from previously retrieved ones. As pages are fetched, their outward links can be added to a list of unvisited pages, which is refer to as the crawl frontier. To judge one page is relevant to the specific topic or not, a domain-specific ontology is used to compute the relevance score. After preprocessing, entities (words occurring in the ontology) are extracted from the visited pages and counted. In this way, frequencies are calculated for the corresponding entities in visited web pages. With the entity distance being defined as the links between an entity in the ontology and the crawling topic, concept weights are calculated by a heuristically predefined discount factor raised to the power of the entity distance. The relevance score is the summation of concept weights multiplied by the frequencies of corresponding entities in the visited web pages. With the relevance score, a candidate list of web pages in order of increasing priority is maintained in the priority queue. Based on the priority queue, the crawler can crawl the relevant web pages and store them into the topic specific web pages repository.

Ehrig's approach, the ontology-focused crawling approach, uses the ontology as background knowledge and applies the weights of concepts in the ontology to compute the relevance score. Since the concept weights are manually predefined, the relevance score can not optimally reflect relevance of concepts with the crawling topics during the whole crawling process. Compared to these approaches discussed above, our method not only provides a way to apply the ontology as background knowledge for focused crawling, but also use an ANN, which quantifies the concept weights based on a set of training examples, to classify the visited web pages. The idea of combining ontology and the ANN for focused crawling has not been researched in detail until now.

3 Ontology-Based Learnable Focused Crawling

In this section, we propose the framework of our learnable focused crawling approach. Based on the framework, we describe each step in the crawling process in detail. Then, the mechanism of relevance computation is elaborated by an example.

3.1 Framework of Learnable Focused Crawling

The proposed learnable focused crawling approach consists of three stages according to their distinct functions. Figure 2 depicts these three stages: data preparation stage, training stage, and crawling stage. The data preparation stage is responsible for preparing training examples that are used for the ANN construction. In the training stage, an ANN is trained by the training examples. In the crawling stage, web pages are visited and judged by the ANN as to whether they will be downloaded or not. Finally, the downloaded web pages will be stored in a topic specific web pages repository.

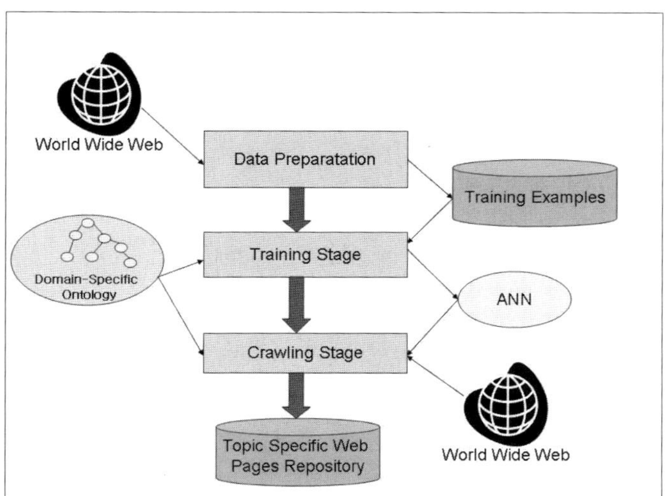

Fig. 2. Framework of learnable focused crawling

Data Preparation Stage. In the first stage, the data preparation stage, we construct a set of training examples by crawling web pages. The weblech [1] crawler is employed to crawl web pages. The weblech crawler is a breadth-first crawler that supports many features required to download web pages and emulate standard web-browser behavior as much as possible. Given a seed URL, the weblech crawler will crawl all the web pages near to this URL without bias. After crawling a collection of web pages, we classify the web pages relevant to the topic

[1] http://weblech.sourceforge.net/

manually. As a result, a set of training examples is composed of positive training examples and negative training examples.

Training Stage. All the training examples constructed in the first stage are fed to the second stage, the training stage, for training an ANN. In the training stage, a set of concepts is selected from a given domain-specific ontology to represent the background knowledge of crawling topics. Those concepts are selected because they are considered to have close relationships with crawling topics. Those selected concepts are called relevant concepts of crawling topics. For each web page in the training examples, we can get a list of entities (words occurring in the set of relevant concepts) by preprocessing it.

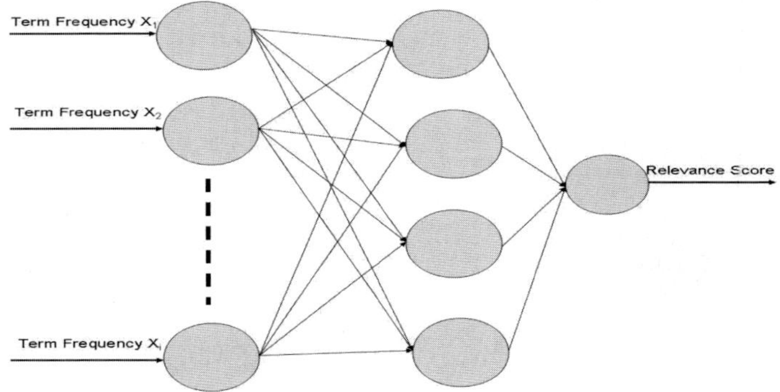

Fig. 3. The structure of artificial neural network

A preprocessing module includes functions of parsing web pages from html to text, making POS (Part-Of-Speech) tagging for web pages, stop words removal, and word stemming. HTML Parser [2] is employed to process the html by removing all the html tags. QTag is used to make POS tagging [3]. After preprocessing of the visited web pages, the entities (words occurring in the set of relevant concepts) are extracted and counted. We get the term frequencies of relevant concepts in the visited web pages.

After calculating the term frequencies of relevant concepts, we use them to train an ANN. The ANN is initialized as a feed forward neural network with three layers (Fig. 3). The first layer is composed of a linear layer with transfer function $y = \beta x$. The number of input nodes is decided by the number of relevant concepts. The hidden layer is composed of a sigmoid layer with transfer function $y = \frac{1}{1+e^{-x}}$. There are four nodes in the hidden layer. Considering the ANN will be used for binary classification, the output layer is also composed of a sigmoid layer. The ANN outputs are the relevance scores of corresponding web pages. We applied the well-known Backpropagation algorithm to train the ANN.

[2] http://htmlparser.sourceforge.net/

[3] http://web.bham.ac.uk/O.Mason/software/tagger/

The training process will not stop until the root mean squared error (RMSE) is smaller than 0.01. A more detailed description of the Backpropagation algorithm can be found in [9].

Crawling Stage. The third stage, the crawling stage, is concerned with the actual crawling and the use of the constructed ANN for classification. First, we retrieve the robots.txt information for the host either from the metadata store or directly from the host. Then we determine if the crawler is allowed to crawl the page or not. If the URL passes the check, we begin to crawl web pages and judge them relevant to the crawling topics or not. Given a domain-specific ontology, a set of concepts, which are called relevant concepts, is selected to represent the background knowledge of crawling topics. When the crawler visits web pages, the term frequencies of relevant concepts are calculated after preprocessing. Then the term frequencies of relevant concepts are acquired as inputs for the ANN. The ANN will determine the relativity of the visited web pages to the crawling topics by computing the relevance scores. The mechanism of relevance computation will be described in detail in the next section. If the web pages are classified as relevant to the crawling topics, we download them and save them into the topic specific web pages repository.

3.2 Mechanism of Relevance Computation

The most important component within the framework is the relevance computation component in the crawling stage. First, a domain-specific ontology was given as a background knowledge base in which the relevant concepts are selected to calculate the relevance of the web pages.

Definition 1. *(Domain-Specific Ontology) Ontology O:={ C, P, I, is-a, inst, prop }, with a set of classes C, a set of properties P, a set of instances I, a class function is-a:* $C \rightarrow C$ *(is-a(*C_1*) =* C_2 *) means that* C_1 *is a sub-class of* C_2 *), a class instantiation function inst:* $C \rightarrow 2^I$ *(inst(C) = I means I is an instance of C), a relation function prop:* $P \rightarrow C \times C$ *(prop(P) = (*C_1*,* C_2 *) means property P has domain* C_1 *and range* C_2*).*

Given the domain-specific ontology, those relevant concepts in the ontology will be selected to calculate the relevance score according to the distance between those concepts and the crawling topics.

Definition 2. *(Distance between Concepts) Distance between concepts* $d(t, c_i)=k$ *, where* $k \in N$ *being the number of links between t and* c_i *in ontology O; t being the crawling topic contained in ontology O;* c_i *being concepts in ontology O,* $i \in \{1,\ldots,n\}$*, n being the amount of concepts of ontology O.*

The distance of concepts reflects the relativity between concepts in ontology and the crawling topic. If $d(t, c_i)$ is too large, the concept c_i will have low relativity with the crawling topic t. Moreover, the number of concepts that need to select from the ontology will increase exponentially when the distance between concepts

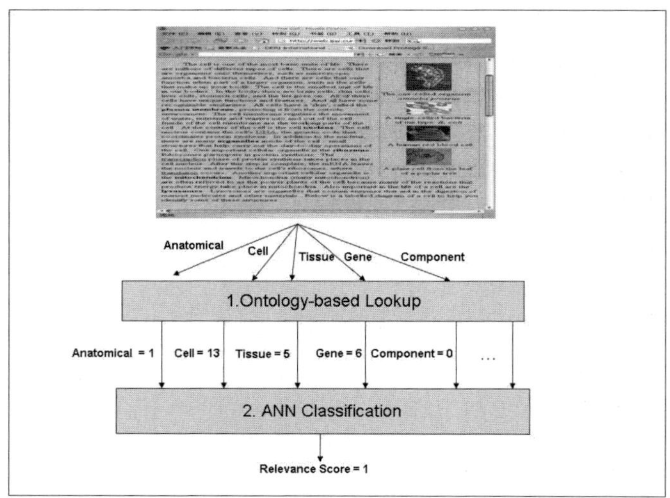

Fig. 4. An example of relevance computation

is becoming large. Because of that, the time complexity of crawling will become higher. To maintain a high crawling efficiency, we need to judge a suitable $d(t, c_i)$ for selecting a set of relevant concepts.

According to the distance between concepts, given a crawling topic t, a set of relevant concepts c_i will be selected from the domain-specific ontology. When visiting web pages, the preprocessing module parses the web pages from html to text, performs POS (Part-Of-Speech) tagging, removes stop words, and processes word stemming. After getting a list of entities (words occurring in the relevant concepts) from the visited web pages, we calculate the term frequencies of relevant concepts based on their occurrence in the visited web pages. The term frequencies are input into the ANN. The ANN will calculate the relevance score of each visited web page and decide whether or not to download the web page.

Figure 4 shows an example of the relevance computation process. The relevance is calculated through several steps, starting with preprocessing such as word stemming and word counting etc., followed by ANN classification and finished with a relevance judgment.

The crawling topic in the example is "Cell". The ontology used here is UMLS (Unified Medical Language System) ontology (Fig. 5) [2]. The UMLS integrates over 2 million names for some 900 000 concepts from more than 60 families of biomedical vocabularies, as well as 12 million relations among these concepts. The concepts relevant to the crawling topics are measured in terms of the distance of the concepts, i.e., $d(t, c_i) <= 3$. Based on the distance of concepts, we select 38 concepts in the UMLS ontology that are relevant to the crawling topic—"Cell" and apply them to construct the ANN with training examples.

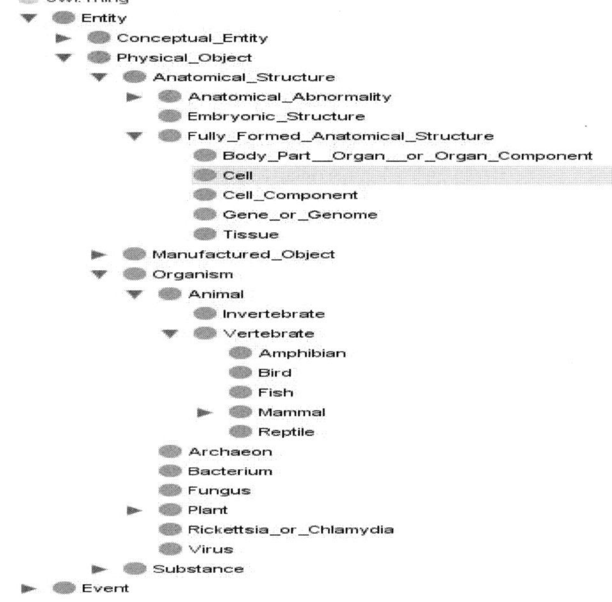

Fig. 5. An UMLS ontology

Note that some of the class concepts in the UMLS ontology are composite words such as "Gene or Genome". With respect to these concepts, we decompose them into single concepts. For example, we decompose the composite concept "Gene or Genome" to concept "Gene" and concept "Genome". When the crawler visits web pages, the term frequencies of "Gene" and "Genome" are calculated.

After preprocessing the visited web pages, entities contained in relevant concepts such as "Cell", "Tissue", and so on, are searched for in the visited web pages. For each entity found, the corresponding relevant concept term frequency is counted. In this example, the concept "Gene" has term frequency 6, the concept "Tissue" has term frequency 5, the concept "Cell" has term frequency 13, the concept "Anatomical" has term frequency 1 and so on. The concept "Component" has term frequency 0, which means there is no such entity in the visited web page. All these calculated term frequencies are input into the ANN. The ANN calculate the relevance score of the visited web pages. Since the result is a "positive" relevance judgement with the relevance score being 1, the crawler decides to download the page.

4 Experiment Results

We compare our approach with the standard breadth-first search crawling approach, the focused crawling approach with simple keyword spotting, and Ehrig's ontology-focused crawling approach [6].

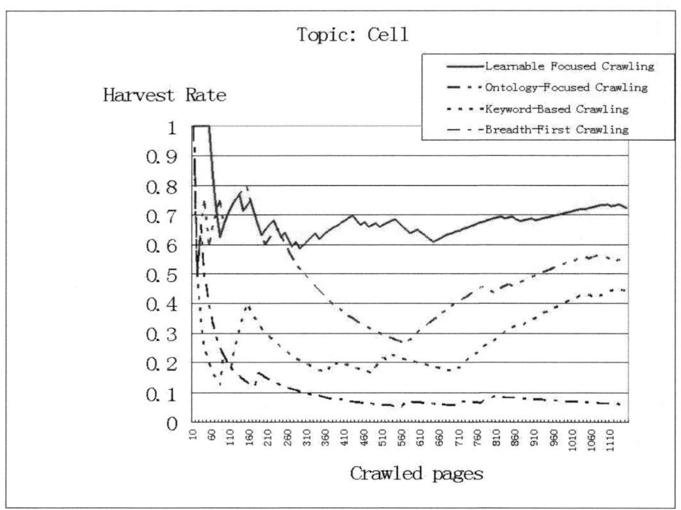

Fig. 6. Harvest rate when the crawling topic is "Cell"

Our experiment focuses on two crawling topics: "Cell" and "Mammal". Because the UMLS ontology [2] (Fig. 5) contains abundant medical information relevant to topic "Cell" and "Mammal", we use it as the background knowledge for focused crawling. For evaluating our approach, we refer to the biology definition of "Cell" [4] and "Mammal" [5] in Wikipedia to judge the web pages whether or not relevant to the crawling topics.

The rates at which relevant pages are effectively filtered are the most crucial evaluation criteria for focused crawling. [1,4] provides a well-established harvest rate metric for our evaluation.

$$hr = \frac{\sharp r}{\sharp p}, hr \in [0, 1] \tag{1}$$

The harvest rate represents the fraction of web pages crawled that satisfy the crawling target $\sharp r$ among the crawled pages $\sharp p$. If the harvest ratio is high, it means the focused crawler can crawl the relevant web pages effectively; otherwise, it means the focused crawler spends a lot of time eliminating irrelevant pages, and it may be better to use another crawler instead. Hence, a high harvest rate is a sign of a good crawling run.

With respect to the crawling topic "Cell", the URL "http://www.biology.arizona.edu/" is used to crawl the web pages in the data preparation stage. 100 training examples are filtered to train an ANN with 64 positive example and 36 negative examples.

In the training stage,, with the distance $d(t, c_i) <= 3$, 38 relevant concepts c_i are selected from the UMLS ontology, such as "Tissue", "Gene", and so on. In

[4] http://en.wikipedia.org/wiki/Cell_(biology)

[5] http://en.wikipedia.org/wiki/Mammal

Table 1. Average harvest rate

Method	Topic "Cell"	Topic "Mammal"
Learnable focused crawling	0.6952	0.8215
Ontology-focused crawling	0.4908	0.6892
Keyword-based crawling	0.3715	0.5641
Breadth-first crawling	0.1132	0.3252

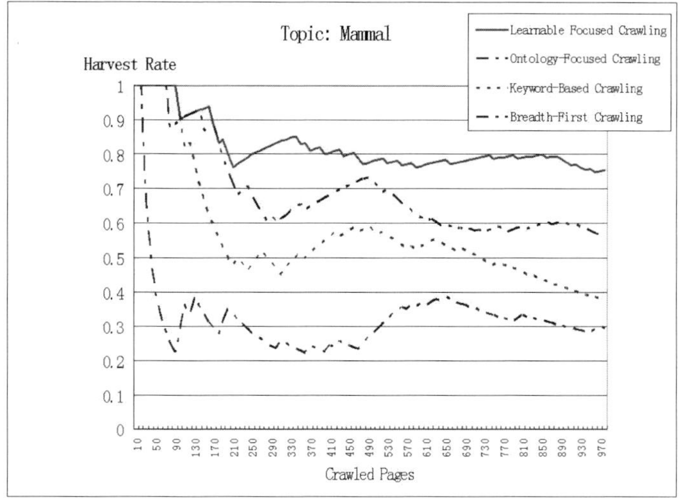

Fig. 7. Harvest rate when the crawling topic is "Mammal"

the crawling stage, we start crawling with the seed URL : "http://web.jjay.cuny.edu/acarpi/NSC/13-cells.htm". Also, the other four crawling approached are applied to crawl web pages with the same seed URL. After crawling 1150 web pages, we get the harvest rates of corresponding crawling approaches shown in figure 6.

With respect to the crawling topic "Mammal", the URL "http://www.enchantedlearning.com/" is used to crawl the web pages in the data preparation stage. 120 training examples are filtered to train an ANN with 73 positive examples and 47 negative examples.

In the training stage, with distance $d(t, c_i) <= 4$, 56 relevant concepts c_i, are selected from the UMLS ontology such as "Vertebrate", "Whale", and so on. We apply the five crawling approaches to start crawling with the seed URL: "http://www.ucmp.berkeley.edu/mammal/mammal.html". After crawling 980 web pages, we get the harvest rates shown in figure 7.

Table 1 shows the average harvest rate of topic "Cell" and topic "Mammal". When the crawling topic is "Cell", our method's average harvest rate is 41.6% higher than the ontology-focused crawling, 87.1% higher than the keyword-based crawling, and 514.1% higher than the breadth-first crawling.

For the crawling topic “Mammal”, our method’s average harvest rate is 19.2% higher than the ontology-focused crawling, 45.6% higher than the keyword-based crawling, and 152.6% higher than the breadth-first crawling.

In the crawling process, harvest rate of the breadth-first crawling was lowest because it does not consider the crawling topics. The keyword-based crawling also performs low because it only considers the crawling topic during the crawling process.

The ontology-focused crawling and our approach have better harvest rates because they relate the crawling topics to the background knowledge base to filter irrelevant web pages. Note that all the crawling processes begin with high harvest rates, but decrease as more pages are crawled. The main reason is that the seed URL has high relevance with the crawling topics. When more pages are crawled, the ratio of irrelevant web pages get higher.

Compared to the ontology-focused crawling, our approach has better harvest rates. It is because the ontology-focused crawling calculates the relevance score based on heuristically predefined concept weights. These predefined concept weights are determined by humans subjectively. It is difficult to optimally scale the contributions of those relevant concepts in real-time. In our approach, concept weights are determined objectively to describe the contribution of each relevant concept. Therefore, the harvest rate of our approach is higher during the entire crawling process. Based on the experiment results, we believe that our approach will play a promising role for focused crawling.

5 Conclusion and Future Work

In order to maintain a high harvest rate during the crawling process, we propose an ontology-based learnable focused crawling approach. The framework of our approach includes three stages: the data preparation stage, the training stage, and the crawling stage. Based on a domain-specific ontology, an ANN is trained by filtered training examples. The ANN is used to calculate the relevance score of visited web pages. The empirical evaluation shows that our approach outperformed the ontology-focused crawling, breadth-first crawling, and keyword-based crawling. Based on the experiment results, we believe that our approach improves the efficiency of the focused crawling approach, which can be employed to build a comprehensive data collection for a given domain.

In the future, the crawler will be applied in a larger application scenario including document and metadata discovery. In particular, our efforts will be focused on building an effective focused crawler with high crawling precision for the medical domain.

Acknowledgements

This work was supported by the IT R&D program of MIC/IITA [2005-S-083-02, Development of Semantic Service Agent Technology for the Next Generation Web].

References

1. Aggarwal, C.C., Al-Garawi, F., Yu, P.S.: Intelligent crawling on the world wide web with arbitrary predicates. In: WWW 10: Proceedings of the 10th international conference on World Wide Web, pp. 96–105. ACM Press, New York (2001)
2. Bodenreider, O.: The unified medical language system (umls): integrating biomedical terminology. Nucl. Acids Res. 32(suppl. 1), D267–270 (2004)
3. Can, A.B., Baykal, N.: Medicoport: A medical search engine for all. Comput. Methods Prog. Biomed. 86(1), 73–86 (2007)
4. Chakrabarti, S., van den Berg, M., Dom, B.: Focused crawling: a new approach to topic-specific Web resource discovery. Computer Networks 31(11–16), 1623–1640 (1999)
5. Diligenti, M., Coetzee, F., Lawrence, S., LeeGiles, C., Gori, M.: Focused crawling using context graphs. In: 26th International Conference on Very LargeDatabases, Cairo, Egypt, pp. 527–534 (2000)
6. Ehrig, M., Maedche, A.: Ontology-focused crawling of web documents. In: SAC 2003: Proceedings of the 2003 ACM symposium on Applied computing, pp. 1174–1178. ACM Press, New York (2003)
7. Hsu, C.-C., Wu, F.: Topic-specific crawling on the web with the measurements of the relevancy context graph. Inf. Syst. 31(4), 232–246 (2006)
8. Maedche, A., Ehrig, M., Handschuh, S., Stojanovic, L., Volz, R.: Ontology-focused crawling of documents and relational metadata. In: Proceedings of the Eleventh International World Wide Web Conference WWW-2002, Hawaii (2002)
9. Mitchell, T.: Machine Learning. McGraw-Hill Science Engineering, New York (1997)
10. Rennie, J., McCallum, A.K.: Using reinforcement learning to spider the Web efficiently. In: Bratko, I., Dzeroski, S. (eds.) Proceedings of ICML 1999, 16th International Conference on Machine Learning, Bled, SL, pp. 335–343. Morgan Kaufmann, San Francisco (1999)
11. Su, C., Gao, Y., Yang, J., Luo, B.: An efficient adaptive focused crawler based on ontology learning. In: Hybrid Intelligent Systems, 2005. HIS 2005. Fifth International Conference, November 6-9 (2005)

Gram-Free Synonym Extraction Via Suffix Arrays

Minoru Yoshida[1], Hiroshi Nakagawa[1], and Akira Terada[2]

[1] Information Technology Center, University of Tokyo
7-3-1, Hongo, Bunkyo-ku, Tokyo 113-0033
{mino,nakagawa}@r.dl.itc.u-tokyo.ac.jp
[2] Japan Airlines
3-2, Haneda Airport 3-Chome, Ota-ku, Tokyo 144-0041 Japan
akira.terada@jal.com

Abstract. This paper proposes a method for implementing real-time synonym search systems. Our final aim is to provide users with an interface with which they can query the system for any length strings and the system returns a list of synonyms of the input string. We propose an efficient algorithm for this operation. The strategy involves indexing documents by suffix arrays and finding adjacent strings of the query by dynamically retrieving its contexts (i.e., strings around the query). The extracted contexts are in turn sent to the suffix arrays to retrieve the strings around the contexts, which are likely to contain the synonyms of the query string.

1 Introduction

This paper considers a problem of extracting synonymous strings of a query given by users. Synonyms, or paraphrases, are words or phrases that have the same meaning but different surface strings. "HDD" and "Hard Drive" in documents related to computers and "BBS" and "Message Boards" in Web pages are examples of synonyms. They appear ubiquitously in different types of documents because, often, the same concept can be described by two or more expressions, and different writers may select different words or phrases to describe the same concept. Therefore, being able to find such synonyms significantly improves the usability of various systems. Our goal is to develop the algorithm that can find synonymous strings to the user input. Applications of such an algorithm include augmenting queries with synonyms in IR or text mining systems, assisting input systems by suggesting expressions similar to users' input, etc. One main problem in such tasks is that we do not know what kind of queries will be posted by users. For example, consider a system that calculates similarities between all the pairs of words or noun phrases in a corpus and provide a list of synonyms of a given query based on such similarity. Such systems can not return any output for queries that are neither words nor noun phrases, such as prepositional phrases like "on the other hand." Of course, this problem can be solved if we have similarities between every substrings in the corpus. However, considering all strings

H. Li et al. (Eds.): AIRS 2008, LNCS 4993, pp. 276–285, 2008.

(or n-grams) in the corpus as synonym candidates greatly increases the number of synonym candidate pairs, which makes computation of similarities between them very expensive in time and space.

We avoid this problem by abandoning extraction of synonym candidates in advance. Instead, we provide an algorithm to retrieve synonyms of user's queries *on the fly*. This goal is achieved by utilizing *suffix arrays*. Suffix arrays are efficient data structures that can index all substrings of a given string. By using them, the system can extract dynamically contexts to calculate similarities between strings. By extracting contexts for the query string, and subsequently by extracting strings that are surrounded by the contexts, synonym candidates can be retrieved in reasonable time. As a result, the system that allow many types of queries, such as "on the other hand", "We propose that", ":-)", "E-mail:", etc., is realized.

Our task is to extract synonymous expressions regardless of whether strings have similar surfaces (*e.g.,* having many characters in common,) or not. In such a situation, similarity calculation is typically done by using *contexts*. The strategy is based on the assumption that "similar words appear in similar contexts." Some previous systems used contexts based on syntactic structures like dependencies [1] or verb-object relations [2][3], but we do not use this type of contexts for the simplicity of modeling and language independency, as well as the fact that our goal is to develop a system that accepts any kinds of queries (*i.e.,* independent from grammatical categories), although incorporating such kinds of contexts into our suffix-array based algorithm is an interesting issue for future work. There also exist studies on the use of other resources such as dictionaries or bilingual corpuses [4] [5], but we assume no such outside resources to make our system available to various kinds of topics and documents. Another type of contexts is *surrounding strings* (i.e., strings that appear near the query). [6] reported that surrounding strings (which they call *proximity*) are effective features for synonym extraction, and combining them with other features including syntactic features stabilized the performance. Using long surrounding strings [7] can specify paraphrases with high precision but low recall, while using short surrounding strings [8] extracts many non-synonyms (i.e., low precision), which make systems to require other clues such as comparable texts for accurate paraphrase detection. We use such surrounding strings *i.e.,* the preceding and following strings, as contexts in our system. In addition, the contexts in our system can be any length to achieve a good precision-recall balance.

The remainder of this paper is organized as follows. Section 2 introduces notations used in this paper and makes a brief explanation of suffix arrays. Section 3 describes our algorithm and Section 4 and 5 reports the experimental results. In Section 6, we concludes this paper and discuss future work.

2 Preliminaries

The input to the system is a corpus S and a query q. Corpus S is assumed to be one string. For a set of documents, S is a result of concatenating those documents into one string. The system finds synonyms of q from S.

In this paper, *context* of the string s is defined as the strings adjacent to s, *i.e.*, $s.x \in S$ or $x.s \in S$ where . represents the concatenation operation and $x \in S$ means x appears in corpus S. If $s.x \in S$, x is called *right context* of s. On the other hand, if $x.s \in S$, x is called *left context* of s.

Suffix arrays [9] are data structures that represent all the suffixes of a given string. It is a sorted array of all suffixes of the string. By using the suffix array constructed on the corpus S, all the positions of s in S can be obtained quickly (in $O(logN)$ time, where N is the length of S) for any s. They require $4N$ bytes[1] of additional space to store indexes and even more space for construction. We assume that both the corpus and the suffix array are on memory.[2]

The algorithm uses two suffix arrays: $\mathcal{A}$ and $\mathcal{A}_r$. The former is constructed from S, and the latter is constructed from $rev(S)$, where $rev(x)$ is a reverse operation on string x. Right contexts are retrieved by querying $\mathcal{A}$ for q and left contexts are retrieved by querying $\mathcal{A}_r$ for $rev(q)$.

We define two operations $nextGrams(\mathcal{A}, x)$ and $freq(\mathcal{A}, x)$. The former returns the set of strings in $\mathcal{A}$ whose prefix is x and whose length is one larger than x, and the latter returns the number of appearance of x in $\mathcal{A}$. We also write them as $nextGrams(x)$ and $freq(x)$ if $\mathcal{A}$ is obvious from contexts.

We use a sorted list *cands* and a fixed-size sorted list $results$.[3] *Cands* retains strings that are to be processed by the algorithm, and $results$ retains a current top-n list of output strings. Elements in *cands* are ranked according to a priority function $priority(x)$, and elements in $results$ are ranked according to a score function $sc(x)$. We define $priority(x)$ to be smaller if x has larger priority (*i.e.*, more important), and $sc(x)$ to be larger if x is more important (*i.e.*, relevant as synonyms). Note that elements in both lists are sorted in the ascending order. $getFirst$ operations therefore return the most important element for *cands* and the least important element for $results$. This means that $getFirst$ operation of $results$ returns the N_2-th ranked element, where N_2 is a size of $results$.

3 Algorithm

The algorithm is divided mainly into two steps: context retrieval and candidate retrieval. The context retrieval step finds top-N_1 (ranked by the score defined below) list of left contexts and right contexts.[4] After that, the candidate retrieval step extracts top-N_2 list of candidates for synonyms.

[1] If each index is represented by four bytes.

[2] We used naive implementation (by ourselves) for suffix arrays. Using more sophisticated implementations like compressed suffix arrays (CSAs) can reduce the memory size required for our algorithm, which is an important future work.

[3] These are implemented by using Java TreeSet data structures, which is the Java class of red-black tree implementation, retains a sorted-list and allows $O(\log(N))$-time *add*, *remove*, and *getFirst* operations. Fixed-sized lists are implemented by replacing the add operation of TreeSets by an add operation of a new element and a subsequent remove operation of the first element.

[4] Among strings with the same score, the ones found earlier are ranked higher.

```
cands ← ‘’
while(cands ≠ ∅){
x = getFirst(cands);
N = nextGrams(A, q.x);
foreach (n ∈ N){
if (sc'_c(n) > sc_c(getFirst(results))){
cands ← cut(n, q);
results ← cut(n, q);
}
}
}
```

Fig. 1. Algorithm: Context Retrieval

3.1 STEP-1: Context Retrieval

The context retrieval step harvests the contexts, *i.e.*, strings adjacent to the query. For example, a left context list for the word *example* might include the string *in the following*. We set parameter N_1 that indicates how many contexts are harvested. We only explain how to extract right contexts, but left contexts can be extracted in a similar manner.

Figure 1 shows the context retrieval algorithm for right contexts. Note that ‘’ indicate a length-zero string. We also write removing context string c from string s as $cut(s, c)$. Starting from a set $cands$ = { ‘’ }, the search proceeds by expanding the length of strings in *cands*. (For example, element *bye* may be added to *cands* when *by* is in *cands*.) Note that this strategy causes search spaces very large because string lengths possibly increase to the end of the corpus. Our idea to avoid this problem is to cut off unnecessary search spaces by terminating string search if the score of current string and their children (i.e., the strings generated by adding suffixes to the current strings,) must be lower than the current N_1-th score. We call such termination *pruning* of search spaces.

The score for context strings are defined as follows, by analogy with tf-idf scoring functions.

$$sc_c(x) = freq(q.x) \log \frac{|S|}{freq(x)}$$

where $|S|$ is a size of corpus S. We also define the score for pruning as

$$sc'_c(x) = freq(q.x) \log |S|$$

Note that $sc'_c(x)$ can be used as the upper bound of $sc_c(x.y)$ for any y because $\log |S| \geq \log \frac{|S|}{freq(x.y)}$ and $freq(q.x) \geq freq(q.x.y)$, resulting in $sc'_c(x) \geq sc_c(x.y)$.

Threshold Values. We introduce the parameter F_1 to reduce execution time of our algorithm. F_1 is set not to include contexts that appears too frequently in the corpus. If a context $freq(c)$ is over F_1, c is not added to *results*.

```
forall(c ∈ C_l){ cands ← (c,“) }
while(cands ≠ ∅){
D = ∅;
while (x does not change){
(c, x) = getFirst(cands);
Y = nextGrams(A, c.x);
foreach(y ∈ Y){
y' = cut(y, c);
D = D ∪ {(c, y')}
sc_l(y') = sc_l(y') + 1; } }
forall ((c, d) ∈ D){
if (sc_l(d) > sc_l(getFirst(results))){
cands ← (c, d);
results ← d; } } }
```

Fig. 2. Algorithm: Candidate Retrieval

3.2 STEP-2: Candidate Retrieval

After context strings are obtained, the algorithm extracts strings adjacent to the contexts. We refer to the strings as *synonym candidates*, or simply *candidates*. We set the parameter N_2 that indicates the number of candidates to be retrieved.

The algorithm proceeds in the following way.

Stage-1: obtain the N_2-best candidates by using left contexts only, according to score function $sc_l(c)$.

Stage-2: obtain the N_2-best candidates by using right contexts only, according to score function $sc_r(c)$.

Stage-3: rerank all obtained candidates according to score function $sc(c)$ and obtain a new top-N_2 list.

Roughly speaking, in stage-1 and 2, the algorithm searches for top-N_2 synonym candidates by using the score which is relatively simple but useful for pruning of search spaces. After that, in stage-3, the algorithm re-ranks these top-N_2 results by using a more complex scoring function. Here, we only explain stage-1, because stage-3 is straightforward and stage-2 can be performed in a similar manner to stage-1.

Figure 2 shows the algorithm. Here, C_l represents a left context set. Note that elements in *cands* are pair (c, x), where the list is sorted according to $priority(x)$. $priority(x)$ is defined to rank strings with the highest score come to the first. The algorithm first makes a set D by expanding the current best candidate x. After that, for each $y \in D$, if $sc_l(y)$ is larger than the current N_2-th score, y is newly added to *cands*.

The score $sc_l(x)$ is defined as the number of $c \in C_l$ for which $c.x \in S$, *i.e.*, how many types of left contexts appearing adjacent to x. The good point of this score is that it is decreasing function of the length of x, *i.e.*, $sc_l(x) \geq sc_l(x.y)$.

This means that if the $sc_l(x)$ is lower than the current n-th best score, there is no need for searching for $x.y$ for any y.

On the other hand, the score in stage-3 is defined as

$$sc(x) = \sum_{c \in C_l} \log \frac{freq(c.x)}{freq_{exp}(c.x)} + \sum_{c \in C_r} \log \frac{freq(x.c)}{freq_{exp}(x.c)}$$

where C_r represents a set of right contexts extracted in step-2, and $freq_{exp}(x)$ is the frequency of x expected from the context frequency and the number of x appearing in the whole-corpus, defined as

$$freq_{exp}(x) = freq(x) \cdot \frac{freq(c)}{|S|}$$

where $|S|$ is the size of corpus S.

List Cleaning. Obtained lists of contexts often contain redundant elements because it contains strings of any length. For example, "have to do" and "have to do it" can be in the same list. To remove such redundancy, list cleaning is performed on each context list. If the n-th element is a substring of the m-th element or m-th element is a substring of the n-th element for $m < n$, the n-th element is removed from the list. Not only it reduces the execution time by reducing the number of contexts, but also we observed that it generally improves the quality of extracted results mainly because it prohibits similar contexts from appearing repeatedly in the same list. List cleaning is also performed on candidates lists. We observed that it also improved the quality of candidate lists. Note that list cleaning operations make the size of resulting lists smaller than N_1 or N_2.

4 Output Examples

We applied our algorithm to the web documents crawled from the web-site of University of Tokyo.[5][6] The size of corpus was about 800 Mbytes and parameters were set to $F_1 = N_1 = N_2 = 1000$. The system was run on an AMD Opteron 248 (2.2GHz) machine with 13Gbytes memory. Figure 3 shows some example results of synonym extraction.[7] Both results were obtained in a few seconds. We observed that phrases like "Natural Language Processing" were correctly associated with the one word string "NLP" without any preprocessing like NP chunkers. In addition, phrases like "We propose" that are not in one phrase structure category (like NP or VP), which are difficult to chunk, were able to be processed thanks to the property of our method that takes into account every-length string.

[5] http://www.u-tokyo.ac.jp/

[6] Due to the noisiness of the data (*e.g.*, it includes very long substrings that appear twice or more because of duplicated web pages), we set a threshold value (50) for the length of strings in context retrieval and candidate retrieval on this data.

[7] The first answer to the first query is a noise caused by variations of white spaces.

Query: Natural Language Processing
Results: 1.Natural LanguageP / 2.NLP / 3.Artificial Intelligence / 4.ning / 5.Retrieval

Query: We propose
Results: 1.We present / 2.we propose / 3.we have develope / 4.we present / 5.This p

Fig. 3. Synonym Extraction Examples from Web documents of University of Tokyo

Query: likely to
Results: 1.will / 2.expected to / 3.going to / 4.ould / 5.able to

Query: U.S.
Results: 1.US / 2.Australian / 3.United States / 4.American / 5.Canadian

Query: doesn't
Results: 1.does not / 2.didn't / 3.would / 4.did not / 5.could

Fig. 4. Synonym Extraction Examples from the Reuter Corpus

We also ran the system with a part of the Reuter corpus (487 MBytes) on the same machine with the same parameter settings. Figure 4 shows the extraction results for some example queries. We observed that the results for acronym query "U.S." correctly included "United States", and the results for contraction query "doesn't" correctly included "does not".

5 Experiments

We used the JAL (Japan Airlines) pilot reports which had been de-identified for data security and anonymity. The reports were written in Japanese except for some technical terms written in English. The size of concatenated documents was 7.4 Mbytes. The system was run on a machine with an Intel Core Solo U1300 (1.06GHz) processor and 2GBytes memory.[8]

Many expressions that have their synonymous variants are found in this corpus, such as loan terms that can be written in both Japanese and English, and long words/phrases that have their abbreviated forms (e.g., "LDG" is an abbreviation of "landing"), etc.

In order to evaluate the performance of the system, we used a thesaurus for this corpus that are manually developed and independent of this research. The thesaurus consists of $(t, S(t))$ pairs where t is a term and $S(t)$ is a set of synonyms of t, such as (CAPT, {Captain}) and (T/O, {Takeoff}), etc. In the experiment, t

[8] Note that the machine is different from the one mentioned in the previous section.

is used as a query to the system, and $S(t)$ is used as a collect answer to evaluate the synonym list produced by the system. The number of queries was 404 and the average number of synonyms was 1.92.

The system performance was evaluated using *average precision* [10]. We provided each query in the test set to the system, which in turn returns a list of synonym candidates $\langle c_1, c_2, ..., c_n \rangle$ ranked on the basis of their similarity to the query. Given this list and synonym set $S = \{s_1, s_2, ..., \}$, the average precision of the result list is calculated as

$$\frac{1}{|S|} \sum_{1 \leq k \leq n} r_k \cdot precision(k),$$

where $precision(k)$ is the accuracy (i.e., ratio of correct answers to all answers) of the top k candidates, and r_k represents whether the k-th document is relevant (1) or not (0). (In other words, $r_k = 1$ if $c_k \in S$, and $r_k = 0$ otherwise.)

We investigated the execution time of the algorithm and average precision values for various parameter values. N_1 parameter was set to 1000. Table 1 shows the results. We observed that setting F_1 threshold value contributed to improvement both of execution time and average precision. Among them, larger parameter values contributed to improvement of output quality, at the expense of execution time. The best result was obtained when $F_1 = N_2 = 1000$. The execution time for that setting was 1.96 seconds.

Table 1. Execution time (in msec) per query for various parameter values

Execution Time				Average Precision (%)			
$F_1 \setminus N_2$	100	300	1,000	$F_1 \setminus N_2$	100	300	1,000
100	725.5	847.0	1070.0	100	32.02	31.71	30.60
300	863.7	1034.1	1359.0	300	34.52	37.18	36.31
1,000	1098.6	1403.4	1963.4	1,000	33.69	38.73	40.32
∞	2635.9	3849.4	6550.6	∞	22.01	28.47	31.40

To analyze quality of outputs of our method, we compared them with the output by the standard vector space model algorithm (VSMs) with cosine similarity measure. Two major features for synonym extraction, namely, *surrounding words* (or *proximity*) [6], and *dependency relations* [3], were used for the vector space model. We defined three types of weighting schemes for vector values: term frequency (TF), tf-idf values (TF-IDF), and logarithm of TF (logTF).[9] The window size for proximity was set to 3.

We compared our algorithm with VSMs on the task of *candidate sorting*, where the task is to make a ranked list of $c \in T$, where T is a set of words in the thesaurus, according to the similarity to the query q. Ranked lists were made from outputs of our method by filtering out elements in output candidate list if they were not T.[10] Note that this task is slightly different from the synonym

[9] Note that logTF is defined as $log(1 + tf)$ to handle the case where $tf = 1$.

[10] In this experiment, we did not perform list cleaning on candidate lists.

Table 2. Candidate ranking results shown in averaged precision values (%)

Algorithm			
Our Method	52.20		
	Dependency	Proximity	Prox + Dep
VSM (logTF)	34.43	55.61	57.35
VSM (TF-IDF)	24.72	37.85	39.19
VSM (TF)	23.25	40.12	40.87

extraction because candidates outside of T is ignored, and therefore the average precision values are higher than the values in Table 1. Parameters of our algorithm was set to $F_1 = 1000, N_2 = 1000$.

Table 2 shows the result. Among VSMs, we observed that proximity features were effective for synonym extraction and the performance was improved by using dependency features. This result agreed with the results reported in [6].

Among three weighting schemes, logTF weighting performed much better than other two schemes. We think that it is because logTF emphasizes the number of types of context words than their frequency, and the number of types of context words shared by the query is important for synonym extraction.

The performance of our method was slightly inferior to the best results among VSMs. We think that it was partly because our context features can not handle relations among strings separated by various strings, and partly because our method does not retrieve all the possible strings which cause some answers in T to be not contained in resulting candidate lists.

6 Conclusions and Future Work

We proposed a method to extract synonymous expressions of a given query on the fly by using suffix arrays. Experimental results on 7M bytes corpus showed that our method was able to extract synonyms in 0.7 – 7.0 seconds. However, qualitative performance of our method was slightly worse than standard vector space model methods. It suggests that our method still leaves room for improvement by, for example, extracting synonymous expressions of context strings themselves and using them as new contexts for synonym extraction. Future work also includes exploring possibility of use of other kinds of features like dependency structures in our suffix-array based retrieving method. Use of more efficient suffix-array implementations like compressed suffix arrays is also an important issue for future work.

References

1. Yamamoto, K.: Acquisition of lexical paraphrases from texts. In: Proceedings of COMPUTERM 2002, pp. 1–7 (2002)
2. Gasperin, C., Gamallo, P., Agustini, A., Lopes, G.P., de Lima, V.: Using syntactic contexts for measuring word similarity. In: Proceedings of the ESSLLI 2001 Workshop on Semantic Knowledge Acquisition and Categorisation (2001)

3. Murakami, A., Nasukawa, T.: Term aggregation: mining synonymous expressions using personal stylistic variations. In: Proceedings of COLING 2004 (2004)
4. Wu, H., Zhou, M.: Optimizing synonym extraction using monolingual and bilingual resources. In: Proceedings of IWP 2003 (2003)
5. van der Plas, L., Tiedemann, J.: Finding synonyms using automatic word alignment and measures of distributional similarity. In: Proceedings of COLING/ACL 2006 Main conference poster sessions, pp. 866–873 (2006)
6. Hagiwara, M., Ogawa, Y., Toyama, K.: Selection of effective contextual information for automatic synonym acquisition. In: Proceedings of COLING/ACL 2006, pp. 353–360 (2006)
7. Pasca, M., Dienes, P.: Aligning needles in a haystack: Paraphrase acquisition across the web. In: Proceedings of IJCNLP 2005, pp. 119–130 (2005)
8. Shimohata, M., Sumita, E.: Acquiring synonyms from monolingual comparable texts. In: Proceedings of IJCNLP 2005, pp. 233–244 (2005)
9. Manber, U., Myers, G.: Suffix arrays: A new method for on-line string searches. In: Proceedings of the first ACM-SIAM Symposium on Discrete Algorithms, pp. 319–327 (1990)
10. Chakrabarti, S.: Mining the Web: Discovering Knowledge from Hypertext Data. Morgan-Kaufmann Publishers, San Francisco (2002)

Synonyms Extraction Using Web Content Focused Crawling

Chien-Hsing Chen and Chung-Chian Hsu

National Yunlin University of Science and Technology, Taiwan
{g9423809,hsucc}@yuntech.edu.tw

Abstract. Documents or Web pages collected from the World Wide Web have been considered one of the most important sources for information. Using search engines to retrieve the documents can harvest lots of information, facilitating information exchange and knowledge sharing, including foreign information. However, to better understand by local readers, foreign words, like English, are often translated to local language such as Chinese. Due to different translators and the lack of translation standard, translating foreign words may pose a notorious headache and result in different transliterations, particularly in proper nouns like person names and geographical names. For example, "Bin Laden" is translated into terms "賓拉登"(binladeng) or "本拉登"(benladeng). Both are valid synonymous transliterations. In this research, we propose an approach to determining synonymous transliterations via mining Web pages retrieved by a search engine. Experiments show that the proposed approach can effectively extract synonymous transliterations given an input transliteration.

Keywords: Transliteration, Associated Word, Unknown Words, Focused Crawling, Speech Sound Comparison.

1 Introduction

A foreign word is usually translated to a local word according to their phonetic similarity in the two languages. Such translated words are referred to as transliterations. Due to the lack of translation standard, a foreign word might be transliterated into different Chinese words, referred to as synonymous transliterations, especially proper names and geographical names, such as terrorist Bin Laden is translated into two different Chinese transliterations as "賓拉登" (binladeng) and "本拉登" (benladeng). Note that we use Hanyu pinyin for the Romanization of Chinese transliterations.

Synonymous transliterations would raise obstacles in reading articles as well as communicating with other people. More importantly, search engines will generate incomplete search results. For example, using "本拉登" as the input keyword cannot retrieve the Web pages that use "賓拉登" instead as the transliteration.

This research aims at mining synonymous transliterations from the Web. Our research results can be applied to construct a database of synonymous transliterations, which can then be used to alleviate the problem of incomplete search of Web pages.

Three major difficulties in extracting synonymous transliterations from huge Web pages are concerned. The first question is *target documents collection*: how to collect

H. Li et al. (Eds.): AIRS 2008, LNCS 4993, pp. 286–297, 2008.

appropriate pages which contain synonymous transliterations from the huge collection of Web pages [1]. We exploit a method, content focused crawling [2-3], to crawl a small number of, but related, documents which contain probable synonymous transliterations from the Web. The second question is *term segmentation*: how to segment words as synonymous transliteration candidates from noisy search-result pages; and the last is *confirming synonymous transliterations*: how to recognize those identified candidates as true synonymous transliterations.

2 The Proposed Approach

2.1 Observation

Chinese transliterations may come with their original foreign words. One approach to retrieving synonymous transliterations is to query the foreign word in Chinese Web pages. However, most commercial search engines limit the number of Web pages returned to the user. For example, Google, Altavista, and Goo return only 1,000 pages even though more pages match the query [4]. Due to this limitation, a user inputs an original foreign word with the search scope limited to Chinese Web pages. The user may not be able to retrieve many distinct Chinese transliterations. Especially when there are a few dominant transliterations which appear more frequent than the others, the retrieved pages may contain only those dominant transliterations. For instance, the transliteration "本拉登' appear less frequent than "賓拉登'.

Another alternative is to use *associated words* which are in terms of context highly relevant to its original word. For example, we used the term "恐怖分子' (terrorist) to submit to search engines, and then we got some Web pages containing the term "本拉登' (Bin Laden) as shown in Fig. 1. Therefore, identifying appropriate associated words for submitting to search engines is a key step to retrieving synonymous transliterations.

2.2 Collecting Candidate Web Pages

Fig. 2 depicts the process of identifying candidate pages which may contain synonymous transliterations. First, some seed Web pages are downloaded by the input transliteration, e.g. "賓拉登" (binladeng). Second, we extract its associated words and assign their weights which represent the degree of significance between the transliteration and the associated words. For instance, "恐怖分子" (terrorist), "阿富汗" (Afghanistan) and "恐怖主義" (terrorism) all have significant weights in terms of their association with "賓拉登". The higher weighted words would not only be treated as keywords to be submitted to a search engine for candidate pages, but also be regarded as a set of characteristics of this input transliteration. The method of weight assignment will be described in Section 2.3.

Third, for each significant associated word which is used as a search keyword, we download a fixed number of Web pages. Four, because the Web consists of pages on diverse topics, naïve queries by users may find matches in many irrelevant pages [4].

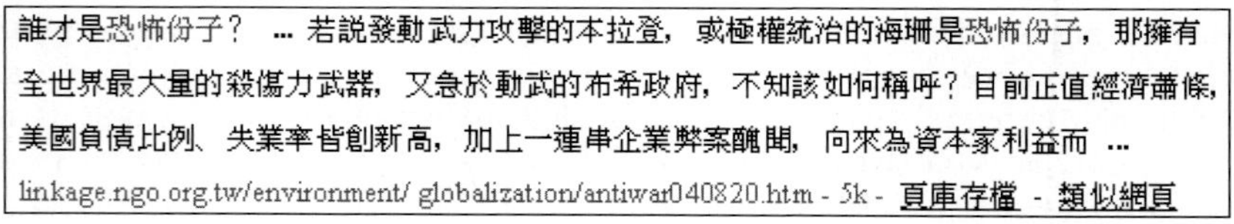

誰才是恐怖份子？ ... 若說發動武力攻擊的本拉登，或極權統治的海珊是恐怖份子，那擁有全世界最大量的殺傷力武器，又急於動武的布希政府，不知該如何稱呼？目前正值經濟蕭條，美國負債比例、失業率皆創新高，加上一連串企業弊案醜聞，向來為資本家利益而 ...
linkage.ngo.org.tw/environment/ globalization/antiwar040820.htm - 5k - 頁庫存檔 - 類似網頁

Fig. 1. The search-result page of query term "恐怖分子" (terrorist) in Google search engine

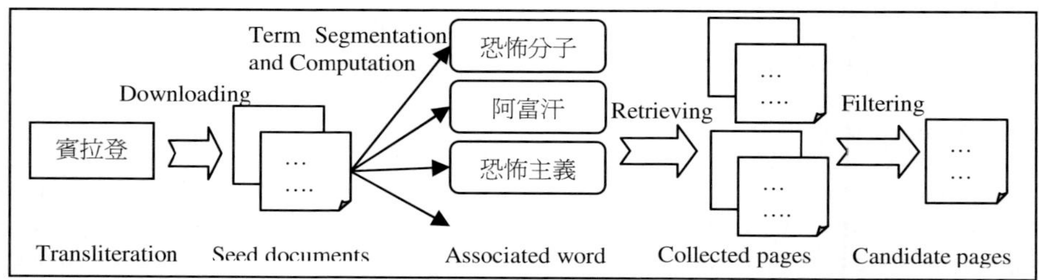

Fig. 2. Illustration of identifying candidate pages with content focused crawling by the term "賓拉登" (binladeng)

So, we use a method to filter out irrelevant Web pages. Specifically, if the Web pages contain many highly weighted associated words of the transliteration (i.e., the characteristics of the transliteration), we regard that Web page a relevant page, such as the search-result page of Fig. 3. In contrast, we can discard the Web page which doesn't contain significant associated terms, such as that of Fig. 4.

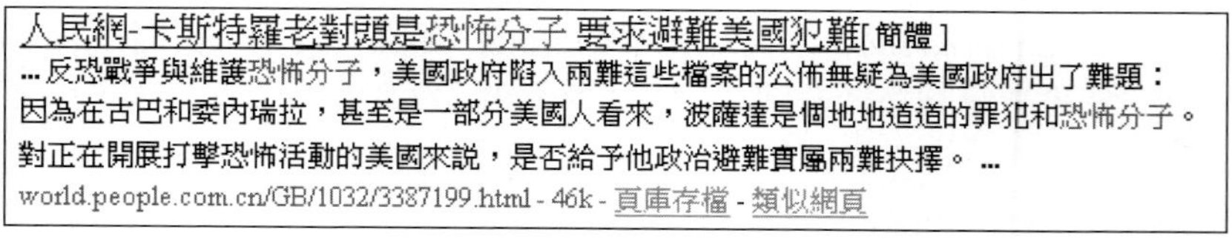

人民網-卡斯特羅老對頭是恐怖分子 要求避難美國犯難[簡體]
... 反恐戰爭與維護恐怖分子，美國政府陷入兩難這些檔案的公佈無疑為美國政府出了難題：因為在古巴和委內瑞拉，甚至是一部分美國人看來，波薩達是個地地道道的罪犯和恐怖分子。對正在開展打擊恐怖活動的美國來說，是否給予他政治避難實屬兩難抉擇。 ...
world.people.com.cn/GB/1032/3387199.html - 46k - 頁庫存檔 - 類似網頁

Fig. 3. A search-result page of query term "恐怖分子" (terrorist) in Google search engine

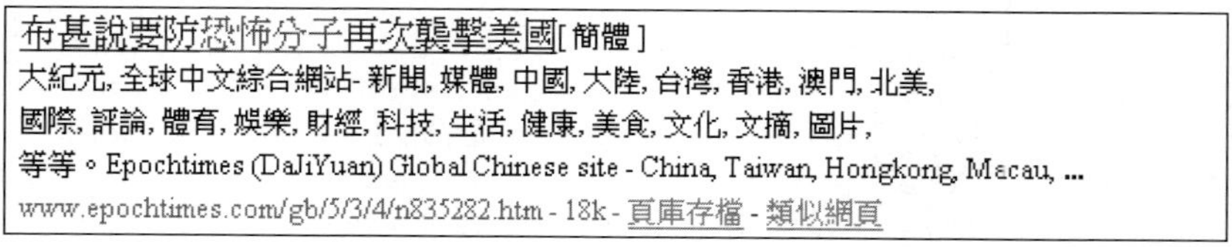

布甚說要防恐怖分子再次襲擊美國[簡體]
大紀元, 全球中文綜合網站- 新聞, 媒體, 中國, 大陸, 台灣, 香港, 澳門, 北美, 國際, 評論, 體育, 娛樂, 財經, 科技, 生活, 健康, 美食, 文化, 文摘, 圖片, 等等。Epochtimes (DaJiYuan) Global Chinese site - China, Taiwan, Hongkong, Macau, ...
www.epochtimes.com/gb/5/3/4/n835282.htm - 18k - 頁庫存檔 - 類似網頁

Fig. 4. Another search-result page of query term "恐怖分子" (terrorist) in Google search engine

2.3 Associated Words Computation

Several statistic methods have been proposed for estimating term association based on co-occurrence analysis, including mutual information, DICE coefficient, chi-square test, and log-likelihood ratio. For chi-square test, the required parameters can be obtained by submitting Boolean queries to a search engine and utilizing the returned page count indicating the estimated number of qualified pages containing the query term [5]. In such a way, we do not need to perform time-consuming page download and text processing for estimating the counts and thus dramatically reduce computation time.

We take the initial step of downloading a fixed number of Web documents for the input transliteration. Then we extract all known words w_i from the set of downloaded pages Q using term segmentation. The vector space model VSM is employed, in which each element represents a word with a value of 1 when the document contains this word or a value of 0 otherwise. We estimate the probability $\Pr(w_i)$ of the documents containing the word w_i .

$$Pr(w_i)=\frac{\sum_{j}^{|Q|} m_{ij}}{|Q|},\ \text{where } m_{ij}=\begin{cases}0, & w_i \notin d_j\\ 1, & w_i \in d_j\end{cases}\quad i=1,2,\ldots,W \text{ and } j=1,2,\ldots,|Q| \tag{1}$$

W represents the size of the extracted vocabulary from the downloaded documents Q. $|Q|$ represents the size of Q. m_{ij} represents a vector element whose value is 1 when the document d_j contains w_i or 0 otherwise.

We consider a word w_i is more important if w_i co-occurs frequently with the transliteration TL. Therefore, the significance of an associated word is proportional to the degree of co-occurrence with the transliteration. The significance (or weight) can be defined by Eq. (2) which can be further written such that it can be estimated by the frequencies of the words. It is worth noting that we just query a search engine and obtain the returned page counts without downloading and processing the Web pages.

$$weight(w_i) = \Pr(w_i) \times \frac{\Pr(TL \mid w_i)}{\Pr(TL' \mid w_i)} = \frac{\sum_{j}^{|Q|} m_{ij}}{|Q|} \times \frac{f(TL, w_i)}{f(w_i) - f(TL, w_i)} \tag{2}$$

$\Pr(TL|wi)$ and $\Pr(TL'|wi)$ represent the probabilities of observing TL or not given a page containing w_i. $f(w_i)$ represents the number of documents queried by w_i. w_i gets higher strength if w_i occurs frequently in documents containing TL, but infrequently in documents without TL.

2.4 Sifting the Collected Pages

Each of significant associated words is submitted to obtain a small set of Web pages. Due to the diversity of the Web, some of the collected pages might not be relevant to the content of the input transliteration. A filtering step is needed to discard irrelevant pages.

We sift the Web pages via the set of the characteristic associated words of the transliteration. Let $AS_1, AS_2, AS_3, \ldots, AS_n$ are the associated words of the transliteration mentioned in Section 2.3. d_i is a crawled Web page. $Wd_{i,k}$ represents a word in d_i. and $weight(Wd_{i,k}) = weight(AS_i)$ if $Wd_{i,k} = AS_i$. The score of d_i with respected to TL is defined as Eq. (3), the aggregated weight of the top-θ weighted words $Top\_Wd_{i,j}$. We consider documents with high scores as candidate pages which might contain synonymous transliterations and discard the other pages.

$$Score\ (d_i) = \prod_{j=1}^{\theta} Weight\ (Top\_Wd_{i,j}) \tag{3}$$

3 Recognizing and Confirming Synonymous Transliterations

For those candidate Web pages, some preprocess is performed to extract unknown words. We first identify and eliminate known words with the help of a dictionary. We then apply an N-gram approach [6] to extract N-grams units on the remaining text strings. N is set between the length of TL minus and plus one, i.e., $|TL| - 1 \leq N \leq |TL| + 1$ since most of synonymous transliterations has a length discrepancy less than or equal to one.

The number of the extracted N-grams units is usually quite large. We use SPLR algorithm [7] to help reduce the size. The SPLR algorithm is effective to detect an unknown word which is a subsequence of an N-grams unit. For instance, the unknown word "本拉登" (Bin Laden) is a subsequence of the 4-characters unit "本拉登的" (of Bin Laden). Once the subsequence is determined an unknown word, its supersequences can be discarded.

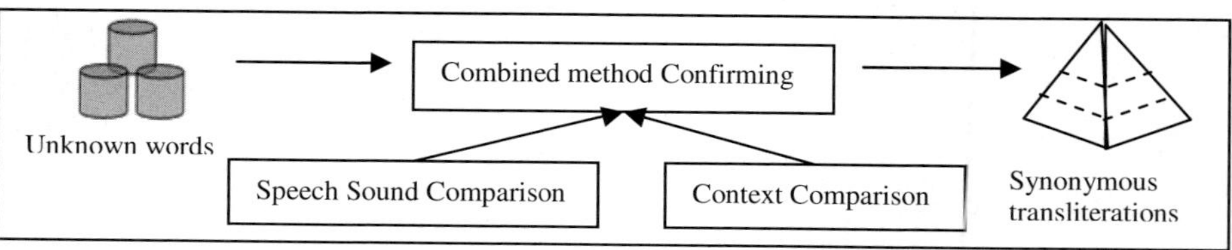

Fig. 5. Process of synonymous transliterations confirmation

The process of recognizing and confirming the unknown words as the synonymous transliterations is illustrated in Fig. 5. Based on our observation, a transliteration has similar speech sounds with its synonymous transliterations, for instance the terms "賓拉登" (binladeng) and "本拉登" (benladeng). Therefore, we use speech sound comparison to identify synonymous transliterations. However, a candidate unknown word which scores high in speech sound comparison is not always a synonymous transliteration. To cope with this problem, a further confirmation step based on context matching is performed.

3.1 Speech Sound Comparison

In [8], we have given a description of the Taiwan's phonetic system and devised an approach to comparing the phonetic similarity of two Chinese words. Our approach uses two similarity matrices of digitalized speech sounds, one for 412 basic Chinese character sounds and the other for the 37 Chinese phonetic symbols used in Taiwan. The advantage of this digitalized sound comparison approach over the conventional statistic-based approaches is that it does not require a corpus for computing the phonetic similarity between two words and thus avoids potential bias inherent in the collected corpus.

We employee the speech sound comparison approach to compare the similarity between the extracted N-grams unknown word and TL. Highly similar pair is likely synonyms. The approach looks at the patterns of similarity in Chinese phonemes as a basis for comparing the degree of similarity between the pronunciations of different characters, and thus for comparing the degree of similarity between two words. We carry out numeric analysis to quantify the similarities among the digitalized sounds. For example, the Taiwan's phonetic symbols ㄅ(p) and ㄆ(b) are both initials and are highly similar in pronunciation. In particular, we first recorded in digitalized form the two sets of sounds including the 412 basic Chinese character sounds and the 37 Taiwan's phonetic symbols. We then adopted speech recognition technique to compare the similarity between every pair of two sounds in each set of the digitalized sounds. As a result, we got two similarity matrices for the 412 character sounds and the 37 phonetic symbols, respectively. Note that the complete set of Chinese character sounds is about 5*412 which considers character's tone. However, the tone does not affect much in terms of comparing phonetic similarity between two Chinese words [8]. Therefore, the tone is not taken into account in the set of character sounds to reduce the size of the set.

Given the two sound similarity matrices, a dynamic programming based algorithm is used to measure the similarity between the N-grams word and TL. Specifically, the algorithm aligns the phonetically similar characters and aggregates their similarity scores. Since the final phoneme of a character has heavy influence during the sound recording, we include a weight on the initial consonant of the character to balance the bias in the computation. The recursive formula for the dynamic programming algorithm to compute the similarity between two Chinese character strings is defined as follows.

$$T(i,j) = \max\begin{Bmatrix} T(i-1,j-1) + sim(S_1(i), S_2(j)), \\ T(i-1,j) + sim(S_1(i), '\ '), \\ T(i,j-1) + sim('\ ', S_2(j)) \end{Bmatrix} \tag{4}$$

where $sim(S_1(i), S_2(j)) = w \times sim_{s37}(S_1(i)IC, S_2(j)IC) + (1-w) \times sim_{s412}(S_1(i), S_2(j))$ $sim_{s412}(\cdot,\cdot)$ gives the similarity between two Chinese character sounds by retrieving the correspond ing entry in the 412-character-sound similarity matrix and $sim_{s37}(\cdot,\cdot)$ give the similarity between the two initial consonants of the two characters by retrieving the entry in the 37 phonetic-symbol similarity matrix. The weight w can be set to 0.5, which experimentall y leads to a good result.

The comparison with the NULL character is $sim(S_1(k), ‘\ ’) = 0$ and $sim(‘\ ’, S_2(k)) = 0$. The initial conditions are $T(i,0) = 0$ and $T(0,j) = 0$.

For an m-length unknown word UW and an n-length TL, the aggregate similarity score is $T(m, n)$ and the score is further normalized by the average length of the two words. In other words, the similarity between UW and TL is defined by $sim(\text{UW, TL}) = 2/(m+n) * T(m, n)$.

3.2 Context Comparison

Some unknown words which are not a synonym of the transliteration might happen to have a high similarity score with respected to TL by sound comparison. For instance the unknown word "九八七年" (jiubaqinian, meaning 987 years) has high similarity score when compared to the transliteration "戈巴契夫" (gebaqifu) of Gorbachev but clearly it is not a synonymous transliteration. Therefore, a confirmation step is needed to reduce false positives by the sound comparison approach in the previous step.

The basic idea of the confirmation technique is that a true synonymous transliteration shall share similar context. Therefore, it shall share common associated words with the input transliteration in its search-result pages. The confirmation step is thus performed as follows. We retrieve some Web pages by using the candidate synonymous transliteration, which is an N-grams unknown word having high similarity score with the input transliteration. If the retrieved pages have similar content with those of a transliteration, the candidate synonym is judged as a synonymous transliteration. To compare the content similarity of the pages, the cosine measure can be used. Again, the previously extracted associated words of TL can be used in this similarity computation.

For example, given a transliteration "戈巴喬夫' (gebaqiaofu) of Gorbachev and its candidate synonymous transliteration N-grams words "戈巴契夫' (gebaqifu) and "九八七年' (jiubaqinian), a search-result page is shown respectively in Fig. 6 and Fig. 7. Although both "戈巴契夫' and "九八七年' have high similarity scores to "戈巴喬夫' in speech sound comparison to "戈巴契夫', we can see that the snippet in Fig. 6 by "戈巴契夫' (gebaqifu) are relevant to "戈巴喬夫', but the snippet in Fig. 7 tells that the page of "九八七年' (jiubaqinian) has nothing to do with "戈巴喬夫'.

3.3 The Combined Approaches

The speech sound comparison measures the similarity of pronunciation whereas context comparison measures the similarity of context between Chinese terms. It might be beneficial to take the advantages of both approaches. Thus, an alternative is to combine these two methods. We use a combined ranking scheme to determine the similarity between a transliteration *TL* and its synonymous transliteration candidate *t* as follow:

$$Rank(TL,t) = \sum_{mi} \alpha_{mi} R_{mi}(TL,t) \quad \text{subject to} \sum_{mi} \alpha_{mi} = 1 \tag{5}$$

where *mi* represents the different methods. α_{mi} is a weight for each *mi*, and $R_{mi}(TL,t)$ represents the rank of the similarity between the pair (*TL*, *t*) with respected to all pairs under the method *mi*. The combined approach is expected to raise the ranking of true synonymous transliterations among other noise terms.

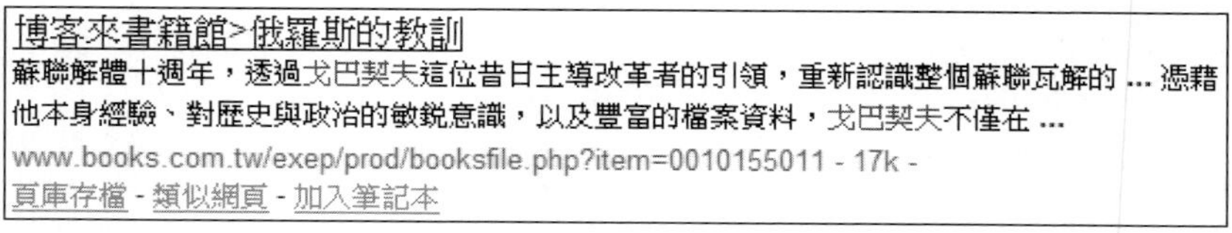

博客來書籍館>俄羅斯的教訓
蘇聯解體十週年，透過戈巴契夫這位昔日主導改革者的引領，重新認識整個蘇聯瓦解的 ... 憑藉他本身經驗、對歷史與政治的敏銳意識，以及豐富的檔案資料，戈巴契夫不僅在 ...
www.books.com.tw/exep/prod/booksfile.php?item=0010155011 - 17k -
頁庫存檔 - 類似網頁 - 加入筆記本

Fig. 6. A search-result page of query "戈巴契夫' (gebaqifu) in Google search engine

[PDF] 香港立法局一 ˉ 九八七年十月 ˉ 十 ˉ 日73 立法局會議過程正式紀錄
檔案類型: PDF/Adobe Acrobat - HTML 版
... 九八七年十月 ˉ 十 ˉ 日. 77. 不過，我已接到通知，政府將於十日左右便可收到報告書，
屆時我們會印製副本，交給立法 ... 九八七年十月 ˉ 十 ˉ 日. 於現時提供的補救措施是否
足夠，市民都表示關注。在這個情形下 ，財政司可否解釋他為何說， ...
www.legco.gov.hk/yr87-88/ chinese/lc_sitg/hansard/h871021.pdf - 類似網頁

Fig. 7. A search-result page of query "九八七年' (jiubaqinian) in Google search engine

4 Experiments

4.1 Associated Words Extraction

Our experimental data are the Web pages collected by the use of Google search engine. Determining the associated words of the transliteration is the core of the quality of content focused crawling. We limit the initial downloaded Web data to 800 pages for each transliteration, and then extract the associated words for each transliteration. We use the 80K CKIP dictionary from Academia Sinica of Taiwan to determine known words in the downloaded Web pages and to extract the associated words. Some of experimental results of extracted associated words are shown in Tables 1. As the results show, most of the extracted Chinese associated words are

Table 1. Some examples of personal names and geographical names and their extracted Chinese associated words

Initial Word	Extracted Associated Words
柯林頓	柯江會, 總統, 柳思基, 美國, 柯江高峰會, 白宮
賓拉登	恐怖份子, 攻擊, 美國, 阿富汗, 五角大廈, 恐怖主義
戈巴契夫	蘇聯, 雷根, 史達林, 柏林圍牆, 冷戰, 總統
麥可喬登	空中飛人, 比爾蓋茲, 籃球, 球鞋, 公牛
莎朗史東	地動天驚, 史蒂芬席格, 金貝辛格, 好萊塢, 第六感
阿諾史瓦辛格	如癡如醉, 複製人, 無異議通過, 換跑道, 合十
陳水扁	總統, 台獨, 大選, 兩岸, 民進黨, 台灣
布魯斯威利	終極密碼戰, 沈痛, 恩主, 兒童心理學, 海倫杭特
查爾斯	安娜, 勝地, 身穿, 穿著, 節省, 水晶, 報業
約翰保羅	教皇, 羅馬, 天主教, 梵蒂岡, 新聞, 去世
羅納多	世足賽, 鐵面人, 雨勢, 金靴獎, 準決賽
海珊	政權, 伊拉克, 總統, 巴格達, 美國, 戰爭, 聯軍
梅爾吉勃遜	搶救雷恩大兵, 致命武器, 羅素, 李奧納多, 海倫杭特
俠客歐尼爾	小飛俠, 中鋒, 鄧肯, 摔角, 公牛, 球季, 士官長, 球員
達賴喇嘛	轉世, 音譯, 汪洋, 流亡, 查理, 印度人, 大汗, 尊稱
雪梨	墨爾本, 澳大利亞, 歌劇院, 澳洲, 坎培拉, 資訊, 國際
悉尼	澳大利亞, 墨爾本, 歌劇院, 奧運會, 坎培拉, 奧運

well relevant to the initial query words, indicating that our approach to determining the associated words works well.

4.2 Synonymous Transliterations Extraction

We show some experiments about synonymous transliterations extraction from the Web. We manually collected 18 transliterations and perform the approach to identify synonymous transliterations via searching and retrieving only a subset of the WWW that pertains to those transliterations. Via content focused crawling, we downloaded related Web pages, and then filter less relevant pages in order to keep the quality of crawling and to reduce the size of pages from which we expected to obtain synonymous transliterations. We set the filtering ratio to 0.7. It means that we downloaded 3,000 pages, and then retain only 900 pages after the filtering.

We perform the context comparison step as mentioned in Section 3.2 to top-200 ranked words after speech sound comparison. Table 2 shows the results.

In Table 2, the first column shows the 18 transliterations that we collected from newspapers. Column 2 shows the extracted synonymous transliterations from the Web. Column 3 SSC (speed sound comparison) shows the ranking of the identified synonymous transliterations among other segmented N-grams words via the speech sound comparison approach while Column 4 SSC+CC (speed sound comparison followed by context comparison) shows the ranking by further comparing the context of extracted synonymous transliterations with those of the input transliteration.

Table 2. The extracted results with the ranking of synonymous transliterations with the SPLR threshold set to 0.1

Transliteration	Extracted Synonymous Transliterations	SSC	SSC +CC
戈巴契夫 (Gorbachev)	戈爾巴喬夫	2	6
	戈巴卓夫	7	4
	戈巴喬夫	2	6
	戈爾喬夫	6	1
布希(Bush)	布什	56	38
	布甚	111	22
布魯斯威利 (Bruce Willis)	布魯斯維利	1	3
弗羅倫斯(Firenze)	翡冷翠	34	16
貝克漢 (Beckham)	貝克厄姆	2	10
	貝克漢姆	1	4
妮可基嫚 (Nicole Kidman)	妮可基曼	1	1
阿諾史瓦辛格 (Arnold Schwarzenegger)	阿諾舒華辛力加	10	4
	安諾德施瓦辛格	2	27
哈珊(Hisun)	海珊	6	25
柯林頓(Clinton)	克林頓	1	29
查爾斯(Charles)	查理	64	66

Transliteration	Extracted Synonymous Transliterations	SSC	SSC +CC
珍妮佛安妮絲頓 (Jennifer Aniston)	詹妮弗安妮斯頓	1	6
	珍妮花安妮斯頓	2	1
雪梨(Sydney)	悉尼	88	2
梅爾吉勃遜 (Mel Gibson)	梅爾吉普森	9	3
	梅爾吉布森	6	8
	梅爾吉布生	7	8
	米路吉勃遜	8	84
	米路吉遜	10	12
莎朗史東 (Sharon Stone)	莎朗斯通	5	17
麥克傑克森 (Michael Jordan)	麥可傑克森	1	1
	邁克杰克遜	3	31
賓拉登 (Bin Laden)	本拉登	1	4
羅納多 (Ronaldo)	羅納度	2	22
	羅納爾多	1	30
約翰保羅 (John Paul)	約望保祿	2	11
	若望保祿	1	27

The experimental results verify that we can get synonymous transliterations well even thought we only retain 900 pages via content focused crawling for a transliteration. It also reveals that context comparison help greatly raise the ranking of some of synonymous transliterations, such as 布甚 (bushe) and 悉尼 (xini). However, the method will reduce the ranking of some synonyms a little. The combined approach shown later will alleviate the problem.

More analysis on the SPLR values is followed. The parameter setting of SPLR often affects the quantity of synonymous transliterations extraction. A loose parameter value will result in generating a large number of unknown words and consequently has less possibility of missing synonymous transliterations. On the contrary, a stricter parameter for unknown words extraction will cause damage, leading to significant reduction of recall rate. Fig. 8 shows the effect of the SPLR threshold value on the number of identified synonymous transliterations.

As mentioned in Section 3.3, the combined ranking from the results of the speech sound comparison and the speech sound comparison followed by context comparison would be better than those of using a single approach. The resultant average rank of the synonymous transliterations by the combined approach is shown in Fig. 9, where different weights are placed to the individual comparison methods. The three curves represent the different SPLR values of 0.1, 0.5 and 0.9, respectively. Naturally, the stricter SPLR value generates better ranking since it eliminate more unknown words.

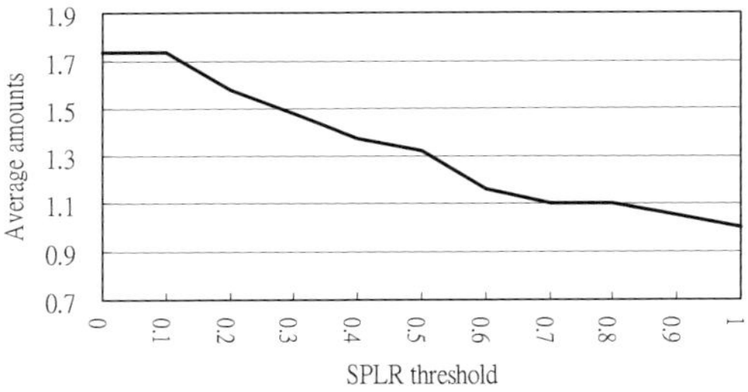

Fig. 8. SPLR threshold influences on harvesting the number of collected synonymous transliterations

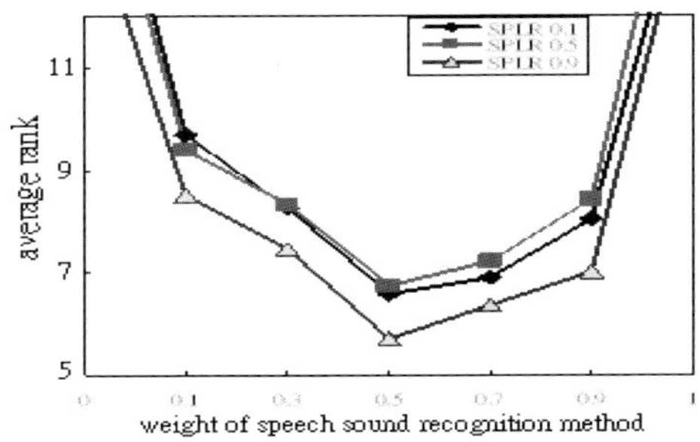

Fig. 9. The average rank of the combined approach with different weighting on individual methods

As we can see in Fig. 9, equal weights on each of the two approach results in the best result in which the average rank of synonymous transliterations is lowest. That is, we shall place a weight 0.5 on the speech sound comparison as well as on the speech sound comparison plus context comparison. Therefore, the speech sound comparison method and the context comparison method are quite complementary to each other. The best average rank of a synonymous transliteration is around six.

Table 3. The inclusion rate of synonymous transliterations

SPLR /Avg. NO. of S.T.	Method	Top 1	Top 3	Top 5	Top 10
	SSC	27%	48%	52%	82%
0.1 / 1.74	SSC+CC	12%	21%	33%	52%
	Combined	27%	45%	64%	85%
	SSC	16%	36%	40%	76%
0.5 / 1.32	SSC+CC	8%	20%	32%	52%
	Combined	20%	48%	68%	80%
	SSC	20%	30%	45%	75%
0.9 / 1.05	SSC+CC	10%	20%	30%	65%
	Combined	20%	55%	75%	85%

Table 3 shows the inclusion rate of the synonymous transliterations, which indicates the percentage of synonymous transliterations, acquired under different SPLR values, being included in Top N terms. The experiments are conducted under three SPLR threshold values, 0.1, 0.5, and 0.9, respectively. '*Avg. NO. of S.T.*' represents the average amount of collected synonymous transliterations for each input data. When SPLR is set to 0.1, an average count of 1.74 synonymous transliterations can be retrieved. Among these retrieved, 85% are ranked within top 10 under the combined approach. The results show that the linear combination of SSC and SSC+CC can help to improve the inclusion rate.

5 Conclusion

The preliminary results showed the approach is appealing. However, more experiments, especially on a large set of test data, is needed to further verify and fine-tune the approach, especially to determine the optimal parameter values involved in the approach. The result of this research has many practical values, can be utilized in many applications, including help to construct a database of synonymous transliterations which can alleviate the incomplete search problem, and help to detect and track news events, in which different transliterations may be used by different editors of news articles.

Acknowledgement

This research is supported by National Science Council, Taiwan under grants NSC94-2416-H-224-007 and NSC 96-2416-H-224-004-MY2.

Reference

1. Netcraft, How many Web sites are there? http://www.boutell.com/newfaq/misc/sizeofWeb.html
2. Han, J., Kamber, M.: Data Mining Concepts and Techniques. Morgan Kaufmann, San Francisco (2001)
3. Qin, J., Zhou, Y., Chau, M.: Building Domain-Specific Web Collections for Scientific Digital Libraries: A Meta-Search Enhanced Focused Crawling Method. In: Proceedings of the 2004 Joint ACM/IEEE Conference on digital Libraries, pp. 135–141 (2004)
4. Oyama, S., Kokubo, T., Ishida, T.: Domain-Specific We Search with Keyword Spices. IEEE Transactions on Knowledge and Data Engineering 16(1), 17–27 (2004)
5. Cheng, P.J., Teng, J.W., Chen, R.C., Wang, J.H., Lu, W.H., Chien, L.F.: Cross-Language Information Retrieval: Translating Unknown Queries with Web Corpora for Cross-Language Information Retrieval. In: Proceedings of the 27th annual international ACM SIGIR conference on Research and development in information retrieval, pp. 146–153 (2004)
6. Gao, J., Zhang, J., Zhou, M.: On the use of Words and N-grams for Chinese Information Retrieval. In: Proceedings of the fifth International Workshop on Information Retrieval with Asian Languages, Beijing, China, pp. 141–148 (2000)
7. Chang, T.H., Lee, C.H.: Automatic Chinese Unknown Word Extraction using Small-Corps-based Method. In: Proceedings of the 2003 International Conference on Natural Language Processing and Knowledge Engineering, pp. 459–464 (2003)
8. Hsu, C.C., Chen, C.H., Shih, T.T., Chen, C.K.: Measuring similarity between transliterations against noise data. ACM Transactions on Asian Language Information Processing 6(1) (2007)

Blog Post and Comment Extraction Using Information Quantity of Web Format

Donglin Cao[1,2,3], Xiangwen Liao[1,2], Hongbo Xu[1], and Shuo Bai[1]

[1] Institute of Computing Technology, Chinese Academy of Sciences, Beijing 100080
[2] Graduate School, the Chinese Academy of Sciences, Beijing 100039
[3] Dept. of Cognitive Science, Xiamen University, Xiamen, 361005, P.R. China
caodonglin@software.ict.ac.cn

Abstract. With the development of the research on blogosphere, acquiring the post and comment from blog page becomes more important in improving the search performance. In this paper, we present a two-stage method. First, we combine the advantage of the vision information and the effective text information to locate the *main text* which represents the theme of blog page. Second, we use the *information quantity of separator* to detect the boundary between the post and comment. According to our experiments, this method achieves a good performance in extraction and improves the performance of blog search.

1 Introduction

As we know, blog is one of the core applications of web 2.0. It is composed by blogrolls, permalinks, comments, trackbacks and posts. With the development of blogosphere, searching information from blog pages becomes more and more important. However, blog pages inevitably include some noises which affect the precision of the information retrieval system. In order to improve the performance of information retrieval system, it is necessary to acquire the post and comment from blog page.

There are some state-of-the-art methods in information extraction. One of the most common methods is link/text removal ratio [2,3]. This method is useful in removing useless links in news page. From the experiment in [2,3], this method shows a good performance in extracting news from news page. However, this method isn't good at differentiating other useless texts from news texts. Besides the research in text extraction, there are a lot of research in extracting data record from the web page. Handcrafted rules, like NLP based, wrapper [1,4,5,6,7,13] and Html-aware [12], are useful in locating the special data record in web page, such as author name, publish time, etc. In [6], Bing Liu uses MDR (Mining Data Region) to automatically find the data record. This method is useful in mining both contiguous and noncontiguous data records. A Partial Tree Alignment based method is presented in [12]. This method is used to align similar data items from multiple data fields. In some specific web sites, the above methods achieve a high precision. However, these methods need to write some polished rules by hand. These rules aren't easy to achieve and they are language

H. Li et al. (Eds.): AIRS 2008, LNCS 4993, pp. 298–309, 2008.

sensitive. We have to write rules for each kind of nature language. Although there are some kinds of methods to automatically acquire rules from the corpus, all these methods require a well-formed corpus annotation which needs a great effort.

Because the characters of the post and comment are different, it is necessary to separate the post and comment. This work is similar to the topic segmentation. Some papers were published in this field. Reynar provides an extensive discussion of algorithms for topic segmentation [10]. Yan Qi introduces a CUrvature-Based segmentation method [8]. Although the above topic segmentation methods achieve very good performance in identifying boundaries in text streams, the topics of the post and comment are often the same.

In this condition, it is necessary to do some research in blog extraction. In this paper, we try to use html format information of blog page from the perspective of information theory. Therefore, we propose a two-stage method which computes the information quantity of the html format information to extract the post and comment in blog page. The main contributions of this paper are as follows.

1. We transform the problem of detecting the boundary between the post and comment into a problem of detecting the redundancy of web format information in the post and comment. By using the information quantity of separator, it is easy to detect the boundary.
2. We combine the advantage of the vision information and the effective text information to locate the main text of blog page which includes the post and comment.

The structure of this paper is as follows. We discuss the framework of our two-stage extractor in section 2. Section 3 describes the algorithm of locating main text and presents an example. Finding separator between the post and comment will be detailed in section 4. The results of experiments will be shown in section 5.

2 Framework of Blog Extraction

A typical function of web page extraction is to locate some useful texts and filter the noises. Here, in blog page, we call these useful texts '*main text*'. Main text is some of object texts which represent the theme of the web page. The main text of blog page includes two parts of text. The first part is the post which is written by the author. The second part is the comment which is written by the reader who is interested in the post. A example of main text is shown in Figure 1. In this figure, all of the texts in cell $C2$ are the main text of the example blog page. Both of these two parts are important in representing the main idea of the document and they have different priority. If we want to find the opinion of the author, it is obvious that the post is the first thing to be concerned. Based on this consideration, we have to separate the post and comment. The separating result of example page is shown in Figure 2.

With the above discussion, the framework of our blog extraction includes two stages, locating the main text and finding separator between the post and comment. Because the format of html can form a DOM (Document Object Model) tree and the main text exists in one of the subtree, in the DOM model, the object of locating the main text is equal to find the minimal subtree which contains the main text. After the locating main text stage, the html format information is used in finding separator between the post and comment. Because of different html format distribution in the post and comment, the object of finding separator is to make a suitable division which can partition the different format of the post and comment.

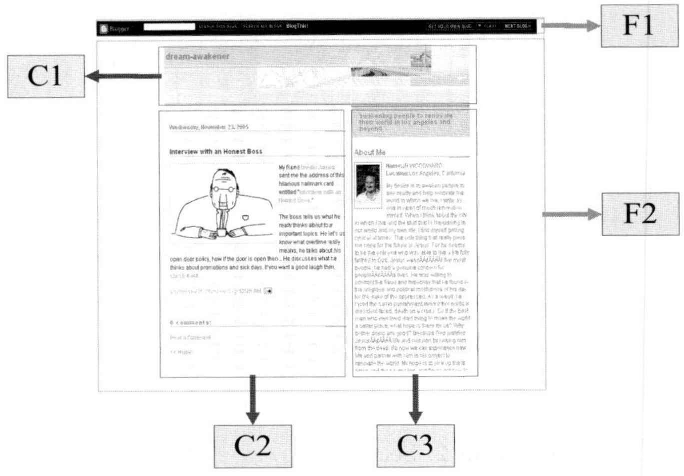

Fig. 1. A typical blog page

3 Locating Main Text

Based on the DOM tree structure, our task of locating main text is to find the minimal subtree which contains the main text. We call this minimal subtree '*minimal main text subtree*'. Because the main text of the blog page only includes the post and comment, the minimal main text subtree is the minimal subtree which contains the post and comment. Here, any other texts and links except the main text are treated as noises. There are three kinds of noises in blog pages. The first kind of noise is some advertisements which have no contribution to the information retrieval system. The second kind of noise is some useful links. Such as the blogrolls which contain a list of other weblogs that the author reads regularly. These links are useful in link analysis, but they are useless in text analysis. The third kind of noise is some *routine texts* which represent some status of the blog. Such as 'copyright' text and 'about author' text. For example, in Figure 1, there are some 'about author' texts in cell $C3$. Although link/text removal ratio is useful in removing the useless links in news page, there are two

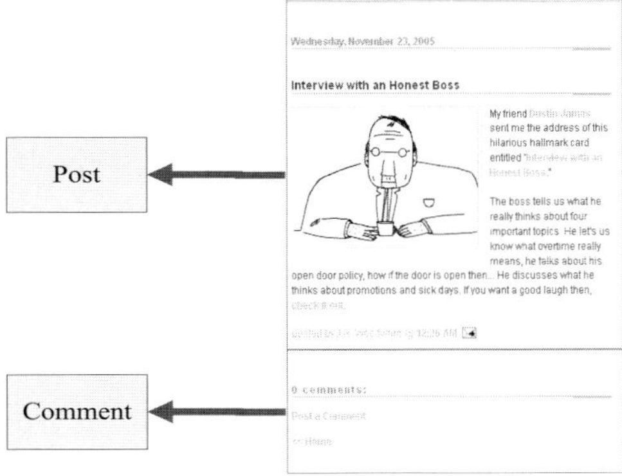

Fig. 2. The separating result of sample page

limitations when it is used in blog page. First, some authors write some useful links in the post. Second, it is difficult to distinguish the routine text and the main text because they are natural language expressions appearing in text.

Based on our experiments, we find two important features of the main text. The first feature is that most of the main texts of blogs hold the largest vision space in comparing with their siblings in the DOM tree. we call this vision space '*vision information*'. Vision information refers to the visual position of each html block, such as width, height, etc. But in this paper, we only use the width in vision information. In [11], it shows that the vision information is useful in differentiating segments in the web page. The second feature is that most of the main texts of blogs contain more words than other routine texts. Therefore, for the first feature, we use css (Cascading Style Sheets) style in html to acquire vision information, for the second feature, we calculate the effective text information in each node of the DOM tree. We use the following formula to calculate *the effective text information* .

$$I_e = \frac{W_e}{W_a} \times W_e \qquad (1)$$

Where W_e is the number of words without links in the text, W_a is the number of words in the text. $\frac{W_e}{W_a}$ is the *effective text information ratio* which represents how much effective text information each word of the text has.

With the above consideration, we build the effective text information based locating main text algorithm which is described as follows.

Locating main text algorithm

1. Build an html DOM tree.
2. Calculate the effective text information of each node.
3. Use the css style to get the visual width of each node.
4. From the root of html DOM tree, do the following steps.

4.1 For the current root node, find the immediate child which has the largest visual width and its number of words exceeds threshold (the value of threshold is 10 in experiment). If the child exists, choose it as root node.
4.2 If all the immediate children have the same visual width, find the immediate child whose effective text information is the biggest and choose it as root node.
4.3 If the *loss ratio* is out of the range of threshold, then go to step 5. Otherwise go to step 4.1. The formula of loss ratio is shown as follows.

$$LossRatio = \frac{I_e(P)}{I_e(C)} \tag{2}$$

Where $I_e(P)$ is the effective text information ratio in the parent node, $I_e(C)$ is the effective text information ratio in the selected child node.
5. Use the range of chosen subtree as the range of main text.

In this algorithm, we define the concept of loss ratio. The main text and other parts of blog page have different distribution of effective text information ratio. If loss ratio changes greatly, it is in a great possibility that we have located the boundary of main text. The value of loss ratio threshold is difficult to set because the value of loss ratio threshold would be different for different blog pages. Therefore, we use the average loss ratio as the value of loss ratio threshold.

According to our algorithm, we divide page into cells and it is easy to find the correct cell $C2$. First, we compare two cells ($F1$ and $F2$). We will find that the width of $F2$ is equal to the width of $F1$. Then we compare the effective text information and choose cell $F2$ as root. Second, we compare three cells ($C1$, $C2$ and $C3$). It is easy to find that $C1$ has the longest width, but the number of its words is fewer than the threshold. Although the effective text information of $C3$ is bigger than $C2$, we select $C2$ because the width of $C2$ is bigger than $C3$. So we select $C2$ as root. And we will find that loss ratio is out of the range of threshold. This is because from cell $F2$ to cell $C2$, the distribution of effective text information ratio changes greatly. As a result, we stop here and choose cell $C2$ as the range of main text.

4 Finding Separator

4.1 Theory Analysis

In tree structure, suppose that the main text is correctly located and the web format information of post and comment exists. We can partition the html tree of main text into three parts. The first part contains a part of the post and the last part contains a part of the comment. Therefore, the problem of finding the separator between the post and comment is to find which part the middle part is similar to. From this purpose, we have to compare two trees to compute their similarity [9]. This kind of method greatly depends on the formula of computing similarity. It is difficult to define a suitable formula.

From another point of view, computing the similarity of tree structure is to detect the redundancy of html format. If a block is redundant to the post (or the comment), it will be similar to the format of the post (or the comment). From data compression and information communication, the information theory shows a great power in detecting the redundancy of information. In the information theory philosophy, for three strings (A, B and C), if C is more similar to A than B, then the increment of information quantity of adding C to A will be smaller than the increment of information quantity of adding C to B. Based on this consideration, the html tag sequence is mapped into string to compute the redundancy of format information.

Without loss of generality, we assume that there are two possible separators (S_1 and S_2) in the main text. Our goal is to find which one is the right one. The whole html tag sequence is separated into three parts (M_1, M_2, M_3). This is shown in Figure 3. We treat the html tags in M_3 as a whole block which is named as D. p_1 is the probability of D in M_1. p_2 is the probability of D in M_2. If $p_1 > p_2$, then $-\log_2(p_1) < -\log_2(p_2)$. Because the number of D in M_3 is 1, the information quantity increment of combining the M_1 and M_3 is $-\log_2(p_1)$ and the information quantity increment of combining the M_2 and M_3 is $-\log_2(p_2)$. It is obvious that the format in M_3 is more close to the format in M_1 than in M_2, and we should choose S_2 as the right separator. If there are more than two possible separators, we can compare each pair of separators to find which one is the best.

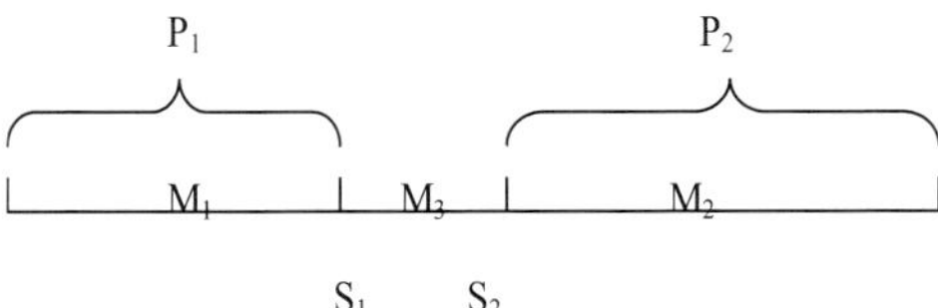

Fig. 3. A simple status of two possible separators

In the above discussion, we treat the html tags in M_3 as a whole block. In unigram model, we consider each html tag as an independent block. Assume that there are m kinds of html tags in blog page. Because each possible separator divides the html tag sequence into two parts (post and comment), we define the *information quantity of separator* as follows.

$$\begin{aligned} IS &= \sum_{i=1}^{m}((-n_{1i}\log_2(p_{1i}))) + \sum_{i=1}^{m}((-n_{2i}\log_2(p_{2i}))) \\ &= \sum_{j=1}^{2}(\sum_{i=1}^{m}((-n_{ji}\log_2(p_{ji})))) \end{aligned} \tag{3}$$

Where j is the jth part which is divided by the separator. For the ith kind of tag in the jth part, its probability is p_{ji} and its number is n_{ji}.

We can use this equation to find the suitable separator. For example, in Figure 3, assume that $p_{M_j i}$ is the probability of the ith kind of tag in M_j. And the number of the ith kind of tags in M_j is $n_{M_j i}$. The information quantity of M_1 and M_2 is $\sum_{i=1}^{m}(-n_{M_1 i}\log(p_{M_1 i}) - n_{M_2 i}\log(p_{M_2 i}))$. Adding M_3 to M_1 (or

M_2) will increase the information quantity. If we choose S_1 as the right separator (adding M_3 to M_2), the information quantity will increase Δ_1. If we choose S_2 as the right separator (adding M_3 to M_1), the information quantity will increase Δ_2. Then we can get the following result.

$$\begin{aligned}
&\Delta_1 - \Delta_2 \\
&= \sum_{i=1}^{m}(-(n_{M_2 i} + n_{M_3 i}) \log(p'_{M_2 i}) + n_{M_2 i} \log(p_{M_2 i}))- \\
&\sum_{i=1}^{m}(-(n_{M_1 i} + n_{M_3 i}) \log(p'_{M_1 i}) + n_{M_1 i} \log(p_{M_1 i})) \\
&= \sum_{i=1}^{m}(-(n_{M_2 i} + n_{M_3 i}) \log(p'_{M_2 i}) - n_{M_1 i} \log(p_{M_1 i}))- \\
&\sum_{i=1}^{m}(-(n_{M_1 i} + n_{M_3 i}) \log(p'_{M_1 i}) - n_{M_2 i} \log(p_{M_2 i})) \\
&= IS_1 - IS_2
\end{aligned} \tag{4}$$

Where $p'_{M_1 i}$ is the probability of the ith kind of tag in M_1 and M_3 when S_2 is separator. $p'_{M_2 i}$ is the probability of the ith kind of tag in M_2 and M_3 when S_1 is separator. IS_i is the information quantity of S_i.

From the above equation, we can find that the separator which has the minimal increment will also be the separator which has the minimal information quantity. So our target becomes to find the separator which has the minimal information quantity.

4.2 Basic Algorithm

Because all the nodes in unigram model are equal, we use the preorder traversal method to map the tree structure into a linear structure. Then we calculate the information quantity of separator to find the suitable separator. The algorithm is shown as follows.

Finding separator algorithm

1. Build an html DOM tree in the range of the main text.
2. Eliminate all the non-tag nodes in DOM tree and build an html tag tree.
3. For each immediate child node of the root, do the following steps.
   3.1 Separate the tree into two parts. The first part includes the current selected immediate child and all siblings which are on the left of it. The second part includes all siblings which are on the right of it.
   3.2 Use the preorder traversal method to map two parts of subtree structure into a linear structure.
   3.3 Calculate the information quantity of the separator.
   3.4 If all immediate child nodes are visited, then go to step 4. Otherwise, go to step 3.1.
4. Choose the separator which has the minimal information quantity.

In this algorithm, we only check the immediate child node of the root node of minimal main text subtree. This is because the post node and comment node exist in the different immediate child node, otherwise the root node won't be the root node of the minimal main text subtree which contains the post and comment. So if we locate wrong minimal main text subtree, the chosen separator in this algorithm will be wrong.

5 Experiment

5.1 Corpus Processing

The goal of our experiment is to test the performance of our information quantity based extracting algorithm. We use the standard blog corpus which comes from the blog track in TREC2006. According to our algorithm, we process the corpus in the following steps. First, we choose all the permalinks from the data which were crawled in December 7, 2005. Second, according to the domain name of each blog page, we count the number of pages in each domain and select the blog pages in the top 100 domains as test data. Third, because our algorithm uses css style to acquire the visual width of each html tag, we download css style file of each page. At last, after eliminating the pages without css style, we get 25910 blog pages and label them in manual. The status of these pages are shown in Table 1. Because a lot of domain names of blog pages are from the same blog site, the final 25910 blog pages are distributed in 9 blog sites. We manually label these pages and the annotation of the corpus denotes the root node of the minimal main text subtree and immediate child node which contains the comment in the minimal main text subtree. Labelling blog page one by one isn't an easy work. Fortunately, although there are hundreds of format styles in these pages, we find that the blog site builders like to use some comprehensible words in html tag to identify the main text and comment. Such as 'main', 'post', 'comment', 'reply', etc. So we use some heuristic words in html tag to group these 25910 pages and get 84 groups. These groups were checked one by one. In each group, we use these heuristic words to manually label the root node of the minimal main text subtree and immediate child node which contains the comment. Finally, we write only 84 templates instead of 25910 labelled pages.

Table 1. Corpus distribution

Site	# pages
www.livejournal.com	16944
blogspot.com	7401
nospeedbumps.com	382
www.plogress.com	340
ipunkrock.com	318
www.blogespierre.com	215
weblogs.java.net	110
redjar.org	100
www.sff.net	100

Table 2. Experiment precision

Type	Precision
$MainText$	85.38%
$PostComment$	77.70%
$PostComment \mid MainText$	91.00%
$MainText + PostComment$	77.70%

5.2 Experiment Result

Experiment evaluation

For the performance evaluation, we define four kinds of precisions as follows.

$Precision(MainText) = \frac{NL}{NCorpus}$
$Precision(PostComment) = \frac{NS}{NCorpus}$
$Precision(PostComment|MainText) = \frac{NSL}{NL}$
$Precision(MainText + PostComment) = \frac{NSL}{NCorpus}$
Where NL is the number of pages which is correct in locating main text. $NCorpus$ is the number of pages in corpus. NS is the number of pages which is correct in finding separator. NSL is the number of pages which is both correct in locating main text and finding separator.

Performance test
In our experiment, the overall performance of our algorithm and the performance of each stage is tested. The results of our algorithm are shown in Table 2.

In Table 2, we can see that both Precision(MainText) and Precision(Post Comment | MainText) are high. It shows that our algorithm can locate main text and find separator precisely. We can also see that Precision(PostComment) and Precision(MainText+PostComment) are so close. As we have mentioned in section 4.2, if the minimal main text subtree is wrong, the chosen separator will be wrong in a great probability. In other words, its contrapositive, if the post and comment is separated correctly (Precision(PostComment)), then the minimal main text subtree is correct in a great probability (Precision(MainText+Post Comment)).

Because our extracting framework includes two stages, we have to test the performance of each stage. In the following experiments, we first compare our locating main text algorithm with other two methods, then we show the performance of finding separator in all kinds of *post/comment* ratio.

In Figure 4, we compare the performance of three kinds of locating methods, our effective text information based method, link/text removal ratio based method and text based method. The x axis is the $MainText/WebText$ ratio, and the y axis is Precision(MainText). Because we consider that other texts and links except the main text as noises, the x axis shows the sequence of ratio of main text and noise in blog page. If the $MainText/WebText$ ratio is small, there will be a lot of noises in blog page. Three curves in this figure show the performance trend of three methods. It is clear that our effective text information based method outperforms other two methods. The link/text removal ratio isn't easy to precisely locate the main text. The text based method achieves a good performance when the $MainText/WebText$ ratio is over 7/10. But the text based method isn't easy to differentiate the main text and other texts when they have the similar number of bytes.

Some blog pages contain few comments (or no comments). Our finding separator algorithm is based on the condition that the web format information of post and comment exists. In our corpus processing, we treat the message text '0 comments' as a part of comment text. For example, in Figure 2, although there are 0 comment, we treat the message text '0 comments' and the web format framework of comments as the comment part. In this condition, the length of post text will be bigger than the length of comment text. It is important to exam

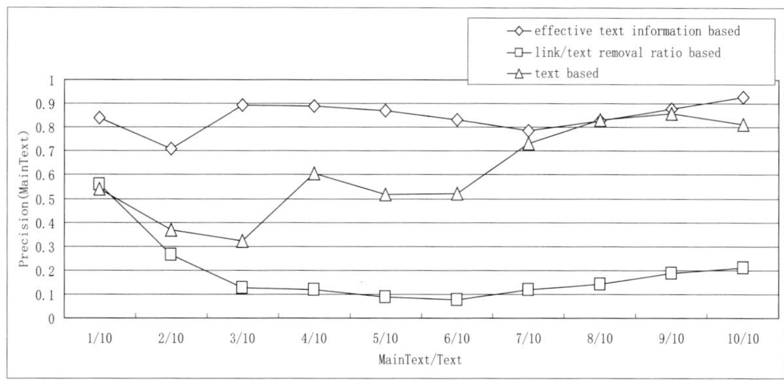

Fig. 4. Performance comparison of three locating main text algorithms

the performance of our information quantity based finding separator method in all kinds of *post/comment* ratio. The result is shown in Figure 5. The x axis is the *post/comment* ratio, and the y axis is Precision(PostComment | MainText). In the result, our method achieves a high precision (over 90%) when the length of comment text is bigger than the length of post text. When the length of post text becomes bigger than the length of comment text, the precision decreases slightly. This is because there are more redundant html formats in the comment than in the post. Our finding separator algorithm is easy to detect the similar (or redundant) html tag node. The trend of curve shows that even in the condition of few comments, our algorithm also achieves a good precision (over 78%).

Blog search test

Based on our blog extraction method, we can extract the post and comment from blog to improve the performance of blog search. In order to test the contribution of our extraction method, we use the 88.8G blog corpus and 50 topics in

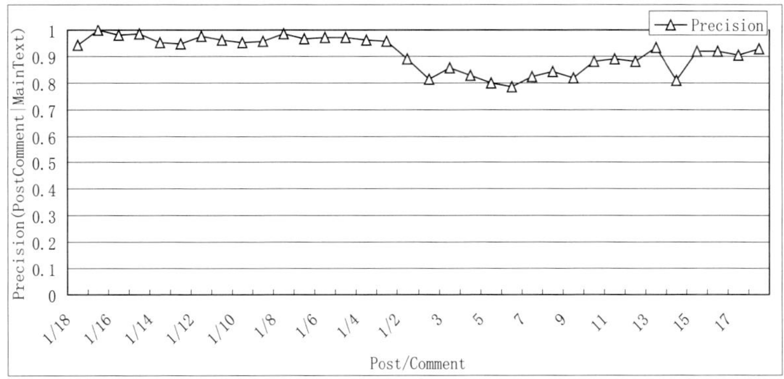

Fig. 5. Performance of finding separator algorithm

Table 3. Blog search performance

Model	Map	R-Prec	bPref	P@10
Language Model (before Extraction)	0.2987	0.4030	0.4039	0.6460
Language Model (after Extraction)	**0.3224**	**0.4335**	**0.4439**	0.6800
Best results in Trec2006	0.2983	0.3925	0.4272	**0.706**

TREC2006 blog track to retrieval the relevant blog pages. The retrieval model is the classical language model with Dirichlet smoothing. Retrieval results are shown in Table 3. The results show that our extraction method improves all four metrics and our results even better than the best results in TREC2006 blog track in three metrics.

6 Conclusions

Extracting the information from the web page is one of the challenging works in information retrieval. In order to acquire the blog post and comment, we build a two-stage method. First, we use the effective text information based method to locate the main text. This method combines the advantage of effective text information and vision information. Second, in finding the separator between the post and comment, we choose the separator which has the minimal information quantity. Both in the theory analysis and experiment, we find this method is very useful. From the results of experiments, our method achieves a good performance.

Although our two-stage method achieves a good performance, there are still some spaces to improve. In finding separator, we only use the unigram model to calculate the information quantity. Compare with bigram and trigram model, unigram model is more simple in representing the information. In fact, there are some relations between the html tags, especially the hiberarchy structure relation between the html tags.

Acknowledgments

The work is supported by the National Fundamental Research Program of China (Project No 2004CB318109). And we would like to thank to Yan Guo, Gang Zhang, Yu Wang and Ruijie Guo for their contributions to the related research work and preparation of this paper.

References

1. Crescenzi, V., RoadRunner, G.M.: Towards automatic data extraction from large web site. In: Proceeding of the 26th International Conference on very Large Database Systems, pp. 109–118 (2001)
2. Gupta, S., Kaiser, G., Neistadt, D., Grimm, P.: Dom-based content extraction of html documents. In: 12th International World Wide Web Conference (May 2003)

3. Gupta, S., Kaiser, G.E., Grimm, P., Chiang, M.F., Starren, J.: Automating content extraction of html documents. World Wide Web 8(2), 179–224 (2005)
4. Irmak, U., Suel, T.: Interactive wrapper generation with minimal user effort. In: WWW 2006, pp. 553–563–224 (2006)
5. Ling, Y., Meng, X., Meng, W.: Automated extraction of hit numbers from search result pages. In: Yu, J.X., Kitsuregawa, M., Leong, H.-V. (eds.) WAIM 2006. LNCS, vol. 4016, pp. 73–84. Springer, Heidelberg (2006)
6. Liu, B., Grossman, R., Zhai, Y.: Mining data records in web pages. In: ACM SIGKDD International Conference on Knowledge Discovery and Data Mining (August 2003)
7. Lu, Y., Meng, W., Zhang, W., Liu, K.-L., Yu, C.T.: Automatic extraction of publication time from news search results. In: ICDE Workshops (2006)
8. Qi, Y., Candan, K.S.: Blogs, wikis and rss: Cuts: Curvature-based development pattern analysis and segmentation for blogs and other text streams. In: Proceedings of the seventeenth conference on Hypertext and hypermedia HYPERTEXT 2006 (August 2006)
9. Reis, D.C., Golgher, P.B., Silva, A.S., Laender, A.F.: Automatic web news extraction using tree edit distance. In: Proceedings of the 13th international conference on World Wide Web WWW 2004 (May 2004)
10. Reynar, J.C.: Topic segmentation: Algorithms and applications. PhD thesis (1998)
11. Song, R., Liu, H., Wen, J., Ma, W.: Learning block importance models for web pages. In: Proceedings of the 13th international conference on World Wide Web WWW 2004 (May 2004)
12. Zhai, Y., Liu, B.: Structured data extraction from the web based on partial tree alignment. IEEE Transactions on Knowledge and Data Engineering (2006)
13. Zhao, H., Meng, W., Yu, C.T.: Automatic extraction of dynamic record sections from search engine result pages. In: VLDB 2006, pp. 989–1000 (2006)

A Lexical Chain Approach for Update-Style Query-Focused Multi-document Summarization

Jing Li[1,2] and Le Sun[1]

[1] Institute of Software, Chinese Academy of Sciences,
No.4, South Fourth Street, Zhongguancun, Haidian District, 100190, Beijing, China
[2] Graduate University of Chinese Academy of Sciences,
No.19, Yu Quan Street, Shijinshan, 100049, Beijing, China
{lijing05,sunle}@iscas.ac.cn

Abstract. In this paper we propose a novel chain scoring (chain selecting) method to enhance the performance of Lexical Chain algorithm in query-focused summarization and present an information filtering strategy to adapt Lexical Chain method to update-style summarization. Experiments on DUC2007 datasets demonstrate the encouraging performance.

Keywords: Multi-document summarization, query-focused, update-style, lexical chain.

1 Introduction

Automatic text summarization is the process of automatically producing a short version of source document's or document-set's main topics. Text summarization is a valid and efficient way to compress, filter, and find information. With the development of WWW, text summarization has drawn more and more attention, such as in Question Answering (QA) and Information Retrieval (IR) tasks. Query-focused summarization, a new branch of automatic text summarization, is introduced in 2005 at DUC conference.

There is a new pilot task in DUC2007 called update-style summarization. In this task there are 10 document-sets with 25 documents and a single query in each. Every document-set has been divided into three subsets according to time. Assuming that readers read documents in each document-set in sequence, first time read the first subset and then the second subset and finally the third, participants are required to generate summaries for the subsets in each document-set conforming to the query and the summary generated for one subset cannot contain information from previous subset(s). There are three levels of task: 1) assuming reader has read nothing, generate a summary for the first document-set; 2) assuming reader has already read documents in the first document-set, generate a summary for the second document-set; 3) assuming reader has already read the former two document-sets, generate a summary for the third document-set [4]. This pilot task shows the update-style summarization.

The precondition of update-style summarization is that readers are reading a series of documents about the same topic or topics related closely and they need a piece of summary for the documents they want to read, which we call candidate documents,

H. Li et al. (Eds.): AIRS 2008, LNCS 4993, pp. 310–320, 2008.

without information they have already known from documents they read before, which we call previous documents. The update-style summarization can filter redundancies while compressing information and thus make information retrieving more efficient.

Our summarization system was initially designed for DUC2005 and was further developed for DUC 2006. For DUC2007 we carried out a number of necessary improvements aimed at enhancing its efficiency and performance. After DUC2007 we attempted to make further reforming on our summarizer especially at chain scoring strategy and update-style summary generation.

We use the existing lexical chain algorithm optimized by Barzilay and Elhadad in [1] as our major means for intermediate representation construction. Several modifications have been made, for ameliorating its efficiency and adapting it to general and update-style query-focused multi-document summarization.

The remaining sections of this paper are organized as follows: in section 2, the related work will be shown, and then the system architecture will give the generalization of whole summarization process in section 3. We describe the focus of our study in this paper, chain scoring strategy and update-style summary generation, in section 4 and section 5 and discuss the evaluation in Section 6. Lastly we conclude this paper in Section 7.

2 Related Works

Automatic text summarization is a difficult task. "This requires semantic analysis, discourse processing, and inferential interpretation. Those actions are proved to be highly complex by the researchers and their systems in the last decade." [14]

The process of summarization has been generally divided into two steps in current research. The first step is to extract the important concepts from the source text and construct some form of intermediate representation. The second one is to use the intermediate representation to generate a coherent summary of the source text. [8]

In the first step most summarization approaches can be generally classified into the following three categories according to the semantic analysis level: [19]

1. Based on extraction. These methods analyze global statistical feathers (word frequency, sentences similarity, etc.) and extract the most important sentences to generate the summary, for example, MEAD [15].
2. Based on simple semantic analysis. Take Lexical Chain for example, it first constructs a tree structure of the original document, and then scores every chain to select the strongest chains for summary generating.
3. Based on deep semantic analysis. For example, Marcu [11] proposed an approach based on the construction of a rhetorical tree which uses heuristic rules and explicit discourse markers to find out which the best rhetorical tree for a given document is.

Obviously methods based on deep semantic analysis can offer the best opportunity to create an appropriate and fluent summary. The problem with such methods is that they rely on detailed semantic representations and domain specific knowledge bases.

And the major problem with methods based on extraction is that they do not take context into account. Specifically, finding the about-ness of a document relies largely on identifying and capturing the existence of duplicate terms and related terms. This concept, known as cohesion, links semantically related terms which is an important component in a coherent text [5].

The simplest form of cohesion is lexical cohesion, namely lexical chain. In 1991, Morris and Hirst first gave a logical description of the implementation of lexical chain using Roget dictionary [13]. It has been used in a variety of IR and NLP applications including summarization in which it is used as an intermediate text representation. Afterwards Hirst and Onge used WordNet for lexical chain construction and adopted a strategy to choose word's sense with respect to those words occurring ahead of it [7]. An optimized strategy was put forward by Barzilay and Elhadad in [1] to insure that all senses of a candidate word be properly considered. It was applied to generate coherent summaries for single document summarization. Barzilay and Elhadad also develop the first summarizer using lexical chain. In their summarizer lexical chains are used to weigh the contribution of sentences to the main topic of a document. Brunn et al. [2] suggests "calculating the chain scores with the pair-wise sum of the chain word relationship strengths in the chains". Then, "sentences are ranked based on the number of 'strong' chain words they contain." [3]

3 System Architecture

In general, there are three main steps to build a summary using lexical chain: 1) selecting candidate words for chain building, 2) constructing and scoring chains to represent the original document and 3) selecting the "strongest" chains to generate a summary. And before main steps we need preprocessing to transform raw documents from plain text into sets of words and to extract query terms as well. We employ GATE and Stanford Tagger to tag words with POS and identify named entities. Figure 1 shows our system architecture.

We utilize the optimized lexical chain algorithm described in [9] and [20] to select candidate words and build lexical chains (single chains and multi chains). Thus, in this paper we mainly focus on chain scoring step and update-style summary generation step.

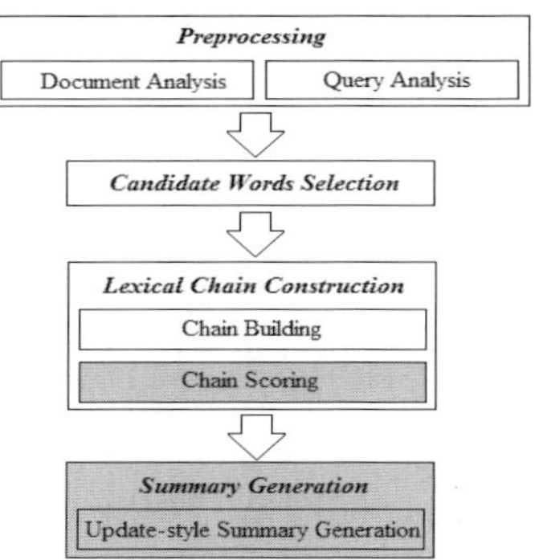

Fig. 1. System Architecture

4 Chain Scoring Strategy

General chain scoring strategy in lexical chain algorithm calculates chain's score according to frequencies of all words in it and the number of distinct words. This empirical strategy works fine with general single document summarization and even multi-document summarization, but it doesn't suit query-focused summarization well. Chain with highest score calculated by this strategy means that the theme represented is most similar with the main theme of the whole document but not the theme response query best. In other words this strategy does not take query into account. Therefore we propose a new chain scoring strategy described below.

1. Select all noun words from query and retrieve their senses using WordNet.
2. For the next noun word in query find an element of chain which is most similar with this word (by calculating similarities between all elements and all senses of the word and choosing the highest one). Record this similarity score as chain score on this query word.
3. Repeat 2) until all noun words in query have been calculated.
4. Accumulate chain scores on all query words as the final score of the chain.
5. Repeat 2), 3) and 4) until all chains have been calculated.

Step 1) is just like candidate words selection procedure. Step 2), 3) and 4) calculate and extract the highest similarities between chain and each query term. Step 5) ensures all chains are calculated. We evaluate both strategies in our experiments to give a comparison.

5 Update-Style Summary generation

The most important task in update-style summary generation step is distinguishing between new information and information already known. We have experimented on two different summary generation strategies, simple strategy and chain filtering strategy.

For convenience we call chain set from candidate document-set the candidate chain set and chain set from previous document-set the previous chain set.

5.1 Simple Strategy

Assuming in candidate document-set information already known, namely candidate chain set, can be ignored we build and merge all chains come from candidate document-set and previous document-set(s), but only extract sentences from candidate document-set. In other words we process candidate documents and previous documents together while constructing chains but separately while generating summaries.

Take the second level task of DUC2007 for example, we build chain set for both the former two document-sets but we extract sentences only from the second document-set, which ought to be summarized, to generate the summary.

By this way we treat update-style summary generation as general summarization having external chains. We use previous document-set (previous chain set), to help our intermediate representation construction without filtering information they contain.

We implemented this strategy in our summarizer when we participated in DUC2007.

5.2 Chain Filtering Strategy

In order to filter information already known from the candidate chain set we need to find out where the information is. Assuming that two kinds of information can be identified by themes of document-set which are expressed by chains we divided the candidate chain set into two parts by comparing chains from the candidate chain set to the previous chain set. In the first part all chains are new and considered containing information new for reader. In the other part chains are related to the previous chain set and considered containing information reader has already read. We give a bounty (5.0 by default) on score to chains in the first parts while extracting sentences to generate summaries. It is carried out by the following procedure.

1. Select the first chain not been processed from the candidate chain set.
2. For the first chain in the previous chain set calculate similarity between it and the chain selected from candidate chain set in 1).
3. For the next chain in the previous chain set calculate similarity between it and the chain selected in 1)
4. Repeat 3) until all chains in the previous chain set have been calculated. Record these similarities and choose the highest one as the similarity between the chain selected in 1) and the previous chain set.
5. Select the next chain not been processed from the candidate chain set and repeat 2), 3) and 4).
6. Repeat 5) until all chains in the candidate chain set have been processed.
7. Choose a critical value (default 0.5) as the criteria to divide the candidate chain set into two parts. Chain has a similarity to the previous chain set less than the criteria will take the bounty.

Step 2), 3) and 4) traverse all chains in the previous chain set to calculate the highest similarity between the chain from the candidate chain set and chains from the previous chain set, which will be considered as similarity between chain and chain set. Step 1), 5) and 6) ensure all chains in the candidate chain set have been calculated. Step 7) divides the candidate chain set into two parts for further process.

The method calculating similarity between chains mentioned in step 2) implemented as follows.

1. Prepare two chains to be calculated, chain A and chain B.
2. For the first element in chain A calculate similarities between it and each element in chain B by WordNet. Take the highest one.
3. For the next element in chain A calculate similarities between it and each element in chain B. Take the highest one.
4. Repeat 3) until all elements in chain A have been calculated.
5. Calculate similarities between chain A and each element in chain B in the same way.

6. Calculate average of highest values of all elements in chain A and chain B taken in previous steps and take that average as the final similarity between chain A and chain B.

Step 2), 3) and 4) calculate similarities between chain B and each element in chain A. Step 5) calculates reverse similarities. Step 6) calculates the final similarity between two chains utilizing results from previous steps.

5.3 Summary Generation

After dividing chain set and distributing bounties we use the following formulas to scoring sentences and generate summaries with the aid of query terms and named entities extracted in preprocessing.

$$Score = \alpha \cdot S(chain) + \beta \cdot S(query) + \gamma \cdot S(namedentity) \tag{1}$$

$$Score = \alpha \cdot S(chain) + \beta \cdot S(namedentity) \tag{2}$$

Where S(chain) is the sum of the scores of the chains whose words come from the candidate sentence, S(query) is the sum of the co-occurrences of key words in a query and the sentence, and S(named entity) is the number of name entities existing in both the query and the sentence.

The upper formula is used for calculating sentence score with chains built by general chain scoring strategy and the lower one with chains built by new chain scoring strategy proposed above. Because new chain scoring strategy calculates chain score with query terms, there is no need to calculate query score again in sentence scoring procedure.

Each score is normalized first. We select the sentence with the next highest score until reaching the word number limit for a summary.

In our experiments, S(query) and S(named entity) are found to affect the system's performance remarkably. Empirically, for the upper formula the three coefficients α, β and γ are set to 0.2, 0.3 and 0.5 and for the lower formula the two coefficients α and β are set to 0.5 and 0.5, respectively.

6 Evaluation

6.1 Data Set

We use the DUC2007 update task dataset for evaluation in the experiments. The update task of DUC2007 aims to evaluate update-style query- focused multi-document summaries with a length of approximately 100 words or less. The dataset has already been described simply in Section 1.1. Table 1 and Table 2 give a short summary of the dataset. Documents in each document-set in this dataset are topic-related and each document-set has been given a query for summarization. The articles all come from news reports and are in XML format.

Table 1. Summary of datasets

	DUC2007
Task	Update task
Number of document-set	10
Number of document	250
Document per set	25
Subset per set	3
Document length (words)	100~1900
Summary length (words)	3×100

Table 2. Summary of subsets in document-set

Subset	Set A	Set B	Set C
Number of document	9~10	8	7~8
Summary length (words)	100	100	100

6.2 Evaluation Toolkit

In our experiments we use the ROUGE toolkit for evaluation, which is widely adopted by DUC for automatic summarization evaluation. ROUGE stands for Recall-Oriented Understudy for Gisting Evaluation. It measures summary quality by counting overlapping units such as the n-gram, word sequences and word pairs between the candidate summary and the reference summary (generated manually).

ROUGE toolkit reports separate scores for 1, 2, 3 and 4-gram, and also for longest common subsequence co-occurrences. Among the scores, bi-gram (ROUGE-2), 4-gram (ROUGE-4) and skip-4-gram co-occurrence (ROUGE-S4 & ROUGE-SU4) perform best for multi-document summarization according to Lin's conclusion in [10]. And DUC takes ROUGE-2 and ROUGE-SU4 for evaluation criteria. For convenience and impartiality we use the same criteria as DUC does.

6.3 General Summary Evaluations

In the first level in DUC2007 update task participants need to generate summaries for the first document-set and there isn't any pervious document. Thus this level of task can be considered as general query-focused multi-document summarization and we can evaluate our chain scoring strategy without the influence of update-style summary generation.

Table 3. General summary ROUGE results

	ROUGE-2	ROUGE-SU4
DUC Best	0.12586	0.15592
New Strategy	0.10171	0.13541
DUC Baseline2	0.08343	0.11479
DUC Ours	0.07734	0.11462
DUC Baseline1	0.04614	0.07830

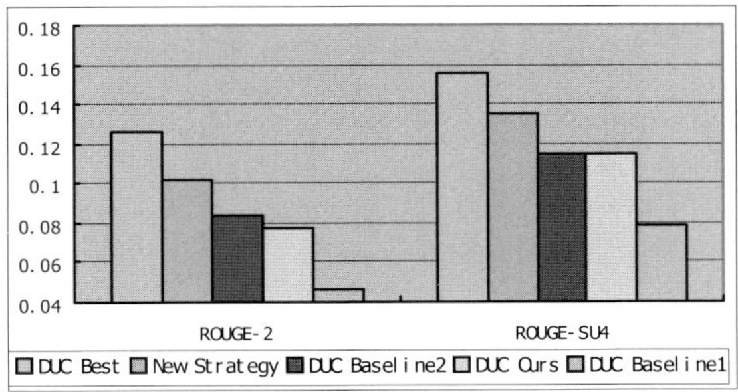

Fig. 2. General Summary ROUGE results comparison

Table 3 shows resulting scores of our summarizer with new scoring strategy generated by ROUGE together with our evaluation result of first level task (with former chain scoring strategy), the best system's performance and baselines in DUC2007 extracted from DUC 2007 update task evaluation results table. And they are visually compared in Figure 2.

From the table and figure above we can see clearly that new chain scoring strategy improves system's performance in evidence.

6.4 Update-Style Summary Evaluation

There are two groups of parameters in update-style experiments, the critical value in dividing chain set (CV for short) with a default value of 0.5 and the critical sense similarity (CSS for short) in building chains. The CV decides whether one chain should take bounty. Lower CV means less chains take bounty and higher means more (details described in Section 2.4.2). The CSS decides whether one sense of a word should be inserted into chain. Lower CSS means longer chains and higher mean shorter (details described in Section 2.2.1).

We have experimented with CV 0.1, 0.3, 0.5, 0.67 and 0.8 respectively, and CSS 0.3, 0.5, 0.67 and 0.8. We abstract the highest one and the mean from these twenty groups

Table 4. Update-style summary ROUGE results

	ROUGE-2	ROUGE-SU4
DUC Best	0.11189	0.14306
Machine Reading [6]	0.11189	0.14306
New Strategy (Highest)	0.09398	0.12718
New Strategy (Average)	0.08810	0.12361
DUC Baseline2	0.08501	0.12247
DUC Ours	0.08068	0.11577
Feature-based Relevance measures [17]	0.06801	0.11143
Fuzzy Co-reference Cluster Graphs [18]	0.05302	0.09560
DUC Baseline1	0.04543	0.08247
Term Frequency Distribution [16]	0.04205	0.07809

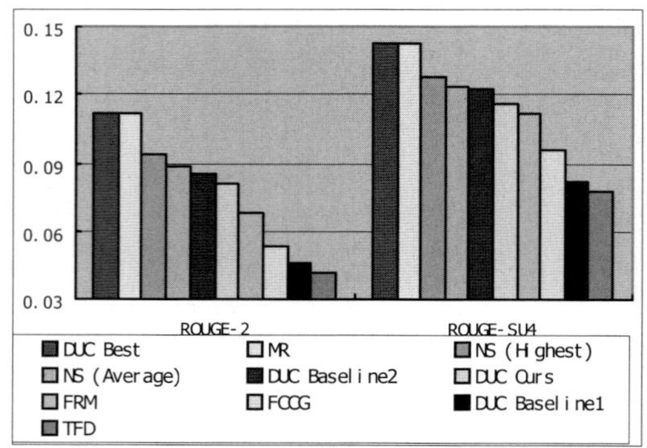

Fig. 3. Update-style summary ROUGE results comparison

of ROUGE resulting scores and compare them to DUC baselines, our evaluation results in DUC, best system's evaluation results and other four participants' system performances according to their workshop papers in DUC2007 [6][16][17][18]. Table 4 and Figure 3 show the comparison.

The table and figure above state that the chain filtering strategy does work with update- style summarization. But the performance improvement enhanced by this new strategy is not as large as we thought. The reason may be that lexical chains can't divide document into themes accurately enough, which causes extra noise during implementing the new strategy.

Experiment results on different parameters are shown in Figure 3. The left figure in Figure 3 shows system's ROUGE-2 and ROUGE-SU4 scores with respect to different CVs (taking default CSS of 0.67) and the right one shows system's ROUGE scores with respect to different CSS (taking CV of 0.1).

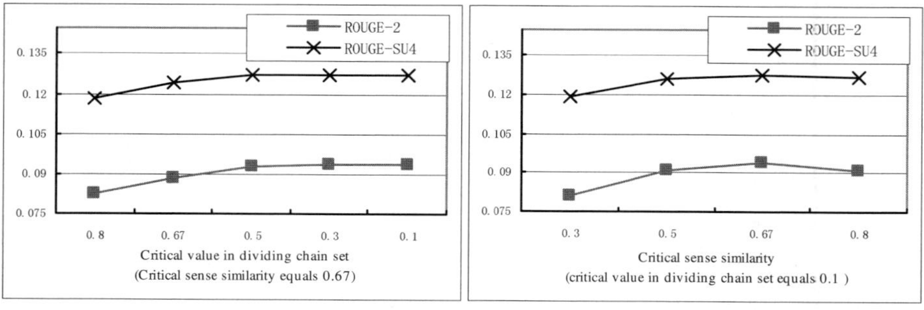

Fig. 4. Rouge score vs. critical value in dividing chain set and vs. critical sense similarity

Seen from Figure 4, while critical value going higher, the ROUGE scores increase. It means the stricter filtering criteria can generate better summaries for update purpose. And while critical sense similarity going higher, the scores of system first increase and then decrease and the best performances are achieved at CSS 0.67. CSS too low leads longer chains being built and more themes being contained in one chain, which brings more noise. CSS too high leads shorter chains being built by contrast and one theme being broken up and distributed into several chains, which brings more noise, too.

7 Conclusion and Future Work

In this paper we proposed a novel chain scoring method and an update-style summarization strategy based on lexical chain. In the chain scoring method we import query information to relate chains to query while scoring chains. In the update strategy we divide chain set built from document-set into two parts, the part containing previous information and the part containing new information, by calculating similarity between chains, and then treat them separately during summary generating. Experimental results on the DUC 2007 update task dataset demonstrate the performance improvement of the new method and strategy. Taking query information into account while scoring chains can adapt Lexical Chain algorithm to query-based summarization better. And by introducing chain filtering summarizer based on Lexical Chain is adapted for update-style summarization.

There are more implementations than chain filtering strategy in this study. In future work, we will explore more summarization methods for update-style to validate the robustness of our summarizer. In addition Lexical Chain algorithm as method based on simple semantic analysis has its own limitation and cannot exceed method based on deep semantic analysis such as Machine Reading [6]. We will attempt to implement some kind of complex method for better performance.

Acknowledgments. This work is partially supported by National Natural Science Foundation of China with the contract No. 60773027 and No. 60736044 and by the 863 project of China with the contract No. 2006AA010108.

References

1. Barzilay, R., Elhadad, M.: Using lexical chains for summarization. In: ACL/EACL 1997 Summarization Workshop, Madrid, pp. 10–18 (1997)
2. Brunn, M., Chali, Y., Pancake, C.J.: Text summarization using lexical chains. In: Workshop on Text Summarization in conjunction with the ACM SIGIR Conference 2001, New Orleans, Louisiana (2001)
3. Doran, W., Stokes, N., Carthy, J., Dunnion, J.: Comparing Lexical Chain-based Summarization Approaches Using an Extrinsic Evaluation (2004)
4. DUC 2005/2006/2007, `http://duc.nist.gov`
5. Halliday, M.A.K., Hasan, R.: Cohesion in English. Longman, London (1976)
6. Hickl, A., Roberts, K., Lacatusui, F.: LCC's GISTexter at DUC 2007: Machine Reading for Update Summarization. In: DUC 2007 (2007)
7. Hirst, G., Onge, D.S.: Lexical chains as representation of context for the detection and correction of malapropisms. In: Fellbaum, C. (ed.) WordNet: An electronic lexical database, pp. 305–332. MIT Press, Cambridge, MA (1998)
8. Jones, K.S.: What might be in summary? Information Retrieval (1993)
9. Li, J., Sun, L., Kit, C., Webster, J.: A Query-Focused Multi-Document Summarizer Based on Lexical Chain. In: DUC 2007 (2007)

10. Lin, C.Y.: ROUGE: A Package for Automatic Evaluation of Summaries. In: Proceedings of the ACL 2004 Workshop on Text Summarization, Spain, pp. 74–81 (2004)
11. Marcu, D.: From Discourse Structures to Text Summaries. In: The Proceedings of the ACL 1997/EACL 1997 Workshop on Intelligent Scalable Text Summarization, pp. 82–88 (1997)
12. Maynard, D., Ananiadou, S.: TRUCKS: A model for automatic multi-word term recognition. Journal of Natural Language Processing 8(1), 101–126 (2000)
13. Morris, J., Hirst, G.: Lexical cohesion computed by thesaural relations as an indicator of the structure of text. Computational Linguistics 17(1), 21–48 (1991)
14. Radev, D.R.: Text summarization. ACM SIGIR tutorial (2004)
15. Radev, D.R., Otterbacher, J., et al.: MEAD ReDUCs: Michigan at DUC 2003 (2003)
16. Reeve, L.H., Han, H.: A Term Frequency Distribution Approach for the DUC 2007 Update Task. In: DUC 2007 (2007)
17. Stokes, N., Rong, J., Laugher, B., Li, Y., Cavedon, L.: NICTA's Update and Question-based Summarisation Systems at DUC 2007. In: DUC 2007 (2007)
18. Witte, R., Krestel, R.: Generating Update Summaries for DUC 2007. In: DUC 2007 (2007)
19. Zhou, Q., Sun, L., Nie, J.Y.: IS_SUM: A multi-document summarizer based on document index graphic and lexical chains. In: DUC 2005 (2005)
20. Zhou, Q., Sun, L., Lv, Y.: ISCAS at DUC 2006. In: DUC 2006 (2006)

GSPSummary: A Graph-Based Sub-topic Partition Algorithm for Summarization

Jin Zhang, Xueqi Cheng, and Hongbo Xu

Institute of Computing Technology, Chinese Academy of Sciences,
Beijing, P.R. China
zhangjin@software.ict.ac.cn,
{cxq,hbxu}@ict.ac.cn

Abstract. Multi-document summarization (MDS) is a challenging research topic in natural language processing. In order to obtain an effective summary, this paper presents a novel extractive approach based on graph-based sub-topic partition algorithm (GSPSummary). In particular, a sub-topic model based on graph representation is presented with emphasis on the implicit logic structure of the topic covered in the document collection. Then, a new framework of MDS with sub-topic partition is proposed. Furthermore, a novel scalable ranking criterion is adopted, in which both word based features and global features are integrated together. Experimental results on DUC2005 show that the proposed approach can significantly outperform existing approaches of the top performing systems in DUC tasks.

Keywords: Multi-document Summarization, Sub-topic, Graph Representation.

1 Introduction

With the rapid increasing of online information and fast development of science and technology, a lot of research efforts have been made on web mining, text mining, information extraction, and information retrieval (IR). However, the conventional IR technologies are becoming more and more insufficient for obtaining useful information effectively. Which makes how to summarize documents with all kinds of information increasingly urgent. Therefore, MDS - capable of summarizing either complete documents sets, or a series of documents in the context of previously ones - is likely to be essential in such situations. The goal of text summarization is to take an information source, extract content from it, and present the most important content to the user in a condensed form and in a manner sensitive to the user's application needs [3]. If the summarization system can make an effective summary, which can be a substitute of the original documents, the retrieval effectiveness or efficiency can be improved and the user can save the reading time.

Usually, the topic of a document collection is composed of some aspects of information, each aspect is named sub-topic of the document collection. In order to

H. Li et al. (Eds.): AIRS 2008, LNCS 4993, pp. 321–334, 2008.

model the sub-topics, many cluster-based approaches have been proposed. These approaches employ a clustering method to model the logic structure of the topic based on the structure of the topic covered in the document collection in the first, follows by a sentence selection method in a a specified cluster. However, the implicit logic structure of the topic covered in the document collection is not only represented by the explicit distribution of features (statistical features in usual), but also represented by the implicit distribution of features (centrality, etc).

In this paper, we argue that information selection in a MDS can be based on the implicit logic structure of the topic covered in the document collection. Using the relationship information with graph representation, we investigate the use of sub-topics as a model of the document collection for the purpose of producing summaries. Furthermore, unlike the two-step cluster-based approaches, we aim to obtain an approach can select important information when modeling sub-topics.

It would be worthwhile to highlight several aspects of our proposed algorithm here:

1. Presenting a new framework of MDS with sub-topic model, according to the implicit logic structure of the topic covered in the document collection.
2. Proposing a scalable criterion to rank the salience of sentences, in which both the word based and global features are modeled explicitly and effectively.
3. Proposing a novel MDS algorithm to determine the sub-topics in global space of a document collection.

The rest of this paper is organized as follows. Section 2 relates a review of the previous work. In section 3, we present the proposed graph-based summarization approach using sub-topic partition. The experimental methodologies and results are reported in section 4 and 5, followed by the conclusion and future work in section 6.

2 Related Work

Generating an effective summary requires the summarizer to select, evaluate, order and aggregate items of information according to their relevance to a particular subject or purpose. These tasks can either be approximated by IR techniques or done in great depth with full natural language processing (NLP). Most previous work in summarization has attempted to deal with the issues by focusing more on a related, but simple problem. Most of the work in sentence extraction applied statistical techniques (frequency analysis, variance analysis, etc.) to linguistic units such as tokens, names, anaphora, etc. (e.g., [9]). Other approaches include the utility of discourse structure [10], the combination of information extraction and language generation [1], and using machine learning to find patterns in text [6][7].

Several researchers have extended various aspects of the single document summarization approach to look at MDS [12][13]. These include comparing templates filled in by extracting information - using specialized, domain specific knowledge sources - from the document, and then generating natural language summaries from the templates, comparing named-entities - extracted using specialized lists - between documents and selecting the most relevant section, finding co-reference chains in the document collection to identify common sections of interest, or building activation networks of related lexical items (identity mappings, synonyms, hypernyms, etc.) to extract text spans from the document collection [13].

Many of recent researches put emphasis on the comprehensiveness while keeping readability of summaries or maximizing the coverage and the anti-redundancy to keep the comprehensiveness and readability to some extent. For example, Radev et al. [14] proposed a method that classifies given documents into some clusters and made one sub-summary for each cluster, then placed them in an order.

Carbnell [1] proposed the Maximal Marginal Relevance (MMR) criterion for combining query relevance with information novelty in the context of text retrieval and summarization. Goldstein et al. [11] proposed a method called MMR-MD (Maximal Marginal Relevance - Multi-Document), which collects passages related to the query from newspaper articles retrieved by an IR system and arranged them into one summary.

As first proposed in [17], the central to the MDS approach has been gained a lot of interest. In 2005, Harabagiu et al. [15] proposed a topic themes method that a MDS can be based on the structure of the topic covering in the document collection.

3 Graph-Based Sub-topic Partition Algorithm

Although the document collection used to generate a summary may be relevant to the same general topic, they do not necessarily include the same information. Extracting all similar sentences would produce a verbose and repetitive summary, while extracting some similar sentences could produce a summary biased towards some sources, as it was noted in [8]. However, the graph-based extractive summarization algorithm succeeds in identifying the most important sentences in a document collection based on information exclusively drawn from the collection itself. In this section, we propose a graph-based algorithm - GSPSummary - to obtain the important sub-topics. GSPSummary starts from the assumption that capturing sub-topic structure of document collection is essential for summarization. It firstly creates a graph representation of the document collection, then selects the salient (or more central) sentences with GSPRank and obtains the most important sub-topics in global graph space iteratively, finally forms the summary supported by the salient sentences of different sub-topics. We will give the definition of graph-based sub-topic representation, the GSPRank criterion, and the GSPSummary algorithm in more details in the subsections below.

3.1 Problem Formalization

Let $G = (V, E)$ be an undirected graph with the set of nodes V and set of edges E, where E is a subset of $V \times V$. Then a graph G of a related document collection can be represented by the set of sentences V, and the similarities to each other

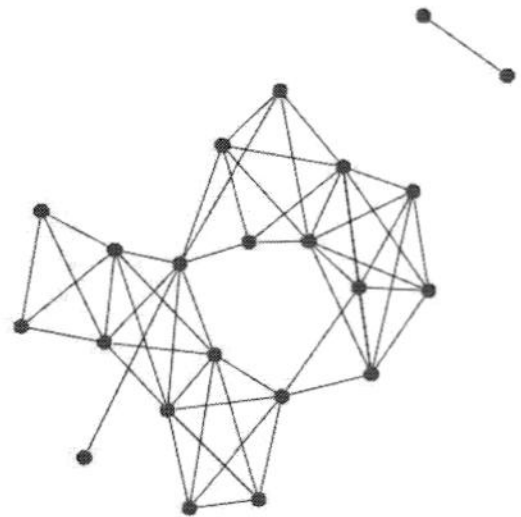
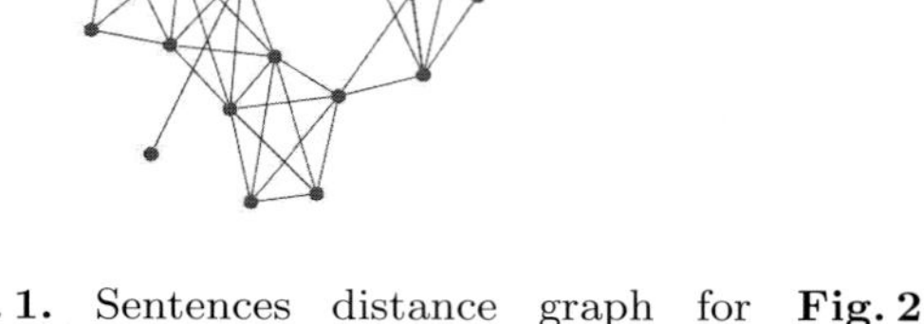

Fig. 1. Sentences distance graph for a small document collection with 22 sentences

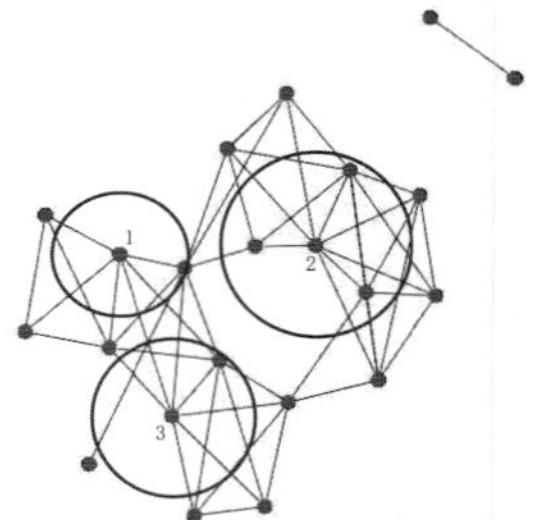

Fig. 2. Sentences distance graph with three sub-topics marked by three circles around node 1, 2, 3 respectively

E. Figure 1 is a sentences distance undirected graph representing a document collection with 22 sentences (the nodes in Figure 1), the edge stands for the distance $dist(u_i, u_j)$ of every pair of sentences u_i, u_j in the document collection. A threshold is defined to eliminate the edges whose distance is higher, since we are interested in significant similarities. This reduction in the graph also provides us with computational savings. To define distance, we use the bag-of-words model to represent each sentence as an N-dimension vector, where N is the number of all possible words. Formally, the distance between two sentences is then defined by the following equation:

$$dist(u_i, u_j) = 1.0 - \frac{\boldsymbol{u_i}.\boldsymbol{u_j}}{|\boldsymbol{u_i}||\boldsymbol{u_j}|} \tag{1}$$

where $\frac{\boldsymbol{u_i}.\boldsymbol{u_j}}{|\boldsymbol{u_i}||\boldsymbol{u_j}|}$ is the traditional cosine similarity between two sentences u_i and u_j using $tfidf$ weighting method. The effectiveness and robustness of the above measure has been proven in IR and NLP.

Suppose there are three sub-topics in this document collection (the three circles marked in Figure 2), and node 1, node 2 and node 3 are the salient sentences of the three sub-topics.

In order to generate the sub-topics set, we need a ranking method which can generate sub-topics using combined criterion of relevance to the given topic and its centrality. Here, relevance and centrality are not two conflicting concepts while belong to two different dimension. Relevance is the relationship of the topic with retrieved sentences set, centrality is based on the relationship among its similar sentences.

We used the following notation throughout this paper:

- $subtopic(S|T)$: the sub-topic coverage of the document collection corresponding to a topic. Given a certain topic T, we may substitute the notation by $subtopic(S)$ which is clear under certain context.
- u: a single node in the document collection, in usual, is a sentence.
- $p_c(u)$: the salient node u of a certain sub-topic S.
- $neighbor(u)$: the nodes near to the salient node u, also these nodes belong to the same sub-topic S.

More precisely, we define the sub-topics $subtopic(S|T)$ as:

$$subtopic(S|T) = p_c(u) + neighbor(u) \tag{2}$$

Then the summarization problem can be formulated as a graph partition problem:

Given a sentences distance graph G of a document collection of a certain topic T, composed of a set of N nodes $U = u_1, u_2, ..., u_N$, and a length l, partition K sub-graphs of nodes $S_i \subseteq U$ as K sub-topics such that: (1) each sub-graph has a salient node u and its neighborhood neighbour(u) and (2) using u as a representative node of S_i; (3) sum of the length of all the K salient nodes should not be more than l.

The key for our task here is to find the appropriate salient node $p_c(u)$ and its neighborhood $neighbour(u)$ in G.

3.2 GSPRank Criterion

Many existing approaches explore the most important units (clauses/ sentences/ paragraphs) in texts [4] with statistics scoring methods and other higher semantic/syntactic structure such as rhetorical analysis, lexical chains, co-reference chains [6]. Unfortunately, these methods are still hard to obtain the really important units, for the important units are not only decided by the statistical features, but also decided by the semantic features and other fields' features. To explore the most important units or assess the salient nodes in graph, we propose a new sentence ranking criterion - GSPRank - served as basis for our GSPSummary method. This criterion has inspired by the ideas in information retrieval and feature selection. Since the summarization is controlled by choosing the central sentences, which we call "salient sentences", it is in principle possible for the salient sentences to be scored according to the word based features - the statistical features or semantic features according to words or phrases - and the global features.

$$g(u) = f_1(u) \cdot f_2(u) \tag{3}$$

where $g(u)$ is the salience score of sentence u, $f_1(u)$ is the score of word based features, and $f_2(u)$ is the score of global features. We can use the product of the two classes of features to assess the salience of sentence u, for they belong to two different feature spaces.

Word Based Features Metrics. Among the word based features proposed previously, the $tfidf$ score of word is the most widely used approach. In the course of our investigation, the word based features can be presented with a linear combination as the following:

$$f_1(u) = \sum_{i=1}^{m} \lambda_i f_{wi}(u) \tag{4}$$

s.t.

$$\lambda_i \geq 0$$

where $f_{wi}(u)$ is a single word based feature, and the parameter λ_i is the factor to adjust different word based features. Normally, we can express these m word based features with a linear combination. In practice, we use the following word based features:

$$f_1(u) = \lambda_1 f_{w1}(u) + \lambda_2 f_{w2}(u, T) + \lambda_3 f_{w3}(D(u), T) \tag{5}$$

That is, $f_{w1}(u)$ is the centrality score of sentence u, $f_{w2}(u, T)$ is the relevance score between sentence u and the document collection's topic T, and $f_{w3}(D(u), T)$ is the relevance score between the document $D(u)$ where sentence u located and the topic T.

To compute the overall centrality $f_{w1}(u)$ of a sentence given to other sentences, Radev et al. [5] proposed a LexRank approach based on the concept of graph-based centrality. The LexRank value of a sentence gives the limiting probability that such a random walk will visit that sentence in the long run. By LexRankthe score of sentence u can be computed as:

$$f_{w1}(u) = l(u) = \frac{d}{N} + (1-d) \sum_{v \in adj[u]} \frac{w(u,v)}{\sum_{z \in adj[v]} w(z,v)} l(v) \tag{6}$$

where $l(u)$ is the LexRank value of sentence u, N is the total number of nodes in the graph, d is a damping factor for the convergence of method, and $w(u, v)$ is the weight of the link form sentence u to sentence v. Equation 6 can be written in the matrix form as

$$l = [d\mathbf{U} + (1-d)\mathbf{B}]^T l \tag{7}$$

where U is a square matrix with all elements being equal to $1/N$. The transition kernel $[d\mathbf{U} + (1-d)\mathbf{B}]$ of the resulting Markov chain is a mixture of two kernels U and B.

Global Features Metrics. Here, global features mainly consider the length, the position, the text pattern of a sentence, and so on. A simple fact is that short sentences cannot carry enough information corresponding to the topic. Thus, they are not appropriate candidates of summary sentences. And due to the constraint of summary length, too long sentences are not appropriate, either. There are some patterns which are unsuitable for appearing in the summary. The sentences which have these patterns will be discounted for summary sentence. Normally, we can consider the global features are independent, then the global features can illustrated in a form of conditional probability in Equation 9.

$$f_2(u) = P(F_g|u) = \prod_{i=1}^{k} p(f_{gi}|u) \tag{8}$$

where F_g are the global features, and $P(F_g|u)$ is the probability of sentence u in global features space, and $P(F_g|u)$ equals to the product of k global features. In our work, global feature space involves three salient phases: the sentence length, sentence position, and sentence pattern.

$$\begin{cases} p(f_{g1}|u) = p(length|u) \\ p(f_{g2}|u) = p(position|u) \\ p(f_{g3}|u) = p(pattern|u) \end{cases} \tag{9}$$

That is, $p(f_{g1}|u)$ is the probability that the observation of length feature was generated by the training data set, $p(f_{g2}|u)$ is the probability of position feature of u, and $p(f_{g3}|u)$ is the probability of sentence pattern of u. What's more, the global features can be exploited from a supervised way by using a machine learning method based on a training corpus of documents, such as HMM.

GSPRank. As mentioned above, we can obtain the new sentence ranking criterion - GSPRank - combining with word based features and global features. From the Equation 3, 5, 6, and 9, we can induce

$$GSPRank(u) = g(u) = j(u) \cdot l(u) \tag{10}$$

s.t.

$$j(u) = (1 + \lambda_1' \frac{f_{w2}(u,T)}{l(u)} + \lambda_2' \frac{f_{w3}(D(u),T)}{l(u)}) \prod_{i=1}^{k} p(f_{gi}|u)$$

where $g(u)$ is the salience score of sentence in Equation 3, $j(u)$ is a feed function for sentence u, and $l(u)$ is the centrality score of sentence u, which is same to $f_{w1}(u)$ noted in Equation 5. As the Equation 6 mentioned, $l(u)$ can be calculated as a Markov chain model. The convergence property of Markov chains provides a simple iterative algorithm, called Power Method[1], to compute the stationary distribution. The algorithm starts with a uniform distribution. At each iteration, the eigenvector is updated by multiplying with the transpose of the stochastic matrix. Since the Markov chain is irreducible and aperiodic, the algorithm is guaranteed to terminate.

Based on these, we can write Equation 3 in the matrix notation as Equation 11. Here, the salience scores of the sentences set U can be formulated with the product of a feed matrix J and a vector L as the following equation.

$$R = J \cdot L \tag{11}$$

where $\mathbf{R}$ is the vector of GSPRank scores of the sentences set U, J is the feed matrix corresponding to U, each diagonal element in J is a feed function for sentence u in Equation 12, and L is the centrality score vector of U, which can be calculate with the Power Method. Since the procedure of calculating L is iterative, the procedure of calculating R can also be presented as an iterative method with the Markov model.

[1] http://math.fullerton.edu/mathews/n2003/PowerMethod- Mod.html

$$\mathbf{J} = \begin{pmatrix} j(u_1) & \dots & 0 & \dots & 0 \\ \dots & \dots & \dots & \dots & \dots \\ 0 & \dots & j(u_i) & \dots & 0 \\ \dots & \dots & \dots & \dots & \dots \\ 0 & \dots & 0 & \dots & j(u_n) \end{pmatrix} \tag{12}$$

3.3 GSPSummary Algorithm

In section 3.2, we proposed a novel ranking criterion - GSPRank - to assess salience of sentence. The GSPRank can be expressed as an iterative way. Equivalently, our procedure of sub-topic partition algorithm can be described iteratively. This way, a GSPSummary algorithm (in Algorithm 1) should include the following stages as the Fig.3 illustrates:

1. Partition a sub-topic: Generate a ranked list G of U with GSPRank, select the most salient node $p_c(u)$, then obtain $neighbor(u)$ with graph searching or graph partition algorithms.
2. Modify adjacency matrix for next partition: Reduce all the nodes of sub-graph S from M (in Fig.3(3)), and generate the next salient node $p_c(u')$ and its neighborhood $neighbor(u')$ until the algorithm can be terminated.

A brief sketch of our GSPSummary algorithm by looking at the graphs in Fig.3 is to find the salient node $p_c(u)$ and its neighborhood $neighbor(u)$ in graph G of the document collection based on sentences relation. Suppose M is the adjacency matrix of G (in Fig.3(1)), M_0 is the initial matrix of G, and each circle is an element. As seen in Fig.3(2), we can use GSPRank to obtain the salient node $p_c(u)$ in the global space of M_0, then the neighborhood of $p_c(u)$ can

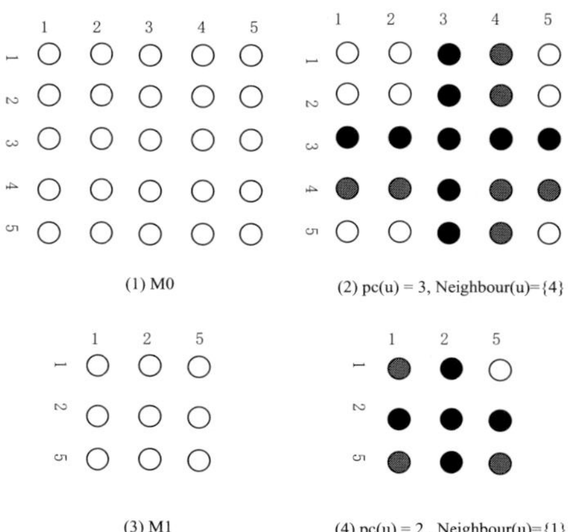

Fig. 3. The procedure of GSPSummary algorithm, (1) illustrates a adjacency matrix of graph, which has 5 nodes, and a circle means an element of the matrix

be generated with graph partition or searching algorithms - e.g, BFS - with a specified neighborhood threshold. At the first iteration, the salient node 3 and its neighborhood node 4 denote the first sub-topic S_1. After eliminated the elements corresponding to S_1, the matrix is be adjusted into a lower dimension one M_1 in Fig.3(3), follows by the next iteration to find the next sub-topic in Fig.3(4). In order to rank the scores of salient nodes, we used the GSPRank method (in Algorithm 2) in each iteration. ζ is the convergence factor for Power Method.

```
Input: A document collection D about the topic T
Output: An array of summary sentences S
repeat
InitGraphMatrix(&M,D);
ArrayRS;
i = GSPRankMethod(M, DistThre, ζ, RS);
ArrayNeighbours = NeighbourSearch(i, M, NeighbourThre);
iLen = LengthOfSentence(i);
if ((iLen + iSummaryLen) > SelectThre) then
    break;
end
InsertIntoSelectedArray(S, i);
iSummaryLen+ = iLen;
UpdateRemainGraph(RS);
until (RS.size() >= MINGRAPHSIZE);
```

Algorithm 1. GSPSummary Algorithm

The following GSPRank Method (Algorithm 2) describes how to select a salient sentence for a given set of sentences with GSPRank. Note that the centrality score vector L is also computed as a side product of the algorithm, and ϵ is the distance threshold used to eliminate some high distance.

4 Experimental Setup

In order to evaluate our GSPSummary approach, we use the ROUGE[2] metrics on DUC2005 data sets for comparison. And the ROUGE score of the DUC2005 start-of-the-art systems came from Hoa's overview of DUC2005 in [2].

4.1 DUC Task Description

Every year, Document Understanding Conferences (DUC[3]) evaluates competing research group's summarization systems on a set of summarization tasks. In DUC2005, the task is to produce summaries of sets of documents in response to short topic statements that define what the summaries should address. The

[2] ROUGE stands for Recall-Oriented Understudy for Gisting Evaluation, http://haydn.isi.edu/ROUGE/

[3] Document Understanding Conferences, http://duc.nist.gov/

```
Input: An array S of n sentences, distance threshold ϵ, a size N feed matrix J
Output: The ID of the salient sentence of S
Array DistMatrix[n][n];
Array Degree[n];
maxDist = −INFINITE;
for i ← 0 to n do
    for j ← 0 to n do
        DistMatrix[i][j]=dist(S[i],S[j]);
        if DistMatrix[i][j] < ϵ then DistMatrix[i][j]=1;
        Degree[i]++;
        else DistMatrix[i][j]=0;
    end
end
Normalization of matrix DistMatrix[i][j];
L = PowerMethod(DistMatrix);
R = J · L;
return the ID with maximal score from R;
```

Algorithm 2. GSPRank Method for obtaining salient sentence in global space

summaries are limited to 250 words in length. The DUC 2005 task was a complex question-focused summarization task that required summaries to piece together information from multiple documents to answer a question or set of questions as posed in a DUC topic. NIST[4] Assessors developed a total of 50 DUC topics to be used as test data. For each topic, the assessor selected 25-50 related documents from the Los Angeles Times and Financial Times of London and formulated a DUC topic statement, which was a request for information that could be answered using the selected documents. The topic statement could be in the form of a question or set of related questions and could include background information that the assessor thought would help clarify his/her information need. The assessor also indicated the granularity of the desired response for each DUC topic. That is, they indicated whether they wanted the answer to their question(s) to name specific events, people, places, etc., or whether they wanted a general, high-level answer. Only one value of granularity was given for each topic, since the goal was not to measure the effect of different granularity on system performance for a given topic, but to provide additional information about the user's preferences to both human and automatic summarization.

4.2 ROUGE

Automatic text summarization has drawn a lot of interest in the NLP and IR communities in the past years. Recently, a series of government-sponsored evaluation efforts in text summarization have taken place in both the United States and Japan. The most famous DUC evaluation is organized yearly to compare the summaries created by systems with those created by humans. Following

[4] National Institute of Standard and Technology, http://www.nist.gov/

the recent adoption of automatic evaluation techniques by the machine translation community, a similar set of evaluation metrics - known as ROUGE [16] - were introduced for both single and multi-document summarization. ROUGE includes four automatic evaluation methods that measure the similarity between summaries: ROUGE-N, ROUGE-L, ROUGE-W, and ROUGE-S.

5 Experimental Results

5.1 Threshold Selection Results

In order to obtain an appropriate sub-topic, the threshold selection of neighborhood is also significant. The higher the threshold, the less the informative; while the lower the threshold, the higher redundancy. On the extreme point where we have a very high threshold or a very low threshold, the GSPSummary algorithm would be of no expected use. Fig.4 demonstrates the effect of threshold for GSPSummary on DUC2005 data set with ROUGE-2 and ROUGE-SU4 metrics. We have experimented with 13 different thresholds - from 0.09 to 0.81 with step 0.06. It is apparent in the figure that the threshold of 0.21 can produce the best summaries together. When the threshold is too lower, the ROUGE scores are decreased for no node in the neighborhood. Similarly, when the threshold is too higher, the ROUGE scores are rapidly decreased for too many nodes in the neighborhood. Therefore, the curves less than 0.08 and higher than 0.81 were not plotted in Fig.4.

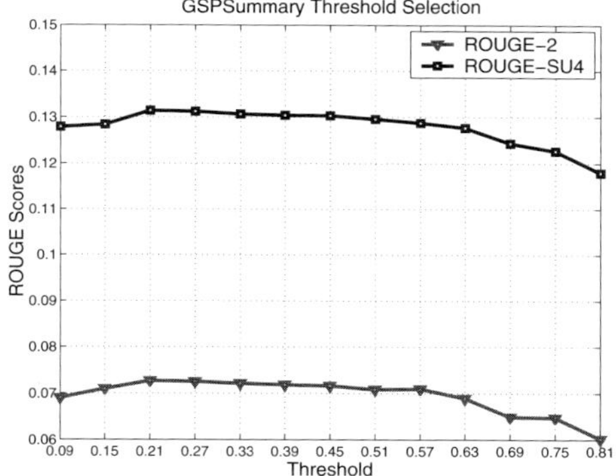

Fig. 4. The thresholds selection of GSPSummry, and the ROUGE scores change when the neighborhood threshold varies

5.2 Performance Comparison

We evaluated the performance of our system in terms of both comparison with LexRank and comparison with DUC2005 results. These comparisons indicate their applicability for real data, DUC2005.

Comparison of GSPSummary and LexRank. In the experiment, the proposed approach was compared with LexRank. With a unit matrix replaced the feed matrix J in Equation 12, our system will degenerate to a hierarchical LexRank system. In practice, we can introduce a hierarchical LexRank method to obtain the most important sub-topics. Unfortunately, for centrality only in LexRank, it is hard to measure the real salience of a sub-topic. Table 1 shows the results of two systems with two different ranks - LexRank and GSPRank. The ROUGE scores of Table 1 illustrates that our system with GSPRank can be quite more effective than the system with LexRank.

Table 1. Performance comparison between systems with LexRank and GSPRank. Columns 2-3 are the ROUGE-2 and ROUGE-SU4 scores of these two systems, and compared with LexRank, the results of GSPRank increase by 48% in ROUGE-2, 26% in ROUGE-SU4 respectively.

Approach	ROUGE-2	ROUGE-SU4
LexRank	0.04943	0.10541
GSPRank	0.07311	0.13231

Table 2. Evaluation results on DUC2005 dataset, IIITH-Sum, PolyU, NUS3 are the state-of-the-art systems competing in DUC2005, PolyU is the rank 2 system in ROUGE-2 and ROUGE-SU4, and NUS3 is the best system in ROUGE-2 and ROUGE-SU4. The last two rows are our two systems GSP-S1 and GSP-S2, and the ROUGE scores of DUC2005 can be highly improved with our GSP-Summary algorithm GSP-S2.

MDS Systems	ROUGE-2	ROUGE-SU4
Baseline	0.04160	0.08946
IIITH-Sum	0.06963	0.12525
PolyU	0.07174	0.12973
NUS3	0.07251	0.13163
GSP-S1	0.06964	0.12923
GSP-S2	**0.07311**	**0.13231**

Comparison of GSPSummary and DUC2005 Results. Table 2 shows the results of our two summarization systems GSP-S1, GSP-S2 on the data set of DUC2005 with ROUGE-2, ROUGE-L, and ROUGE-SU4. The baseline is the result provided by NIST, NUS3 is the best system in the two NIST official ROUGE scores: ROUGE-2 and ROUGE-SU4 recall. The GSP-S1 is used the GSPRank without consideration global features, while the GSP-S2 is used the GSPRank with the global features consideration. The score of our GSP-S1 in ROUGE-2 can obtain the 3rd rank, and the score of ROUGE-SU4 can obtain the 3rd place in DUC2005. Furthermore, comparing with IIITH-Sum - the third ranked system in ROUGE-2 and the 5th ranked system in ROUGE-SU4 - our GSP-S1 system has significant superiority in performance. The scores of our GSP-S2 can both obtain the 1st place in DUC2005. In comparison with the scores in GSP-S1, the ROUGE-2 score and the ROUGE-SU4 score increase 5.0%

and 2.4% respectively, which demonstrates the influence of the global features in the proposed approach. The results confirm that our graph-based sub-topic partition summarizer performs well as comparing to the state-of-the-art systems competing in DUC.

6 Conclusions and Future Work

Summarization is a product of electronic document explosion, and can be seen as the condensation of the document collection. As summary is concise, accurate and explicit, it became more and more important. In this paper, we present a new sub-topic representation model for MDS, and a new rank criterion is presented to obtain sub-topics. Furthermore, a new procedure and algorithm for generic and topic-oriented summarization is proposed. With the representation of graph, our algorithm can obtain the appropriate sub-topic with an iterative procedure in global space. We test our algorithm with DUC2005 data set, and the results suggest that our algorithm is effective in MDS.

The study has three main contributions: (1) we propose a new framework of MDS with sub-topic representation model, according to the logic structure of the topic covered in the document collection. (2) we propose a new ranking criterion GSPRank, in which both the word based and global features are modeled explicitly and effectively. (3) we present a new graph-based sub-topic partition algorithm GSPSummary for MDS.

As future work, we plan to explore in how to generate neighborhood with some other graph searching and partition algorithms. To some extent, our GSPSummary approach can be viewed as a simple version of hierarchical Markov model, with the scalable ranking criterion GSPRank, our algorithm can be further improved. Thus, in future work, we will study how to deal with such issues, and use fitful neighborhood searching or partition algorithms to model sub-topics.

Acknowledgments. The work is supported by the National Grand Fundamental Research 973 Program of China "Large-Scale Tex Content Computing" under Grand NO.2004CB318109 an Grand NO.2007CB311100.

References

1. Carbonell, J., Goldstein, J.: The Use of MMR, Diversity-Based Reranking for Reordering Documents and Producing Summaries. In: Proceedings of SIGIR 1998 (August 1998)
2. Dang, H.T.: Overview of DUC 2005 (2005), http://duc.nist.gov/pubs/2005papers/
3. Mani, I.: Recent developments in text summarization. In: Proceedings of CIKM 2001, Atlanta, Georgia, USA, pp. 529–531 (2001)
4. Mani, I., Maybury, M.T.: Advances in Automatic Text Summarization. MIT Press, Cambridge (1999)
5. Erkan, G., Radev, D.R.: LexRank: Graph-based Lexical Centrality as Salience in Text Summarization. Journal of Artificial Intelligence Research (JAIR) (July 2004)

6. Barzilay, R., Elbadad, M.: Using Lexical Chains for Text Summarization. In: Proceedings of the ACL Intelligent Scalable Text Summarization Workshop, pp. 86–90 (1997)
7. Teufel, S., Moens, M.: Sentence Extraction as a Classification Task. In: Proceedings of the ACL Intelligent Scalable Text summarization Workshop (July 1997)
8. Barzilay, R., McKeown, K.R., Elhadad, M.: Information Fusion in the Context of Multi-Document Summarization. In: Proceedings of ACL 1999, June 16-20 (1999)
9. Mitra, M., Singhal, A., Buckley, C.: Automatic text summarization by paragraph extraction. In: ACL/EACL-1997 Workshop on Intelligent Scalable Text Summarization, July 1997, Madrid, Spain (1997)
10. Marcu, D.: From discourse structures to text summaries. In: Proceedings of the ACL 1997/EACL 1997 Workshop on Intelligent Scalable Text Summarization, Madrid, Spain (1997)
11. Goldstein, J., Mittal, V.O., Carbonell, J.G., Callan, J.P.: Creating and Evaluating Multi-Document Sentence Extract Summaries. In: Proceedings of CIKM 2000 (2000)
12. Radev, D.R., McKeown, K.R.: Generating natural language summaries from multiple online sources. Computational Linguistics 24(3) (1998)
13. Mani, I., Bloedern, E.: Multi-document summarization by graph search and merging. In: Proceedings of AAAI-1997, pp. 622–628 (1997)
14. Radev, D.R., Jing, H., Budzikowska, M.: Summarization of multiple documents: clustering, sentence extraction, and evaluation. In: Proceedings, ANLP-NAACL Workshop on Automatic Summarization, April 2000, Seattle, WA (2000)
15. Harabagiu, S., Lacatusu, F.: Topic themes for multi-document summarization. In: Proceedings of SIGIR 2005 (2005)
16. Lin, C.-Y.: ROUGE: a Package for Automatic Evaluation of Summaries. In: Proceedings of the Workshop on Text Summarization Branches Out, Barcelona, Spain (2004)
17. Lin, C.-Y., Hovy, E.: The automated acquisition of topic signatures for text summarization. In: Proceedings of the 18th COLING Conference, Saarbrucken, Germany (2000)

An Ontology and SWRL Based 3D Model Retrieval System

Xinying Wang[1,2], Tianyang Lv[1,2,3], Shengsheng Wang[1,2], and Zhengxuan Wang[1,2]

[1] College of Computer Science and Technology, Jilin University, Changchun, China
[2] Key Laboratory for Symbolic Computation and Knowledge Engineering of Ministry of Education, Jilin University, Changchun, China
[3] College of Computer Science and Technology, Harbin Engineering University, Harbin, China
xinying_wang2005@tom.com

Abstract. 3D model retrieval is an important part of multimedia retrieval. However, the performance of traditional retrieval method of text-based and content-based is not satisfying because the keyword and the shape feature do not include the semantic information of the 3D models. The paper explores an ontology and SWRL-based 3D model retrieval system Onto3D. It can infer 3D models' semantic property by rule engine and retrieve the target models by ontology. The experiments on Princeton Shape Benchmark show that Onto3D achieves good performance not only in text retrieval but also in shape retrieval.

Keywords: Ontology; SWRL; 3D Model Retrieval; Clustering.

1 Introduction

With the proliferation of 3D models and their wide spread through internet, 3D model retrieval emerges as a new field of multimedia retrieval and has great value in industry, military etc. [1]. Nowadays, two kinds of approaches are widely applied in 3D model retrieval to obtain the desired models, the text-based method (TBR) and the content-based method (CBIR) [2].

The method of text-based retrieval looks on the model as an object in the database. It describes the models with keywords and text. Although the text-based approach may be convenient for user, it relies on artificial annotation and has many shortcomings such as time-consuming and hard-sledding. Moreover, the retrieval results need match strictly in query keyword, so its performance is not always satisfying.

Therefore, the content-based retrieval way becomes the research focus in 3D model retrieval, especially the shape-based retrieval [1]. It is widely accepted that the key problem of shape-based retrieval is extracting model's shape feature with good properties, such as rotation invariant. Researches of this kind concentrate on improving the describing ability of the shape feature. However, model's shape feature only reflects its physics information and can not represent its semantics. Due to the influence of Semantic Gap [3], the shape-based method doesn't perform quite well.

The root of the problem is keyword and feature have not any meaning for computer, namely they do not have any semantics. So the retrieval results cannot satisfy

H. Li et al. (Eds.): AIRS 2008, LNCS 4993, pp. 335–344, 2008.

the user's intention, then we consider introducing ontology technology to the field of 3D model retrieval. On the one hand, it can instruct retrieval intelligently according to the semantic relation among the models; on the other hand ontology can also be used to link the model's low-level features and its high-level semantic information.

The paper explores an ontology and SWRL-based 3D model retrieval system Onto3D. It can infer 3D models' semantic property by rule engine and retrieve the target models by ontology. The experiments on Princeton Shape Benchmark [4] show that Onto3D achieves good performance not only in text retrieval but also in shape retrieval.

The rest of the paper is organized as follows: Section 2 introduces the way to construct the ontology for 3D models; Section 3 states the system realization; Section 4 gives the experimental result; finally, Section 5 concludes the paper.

2 Knowledge Base and Semantic Web Rule Language

2.1 Ontology Structure

The Knowledge Base of Onto3D contains a general ontology. Since our purpose is to create a large ontology of portrayable objects, we consider using the English lexical database *WordNet* [5]. *WordNet* covers both the abstract and concrete vocabulary of English, so we must prune it and extract only those branches containing these objects. In our experiment, the basis of pruning is the manual classifications of 1,814 models in the Princeton Shape Benchmark (PSB) [4]. By classification names, we can search their synonymy, hypernymy, hyponymy and meronymy in *WordNet*, and then supply their correlated concept as well as the attribute relations to our ontology. After pruning, final 3D model ontology partial structure is shown in Fig. 1.

In Fig. 1, ontology is represented by a directed acyclic graph and every node in the graph represents a concept. Concepts are interconnected by means of interrelationships. If there is an interrelationship R between concepts C_i and C_j, then there is also an interrelationship R ' (R inverse) between concepts C_j and C_i. In Fig. 1, interrelationships are represented by labeled arcs/links. Three kinds of interrelationships are used to create our ontology: specialization (*Is-a*), instantiation (*Instance-of*), and component membership (*Part-of*).

Is-a interrelationship is used to represent specialization (concept inclusion). A concept represented by C_j is said to be a specialization of the concept represented by C_i if C_j is kind of C_i. For example, bed is a kind of furniture .In Fig. 1, *Is-a* interrelationship between C_i and C_j goes from generic concept C_i to specific concept C_j represented by a real line with an arrowhead.

The *Instance-Of* relationship denotes concept instantiation. If a concept C_j is an example of concept C_i, the interrelationship between them corresponds to an *Instance-Of* denoted by a broken line with an arrowhead. For example, model m118 is an instance of a concept human; *Ins_hand* and *Ins_face* are instances of the concepts hand and face respectively. In our ontology, the instance which has the 3D model's serial number (such as m118) is the actual existence, and the model's portion (such as *Ins_hand*) is the hypothesized instance, it only used to annotate the model's portion.

A concept is represented by C_j is *Part-Of* a concept represented by C_i if C_i has a C_j (as a part) or C_j is a part of C_i. For example, the concept hand is *Part-Of* the concept human. The *Part-of* relationship is obtained by meronymy relation in *WordNet*, and in order to consistent with the *WordNet* expression, we use its inverse relation *hasPart* express the model whole/partial relationship, see Fig.1 model m118 *hasPart Ins_hand* and *Ins_face*.

In ontology, the relationship is expressed by concept property. *hasPart* is a kind of object property. Moreover, 3D model's name has *SameAs* property, it is a kind of datatype property, and used to express the concept synonym relationship, see Fig.1.

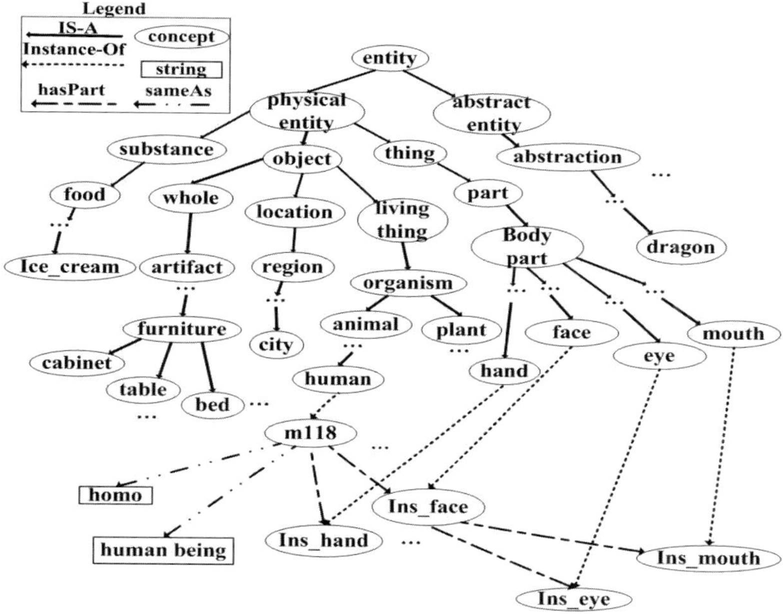

Fig. 1. Portion of 3D Model Ontology

2.2 OWL

Each of the 3D models is an instance of a certain concept in the ontology. In the experiment, we use Protégé 3.3 as the ontology modeling tool and OWL [6] as ontology modeling language. Fig.2 lists part of the OWL of the ontology about the concept human.

Fig. 2 lists a portion of the ontology expressed in OWL about the concept human. It is the description about the 3D model m118 of PSB. It also contains some annotation information about the model. For example m118 is a human and it can be described by some synonyms, such as homo, man, human_being etc.. The model has two parts: face and hand.

```
<owl:Class rdf:about="#human">
    <rdfs:subClassOf rdf:resource="#hominid"/>
    <rdfs:subClassOf>
      <owl:Restriction>
        <owl:someValuesFrom rdf:resource="#face"/>
        <owl:onProperty>
          <owl:ObjectProperty rdf:ID="hasPart"/>
        </owl:onProperty>
      </owl:Restriction>
    </rdfs:subClassOf>
    <rdfs:subClassOf>
      <owl:Restriction>
        <owl:someValuesFrom rdf:resource="#hand"/>
        <owl:onProperty>
          <owl:ObjectProperty rdf:about="#hasPart"/>
        </owl:onProperty>
      </owl:Restriction>
    </rdfs:subClassOf>
 </owl:Class>
 <owl:ObjectProperty rdf:about="#hasPart">
    <rdfs:range rdf:resource="#entity"/>
    <rdf:type rdf:resource="http://www.w3.org/2002/07/owl#TransitiveProperty"/>
    <rdfs:domain rdf:resource="#entity"/>
  </owl:ObjectProperty>
  <owl:DatatypeProperty rdf:ID="sameAs">
    <rdfs:domain rdf:resource="#entity"/>
  </owl:DatatypeProperty>
<human rdf:ID="m118">
    <hasPart rdf:resource="#Ins_face"/>
    <hasPart rdf:resource="#Ins_hand"/>
    <sameAs rdf:datatype="http://www.w3.org/2001/XMLSchema#string"
    >homo</sameAs>
    <sameAs rdf:datatype="http://www.w3.org/2001/XMLSchema#string"
    >human_being</sameAs>
</human>
```

Fig. 2. Portion of 3D Model Ontology in OWL

2.3 SWRL

SWRL [7] (Semantic Web Rule Language) is based on a combination of the OWL DL and OWL Lite sublanguages of the OWL Web Ontology Language the Unary/Binary Datalog sublanguages of the Rule Markup Language. SWRL allows users to write Horn-like rules expressed in terms of OWL concepts to reason about OWL individuals. The rules can be used to infer new knowledge from existing OWL knowledge bases.

In our ontology we need a SWRL rule to express the transfer relationship of the *hasPart* properties:

$$\text{hasPart}(?x,?y) \wedge \text{hasPart}(?y,?z) \rightarrow \text{hasPart}(?x,?z)$$

Furthermore, for every type of model we also need a SWRL rule to express its whole/partial relationship, for example:

$$\text{table}(?x) \rightarrow \text{hasPart}(?x,\text{Ins\_tabletop}) \wedge \text{hasPart}(?x,\text{Ins\_leg})$$

$$\text{bed}(?x) \rightarrow \text{hasPart}(?x,\text{Ins\_mattress}) \wedge \text{hasPart}(?x,\text{Ins\_bedstead})$$

We can create, edit, and read/write SWRL rules in the Protégé SWRL Editor. The Protégé SWRL Editor is an extension to Protégé-OWL that permits interactive editing of SWRL rules. Figure 3 shows the part of the SWRL Editor tab in Protégé-OWL.

Rule-1	→ hasPart(?x, ?y) ∧ hasPart(?y, ?z) → hasPart(?x, ?z)
Rule-2	→ table(?x) → hasPart(?x, Ins_tabletop) ∧ hasPart(?x, Ins_leg)
Rule-3	→ bed(?x) → hasPart(?x, Ins_mattress) ∧ hasPart(?x, Ins_bedstead)
Rule-4	→ face(?x) → hasPart(?x, Ins_eye) ∧ hasPart(?x, Ins_mouth)
Rule-5	→ human(?x) → hasPart(?x, Ins_face) ∧ hasPart(?x, Ins_hand) ∧ hasPart(?x, Ins_arm) ∧ hasPart(?x, Ins_foot)

Fig. 3. Part of the SWRL Editor tab in Protégé-OWL

In the Protégé SWRL Editor, the interaction between OWL and the Jess rule engine is user-driven. After defining the SWRL rules, we can make OWL knowledge and SWRL rules transfer to Jess, and perform the inference using those knowledge and rules, and then make the resulting Jess facts be transferred back to Protégé-OWL as OWL knowledge.

Figure 4 shows part of the Jess Asserted Properties Tab after jess inference in the Protégé SWRL Editor.

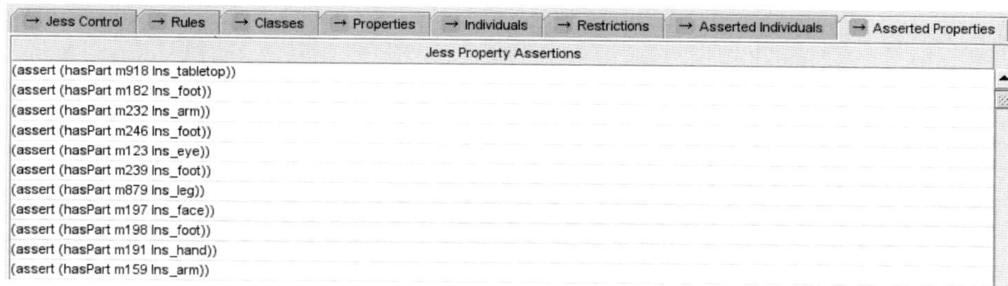

Fig. 4. Jess Asserted Properties Tab in the Protégé SWRL Editor

3 System Realization

3.1 Working Principle

We can use the two methods of text-based and content-based retrieval to retrieval the 3D models in Onto3D.

The user who used the method of text-based retrieval can make retrieval of 3D objects by textual query or browse the ontology and select the appropriate concept. The concrete working is as follows:

1. System needs to establish an ontology knowledge-base which is connected with the PSB. The knowledge-base consists of SWRL.OWL files which contain the 3D models instance data, metadata and the inference rules.
2. User can input keyword or browse the ontology tree and select the model name to retrieval.
3. The Knowledge Base can infer and obtain the retrieval results by query request of user, then make the satisfied result return to the user by 2D images.

About the method of content-based retrieval, the model is based on the hypothesis that similar models may partially share the same semantic. So we use the method of clustering and relevance feedback for correlating 3D shape low-level feature with high-level semantic of the models. The concrete working is as follows:

1. System also needs an ontology Knowledge-base which is connected with PSB. We use the same ontology Knowledge-base as the method of text-based retrieval used.
2. User uploads a sample 3D model.
3. Extracting the sample model's features.
4. Cluster the features of all the models in KB and the sample model. After that, we can obtain many of clusters of the models.
5. System returns all of the models of the cluster which the sample model belong to and waits for user to do information feedback. There are three kinds of selection of the feedback: the completely correct models (or completely relevant models), the category relevant models and the completely wrong models (or completely irrelevant models).
6. If the first returned results have some completely relevant 3D models or category relevant models, then the knowledge-base can infer the sample model's category or the category of its super-class according to the result model's annotation. Finally, system returns all of the results to user. If the first returned results are all completely irrelevant, then system returns to user the nearest cluster of the cluster which sample model belongs to and waits for users to do relevance feedback again.

The system architecture is depicted in Fig.5.

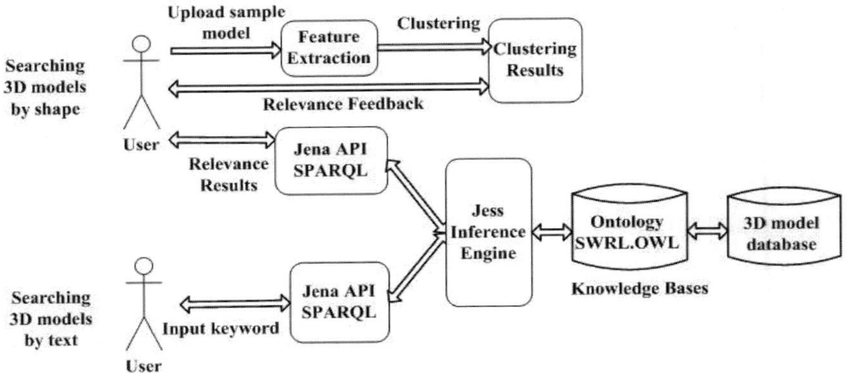

Fig. 5. Working Principle of Onto3D

3.2 Shape Feature Extraction and Clustering

The method of shape retrieval is mainly to search the models which are correlated with the 3D model that user provided.

At present, there are many kinds of feature extraction methods [8]. In our system we used the method of shape feature extracting based on depth-buffer [9] to extract features of 3D models.

We use X-means [10] algorithm to cluster the 3D models in the PSB and utilize the cluster results to simply classify the models. X-means is an important improvement of the well known method K-means. To overcome the highly dependency of K-means on the pre-decided number k of clusters, X-means requires but not restricts to the parameter k. Its basic idea is that: in an iteration of clustering, it splits the center of selected clusters into two children and decides whether a child survives. During clustering, the formula (BIC) is used to decide the appropriate opportunity to stop clustering, where L(x | c) is the log-likelihood of dataset x according to the model c, and m is the dimensionality of the data. In this way, the appropriate number of clusters can be automatically decided.

$$(BIC): BIC(c \mid x) = L(x \mid c) - \frac{k(m+1)}{2} \log n$$

In our experiment, we assigned X-means parameter range is [130, 150] which is according to the count of the basic manual classification in PSB. Finally, we can obtain 130 clusters after executing the X-means clustering algorithm. In the course of retrieval, the system can return all the models in the cluster which the sample model belongs to. Meanwhile, the system can calculate the nearest N neighbors according to the distances between their centers, and saved the result to the temp document for user feedback later.

3.3 Retrieval Realization and Relevance Feedback

After annotating the 3D models with ontology concepts users have two possibilities. They can make retrieval of 3D objects by textual query. A query can be typed by the user or can be formed by ontology browsing. For example a user interested in table models can input the concept in a text box. Alternatively he can browse the ontology and select the appropriate concept. The system will answer the user query by returning all the models annotated with the input concept or with a sub-concept of the input concept. Specially, the system provides the synonyms query. For example a user wants to query human, he/she can input man, homo or human being.

The second possibility is to query by a sample model. Here a user can upload a new model. In the method of shape retrieval, the Knowledge-Base's inference work starts after user's relevance feedback, the selection of feedback can be divided into three kinds: the completely relevant models, the category relevant models and the completely irrelevant models.

The annotation information of the completely relevant models contains all the semantic of the target models, so we can take such model's property of *http://www.w3.org/1999/02/22-rdf-syntax-ns#type* as the property of target model and retrieve all the models having the same property in the knowledge-base.

The category relevant models are those models that they and sample model should belong to the identical category such as table and bed. Although they are not same models, they both are the subclass of the concept furniture, namely they have the same hypernymy (see Fig.1). For this kind of the model, we may infer its upper level concept according to its concept, and then return all of the instances of the upper level concept.

If all of the results are completely irrelevant models, we should use the X-means algorithm to find the nearest neighbor cluster, and return all of the models in the nearest neighbor cluster.

In addition, the completely relevant models have the highest priority, the category relevant models are the next high and the completely irrelevant models have the lowest priority. For the unknown model which was not in the knowledge-base, according to the retrieval result, the system may annotate it automatically by the relevant model's annotation and supply the unknown model as the instance of the certain concept to the knowledge-base.

We use Jena [11] as the reasoning tool in the model retrieval process and use SPARQL [12] to assist Jena perform query. The Protégé-OWL Plugin was used to generate the SWRL.OWL files and to retrieve the rules from the SWRL.OWL files for Jess to process.

Jena is a Java frame which is used to construct application programs of semantic web, and it was developed by HP laboratory. Jena has the function of inference and retrieval knowledge-base. SPARQL (Simple Protocol and RDF Query Language) is a kind of RDF query language which is proposed by W3C. Jess (Java Expert System Shell) is a rule engine and scripting environment written entirely in Sun's Java language by Ernest Friedman-Hill at Sandia National Laboratories in Livermore, CA. It is not open source but can be downloaded free for a 30-day evaluation period and is available free to academic users.

4 Experiment and Analysis

In experiment, we use the models of Princeton Shape Benchmark. The main page of the system is showed in Fig.6. Fig.6 (a), (c) and (d) shows the first retrieval results through the clustering arithmetic after user uploads a sample model in shape-retrieval. System also provides the model's annotation information under the each result model. The annotation information is used to assist users to do relevant feedback. Fig.6 (b) shows the final feedback result of the query in the Fig.6(a).

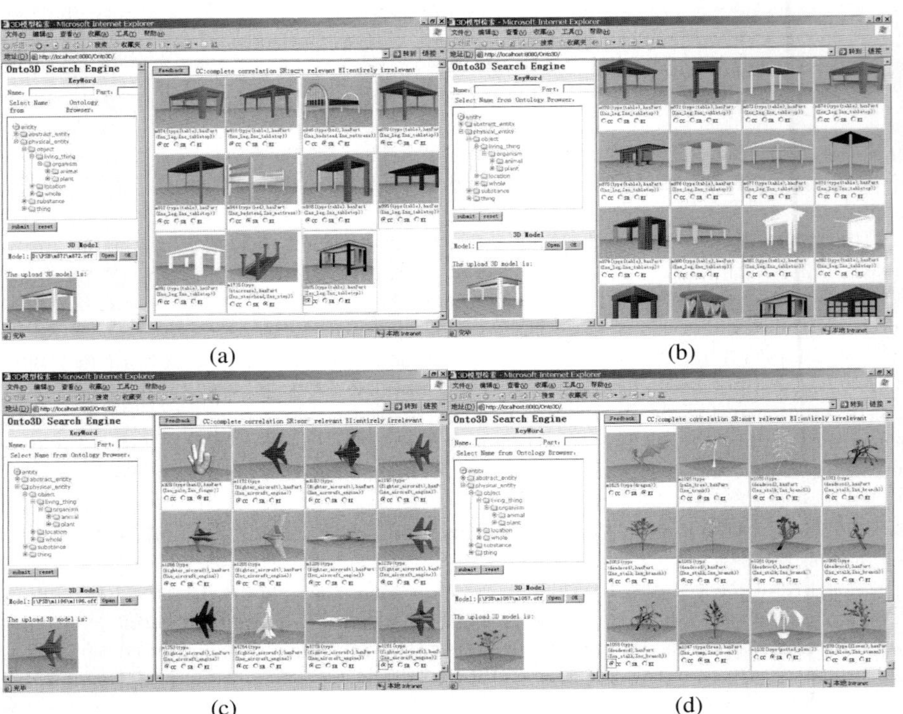

Fig. 6. Main Page of Onto3D System

In shape retrieval, we chose 20 kinds of the representative models, and selected one model in each kind respectively as sample model to perform experiment. We delete all the sample models from the PSB to improve the accuracy of the experiment. In the 20 tests, 14 tests retrieved semantic completely relevant models at first time, then after user's first relevance feedback the system can return all of the correct models according to the relevant models' annotation. But in other 6 tests, we cannot retrieve any semantic relevant or category relevant models at first time. So the system needs to return all the models in the nearest neighbor cluster of the sample model as the query results. In order to compare the precision of the retrieval, we used the file compare function of the Princeton 3D search engine. The Princeton 3D search engine has about 36000 models in its database and it can return 100 results after user upload a model file. Table 1 shows the retrieval efficiency of Onto3D.

From table 1 we can see that Onto3D system's precision of the first time retrieval is satisfying in the majority of retrieval experiment, then the system can retrieve all the relevant models after user's first time relevance feedback. For those sample models that had not retrieved any relevant models at first time, the experiment shows that the system can retrieve the semantic relevant models in 2 or 3 times user feedbacks.

Table 1. Precision Comparison of Princeton 3D Search Engine and Onto3D (%)

3D Model Type (sample model)	Precision of Princeton 3D Search Engine(%)	Precision of Onto3D Search Engine			
		System First returned results	Users' First feedback results	Users' Second feedback results	Users' Third feedback results
human(m185)	81	93.3	100	-	-
biplane(m1134)	78	61.9	100	-	-
horse(m107)	11	22.7	100	-	-
table(m872)	58	72.7	100	-	-
chair(m819)	90	93.8	100	-	-
face(m320)	37	33.3	100	-	-
ship(m1433)	19	57.1	100	-	-
car(m1518)	69	45.5	100	-	-
bicycle(m1472)	5	27	100	-	-
desk_lamp(m612)	4	40	100	-	-
hot_air_balloon(m1343)	1	11.6	100	-	-
door(m1719)	28	50	100	-	-
guitar(m632)	3	20	100	-	-
sword(m707)	50	47.9	100	-	-
staircase(m1735)	2	0	0	6.9	100
flower(m1570)	3	0	5	100	-
ant(m3)	2	0	3.5	100	-
sailboat(m1448)	7	0	3.8	100	-
flower(m975)	1	0	10	100	-
shoe(m1741)	21	0	23	100	-

But the system still has some shortcomings. Firstly, text-based retrieval relies on the words appeared in *WordNet*. For the words do not in *WordNet* or the non-entity words, the system could not find any results. Secondly, content-based retrieval relies on the cluster results. For the models which could not find any relevant result after feedback many times, retrieval efficiency of the system must be reduced greatly.

5 Conclusions

The paper explores an ontology and SWRL-based 3D model retrieval system Onto3D. It can infer 3D models' semantic property by rule engine and retrieve the target models by ontology. The experiments on Princeton Shape Benchmark show that Onto3D achieves good performance not only in text retrieval but also in shape retrieval. Our future work will concentrate on improve the system and promote the Onto3D system to adopt more of the widespread 3D model database.

Acknowledgements

This work is sponsored by Foundation for the Doctoral Program of the Chinese Ministry of Education under Grant No.20060183041 and the National Natural Science Foundation of China under Grant Nos. 60773096, 60603030.

References

1. Iyer, N., Lou, K., Janyanti, S., et al.: Three Dimensional Shape Searching: State-of-the-art Review and Future Trends. J. Computer-Aided Design 37(5), 509–530 (2005)
2. Yubin, Y., Hui, L., Qing, Z.: Content-Based 3D Model Retrieval: A Surver. Chinese journal of computers 27(10), 1297–1310 (2004)
3. Zhao, R., Grosky, W.I.: Negotiating the semantic gap: from feature maps to semantic landscapes. J. Pattern Recognition. 35(3), 51–58 (2002)
4. Shilane, P., Min, P., Kazhdan, M., Funkhouser, T.: The Princeton Shape Benchmark. In: Proceedings of the Shape Modeling International 2004 (SMI 2004), Genova, Italy, pp. 388–399 (2004)
5. Felbaum, C.: WordNet: an Electronic Lexical Database. MIT Press, Cambridge (1998)
6. OWL Web Ontology Language Guide, `http://www.w3.org/TR/2004/REC-owl-guide-20040210/`
7. `SWRL: A Semantic Web Rule Language Combining OWL and RuleML, http://www.daml.org/rules/proposal/`
8. Chenyang, C., Jiaoying, S.: Analysis of Feature Extraction in 3D Model Retrieval. Journal of computer-aided design & computer graphics 16(7), 882–889 (2004)
9. Heczko, M., Keim, D., Saupe, D., Vranic, D.: Methods for similarity search on 3D databases (in German). Datenbank-Spektrum 2(2), 54–63 (2002)
10. Pelleg, D., Moore, A.: X-means: Extending K-means with efficient estimation of the number of clusters. In: Proc. 17th ICML, Stanford University, pp. 89–97 (2000)
11. Carroll, J.J., et al.: Jena: implementing the semantic web recommendations. In: Proceedings of the 13th international World Wide Web conference on Alternate track papers & posters, pp. 74–83. ACM Press, New York (2004)
12. SPARQL Query Language for RDF, `http://www.w3.org/TR/2006/WD-rdf-sparql-query-20061004/`

Multi-scale TextTiling for Automatic Story Segmentation in Chinese Broadcast News

Lei Xie[1], Jia Zeng[2], and Wei Feng[3]

[1] Audio, Speech & Language Processing Group (ASLP),
School of Computer Science,
Northwestern Polytechnical University, Xi'an, China
[2] Department of Electronic Engineering,
City University of Hong Kong, Hong Kong SAR
[3] School of Creative Media,
City University of Hong Kong, Hong Kong SAR
lxie@nwpu.edu.cn, {j.zeng, wfeng}@ieee.org

Abstract. This paper applies Chinese subword representations, namely character and syllable n-grams, into the TextTiling-based automatic story segmentation of Chinese broadcast news. We show the robustness of Chinese subwords against speech recognition errors, out-of-vocabulary (OOV) words and versatility in word segmentation in lexical matching on errorful Chinese speech recognition transcripts. We propose a multi-scale TextTiling approach that integrates both the specificity of words and the robustness of subwords in lexical similarity measure for story boundary identification. Experiments on the TDT2 Mandarin corpus show that subword bigrams achieve the best performance among all scales with relative f-measure improvement of 8.84% (character bigram) and 7.11% (syllable bigram) over words. Multi-scale fusion of subword bigrams with words can bring further improvement. It is promising that the integration of syllable bigram with syllable sequence of word achieves an f-measure gain of 2.66% over the syllable bigram alone.

Key words: story segmentation, topic segmentation, spoken document segmentation, TextTiling, multi-scale fusion, spoken document retrieval, multimedia retrieval

1 Introduction

Story segmentation is the task of dividing text, audio or video into homogenous regions, each addressing a single central topic. It is a necessary pre-processing step for a wide range of speech and language processing tasks, namely topic tracking, summarization, information extraction, indexing and retrieval. These tasks usually assume the presence of individual 'documents'. For broadcast news (BN), a major media channel in the information era, the objective of story segmentation is to segment continuous audio/video streams into news stories. Manual segmentation requires annotators to work through the whole audio/video,

H. Li et al. (Eds.): AIRS 2008, LNCS 4993, pp. 345–355, 2008.

which takes a large amount of time that makes it an intractable task. To perform automatic segmentation, three kinds of cues have been explored, namely acoustic/prosodic cues from audio, lexical cues from speech recognition transcripts or video captions and video cues such as anchor face and color histograms.

Borrowed from traditional text segmentation techniques, lexical-based story segmentation in BN is mainly performed on errorful text transcribed from audio stream via a large vocabulary continuous speech recognizer (LVCSR). Main approaches include word cohesiveness [1], [2], use of cue phrases and modeling (e.g. hidden Markov model [3]). *TextTiling* [1], a classical word-cohesiveness-based approach, has been recently introduced to segmenting spoken documents such as BN [4] and meetings [5] due to its simplicity and efficiency. This approach is based on a straightforward argument that different topic usually employs different set of words. Shifts in word usage are indicative of changes in topic. Therefore, lexical similarity measure between consecutive word chunks is performed across the text and a local similarity minimum indicates a possible topic shift.

Despite of receiving a considerable amount of attention from TREC SDR, TDT and TRECVID evaluation programs, state-of-the-art story segmentation error rates on BN transcripts remain fairly high. This is largely because of the high level of speech recognition error rates, e.g. about 30% for English and about 40% for Chinese Mandarin and Arabic in terms of the word error rate (WER) reported in TRECVID 2006. Besides the errors caused by harsh acoustic conditions (especially for 'field' speech) and diverse speaking styles from various speakers (anchors, reporters and interviewees), the out-of-vocabulary (OOV) problem remains the major obstacle for broadcast news segmentation. New words are introduced continuously from growing multimedia collections and words outside the speech recognizer vocabulary cannot be correctly recognized. Specifically for Chinese BN, the OOV words are largely named entities (e.g. Chinese person names and transliterated foreign names) that are highly related to topics. These OOV words induce incorrect lexical similarity measures across the word stream and thus decrease the segmentation performance greatly.

Recently, the use of *subword* indexing units (e.g. phonemes, syllables and sub-phonetic segments) has been shown to be very helpful to alleviate problems of speech recognition errors and OOV words in the spoken document retrieval (SDR) task [6]. Especially for Chinese, retrieval based on character or syllable indexing units is superior to words due to the special features of Chinese, namely character-based, monosyllabic and flexible wording structure [7], [8]. In this paper, we propose to apply subword units (characters and syllables) in automatic story segmentation of Chinese broadcast news. We present a *multi-scale* TextTiling approach, in which lexical similarities are measured in multiple scales (word and subword scales) and integrated in two schemes, namely representation fusion and score fusion. We aim at fusing the specificity of words and the robustness of subwords to improve story segmentation performance on errorful speech recognition transcripts.

2 Corpus

We experiment with the TDT2 Mandarin corpus[1] that contains about 53 hours of Voice of America (VOA) Mandarin Chinese broadcast news with time span from February to June, 1998. The 177 audio files are accompanied with manually annotated story boundary files and word-level speech recognition transcripts. In TDT2, speech recognition was performed by the Dragon LVCSR with word, character and syllable error rates of 37%, 20% and 15%, respectively. We adopt a home-grown Pinyin lexicon to get the syllable sequences of words. The corpus covers about 2,907 news stories, and the average story duration is 65 seconds (142 words, 248 characters). We separate the corpus into three parts: a quarter as the training set, a quarter as the development set and another half as testing set for evaluation. According to the definition of TDT2, a detected story boundary is considered correct if it lies within a fifteen-second tolerant window on each side of a hand-annotated reference boundary (ground truth).

3 Robustness of Chinese Subwords

Robustness to Speech Recognition Errors. Chinese language is fundamentally different from western languages. Chinese is *character-based* while English is alphabetical. Each Chinese character is pronounced as a tonal syllable (known as the *monosyllabic* feature) and almost every Chinese character is a morpheme with its own meaning. A 'word' is made up of one or several characters. Chinese syllables are tonal because syllables with different lexical tones convey different meanings[2]. There are four lexical tones and a neutral tone in Mandarin Chinese. About 1200 phonologically allowed tonal syllables correspond to over 6500 commonly used simplified Chinese characters. When tones are disregarded, the 1200 tonal syllables are reduced to only about 400 base syllables. This means that the small number of syllables implies large number of *homonym* characters sharing the same syllable. Therefore in errorful Chinese ASR transcripts, it is common that a word is substituted by another character sequence with the same or similar pronunciations, in which homophone characters are the probable substitutions. There are considerable possibilities that a word (or part of a word) is substituted with their homophones in the ASR transcripts. Matching at syllable scale is robust to this kind of recognition errors.

Table 1 shows some word matching failures extracted from the TDT2 corpus. In Table 1, (a) illustrates the case that the foreign person name 哈里斯 is substituted by another person name 哈里森 in its neighborhood in an ASR transcript. Although the first two characters (哈 and 里) are correctly recognized, a single mis-recognized character (the third character 斯 was replaced by 森) will make the word level match fail. In story segmentation approaches based

[1] http://projects.ldc.upenn.edu/TDT2/

[2] Lately, we have shown that the tonality of Mandarin Chinese affects the pitch reset cue for prosody-based story segmentation in Chinese broadcast news. Please refer to [9] for details.

Table 1. Samples of word matching failures due to speech recognition errors

#	Character sequence	Syllable sequence	English translation
(a)	哈里斯	/ha-li-si/	*Harris*
	哈里森	/ha-li-sen/	*Harrison*
(b)	阿尔及利亚	/a-er-ji-li-ya/	*Algeria*
	鲍尔 激励 要	/bao-er ji-li yao/	*Bauer inspire want*
(c)	股市	/gu-shi/	*stock exchange*
	故事	/gu-shi/	*story*

on word cohesiveness, e.g. TextTiling, this failure will induce incorrect lexical similarity measures. However, if character or syllable matching is adopted, the first two characters 哈 and 里 in the two different words still can be matched. Another failure in word matching is shown in (b), where the country name 阿尔及利亚 was mis-recognized as a sequence of three distinct words {鲍尔 激励 要} with similar pronunciations as word 阿尔及利亚. In this case, syllable matching can recall largely of the country name because character sequence 尔激励 has the same syllable representation with 尔及利 in the correct word 阿尔及利亚. (c) shows another failure, where the word 股市(*stock exchange*) was substituted by another word 故事(*story*) with character homophones. In this case, syllable matching can still link the two words together.

Robustness to OOV words. The limited number of Chinese characters with different meanings can be combined to produce unlimited Chinese words. As a result, there does not exist a commonly accepted lexicon for Chinese since new words are born everyday. Therefore, the monosyllabic feature makes the OOV problem more pronounced in Chinese spoken document segmentation, especially in the broadcast news domain. An OOV word distributed in different places of a spoken document may be substituted by several totally different character strings with the same (or partially same) syllable sequence. For example, foreign proper names are common OOV words in Chinese spoken documents as they are transliterated to Chinese character sequences based on the pronunciations (i.e. phonetic transliteration). As a result, speech recognizer may return different character sequences with the same or similar pronunciations.

Table 2 shows two OOV words from the TDT2 corpus. In (a), the OOV word, 尼姆佐夫 (*Nimzov*, a Russian person name), was substituted by four different character sequences (mainly come as singletons) within a news story. Lexical similarity measure at syllable level can partially recover this highly-topical-related OOV word because the last two syllables are the same for the four recognition outputs (all /*zuo fu*/). In (b), the former Korean president 金大中 (*Kim Dae-Jung*) was an OOV word and mis-recognized as three singleton sequences. Lexical similarity measure at syllable scale will successfully recover this OOV word since the three recognition outputs have the same syllable sub-sequence /da zhong/.

Table 2. Samples of word matching failures due to the OOV problem

#		Character sequence	Syllable sequence
(a)	OOV word	尼姆佐夫 (Nimzov)	/ni-mu-zuo-fu/
	Recognizer output	名 模 作 福	/ming mo zuo fu/
		你 目 作 赋	/ni mu zuo fu/
		英国 作 赋	/ying-guo zuo fu/
		你 没 坐 夫	/ni mei zuo fu/
(b)	OOV word	金大中 (Kim Dae-Jung)	/jin-da-zhong/
	Recognizer output	金 大 中	/jin da zhong/
		竟 达 中	/jing da zhong/
		近 大众	/jin da-zhong/

Robustness to Versatility in Word Segmentation. Words and sentences in Chinese text appear as a sequence of characters. Different from English, there are no blanks in Chinese text serving as word boundaries. As a result, 'word' is not clearly defined in Chinese. Consequently, word segmentation in Chinese texts is an ambiguous process and definitely not unique. For LVCSR, the same character sequence in different places might be recognized as several different word sequences that both syntactically valid and semantically meaningful. Therefore, the flexible wording structure of Chinese may contribute considerably to the word recognition error. For example, 北京市领导 (Translation: *leaders of the Beijing city*) can be segmented to words by the speech recognizer as:

a). 北京市 领导 (Translation: *Beijing city leaders*);
b). 北京 市 领导 (Translation: *Beijing city leaders*);
c). 北京 市领导 (Translation: *Beijing city leaders*).

Although they are indicating the same, word matching cannot link the three together. Specifically, matching between 北京市 (Translation: *Beijing city*) in a) and 北京 (Translation: *Beijing*) in b) leads to a failure. This problem can be solved easily by character matching.

4 Multi-scale TextTiling

We have demonstrated that Chinese subword units are robust to speech recognition errors, OOV words and versatility in word segmentation in imperfect Chinese ASR transcripts. This motivates the use of characters and syllables in TextTiling for story segmentation of Chinese BN. In this section, we first describe our TextTiling-based story segmentation algorithm, and then define the subword overlapping n-grams used in lexical similarity measure in subword scales. Finally, we explore the fusion schemes of multiple lexical representations (words and subwords n-grams) for further improving the story segmentation performance.

4.1 TextTiling-Based Story Segmentation

The classical TextTiling algorithm is composed of three steps: tokenization, lexical score determination and boundary identification [1]. The tokenization step splits a character stream into words, i.e. word segmentation. As the speech recognition outputs are word-level text transcripts in the TDT2 corpus, we thus bypass the tokenization step and jump to the rest two steps.

Lexical Score Determination. The Text Tiling algorithm first divides the text document into sentences (or pseudo-sentences). At each inter-sentence gap along the text stream, adjacent windows of fixed number of sentences are compared in terms of lexical similarity. For ASR transcripts of spoken documents, sentence boundaries are not readily available. A possible way for story boundary investigation is to perform lexical similarity measure at places with significant pauses (a useful prosodic cue). To evaluate story segmentation performance *purely* based on the lexical information, we divide the ASR transcripts into blocks/windows of fixed number of words (W). The *lexical score* between adjacent windows (at inter-window gap g) is calculated by cosine similarity:

$$lexscore(g) = \cos(\mathbf{v}_l, \mathbf{v}_r) = \frac{\sum_{i=1}^{I} v_{i,l} v_{i,r}}{\sqrt{\sum_{i=1}^{I} v_{i,l}^2 \sum_{i=1}^{I} v_{i,r}^2}} \quad (1)$$

where $\mathbf{v}_l$ and $\mathbf{v}_r$ are the term frequency vectors for the two adjacent windows (left and right windows to the inter-window gap). $v_{i,l}$ is the ith element of $\mathbf{v}_l$, i.e., the term frequency of word i registered in the vocabulary (with size of I).

Since story boundaries are searched at inter-window gaps, we increase the boundary hypotheses by sliding. The lexical scores are calculated at positions of $J, 2J, 3J \cdots$ word positions along the ASR transcript, where J is the sliding length and $J \leq W$.

Boundary Identification. Boundary identification can be carried out directly on the time trajectory of lexical score. However, TextTiling adopts the relative score information instead of the absolute values. The inter-window gaps whose similarity valleys represent a 'valley' are considered for boundary identification. Specifically, *depth score* is calculated for each valley point on the lexical score time trajectory. Denote the valley point as v, and the nearest left and right peaks, p_l and p_r, around the valley points, the depth score of valley v is

$$depthscore(v) = (lexscore(p_l) - lexscore(v)) + (lexscore(p_r) - lexscore(v)). \quad (2)$$

Note that every non-valley point is given a depth score of 0. The depth score considers a sharp drop in lexical similarity as more indicative of a story boundary than a gentle drop. This helps make decisions in the cases in which a inter-window gap's lexical score falls into the middle of the lexical score range, but is flanked by tall peaks on either side. This situation happens commonly enough to be possible story boundaries. Boundary identification is performed on the depth score time trajectory, in which a point whose depth score exceeds a pre-set threshold θ is considered as a story boundary.

4.2 Subword Overlapping N-grams

We perform lexical similarity measures in subword scales by subword overlapping n-grams. For a sequence of m subword units (characters or syllables) $\{S_1S_2S_3\cdots S_m\}$, the subword overlapping bigram and trigram are formed as

$$\{S_1S_2\ S_2S_3\ S_3S_4\cdots S_{m-1}S_m\} \text{ and} \tag{3}$$

$$\{S_1S_2S_3\ S_2S_3S_4\ S_3S_4S_5\cdots S_{m-2}S_{m-1}S_m\} \tag{4}$$

respectively. Other n-grams can be listed accordingly. To reduce the possibility of missing any useful information embedded in the subword sequence, overlapping between subwords is used. Term frequency vectors, lexical scores and depth scores are calculated on sequences of subword overlapping n-gram units transformed from ASR word transcripts.

4.3 Fusion of Multi-scale Representations

We propose to combine multiple lexical scales (words and subwords) for improving story segmentation performance by taking advantage of different scales. We present two different fusion strategies: *representation fusion* and *score fusion.*

Representation Fusion. Representation fusion merges lexical representations from different scales before the lexical similarity measure. The term frequency vectors for all scales are combined to form a concatenated vector with $\sum_{k=1}^{K} I^k$ dimensions, where I^k denotes the dimension of scale k. The concatenated vector is formulated as

$$\mathbf{v} = [w_1 \cdot \mathbf{v}^1, w_2 \cdot \mathbf{v}^2, \cdots, w_K \cdot \mathbf{v}^K], \tag{5}$$

where w_k denotes the fusion weight for scale k and $\sum_{k=1}^{K} w_k = 1$. The fusion weight is used to reflect the importance of each lexical scale. The lexical score and the depth score are thus calculated on the concatenated vector, and boundary identification is performed on the time trajectory of depth score.

Score Fusion. In score fusion, the lexical scores are calculated for all scales separately, and then linearly integrated to a combined score by

$$lexscore = \sum_{k=1}^{K} w_k \cdot lexscore^k, \tag{6}$$

where $lexscore^k$ is the lexical score for scale k calculated on the term frequency vector $\mathbf{v}^k$ and $\sum_{k=1}^{K} w_k = 1$. Subsequently the depth scores are calculated and story boundary hypotheses are made.

5 Evaluations

We have carried out story segmentation experiments on the TDT2 corpus described in Section 2. We evaluate our story segmentation approach by comparing

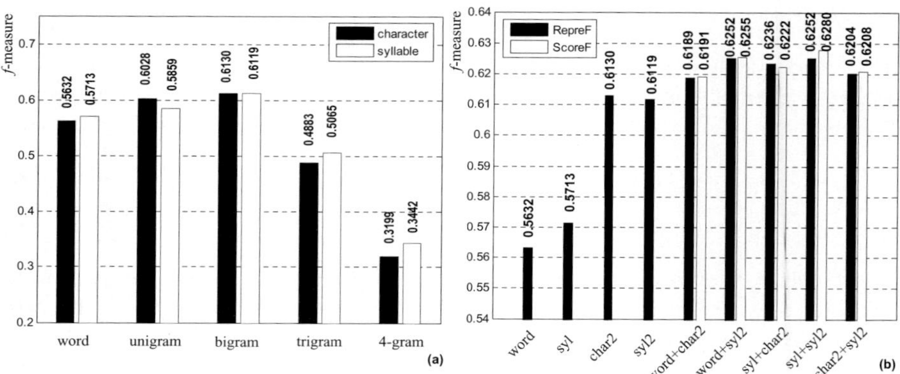

Fig. 1. Experimental results on (a) individual scales and (b) multi-scale fusion. Syl: syllable sequence of word; char2: character bigram; syl2: syllable bigram.

the automatically detected boundaries to the human-labeled boundaries. Recall, precision and their harmonic mean, f-measure, are adopted as the evaluation metrics. The training set is used to optimize the window size (W), the sliding length (J) and the boundary identification threshold (θ). An empirical selection procedure is adopted on the testing set to choose the best W, J and θ maximizing the f-measure for the word scale. Experiment shows that the combination of $W = 50$ and $J = 10$ achieves the best f-measure. The selected W and J are then fixed in the empirical searching for the best θ on each subword scale. For multi-scale fusion experiments, we use the development set to find the optimal fusion weights, w_k for each fusion form. We iterate through weights from 0 to 1.0 at intervals of 0.1 to seek the best weights that maximize the f-measure on the development set. Story segmentation results are reported for the testing set.

Experimental Results on Individual Scales. Story segmentation results for individual lexical scales are shown in Fig. 1 (a). We observe that subword bigrams perform the best as compared with other lexical scales in story segmentation of Chinese BN. The f-measure for character bigram and syllable bigram are 0.6130 and 0.6119, respectively. Promisingly, the relative improvement of subword bigrams over words are as high as 8.84% (characters) and 7.11% (syllables). This improvement is mainly because of the robustness of bigrams against speech recognition errors, OOV words and versatility in word segmentation, as illustrated in Section 3. This observation also accords with Chinese SDR using subword units, where subword bigrams are also superior to other scales [7], [8].

Another observation is that as n increases from one to four for the subword n-grams, performance reaches the peak at bigram and gradually drops from bigram to 4-gram. It can also be seen that subword unigrams also bring considerable improvement over words, performing the second best. The subword trigram and 4-gram perform worse even than the word scale. This can be explained by the fact that most frequently used words in Chinese are bi-character, and the proba-

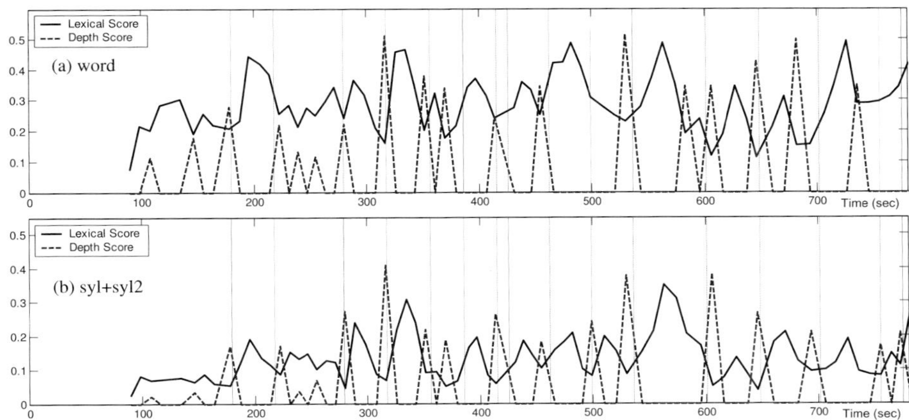

Fig. 2. Lexical score and depth score time trajectories for a broadcast news audio file (partial) in the testing set, calculated using (a) word scale alone and (b) fusion between syllable sequence of words and syllable bigrams (*syl+sy2*, score fusion). The vertical lines (in red) show the positions of human-labeled boundaries.

bility of long sequences with correctly recognized characters is smaller than two character units.

Experimental Results on Multi-Scale Fusion. Since subword bigrams show superior performance among other scales, we run experiments on multi-scale fusion that integrates both subword bigrams and words for TextTiling-based story segmentation. Results in Fig. 1 (b) shows that multi-scale fusion can bring further improvement to the story segmentation performance. All multi-scale integration forms under investigation outperform the single scales. The best performance is achieved when syllable bigrams are combined with syllable sequence of words (*syl+syl2*) in terms of score fusion. The f-measure is as high as 0.6280 and the performance gain of 2.66% is achieved over using the syllable bigram alone. The integration of words and syllable bigrams (*word+syl2*) offers the second best performance with f-measures of 0.6255 for score fusion and 0.6252 for representation fusion. When syllable bigrams and character bigrams are integrated (*char2+syl2*), the improvement is not as salient as the integration between words and subword bigrams. This is probably because the two subword bigram scales carry similar information (which can be seen from their very close performance in Fig. 1 (a)) and neither have the specificity of words. Results also show that score fusion marginally outperforms representation fusion.

Fig. 2 illustrates the lexical score and depth score time trajectories for a broadcast news clip in the testing set, where (a) is calculated using the word scale alone and (b) the fusion of syllable sequence of word and syllable bigrams (*syl+syl2*, score fusion). We can clearly see that some of the story boundary misses and false alarms in (a) are corrected in (b). This performance gain is mainly due to the robustness of the syllable bigram.

6 Conclusions and Future Work

In this paper, we have applied subword representations (characters and syllables) into story segmentation of Chinese broadcast news. We have shown the robustness of Chinese subwords in lexical matching in errorful speech recognition transcripts. This have motivated us to propose a multi-scale TextTiling story segmentation approach to integrate both the word and subword scales. This approach aims at combining the specificity of words and the robustness of subwords. Two different fusion schemes have been adopted, namely representation fusion and score fusion. Experimental results on the TDT2 Mandarin corpus show that subword bigrams achieve the best performance among all scales with relative f-measure improvement of 8.84% (character bigram) and 7.11% (syllable bigram) over words. Multi-scale fusion experiments demonstrate that integration of subword bigrams with words can bring further improvement for both fusion schemes. The integration between syllable bigram and syllable sequence of words achieves an f-measure gain of 2.66% over the syllable bigram alone.

There is still substantial work to be done. The overall story segmentation performance of the TextTiling-based approach is not high. This is mainly because the TextTiling algorithm is based on local lexical similarity measure, and thus sensitive to sub-topics changes. This motivates us to integrate both local and global measurements in the future work. In addition, we will experiment with the fusion between lexical cues and prosodic cues to further improve the story segmentation performance in Chinese broadcast news.

Acknowledgements

This work is supported by the Research Fund for the Doctoral Program of Higher Education in China (Program No. 20070699015), the Natural Science Basic Research Plan in Shaanxi Province of China (Program No. 2007F15), NPU Aoxiang Star Plan (Program No. 07XE0150) and NPU Foundation for Fundamental Research.

References

1. Hearst, M.A.: TexTiling: Segmenting text into multiparagraph subtopic passages. Computational Linguistics 23(1), 33–64 (1997)
2. Chan, S.K., Xie, L., Meng, H.: Modeling the statistical behavior of lexical chains to capture word cohesiveness for automatic story segmentation. In: Proc. Interspeech, pp. 2851–2854 (2007)
3. Yamron, J., Carp, I., Gillick, L., Lowe, S., van Mulbregt, P.: A hidden Markov model approach to text segmentation and event tracking. In: Proc. ICASSP, pp. 333–336 (1998)
4. Rosenberg, A., Hirschberg, J.: Story segmentation of broadcast news in English, Mandarin and Arabic. In: Proc. HLT-NAACL, pp. 125–128 (2006)
5. Banerjee, S., Rudnicky, I.A.: A TextTiling based approach to topic boundary detection in meetings. Proc. Interspeech (2006) 57–60

6. Ng, K.: Subword-based approaches for spoken document retrieval. Ph.D. Thesis of MIT (2000)
7. Chen, B., Wang, H.M., Lee, L.S.: Discriminating capabilites of syllable-based features and approaches of utilizing them for voice retrieval of speech information in Mandarin Chinese. IEEE Transactions on Speech and Audio Processing 10(5), 202–314 (2002)
8. Lo, W.K., Meng, H., Ching, P.C.: Multi-scale spoken document retrieval for Cantonese broadcast news. International Journal of Speech Technology 7(2-3), 1381–2416 (2004)
9. Xie, L., Liu, C., Meng, H.: Combined Use of Speaker- and Tone-Normalized Pitch Reset with Pause Duration for Automatic Story Segmentation in Mandarin Broadcast News. In: Proc. HLT-NAACL, pp. 193–196 (2007)

Improving Spamdexing Detection Via a Two-Stage Classification Strategy

Guang-Gang Geng, Chun-Heng Wang, and Qiu-Dan Li

Key Laboratory of Complex System and Intelligent Science, Institute of Automation
Chinese Academy of Sciences, Beijing 100080, P. R. China
{guanggang.geng,chunheng.wang,qiudan.li}@ia.ac.cn

Abstract. Spamdexing is any of various methods to manipulate the relevancy or prominence of resources indexed by a search engine, usually in a manner inconsistent with the purpose of the indexing system. Combating Spamdexing has become one of the top challenges for web search. Machine learning based methods have shown their superiority for being easy to adapt to newly developed spam techniques. In this paper, we propose a two-stage classification strategy to detect web spam, which is based on the predicted spamicity of learning algorithms and hyperlink propagation. Preliminary experiments on standard WEBSPAM-UK2006 benchmark show that the two-stage strategy is reasonable and effective.

1 Introduction

The exploitation of search engines today is a serious issue, but, like it or not, most businesses see it as something that must be done–an online business imperative. For many commercial web sites, an increase in search engine referrals translates to an increase in sales, revenue and profits. The practices of crafting web pages for the sole purpose of increasing the ranking of these or some affiliated pages, without improving the utility to the viewer, are called spamdexing(web spam) [6]. Web spam seriously deteriorates search engine ranking results, makes it harder to satisfy information need, and frustrates users' search experience. Combating spamdexing has become one of the top challenges for web search.

The relation of spamdexing and spam detection has been likened to an "arm race". Once an anti-spam technique is developed, new spamdexing techniques will be implemented to confuse search engine. Machine learning based methods have shown their superiority for being easy to adapt to newly developed spam techniques [2] [17] . In most machine learning based spam detection algorithms, spam identification is treated as a binary classification problem. After extracting content based or hyperlink relevant features, classification algorithms are employed to detect web spam.

In this paper, our focus is placed on finding more effective learning strategy by exploiting the hyperlinks dependent relation. Generally, more training samples will provide more discriminant information for classification. Based on such consideration, a two-stage classification strategy based on predicted spamicity(PS) is proposed. Self-learning and label propagation through hyperlink graph are used to expand the training set in this strategy. The samples with high classification confidence or propagation confidence will be put into the training set for the next stage learning. The strategy provides

H. Li et al. (Eds.): AIRS 2008, LNCS 4993, pp. 356–364, 2008.

a well-founded mechanism to integrate the existing learning algorithms for spam detection. Experiments on WEBSPAM-UK2006 show that the strategy can improve the web spam detection performance effectively.

The remainder of this paper is organized as follows. We review the previous research work in Section 2. In section 3, we describe the proposed two-stage classification strategy in detail. Experimentally evaluation of the proposed strategy on the standard collection is presented in section 4. At last, we give the concluding remarks and future research directions.

2 Related Works

Generally, search engine ranks a web page according to the content relevance and the page importance which is computed with link analysis algorithm, such as PageRank. Accordingly, web spam can be broadly classified into two categories: content spam and link spam[11]. Previous work on web spam identification mostly focused on these two types of spam.

Content spam is created for obtaining a high relevance score, which refers to the deliberate changes in the content of the pages, including inserting a large number of keywords, content hiding and cloaking. Similar with the methods used in fighting email spam, [5] detected content spam via a number of statistical content-based attributes. These attributes are parts of the features in our experiments.

Link spam tries to unfairly gain a high ranking on search engines for a web page without improving the user experience, by mean of trickily manipulating the link graph [14]. Based on the assumption that good pages seldom point to spam pages, but spam pages may very likely point to good pages. Gyongyi et al. [7] proposed a method to separate reputable pages from spam. The method consists of the following processes, firstly, a small set of seed pages were evaluated by experts; then, the link structure of the web was used to discover other pages that are likely to be good.

In recent years, machine learning based detection methods have received much attention, which shows its superiority for being easy to adapt to newly developed spam techniques. According to the characteristic of web spam, many link-based features have been extracted to construct automatic classifiers [1] [2] [8], such as transformed PageRank [3],TrustRank [7], and Truncated PageRank[1], etc. [19] used a number of features for Web spam filtering based on the occurrence of keywords that are either of high advertisement value or highly spammed. [17] designed several heuristics to decide if a node should be relabeled based on the preclassified result and knowledge about the neighborhood.

Many studies show that properties of neighboring nodes are often correlated with those of a node itself [2] [8] [17]. Literature [2] computed the topological dependencies of spam nodes on standard WEBSPAM-UK2006, the experiment shows that: "Non-spam nodes tend to be linked by very few spam nodes, and usually link to no spam nodes" and "Spam nodes are mainly linked by spam nodes". Based on these facts, we plan to increase the samples in the train set through link topology learning.

In our previous work[9][18], we proposed a predicted spamicity-based ensemble under-sampling strategy for spamdexing detection. Within this strategy, many existing learning algorithms, such as C4.5, bagging and adaboost, can be applied in this strategy; distinguishing information involved in the massive reputable websites are fully explored and solves the class-imbalance problem well. Based on the strategy, we will probe new classification method to improve spamdexing detection performance in this paper. A two-stage classification strategy is proposed, which integrates the Self-learning and Link-learning. This method takes full advantage of the information involved in the web topology and expands the training data sets for the ultimate spam detection.

3 Proposed Two-Stage Classification Strategy

In this section, we will propose an two-stage classification strategy, which is based on the predicted spamicity(PS)[18]. PS is defined as the probability of a website belonging to web spam, which is based on the predicted confidence of classifiers.

The flow chart of the proposed two-stage classification strategy is depicted in Figure 1. The arrows indicate the direction of data flow, and the figures on the line represent the order of the execution process.

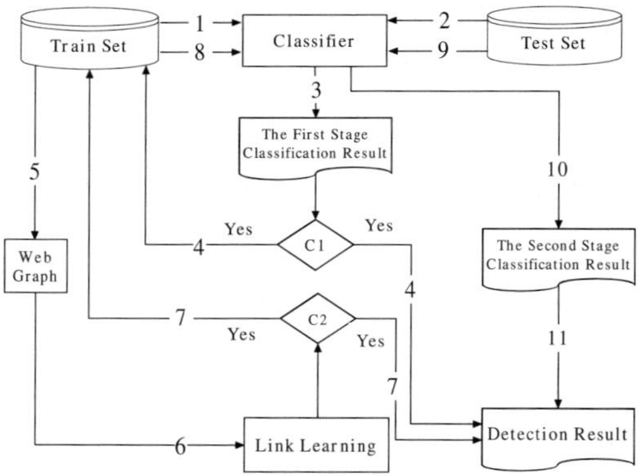

Fig. 1. Flow chart of the proposed two-stage classification strategy

Figure 1 illustrates the general bootstrapping process. The bootstrapping algorithm aims to improve the classification performance, by integrating examples from the unlabeled data into the labeled data set. In the first stage, we implement a preliminary classification, then converts the most confidently predicted examples of each class into a labeled training example. After that, link learning is carried out, and the samples with high propagation confidence are converted into labeled ones, too. In the second stage, a new classifier is trained with the expanded train set, then the the remaining unlabeled examples are tested with the learned machine. In the next part, we will describe the algorithm processes in detail.

The strategy starts with a set of labeled data(train set), and builds a classifier, which then applied on the test set. The instances with a PS value exceeding a certain threshold are added to the labeled set. Taking into account the dependence between samples, we use link learning to further expand the labeled samples set.

As the intention of link learning is to label more unlabeled samples for next step learning, we adopt a more stringent restrictions, ie. both inlink and outlink nodes properties are taking into account. In order to remove the noise hyperlinks, not all the hyperlinks are taken into account. If the number of hyperlinks from one node(website) to another one is more than 5, we think that there exist a valid link relation between the two nodes. The propagated confidence of node is computed as follows:

$$Conf_{normal}(i) = K_1 \cdot \frac{out_n(i) + in_n(i) + \epsilon}{out_s(i) + in_s(i) + \epsilon} \tag{1}$$

$$Conf_{spam}(i) = K_2 \cdot \frac{out_s(i) + in_s(i) + \epsilon}{out_n(i) + in_n(i) + \epsilon} \tag{2}$$

where $out_s(i)$ and $in_s(i)$ are the spam nodes number among inlink and outlink of node i, respectively. Correspondingly, $out_n(i)$ and $in_n(i)$ are the non-spam nodes number among inlink and outlink of node i, respectively. ϵ, is the smoothing factor, whose value belongs to interval $(0, 1)$. The value of K_1 and K_2 are decided by the following algorithm:

Algorithm 1. Computing the Value of K_1 and K_2

Require: $M_1 > M_2 > M_3 \geq 1$,$N_1 > N_2 > N_3 \geq 1$
if (($out_n(i) > 0$) and ($in_n(i) > 0$)) **then**
  if (($in_s(i) == 0$) and ($out_s(i) == 0$)) **then**
    $K_1 = M_1$
  else
    $K_1 = M_2$
  end if
else
  $K_1 = M_3$
end if
if (($out_s(i) > 0$) and ($in_s(i) > 0$)) **then**
  if (($in_n(i) == 0$) and ($out_n(i) == 0$)) **then**
    $K_2 = N_1$
  else
    $K_2 = N_2$
  end if
else
  $K_2 = N_3$
end if

The condition C1 is the boolean conditions for selecting the samples with high predicted confidence and condition C2 is the conditions for selecting the instances with high propagated confidence. For keeping consistency with original data, selected samples must have the same class ratio as the train set.

Algorithm 2 is the detailed implementation of our proposed two-stage classification strategy.

Algorithm 2. Two-stage classification strategy for web spam detection

Input $TrainSet$: Train set
$TestSet$: Test set
$Classifer$: Classifier
Output The spamdexing detection results

```
Train Classifier with TrainSet
Save the learned model Model_1
for all x such that x ∈ TestSet do
    Test x with model Model_1
end for
for ∀x such that x ∈ TestSet do
    if condition C1 is met then
        TrainSet+={x}
        TestSet-={x}
        Put x and predicted label to result set
    end if
end for
Label the web graph with TrainSet samples
Link learning on the labeled web graph
for ∀x such that x ∈ TestSet do
    if condition C2 is met then
        TrainSet+={x}
        TestSet-={x}
        Put x and predicted label to result set
    end if
end for
Train Classifier with TrainSet
Save the learned model Model_2
for all x such that x ∈ TestSet do
    Test x with model Model_2
    Put x and the predicted label into result set
end for
```

4 Experiments

4.1 Data Collection

The WEBSPAM-UK2006 [6] collection is used in our experiments, which is a publicly available data set. The collection includes 77.9 million pages, corresponding to roughly 11400 hosts, among which over 8000 hosts have been labeled as “spam”, “non-spam(normal)” or “borderline”.

We use all the labeled data with their homepage in the summarized samples[6], where 3810 hosts are marked normal and 553 hosts are marked spam[1].

[1] The original human labeled data are in webspam-uk2006-labels-DomainOrTwoHumans.txt, which were downloaded from http://www.yr-bcn.es/webspam/datasets/webspam-uk2006-1.3.tar.gz

4.2 Features

The features used for classification involve transformed link-based features[2] and content-based features[5]. All the link-based features are computed with the whole web graph; and the content based features used in the rest of the paper are extracted from the summarized samples containing 3.3 million pages[2].

Similar with [2][18], most of the link-based features are computed for the home page and page in each host with the maximum PageRank. The content-based features include the number of words in page, amount of anchor text, average length of words, fraction of visible content and compressibility etc [2][5].

The following experiments were run with a combination of all the features mentioned above, and preprocessed with information gain. After feature selection, each sample has 198 dimensions attributes.

4.3 Classification Algorithms

Most of the existing classification algorithms are applicable in our proposed two-stage strategy. In [18], we proposed an ensemble random under-sampling classification strategy, which exploits the information involved in the large number of reputable websites to full advantage. In the ensemble method, several subsets $S_1, S_2, ..., S_n$ are independently sampled from spam set S. For each subset $S_i(i \in N)$, a classifier C_i is trained using S_i and non-spam set M. All the results generated by the sub classifiers are combined for the final decision. Experimental results on standard WEBSPAM-UK2006 collection show that the proposed learning strategy is robust, and can improve the web spam detection performance effectively. Especially when choose adaboost as the base classifier, system achieved the best results. Thereby, we chose ensemble adaboost as base classifier for our two-stage classification test. Decision stump is used as the weak classifier for adaboost.

4.4 Experimental Results

5 times 2-fold cross-validation is run on the data set. The precision, recall, true positive rate(TP), false positive rate(FP), area under ROC curve (AUC) and F-measure were used to measure the performance of the learning algorithms.

We use the usual definition for ROC Area (AUC). An ROC plot is a plot of true positive rate vs. false positive rate as the prediction threshold sweeps through all the possible values. AUC is the area under this curve. AUC of 1 is perfect prediction – all positive cases sorted above all negative cases. AUC of 0.5 is random prediction – there is no relationship between the predicted values and truth.

All the results were obtained with $M_1 = 3 > M_2 = 2 > M_3 = 1$, $N_1 = 1 > N_2 = 1 > N_3 = 1$, the resampling ratio $K = 3$, resampling times $n = 15$ in ensemble under-sampling classification.

Figure. 2 shows the performance of AUC with different feedback sample ratio R in step 4 and 7. The X-axis represents the ratio of annotating unlabeled samples ratio in test set, and the Y-axis is the AUC values. The baselines of AUC were computed with one-stage classification.

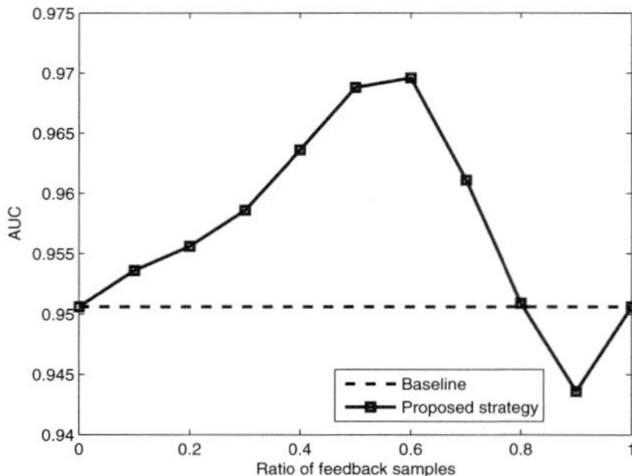

Fig. 2. AUC value with proposed strategy

From figure. 2, we can find that two-stage classification boosts the detection performance. The classification strategy is robust, especially, when $R \in [0.1, 0.7]$, the improvement is notable. It can be seen from figure. 2 that the AUC is poor with $R > 0.8$. Actually, the phenomenon is understandable because bigger R value means more unlabeled samples are put into train set according to the predicted confidence, then there will be more samples are incorrectly labeled. When the ratio of wrong samples becomes more and more in the new train set, the performance decline is inevitable.

To gain more understanding of the performance comparison, we further list in Table 1 all the assessment criteria with two-stage classification and one-stage classification. All the data is computed with $R = 0.5$.

Table 1. Performace of Proposed Two-Stage Classification Strategy

Evaluation criteria	TP	FP	Precision	Recall	F-measure	AUC
Baseline	0.7996	0.0537	0.6845	0.7996	0.7372	0.9506
Proposed Method	0.8402	0.0554	0.6876	0.8402	0.7561	0.9688

From the table, we can find that our proposed method can boost the detection performance effective, especially TP(Recall), F-measure and AUC. For more conservative spam filtering application, cost sensitive classification can be employed. The false positive changes with different cost matrix.

5 Conclusions

In this paper, a two-stage classification strategy is proposed. Self-learning and link learning through hyperlink graph are used to expand the training set in this strategy. The samples with high classification confidence or propagation confidence are put into training set for the next stage learning. Experimental results on standard WEBSPAM-UK2006 collection show that the proposed learning strategy is feasible and effective.

Future work involves testing more learning algorithms with the proposed strategy, taking the weight information of hyperlinks into account and improving the strategy from two-stage classification to multi-stage classification.

Acknowledgments

This work is supported by the National High Technology Research and Development Program of China (863 Program) under Grant No 2006AA010106.

References

1. Becchetti, L., Castillo1, C., Donato1, D., Leonardi, S., Baeza-Yates, R.: Using Rank Propagation and Probabilistic Counting for Link Based Spam Detection. In: Proc. of WebKDD 2006 (August 2006)
2. Castillo, C., Donato, D., Gionis, A., Murdock, V., Silvestri, F.: Know your Neighbors: Web Spam Detection using the Web Topology. Technologies Project (November 2006)
3. Page, L., Brin, S., Motwani, R., Winograd, T.: The PageRank Citation Ranking: Bringing Order to the Web. Stanford Digital Library Technologies Project (1998)
4. Benczúr, A.A., Csalogány, K., Sarlós, T., Uher, M.: Spamrank: Fully Automatic Link Spam Detection. In: Proc. of AIRWeb 2005, May 2005, Chiba, Japan (2005)
5. Ntoulas, A., Najork, M., Manasse, M., Fetterly, D.: Detecting Spam Web Pages through Content Analysis. In: Proc. of the World Wide Web conference (May 2006)
6. Yahoo! Research: Web Collection UK-2006, `http://research.yahoo.com/` Crawled by the Laboratory of Web Algorithmics, University of Milan (retrieved Febrary 2007), `http://law.dsi.unimi.it/`
7. Gyöngyi, Z., Molina, H.G., Pedersen, J.: Combating Web Spam with TrustRank. In: Proc. of the Thirtieth International Conference on Very Large Data Bases, August 2004, Toronto, Canada (2004)
8. Benczúr, A., Csalogány, K., Sarlós, T.: Link-based Similarity Search to Fight Web Spam. In: Proc. of AIRWeb 2006 (2006)
9. Geng, G.G., Wang, C.H., Jin, X.B., Li, Q.D., Xu, L.: IACAS at Web Spam Challenge 2007 Track I, Web Spam Challenge (2007)
10. Wu, B.N., Davison, B.: Cloaking and Redirection: a Preliminary Study. In: Proc. of the 1st International Workshop on Adversarial Information Retrieval on the Web (May 2005)
11. Gyöngyi, Z., Garcia-Molina, H.: Web Spam Taxonomy. In: Proc. of First Workshop on Adversarial Information Retrieval on the Web (2005)
12. Weiss, G.M.: Mining with Rarity - Problems and Solutions: A Unifying Framework. In: SIGKDD Explorition (2004)
13. Preund, Y., Schapire, R.E.: A Decision-theoretic Generalization of on-line Learning and an Application to Boosting. Journal of Computer and System Sciences 55(1), 119–139 (1997)
14. Gyöngyi, Z., Molina, H.G.: Link Spam Alliances, Technical Report (September 2005)
15. Witten, I.H., Frank, E.: Data Mining: Pratical Machine Learning Tools and Techniques. 2nd edition. Morgan Kaufmann (2005)
16. Henzinger, M., Motwani, R., Silverstein, C.: Challenges in web search engines. SIGIR Forum (2002)

17. Gan, Q.Q., Suel, T.: Improving Web Spam Classifiers Using Link Structure. In: AIRWeb 2007, May 2007, Banff, Canada (2007)
18. Geng, G.G., Wang, C.H., Li, Q.D., Xu, L., Jin, X.B.: Boosting the Performace of Web Spam Detection with Ensemble Under-Sampling Classification. In: Proc. of the 4th International Conference on Fuzzy Systems and Knowledge Discovery, FSKD 2007 (August 2007)
19. Benczúr, A., Biró, I., Csalogány, K., Sarlós, T.: Web Spam Detection via Commercial Intent Analysis. In: Proc. of the 3rd International Workshop on Adversarial Information Retrieval on the Web, May 2007, Banff, Canada (2007)

Clustering Deep Web Databases Semantically

Ling Song[1,2], Jun Ma[1], Po Yan[1], Li Lian[1], and Dongmei Zhang[1,2]

[1] School of Computer Science &Technology, Shandong University, 250061, China
[2] School of Computer Science & Technology, Shandong Jianzhu University, 250101, China
song_ling@sdjzu.edu.cn

Abstract. Deep Web database clustering is a key operation in organizing Deep Web resources. Cosine similarity in Vector Space Model (VSM) is used as the similarity computation in traditional ways. However it cannot denote the semantic similarity between the contents of two databases. In this paper how to cluster Deep Web databases semantically is discussed. Firstly, a fuzzy semantic measure, which integrates ontology and fuzzy set theory to compute semantic similarity between the visible features of two Deep Web forms, is proposed, and then a hybrid Particle Swarm Optimization (PSO) algorithm is provided for Deep Web databases clustering. Finally the clustering results are evaluated according to Average Similarity of Document to the Cluster Centroid (ASDC) and Rand Index (RI). Experiments show that: 1) the hybrid PSO approach has the higher ASDC values than those based on PSO and K-Means approaches. It means the hybrid PSO approach has the higher intra cluster similarity and lowest inter cluster similarity; 2) the clustering results based on fuzzy semantic similarity have higher ASDC values and higher RI values than those based on cosine similarity. It reflects the conclusion that the fuzzy semantic similarity approach can explore latent semantics.

Keywords: Semantic Deep Web clustering, Fuzzy set, Ontology, PSO, K-Means.

1 Introduction

The Deep Web usually means the databases available through the HTML pages in the Internet. Unlike the surface web, the Deep Web refers to the collection of web data that is accessible by interacting with a web-based query form, and not through the traversal of hyperlinks. Data mining in Deep Web sources has aroused a lot of interesting in the recent researches. Approaches have been proposed for both clustering and classifying Deep Web sources [1-3].

As an important technique of data mining, document clustering is the operation of grouping similar documents together into clusters which are coherent internally but clearly different from each other. There have been a lot of researches focused on clustering algorithm, such as K-Means algorithm [4], Particle Swarm Optimization (PSO) and hybrid PSO approaches [5-9]. Compared with common data clustering, Deep Web clustering faces challenges due to a very wide variation in the way of web-site designers' model that is not possible to assume certain standard form field names and structures [10-11].

H. Li et al. (Eds.): AIRS 2008, LNCS 4993, pp. 365–376, 2008.

Given a large scale of searchable forms of Deep Web databases, in order to group together forms that correspond to similar databases, pre-query and post-query approaches can be used [12]. Post-query techniques issue query probing and the retrieved results are used for clustering purposes [3]. Pre-query techniques rely on visible features of forms such as attribute labels and page contents [13-14]. In these approaches, visible features are represented as vectors and cosine similarity is used as the basis for similarity comparison. In classical VSM, measure of cosine similarity between two vectors d_i and d_j is as follows:

$$sim(d_j, d_k) = \frac{\vec{d}_j \cdot \vec{d}_k}{\left|\vec{d}_j\right| \times \left|\vec{d}_k\right|} = \frac{\sum_{i=1}^{n} w_{ji} \times w_{ki}}{\sqrt{\sum_{i=1}^{n} w^2_{ji}} \times \sqrt{\sum_{i=1}^{n} w^2_{ki}}} \quad (1)$$

Where: w_{ji} and w_{ki} are weights of the ith term in d_j and d_k respectively. However, the representation of document as vector of bags of words suffers from well-known limitations: its inability to represent semantics. The main reason is that VSM is based on lexicographic term matching. Two terms can be semantically similar although they are lexicographically different. Therefore, such lexicographic term matching results is inability to exploit semantic similarity between two Deep Web features, which affects the quality of clustering at last.

Ontology is a specification of a conceptualization of a knowledge domain, which is a controlled vocabulary that describes concepts and the relations between them in a formal way, and has a grammar for using the concepts to express something meaningful within a specified domain of interest. Researchers have focused on how to integrate domain ontology as background knowledge into document clustering process and shown that ontology could improve document clustering performance [15-18]. In these measures, the basic idea is to re-weight terms and assign more weight to terms that are semantically similar with each other and cosine measure is used to measure document similarity. However, these term re-weighting approaches that ignore some of terms might cause serious information loss. Zhang's experiments showed that term re-weighting might not be an effective approach when most of the terms in a document are distinguish core terms [19].

Facing the above problems, in this paper we firstly present a fuzzy semantic measure, which integrates semantics of ontology and fuzzy set theory to compute similarity between visible features of Deep Web forms. Then we present a hybrid PSO algorithm for Deep Web clustering. Main contributions of this paper are:

1. Proposes a semi-automatic approach of building domain ontology.
2. Defines a similarity matrix of concepts based on domain ontology.
3. Defines a fuzzy set to represent a conceptual vector of the Deep Web form, which can explore semantics in domain ontology.
4. Necessity degree of matching between fuzzy sets is used in comparing Deep Web forms.
5. A hybrid PSO algorithm is proposed for Deep Web databases clustering.

The organization of this paper is as follows: After giving an approach of building domain ontology in Sec. 2, a fuzzy semantic similarity measure with respect to ontology is proposed in Sec. 3, which is necessary for the work of Sec. 4, where it is applied in Deep Web databases clustering with hybrid PSO. In Sec. 5, the efficacy of our approach is demonstrated through relevant experiments. Conclusions are given in Sec.6.

2 Core Ontology and Domain Ontology

A frame system for the core ontology is given in [20]:

Definition 1 (Core Ontology). An core ontology is a sign system *O:=(L;F;C; H;ROOT;)*, which consists of: A lexicon *L* consists of a set of terms ; A set of concepts *C*, for each $c \in C$, there exists at least one statement in the ontology; A reference function *F*, with $F{:}\ 2^L \rightarrow 2^C$. *F* links sets of lexical entries $\{L_i\} \subset L$ to the set of concepts they refer to. The inverse of *F* is F^{-1}; Concept hierarchy structure *H:* Concepts are taxonomically related by the directed, acyclic, transitive relation *H*, $(H \subset C \times C)$. $\forall c_1, c_2 \in C$, $H(c_1,c_2)$ means that c_1 is a hierarchy relation of c_2; A top concept $ROOT \in C$. For all $c \in C$ it holds *H(c, ROOT)*.

Based on the frame system of above core ontology, a semi-automatic approach to build domain ontology from a given set of query forms is proposed. Firstly attribute features of the Deep Web forms are parsed. The OntoBuilder project supports the extraction of attributes from web search forms and saves as XML format [21]. From these XML files concepts and instances of concepts are then extracted, with which we can build domain ontology. Fig. 1 and fig. 2 are extracted form attributes and its XML file of a search form, respectively.

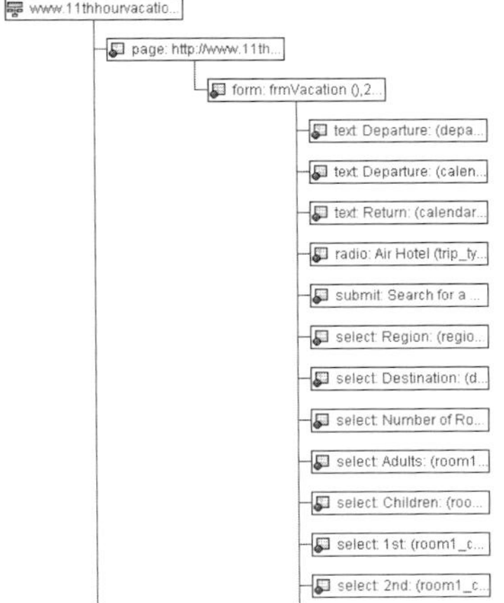

Fig. 1. A fragment of extracted attributes

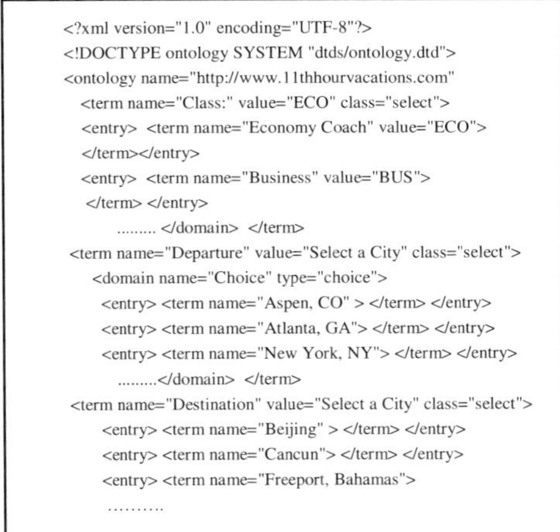

```
<?xml version="1.0" encoding="UTF-8"?>
<!DOCTYPE ontology SYSTEM "dtds/ontology.dtd">
<ontology name="http://www.11thhourvacations.com"
  <term name="Class:" value="ECO" class="select">
  <entry> <term name="Economy Coach" value="ECO">
  </term></entry>
  <entry> <term name="Business" value="BUS">
   </term> </entry>
       ......... </domain> </term>
 <term name="Departure" value="Select a City" class="select">
    <domain name="Choice" type="choice">
     <entry> <term name="Aspen, CO" > </term> </entry>
     <entry> <term name="Atlanta, GA"> </term> </entry>
     <entry> <term name="New York, NY"> </term> </entry>
       .........</domain> </term>
 <term name="Destination" value="Select a City" class="select">
     <entry> <term name="Beijing" > </term> </entry>
     <entry> <term name="Cancun"> </term> </entry>
     <entry> <term name="Freeport, Bahamas">
      ..........
```

Fig. 2. A fragment of XML file

Our building process has the following components: (1) *Concepts:* e.g. "departure_city", "destination_city"-are concepts of airfares domain. (2) *Instances of concept:* e.g. - ----- - -and - -- ~ --- are instances of the concept "departure_city". (3)

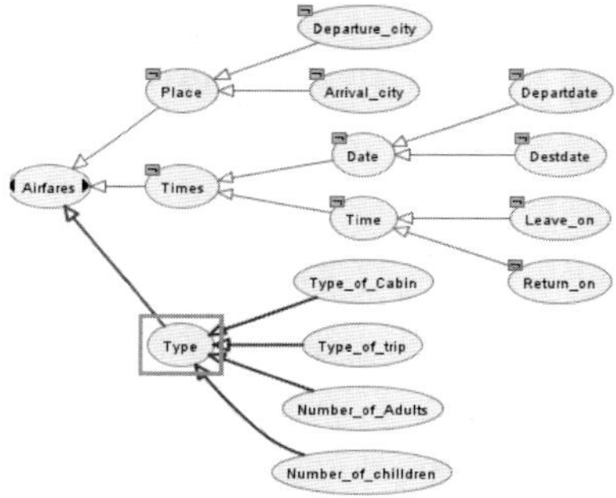

Fig. 3. A Graph of Airfares ontology

Mapping: e.g. the concept "departure_city"-may also be denoted by a set of terms such as "vacation from", "departure", "depart", "leave from", "Origin", "leaving from", "depart from", "departure from", etc, which should be mapped to concept " departure_city". (4) *Concepts relationships: A-Kind-Of relation.* (5) Slot and facet.

The main characteristics of a domain ontology based on query forms are: firstly, it orients to a domain, such as books, airfares, movies etc.; secondly, its semantic relation is very simple, only exists A-Kind-Of semantic relation; thirdly, the depth of the ontology isn't deep. A domain ontology *O* can be represented by a graph *G=(V, E)*, vertices and edges represent concepts and semantic relations respectively. All concepts are semantic related by way of the transitivity relationship over the domain ontology.

Protégé is a free, open source ontology editor and knowledge base framework [22]. We use protégé to edit domain ontology according to the above steps. Fig. 3 is a graph of airfares ontology.

3 Our Work: Semantic Similarity with Respect to Ontology

3.1 Similarity Matrix of Concepts

Conceptual similarity in ontology plays an important role in mining semantics among concepts. To represent the role of semantic relations, weights are assigned to edges. In order to make the implicit membership relations explicit, we represent the graph structure by means of adjacency matrices and apply a number of operations to them. Adjacency matrix *T*, which is symmetry, is used to express the immediate weight between two concepts. Assuming that α express immediate weight of semantic relation, adjacency semantic matrix *T* is defined as:

$$T=\begin{cases} 1 & \text{if } i=j \\ \alpha & \text{if } i\neq j \text{ and } (i, j)\in \text{A-Kind-Of} \\ 0 & \text{otherwise .} \end{cases} \quad (2)$$

In order to represent transitivity of semantic relations, we define matrix composition operation $\otimes$ on matrices as follows: $[A\otimes B]_{ij}=max(A_{ik}\cdot B_{kj})$, where *A* and *B* are two adjacency matrices respectively. Semantic relation is reflexivity, symmetry and transitivity, so the transitivity closure of *T*, denoted T^+ is defined as follows:

$$T^{(0)} = T, T^{(r+1)} = T^{(0)} \otimes T^{(r)}, then\ \ T^+ = \lim_{r \to \infty} T^{(r)}. \tag{3}$$

Note that the computation of the closure T^+ converges in limited number of steps that are bounded by maximum depth of the ontology.

3.2 Fuzzy Set to Represent a Concept

Unlike the case of a classic crisp set, an element may belong partially to a fuzzy set.

Definition 2. A fuzzy set *A* on a domain *X* is defined by a membership function μ_A from *X* to [0, 1] that denotes the degree to which *x* belongs to *A* with each element *x* of *X*.

In our work, a set of concepts in domain ontology is regarded as domain *X*. Fuzzy set, which is more generally used to represent a concept in ontology whose border is not strictly delimited, can be used to define the granularity of a concept by associating a degree with every candidate value in domain. Semantic similarity matrix in Sec. 3.1 can be generalized to a fuzzy measure if the weights for the relation links are replaced by membership degrees indicating the strength of the relationships between the parent and the child concepts. With similarity matrix T^+, we import the fuzzy set theory into a concept. Suppose Fig. 4 is a domain ontology graph.

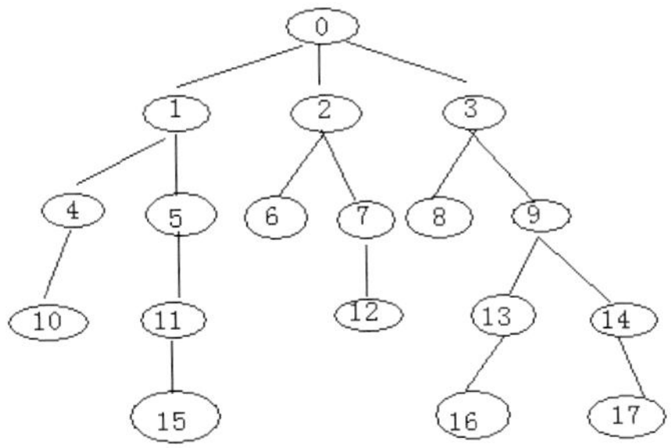

Fig. 4. An Example of Domain Ontology Graph

Example 1. In fig. 4, let α=0.8, for concept c_{14}, its fuzzy set can be described as:$\mu_{c14}=\{0.51/c_0, 0.41/c_1, 0.41/c_2, 0.64/c_3, 0.33/c_4, 0.33/c_5, 0.33/c_6, 0.33/c_7, 0.51/c_8, 0.8/c_9, 0.26/c_{10}, 0.26/c_{11}, 0.26/c_{12}, 0.64/c_{13}, 1/c_{14}, 0.16/c_{15}, 0.51/c_{16}, 0.8/c_{17}\}$

For concept 17, its fuzzy set can be described as:$\mu_{17}=\{0.41/c_0, 0.33/c_1, 0.33/c_2, 0.51/c_3, 0.26/c_4, 0.26/c_5, 0.26/c_6, 0.26/c_7, 0.41/c_8, 0.64/c_9, 0.21/c_{10}, 0.21/c_{11}, 0.21/c_{12}, 0.51/c_{13}, 0.8/c_{14}, 0.16/c_{15}, 0.41/c_{16}, 1/c_{17}\}$.

Where 0.51/c_0 means the degree that c_0 contributes to concept c_{14} is 0.51, *0.41/*c_1 means the degree that c_1 contributes to concept c_{14} is 0.41, 0.51 and 0.41all come from T^+.

3.3 Fuzzy Set to Represent a Vector of Concepts

Fuzzy set can also be used to define the granularity of a documents' conceptual vector by associating a degree with every candidate value in ontology. In order to explore the underlying semantics, a conceptual vector with respect to a document can be endowed with the structure of domain ontology by the fuzzy set operation. Here we use the standard min-max system proposed by Zadeh [23]:

$$\mu_{\neg A}(x)=1-\mu_A(x). \tag{4}$$

$$\mu_{A\cap B}(x)=\min(\mu_A(x), \mu_B(x)). \tag{5}$$

$$\mu_{A\cup B}(x)=\max(\mu_A(x), \mu_B(x)). \tag{6}$$

Example 2. With domain ontology in Fig.4, if the conceptual vector *CV* of document *A* is represented as *CV* =(0,0,0,0,0,0,0,0,0,0,0,0,0,0,1,0,0,1), it is described to be a fuzzy set with fuzzy set operations (equation (6)):
$\mu_A=\mu_{14\cup 17}=\{0.51/c_0, 0.41/c_1, 0.41/c_2, 0.64/c_3, 0.33/c_4, 0.33/c_5, .33/c_6, 0.33/c_7, 0.51/c_8, 0.8/c_9, 0.26/c_{10}, 0.26/c_{11}, 0.26/c_{12}, 0.64/c_{13}, 1/c_{14}, 0.16/c_{15}, 0.51/c_{16}, 1/c_{17}\}$
From above discussion, a conceptual vector with respect to a document can be described as a fuzzy set.

3.4 Fuzzy Similarity Measure between Documents

Necessity degree of matching is used in fuzzy pattern matching to evaluate fuzzy semantic similarity between two fuzzy sets [24-25]:

Definition 3. Let *A* and *B* be two fuzzy sets defined on a domain *X*:
The necessity degree of matching between *A* and *B*, denoted *N(A;B)*, is a "pessimistic" degree of inclusion that estimates the extent to which it is certain that *A* is compatible with *B*, and is defined by:

$$N(A;B)=\inf_{x\in X}\max(\mu_A(x), 1-\mu_B(x)). \tag{7}$$

Example 3. Assume the conceptual vector of document *A* is (0,0,0,0,0,0,0,0,0,0,0,0,0,0,1,0,0,1), and the conceptual vector of document *B* is (0,0,0,0,0,0,0,0,0,0,0,0,0,1,0,0,1,0), their fuzzy semantic similarity is computed as following: Firstly, according to the computing in Sec. 3.3, the fuzzy sets of document *A* and document *B* are described as below:

$\mu_A=\{0.51/c_0, 0.41/c_1, 0.41/c_2, 0.64/c_3, 0.33/c_4, 0.33/c_5, 0.33/c_6, 0.33/c_7, 0.51/c_8, 0.8/c_9, 0.26/c_{10}, 0.26/c_{11}, 0.26/c_{12}, 0.64/c_{13}, 1/c_{14}, 0.16/c_{15}, 0.51/c_{16}, 1/c_{17}\}$
$\mu_B=\{0.51/c_0, 0.41/c_1, 0.41/c_2, 0.64/c_3, 0.33/c_4, 0.33/c_5, 0.33/c_6, 0.33/c_7, 0.51/c_8, 0.8/c_9, 0.26/c_{10}, 0.26/c_{11}, 0.26/c_{12}, 1/c_{13}, 0.64/c_{14}, 0.21/c_{15}, 0.8/c_{16}, 0.51/c_{17}\}$

With equation (7), the necessity degree of matching is $N(A; B)$=0.49. But with cosine similarity measure, similarity between document A and B is 0, because there are no co-occur lexicographic terms in document A and B.

4 The Hybrid Particle Swarm Optimization (PSO) Clustering Algorithm

Clustering problem is to find the partition C^* that has optimal objective function with respect to all other feasible solutions. It has shown that the clustering problem is an NP-hard problem [26]. PSO is a population-based stochastic search algorithm, where each particle represents a potential solution to an optimization problem. In the PSO algorithm of document clustering, the whole document collection can be represented as a d-dimension problem space with n dots, where each concept feature represents one dimension and each dot represents a document. One particle in the swarm represents one possible solution for clustering the document collection, $X_i = (C_1, C_2, \ldots, C_i, .., C_k)$, where C_i represents the ith cluster centroid vector and k is the number of clusters. When a particle moves to a new location, a different clustering solution is generated. Therefore, a swarm represents a number of candidate clustering solutions for the document collection. PSO must also have a fitness evaluation function that takes the particle's position. $x_{id}(t)$ is the position of ith particle in a d-dimensional space at time step t, and $v_{id}(t)$ is the velocity of ith particle in a d-dimensional space at time step t. When a particle takes all the population as its topological neighbors, the best value is a global best and is called p_{gd}. Each particle also keeps track of its highest fitness value. The location of this value is called its personal best p_{id}. At each iteration, every particle adjusts its velocity vector, basing on its momentum and the influence of both its best solution and the global best solution of its neighbors, and then computes a new point to examine. The basic PSO document-clustering algorithm is:

(1) Initialize locations and velocities: here each particle randomly chooses k different document conceptual vectors from the document collection as the initial cluster centroid vectors;

(2) For each particle:
   - (a) Assign each conceptual vector in the document collection to the closest centroid vector. Evaluate the fitness value (equation 10);
   - (b) Compare particle's fitness evaluation with particle's best solution p_{id}. If current value is better than p_{id}, then set p_i value equal to the current value, and the p_{id} position equal to the current position in d-dimensional space;

(3) Compare fitness evaluation with the population's overall previous best. If current value is better than the p_{gd}, then reset p_{gd} to the current particle's value and position;

(4) Modify velocities and position with equation (8) and (9);

(5) Repeat Step (2)-Step (5) until the average change in centroid vectors is less than a predefined value or a predefined number of iterations are completed.

The particle updates its velocity and positions with following equation (8) and (9):

$$v_{id}(t+1)= \omega\cdot v_{id}(t)+c_1\cdot rand_1\cdot(p_{id}(t)-x_{id}(t))+c_2\cdot rand_2\cdot(p_{gd}(t)-x_{id}(t)). \quad (8)$$

$$x_{id}(t+1)=x_{id}(t)+ v_{id}(t+1) \quad . \quad (9)$$

Where ω denotes the inertia weight factor that controls the convergence of the particles; c_1 and c_2 are constants and known as acceleration coefficients; $rand_1$ and $rand_2$ are used to make sure that particles explore a wide search space before converging around the optimal solution.

In this paper we use similarity measure between documents to compute fitness value (document function). The fitness value, whose goal is to attaining high intra-cluster similarity and low inter-cluster similarity, is measured by the equation below:

$$f = \frac{\sum_{i=1}^{k} \left\{ \frac{\sum_{j=1}^{P_i} sim(C_i, m_{ij})}{|C_i|} \right\}}{k} \quad . \quad (10)$$

Where m_{ij} denotes the *j*th document conceptual vector, which belongs to cluster *i*; C_i is the centroid vector of the *i*th cluster; $sim(C_i, m_{ij})$ is the similarity between document m_{ij} and the cluster centroid C_i; $|C_i|$ stands for the number of documents in cluster C_i; and *k* stands for the number of clusters. The larger a fitness value is, the better quality of clustering is.

PSO can conduct a globalized searching for the optimal clustering, but requires more iteration numbers than K-Means algorithm. However the K-Means algorithm tends to converge faster than the PSO algorithm, but usually can be trapped in a local optimal area. The hybrid algorithm combines the ability of globalized searching of the PSO algorithm and the fast convergence of the K-means algorithm and avoids the drawbacks of both algorithms. In this paper our hybrid algorithm firstly executes the PSO process until the maximum number iterations are exceeded, and inherits clustering result from PSO as the initial centroid vectors of K-Means module, then starts K-Means process until maximum number of iterations is reached.

5 Experiments

TEL-8 Query Forms [27] are used as knowledge acquisition database to build 8 domain ontologies. These 8 domains are airfares, automobiles, books, car rentals, hotels, jobs, movies and music records. We also gather 431 form pages of above 8 categories from CompletePlanet [28] and invisible-web [29] as test set to cluster. Attributive terms are extracted from web search forms of test set with OntoBuilder[21]. Each Deep Web search form is represented by attributive term vector. According to the definition of core ontology, attributive terms should be mapped to concepts with the help of domain ontology. Table 1 is statistical data about number of attributive terms and concepts in 8 categories.

Table 1. Numbers of attributive terms and concepts in each category about TEL-8 query forms

	Airfares	Automo-biles	Books	Car Rentals	Hotels	Jobs	Movies	Music Records
Number of terms	68	169	104	72	158	153	122	130
Number of concepts	58	122	68	64	141	133	78	100

After mapping attributive terms to concepts, a conceptual vector can be represented by a fuzzy set. Necessity degree of matching N (equation (7)) is used to compute similarity between two fuzzy sets. Hybrid PSO clustering algorithm in Sec. 4 is used to perform Deep Web resources clustering.

Table 2. ASDC value in K-Means

Simulation	Cosine Similarity	Fuzzy Semantic Similarity
1	0.4527	0.5489
2	0.4483	0.5659
3	0.4238	0.5625
4	0.4301	0.5269
5	0.4687	0.5393
6	0.4708	0.4708
7	0.3741	0.5627
8	0.4450	0.5725
9	0.4468	0.5721
10	0.4836	0.5337
Average	0.44439	0.55561

Table 3. ASDC value in PSO

Simulation	Cosine Similarity	Fuzzy Semantic Similarity
1	0.4625	0.5844
2	0.4768	0.5859
3	0.4758	0.5969
4	0.4771	0.5822
5	0.4700	0.5893
6	0.4658	0.5975
7	0.4757	0.6085
8	0.4721	0.5856
9	0.5037	0.5730
10	0.4762	0.5893
Average	0.47299	0.58926

Table 4. ASDC value in hybrid PSO

Simulation	Cosine Similarity	Fuzzy Semantic Similarity
1	0.4633	0.5887
2	0.4799	0.5859
3	0.4780	0.6061
4	0.4778	0.5840
5	0.4711	0.5902
6	0.4697	0.5994
7	0.4757	0.6108
8	0.4727	0.5892
9	0.5045	0.5739
10	0.4935	0.5911
Average	0.47862	0.59193

In our experiment, our emphasis is to compare the clustering's performance of the fuzzy semantic similarity and cosine similarity with K-means, PSO and hybrid PSO clustering approaches. The fitness equation (10) is used not only in the PSO algorithm for fitness value calculation, but also in the evaluation of the cluster quality. It indicates the value of the average similarity of document to the cluster centroid (ASDC).

For an easy comparison, the K-means and PSO approaches run 50 iterations in each experiment. Parameter k in K-means is set to 8. Because the attribute features of forms are not a high dimensional problem space, particle number in PSO is set to 20. In the PSO algorithm, the inertia weight w is initially set to 0.7 and the acceleration coefficient constants c_1 and c_2 are set to 2. The inertia weight w is reduced by 2% at each generation to ensure good convergence. Table 2, Table 3 and Table 4 demonstrate the experimental results by using the K-Means, PSO, hybrid PSO respectively. Ten simulations are performed for each algorithm.

ASDC is internal criterion for the quality of high intra cluster similarity and low inter cluster similarity. We use The Rand Index (RI) measure as an external criterion that evaluates how well the clustering matches a set of gold standard classes [4]. We can view clustering as a series of decisions, one for each of the N(N−1)/2 pairs of documents in the collection. We want to assign two documents to the same cluster if and only if they are similar. A true positive (TP) decision assigns two similar documents to the same cluster, and a true negative (TN) decision assigns two dissimilar documents to different clusters. There are two types of errors we might commit. A false positive (FP) decision assigns two dissimilar documents to the same cluster. A false negative (FN) decision assigns two similar documents to different clusters. The Rand Index (RI) measures the percentage of RI decisions that are correct.

$$RI = \frac{\mathrm{TP} + \mathrm{TN}}{\mathrm{TP} + \mathrm{FP} + \mathrm{FN} + \mathrm{TN}} \times 100\% \cdot \tag{11}$$

Figure 5 is a comparing graph of RI value about hybrid PSO clustering with cosine similarity and fuzzy semantic similarity.

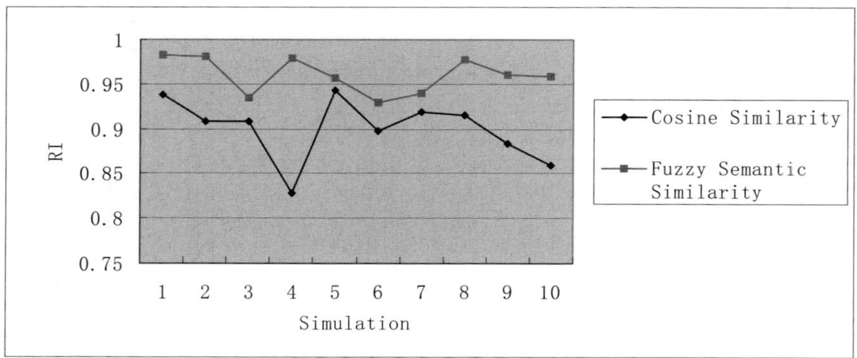

Fig. 5. RI of hybrid PSO clustering with cosine similarity and fuzzy semantic similarity

From the experiment, we can conclude:

(1) Dimension reduces in feature space with ontological mapping approach (See table 1).

(2) As shown in table 2~4, each algorithm generates the clustering results with a higher ASDC value for fuzzy semantic similarity than cosine similarity. These results show that the fuzzy semantic method that we propose in the paper performs better than traditional cosine similarity method.

(3) As shown in table 2~4, the hybrid PSO approach generates the clustering results with the highest ASDC value than PSO approach and K-Means approach.
(4) As shown in fig. 3, fuzzy semantic similarity performs higher RI value than cosine similarity in hybrid PSO clustering.

6 Conclusions

Facing the problem of how to organize large amount of Deep Web databases, we propose a semantic clustering approach of Deep Web databases. Firstly domain ontology is built semi-automatically. Then the database form is represented by fuzzy set with the semantic help of domain ontology. And necessity degree of matching is used to compute fuzzy semantic similarity between two forms. A hybrid PSO is provided for Deep Web databases clustering. Finally experiments were carried out to evaluate the clustering results according to ASDC and RI. The experimental results show the clustering results based on fuzzy semantic similarity have higher ASDC values and higher RI values than those based on cosine similarity. The reason is that cosine similarity between documents is 0 if there are no co-occur lexicographic terms, but its fuzzy similarity measure couldn't be 0 if there exist conceptual semantics. It reflects the conclusion that the fuzzy semantic similarity approach can explore latent semantics.

References

[1] Hedley, Y.-L., Younas, M., James, A.: The categorisation of hidden web databases through concept specificity and coverage. In: Advanced Information Networking and Applications, 2005. 19th International Conference on AINA 2005, March 28-30, 2005, vol. 2(2), pp. 671–676 (2005)

[2] Peng, Q., Meng, W., He, H., Yu, C.T.: WISE-cluster: clustering e-commerce search engines automatically. In: Proceedings of the 6th ACM International Workshop on Web Information and Data Management, Washington, pp. 104–111 (2004)

[3] He, B., Tao, T., Chang, K.C.-C.: Organizing structured web sources by query schemas: a clustering approach. In: CIKM, pp. 22–31 (2004)

[4] Manning, C.D., Raghavan, P., Schütze, H.: An Introduction to Information Retrieval. Cambridge University Press, Cambridge (2006)

[5] Cui, X., Potok, T.E., Palathingal, P.: Object Clustering using Particle Swarm Optimization. In: Proceedings of the 2005 IEEE Swarm Intelligence Symposium, Pasadena, California, USA, June 2005, pp. 185–191 (2005)

[6] Shan, S.M., Deng, G.S., He, Y.H.: Data Clustering using Hybridization of Clustering Based on Grid and Density with PSO. In: IEEE International Conference on Service Operations and Logistics, and Informatics, Shanghai, June 2006, pp. 868–872 (2006)

[7] Van der Merwe, D.W., Engelbrecht, A.P.: Data Clustering using Particle Swarm Optimization. In: The 2003 Congress on Evolutionary Computation, vol. 1, pp. 215–220 (2003)

[8] Srinoy, S., Kurutach, W.: Combination Artificial Ant Clustering and K-PSO Clustering Approach to Network Security Model. In: ICHIT 2006. International Conference on Hybrid Information Technology, Cheju Island, Korea, vol. 2, pp. 128–134 (2006)

[9] Chen, C.-Y., Ye, F.: Particle Swarm Optimization Algorithm and Its Application to Clustering Analysis. In: Proceedings of the 2004 IEEE international Conference on Networking, Sensing Control, Taipei, Taiwan, March 2004, vol. 2, pp. 789–794 (2004)

[10] http://www.11thhourvacations.com
[11] Halevy, A.Y.: Why your data don't mix. ACM Queue 3(8) (2005)
[12] Ru, Y., Horowitz, E.: Indexing the invisibleWeb: a survey. Online Information Review 29(3), 249–265 (2005)
[13] Caverlee, J., Liu, L., Buttler, D.: Probe, Cluster, and Discover:Focused Extraction of QA-Pagelets from the Deep Web
[14] Barbosa, L., Freire, J., Silva, A.: Organizing hidden-web databases by clustering visible web documents. In: Data Engineering, 2007. IEEE 23rd International Conference on ICDE 2007, April 15-20, 2007, pp. 326–335 (2007)
[15] Bloehdorn, S., Cimiano, P., Hotho, A.: Learning Ontologies to Improve Text Clustering and Classification. In: Data and Information Analysis to Knowledge Engineering, pp. 334–341. Springer, Heidelberg (2006)
[16] Castells, P., Fernańdez, M., Vallet, D.: An Adaptation of the Vector-Space Model for Ontology-Based Information Retrieval. IEEE Transactions on Knowledge and Data Engineering 19(2), 261–272 (2007)
[17] Shamsfard, M., Nematzadeh, A., Motiee, S.: ORank: An Ontology Based System for Ranking Objects. International Journal Of Computer Science 1(3), 1306–4428 (2006)
[18] Varelas, G., Voutsakis, E., Raftopoulou, P.: Semantic Similarity Methods in WordNet and their Application to Information Retrieval on the Web. In: Proceedings of the 7th annual ACM international workshop on Web information and data management, Bremen, Germany, pp. 10–16 (2005)
[19] Zhang, X., Jing, L., Hu, X., Ng, M., Zhou, X.: A Comparative Study of Ontology Based Term Similarity Measures on PubMed Object Clustering, http://www.pages.drexel.edu/~xz38/pdf/209_Zhang_DASFAA07.pdf
[20] Chaudhri, V.K., Farquhar, A., Fikes, R., Karp, P.D., Rice, J.P.: OKBC: A Progammatic Foundation for Knowledge Base Interoperability. In: Proceedings of the fifteenth national/tenth conference on Artificial intelligence/Innovative applications of artificial intelligence, Madison, Wisconsin, United States, pp. 600–607 (1998)
[21] http://iew3.technion.ac.il/OntoBuilder
[22] http://protege.stanford.edu
[23] Zadeh, L.A.: Similarity Relations and Fuzzy Orderings. Information Science 3, 177–200 (1971)
[24] Thomopoulos, R., Buche, P., Haemmerle, O.: Fuzzy Sets Defined on a Hierarchical Domain. IEEE Transaction on knowledge and engineering 18(10), 1397–1410 (2006)
[25] Zadeh, L.A.: Fuzzy sets as a basis for a theory of possibility. Fuzzy Sets and Systems 100(supp. 1), 9–34 (1978)
[26] Brucker, P.: On the complexity of clustering problems. In: Beckmenn, M., Kunzi, H.P. (eds.) Optimization and Operations Research. Lecture Notes in Economics and Malhemorical Sysrem, vol. lS7, pp. 45–54. Springer, Berlin (1978)
[27] http://metaquerier.cs.uiuc.edu/repository/datasets/tel-8/index.html
[28] http://aip.completeplanet.com
[29] http://www.invisible-web.net

PostingRank: Bringing Order to Web Forum Postings*

Zhi Chen[1], Li Zhang[2], and Weihua Wang[3]

[1] School of Software, Tsinghua University, Beijing 100084, China
chenzhi05@mails.tsinghua.edu.cn
[2] School of Software, Tsinghua University, Beijing 100084, China
lizhang@mail.tsinghua.edu.cn
[3] School of Software, Tsinghua University, Beijing 100084, China
wwh05@mails.tsinghua.edu.cn

Abstract. Web forum is an important information resource. Each day innumerable postings on various topics are created in thousands of web forums in internet. However, only a small part of them are utilized for the reason that it is difficult to rank the postings importance. Unlike general web sites with hyperlinks in web pages created by editors, links in web forums are automatically generated, therefore, traditional link-based methods, such as famous PageRank are useless to rank postings. In this paper, we propose a novel algorithm named PostingRank to rank postings. The main idea of our method is to exploit the common repliers between postings and leverage the relationship between these common repliers. We build implicit links based on that co-repliers relation and construct a link graph. In the way of iterative calculation, each posting's importance score can be obtained. The experimental results demonstrate that our method can improve retrieval performance and outperforms traditional link-based methods.

1 Introduction

Web forum is a special kind of online portal for the internet users to discuss on specific topics. Nowadays, web forum is quite popular. Everyday thousands of hundreds of postings are created by millions of users in web forums. These postings cover very broad topics and they are also great knowledge sources for people. For example, someone who is planning to buy a car will deliver a posting to ask for recommendation in an auto forum. Other users of this forum, who are familiar with this area, probably give a reply. This kind of information from web forums is more comprehensive and objective than that from company website. Commercial search engines, such as Google [1]and Yahoo! [2], have indexed some part of webpages in web forums. We find that if the submitted queries have the keywords "how to", the search results returned always contains many webpages

* This work is sponsored by National Basic Research Program of China Project NO.2002CB312006, National Natural Science Funds of China Project NO.60673008 and National High Technology Research and Development Program of China Project NO.2007AA04Z135.

H. Li et al. (Eds.): AIRS 2008, LNCS 4993, pp. 377–384, 2008.

from web forums. There is no doubt that the huge volume of information hidden in the forum website should not be neglected, and how to effectively utilize these information has become a valuable issue.

Link analysis is one of the most common methods to rank webpages. The famous PageRank, a hyperlink based link analysis algorithm, has been proved very effective by Google and other commercial search engine for ranking webpages. In link analysis algorithm, there are two basic assumptions [3]:(1) A link from page A to page B is a recommendation of page B by the author of page A.(2) If page A and page B are connected by a link, the probability that they belong to the same topic is higher than not connected. However, these assumptions are not so reliable. The navigational and ad links don't mean any recommendation. The webpages connected by those kinds of links are often irrelevant [4]. Therefore, they are noisy hyperlinks and should not be used for link analysis to rank. In web forums, the situation is worse. There are few links created by users or editors. Most postings are generated automatically by forum system. The purpose of this kind of links is to provide the function of navigation or other operations like creating posting or replying posting. Therefore, the link analysis algorithm based on these noisy links is not reasonable.

In this paper, a novel posting rank algorithm named PostingRank is proposed. The repliers, who reply postings, are taken into consideration in posting importance calculation. Because most hyperlinks in the web forum are noisy, those hyperlinks can't be used for postings importance score calculation. We utilize the relationship of postings' repliers to build an implicit link graph. The behavior of users' reply shows their interests or concerns about postings, which can be regarded as recommendation. One posting may be replied by many different users and the users may reply many postings. The Postings replied by the same repliers have correlation. The more common repliers, the more related the two posting are. That the postings have common users means they are recommended by same users, which can be considered as mutual recommendation through those co-repliers. We exploit this relation between the repliers, build implicit links between postings which have common repliers, and the weight of link is proportioned to the ratio of the common repliers among all repliers. After that, a link graph can be obtained. Then we can use iterative algorithm like traditional hyperlink based link analysis algorithm to rank postings.

The rest of this paper is organized as follows. In Section 2, we review the recent works on link analysis and forum search. In Section 3, we present our PostingRank algorithm, including implicit link building and link graph construction. Our experimental results are given in Section 4. Finally, we summarize our main contributions in Section 5.

2 Previous Work

2.1 Previous Work on Link Analysis

The link analysis technology has been widely used to analyze the web pages importance, such as HITS [5] and PageRank [6, 7].

In HITS, a two level weight propagation is proposed. Every page is thought to have two important properties, called hub and authority. The good hub pages have links pointing to many good authority pages, and good authority pages have links from many good hub pages. Thus the score of a hub (authority) is the sum of the score of connected authorities (hubs):

$$A(p) = \sum_{q:q \rightarrow p} H(q) \tag{1}$$

$$H(p) = \sum_{q:p \rightarrow q} A(q) \tag{2}$$

PageRank models the users' browsing behaviors as a random surfing model, which assumes that a user either follows a link from current page or jumps to a random page in the graph. The PageRank of a page p_i is computed by the following equation:

$$PR(p_i) = \frac{\alpha}{n} + (1 - \alpha) \times \sum_{j:j \rightarrow i} \frac{PR(p_j)}{outdegree(p_j)} \tag{3}$$

where α is a damping factor, which is always set as 0.15; n is the number of nodes in whole web graph; and $outdegree(p_j)$ is the number of hyperlinks in page p_j. Both of above algorithm could be computed in an iterative way.

2.2 Previous Work on Forum Search

Xu [8] noticed the problem that the traditional link-based ranking algorithm is ineffective for forum search and proposed a ranking algorithm named *FGRank*, which introduces the content similarities into the link graph to build a topic hierarchy, and calculates the postings score based on the topic hierarchy. To generate the topic hierarchy, a method named hierarchical co-clustering is used. However, the result of clustering may be not satisfying on large scale. Moreover, the quality of postings content is not reliable. Many postings only have a few words and it's difficult to cluster these kind of postings simply based on their content. Our work is different in introducing the reply information into algorithm framework.

3 Algorithm

Formally, in our framework, a set of web forum users u_i, $i = 1, ..., n$ and a set of postings p_j, $j = 1, ..., m$ are dealed with. The goal of PostingRank is to compute the postings importance. In this section, we present how to form a data model and how to rank postings based on that model.

3.1 Data Model: Reply Graph

To use PostingRank, the implicit link graph should be constructed first. The link between posting p_i and posting p_j is built if p_i and p_j have common users and the weight can be calculated as follows:

$$w_{ij} = \begin{cases} |U_i \cap U_j| & i \neq j \\ 0 & i = j \end{cases} \tag{4}$$

where U_i denotes the set of repliers of p_i and $|\cdot|$ denotes the cardinality of a set. A link graph can be obtained after building all links between posting pairs, which we name *Reply Graph.* Now we compute the $(|P| \times |P|)$ matrix R (P denotes the set of postings). R contains the weight of link(p_i, p_j),

$$R_{ij} = w_{ij} \tag{5}$$

. Noticing that w_{ij} equals to w_{ji}, and $\forall i, R_{ii} = 0$, so R is a symmetrical matrix. Then We normalize the matrix R as follows

$$\tilde{R} = D^{-\frac{1}{2}} R D^{-\frac{1}{2}} \tag{6}$$

where $\tilde{R}$ is the reply matrix reflecting the relationship of co-reply between postings and D is the diagonal matrix with D_{ii} equal to the sum of the ith row of R.

3.2 PostingRank Algorithm

The idea underlying PostingRank is spreading the repliers recommendation through the reply graph. The spreading scores should decay as propagation, because after many times spreading, the recommendation effect is decreased. So the algorithm we apply should have two key properties: propagation and attenuation. These properties reflect two key assumptions. First, the good postings are always related to other good postings. Through the *Reply Graph* we can find the relation between the postings. If p_i is a good postings, the p_j related to p_i is probably a good postings. Second, the postings replied by many user should rank highly. The postings with more repliers are more likely to share larger number of repliers with other postings. SALSA [9] has both propagation and attenuation properties we need. Considering the generic graph $G = (V, E)$, the authorities of all nodes $n \in V$ can be calculated by iterating following equation:

$$\boldsymbol{a}(n+1) = \boldsymbol{a}(n) W_c^T W_r \tag{7}$$

where $\boldsymbol{a}$ is the authority vector and W_r denote the matrix derived from affinity matrix W for graph G by normalizing the entries such that, for each row, the sum of the entries is 1, and let W_c denote the matrix derived from affinity matrix W for graph G by normalizing the entries such that, for each column, the sum of the entries is 1. Then the stationary distribution of the SALSA algorithm is the principal left eigenvector of matrix $W_c^T W_r$ [9]. We replace $W_c^T W_r$ with normalized matrix $\tilde{R}$ to calculate the $\boldsymbol{a}$. Finally, we combine the postings importance score obtained in the iterate procedure with the forum sites importance score given by Google. p_i's postingRank score $a_i' = a_i \cdot \beta_i$, where β_i is the importance coefficient of the forum site where p_i is from.

We define PostingRank procedure as follows:

1. Calculate the link weight w_{ij} for each pair of (p_i, p_j)

2. Form a matrix R, the entry R_{ij} equals to w_{ij}. Symmetrically normalize R by $\tilde{R} = D^{-\frac{1}{2}} R D^{-\frac{1}{2}}$
3. Initialize all elements in $\boldsymbol{a}(0)$ to 1.
4. Iterate the following equation until $\boldsymbol{a}$ converge:
   (a) $\boldsymbol{a}(n+1) = \boldsymbol{a}(n)\tilde{R}$

   (b) Normalize $\boldsymbol{a}(n+1)$
5. Combine $\boldsymbol{a}$ with forum sites importance to generate $\boldsymbol{a'}$

3.3 Convergence Issue

We now prove that the iterate procedure converges as n increases arbitrarily. It is easily derived that the iterate procedure can be written as $\boldsymbol{a}(n) = \boldsymbol{a}(0)\tilde{R}^n$. A result from linear algebra [10] states that if M is a symmetrical $n \times n$ matrix, and $\boldsymbol{v}$ is a vector not orthogonal to the principal eigenvector $w_i(M)$, then the vector $\boldsymbol{v}$ in the direction of $M^k\boldsymbol{v}$ converges to $w_i(M)$ as k increases without bound. If M has only nonnegative entries, the the principal eigenvector of M has only nonnegative entries. The matrix $\tilde{R}$ is a symmetrical matrix, so $\boldsymbol{a}$ will converges to the principal eigenvector of $\tilde{R}$ [5].

3.4 Improved PostingRank Algorithm

According to our observation, we find that it is likely that the replies at the end of reply list are not relevant to the first posting in the group when there are too many replies. For instance, some user, who wonder which is better between Chevy and Ford, would post his question in an auto related board in web forum. Other users reply this posting and discuss on this topic. After some replies, someone proposes that GM is better. Then the discussion shifts to the topic of which is the best among GM, Chevy and Ford. The later replier may not reply to the topic starter, but to other former repliers. We name this phenomenon "topic drift". This kind of replies may not mean that the repliers are interested in original topic. As the number of replies increases, the probability of the topics drift become larger. We should punish this kind of replies. To model the phenomenon above, we improve our method to calculate the link weight. To handle the irrelevant replies problem, we set a threshold t. We also take the replies order into consideration. We prune the common repliers if they reply the posting after t replies. We introduce a decaying factor to decrease the influence of topic drift. The formula to calculate the weight w_{ij} can be re-written as follows:

$$w_{ij} = \sum_{k=2}^{t} \frac{c_i(k)}{\log_b k} + \sum_{k=2}^{t} \frac{c_j(k)}{\log_b k} \tag{8}$$

$c_i(k) = 1$ when the kth replier of p_i exists in $U_i \cap U_j$, otherwise, $c_i(k) = 0$. k start from 2, because $k = 1$ is the topic starter. $\log_b k$ is the decaying factor, and t is for tuning. Based on the new weight, a new *Reply Graph* can be obtained. Other procedure is the same as original PostingRank.

4 Experiments

To evaluate our proposed algorithm, we compare the retrieval performance of well-known ranking algorithms versus the proposed algorithms.

4.1 Experiments Setting

The data for experiment are crawled from web forums, which covered 8 topics,including auto,computer,and etc. More than 2 million webpages are crawled. In this data set, postings are divided into groups according to their reply relation, which means that the postings regarding the same detailed topic, like "which is better CHEVY or FORD", will be grouped together. After preprocess, we get about 538,691 groups. Among all groups, 192,727 groups have not any replies and 5,288 groups have replies but not common replies with other groups. In this data set, there are 319,991 unique usernames and no anonymous user among them, so each username will be considered as a unique user. On average, each posting has about 9.29 repliers.The largest number of repliers in a group is 21,131.

To evaluate the algorithms, we need to collect the queries. There are two query sources. Some queries are collected by refining the frequently asked questions from the forum sites into query terms as [8]. Other queries are long queries about the very 8 topics manually selected from a commercial search engine query log. Finally, we obtain 25 representative queries (e.g. "FORD"). we manually judge the relevance of returned search results. Each returned posting is labeled as relevant or irrelevant result.

We compare our proposed algorithm with BM2500 score [11] and PageRank [6, 7]. Here we don't use *FGRank* [8] as comparison candidate since the features used for link analysis are quite different. *FGRank* introduces content to link building but we don't. The replier information are utilized in our algorithm. We use the content of postings to calculate BM2500 score. We linearly combine BM2500 score with the link-based importance score ,namely PageRank and PostingRank, as follows:

$$s_i = \lambda \cdot f_{importance}(p_i) + (1 - \lambda) \cdot f_{relevance}(p_i) \tag{9}$$

where s_i is the overall score of p_i and λ is the combination parameter

To evaluate the algorithms, the parameters are tuned to achieve the best performance. For PageRank, the damping factor is set as 0.15. For improved PostingRank, t is set as 50 and b is set as 2 according to tuning.

4.2 Evaluation and Results

Here we use P@10 and Means Average Precision (MAP) as the evaluation metrics for they are widely used in TREC and the details can be found in [12].

After about 40 times iteration, the algorithm converges. The PostingRank scores of the postings without replies are zero. It is reasonable that if nobody is interested in that posting or nobody can answer the question in the posting, the posting is useless.

Fig. 1 shows the performance comparison between three algorithms. Improved PostingRank get the best performance, outperforms PageRank and BM2500 by 5.4% and 11.2% on P@10 and by 4.1% and 8.6% on MAP,respectively. It also outperforms original PostingRank by 2.4% on P@10 and 1.9% on MAP. As the postings in web forums are topics-oriented, the performance is higher than the general web search task. To understand whether these improvements are statistically significant, we performed t-tests. It shows that our algorithm significantly improve the search result based on MAP measure. But our improvements on P@10 failed to pass t-test.

Fig. 2 shows the performance of improved PostingRank changes with parameter t. At first, as t increases, the performance is improving. The reason is that more useful replies are included. But when t is larger than 50, the performance is deteriorated, since too many noisy replies are included.

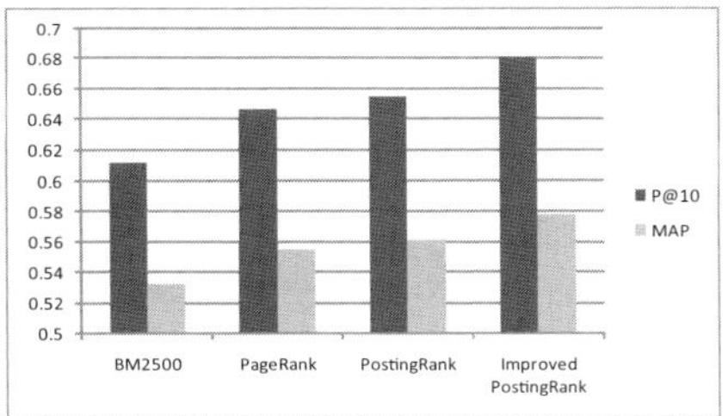

Fig. 1. The performance comparison between BM2500, PageRank, PostingRank and improved PostingRank

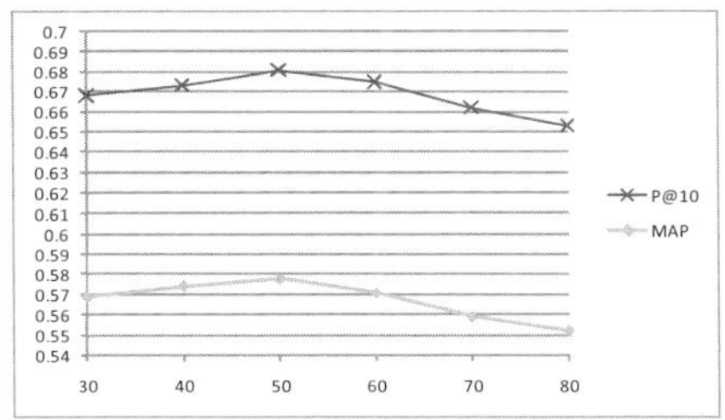

Fig. 2. The performance of improved PostingRank changes with the parameter t

5 Conclusion

Because of the noisy link environment in web forums, the traditional linked-base algorithm like PageRank is not suitable for ranking postings. In this paper, we introduce the replier information to rank postings. We exploit the relation of co-repliers between each pair of postings, build implicit link between postings and generate a reply graph for our PostingRank. As the experiments demonstrated, PostingRank and improved PostingRank algorithm proposed in this paper, outperform other ranking methods.

In this work, we conduct the experiments on the dataset we crawled. In future, we hope to make a standard dataset for forum search. We will try to extract more features from web forum to improve the performance of ranking method for postings.

References

1. Google search engine, `http://www.google.com`
2. Yahoo!search engine, `http://search.yahoo.com`

3. Henzinger, M.: Link Analysis in Web Information Retrieval. IEEE Data Engineering Bulletin (2000)
4. Bergmark, D., Lagoze, C., Sbityakov, A.: Focused crawls, tunneling, and digital libraries. In: Proceeding of the 6th European Conference on Digital Libraries, pp. 91–106 (2002)
5. Kleinberg, J.: Authoritative sources in a hyperlinked environment. In: Proceedings of the Ninth Annual ACM-SIAM Symposium on Discrete Algorithms (1998)
6. Page, L., Brin, S., Motwani, R., Winograd, T.: The PageRank Citation Ranking: Bringing Order to the Web. Technical report, Stanford University (January 1998)
7. Brin, S., Page, L.: The Anatomy of a Large-Scale Hypertextual Web Search Engine. In: Proceedings of the 7th International World Wide Web Conference, pp. 107–117 (1998)
8. Xu, G., Ma, W.Y.: Building Implicit Links from Content for Forum Search. In: Proceedings of the 29th International Conference on Research and Development in Information Retrieval (2006)
9. Lempel, R., Moran, S.: The stochastic approach for link-structure analysis (SALSA) and the TKC effect. In: Proceedings of the 9th International World Wide Web Conference (2000)
10. Golub, G., Vanloan, C.G.: Matrix Computations. Johns Hopkins University Press
11. Robertsom, S.E.: Overview of the OKAPI projects. Journal of Documentation 53, 3–7 (1997)
12. Baeza-Yates, R., Ribeiro-Neto, B.: Modern Information Retrieval. Addisom-Wesley Longman Publishing Co., Inc., Boston (1999)

A Novel Reliable Negative Method Based on Clustering for Learning from Positive and Unlabeled Examples

Bangzuo Zhang[1,2] and Wanli Zuo[1]

[1] College of Computer Science and Technology, Jilin University, ChangChun, 130012, China
[2] College of Computer, Northeast Normal University, ChangChun, 130024, China
zhangbz@nenu.edu.cn, wanli@mail.jlu.edu.cn

Abstract. This paper investigates a new approach for training text classifiers when only a small set of positive examples is available together with a large set of unlabeled examples. The key feature of this problem is that there are no negative examples for learning. Recently, a few techniques have been reported are based on building a classifier in two steps. In this paper, we introduce a novel method for the first step, which cluster the unlabeled and positive examples to identify the reliable negative document, and then run SVM iteratively. We perform a comprehensive evaluation with other two methods, and show experimentally that it is efficient and effective.

Keywords: Semi-Supervised Learning, Text Classification, Bisecting k-means Clustering, Learning from Positive and Unlabeled Examples (LPU).

1 Introduction

With the ever-increasing volume of text data from various online sources, it is an important task to categorize or classify these text documents into categories that are manageable and easy to understand. Text categorization or classification aims to automatically assign text documents to pre-defined classes. In supervised learning, text classifier relies on labeled training examples. For binary problems, positive and negative examples are mandatory for machine learning. The main bottleneck of building such a classifier is that a large, often prohibitive, number of labeled training documents are needed. But, for many learning task, labeled examples are rare while numerous unlabeled examples are easily available.

Recently, semi-supervised learning algorithms from a small set of labeled data with the help of unlabeled data have been defined. These techniques alleviate some labor-intensive effort. Semi-supervised learning includes two main paradigms: (1) learning from a small set of labeled examples and a large set of unlabeled examples; and (2) learning from positive examples and unlabeled examples (with no labeled negative examples). Many researchers have studied learning in the first paradigm [1]. In learning from positive and unlabeled examples, some theoretical studies and practical algorithms have been reported in [2-9].

In this paper, we study learning from positive data with the help of unlabeled data, which is also common in practice. For instance, in many text mining tasks, such as document retrieval and classification, one goal is the efficient classification and re-

H. Li et al. (Eds.): AIRS 2008, LNCS 4993, pp. 385–392, 2008.

trieval of interests of some users. Positive information is readily available and unlabeled data can easily be collected. One example is learning to classify web page as "interesting" for a specific user. Documents pointed by the user's bookmarks defined a set of positive examples because they correspond to interesting web pages for him and negative examples are not available at all. Nonetheless, unlabeled examples are easily available on the World Wide Web.

Theoretical results show that in order to learn from positive and unlabeled data, it is sometimes sufficient to consider unlabeled data as negative ones [2-3]. Recently, a few algorithms were proposed to solve the problem. One class of algorithms is based on a two-step strategy as follow. These algorithms include Roc-SVM [7], S-EM [8], PEBL (Positive Examples Based Learning) [9].

Step 1: Identifying a set of reliable negative documents from the unlabeled set. In this step, S-EM uses a Spy technique, PEBL uses a technique called 1-DNF, and Roc-SVM uses the Rocchio algorithm.

Step 2: Building a set of classifiers by iteratively applying a classification algorithm and then selecting a good classifier from the set. In this step, S-EM uses the Expectation Maximization (EM) algorithm with a NB (Naive Bayesian) classifier, while PEBL and Roc-SVM use SVM (Support Vector Machine). Both S-EM and Roc-SVM have some methods for selecting the final classifier. PEBL simply uses the last classifier at convergence.

The underlying idea of these two-step strategies is to iteratively increase the number of unlabeled examples that are classified as negative while maintaining the positive examples correctly classified. This idea has been justified to be effective for this problem [8].

In this paper, we first introduce another method for the first step, i.e. cluster the positive and unlabeled examples to identify the reliable negative document, and evaluate our method with other two methods (PEBL, and Roc-SVM).

The remainder of this paper is organized as follow: We would like to first review the existing reliable negative methods to this problem in section 2; propose a novel clustering based approach in section 3; and comparative experiments have been made in section 4; finally make conclusion in section 5.

2 Related Works

In this section, we introduce algorithms for the first step that based on the two-step strategy. The techniques of the Roc-SVM, the S-EM and the PEBL have been reported in [7], [8], [9] respectively.

In this paper, we use *P* to denote the positive examples set, *U* for unlabeled examples set, and *RN* for reliable negative examples set that produced from the unlabeled examples set *U*.

Li, X.L. et al. report the Spy technique in the S-EM [7]. It first randomly selects a set *S* of positive documents from *P* and put them in *U*. Documents in *S* act as "spy" documents. The spies behave similarly to the unknown positive documents in *U*. Hence they allow the algorithm to infer the behavior of the unknown positive documents in *U*. In step 2, it then run EM to build the final classifier. Since NB is not a strong classifier for text classification, so we do not compare with it. This algorithm performs stably when the positive examples set is very small. When the positive examples set is large, it is worse than others.

The Roc-SVM algorithm uses the Rocchio method to identify a set *RN* of reliable negative documents from *U*. Rocchio is an early text classification method. In this method, each document is represented as a vector, Let *D* be the whole set of training documents, and C_j be the set of training documents in class *j*. Building a Rocchio classifier is achieved by constructing a prototype vector $\vec{C}_j$ for each class *j*. In classification, for each test document *td*, it uses the cosine similarity measure to compute the similarity of *td* with each prototype vector. The class whose prototype vector is more similar to *td* is assigned to *td*.

$$\vec{Cj} = \alpha \frac{1}{|Cj|} \sum_{\vec{d} \in Cj} \frac{\vec{d}}{\|\vec{d}\|} - \beta \frac{1}{|D - Cj|} \sum_{\vec{d} \in D - Cj} \frac{\vec{d}}{\|\vec{d}\|}. \quad (1)$$

When use this method, the amount of *RN* is so big that biased the classifier of step 2 and poor performance, especially when the *P* set is small.

The PEBL uses the 1-DNF method, first builds a positive feature set *PF* which contains words that occur in the positive examples set *P* more frequently than in the unlabeled examples set *U*. Then it tries to filter out possible positive documents from *U*. A document in *U* that does not have any positive feature in *PF* is regarded as a strong negative document. In this algorithm, the amount of *RN* set is always small and sometimes is short text examples. Its performance is poor when the number of positive examples set is small. When the positive examples set is large, it becomes more stable.

3 The Proposed Technique

In this section, we introduce a new method for the first step that use clustering to identify a set *RN* of reliable negative documents from the unlabeled examples set *U* and positive examples set *P*.

For information retrieval and text mining, a general definition of clustering is the following: given a large set of documents, automatically discover diverse subsets of documents that share a similar topic. Clustering provides unique ways of digesting large amounts of information. Clustering algorithms divide data into meaningful or useful groups, called clusters, such that the intra-cluster similarity is maximized and the inter-cluster similarity is minimized. These discovered clusters could be used to explain the characteristics of the underlying data distribution and thus serve as the foundation for various data mining and analysis techniques.

The standard clustering algorithms can be categorized into partitioning algorithms such as k-means and hierarchical algorithms such as Single-Link or Average-Link. Many variants of the k-means algorithm have been proposed for the purpose of text clustering. A recent study has compared partitioning and hierarchical methods of text clustering on a broad variety of test datasets. It concludes that k-means clearly outperforms the hierarchical methods with respect to clustering quality. A variant of k-means called bisecting k-means [10] is introduced, which yields even better performance. Bisecting k-means uses k-means to partition the dataset into two clusters. Then it keeps partitioning the currently largest cluster into two clusters, again using k-means, until a total number of k clusters has been discovered.

We propose a novel method for the first step as shown in fig. 1. First, set *RN* to null, and then run bisecting k-means clustering algorithm with the union of positive examples set and unlabeled examples set with parameter *k*. Last, if proportion of positive examples in each cluster is lower than the threshold that given, and then add this cluster to *RN*.

```
Algorithm: Exploiting Reliable Negative by Clustering
Input: P positive examples set
       U unlabeled examples set
       K number of cluster
       T threshold
Output: RN (reliable negatives set)
Steps: 1.RN ={};
       2.Clustering set E = P •U;
       3.run bisecting k-means with parameter k on E,
         and divide into E1, E2, … Ek, in each Ei(i = 1,
         2, …, k), the positive examples in it is Pi;
       4.for each Ei (i = 1, 2, … , k )
             if |Pi|/|Ei|<T then RN = RN •Ei.
```

Fig. 1. The algorithm of exploiting reliable negative by clustering

We use the CLUTO toolkit package [12] for clustering, which use bisecting k-means algorithm. The parameter *T* generally is small, usually set to zero, i.e. the cluster that has no positive examples can be used as reliable negative examples set. Yang, Y. suggests that the numbers of text clustering impacts the resulting difference in F_1 scores are almost negligible [11]. From our experiments in section 4, we also observed that the choice of *k* does not affect classification results much as long as it is not too small. So we set *k* as 20.

```
Algorithm: Iterative SVM
Input: P  positive examples set
       RN reliable negative examples set by step 1
       Q  the remaining unlabeled examples set, U -RN;
Output: The final classifier S;
Steps:
    1. Assigned the label 1 to each document in P;
    2. Assigned the label -1 to each document in RN;
    3. While(true)
    4.     Training a new SVM classifier Si with P and
RN;
    5.     Classify Q using S;
    6.     Let the set of documents in Q that are
           classified as negative be W;
    7.     If W ={} then break;
    8.     Else Q = Q - W; RN = RN •W;
    9.     End if
   10. End while
```

Fig. 2. The algorithm of iterative SVM

For step 2, we run SVM iteratively as shown in fig. 2. This method is similar to the step 2 of PEBL technique and Roc-SVM technique except that we do not use an additive classifier selection step. The basic idea is to use each iteration of SVM to exact more possible negative examples from *Q* (*U* – *RN*) and put them in *RN*. The iteration converges when no document in *Q* is classified as negative. Our technique does not select a good classifier from a set of classifiers built by SVM, and use the last SVM classifier at convergence. For Roc-SVM, the reason for selecting a classifier is that there is a danger in running SVM repetitively, since SVM is sensitive to noise. However, it is hard to catch the best classifier [6].

4 Experiments and Results

We now evaluate our proposed method with the Roc-SVM technique [7] and the PEBL technique [9]. We do not compare with the S-EM technique [8], because it uses the Naïve Bayesian method, which is a weaker classifier than the SVM, and our proposed technique is much more accurate than S-EM. Liu, B. et al. [6] have surveyed and compared these three methods, and our experiments on the dataset are with the same setting as [6] in order to allow comparison on the square.

4.1 Experiments Setup and Data Preprocess

We use Reuters-21578, the popular text collection in text classification experiment, which has 21578 documents collected from the Reuters newswire. Among 135 categories, only the most populous 10 are used. 9980 documents are selected to use in our experiment, as shown in Table 1. Each category is employed as the positive examples class, and the rest as the negative examples class. This gives us 10 datasets.

Table 1. The most popular 10 categories on Reuters-21578 and their quantity

Acq	Corn	Crude	Earn	Grain	Interest	Money-fx	Ship	Trade	Wheat
2369	**237**	**578**	**3964**	**582**	**478**	**717**	**286**	**486**	**283**

In data preprocessing, we use the Bow toolkit [13]. We applied stopword removal, but no feature selection or stemming was done. The tf-idf value is used in the feature vectors. For each dataset, 30% of the documents are randomly selected as test documents. The rest (70%) are used to create training sets as follows: γ percent of the documents from the positive examples class is first selected as the positive examples set *P*. The rest of the positive and negative documents are used as unlabeled examples set *U*. We range γ from 10%-90% to create a wide range of scenarios.

4.2 Evaluation Measures

In our experiments, we use the popular F_1 score on the positive examples class as the evaluation measure. F_1 score takes into account of both recall and precision. Precision, recall and F_1 defined as:

$$Precision = \frac{\#\ of\ \ correct\ \ positive\ \ predictions}{\#\ of\ \ positive\ \ predictions} \qquad (2)$$

$$Recall = \frac{\#\ of\ \ correct\ \ positive\ \ predictions}{\#\ of\ \ positive\ \ examples} \qquad (3)$$

$$F_1 = \frac{2 \times precision \times recall}{precision + recall} \qquad (4)$$

For evaluating performance average across categories, there are two conventional methods, namely macro-average and micro-average. Macro-averaged performance scores are determined by first computing the performance measures per category and then averaging those to compute the global means. We use macro-averaging.

4.3 Experiment Results

In our experiments, we implemented the 1-DNF method used in PEBL and the Rocchio method in the Roc-SVM. We use the CLUTO toolkit package [12] for clustering, and set k=20. For SVM, we use the SVM^{light} system [14] with linear kernel, and do not tune the parameters. For Roc-SVM, we use α=16 and β=4 in formula (1).

We first compare the quantity of reliable negative examples produced by three methods. Table 2 shows the averaged quantity on the Reuters collection. The γ denotes the percent of the document from the positive examples class is selected as positive examples set P. For the PEBL, the quantity of initial negative examples is so small; by browsing the initial negative examples, we found these examples sometimes are short paper, and the quality is poor too. For the Rocchio method, the quantity of reliable negative examples is so big that near the two third of training data. For clustering method, the quantity is moderate, and sometimes balanced the training set.

Table 2. Averaged reliable negative quantity of three methods on the 10 Reuters collection

γ	PEBL	Rocchio	Clustering
10	**394.1**	**6253.9**	**3760.5**
20	**301.5**	**6894.7**	**3462.2**
30	**201.6**	**6642.3**	**3248.6**
40	**224.6**	**5845.0**	**3334.0**
50	**227.6**	**6793.0**	**3109.9**
60	**218.5**	**6779.1**	**2931.3**
70	**207.0**	**6802.5**	**2909.7**
80	**186.0**	**6816.3**	**2689.8**
90	**208.2**	**6863.6**	**2496.9**

Then we compare the F_1 score of our method with other two methods. The results of the PEBL method and the Rocchio method are extract from the experiment of Bing Liu et al. [6]. Fig.3 shows the macro-averaged F_1 score on the 10 Reuters datasets for each γ setting. When γ is smaller (<50), our method outperforms than other two. When γ is bigger, our method is as good as other methods. But there is still room for further improvement.

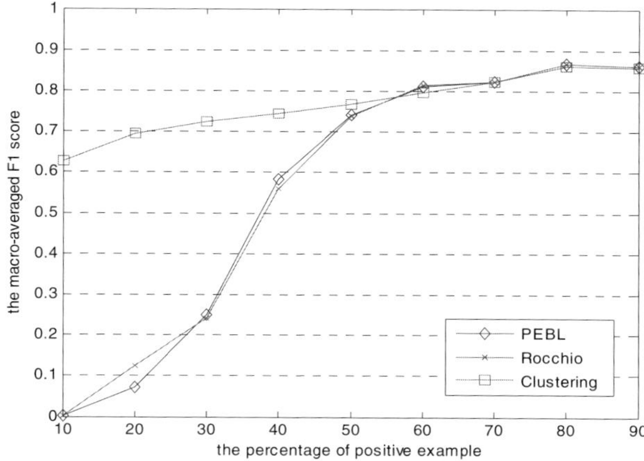

Fig. 3. The macro-averaged F1 scores on the Reuters 10 collection

The poor quality and quantity of the reliable negative examples by PEBL increase the number of the iterations of SVM, which ends up longer training time. The quantity of negative examples of Rocchio method is so big that biased the training set. Our proposed method produces the moderate quantity reliable negative examples.

5 Conclusion

In this paper, we discussed the two-step strategies for learning a classifier from positive examples and unlabeled examples data. The clustering method was added to the existing techniques. A comprehensive evaluation was conducted to compare their performances. Our method produces the moderate quantity reliable negative examples and sometimes balanced the training set. Our experiment shows that our method is efficient and effective. In particular, when positive examples are small, our method outperforms than other two; when γ is bigger, our method is as good as other methods.

Acknowledgments

The work was supported by the National Natural Science Foundation of China under Grant No. 60373099, the Project of the Jilin Province Science and Technology Development Plan under the Grant No.20070533, and the Science Foundation for Young Teachers of Northeast Normal University (No.20070602). We would like to thank the anonymous reviewers for their comments and suggestions.

References

1. Nigam, K., McCallum, A., Thrun, S., Mitchell, T.: Learning to Classify Text from Labeled and Unlabeled Documents. In: AAAI-98, pp. 792–799. AAAI Press, Menlo Park (1998)
2. Denis, F.: PAC Learning from Positive Statistical Queries. In: Richter, M.M., Smith, C.H., Wiehagen, R., Zeugmann, T. (eds.) ALT 1998. LNCS (LNAI), vol. 1501, pp. 112–126. Springer, Heidelberg (1998)

3. Letouzey, F., Denis, F., Gilleron, R.: Learning From Positive and Unlabeled Examples. In: Proceedings of 11th International Conference on Algorithmic Learning Theory (2000)
4. Denis, F., Gilleron, R., Tommasi, M.: Text Classification from Positive and Unlabeled Examples. In: Conference on Information Processing and Management of Uncertainty in Knowledge-Based Systems (2002)
5. Denis, F., Gilleron, R., Laurent, A., Tommasi, M.: Text Classification and Co-Training from Positive and Unlabeled Examples. In: Proceedings of the ICML 2003 Workshop: The Continuum from Labeled to Unlabeled Data (2003)
6. Liu, B., Dai, Y., Li, L.X., Lee, W.S., Yu, P.: Building Text Classifiers Using Positive and Unlabeled Examples. In: Proceedings of the Third IEEE International Conference on Data Mining (2003)
7. Li, X.L., Liu, B.: Learning to Classify Text using Positive and Unlabeled Data. In: Proceedings of Eighteenth International Joint Conference on Artificial Intelligence (2003)
8. Liu, B., Lee, W.S., Yu, P., Li, X.L.: Partially Supervised Classification of Text Documents. In: Proc. 19th Intl. Conf. on Machine Learning (2002)
9. Yu, H., Han, J., Chang, K.C.C.: PEBL: Web Page Classification Without Negative Examples. J. IEEE Transactions on Knowledge and Data Engineering (Special Issue on Mining and Searching the Web) 16(1), 70–81 (2004)
10. Zhao, Y., Karypis, G.: Hierarchical Clustering Algorithms for Document Datasets. J. Data Mining and Knowledge Discovery 10(2), 141–168 (2005)
11. Yang, Y.: An Evaluation of Statistical Approaches to Text Categorization. J. of Information Retrieval 1(1/2), 67–88 (1999)
12. The CLUTO toolkit package, `http://glaros.dtc.umn.edu/gkhome/cluto/cluto/`download
13. Bow, A.: Toolkit for Statistical Language Modeling, Text Retrieval, Classification and Clustering, `http://www.cs.cmu.edu/~mccallum/bow/`
14. Joachims, T.: Making large-Scale SVM Learning Practical. In: Advances in Kernel Methods - Support Vector Learning, MIT Press, Cambridge (1999)

Term Weighting Evaluation in Bipartite Partitioning for Text Clustering

Chao Qu[1,2], Yong Li[1], Jun Zhu[1,2], Peican Huang[1], Ruifen Yuan[1], and Tianming Hu[1,3]

[1] Dongguan University of Technology, China
[2] Zhongshan University, China
[3] East China Normal University, China
tmhu05@gmail.com

Abstract. To alleviate the problem of high dimensions in text clustering, an alternative to conventional methods is bipartite partitioning, where terms and documents are modeled as vertices on two sides respectively. Term weighting schemes, which assign weights to the edges linking terms and documents, are vital for the final clustering performance. In this paper, we conducted an comprehensive evaluation of six variants of *tf*/*idf* factor as term weighting schemes in bipartite partitioning. With various external validation measures, we found *tfidf* most effective in our experiments. Besides, our experimental results also indicated that *df* factor generally leads to better performance than *tf* factor at moderate partitioning size.

1 Introduction

The high dimension in text clustering is a major difficulty for most probabilistic methods such as Naive Bayes [1] and AutoClass [2]. To circumvent this problem, graph-theoretic techniques have been considered for clustering [3]. They model the document similarity by a graph whose vertices correspond to documents and weighted edges give the similarity between vertices. Graphs can also model terms as vertices and similarity between terms is based on documents in which they co-occur. Partitioning the graph yields a clustering of terms, which is assumed to be associated with similar concepts [4]. The duality between document and term clustering can also be naturally modeled using a bipartite, where documents and terms are modeled as vertices on two sides respectively and only edges linking different types of vertices are allowed in the graph [5]. Finding an optimal partitioning in such a bipartite gives a co-clustering of documents and terms, with the expectation that documents and terms in the same cluster are related to the same topic. In addition, bipartite graphs have also been used to model other relationships, such as (documents, concepts)[6] and (authors, publications)[7].

Since the general partitioning goal is to minimize the edge cut, term weighting schemes, which assign weights to edges linking term and document vertices, is a vital step to the final clustering performance. Many researchers have studied text clustering based on different term weighting schemes or different criterion

H. Li et al. (Eds.): AIRS 2008, LNCS 4993, pp. 393–400, 2008.

functions using conventional probabilistic methods [8,9,10,11]. For instance, the authors of [9] pointed out that it is the text representation schemes that dominate the clustering performance rather than the kernel functions of support vector machines (SVM). In other words, choosing an appropriate term weighting scheme is more important than choosing and tuning kernel functions of SVM for text categorization. However, to the best of our knowledge, there is little work on comparing weighting schemes for graph based text clustering, not to mention bipartite partitioning. For this purpose, we concentrate on the comparison of various term weighting schemes in bipartite based text clustering. Specifically, we provide a comprehensive experimental evaluation with real world document datasets from various sources, and with various external validation measures.

Overview. The rest of this paper is organized as follows. Section 2 describes the necessary background. Section 3 provides the results of extensive experiments. Finally, we draw conclusions in Section 4.

2 Background

In this section, we first describe the overview of bipartite based text clustering. Then we introduce the term weighting schemes that will be compared in this paper.

2.1 Bipartite Generation

To apply clustering algorithms, a document data set is usually represented by a matrix. First we extract from documents unique content-bearing words as features, which involves removing stopwords and those with extreme document frequencies. More sophisticated techniques use support or entropy to filter words further. Then each document is represented as a vector in this feature space. With rows for documents and columns for terms, the matrix A's non-zero entry A_{ij} indicates the presence of term w_j in document d_i, while a zero entry indicates an absence.

A graph $G = (V, E)$ is composed of a vertex set $V = \{1, 2, ..., |V|\}$ and an edge set $\{(i, j)\}$ each with edge weight E_{ij}. The graph can be stored in an adjacency matrix M, with entry $M_{ij} = E_{ij}$ if there is an edge (i, j), $M_{ij} = 0$ otherwise. Given the $n \times m$ document-term matrix A, the bipartite graph $G = (V, E)$ is constructed as follows. First we order the vertices such that the first m vertices index the terms while the last n index the documents, so $V = V_W \cup V_D$, where V_W contains m vertices each for a term, and V_D contains n vertices each for a document. Edge set E only contains edges linking different kinds of vertices, so the adjacency matrix M may be written as $\begin{pmatrix} 0, A \\ A^T, 0 \end{pmatrix}$.

2.2 Bipartite Partitioning

Given a weighted graph $G = \{V, E\}$ with adjacency matrix M, clustering the graph into K parts means partitioning V into K disjoint clusters of vertices

$V_1, V_2, ..., V_K$, by cutting the edges linking vertices in different parts. The general goal is to minimize the sum of the weights of those cut edges. Formally, the cut between two vertex groups V_1 and V_2 is defined as $cut(V_1, V_2) = \sum_{i \in V_1, j \in V_2} M_{ij}$. Thus the goal can be expressed as $min_{\{V_1, V_2, ..., V_K\}} \sum_{k=1}^{K} cut(V_k, V - V_k)$. To avoid trivial partitions, often the constraint is imposed that each part should be roughly balanced in terms of part weight $wgt(V_k)$, which is often defined as sum of its vertex weight. That is, $wgt(V_k) = \sum_{i \in V_k} wgt(i)$. The objective function to minimize becomes $\sum_{k=1}^{K} \frac{cut(V_k, V - V_k)}{wgt(V_k)}$. Given two different partitionings with the same cut value, the above objective function value is smaller for the more balanced partitioning.

In practice, different optimization criteria have been defined with different vertex weights. The ratio cut criterion [12], used for circuit partitioning, defines $wgt(i) = 1$ for all vertices i and favors equal sized clusters. The normalized cut criterion [13], used for image segmentation, defines $wgt(i) = \sum_j M_{ij}$. It favors clusters with equal sums of vertex degrees, where vertex degree refers to the sum of weights of edges incident on it.

Finding a globally optimal solution to such a graph partitioning problem is in general NP-complete [14], though different approaches have been developed for good solutions in practice [15,16]. Here we employ Graclus [16], a fast kernel based multilevel algorithm, which involves coarsening, initial partitioning and refinement phases. As for the graph partitioning criterion used in Graclus, we tried both the normalized cut criterion and the ratio cut criterion. We found the former always produces better results, possibly for the following reasons. First, our datasets are highly imbalanced, which makes unreasonable the constraint of equal sized clusters by the ratio cut criterion. Second we find that sometimes it yields clusters of pure word vertices, which makes it impossible to determine the number of document clusters (clusters containing the document vertices) beforehand. Those terms with low frequencies are likely to be isolated together, since few edges linking outside are cut. As for the normalized cut criterion that tries to balance sums of vertex degrees in each cluster, the resultant clusters tend to contain both document and term vertices. So in this paper we only report results from the normalized cut criterion.

2.3 Term Weighting

Term weighting schemes determine the value of non-zero entry A_{ij} in the document-term matrix when term w_j appears in document d_i. Two frequencies are commonly used. Term frequency tf_{ij} denotes the raw frequency of term w_j in document d_i. Inverse document frequency $idf_j = log(n/n_j)$ considers the discriminating power of w_j, where n_j is the number of documents that contain w_j. In this paper, we compared six term weighting schemes listed in Table 1. Most of these term weighting schemes have been widely used in information retrieval and text categorization. The first four term weighting schemes are different variants of tf factor. The last two incorporate idf. According to a recent

Table 1. Term weighting schemes

name	description
binary	1 for presence and 0 for absence
tf	raw tf
$logtf$	$log(1+tf)$
itf	$1-1/(1+tf)$
idf	$log(n/n_j)$
$tfidf$	$tf \times idf$

study of text classification with SVM [11], although the first four schemes relate with term frequency alone, all of them show competitive performance with other sophisticated schemes except *binary*. The *idf* factor, taking the collection distribution into consideration, does not improve the terms discriminating power with SVM. The $tfidf$ factor, combining both term and document frequencies, usually yields best results in query based document retrieval [17].

3 Experimental Evaluation

In this section, we present an extensive experimental evaluation of various term weighting schemes. First we introduce the experimental datasets and cluster validation criteria, then we report comparative results.

3.1 Experimental Datasets

For evaluation, we selected 10 real data sets from different domains used in [10]. The RE0 and RE1 datasets are from the Reuters-21578 text categorization test collection Distribution 1.0. The datasets K1a, K1b and WAP are from the WebACE project; each document corresponds to a web page listed in the subject hierarchy of Yahoo. In particular, K1a and K1b contain the same data but K1a's class labels are at a finer level. The datasets TR31 and TR41 were derived from the TREC collection. The LA1 and LA2 datasets were obtained from articles of the Los Angeles Times that was used in TREC-5. The FBIS dataset is from the Foreign Broadcast Information Service data of TREC-5. For all data sets, we used a stoplist to remove common words, stemmed the remaining words using Porter's suffix-stripping algorithm, and discard those with very low document frequencies. Some characteristics of them are shown in Table 2.

3.2 Validation Measures

Because the true class labels of documents are known, we can measure the quality of the clustering solutions using external criteria that measure the discrepancy between the structure defined by a clustering and what is defined by the class labels. First we compute the confusion matrix C with entry C_{ij} as the number of documents from true class j that are assigned to cluster i. Then we calculate the following four measures: normalized mutual information(NMI), conditional entropy(CE), error rate(ERR) and F-measure.

Table 2. Characteristics of data sets

data	re0	re1	k1a	k1b	wap	tr31	tr41	la1	la2	fbis
#doc	1504	1657	2340	2340	1560	927	878	3204	3075	2463
#word	2886	3758	4707	4707	8460	10128	7454	6188	6060	2000
#class	13	25	20	6	20	7	10	6	6	17
MinClass	11	13	9	60	5	2	9	273	248	38
MaxClass	608	371	494	1389	341	352	243	943	905	506
min/max	0.018	0.035	0.018	0.043	0.015	0.006	0.037	0.29	0.274	0.075
source	Reuters-21578		WebACE			TREC-6,7		TREC-5		

NMI and CE are entropy based measures. The cluster label can be regarded as a random variable with the probability interpreted as the fraction of data in that cluster. Let T and C denote the random variables corresponding to the true class and the cluster label, respectively. The two entropy-based measures are defined as $NMI = \frac{H(T)+H(C)-H(T,C)}{\sqrt{H(T)H(C)}}$, $CE = H(T|C) = H(T,C)-H(C)$, where $H(X)$ denotes the entropy of X and $\log_2$ is used here in computing entropy. NMI measures the shared information between T and C and it reaches the maximal value of 1 when they are the same. CE tells the information remained in T after knowing C and it reaches the minimal value of 0 when they are identical. Error rate $ERR(T|C)$ computes the fraction of misclassified data when all data in each cluster is classified as the majority class in that cluster. It can be regarded as a simplified version of $H(T|C)$.

F-measure combines the precision and recall concepts from information retrieval [17]. We treat each cluster as if it were the result of a query and each class as if it were the desired set of documents for a query. We then calculate the recall and precision of that cluster for each given class as follows: $R_{ij} = C_{ij}/C_{+j}$, $P_{ij} = C_{ij}/C_{i+}$, where C_{+j}/C_{i+} is the sum of jth column/i-th row, i.e., j-th class size /i-th cluster size. Note that C_{+j} could be larger than the true size of class j if some documents from it appear in more than one cluster. F-measure of cluster i and class j is then given by $F_{ij} = \frac{2R_{ij}P_{ij}}{P_{ij}+R_{ij}}$. The overall value for the F-measure is a weighted average for each class, $F = \frac{1}{n}\sum_j C_{+j} max_i\{F_{ij}\}$, where n is the total sum of all elements of matrix C. F-measure reaches its maximal value of 1 when the clustering is the same as the true classification.

3.3 Clustering Results

Let c in cK denote the number of partitions we set. The six term weighting schemes are evaluated in terms of the four validation measures at $5K, 10K, 15K$ and $20K$. The detailed results are shown in Table 3. NMI and F are preferred large while ERR and CE are preferred small. For each setting, the best results are highlighted in bold numbers. One can see in most cases $tfidf$ gives the best results and $binary$ performs worst. Although tf, $logtf$ and itf perform best in certain settings, their gap between $tfidf$ is not significant. The last six rows of Table 3 give the number of wins over all measures at four levels of clustering granularity, respectively. The superiority of $tfidf$ is obvious at three levels.

Table 3. Comparison of clustering results

		5K				10K				15K				20K			
data	wgt	NMI	CE	ERR	F	NMI	CE	ERR	F	NMI	CE	ERR	F	NMI	CE	ERR	F
re0	binary	0.28	1.43	0.48	0.40	0.32	2.15	0.46	0.36	0.32	2.71	0.40	0.34	0.33	2.87	0.39	0.31
	tf	**0.33**	1.40	0.45	0.47	0.31	2.18	0.45	0.36	0.33	2.69	0.39	0.37	0.32	2.95	0.42	0.31
	logtf	0.30	1.28	0.47	0.43	0.30	2.24	0.43	0.35	0.33	2.64	0.40	0.33	0.35	2.82	0.38	0.33
	itf	0.29	1.40	0.51	0.42	**0.35**	**2.07**	**0.39**	**0.40**	**0.37**	**2.43**	**0.37**	**0.39**	**0.39**	**2.77**	**0.35**	**0.36**
	idf	0.29	**1.26**	**0.44**	**0.49**	0.26	2.10	0.48	0.36	0.24	2.72	0.44	0.30	0.27	3.01	0.43	0.29
	tfidf	0.25	1.44	0.47	0.45	0.29	2.15	0.44	0.35	0.31	2.68	0.41	0.33	0.32	2.87	0.43	0.33
re1	binary	0.24	**1.19**	0.66	0.29	0.29	1.67	**0.58**	0.33	0.31	2.36	0.56	0.33	0.33	2.51	0.55	0.34
	tf	0.26	1.46	0.63	0.31	**0.31**	1.68	0.58	**0.36**	0.33	2.37	0.56	0.31	**0.35**	2.40	**0.51**	**0.35**
	logtf	**0.28**	1.36	**0.62**	0.33	0.26	1.93	0.64	0.28	0.31	2.17	0.56	0.32	0.35	2.42	0.53	0.35
	itf	0.25	1.22	0.66	0.30	0.26	1.96	0.60	0.31	0.28	2.28	0.60	0.32	0.33	2.67	0.56	0.33
	idf	0.20	1.27	0.70	0.24	0.26	1.78	0.61	0.31	0.30	2.12	0.57	**0.34**	0.32	2.35	0.55	0.32
	tfidf	0.28	1.34	0.62	**0.36**	0.30	**1.60**	0.60	0.32	**0.33**	**2.12**	**0.56**	0.32	0.35	**2.16**	0.56	0.31
k1a	binary	0.08	0.82	0.75	0.19	0.29	0.93	0.66	0.31	0.24	1.51	0.67	0.32	0.30	1.62	0.64	0.38
	tf	0.11	0.45	0.75	0.21	0.38	0.66	0.63	0.36	0.32	1.18	0.60	0.38	0.39	**1.26**	0.56	0.43
	logtf	0.16	0.39	0.73	0.22	**0.39**	**0.62**	0.63	0.36	0.38	**1.03**	0.58	0.41	0.37	1.32	0.59	0.43
	itf	0.16	**0.37**	0.73	0.22	0.36	0.69	0.62	0.35	0.34	1.10	0.60	0.38	0.37	1.29	0.59	0.42
	idf	**0.31**	0.66	0.65	0.30	0.31	1.19	0.63	0.36	**0.45**	1.32	**0.49**	**0.48**	0.39	1.68	**0.53**	0.44
	tfidf	0.30	0.95	**0.64**	**0.33**	0.35	1.44	**0.59**	**0.39**	0.40	1.30	0.59	0.42	**0.41**	1.56	0.54	**0.44**
k1b	binary	0.07	0.88	0.41	0.49	0.33	1.09	0.30	0.63	0.29	1.64	0.27	0.51	0.34	1.83	0.24	0.54
	tf	0.03	0.59	0.40	0.50	0.39	0.93	0.28	**0.69**	0.29	1.53	0.30	0.60	0.42	1.57	0.22	**0.60**
	logtf	0.06	0.57	0.40	0.49	0.40	**0.89**	0.26	0.68	0.38	**1.38**	0.29	**0.64**	0.42	1.56	0.22	0.58
	itf	0.08	**0.53**	0.40	0.49	0.36	0.96	0.27	0.64	0.33	1.44	0.24	0.57	**0.43**	**1.53**	0.22	0.59
	idf	0.33	0.86	0.33	0.63	0.37	1.34	0.25	0.57	**0.47**	1.71	**0.19**	0.56	0.42	2.04	0.22	0.51
	tfidf	**0.33**	1.14	**0.23**	**0.65**	**0.41**	1.67	**0.20**	0.66	0.45	1.58	0.22	0.60	0.42	1.95	**0.21**	0.50
wap	binary	0.46	0.32	0.61	0.39	0.40	1.40	0.53	0.41	0.39	1.52	0.56	0.38	0.43	1.56	0.49	0.42
	tf	0.46	0.76	0.58	0.45	0.40	1.32	0.56	0.39	0.40	1.37	0.54	0.36	0.43	**1.39**	0.51	0.34
	logtf	0.45	0.82	0.56	**0.46**	0.45	1.17	0.52	0.41	0.41	1.45	0.53	0.38	0.38	1.61	0.55	0.36
	itf	0.48	0.40	0.57	0.43	**0.50**	0.97	0.52	**0.48**	0.43	1.53	0.54	0.42	0.44	1.80	0.46	0.42
	idf	0.47	0.75	**0.55**	0.46	0.50	1.03	**0.50**	0.44	0.50	**1.31**	0.46	**0.48**	0.50	1.62	0.46	**0.48**
	tfidf	**0.48**	**0.23**	0.60	0.42	0.50	**0.73**	0.52	0.44	**0.52**	1.48	**0.44**	0.46	**0.52**	1.65	**0.40**	0.48
tr31	binary	0.23	**0.51**	0.52	0.51	0.21	1.61	0.48	0.49	0.31	1.64	0.39	0.58	0.29	1.90	0.38	0.55
	tf	0.25	0.78	0.51	0.48	**0.52**	**0.68**	**0.28**	**0.72**	**0.49**	**1.02**	**0.28**	**0.67**	0.54	1.27	0.20	0.68
	logtf	0.22	1.00	0.51	0.48	0.44	0.87	0.32	0.69	0.44	1.19	0.29	0.61	0.50	**1.26**	0.22	**0.70**
	itf	0.18	1.00	0.53	0.45	0.41	0.82	0.32	0.66	0.44	1.16	0.29	0.64	0.47	1.34	0.24	0.66
	idf	0.32	0.69	**0.40**	0.54	0.48	1.07	0.29	0.64	0.39	1.75	0.33	0.58	0.51	1.76	0.20	0.60
	tfidf	**0.34**	1.11	0.50	**0.55**	0.51	1.11	0.28	0.64	0.48	1.41	0.28	0.60	**0.56**	1.42	**0.17**	0.69
tr41	binary	0.54	0.89	0.33	0.66	0.57	1.44	0.30	0.63	0.48	1.98	0.32	0.48	0.45	2.25	0.36	0.42
	tf	0.67	0.51	0.28	0.71	0.60	1.36	0.29	0.68	0.61	1.77	**0.20**	0.58	0.59	2.06	0.19	0.55
	logtf	**0.67**	**0.49**	**0.27**	**0.72**	**0.68**	**1.02**	**0.18**	**0.74**	**0.62**	1.67	0.21	**0.58**	0.58	2.04	0.21	0.52
	itf	0.65	0.57	0.29	0.70	0.58	1.32	0.24	0.61	0.57	1.76	0.24	0.57	0.55	2.12	0.24	0.54
	idf	0.64	0.60	0.31	0.67	0.59	1.20	0.30	0.66	0.54	**1.66**	0.28	0.53	0.54	2.00	0.30	0.51
	tfidf	0.58	0.76	0.33	0.64	0.61	1.14	0.26	0.65	0.58	1.67	0.24	0.57	**0.61**	**1.90**	**0.18**	**0.57**
la1	binary	0.01	1.02	0.70	0.32	0.15	1.43	0.60	0.41	0.10	2.00	0.62	0.37	0.15	2.23	0.54	0.42
	tf	0.05	**0.64**	0.68	0.34	0.23	**1.09**	0.58	0.45	0.33	**1.33**	0.42	0.57	0.33	**1.62**	0.42	0.56
	logtf	0.02	0.76	0.70	0.32	0.20	1.17	0.59	0.43	0.27	1.49	0.45	0.54	0.33	1.66	0.43	0.53
	itf	0.02	0.76	0.70	0.32	0.18	1.22	0.61	0.42	0.26	1.51	0.43	0.55	0.32	1.65	0.43	0.53
	idf	0.24	1.14	0.54	0.45	**0.37**	1.22	**0.43**	**0.58**	0.36	1.59	0.40	0.58	0.32	1.99	0.39	0.57
	tfidf	**0.29**	0.91	**0.51**	**0.49**	0.27	1.33	0.53	0.48	**0.39**	1.45	**0.37**	**0.64**	**0.35**	1.82	**0.38**	**0.61**
la2	binary	0.01	0.82	0.70	0.32	0.16	1.28	0.59	0.42	0.10	1.86	0.63	0.38	0.12	2.17	0.58	0.41
	tf	0.03	0.52	0.69	0.33	0.23	1.00	0.57	0.45	0.30	**1.31**	0.44	0.55	0.34	**1.48**	0.41	0.59
	logtf	0.03	0.55	0.69	0.33	0.22	1.04	0.58	0.44	0.17	1.61	0.56	0.42	0.30	1.61	0.44	0.52
	itf	0.02	0.57	0.70	0.33	0.19	1.11	0.59	0.43	0.14	1.68	0.59	0.40	0.21	1.83	0.50	0.49
	idf	0.06	0.94	0.67	0.34	0.39	1.10	0.41	0.60	**0.32**	1.55	**0.41**	**0.57**	0.30	1.92	0.40	0.56
	tfidf	**0.36**	**0.50**	**0.48**	**0.50**	**0.42**	**1.00**	**0.36**	**0.64**	0.32	1.52	0.45	0.52	**0.36**	1.73	**0.40**	**0.60**
fbis	binary	0.38	0.97	0.58	0.44	0.51	1.37	0.41	0.54	0.33	2.59	0.47	0.45	0.32	2.94	0.46	0.40
	tf	0.41	0.81	0.57	0.44	0.51	1.38	0.42	**0.58**	0.50	1.90	0.39	0.52	0.50	2.14	0.37	**0.55**
	logtf	0.32	0.84	0.63	0.39	0.48	1.49	0.44	0.52	**0.51**	**1.68**	0.39	**0.55**	0.50	2.15	**0.37**	0.49
	itf	0.37	0.86	0.58	0.43	0.47	1.51	0.43	0.52	0.48	1.71	**0.38**	0.53	0.49	2.14	0.38	0.52
	idf	**0.45**	**0.77**	**0.49**	**0.48**	0.45	1.36	0.46	0.55	0.46	1.79	0.42	0.51	0.46	2.25	0.40	0.50
	tfidf	0.41	0.86	0.56	0.44	**0.53**	**1.29**	**0.40**	0.55	0.50	1.81	0.41	0.55	**0.51**	**1.91**	0.38	0.53
#wins	binary	2				1				0				0			
	tf	2				9				6				9			
	logtf	7				7				8				3			
	itf	2				6				7				6			
	idf	10				4				**11**				2			
	tfidf	**17**				**13**				8				**20**			

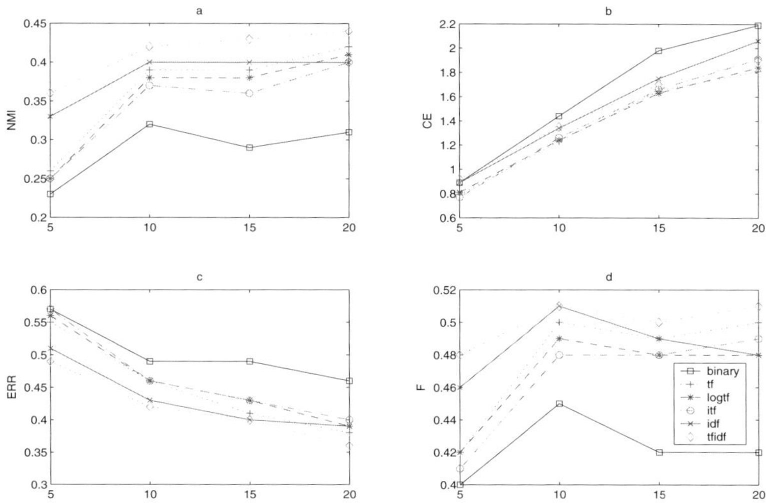

Fig. 1. The average results over the 10 datasets

To give a summary of performance, Figure 1 illustrates the average results over all datasets for each measure. Apparently, $tfidf$ gives the best results on all the measures except CE. Recall that CE is an un-normalized measure, which makes averaging questionable. This indicates that the incorporation of the discriminating factor, *idf*, really makes a difference. It is confirmed again if we compare the performance by *idf* and tf alone. One can see at $5K, 10K, 15K$, *idf* beats tf. It shows that at relatively moderate clustering granularity, the term relevance for the whole document set is more important than that for each individual document. However, as the number of clusters gets larger (e.g., $20K$) and each cluster gets smaller, the term frequency within each document matters.

4 Conclusions

Since the goal of bipartite partitioning is to minimize the edge cut, term weighting schemes are essential for the final clustering performance. In this paper, we provided an extensive comparison of six commonly used schemes. Our experimental results show that $tfidf$ generally yields better performance than other term weighting schemes in terms of various external validation measures. Besides, at moderate clustering granularity, *idf* is more important than tf. Because the graph partitioning is always subject to the balance constraint, the vertex weighting also plays an important role. For the future work, we plan to investigate the impact of vertex weighting schemes. To capture the full semantics that cannot be represented with single words, another direction is to augment the raw vocabulary with word-sets based on frequent itemsets [18,19] or hypercliques [20,21].

Acknowledgments. This work was partially supported by the Scientific Research Foundation for the Returned Overseas Chinese Scholars, State Education Ministry of China, and the Dongguan Foundation of Scientific and Technological Plans.

References

1. Rennie, J.D., Shih, L., Teevan, J., Karger, D.R.: Tackling the poor assumptions of naive bayes text classifiers. In: ICML, pp. 616–623 (2003)
2. Cheeseman, P., Stutz, J.: Bayesian classification (AutoClass): Theory and results. In: KDD, pp. 153–180 (1996)
3. Strehl, A., Ghosh, J., Mooney, R.: Impact of similarity measures on web-page clustering. In: Proc. AAAI: Workshop of Artificial Intelligence for Web Search, pp. 58–64 (2000)
4. Baker, L.D., McCallum, A.: Distributional clustering of words for text classification. In: SIGIR, pp. 96–103 (1998)
5. Dhillon, I.S.: Co-clustering documents and words using bipartite spectral graph partitioning. In: SIGKDD, pp. 269–274 (2001)
6. Yoo, I., Hu, X., Song, I.Y.: Integration of semantic-based bipartite graph representation and mutual refinement strategy for biomedical literature clustering. In: SIGKDD, pp. 791–796 (2006)
7. Sun, J., Qu, H., Chakrabarti, D., Faloutsos, C.: Relevance search and anomaly detection in bipartite graphs. ACM SIGKDD Explorations 7(2), 48–55 (2005)
8. Salton, G., Buckley, C.: Term-weighting approaches in automatic text retrieval. Information Processing and Management 24(5), 513–523 (1988)
9. Leopold, E., Kindermann, J.: Text categorization with support vector machines: How to represent texts in input space? Machine Learning 46(1-3), 423–444 (2002)
10. Zhao, Y., Karypis, G.: Empirical and theoretical comparisons of selected criterion functions for document clustering. Machine Learning 55(3), 311–331 (2004)
11. Lan, M., Tan, C.L., Low, H.B.: Proposing a new term weighting scheme for text categorization. In: AAAI (2006)
12. Hagen, L., Kahng, A.: New spectral methods for ratio cut partitioning and clustering. IEEE Trans. CAD 11, 1074–1085 (1992)
13. Shi, J., Malik, J.: Normalized cuts and image segmentation. IEEE Trans. PAMI 22, 888–905 (2000)
14. Garey, M.R., Johnson, D.S.: Computers and Intractability: A Guide to the Theory of NP-Completeness. W. H. Freeman & Company, New York (1979)
15. Karypis, G., Kumar, V.: A fast and high quality multilevel scheme for partitioning irregular graphs. SIAM J. Scientific Computing 20(1), 359–392 (1998)
16. Dhillon, I.S., Guan, Y., Kulis, B.: A fast kernel-based multilevel algorithm for graph clustering. In: SIGKDD, pp. 629–634 (2005)
17. Baeza-Yates, R., Ribeiro-Neto, B.: Modern Information Retrieval. Addison-Wesley, Reading (1999)
18. Agrawal, R., Imielinski, T., Swami, A.N.: Mining association rules between sets of items in large databases. In: SIGMOD, pp. 207–216 (1993)
19. Agrawal, R., Srikant, R.: Fast algorithms for mining association rules. In: VLDB, pp. 487–499 (1994)
20. Xiong, H., Tan, P.N., Kumar, V.: Mining strong affinity association patterns in data sets with skewed support distribution. In: ICDM, pp. 387–394 (2003)
21. Xiong, H., Tan, P.N., Kumar, V.: Hyperclique pattern discovery. Data Mining and Knowledge Discovery Journal 13(2), 219–242 (2006)

A Refinement Framework for Cross Language Text Categorization

Ke Wu and Bao-Liang Lu*

Department of Computer Science and Engineering, Shanghai Jiao Tong University
800 Dong Chuan Road, Shanghai 200240, China
{wuke,bllu}@sjtu.edu.cn

Abstract. Cross language text categorization is the task of exploiting labelled documents in a source language (e.g. English) to classify documents in a target language (e.g. Chinese). In this paper, we focus on investigating the use of a bilingual lexicon for cross language text categorization. To this end, we propose a novel refinement framework for cross language text categorization. The framework consists of two stages. In the first stage, a cross language model transfer is proposed to generate initial labels of documents in target language. In the second stage, expectation maximization algorithm based on naive Bayes model is introduced to yield resulting labels of documents. Preliminary experimental results on collected corpora show that the proposed framework is effective.

1 Introduction

Due to the popularity of the Internet, an ever-increasing number of documents in languages other than English are available in the Internet, thus creating the need of automatic organization of these multilingual documents. In addition, with the globalization of business environments, for many international companies and organizations, huge volume of documents in different languages need to be archived into common categories. On the other hand, in order to build a reliable model for automated text categorization, we typically need a large amount of manually labelled documents, which cost much human labor. Consequently, in multilingual scenario, how to employ the existing labelled documents written in a source language (e.g. English) to classify the unlabelled documents other than the language has become an important task, as it can be leveraged to alleviate cost of labelling. We refer to the mentioned-above task as cross language text categorization (CLTC).

Cross language information retrieval is highly related to CLTC. Also, the use of bilingual lexicon has been extensively studied in cross language information

* Corresponding author. This work was supported in part by the National Natural Science Foundation of China under the grants NSFC 60375022 and NSFC 60473040, and the Microsoft Laboratory for Intelligent Computing and Intelligent Systems of Shanghai Jiao Tong University.

H. Li et al. (Eds.): AIRS 2008, LNCS 4993, pp. 401–411, 2008.

retrieval [1,2,3]. However, to our knowledge, there is little research on the direction for CLTC. This paper will focus on investigating the use of a bilingual lexicon. Accordingly, we propose a novel refinement framework for CLTC.

The basic idea is that we assume that initial and inaccurate labels from the transferred model can be refined in the original documents into better resulting labels. Specifically, the framework consists of two stages. In the first stage, a cross language model transfer is proposed to generate preliminary labels of documents in target language. In the second stage, expectation maximization algorithm (EM) [4] based on naive Bayes model is introduced to generate resulting labels of documents. Preliminary experimental results on collected corpora show that in the case of sufficient test data, with a small number of training documents, the proposed refinement framework can achieve better performance than monolingual text categorization and with a large number of training documents, it can also obtain promising results close to that of monolingual text categorization.

The remainder of this paper is organized as follows. Section 2 introduces related work. Section 3 presents the refinement framework. Section 4 performs evaluation over our proposed framework. Section 5 is conclusions and future work.

2 Related Work

Cross language text categorization is divided into two cases, which are polylingual training and cross-lingual training [5]. The term poly-lingual training indicates the case that enough training documents available for every language. However, such scenarios are not particularly interesting as they can be handled with separate monolingual solutions. The term cross-lingual training indicates that another case that enough training documents available for a language but no training documents for other languages. Currently, researchers focus their effort on the latter case. In this paper, we also focus on this case.

Typically, some external lexical resources are used for CLTC. Li and Shawe-Taylor [6] applied kernel canonical correlation analysis (KCCA) and latent semantic analysis (LSA) to parallel corpora and induced the semantic space for CLTC. Olsson et al. [7] used the probabilistic bilingual lexicon induced by parallel corpora to ensure that test data is translated into the language of training data. However, a good semantic space or accurate translation probabilities depend on the amount of parallel corpora. Unfortunately, large-scale parallel corpora are not easily obtained. To alleviate the difficulty, Gliozzo and Strapparava [8] exploit comparable corpora to induce a semantic space by LSA. Nevertheless, this method is applicable only for language pairs, which have common words for the same concepts. Furthermore, Fortuna and Shawe-Taylor [9] applied machine translation system to generate pseudo domain-specific parallel corpus. Rigutini et al. [10] used a machine translation system to bridge the gap between different languages. However, there are not machine translation systems for many language pairs and there is still wide gap of statistical characteristics between translated documents and original documents.

Compared with the above lexical resources, bilingual lexicon is a kind of cheap resource, which is readily available. However, there is little research on the use of a bilingual lexicon alone for CLTC. In this paper, we wish to concentrate on the direction.

3 Refinement Framework

Figure 1 shows the overall architecture of our refinement framework. L_1 denotes the source language (i.e. the language in which documents are manually labelled); L_2 denotes the target language (i.e. the language in which documents are to be classified according to the categories from language L_1). The framework consists of two stages. The preliminary labels are generated in the first stage and a refinement with the preliminary labels is performed in the second stage. In the following two sections, we shall explain the two stages in details.

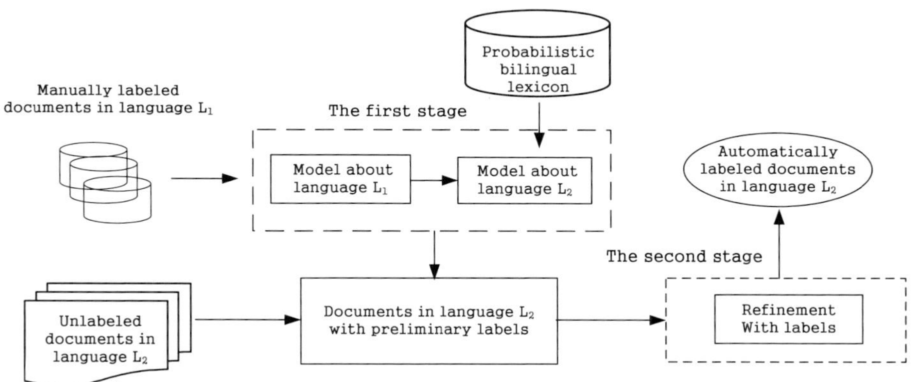

Fig. 1. Refinement framework for cross language text categorization

3.1 The First Stage

In this stage, preliminary labels about documents in L_2 are generated. Accordingly, a learning model about L_2 needs to be generated for label assignment of the documents in L_2. However, a learning model about L_2 can not be derived directly. As a result, we propose an approach which transfers the trained model in L_1 into the new model in L_2 via a bilingual lexicon. This approach is called as `cross language model transfer` (CLMT). In this paper, we choose to investigate the transfer of the naive Bayes model from L_1 to L_2, since naive Bayes model is efficient and effective for multi-class case. The details of naive Bayes model can be referred to [11]. For naive Bayes model in language L_2, we need estimate two parameters, $P(w_f|c)$ and $P(c)$, where $P(w_f|c)$ denotes the probability that word w_f in L_2 occurs, given class c and $P(c)$ denotes the probability that class c occurs in language L_2.

In cross language information retrieval, a unigram language model in one language is combined with a probabilistic bilingual lexicon to yield a unigram language model in another language. Our approach is inspired by this well-known technique. We extend this by using a class-conditional bilingual lexicon. It can be formalized as follows:

$$P(w_f|c) = \sum_{w_e \in \mathcal{V}^{L_1}} P(w_e|c)P(w_f|w_e,c) \tag{1}$$

where $\mathcal{V}^{L_1}$ denotes the vocabulary in L_1; $P(w_e|c)$ denotes the probability that word w_e in L_1 occurs given class c; and $P(w_f|w_e,c)$ denotes the probability that word w_e is translated into word w_f given class c.

$P(w_e|c)$ is derived from the parameter estimation of model about L_1. There are two solutions for $P(w_f|w_e,c)$. First, a naive and direct method is that we simply assume for each class c, a word w_e is translated into w_f with the same probability, which is a uniform distribution on a word's translations. If a word w_e has n translations in our bilingual lexicon $\mathcal{L}$, each of them will be assigned equal probability, i.e. $P(w_f|w_e,c) = \frac{1}{n}$, where w_f is a translation of w_e in $\mathcal{L}$; otherwise $P(w_f|w_e,c) = 0$.

Second, we propose to apply EM algorithm to deduce the conditional translation probabilities given class c, via the bilingual lexicon $\mathcal{L}$ and the training document collection at hand. This idea is inspired by the work of word translation disambiguation [12]. We can assume that given class c, each word w_e in language L_1 is independently generated by a finite mixture model according to $P(w_e|c) = \sum_{w_f} P(w_f|c)P(w_e|w_f,c)$.

Therefore we can use EM algorithm to estimate the parameters of the model. Specifically, $p(w_f|w_e,c)$ is initialized through the first solution and then the following two steps are iterated until $p(w_f|w_e,c)$ remains unchanged.

- E-step:

$$P(w_f|w_e,c) = \frac{P(w_f|c)P(w_e|w_f,c)}{\sum_{w_f \in \mathcal{V}^{L_2}} P(w_f|c)P(w_e|w_f,c)} \tag{2}$$

- M-step:

$$P(w_e|w_f,c) = \frac{N(w_e,c)P(w_f|w_e,c)}{\sum_{w_e \in \mathcal{V}^{L_1}} N(w_e,c)\,P(w_f|w_e,c)} \tag{3}$$

$$P(w_f|c) = \frac{\sum_{w_e \in \mathcal{V}^{L_1}} N(w_e,c)P(w_f|w_e,c)}{\sum_{w_e \in \mathcal{V}^{L_1}} N(w_e,c)} \tag{4}$$

where $N(w_e,c)$ denotes the times of co-occurrence of w_e and c.

For $P(c)$, there are two solutions, too. A simple solution is that we use estimation from the labelled documents in langauge L_1, since we assume that documents from different languages have the same class distribution. Another solution is that we can assume that the class distribution for documents in L_2 conforms to the uniform distribution, i.e. $P(c) = \frac{1}{|\mathcal{C}|}$. The true class priors for documents in language L_2 may be different from those for documents in language L_1. We do not simply estimate $P(c)$ in language L_2 from the documents in language L_1. That is, we have no idea of any information about the class distribution for documents in language L_2. According to principle of maximum entropy, we can assume that the class distribution for documents in L_2 conforms to the uniform distribution.

3.2 The Second Stage

In this stage, preliminary labels of documents in language L_2 from the first stage are used as input and an EM algorithm is introduced to obtain the final labels of document in language L_2. The iterations of EM are a hill-climbing algorithm in parameter space that locally maximizes the entire log likelihood of documents in the collection. In this paper, we use naive Bayes model for the EM, similar to [11]. The algorithm we use is an unsupervised clustering whereas [11] is a semi-supervised learning. The basic idea is that EM is initialized to onto a right hill and then hill-climb the top. Specifically, $P(c|d)$ is initialized based on the preliminary labels of documents in language L_2 and then the following two steps are iterated until $P(c|d)$ stays unchanged, where d denotes a document.

- E-step:

$$P(c|d) = \frac{P(c)P(d|c)}{\sum_c P(c)P(d|c)} \tag{5}$$

- M-step:

$$P(w_f|c) = \frac{1 + \sum_d N(w_f, d)P(c|d)}{|\mathcal{V}^{L_2}| + \sum_c \sum_d N(w_f, d)P(c|d)} \tag{6}$$

$$P(c) = \frac{1 + \sum_d P(c|d)}{|\mathcal{C}| + |\mathcal{D}|} \tag{7}$$

where the calculation of $P(d|c)$ is referred to [11]. The resulting labels of documents in language L_2 are assigned according to the following equation:

$$c = \arg\max_c P(c|d) \tag{8}$$

Notice that in this stage only original documents in language L_2 are involved.

4 Evaluation

4.1 Setting

We chose English and Chinese as our experimental languages, for we can easily setup our experiments and they are quite different languages. Standard evaluation benchmark is not available and thus we developed a test data from the Internet, containing Chinese Web pages and English Web pages. We applied

Table 1. Source of Chinese Web pages

Chinese Web sites	Number of Web pages
people.com.cn	464
sina.com.cn	4814
tom.com	94
xinhuanet.com	408
chinanews.com.cn	41
jfdaily.com	32
voanews.com	18
takungpao.com	140
Total	6011

Table 2. Source of English Web pages

English Web sites	Number of Web pages
abcnews.go.com	232
allafrica.com	110
english.people.com.cn	416
football.guardian.co.uk	191
gradschool.about.com	19
news.bbc.co.uk	794
news.xinhuanet.com	142
soccernet.espn.go.com	237
yahoo.com	963
cbc.ca	81
cnn.com	335
nba.com	353
news.gov.hk	56
nytimes.com	740
soccerway.com	164
sportnetwork.net	246
uefa.com	290
voanews.com	93
Total	5462

RSS reader[1] to acquire the links to the needed content and then downloaded the Web pages. Although category information of the content can be obtained by RSS reader, we still used three Chinese-English bilingual speakers to organize these Web pages into the predefined categories. The data consists of news during December 2005. There are total 5462 English Web pages which are from 18 news Web sites and 6011 Chinese Web pages which are from 8 news Web sites. The details of the sources of Web pages are shown in Table 1 and Table 2. Data distribution over categories is shown in Table 3.

Some preprocessing steps are applied to those Web pages. First we extract the pure texts of all Web pages, excluding anchor texts which introduce much noise. Then for Chinese corpus, all Chinese characters with BIG5 encoding first were

[1] http://www.rssreader.com/

Table 3. Distribution of documents over categories

Categories	English	Chinese
Sports	1797	2375
Business	951	1212
Science	843	1157
Education	546	692
Entertainment	1325	575
Total	5462	6011

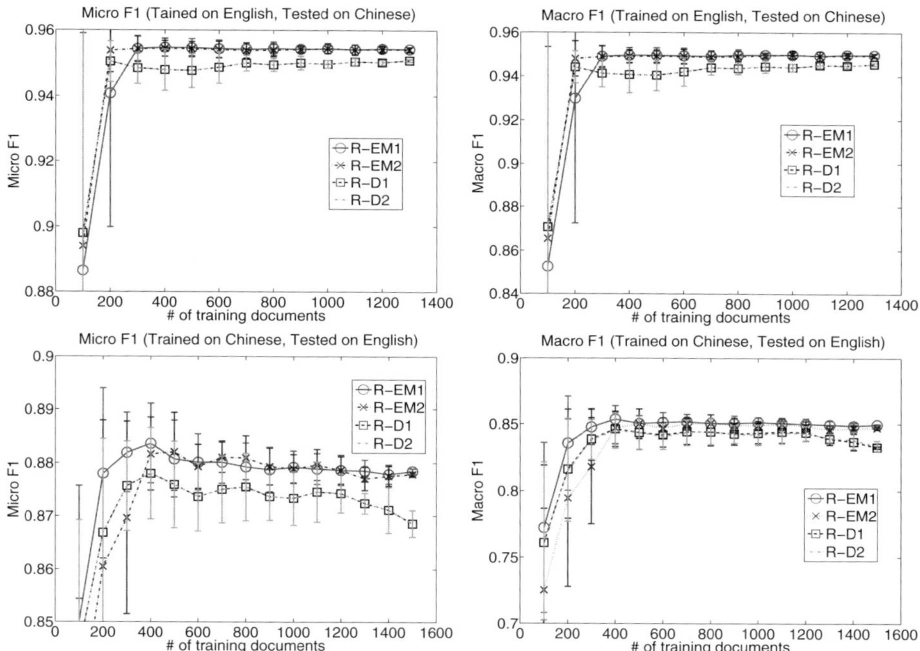

Fig. 2. Performance comparison of our refinement framework with different parameter estimations. The entire test data is used. Each point represents the mean performance for 10 arbitrary runs. The error bars show standard deviations for the estimated performance.

converted into ones with GB2312 encoding, applied a Chinese segmenter tool[2] by Zhibiao Wu from linguistic data consortium (LDC) to our Chinese corpus and removed words with one character and less than 4 occurrences; for English corpus, we used a stop list from SMART system [13] to eliminate common words. Finally, We randomly split both the English and Chinese documents into 2 sets, 25% for training and 75% for test.

[2] http://projects.ldc.upenn.edu/Chinese/LDC_ch.htm

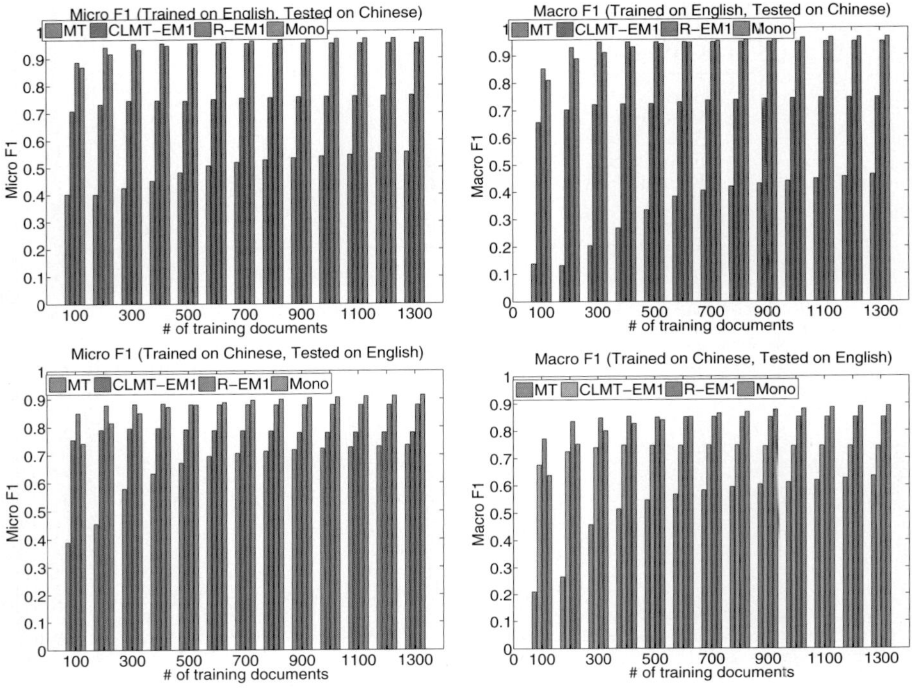

Fig. 3. Performance comparisons of different methods. The entire test data is used. Each value represents the mean performance for 10 arbitrary runs.

We compiled a general-purpose English-Chinese lexicon, which contains 276,889 translation pairs, including 53,111 English entries and 38,517 Chinese entries. Actually we used a subset of the lexicon including 20,754 English entries and 13,471 Chinese entries , which occur in our corpus.

4.2 Evaluation Measures

The performance of the proposed methods was evaluated in terms of conventional precision, recall and $F1$-measures. Furthermore, there are two conventional methods to evaluate overall performance averaged across categories, namely micro-averaging and macro-averaging [14]. Micro-averaging gives equal weight to each document while macro-averaging assigns equal weight to each category. In this paper, it is a multi-class case. Micro F1 and Macro F1 are short for micro-averaging F1 and macro-averaging F1.

4.3 Results

In our experiments, all results are averaged over 10 arbitrary runs. For the proposed CLMT approach for initial labels of documents in language L_2, four

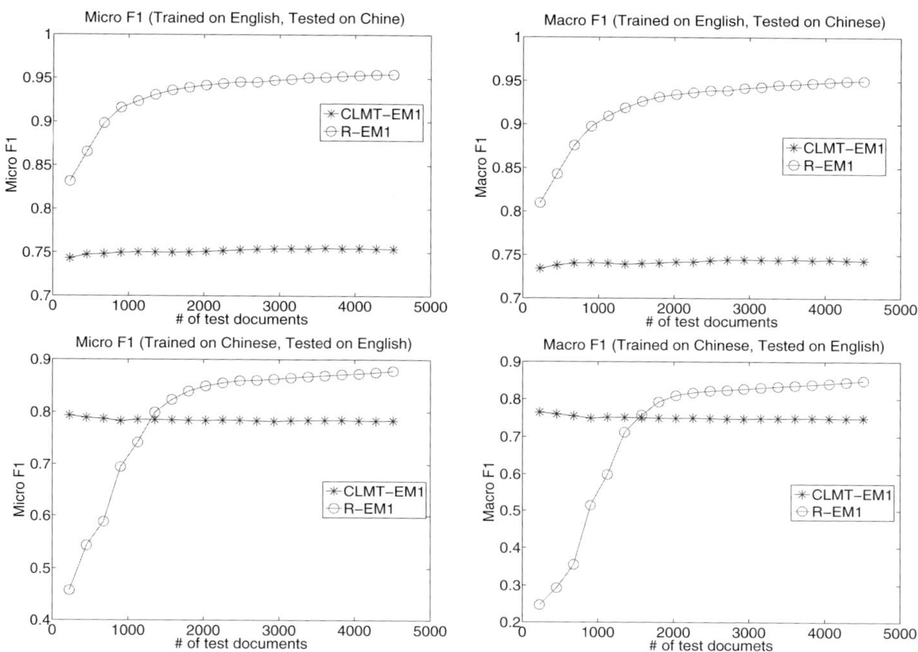

Fig. 4. Performance comparisons of CLMT-EM1 and R-EM1 varying the size of test data. The entire training data is used. Each value represents the mean performance for 10 arbitrary runs.

variants are naturally yielded as different parameter estimations may be used. As a result, we first investigate the impact on resulting performance, through varying different parameter estimations of CLMT. For ease of description, we call them as R-D1, R-D2, R-EM1 and R-EM2, where R indicates refinement framework, D indicates the first solution to estimate $P(w_f|w_e, c)$, EM indicates the second solution to estimate $P(w_f|w_e, c)$, digit 1 denotes the first solution to estimate $P(c)$ in language L_2, and digit 2 denotes the second solution to estimate $P(c)$ in language L_2. Their results on collected corpora are shown in Fig. 2. We can notice that R-EM1 and R-EM2 consistently work better than R-D1 and R-D2 over experiments trained on English documents and tested on Chinese documents or trained on Chinese documents and tested on English documents. In addition, R-EM1 performs slightly better than R-EM2.

For further evaluation of our framework, we compare our approach with the following three baselines. In our experiments, we use Naive Bayes as our classifier for fair comparison.

Mono (Monolingual text categorization). Training and testing are performed on documents in the same language.

CLMT-EM1. It is used to generate preliminary labels for R-EM1. It sets a starting point of refinement for R-EM1, which is used as representative of our methods, since it perform better than other methods.

MT (machine translation). We used Systran premium 5.0 to translate training data into the language of test data, since the machine translation system is one of the best commercial machine translation systems. Then use the translated data to learn a statistical model for classifying the test data.

The results are shown in Fig. 3. We notice that with fewer training documents, R-EM1 works best among all methods and with more training documents, R-EM1 achieves a performance close to monolingual text categorization. In addition, we observe that MT obtains poor performance. This may be because statistical property of the translated documents is quite different from that of the original documents, although human can understand the translated documents produced by Systran premium 5.0.

To examine how the size of test data affects resulting performance, we compare CLMT-EM1 with R-EM1, varying the size of test data. The results are shown in Fig. 4. Experiments show that higher performance benefits from more test data. Meanwhile, we can also notice that when applied on a small portion of English test data set, EM based on naive Bayes model obtains results contrary to what we expect. The EM does not improve the performance of initial labels. On the contrary, it makes resulting performance worse than initial performance. It may be because there are too many parameters to be estimated but few data do not provide potential of accurate parameter estimation.

5 Conclusions and Future Work

This paper proposes a novel refinement framework for cross language text categorization. Our preliminary experiments on the collected data show that our refinement framework is effective for CLTC. This work has the following three main contributions. First, we are apparently the first to investigate the use of a bilingual lexicon alone for cross language text categorization. Second, a refinement framework is proposed for the use of a bilingual lexicon on cross language text categorization. Third, a cross language model transfer approach is proposed for the transfer of naive Bayes models from different languages via a bilingual lexicon and an EM algorithm based on naive Bayes model is introduced for the refinement of initial labels yielded by the proposed cross language model transfer method.

In the future, we shall improve our work from the following three directions. First, our data set is limited and the predefined categories are coarse. we plan to collect larger data collection with finer categories and test our proposed refinement framework on it. Second, different monolingual text categorization algorithms will be explored with the framework and accordingly new cross language model transfer approaches need to be proposed. Third, the EM algorithm is easily trapped into local optima. Therefore, we plan to propose a new refinement approach to avoid this case. Finally, people have recently tried to automatically collect bilingual corpora from web [15,16], and therefore we may benefit by using the translation probabilities trained from the bilingual corpora.

References

1. Gao, J., Xun, E., Zhou, M., Huang, C., Nie, J.Y., Zhang, J.: Improving query translation for cross-language information retrieval using statistical models. In: ACM SIGIR 2001, pp. 96–104 (2001)
2. Gao, J., Nie, J.Y.: A study of statistical models for query translation: finding a good unit of translation. In: SIGIR 2006, pp. 194–201. ACM Press, New York (2006)
3. Liu, Y., Jin, R., Chai, J.Y.: A maximum coherence model for dictionary-based cross-language information retrieval. In: SIGIR 2005, pp. 536–543 (2005)
4. Dempster, A.P., Laird, N.M., Rubin, D.B.: Maximum likelihood from incomplete data via the EM algorithm. Journal of the Royal Statistical Society,Series B 39, 1–38 (1977)
5. Bel, N., Koster, C.H.A., Villegas, M.: Cross-Lingual Text Categorization. In: Koch, T., Sølvberg, I.T. (eds.) ECDL 2003. LNCS, vol. 2769, pp. 126–139. Springer, Heidelberg (2003)
6. Li, Y., Shawe-Taylor, J.: Using KCCA for Japanese-English cross-language information retrieval and document classification. Journal of Intelligent Information Systems 27, 117–133 (2006)
7. Olsson, J.S., Oard, D.W., Hajič, J.: Cross-language text classification. In: Proceedings of SIGIR 2005, pp. 645–646. ACM Press, New York (2005)
8. Gliozzo, A.M., Strapparava, C.: Exploiting comparable corpora and bilingual dictionaries for cross-language text categorization. In: Proceedings of ACL 2006, The Association for Computer Linguistics (2006)
9. Fortuna, B., Shawe-Taylor, J.: The use of machine translation tools for cross-lingual text mining. In: Learning With Multiple Views, Workshop at the 22nd International Conference on Machine Learning (ICML) (2005)
10. Rigutini, L., Maggini, M., Liu, B.: An EM based training algorithm for cross-language text categorization. In: Proceedings of WI 2005, Washington, pp. 529–535. IEEE Computer Society, Los Alamitos (2005)
11. Nigam, K., McCallum, A., Thrun, S., Mitchell, T.: Text classification from labeled and unlabeled documents using EM. Machine Learning 39, 103–134 (2000)
12. Li, C., Li, H.: Word translation disambiguation using bilingual bootstrapping. In: Proceedings of ACL 2002, pp. 343–351 (2002)
13. Buckley, C.: Implementation of the SMART information retrieval system. Technical report, Ithaca, NY, USA (1985)
14. Yang, Y., Pedersen, J.O.: A comparative study on feature selection in text categorization. In: Fisher, D.H. (ed.) Proceedings of ICML 1997, 14th International Conference on Machine Learning, Nashville, US, pp. 412–420. Morgan Kaufmann Publishers, San Francisco (1997)
15. Zhang, Y., Wu, K., Gao, J., Vines, P.: Automatic Acquisition of Chinese–English Parallel Corpus from the Web. In: Lalmas, M., MacFarlane, A., Rüger, S.M., Tombros, A., Tsikrika, T., Yavlinsky, A. (eds.) ECIR 2006. LNCS, vol. 3936, pp. 420–431. Springer, Heidelberg (2006)
16. Resnik, P., Smith, N.A.: The web as a parallel corpus. Comput. Linguist. 29, 349–380 (2003)

Research on Asynchronous Communication-Oriented Page Searching

Yulian Fei, Min Wang, and Wenjuan Chen

Zhejiang Gongshang University
No.18 Xuezheng Str., Xiasha University Town, Hangzhou, Zhejiang, China 310035
fyl@mail.zjgsu.edu.cn, wangmg@mail.zjgsu.edu.cn,
cece1820@hotmail.com

Abstract. Researches on asynchronous communication-oriented page searching aim at solving the new problems for search engine brought about by the adoption of asynchronous communication technology. At present, a full text search engine crawler mostly adopts the algorithm based on a hyperlink analysis. The crawler searches only the contents of the HTML page and ignores the codes in the script region. But it is through the script codes that asynchronous communication is realized. Since a great number of hyperlinks are hidden in the script region, it is necessary to improve the present search engine crawler to search the codes in the script region and extract the hyperlinks hidden in the script region. This paper proposes an approach, which, with the help of script code operation environment, takes advantage of the Windows message mechanism, and employs simulation clicking script function to extract hyperlinks. Meanwhile, in view of the problem that a feedback webpage is not integrated resulting from the asynchronous communication technology, this paper adopts a method that loads in the source page where hyperlinks locate and uses partial refreshing mechanism to save the refreshed page to solve the problem that information cannot be directly stored.

Keywords: asynchronous communication, search engine, crawler.

1 Introduction

With the development of network communication technology, the traditional web application model of client/server and correlative N layer frame construction has increasingly restricted web application[1][2], which is mainly reflected in the following two aspects. Firstly, with the traditional web model, web application is done through synchronous communication. Thus, the client has to wait for the server's response after sending each request, which affects the user's efficiency and experience. Secondly, the traditional web model adopts a whole page refreshing mechanism, by which a lot of data of the server backs out leading to the increase of network delay and load and the waste of network communication resources. As a result, an asynchronous communication model represented by Ajax emerges[3] which has the following strengths[4]: First, asynchronous communication. When the user needs the data from the server, the Ajax engine will transfer synchronously, so the

H. Li et al. (Eds.): AIRS 2008, LNCS 4993, pp. 412–417, 2008.

user can continue the operation on the page with no need to wait for response from the server. And the Ajax will automatically refresh the page showing the result when there is response. Second, data is extracted as required. Since it is a synchronous request from the Ajax, the rest of the template file, the navigation list and the page layout has been sent to the browser with the initial page before the relative data is sent, which reduces redundant requests, lessens the burden on the server and cuts down the gross data downloading. Third, partial refreshing. It only refreshes the necessary part, which enhances user experience. Finally, it increases client function and alleviates the burden on the server.

The adoption of asynchronous communication has promoted the development of web application, which greatly challenges the traditional search engine. Firstly, with the adoption of asynchronous communication the majority of labels will lose the HREF property, and the traditional directly extracting HREF property WebPages crawl technology based on hyperlink analysis will lose its effect. Secondly, it is very difficult and of great importance to extract integrated and accurate URL and to assure ordinal and non-repeating visits, because within the website adopting the asynchronous communication technology, it is impossible to get all URL lists when visiting the WebPages, the URL is extracted while the function is being sent. Thirdly, as asynchronous communication technology employs partial refreshing, the mechanism by which the traditional search engine uses URL as the primary key and stores the server response content in database is not advisable. Based on this, research on asynchronous communication-oriented page searching has become a new field.

At present, a lot of researches have been done on page search engine both at home and abroad. For example, Michael Kohlhase[5] has presented a search engine for mathematical formulae. W. John Wilburwe[6] describes the methodology he has developed to perform spelling correction for the PubMed search engine.Chen Hanshen[7] discusses a new intelligentized and personalized search engine based on server port and client port. Tan Yihong[8] has put forward a method of query expansion based on the association rules-based databases. But none of the above mentioned researches aim at page searching of asynchronous communication website. Up till now, there are no papers on asynchronous communication-oriented page searching. Therefore, the asynchronous communication-oriented page searching approach presented in this paper may serve as an introduction so that other researches may come up with valuable opinions.

2 The Structure of the Asynchronous Communication-Oriented Page Crawler System

There are not many obvious HRFE on the WebPages adopting asynchronous communication technology, for great quantities of URL are hidden in the district of Script code. Therefore, for the WebPages adopting asynchronous communication technology, it is not only necessary to have the traditional page URLextraction, but more necessary to have the WebPages additionally processed to enable the crawler to collect the great number of URL hyperlinks hidden in JavaScript code. Furthermore, because of the adoption of partial refreshing technology, feedback information of the remote server corresponding to the URL hyperlink hidden in the script code is not an integrated HTML WebPages, which cannot store any data unless processed. Thus, compared with the traditional WebPages crawler, the asynchronous communication-oriented WebPages crawler needs additional processing in terms of URL list extraction and WebPages storage. In accordance with the characteristics of asynchronous communication technology, this paper has improved the workflow of the traditional crawler. The main workflow is as follows:

(1) To use page classifier to distinguish the current page. In accordance with the characteristics of the current page, a classifier model is used to get the classification result, which distinguishes an asynchronous communication technology WebPages from a common WebPages. Reprocessing is only aimed at the WebPages, which has adopted asynchronous communication technology so as to enhance work efficiency of the search engine crawler.
(2) For the WebPages adopting asynchronous communication technology, the URL hyperlinks hidden in the district of script codes are extracted and added in the waiting queue for visits. Meanwhile, URL hyperlinks and the WebPages URL under processing are recorded.
(3) When the current URL hyperlink visited by the crawler is logged in, it suggests that the URL is extracted from a WebPages adopting asynchronous communication technology. It is necessary to process correspondingly the storage of the WebPages adopting asynchronous communication technology in order to rightly store the multi-kind, multi-format feedback information from the server.

3 Crawler Hyperlink Extractions and Processing Technology

The common crawler spider can only visit the main code of the page that is the contents of HTML, so it is not able to effectively extract URL hyperlinks of the WebPages adopting asynchronous communication technology. In order to enable the spider to have access to the contents of the script codes, it is necessary to provide the crawler with an environment that may guarantee the function of the script codes. In addition, the crawler should be able to produce corresponding mouse or keyboard events so that it can guarantee JavaScript events. The concrete workflow is as follows.

Step1: To give the WebPages under processing a unique number for registration.
Step2: To analyze the page property, compare page character string according to the characteristics of JavaScript language and the name list of JavaScript events, extract the function list of the JavaScript events and add the event function to the event queue.
Step3: To provide JavaScript with an operation environment to guarantee the well run of JavaScript events and the sending of the page to the operation environment, and to produce the corresponding mouse or keyboard events in proper order in accordance with the prior extraction event list.
Step4: To monitor the port in the local area, intercept the request if it is sent from the port, analyze the request message according to Http request message format; extract URL, and record URL and the contents of the message.
Step5: To output URL hyperlinks and add URL to the request queue.

According to the above extracting flow, this paper divides the whole hyperlink extracting and processing system into the page analyzer, the event producer, the URL interceptor and the HTTP message analyzer. The page analyzer takes charge of reading the page, using JavaScript event list to analyze page characteristics, extracting JavaScript event function list and adding it to the event list. The event producer is responsible for reading the event list and producing mouse or keyboard events in accordance with event type. The URL interceptor is in charge of monitoring the port and intercepting requests before sending them to the HTTP message analyzer. The HTTP message analyzer is in charge of the analysis of the message, the extraction of URL, the registration of the message contents and so on. The crawler's workflow of hyperlink extraction and subsystem processing is shown in Figure 1.

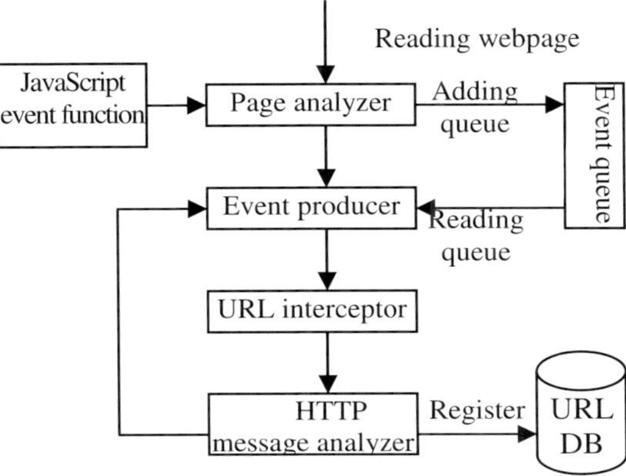

Fig.1. The crawler's workflow of hyperlink extraction and subsystem processing

4 Crawler Page Storage Technologies

The feedback information from the server of a WebPages adopting asynchronous communication technology varies. It's not an integrated page, possibly a part of the page or only an XML file, or even several JavaScript functions. But the search engine indexer requires an integrated page. Therefore, the feedback information must be reprocessed, as the crawler cannot directly store it. Asynchronous communication technology usually adopts callback function making the page refreshed. So the page must be sent to the browser once more, and the feedback data from the server is transmitted so that the callback function could be called and the page refreshed. And then the refreshed page is saved as a page storage cell. The whole page storage flow is as follows:

Step1: To read URL queue and judge whether the URL is extracted from a WebPages adopting asynchronous communication technology, if yes, extract the host computer information from URL and request a link to the host computer. If it is successfully linked, then take the next step; if no, extract the next URL.

Step2: Load in URL and request data from the host computer when the link with the host computer is successful. Delay storing the feedback data flow from the server.

Step3: Refer to the characteristics of HTML page integrity before judge whether the feedback page from the server is integrated or not, if yes, save the page directly; if no, take the next step.

Step4: Look up in the registration list the source page URL corresponding to the current URL.

Step5: Use the homepage URL to request the source page contents from the server.

Step6: Load in both the source page and the delay-stored data to the operation environment, call the callback function, and realize partial refreshing of the source page. Export the integrated page and save in the WebPages database.

5 An Evaluation of the Asynchronous Communication-Oriented Crawler

In order to offer a clear explanation and better contrast, this paper uses the Ajax demonstrative page as a reference to make a simple contrast between the searching effects of the asynchronous communication-oriented crawler and the common full Text search engine crawler, see Table1 for the URL addresses collected by the

Table 1. A list of hyperlinks

(http://openrico.org/rico/demos.page?demo=rico_ajax_inner_HTML)

Id	URLstring	A	B
1	home.page	✓	✓
2	demos.page	✓	✓
3	downloads.page	✓	✓
4	http://Weblog.openrico.org	✓	✓
5	http://forum.openrico.org	✓	✓
6	http://tracking.openrico.org:9004/rico	✓	✓
7	docs.page	✓	✓
8	demos.page?demo=rico_ajax_inner_HTML	✓	✓
9	demos.page?demo=rico_ajax_complex	✓	✓
10	demos.page?demo=rico_drag_and_drop_simple	✓	✓
11	demos.page?demo=rico_drag_and_drop_custom_draggable	✓	✓
12	demos.page?demo=rico_drag_and_drop_custom_dropzone	✓	✓
13	demos.page?demo=rico_effect_position	✓	✓
14	demos.page?demo=rico_effect_size	✓	✓
15	demos.page?demo=rico_effect_size_and_position	✓	✓
16	demos.page?demo=rico_effect_fade_to	✓	✓
17	demos.page?demo=rico_effect_round	✓	✓
18	demos.page?demo=rico_color	✓	✓
19	demos.page?demo=rico_corner	✓	✓
20	demos.page?demo=rico_accordion	✓	✓
21	demos.page?demo=rico_weather	✓	✓
22	livegrid.page	✓	✓
23	yahoo_search	✓	✓
24	ajax_person_info?firstName=Debbie&lastName=Holloman&_=%0D%0A		✓
25	ajax_person_info?firstName=Pat&lastName=Barnes&_=%0D%0A		✓
26	ajax_person_info?firstName=Joan&lastName=Dampier&_=%0D%0A		✓
27	ajax_person_info?firstName=Randy&lastName=Alvarez&_=%0D%0A		✓
28	ajax_person_info?firstName=William&lastName=Neil&_=%0D%0A		✓
29	ajax_person_info?firstName=Kimber&lastName=Hardoway&_=%0D%0A		✓
30	ajax_person_info?firstName=Leslie&lastName=Story&_=%0D%0A		✓
31	ajax_person_info?firstName=Charlie&lastName=Lott&_=%0D%0A		✓
32	ajax_person_info?firstName=Sabrina&lastName=Patton&_=%0D%0A		✓
33	ajax_person_info?firstName=Juan&lastName=Lopez&_=%0D%0A		✓

Note: (1) Host computer is http://openrico.org/rico; (2) A means URL addresses collected by the common full text search engine crawler; (3) B means URL addresses collected by the asynchronous communication-oriented crawler.

common full text search engine crawler or by the asynchronous communication-oriented crawler based on hyperlink analysis.

By contrasting Column A with Column B in Table 1, we can see the items after number 24 in Column B disappear in Column A, it is obvious that the asynchronous communication-oriented mechanism crawler designed and realized in this paper is superior to the traditional full text search engine crawler.

6 Conclusions

This paper has structured a hyperlink extraction and analysis system for the WebPages adopting the asynchronous communication technology. With the aid of script operation environment and windows message response, the hyperlinks hidden in the script code of the WebPages are effectively extracted. In view of the problem that the feedback information from the server is no longer an integrated page that can be directly saved, a crawler page storage system is structured. As is shown by the experiment, the asynchronous communication-oriented page searching technology proposed in this paper is effective. Nevertheless, the asynchronous communication-oriented page searching technology is a very complicated process, I would like to do further research on such problems as how to exploit lightweight JavaScript operation environment and how to optimize saving the WebPages adopting the asynchronous communication technology.

References

1. Mauldin, M.L., Lycos: Design Choices in an Internet Search Service. IEEE Expert 12(1), 8–11 (1997)
2. Gehtland, J., Galbraith, B., Almaer, D.: Pragmatic Ajax, Pragmatic Bookshelf (2006)
3. Merrill, C.L.: Performance Impacts of AJAX Development: Using AJAX to Improve the Bandwidth Performance of Web Applications. Web Performance (2006)
4. Crane, D., Pascarello, E., James, D.: Ajax in Action. Posts & telecom press (2006)
5. Kohlhaseand, M., Sucan, I.: A Search Engine for Mathematical Formulae. Artificial Intelligence and Symbolic Computation, pp. 241–253. Springer, Heidelberg (2006)
6. John Wilbur, W., Kim, W., Xie, N.: Spelling correction in the PubMed search engine. Information Retrieval, 543–564 (September 2006)
7. Hanshen, C., Weizhong, L.: A New Intelligent and Personalized Search Engine Based on C/S. Journal of the China Society for Scientific and Technical Information 25(1), 70–73 (2006)
8. Yihong, T., Xin, W., Tiejun, Z.: Design and Implementation of Concept based Retrieval for Chinese Search Engine. Computer Applications and Software 23(5), 38–40 (2006)

A Novel Fuzzy Kernel C-Means Algorithm for Document Clustering

Yingshun Yin[1], Xiaobin Zhang[1], Baojun Miao[2], and Lili Gao[1]

[1] School of computer science, Xi'an polytechnic university, Shaanxi, China
[2] Schol of mathematical Science, Xuchang University, Henan, China
yinyingshun@yahoo.com.cn, xbzhangcn@gmail.com

Abstract. Fuzzy Kernel C-Means (FKCM) algorithm can improve accuracy significantly compared with classical Fuzzy C-Means algorithms for nonlinear separability, high dimension and clusters with overlaps in input space. Despite of these advantages, several features are subjected to the applications in real world such as local optimal, outliers, the c parameter must be assigned in advance and slow convergence speed. To overcome these disadvantages, Semi-Supervised learning and validity index are employed. Semi-Supervised learning uses limited labeled data to assistant a bulk of unlabeled data. It makes the FKCM avoid drawbacks proposed. The number of cluster will great affect clustering performance. It isn't possible to assume the optimal number of clusters especially to large text corps. Validity function makes it possible to determine the suitable number of cluster in clustering process. Sparse format, scatter and gathering strategy save considerable store space and computation time. Experimental results on the Reuters-21578 benchmark dataset demonstrate that the algorithm proposed is more flexibility and accuracy than the state-of-art FKCM.

Keyword: Text clustering, Semi-supervised Learning, Fuzzy Kernel C-Means, Kernel Validity Index.

1 Introduction

Fuzzy Kernel C-Means (FKCM) algorithm [2] extends kernel methods to Fuzzy C-Means algorithm which is introduced by Bezdek [1]. FKCM algorithm achieves better performance than classical FCM for nonlinearly separable data and clusters with overlaps. FCM algorithm often minimizes the sum of square of Euclidean distance between samples and centroids. The assumption behind this measure is the belief that the data space consist of isolated elliptical regions. However, such an assumption is not always held on real world applications. Mapping the data to higher dimension space satisfies the requirement of the optimization measure.

Though FKCM algorithms have excellent performance in many applications, it suffer from several drawbacks: The c parameter specified in advance and fuzzier m, significant computation time and memory space for introducing kernel function, local optimal and bad convergence speed, These drawbacks restrict application in real world especially for large scale document clustering.

H. Li et al. (Eds.): AIRS 2008, LNCS 4993, pp. 418–423, 2008.

2 Adaptive Semi-supervised Fuzzy Kernel C-Means Algorithm

The major challenges in using FKCM algorithm on text clustering lie not in performing the clustering itself, but rather in choosing the number of clusters and tacking with the high dimensional, sparse document vectors. Worse, kernel functions always consume significant computation time and store space especially for large text collection. More unfortunately, extremely sparse feature vectors and the large difference size of clusters make some vectors be merged into the larger clusters. All these drawbacks are subject to generalize in practical applications.

2.1 Kernelised Validity Index

It's often unfeasible to predefine the number of clusters in advance for large, high-dimensional text data. With an exponential increase in the complexity and volume of data, it is blind to assign labels to document without knowing any information about categories. Many researches have been conducted. Validity indexes find the optimal c cluster that can measure the description of the data structure. It is tradeoff of the compactness and separation [4, 5, 6].

In order to save computation cost, Gauss kernel is extended to the validity index introduced by Bensaid [6]. Then the validity index function can be rewritten as

$$V_{KBszid}(U,V;c)=\sum_{i=1}^{c}\left[\frac{\sum_{j=1}^{n}u_{ij}(1-K(v_i,x_j))}{n_j\sum_{i=1}^{c}(1-K(v_i,v_j))}\right] \qquad \text{where, } n_j=\sum_{j=1}^{n}u_{ij}$$

Let $\sum_{j=1}^{n}u_{ij}(1-K(v_i,x_j))$ be compactness and be separation $n_j\sum_{i=1}^{c}(1-K(v_i,v_j))$

It's readily shown that the following advantages are true. First, it is able to observing the same size of clusters but distinct partitions. Second, by measuring the average compactness sum of each cluster, it's not sensitive to the size of cluster.

2.2 Sparse Format and Scatter-and-Gathering

Each document vector generally has small percentage of nonzero elements. Therefore, storing the data set in sparse format may not only reduce the computational time but also reduce considerable space to store it.

Inherent drawbacks of the kernel do lie in dot production computations consume significant time and kernel function need to marked memory space. To alleviate the expensive kernel computational and store cost, we introduce Scatter-and-Gathering strategy to further enhance performance [9].

The Scatter-and-Gathering strategy is an efficient way of computing vector dot production for sparse vectors. The main idea is to first scatter the sparse vector into a full length vector, then looping through the nonzero element of sparse matrix to evaluate the vector product. The strategy can explore the pipeline effect of the CPU to reduce the number of CPU cycles and lead to significant computing saving. Therefore, the strategy releases the expensive burden for kernel and makes it suitable for large-scale text data.

2.3 Semi-supervised Learning

Text feature vectors are always very high dimensional and extremely sparse, leading to the clusters with rarely data merging into large cluster. So, the performance of clustering suffers from great impact. The key advantage of incorporating prior knowledge into clustering algorithm lies in their ability to enforcing the correlations of feature vectors and enhancing the speed of convergence.

Semi-Supervised clustering falls into two general approaches that we call Constraint-based and Distance-based methods [8, 10].

We introduce the Semi-Supervised Learning into FKCM algorithm. Then, We briefly discusses the problem.

Let labeled vector B=[b_j], $j=1,2,\ldots,n$

$$b_j = \begin{cases} 1, x_j labeled \\ 0, otherwise \end{cases} \quad (1)$$

In the help of labeled vectors, the better the degree of membership F is obtained F=$[f_{ij}]$ $1 \le i \le c, 1 \le j \le n$

For simultaneously obtaining the minimum of distance from clustering data to clustering center and the degree of prior membership, we rewrite

$$J(U,V) = \sum_{i=1}^{c}\sum_{j=1}^{n} u_{ij}^m + \alpha \sum_{i=1}^{c}\sum_{j=1}^{n} (u_{ij}^m - f_{ij}b_j)^m d_{ij}^2$$

$$s.t. \begin{cases} \sum_{i=1}^{c} u_{ij} = 1, j = 1,2,\ldots,n \\ 0 < \sum_{j=1}^{n} u_{ij} < n, i = 1,2,\ldots,c \end{cases} \quad (2)$$

Here, α is coefficient that adjusts the proportion of unsupervised clustering and semi-supervised clustering. The larger α is, the more important of labeled data are. The α is proportional to labeled data. Due to labeled data are far smaller than unlabeled data, we calculate α with the equation $a = \frac{n}{M}$, where n and M represents the total number of objects and the number labeled data respectively.

Minimizing the objective function, we introduce the Lagrange multiplier λ .

$$L(U,V,\lambda) = \sum_{i=1}^{c} u_{ik}^m + \alpha \sum_{i=1}^{c} (u_{ik}^m - f_{ik}b_k)^m d_{ik}^2 - \lambda(\sum_{i=1}^{c} u_{ik} - 1) \quad (3)$$

Setting the partial derivative of the Lagrange λ and variable respectively of equation (3) to zero u_{st} yield

$$\frac{\partial L(U,V,c)}{\partial u_{st}} = mu_{st}^{m-1}d_{st}^2 + 2\alpha m(u_{ik} - f_{ik}b_k)^{m-1}d_{ik}^2 - \lambda(\sum_{i=1}^{c} u_{ik} - 1) = 0 \quad (4)$$

From validity of clustering, we take the optimal value in interval [1.5, 2.5] [4]. Bezdek [1, 6] considered $m = 2$ is the optimal value. So we take $m = 2$, then we have

$$u_{ij} = \frac{1}{1+\alpha}\left\{ \frac{\frac{1+\alpha(1-b_j\sum_{k=1}^{c} f_{kj})}{1-K(v_i,x_j)}}{\sum_{k=1}^{c}\frac{1}{1-K(v_i,x_j)}} \right\} \quad (5)$$

AS^{2}FKCM algorithm

Step1: Initial centroids using labeled data $v_i^{(0)}(i=1,2,\ldots c)$ and termination value ε =0.01

Step2: Compute and update the degree of membership $u_{ij}^{(t)}$ using equation (5)

Step3: Compute and update kernel $K(v_i,x_j)$ using $K(v_i,x_j) = \frac{\sum_{j=1}^{n} u_{ij}^m .K(v_i,x_j)}{\sum_{j=1}^{n} u_{ij}^m}$

Compute kernel using Scatter-and-Gathering strategy. Update $u_{ij}^{(t)}$ to $u_{ij}^{(t+1)}$

Step4: If $\max_{i,j}\left|u_{ij}^{(t)} - u_{ij}^{(t+1)}\right| < \varepsilon$, then stop; otherwise, go to step 3

Step5: If $V_{KBszid}^{t+1} < V_{KBsaid}^{t}$, then $c = c+1$, go to step 2; otherwise, Validity index get the optimal the number of cluster, $c = c-1$ stop.

3 Experiments

3.1 Dataset

Experiments were carried out on the popular dataset which was evaluated performance of text clustering algorithms.

Table 1. The number of labeled and unlabeled documents in five categories

	corn	grain	wheat	sugar	ship
Labeled	10	10	10	1	1
Unlabeled	90	90	90	9	9

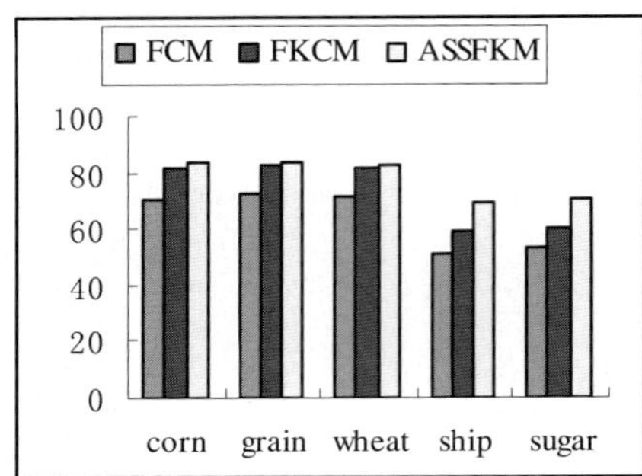

Fig 1. The performances of three algorithms on Reuters-21578

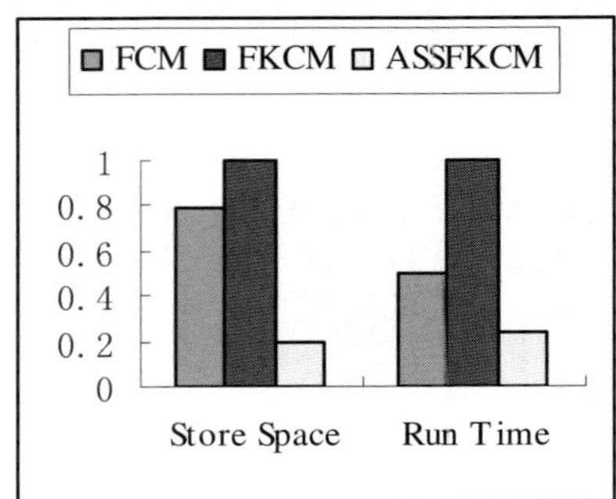

Fig 2. Comparison of store space and run time for three algorithms on the benchmark dataset

Reuters-21578: We constructed the subset of Reuter-21578 by sampling 320 documents from the original dataset. Features were extracted by removing stopwords by stoplist, performing stemming. The 1 percentage documents of each category have been not removed topics. The subset embodies the feature characteristics of a typical text collection, which are high dimensional, sparse, some significantly overlapping and skewed. The briefly describes as follows.

3.2 Results

The results on Reuter-21578 are shown in Figure 1. From the results, we see that the three algorithms perform well regarding balanced dataset with high overlapping. ASSFKCM obtains a small amount of improvement than FKCM for balanced dataset. FKCM gain better results indicate that the boundaries of text clusters are in possession of nonlinear relations. But the FCM and FKCM have no ability to tackle with small datasets overlapped with large dataset. Small datasets always be merged big cluster. Incorporating limited labeled data can marked improve in term of skewed dataset.

A few things need to be noted regarding the run time and the store space in Figure2. Generally, document vectors are all high-dimensional, extremely sparse. Storing the vector in sparse format, we save marked store space. We need not load big matrix into memory time after time. It is important to note that kernel computation and storing consume exponent power in term of the number of objects. Figure 2 indicates that sparse format and scatter-and –gathering strategy achieve sharp decrease.

4 Conclusion

FKCM algorithms which are based on minimizing the objective function have two drawbacks. The first one is sensitivity of algorithms to the initialization of the parameter c . The second is that local optimal and slow convergence speed for skewed clusters. In this paper, we first have introduced a kernelised validity index measures the fitness of clustering algorithms. The objective is to find optimal c clusters that can ensure the best description of the data structures. Then, semi-supervised learning has been explored to enhance the convergence and performance of clustering algorithms. Labeled data efficiently strengthen the correlations of unlabeled data and centroids. Last but least, avoiding the issue of high dimensionality, extremely sparse, we have introduced spare format and scatter-and-gathering strategy. In the end, the experiments on the popular benchmark datasets indicate that the algorithm proposed has high ability to automatic tackling with skewed and pronounced overlapping data clustering.

References

1. Bezdek, J.C.: Pattern Recognition with Fuzzy Objective Function Algorithms. Plenum Press, New York (1981)
2. Wu, Z.-d., Xie, W.-x., Yu, J.-p.: Fuzzy C-means clustering algorithm based on kernel method. In: Proceedings of Fifth International Conference on Computational Intelligence and Multimedia Applications, pp. 49–56 (2003)
3. Shawe-Taylor, J., Cristianini, N.: Kernel Methods for Pattern Analysis, pp. 327–338. Cambridge University Press, Cambridge (2004)
4. Pal, N.R., Bezdek, J.C.: On clustering for the fuzzy c-means model. IEEE Transaction on Fuzzy System 3(3), 370–379 (1995)
5. Xie, X.L., Beni, G.: A validity measure for fuzzy clustering. IEEE Transactions on Pattern Analysis and Machine Intelligence, 841–847 (1999)
6. Bensaid, A.M., Hall, L.O., Bezdek, J.C.: Validity-guided (re)clustering with applications to image segmentation. IEEE Transactions on Fuzzy Systems, 112–123 (1996)
7. Li, K., Liu, Y.: KFCSA:A Novel clustering Algorithm for High-Dimension Data. In: Wang, L., Jin, Y. (eds.) FSKD 2005. LNCS (LNAI), vol. 3613, pp. 531–536. Springer, Heidelberg (2005)
8. Chapelle, O., Schölkopf, B., Zien, A.: Semi-Supervised Learning. MIT Press, Cambridge (2006)
9. Huang, T.-M., Kecman, V., Kopriva, I.: Kernel Based Algorithms for Mining Huge Data Sets: Supervised, Semi-supervised, and Unsupervised Learning. Springer, Berlin (2006)
10. Bouchachia, A., Pedrycz, W.: Data Clustering with Partial Supervision Data Mining and Knowledge Discovery 12, 47–78 (2006)

Cov-HGMEM: An Improved Hierarchical Clustering Algorithm

Sanming Song, Qunsheng Yang, and Yinwei Zhan

Faculty of Computer, Guangdong University of Technology
510006 Guangzhou, P. R. China
SanmSoong@gmail.com, jsjqsy@sina.com, ywzhan@gdut.edu.cn

Abstract. In this paper we present an improved method for hierarchical clustering of Gaussian mixture components derived from Hierarchical Gaussian Mixture Expectation Maximization (HGMEM) algorithm. As HGMEM performs, it is efficient in reducing a large mixture of Gaussians into a smaller mixture while still preserving the component structure of the original mode. Compared with HGMEM algorithm, it takes covariance into account in Expectation-Step without affecting the Maximization-Step, avoiding excessive expansion of some components, and we simply call it Cov-HGMEM. Image retrieval experiments indicate that our proposed algorithm outperforms previously suggested method.

1 Introduction

Content-Based Image Retrieval (CBIR), searching large image repositories according to their content, has emerged as an important area in computer vision and multimedia. A proper image representation and an appropriate distance measure between images are the main issues for an image retrieval framework. We model an image as a set of coherent regions. Each homogeneous region in the image plane is represented by a Gaussian distribution, and the set of all the regions in the image is represented by a Gaussian mixture model (GMM). It is exhaustive to traverse in a large database. Hierarchically searching could be a promising solution. All Gaussian mixture components that belong to the same class were clustered into a simpler GMM. Then, evidently, it makes work easy to navigating from coarse classes to fine images. HGMEM proposed by N.Vasconcelos in [1,2], is an efficient algorithm in reducing a large mixture of Gaussians into a smaller mixture while still preserving the component structure of the original mode, which is achieved by clustering the components. Without taking the differences of covariance between components into consideration, it usually excessively expands those components have big covariance, leading to a final GMM which doesn't reflect the real semantics of the class. Semantic multinomial (SMN) was adopted in [3] to denote the similarity between query image and image classes, which not only result in too complex computation, but also enlarge the time complexity.

In this paper, we propose an improved hierarchical clustering algorithm-Cov-HGMEM. Image retrieval experiments show that our proposed algorithm not only inherit the exceptional computational complexity of HGMEM, but also improves retrieval efficiency and proficiency.

H. Li et al. (Eds.): AIRS 2008, LNCS 4993, pp. 424–429, 2008.

2 Basic Hierarchical Image Retrieval Framework

2.1 Image Modeling

An image is represented as a GMM in feature space. Here, we only take color into consideration. It should be noted that the representation model can be extended to any feature space. We choose to work in (L,u,v) color space [10]. Besides, including the spatial information enabling a localized representation [6], we append the (x, y) position of pixel to the feature vector. Therefore, each pixel is represented by a 5-dimensional feature vector (L,u,v,x,y). Expectation-Maximization (EM) [8] algorithm is adopted to determine the parameters. The number of Gaussian mixtures could be selected by the Minimum Description Length (MDL) principle [7], we simply choose 8 [9] in this paper. Figure 1 shows an example of learning a GMM for a given image. Each Gaussian in the model is displayed as a localized colored ellipsoid.

Fig.1. Sample image and its GMM components

2.2 HGMEM Algorithm and Image Class Modeling

Denote $\left(\pi_i^{l+1},\mu_i^{l+1},\Sigma_i^{l+1}\right)$ the a priori probability, mean vector and covariance of the ith component at child level $l+1$, $\left(\pi_j^l,\mu_j^l,\Sigma_j^l\right)$ the jth component at father level l. In image analysis, it corresponds to components of image and image class respectively. Assume the total number of virtual samples taken from child level l is M. HGMEM algorithm iterates between the following steps:

Expectation-Step: Compute

$$h_{ij} = \frac{\pi_j^l\left(G\left(\mu_i^{l+1},\mu_j^l,\Sigma_j^l\right)e^{-\frac{1}{2}\mathrm{trace}\left[\left(\Sigma_j^l\right)^{-1}\left(\Sigma_i^{l+1}\right)\right]}\right)^{\mathrm{M}_i}}{\sum_k \pi_k^l\left(G\left(\mu_i^{l+1},\mu_k^l,\Sigma_k^l\right)e^{-\frac{1}{2}\mathrm{trace}\left[\left(\Sigma_k^l\right)^{-1}\left(\Sigma_i^{l+1}\right)\right]}\right)^{\mathrm{M}_i}}, \tag{1}$$

where $\mathrm{M}_i = \pi_i^{l+1}\mathrm{M}$ denotes the size of virtual samples from the ith child component, and $G(,,)$ stands for Gaussian mixture component.

Fig.2. Image class “sunset” and the final models using different clustering algorithms ($k = 4$) (a) Samples from “sunset”, (b) Results given by HGMEM, (c) Results given by Cov-HGMEM

Maximization-Step: Set

$$\pi_j^l = \frac{\sum_i h_{ij}}{C^{l+1}}, \qquad \mu_j^l = \frac{\sum_i h_{ij} \mathbf{M}_i \mu_i^{l+1}}{\sum_i h_{ij} \mathbf{M}_i},$$
$$\Sigma_j^l = \frac{1}{\sum_i h_{ij} \mathbf{M}_i} \left[\sum_i h_{ij} \mathbf{M}_i \Sigma_i^{l+1} + \sum_i h_{ij} \mathbf{M}_i \left\| \mu_i^{l+1} - \mu_j^l \right\|^2 \right]. \tag{2}$$

Fig 2(a) shows 12 images belong to class “sunset”, if we extract 8 Gaussian components from each image, there would be 96 in total, we cluster them into a GMM with only 4 components and show the result in Fig.2 (b).

2.3 Similarity Function

Consider a query mixture $P(x) = \sum_{j=1}^{C} G(x, \mu_{q,j}, \Sigma_{q,j})\, p(\omega_j)$ and a database mixture $P_i(x) = \sum_{k=1}^{C_i} G(x, \mu_{i,k}, \Sigma_{i,k})\, p_i(\omega_k)$, where C and C_i represents the number of components of query image and database image respectively. ALA [4] is defined

$$\mathrm{ALA}(P \,\|\, P_i) = \sum_j p(\omega_j) \left\{ \log p_i\left(\omega_{\beta(j)}\right) + \left[\log G\left(\mu_{q,j}, \mu_{i,\beta(j)}, \Sigma_{i,\beta(j)}\right) - \frac{1}{2} trace\left[\Sigma_{i,\beta(j)}^{-1} \Sigma_{q,j} \right] \right] \right\}, \tag{3}$$

where,

$$\beta(j) = k \Leftrightarrow \left\| \mu_{q,j} - \mu_{i,k} \right\|_{\Sigma_{i,k}}^2 - \log p_i(\omega_k) < \left\| \mu_{q,j} - \mu_{i,l} \right\|_{\Sigma_{i,k}}^2 - \log p_i(\omega_l), \forall l \neq k. \tag{4}$$

The plausibility of the two assumptions that ALA has to satisfy [4] grows with dimension of the feature space.

3 Ameliorate HGMEM Algorithm -Cov-HGMEM

Go to Equation (1) and consider the expression

$$G\left(\mu_i^{l+1},\mu_j^l,\Sigma_j^l\right)e^{-\frac{1}{2}\text{trace}\left[\left(\Sigma_j^l\right)^{-1}\left(\Sigma_i^{l+1}\right)\right]} \tag{5}$$

Consider $G\left(\mu_i^{l+1},\mu_j^l,\Sigma_j^l\right)=\dfrac{1}{\sqrt{(2\pi)^D\left|\Sigma_j^l\right|}}e^{-\frac{1}{2}\left\|\mu_i^{l+1}-\mu_j^l\right\|_{\Sigma_j^l}^2}$, where D denotes the dimension of feature space. It is obvious that $0<e^{-\frac{1}{2}\left\|\mu_i^{l+1}-\mu_j^l\right\|_{\Sigma_j^l}^2}\le 1$, so if $\left|\Sigma_j^l\right|>\dfrac{1}{(2\pi)^D}$ holds, expression (5) ranges from 0 to 1 , for $e^{-\frac{1}{2}\text{trace}\left[\left(\Sigma_j^l\right)^{-1}\left(\Sigma_i^{l+1}\right)\right]}$ also lies in 0 and 1 . The condition can be met easily in high dimension spaces. Experiments show it is very small. So, taking the power of M_i will lead to a catastrophic result, it is very inaccurate. But Expectation-Step holds only when M_i goes to infinite.

Generally speaking, when M_i small, component that has a big covariance influenced much. So, adding an affection factor relates to covariance may be a promising choice. Let's return to deduction process in [1]. Adding a factor $a\left|\Sigma_j^l\right|$ to equation [1](9), a being a constant, we obtain

$$\begin{aligned}\log P\left(X_i \mid z_{ij}=1,M_l\right)&=\mathrm{M}_i\left[\frac{1}{\mathrm{M}_i}\sum_{m=1}^{\mathrm{M}_i}\log p\left(x_i^m \mid z_{ij}=1,M_l\right)\right]\\ &\approx \mathrm{M}_i E_{M_{l+1,i}}\left[\log P\left(x \mid z_{ij}=1,M_l\right)-\log\left(a\left|\Sigma_j^l\right|\right)\right]\\ &=\mathrm{M}_i\left\{-\frac{1}{2}\text{trace}\left[\left(\Sigma_j^l\right)^{-1}\left(\Sigma_i^{l+1}\right)\right]+\log G\left(\mu_i^{l+1},\mu_j^l,\Sigma_j^l\right)-\log\left(a\left|\Sigma_j^l\right|\right)\right\},\end{aligned} \tag{6}$$

that is,

$$P\left(X_i \mid z_{ij}=1,M_l\right)=\left(\frac{G\left(\mu_i^{l+1},\mu_j^l,\Sigma_j^l\right)e^{-\frac{1}{2}\text{trace}\left[\left(\Sigma_j^l\right)^{-1}\left(\Sigma_i^{l+1}\right)\right]}}{\left(a\left|\Sigma_j^l\right|\right)}\right)^{\mathrm{M}_i}. \tag{7}$$

Substitute Equation (7) in post probability equation [1](9), Expectation-Step becomes

$$h_{ij}=\frac{P\left(X_i \mid z_{ij}=1,M_l\right)}{\sum_k \pi_k^l P\left(X_i \mid z_{ik}=1,M_l\right)}. \tag{8}$$

We needn't modify Maximization-Step, and we briefly call Equation (8) and (2) Cov-HGMEM iteration steps.

It obviously that modification on the expression of Expectation-Step will not influence Maximization-Step.

Similar to HGMEM algorithm, we model image class as a simpler GMM using Cov-HGMEM algorithm. As shown in Fig. 2(c), Cov-HGMEM performs better than HGMEM algorithm. For, taking covariance into consideration, components with small covariance will not be absorbed into big ones. The final mixture model reflects the semantic of original class more accurately.

4 Experimental Evaluation

To evaluate the efficiency of our proposed Cov-HGMEM Algorithm in comparison to HGMEM, we conducted experiments with image sets selected from Corel database, including sunset, coast scenes, skiing , waterfall, elephant, horse, flowers, and so on.

We start with image modeling and image class modeling as Gaussian mixture in feature space. And the essential step is to find the best matched class. We use probability of class match order (PCMO) to compare the efficiencies of two clustering algorithms. Assume class match order counter (CMOC) is an array with length 20. For each picture in the database, we obtain 20 ALA values by matching to each class. After arrangement, if the ith $(i=1,...20)$ class is the object class, CMOC[*i*] increase by 1. When the traverse finished, we obtain average PCMO by dividing CMOC with the number of images in database.

Fig.3 compares HGMEM and Cov-HGMEM systematically. We plot the probability mass function in Fig.4 (a), as it shows, PMCO[1], the probability that the most similar class is the object class lifts from 37.00% to 57.45%! Due to overlap between image classes, the most similar class doesn't necessarily correspond to the object class. Under that condition, we may search in the most similar classes. The probability that the object class contained in the most similar 4 classes lifts from 72.55% to 86.25%. Fig.4 (b) shows PMCO[1] for each image class. And we found HGMEM performs better. With parameters a in Cov-HGMEM, we evaluate it experimentally. We found algorithm has best retrieval performance when $a = 0.001$,.

As mentioned above, we only adopt spatial and range information as feature, so retrieval performances is sensitive to color. ALA values would be very close if two classes have a similar Gaussian component. For example, as shown in Fig.1, sky in "steamboat and snow in "snow scenes" seems similar, and both prevail a big area respectively.

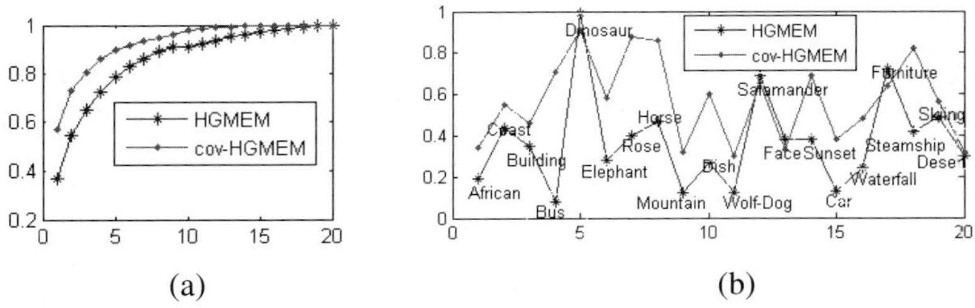

Fig.3. (a) Probability mass function of PMCO, (b) PMCO for each class

5 Conclusions

In this paper, we propose a new and efficient hierarchical clustering algorithm-Cov-HGMEM. Compared with HGMEM algorithm, Cov-HGMEM tries to prevent excessive expansion of those components which have a big covariance, preserving original component structure better. Image retrieval Experiments show that our proposed algorithm performs better than HGMEM.

Acknowledgments

This work was supported in part by the National Natural Science Foundation of China (Grant No. 60572078) and by Guangdong Natural Science Foundation (Grant No. 05006349).

References

1. Vasconcelos, N., Lippman, A.: Learning mixture hierarchies. In: Proc. of Neural Information Processing Systems (1998)
2. Vasconcelos, N.: Image Indexing with Mixture Hierarchies. In: Proceedings of IEEE Conference on Computer Vision and Pattern Recognition, Kauai, Hawai (2001)
3. Rasiwasia, K.N., Vasconcelos, N., Moreno, P.J.: Query by Semantic Example. In: Sundaram, H., Naphade, M., Smith, J.R., Rui, Y. (eds.) CIVR 2006. LNCS, vol. 4071, pp. 51–60. Springer, Heidelberg (2006)
4. Vasconcelos, N.: On the Efficient Evaluation of Probabilistic Similarity Functions for Image Retrieval. IEEE Transactions on Information Theory 50(7), 1482–1496 (2004)
5. Kullback, S.: Information theory and statistics. Dover Publications, NewYork (1968)
6. Goldberger, J., Greenspan, H., Gordon, S.: Unsupervised image clustering using the information bottleneck method. In: The Annual Pattern Recognition Conference DAGM, Zurich (2002)
7. Rissanen, J.: Stochastic Complexity in Statistical Inquiry. World Scientific, Singapore (1989)
8. Dempster, A., Laird, N., Rubin, D.: Maximum-likelihood from incomplete data via the EM algorithm. J. Roy. Statist. Soc. B-39 (1977)
9. Jeong, S., Gray, R.M.: Minimum Distortion Color Image Retrieval Based on Lloyd-Clustered Gauss Mixtures. In: Proceedings of the Data Compression Conference, pp. 279–288 (2005)
10. Wyszecki, G., Stiles, W.S.: Color Science: Concepts and Methods, Quantitative Data and Formulae, 2nd edn. Wiley, Chichester (1982)

Improve Web Image Retrieval by Refining Image Annotations

Peng Huang, Jiajun Bu*, Chun Chen, Kangmiao Liu, and Guang Qiu

College of Computer Science, Zhejiang University, Hangzhou, China
{hangp,bjj,chenc,lkm,qiuguang}@zju.edu.cn

Abstract. Automatic image annotation techniques are proposed for overcoming the so-called semantic-gap between image low-level feature and high-level concept in content-based image retrieval systems. Due to the limitations of techniques, current state-of-the-art automatic image annotation models still produce some irrelevant concepts to image semantics, which are an obstacle to getting high-quality image retrieval. In this paper we focus on improving image annotation to facilitate web image retrieval. The novelty of our work is to use both WordNet and textual information in web documents to refine original coarse annotations produced by the classic Continuous Relevance Model (CRM). Each keyword in annotations is associated with a certain weight, and larger the weight is, more related to image semantics the corresponding concept is. The experimental results show that the refined annotations improve image retrieval to some extent, compared to the original coarse annotations.

1 Introduction

Currently web image retrieval systems usually fall into two main categories: text-based and content-based. In text-based systems, images are first annotated with texts which are produced by people manually, or extracted from image surroundings automatically, then text retrieval techniques are used to performed image retrieval. However, annotating image manually is tedious, time-consuming and subjective, while annotating images automatically with surroundings often involves terms irrelevant to image semantics unavoidably. Content-based image retrieval (CBIR) systems automatically index and images by their low-level visual features. However, it is flawed in the following ways. Firstly, the user query must be provided in the form of a draft of the desired image. Secondly, the images with similar low-level features may have different contents [8]. Finally, there is so-called semantic gap between image low-level features and high-level concepts.

Recently, many approaches have been proposed to automatically annotate images with keywords [7,2,6]. Automatic image annotation is a promising methodology for image retrieval. However it is still in its infancy and is not sophisticated enough to extract perfect semantic concepts according to image low-level features, often producing noisy keywords irrelevant to image semantics. Noisy concepts may be an obstacle to getting high-quality image retrieval. In this paper we propose a novel approach to improve image retrieval, which utilizes *coherence* between coarse concepts and *relatedness* between concepts and web textual information to refine image annotations. The

* Corresponding author.

H. Li et al. (Eds.): AIRS 2008, LNCS 4993, pp. 430–435, 2008.

refinement is based on two basic assumptions. One assumption is that concepts contained in an image should be semantically related to each other. Another assumption is intuitive that the observation of some specific terms in web documents should increase the belief of certain semantically similar concept. For example, if 'tiger' is included in web documents, the annotated concept 'tiger' should be more credible than before the observation. In this paper image is annotated as follows: first we use the classic annotation model CRM [6] to associate an image with a set of keywords (coarse concepts); then these coarse concepts are associated with weights which are calculated from *coherence* and *relatedness* using ontological lexicon WordNet [3]. The model proposed by Jin et al. [5] also uses WordNet to improve image annotations, but there are two important differences comparing to our work. First, Jin et al. only take into account *coherence* to remove noisy concepts, while we use extra text in web documents. Second, Jin et al. focused on eliminating 'noisy' concepts, while we focus on applying these weighted concepts to improving the ranking of image retrieval results.

The remainder of this paper is organized as follows. Section 2 briefly introduces the classic image annotation model, CRM. Section 3 describes how to refine coarse concepts produced by CRM in details, together with a brief introduction to an image retrieval prototype. Section 4 presents experimental results and some discussions. The last section concludes this paper plus some ideas for future work.

2 Continuous Relevance Model

Let V be the annotation vocabulary, T be the training set, J be an element of T. J is partitioned into a set of fixed-size small regions $\mathbf{r}_J = \{r_1, \ldots, r_n\}$, along with corresponding annotation $\mathbf{w}_J = \{w_1, \ldots, w_m\}$ where $w_i \in V$. The Continuous Relevance Model [6] (CRM) assumes that generating J is based on three distinct probability distributions. First, the set of annotation words $\mathbf{w}_J$ is a result of $|V|$ independent samples from underlying multinomial distribution $P_V(\cdot|J)$. Second, each image region r is a sample of a real-valued feature vector g using a kernel-based probability density function $P_G(\cdot|J)$. Finally, the rectangular region r is produced according to some unknown distribution conditioned on g, so $\mathbf{r}_J$ are produced from a corresponding set of vectors $g_1 \ldots g_n$ according to a process $P_R(r_i|g_i)$ which is independent of J. Now let $\mathbf{r}_A = \{r_1, \ldots, r_n\}$ be the feature vectors of certain image A, which is not in the training set T. Similarly, let $\mathbf{w}_B$ be some arbitrary subset of V ($|\mathbf{w}_B| = m$). Then we like to model $P(\mathbf{r}_A, \mathbf{w}_B)$, the joint probability of observing an image defined by $\mathbf{r}_A$ together with annotation words $\mathbf{w}_B$. The observation of $\{\mathbf{r}_A, \mathbf{w}_B\}$ can be supposed to come from the same process that generated one of the image J^* in the training set T. Formally, the probability of a joint observation $\{\mathbf{r}_A, \mathbf{w}_B\}$ is:

$$P(\mathbf{r}_A, \mathbf{w}_B) = \sum_{J \in T} P_T(J) \cdot \prod_{b=1}^{m} P_V(w_b|J) \times \prod_{a=1}^{n} \int_{\mathbb{R}^k} P_R(r_a|g_a) P_G(g_a|J) dg_a \quad (1)$$

So given a new image we can split it into regions $\mathbf{r}_A$, compute feature vectors $g_1 \ldots g_n$ for each region and then use formula 1 to determine what subset of vocabulary w^* is the most likely to co-occur with the set of feature vectors:

$$\mathbf{w}^* = arg \max_{\mathbf{w} \in V} \frac{P(\mathbf{r}_A, \mathbf{w})}{P(\mathbf{r}_A)} \tag{2}$$

Here we only give a brief introduction to CRM, and for details please refer to [6].

3 Refining Image Annotations

In previous section, we have described how to use CRM to assign an a coarse concept sequence $(c_1, \ldots, c_T)$ to image. However, some concepts are possible noisy or incorrect with respect to image semantics. In what follows we will describe how to distinguish these 'noisy' concepts from others by using the notions of *coherence* and *relatedness* based on WordNet. The notion of *coherence* assumes that concepts in annotations should be semantically similar each other, while *relatedness* refers to the semantic similarity between image annotations and terms in web documents.

3.1 Measuring *Coherence* and *Relatedness*

The JCN algorithm [4] is adopted here to measure the similarity between words (concepts) due to its effectiveness, in which the similarity measure of two concepts 'c_1' and 'c_2' is based on the notations of Information Content (*IC*) and concept-distance, defined as:

$$sim_{jcn}(c_1, c_2) = \frac{1}{dist_{jcn}(c_1, c_2)} = \frac{1}{IC(c_1) + IC(c_2) - 2 \times IC(lcs(c_1, c_2))} \tag{3}$$

where $IC(c) = -logP(c)$ and $P(c)$ is the probability of encountering an instance of concept 'c' in WordNet; $lcs(c_1, c_2)$ is the lowest common sub-summer that subsumes both concepts 'c_1' and 'c_2'. Note that all measures are normalized so that they fall within a 0-1 range. For simplicity, normalization factor is omitted, simply assuming that $0 \le sim_{jcn}(c_i, c_j) \le 1$.

Let $C = (c_1, \ldots, c_T)$ be the coarse concepts produced by CRM, $D = (d_1, \ldots, d_n)$ be the terms in page title and image surroundings etc., then the measure of *coherence* of concept c_i is defined as:

$$a_i = \frac{1}{\eta_1} \sum_{j=1 \wedge j \ne i}^{T} sim_{jcn}(c_j, c_i), \quad \eta_1 = \sum_{i=1}^{T} \sum_{j=1 \wedge j \ne i}^{T} sim_{jcn}(c_j, c_i) \tag{4}$$

where η_1 is normalization factor. Similarly, the measure of increased belief for a concept c_i according to the relatedness between concept c_i and textual information D is defined as:

$$b_i = \frac{1}{\eta_2} \sum_{j=1}^{n} sim_{jcn}(d_j, c_i), \quad \eta_2 = \sum_{i=1}^{T} \sum_{j=1}^{n} sim_{jcn}(d_j, c_i) \tag{5}$$

Now we get two variables, a_i and b_i, as the measure of the importance of concept c_i to the semantics of an image. Note that $\sum a_i$and $\sum b_i$ are both 1, that is to say, a_i

and b_i can be regarded as two independent probability distributions of the quantified importance of concept c_i. We combine these two factors linearly as follows:

$$s(c_i) = (a_i + b_i)/2 \tag{6}$$

where $s(c_i)$ is the final score associated with concept c_i. The larger $s(c_i)$ is, the more important it is to the semantics of corresponding image.

3.2 Retrieval Prototype

In the rest of this paper, we refer to $C = \{c_1, \ldots, c_T\}$ as the annotation-set of image *I*, and $Q = \{q_1, \ldots, q_m\}$ as the query. Let *n* be the number of index terms in the system, k_i be a generic index term. $K = \{k_1, \ldots, k_n\}$ is the set of all index terms. A weight $w_{i,c} \geq 0$ is associated with each index term k_i of a annotation-set *C*. For an index term which does not appear in the annotation-set *C*, $w_{i,c} = 0$. With the annotation-set *C* associated, an index term vector $\overrightarrow{c}$ is represented by $\overrightarrow{c} = (w_{1,c}, w_{2,c}, \ldots, w_{n,c})$. Similarly, let $w_{i,q}$ be the weight associated with the pair (k_i, Q) where $w_{i,q} \geq 0$, and $\overrightarrow{q} = (w_{i,q}, \ldots, w_{n,q})$, then we rank the retrieved images as follows:

$$rank(I,Q)=rank(\overrightarrow{c},\overrightarrow{q})=\frac{\overrightarrow{c}\cdot\overrightarrow{q}}{|\overrightarrow{c}|\times|\overrightarrow{q}|}=\frac{\sum_{i=1}^{n}(w_{i,c}\times w_{i,q})}{\sqrt{\sum_{i=1}^{n}(w_{i,c})^2}\times\sqrt{\sum_{i=1}^{n}(w_{i,c})^2}} \tag{7}$$

where the specifications of $w_{i,c}$ and $w_{i,q}$ are as follows::

$$w_{i,c} = \begin{cases} s(k_i), \; k_i \in C \wedge s(k_i) \geq \beta \\ 0, \quad \text{otherwise.} \end{cases}, \; w_{i,q} = \begin{cases} 1, \; k_i \in Q \\ 0, \; \text{otherwise.} \end{cases} \tag{8}$$

where *C* is the annotation-set for image *I* produced by CRM, $s(k_i)$ is the scoring function in formula 6, and β is a threshold for filtering noisy concepts whose scores are below it. In our experiment β was set be 0.1 empirically. For the sake of comparison, we implemented another probability based ranking strategy like in [6]: given a text query Q, we get a conditional probability $P(Q|J)$ for image J according to formula 1, then retrieved images are ranked according to $P(Q|J)$. In short, we use two ranking strategies in this paper, one is the proposed vector-based ranking (VIR) and another is probability-based ranking (PIR).

4 Experiment and Results

The training data set is the Corel Image Dataset, consisting of 5000 images from 50 Stock Photo CDs. Each image is partitioned into 4×6 regions. These images are annotated with words drawn from a vocabulary of size 374, denoted by *V*. In addition, we have previously downloaded 10,000 web pages from the WWW accompanied with images. These images are used as test set in the experiments. We selected top 25 frequent terms in training set as test queries. The CRM was used to annotate each web image with up to 5 keywords. These keywords were used as image indexes for image retrieval.

We used *precision* and *recall* metrics to evaluate the image retrieval results. Given a query *Q* and a set *R* of relevant images for a query *Q*, we obtained a set *A* of relevant

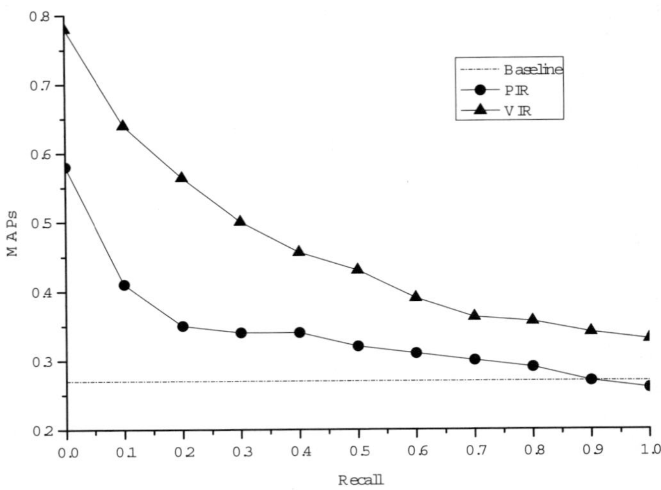

Fig. 1. MAPs of 25 queries using PIR and VIR ranking strategies plus Baseline

images for *Q* through one of above two ranking strategies, then precision is $|A \cap R|/|A|$, and recall is $|A \cap R|/|R|$. To determine the set *R* of relevant images to each of the 25 test queries, we adopted a strategy in [1]: For each test query, we ran out two ranking strategies above. The 40 highest ranked images returned by each of the two ranking strategies were pooled into a set of unique images and then classified by volunteers as relevant or irrelevant with respect to the query term. At the same time, the byproduct of the construction of R is a small set of images which have been labeled be relevant or irrelevant to certain query term by human, denoted by D_R. So we evaluated CRM over D_R as the baseline. Note that all images in $|D_R|$ was associated with the query word, so the precision of CRM over D_R is the number of images correctly annotated with a given word, divided by $|D_R|$. Additionally, it is evident that the recall of CRM over D_R is 100%. We calculated the mean precision for 25 queries and obtained 27%. It was used as baseline and depicted in figure 1.

Usually we want to evaluate average precision at given recall levels. The standard 11-point average precision curve is used for this purpose. It plots precisions at 0 percent, 10 percent, ..., 100 percent of recalls. The mean average precision (*MAP*) is the arithmetic mean of average precision calculated over all queries (here 25) at some specific percent recall. Note that the results of CRM was a straight line, rather than a curve, because they were annotation accuracies rather than retrieval accuracies. The results were depicted in figure 1. As indicated earlier, the baseline is the mean precision of CRM over D_w for 25 query terms w. The curve for baseline in figure 1 reveals the weakness of automatic image annotation technique in image retrieval task without any ranking strategy, only 27 percent precision. In contrast, both of PIR and VIR ranking strategies improved image retrieval. Furthermore, the performance of the proposed approach (VIR) is overall superior to PIR owing to the removal of some noisy concepts and more reasonable weights associated with concepts. Because some noisy concepts were removed, the final precision at recall level 100 percent of VIR is above that of PIR. Especially, in our experiment some retrieved images by using PIR ranking strategy would never be retrieved by using VIR ranking strategy because the correct annotation keyword, i.e. query term, was accidentally removed as noise. For this situation, we simply removed this image from the image pool. This should not affect our final conclusions, since this happened seldom.

5 Conclusions and Future Work

Due to the limitations of current techniques, image annotations have a poor performance in image retrieval systems. To mitigate this problem, we propose an model which scores each annotated concept using semantic similarity measure based on knowledge-based lexicon WordNet. The experimental results show that the precision is improved to some extent. Moreover, after re-ranking, most correctly annotated images are associated with higher rank. In real life it is reasonable since users often are interest to a first couple of retrieval results. However, some problems still need be further researched. The experimental training data has a limited size of vocabulary, so the annotation results have a low coverage over total keyword space. In addition, the evaluation of image retrieval is conducted on a small set of retrieved results using only top 25 terms, since judging relevancy/irrelevancy to test queries requires substantive human endeavors. A wider evaluation on larger data set will be carried out in future work

References

1. Coelho, T.A.S., Calado, P.P., Souza, L.V., Ribeiro-Neto, B.: Image retrieval using multiple evidence ranking. Image 16(4), 408–417 (2004)
2. Duygulu, P., Barnard, K., de Freitas, N., Forsyth, D.: Object recognition as machine translation: Learning a lexicon for a fixed image vocabulary. In: Proceedings of the 7th European Conference on Computer Vision-Part IV, pp. 97–112 (2002)
3. Fellbaum, C.: Wordnet: an electronic lexical database. MIT Press, Cambridge (1998)
4. Jiang, J.J., Conrath, D.W.: Semantic similarity based on corpus statistics and lexical taxonomy. In: Proceedings of International Conference on Research in Computational Linguistics, pp. 19–33 (1997)
5. Jin, Y., Khan, L., Wang, L., Awad, M.: Image annotations by combining multiple evidence & wordnet. In: Proceedings of the 13th annual ACM international conference on Multimedia, pp. 706–715 (2005)
6. Lavrenko, V., Manmatha, R., Jeon, J.: A model for learning the semantics of pictures. In: Proceedings of Advance in Neutral Information Processing (2003)
7. Mori, Y., Takahashi, H., Oka, R.: Image-to-word transformation based on dividing and vector quantizing images with words. In: First International Workshop on Multimedia Intelligent Storage and Retrieval Management (1999)
8. Sheikholeslami, G.C., Zhang, W.A., Syst, C., Jose, C.A.S.: Semquery: semantic clustering and querying on heterogeneous features for visual data. Knowledge and Data Engineering, IEEE Transactions on 14(5), 988–1002 (2002)

Story Link Detection Based on Event Model with Uneven SVM

Xiaoyan Zhang, Ting Wang, and Huowang Chen

Department of Computer Science and Technology, School of Computer,
National University of Defense Technology
No.137, Yanwachi Street, Changsha, Hunan 410073, P.R. China
{zhangxiaoyan,tingwang,hwchen}@nudt.edu.cn

Abstract. Topic Detection and Tracking refers to automatic techniques for locating topically related materials in streams of data. As a core of it, story link detection is to determine whether two stories are about the same topic. Up to now, many representation models have been used in story link detection. But few of them are specific to stories. This paper proposes an event model based on the characters of stories. This model is used for story link detection and evaluated on the TDT4 Chinese corpus. The experimental results indicate that the system using the event model achieves a better performance than that using the baseline model. Furthermore, it shows a larger improvement to the former, especially when using uneven SVM as the multi-similarity integration strategy.

Keywords: story link detection, event model, uneven SVM.

1 Introduction

Topic Detection and Tracking (TDT) [1] refers to a variety of automatic techniques for discovering and threading together topically related materials in streams of data such as newswire or broadcast news. As a core of TDT, story link detection is the task of determining whether two stories are about the same topic. TDT defines “topic” to mean a specific event or activity plus directly related events or activities. An event is "something that happens at some specific time and place".

A number of works have been developed on story link detection, which can be classified into two categories: vector-based methods and probabilistic-based methods.

As the vector-based approaches are widely used in IR and Text Classification research, cosine similarity between documents vectors with *tf*idf* term weighting[2,3] has also been one of the best methods for link detection. We have also examined a number of similarity measures in the link detection task, including weighted sum, language modeling, Kullback-Leibler divergence, Hellinger and Tanimoto, and find that cosine similarity produces the most outstanding results. Furthermore, [4] also confirms this conclusion in its story link detection research. Probabilistic-based methods have been proved to be very effective in several IR applications. One of their attractive features is that it is firmly rooted in the theory of probability, thereby allows the researcher to explore more sophisticated models guided by the theoretical framework. [5,6] both apply probability-based models (language model or relevance model) to

H. Li et al. (Eds.): AIRS 2008, LNCS 4993, pp. 436–441, 2008.

story link detection, and the experimental results indicate that the performances are comparable with those using traditional vector space models, if not better. The story models are all vector-based in this paper. We have concluded in our previous research that the multi-vector model is superior to the single-vector model for news stories. So a multi-vector model is used as the baseline model. However, we know that a story usually describes an event mainly constructed with time, location, person, organization, etc. So a new event model is proposed to represent the news story. The experimental results show that the event model is more proper for stories in TDT.

The paper is organized as follows: Section 2 simply describes the procedure for story link detection; Section 3 describes the baseline multi-vector model and the event model, which share preprocessing, feature weighting, similarity computation and multi-similarity integration except model construction. The experimental results and analysis are given in Section 4; finally, Section 5 concludes the whole paper.

2 Problem Definition

In story link detection, a system is given by a sequence of time-ordered news source files $S=\langle S_1, S_2, \ldots, S_n\rangle$, each S_i $(i=1, 2, \ldots, n)$ includes a set of stories and a sequence of time-ordered story pairs $P=\langle P_1, P_2, \ldots, P_m\rangle$, where $P_i= (s_{i1}, s_{i2})$, $s_{i1}\in S_j$, $s_{i2}\in S_k$, $1\le i\le m$, $1\le j\le k\le n$. The link system is required to make decisions on each story pair to judge if two stories in it are topically linked. The procedure of processing a story pair is as follows. For current story pair $P_i= (s_{i1}, s_{i2})$:

1. Get background corpus B_i of P_i. Normally, according to the supposed application situation, the system is allowed to look ahead N (usually 10) source files when deciding whether the current pair is linked. So,

$$B_i = \{S_1, S_2, \ldots, S_l\}, \text{ where } l = \begin{cases} k+10, & s_{i2} \in S_k \ and \ (k+10) \le n \\ n, & s_{i2} \in S_k \ and \ (k+10) > n \end{cases}.$$

2. Produce the representation models (M_{i1}, M_{i2}) for two stories in P_i. $M= \{(f_s, w_s) \mid s\ge 1\}$, where f_s is a feature extracted from the story and w_s is the weight of the feature in the story. They are computed with some parameters counted from current story or the background.
3. Choose a similarity function F and compute the similarity between two models. If $F(M_{i1}, M_{i2})$ is larger than a predefined threshold, then two stories are decided to be topically linked.

3 Baseline Model and Event Model

First of all, a preprocessing is provided for all the stories. For each story we tokenize the text, tag the tokens, remove stop words, and get a candidate set of features for its vector-based model. In this paper, a token plus its tag is taken as a feature. If two tokens with same spelling are tagged with different tags, they will be taken as different features. The segmenter and tagger are completed by ICTCLAS[1]. The stop word list contains 507 words.

3.1 Model Construction

In the baseline model, we divide the feature set into disjoint subsets according to the tags of tokens in the set. One vector represents a subset. After that ten vectors are picked out to represent the story. The ten corresponding tags are person name, location name, organization name, number, time, noun (including noun and proper noun), verb (including verb, associate verb and nounness verb), adverb, adjective (including adjective, associate adjective and nounness adjective), idiom (including idiom and phraseology). We think that those tokens, which are tagged with any of these ten tags, have comparably more information. So the baseline multi-vector model is actually a ten-vector model.

However, we know that the research objects in TDT are news stories, not usual documents. Stories have their own features besides the usual characters. For example, almost each story describes an event. The time, location, person, and so on, compose the framework of an event. And also the first few sentences often summarize the event and the rest sentences explain what the event is about in detail. Story link detection is to decide whether two stories are topically linked. The topic here is the event described in a story and other event directly related to it. Topic here is event based. Therefore the representation model for stories should reflect the events described in them from both content and structure. The event frame maybe a more proper and natural partition standard for a multi-vector model to represent a story. In this paper we build a new multi-vector model for a story according to the event framework. We call it event model. The later experimental results also verify that just splitting the feature set according to the tag is not the best choice. The event model gets much improvement to the performance of story link detection and has also a larger potentiality for improvement than the baseline model.

The main difference between the event model and the baseline model is the partition standard of the feature set. The baseline model is built according to the tags while the event model is built according to the event framework. The primary component elements are time, number, person, location, organization, abstract and content description in our event framework. The first five will be the same as those in the baseline model. The features in the abstract vector and the content description vector are tokens with their tags individually occurring in the first two sentences and the rest sentences in a story, while they do not belong to the former five kinds of tokens. It is because we find that the first two sentences in a story usually summarize the story and the rest sentences explain what happened in detail. According to this splitting standard, the feature set will be split into seven subsets.

3.2 Other Common System Components

Firstly, the weights of the features in above two models are all based on the *tf*idf* form. Furthermore, it is dynamic, which lies in the dynamic computation of some

---

[1] http://sewm.pku.edu.cn/QA/reference/ICTCLAS/FreeICTCLAS/

parameters in the *tf*idf* form. The story collection used for statistics is incremental, since more story pairs are processed, more source files could be seen, and the story background corpus is bigger. Whenever the size of the story background has changed, the values of some certain parameters will update accordingly. We call this incremental *tf*idf* weighting.

Secondly, another important issue is to determine the right function to measure similarity between two corresponding sub-vectors. The critical property of the similarity function is its ability to separate vectors describing the same information from vectors on different information. We consider the classical cosine function in this paper. This measure is simply an inner product of two vectors, where each vector is normalized to unit length. It represents cosine of the angle between two vectors.

The last important step for the baseline model and the event model is to integrate multiple similarities into a single value to decide whether two stories are topically linked. We do this with a machine learning classifier SVM in this paper. It is because SVM has a good generalization property and has been shown to be a competitive classifier for a variety of other tasks [7]. Firstly, SVM is trained on a set of labeled vector where the features are similarities between two corresponding subvectors and the topically link label in each vector. The generated model is then used to automatically decide whether two stories in a new pair are linked. We use the SVMlight[2] software in this paper. A radial basis function is used in all the reported experiments. In addition to making a decision for whether two stories are linked, SVM also produces a value as the measure of confidence, which is served as input to the evaluation software.

We also notice that the numbers of positive and negative examples in the training data are very different. But the usual SVM treats positive and negative data equally, which may result in poor performance when applying to very unbalanced detection problems. [8] presents a method to solve this problem, where the cost factor for positive examples is distinguished from the cost factor for negative examples to adjust the cost of false positive vs. false negative. This approach is implemented by the SVMlight, in which an optional parameter j ($=C_+/C_-$) is provided to control different weightings of training errors on positive examples to errors on negative examples. Therefore, we also integrate a set of similarities with uneven SVMlight in this paper.

4 Experiment

We use the Chinese subset of TDT4 corpus in this paper. There are totally 12334 story pairs extracted for our experiments. The answers for these pairs are based on 28 topics in TDT 2003 evaluation. The first 9000 pairs are used for training. The rest 3334 pairs are used for testing. The goal of link detection is to minimize the detection cost (C_{det}) due to errors caused by the system, which is a function of the miss probability (P_{miss}) and the false alarm probability (P_{fa}). The cost for each topic is equally weighted and normalized so that the normalized value ($(C_{det})_{norm}$) can be no less than one. The detailed explanation can refer to [1].

[2] http://svmlight.joachims.org/

4.1 Experimental Results

To verify the effectiveness of the event model and the uneven SVM strategy, we have designed four story link detection systems: System1 uses the baseline model with even SVMlight, System2 uses the baseline model with uneven SVMlight, System3 uses the event model with even SVMlight, and System4 uses the event model with uneven SVMlight. All the systems are implemented according to the procedure introduced in the section of problem definition. When the generated SVM classify model is used to automatically decide whether two stories in a new pair are linked, the default optimum threshold is zero. The following table shows the topic-weighted experimental results of these systems.

Table 1. Topic-Weighted Experimental Results of Four Systems

	P_{miss}	P_{fa}	C_{det}	$(C_{det})_{norm}$
System1	0.0429	0.0048	0.0013	0.0664
System2	0.0149	0.0090	0.0012	0.0588
System3	0.0267	0.0052	0.0010	0.0524
System4	0.0193	0.0060	0.0010	0.0487

From the table we can see that story link detection using the event model has a lower detection cost than that using the baseline model whenever using even or uneven SVM. This may be because the event model does no discriminate those features which are not named entities (person, location, organization, time, number) and just splits them into abstract and content description vectors. On one hand it avoids the cost loss caused by exact matching when comparing two sub vectors. On the other hand it reflects the character that the first two sentences in a story are usually summary of what happened and the rest are explanation of what happened.

Relative to even SVM, uneven SVM has made a notable decrement in P_{miss} while a little increment in P_{fa}. This is because uneven SVM is completed under the principle of decreasing the loss errors at the cost of little increment in false alarm errors. We have verified through experiments that the event model is always superior to the baseline model no matter what the optional parameter j is. So we think that the event model is more appropriate to represent news stories in TDT.

Although the event model has got a comparable lower detection cost, we should also notice that the information of relation between features in event model has not yet been abstracted and used. Only using information of tokens and tags may be insufficient. If we could get the relation between the event in a story and the seminal event of its corresponding topic, we should be able to make the right decision at a larger confidence. We will try to exploit relation information and use them properly in our future work.

5 Conclusion

Story link detection is a key technology in TDT research. Though many models have been used, few of them are specific to stories in TDT. After analyzing the characters of stories in TDT corpora, this paper proposes a new event model. It represents the events in a story according to the event framework. The experimental results indicate that story link detection using this model can achieve a better performance than that using the baseline model, especially when using the uneven SVM integration strategy. So we think the event model is more proper for TDT stories. However, we realize that the event model is only used in a simple way in this paper, just looking it as a multi-vector model. How to mine and utilize the relation information between features in the event model will be our future work.

Acknowledgement

This research is supported by the National Natural Science Foundation of China (60403050), Program for New Century Excellent Talents in University (NCET-06-0926) and the National Grand Fundamental Research Program of China (2005CB321802).

References

[1] Allan, J.: Introduction to topic detection and tracking. In: Allan, J. (ed.) Topic Detection and Tracking - Event-based Information Organization, pp. 1–16. Kluwer Academic Publisher, Dordrecht (2002)

[2] Connell, M., Feng, A., Kumaran, G., Raghavan, H., Shah, C., Allan, J.: Umass at tdt 2004. In: TDT 2004 Workshop (2004)

[3] Chen, F., Farahat, A., Brants, T.: Multiple similarity measures and source-pair information in story link detection. In: HLT-NAACL, pp. 313–320 (2004)

[4] Allan, J., Lavrenko, V., Malin, D., Swan, R.: Detections, bounds, and timelines: Umass and tdt–3. In: Proceedings of Topic Detection and Tracking (TDT–3), pp. 167–174 (2000)

[5] Nallapati, R.: Semantic language models for topic detection and tracking. In: HLT-NAACL (2003)

[6] Lavrenko, V., Allan, J., DeGuzman, E., LaFlamme, D., Pollard, V., Thomas, S.: Relevance models for topic detection and tracking. In: Proceedings of Human Language Technology Conference (HLT), pp. 104–110 (2002)

[7] Van Der Walt, C.M., Barnard, E.: Data characteristics that determine classifier performance. In: Sixteenth Annual Symposium of the Pattern Recognition Association of South Africa, pp. 160–165 (2006)

[8] Morik, K., Brockhausen, P., Joachims, T.: Combining statistical learning with a knowledgebased approach - a case study in intensive care monitoring. In: Proceedings of the 16th International Conference on Machine Learning (ICML 1999), pp. 268–277. Morgan Kaufmann, San Francisco, CA (1999)

Video Temporal Segmentation Using Support Vector Machine

Shaohua Teng and Wenwei Tan

Guangdong University of Technology,Guangzhou,P.R. China
shteng@gdut.edu.cn, tanww06@yahoo.com.cn

Abstract. A first step required to allow video indexing and retrieval of visual data is to perform a temporal segmentation, that is, to find the location of camera-shot transitions, which can be either abrupt or gradual. We adopt SVM technique to decide whether a shot transition exists or not within a given video sequence. Active learning strategy is used to accelerate training of SVM-classifiers. We also introduce a new feature description of video frame based on Local Binary Pattern (LBP).Cosine Distance is used to qualify the difference between frames in our works. The proposed method is evaluated on the TRECVID-2005 benchmarking platform and the experimental results reveal the effectiveness of the method.

Keywords: shot boundary detection, temporal video segmentation, video retrieval, support vector machine.

1 Introduction

Temporal video segmentation is the essential first step towards automatic annotation of digital video sequences. Its goal is to divide the video stream into a set of meaningful and manageable segments (shots) that are used as basic elements for indexing. A video shot is defined as a series of interrelated consecutive frames taken contiguously by a single camera and representing a continuous action in time and space. as such, shots are considered to be the primitives for higher level content analysis. According to whether the transition between shots is abrupt or not, the shot boundaries are categorized into two types: cuts and gradual transitions. Abrupt transitions (cuts) are simpler. In the case of shot cuts, the content change is usually large and easier to detect than the content change during a gradual transition [1]. Actually, the content change caused by camera operations, such as object movement, can be of the same magnitude as those caused by gradual transitions. This makes it difficult to differentiate between changes caused by a continuous edit effect and those caused by object and camera motion [2]. Many researchers have exploited solutions extensively in various ways [3-7]. In spite of all the improvement resulted from the works introduced above, there still exists a radical problem which has not been settled properly yet. Distinguishing shot change from motions, whether it is a camera motion or object movement, has been a very hard and still open problem. It's necessary to explore new methods for temporal video segmentation.

H. Li et al. (Eds.): AIRS 2008, LNCS 4993, pp. 442–447, 2008.

2 Proposed Method

We propose a video segmentation method based on support vector machines. Videos can be regarded as continuous frame series. We label the frames as shot boundary and non-boundary. Both of them have their own patterns, so the shot boundary detection can be regarded as a problem of pattern recognition. Meanwhile we introduce a texture descriptor called LBP (Local Binary Pattern) to build a new feature.

2.1 Feature Construction Based on LBP

The LBP operator was originally designed for texture description [8]. The operator assigns a label to every pixel of an image by thresholding the 3x3-neighborhood of each pixel with the center pixel value and considering the result as a binary number. Then the histogram of the labels can be used as a texture descriptor. See Figure 1 for an illustration of the basic LBP operator.

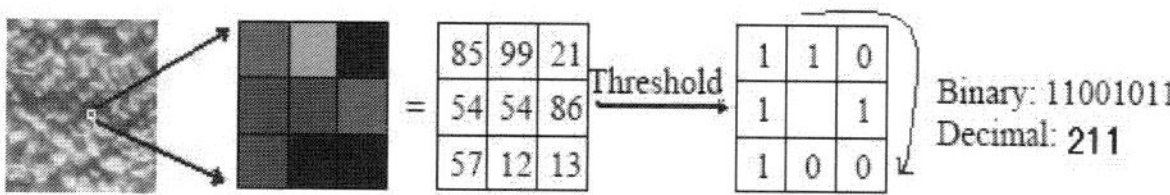

Fig. 1. Basic LBP operator

The video image is divided into local regions and texture descriptors are extracted from each region independently. The descriptors are then concatenated to form a global description of the video image. See Figure 2 for an example of a video image divided into rectangular regions.

Fig. 2. An image from real video is divided into 2×2, 3×3, 4×4 rectangular regions

A histogram of the labeled image $f_l(x, y)$ can be defined as

$$H_i = \sum_{x,y} I\{f_l(x,y) = i\}, i = 0,\ldots,n-1$$

in which n is the number of different labels produced by the LBP operator and $I\{A\} = \{^{1, A\ is\ ture}_{0, A\ is\ false}$. This histogram contains information about the distribution of the local micro patterns, such as edges, spots and flat areas, over the whole image. For efficient image representation, one should also retain spatial information. For this purpose, the image is divided into m regions $R_0, R_1, \ldots, R_{m-1}$ and the spatially enhanced histogram is defined as

$$H_{i,j} = \sum_{x,y} I\{f_l(x,y) = i\} I\{(x,y) \in R_j\}, i = 0,1,...,n-1,\ j = 0,1,...,m-1.$$

Finally these histograms are concatenated as a feature vector: $R = (H_{0,0}, H_{1,0}, ...H_{L-1,0}, H_{0,1}, ..., H_{L-1,m-1})$.

2.2 Dissimilarity Measure

We use aforementioned feature vector to build an effective description of the video image. A dissimilarity measure called Cosine Distance can be exploited from the idea of a spatially histogram. The cosine distance $D_{COS}(A,B)$ between vector A and B can be defined as follows:

$$D_{COS}(A,B) = \frac{\sum_{i=1}^{N} a_i \bullet b_i}{\sqrt{\sum_{i=1}^{N} a_i^2} \bullet \sqrt{\sum_{i=1}^{N} b_i^2}}$$

Where a_i is a component in A and b_i is corresponding component in B. N is the dimension number of the vector. The cosine distance is basically the dot product of two unit vector. Therefore a small value for $D_{COS}(A,B)$ indicates that the frames being considered are quite dissimilar, while a large $D_{COS}(A,B)$ value indicates similarity. Figure 3 shows a time varying curve of the cosine distance in a video clip which involves several gradual transitions, abrupt transitions, and motions.

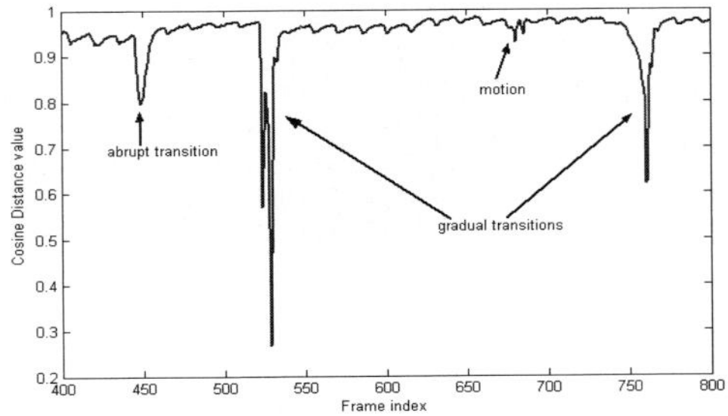

Fig. 3. A time varying curve of the cosine distance between frames

3 Training of SVM

3.1 Feature Selection

Having obtained the aforementioned curve of cosine distance between frames, each boundary corresponds to a valley on the curve. Not every valley is a shot boundary because of some false alarms caused by object or camera motions. In order to accurately classify the potential transitions into the types of CUT and GT,we need to characterize the transitions corresponding with valleys. We can construct the feature vector, which characterizes the shape of the valleys locating on the curve. Let D_k denotes the value of the cosine distance between k and k+1 frame , the feature vector is formally defined by the following equation:

$$V_i = [D_{i-m}, ..., D_{i-1}, D_i, D_{i+1}, ..., D_{i+m}]$$

where D_i is the local minima of a valley. Given a certain value for m, V_i is a vector which characterizes the potential shot boundary at i frame. In our experiment, we choose m=5 in cut detection and m=10 for gradual transition detection. Feature vectors are inputted for classification by SVM.

3.2 Active Learning Strategy

Active learning is a generic approach to accelerate training of classifiers in order to achieve a higher accuracy with a small number of training examples. The active learner can select its own training data. We firstly collect all the valleys under a specified threshold from all the available local minima. The valleys collected, including real boundaries and various false alarms, constitute the training set. Then the real boundaries are labeled as positive examples, and the false alarms are labeled as negative examples, as shown in Figure 4.

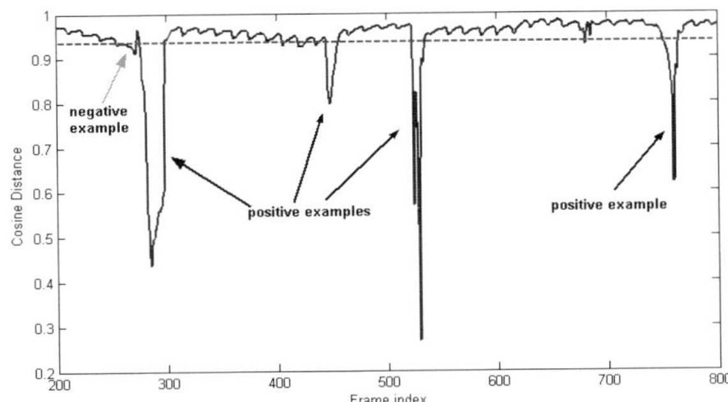

Fig. 4. Selective sampling of the training samples from test videos

Both positive examples and negative constitute the training set. Obviously, all of boundaries will be collected as long as the threshold is low enough.

4 Experiment and Evaluation

To examine the effectiveness of the proposed framework, we evaluate it in the shot boundary task of TRECVID, which is a TREC-style video retrieval evaluation benchmarking platform [9]. Similar to other information retrieval task, the performance

Table 1. Evaluation results of the proposed method

Name	Cut			Gradual transition		
	Recall	Precision	F	Recall	Precision	F
1	0.86	0.97	0.92	0.97	0.77	0.86
2	0.95	0.99	0.97	0.73	0.85	0.79
3	0.94	0.98	0.96	0.89	0.82	0.85
4	0.90	0.98	0.94	0.64	0.88	0.74
5	0.92	0.98	0.95	0.96	0.69	0.80
6	0.84	0.92	0.88	0.98	0.84	0.90
7	0.85	1.00	0.92	0.99	0.81	0.89

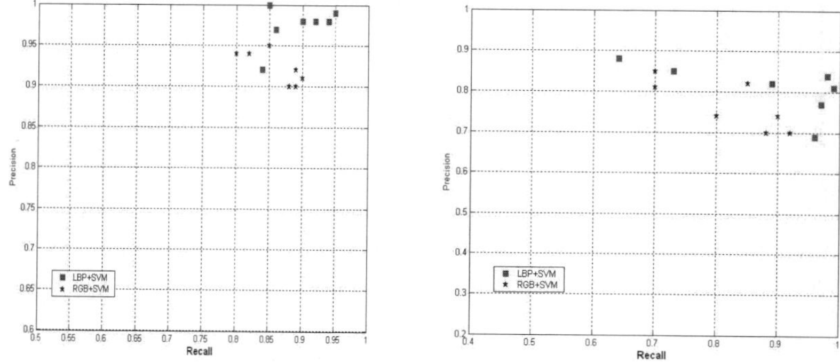

Fig. 5. Recall and precision for cut(left) and Recall and precision for GT(right)

is evaluated by recall and precision criteria. Recall measures the capability in detecting correct shot transitions, while precision measures the ability in preventing false alarms. Test collection of 2005 TRECVID, including 7 videos, is utilized to test the performance of the proposed method. The evaluation results are presented in table 1.

As the previous studies reveal, color histogram is a kind of simple yet effective way for the content representation of videos. Usually, global histogram is extracted to represent the content of the frames. In this paper, we extract RGB color histogram features and then compare the ability of representation to our proposed features. The performances are depicted in Figure 5.

Obviously, our proposed method based on LBP performs better in cut detection and GT detection. New feature based on LBP can describe video frames well. Promising results reveal its effectiveness at the same time.

5 Conclusions

A method for temporal video segmentation is presented. We propose a new feature based on LBP to describe video frame and a measure qualifying the differences between frames. Then a method for cut detection and gradual transitions detection is implemented by SVM. We adopt active learning strategy to improve accuracy of SVM. Promising comparison results encourage our works.

Acknowledgments

This work was supported by Guangdong Provincial Natural Science Foundation (Grant No. 06021484) and (Grant No. 04107411), Guangdong Provincial science and technology project (Grant No. 2005B16001095) and (Grant No.2005B10101077).

References

1. Lienhart, R.: Comparison of automatic shot boundary detection algorithms. In: SPIE Conf. on Storage and Retrieval for Image and Video Databases VII, vol. 3656, pp. 290–301 (1999)
2. Smeaton, A.F.: Techniques used and open challenges to the analysis, indexing and retrieval of digital video. Information Systems 32(4), 545–559 (2007)
3. Ngo, C.-W., Pong, T.-C., Chin, R.T.: Video Partitioning by Temporal Slice Coherency. IEEE Trans. Circuit Syst. Video Technol. 11(8), 941–953 (2001)
4. Zhao, Z.-C., Cai, A.-N.: Shot boundary detection algorithm in compressed domain based on adaboost and fuzzy theory. In: Jiao, L., Wang, L., Gao, X.-b., Liu, J., Wu, F. (eds.) ICNC 2006. LNCS, vol. 4221, Springer, Heidelberg (2006)
5. Yeung, M., Yeo, B., Liu, B.: Segmentation of Videos by Clustering and Graph Analysis, CVIU (1998)
6. Cooper, M., Liu, T., Rieffel, E.: Video segmentation via temporal pattern classification. IEEE Trans. Multimedia 9(3), 610–618 (2007)
7. Yuan, J., Wang, H., et al.: A formal study of shot boundary detection, IEEE Trans. Circuit Syst. Video Technol. 17(2), 168–186 (2007)
8. Ojala, T., Pietikainen, M., Harwood, D.: A comparative study of texture measures with classification based on feature distributions. Pattern Recognition 29, 51–59 (1996)
9. Hauptmann, A.G.: TRECVID: The utility of a content-based video retrieval evaluation. In: Internet Imaging VII - Proceedings of SPIE-IS and T Electronic Imaging (2006)

Using Multiple Combined Ranker for Answering Definitional Questions

Junkuo Cao, Lide Wu, Xuanjing Huang, Yaqian Zhou, and Fei Liu

Fudan University, Shang Hai, China
{jkcao,ldwu,xjhuang,zhouyaqian, liufei}@fudan.edu.cn

Abstract. This paper presents a Multiple Combined Ranker (MCR) approach for answering definitional questions. Generally, our MCR approach first extracts question target-related knowledge as much as possible, then using this knowledge to pick up appropriate question answers. The knowledge includes both online definitions and related terms (RT). In our system, extraction of related terms is different from traditional methods which are largely based on calculating the co-occurred frequency of target words. We adopted the significance of sentences and documents, from which RT were extracted. The MCR approach shows state-in-art performance in handling with increasingly complex definitional questions.

Keywords: Definitional question answering, Multiple Combined Ranker, Related Terms Extraction.

1 Introduction

The objective of question answering task is to focus research on systems that return merely answers, rather than documents containing answers. Related work concerning to definitional question answering are mostly concentrated on Patterns Extraction, Ccentroid-based ranking, as well as utilizing Web knowledge as external source. Patterns Extraction has been extensively adopted in information retrieval tasks. These patterns are often expressed and matched against as regular expressions. Sudo et al. [1] employed TF*IDF to get a set of relevant sentences and built patterns from them. Other approaches employed to extract definitional sentences include various pattern matching methods, in which hand-crafted or machine learned rules are generated to find nuggets[2][3][4]. Moreover, some definitional question answering systems adopt a centroid-based ranking method to identify and select definition sentences [2][5]. For each question target, a series of centroid words were identified and grouped into a centroid vector, which was utilized to rank input sentences using cosine similarity.

Our multiple combined ranker (MCR) approach for answering definitional questions differs from the above in that we perform sentence selection process in a novel and effective way. Instead of using Centroid-based or Pattern-based method, we adopt different rankers, which respectively measures candidate sentences' importance based on AQUAINT corpus, question target expansion, as well as Web knowledge collections. These three rankers act as mutual supplements.

H. Li et al. (Eds.): AIRS 2008, LNCS 4993, pp. 448 – 453, 2008.

2 Multiple Combined Ranker

2.1 Basic Ranker

We use Basic Ranker as the first part of definitional question answering process, it consists of two components: the searching procedure and the refining procedure. A search engine (Lucene) retrieves documents in respond to the target query, and ranks them using some algorithm. We make the assumption that the search engine already produced a good result. Consequently, the sentences in these documents are supposed to have a tight relationship with the question target. The refining procedure considers other possible factors that might make a candidate sentence become an appropriate answer. The assumption is that the sentences containing words, phrases and Name Entities that co-occur frequently with the target are largely possible to be important ones for answering the question. Also, a single sentence could appear in several documents, the total number of these documents is supposed to be in direct relationship to the significance of this sentence concerning the target. Based on the above two procedures, the Basic Ranker scores the candidate sentences so that the relevant sentences receive higher scores.

2.2 Web Ranker

Until recently, Web knowledge bases (Web KBs) are increasingly recognized as a promising way to provide online knowledge, thus we adopt Web KBs as an alternative way for knowledge acquisition and build another ranker called Web Ranker. During this procedure, we calculate the similarity scores between candidate sentence and definitions from different knowledge bases respectively, and merge these scores to rank the candidate sentences. For each target, its candidate sentences are ranked using definitions from Web KBs. Firstly we construct a words vector space, which is based on TF*IDF, for all candidate sentences and Web definitions. Each of them is projected into this vector space. Secondly, the similarity of a particular candidate sentence and Web definition are computed based on the cosine of the two vectors.

Our definitional question answering systems got promising results by employing several external Web knowledge bases during TREC 2003 and TREC 2004. However, this may be due to the intrinsic simplicity of question targets more or less. But unfortunately, definitional questions have shown an increasingly complex characteristic, by way of adopting more complex question types. As a result, we could only find out online definitions for about 65% of the total question targets in TREC2005. Although we still employ Web ranker as one of our strategies to rank the candidate sentences, it is not reliable as before, more details of this approach could be referred to in [6].

2.3 Related Terms Ranker

We construct the Related Terms (RT) Ranker based on the extension of the question targets, for the purpose of obtaining more reliable and target-related information nuggets. At the heart of RT Ranker is the process of identifying and selecting words, phrases, and Name Entities, which are in tight relationship with the question targets. These terms were acquired at the end of preliminary processes like word segmentation and stemming. Also, a Relation Degree is defined to weigh the relationship between extracted terms and the question target In previous work, expansion of terms were adopted in automatic query expansion, as well as open-domain question answering [7][8]. Our approach differ from the above in that 1) Making full use of NE extraction technology which is quite helpful in identifying Related Terms. 2) Taking into account of not only the Relation Degree of the terms, but also their weights, which are related with the Basic Ranker score of the sentence that they belong to. The set of related words, phrases, and Name Entities (naturally named Related Terms) is denoted by $T=\{t_1, t_2, ..., t_n\}$. The process of Relation Degree computing is defined as below:

$$r(t_i) = \sum_j E(t_i, S_j) \times initscore(S_j)$$

$$\text{Where } E(t_i, S_j) = \begin{cases} 1 & t_i \in S_j \\ 0 & t_i \notin S_j \end{cases}$$

Where $initscore(S_j)$ stands for the Basic Ranker score of sentence S_j. After that, $r(t_i)$ is normalized and ranked, top terms were selected as RT for further processing.

Consequently, we rank the candidate sentences based on Relation Degree of RT. Let n_w, n_p, n_e respectively represent the number of words, phrases and Name Entities that a particular sentence S' contains, and $r(w_i)$, $r(p_i)$, $r(e_i)$ denotes the Relation Degree of them, $RT\_score(S')$ is introduced to denote the score of this sentence according to the Related Terms Ranker, which is defined as follows:

$$RT\_score(S') = \alpha \sum_{r_w} r(w_i) \Big/ n_w + \beta \sum_{r_p} r(p_j) \Big/ n_p + \gamma \sum_{r_e} r(e_k) \; / n_e$$

According to experiments and heuristic assumptions, Name Entity should play a relative important role in RT, thus γ received a slightly higher weight than other two parameters. In our system, they are allotted to 0.3, 0.3 and 0.4 respectively to receive the optimal result.

3 Experiments

To evaluate the effectiveness of multiple combined ranker, we utilize the data set from TREC 2006 QA track, which contained 75 series as well as answer judgments. Our system official F(β=3) scores is 0.223, ranked second in all participated systems. To further compare the effectiveness of our MCR approach, we experimented on the TREC 2005 definition question set using our evaluation system, which can keep the rank when evaluates the top 10 submitted result.

The purpose of our first experiment is to judge the effectiveness of the results of document retrieval, which is the foundation of Basic Ranker and Related Terms

Ranker. In the second experiment we evaluate effectiveness of sentence selection. The purpose of the third experiment uses the Basic Ranker as a baseline, and Multiple Combined Ranker is compared with the baseline to show its effectiveness.

3.1 Effectiveness of Document Retrieval

In this part, we utilize Lucene 2.0 as our search engine and judge the returned documents by Vital and Okay nuggets recall, respectively. We vary the number of returned documents from 1 to 200 to study the effect of document number on nuggets recall. The result is listed in Table 1.

As shown from Table 1, Vital nugget recall in all TOP200 documents can achieve up to 90.0% recall and Okay nugget recall reach 81.9%, which are especially high scores. However, this higher score is achieved at the cost of precision score since returning too many sentences for a question target inevitably adds in noise information nuggets. So we also test nuggets recall on top N (1-100) returned documents, experiment results show that the Vital nuggets recall is higher than Okay nuggets recall in TOPN documents. Because the Vital nugget is more important than Okay one, our solution of document retrieval is successful. We can also see from Table 1, R(V)/N and R(O)/N decrease with the increasing of N, which is in accordance with the Basic Ranker hypothesis.

3.2 Candidate Sentences Selection Evaluation

The returned documents always contain some sentences that were not related to the question target. Therefore, discarding the noise sentences is very important. In order to evaluate the process of candidate sentence selection, we use the same method (MCR) for definitional question answering but with different candidate sentences sets. The first set is all candidate sentences without selection. Although all candidate sentences contain 90.0% Vital nuggets and 81.9% Okay nuggets, the system's F-score is only 0.187. In contrast, the other candidate sentence set is selected by Basic Ranker and some manual constructed rules. More candidate sentences were discarded in the selection process, as shown in Table 2. The Vital nugget recall and Okay nuggets recall decreased 30.3% and 45.1% respectively. However, although both Vital recall and Okay recall decreased obviously, the system performance improved 72.0%. In the same time, we try some different candidate sentence sets in our system. These

Table 1. The performance of all candidate sentences from TOPN documents of TREC2005 definitional QA. R(V) denotes Vital Nugget Recall, and R(O) denotes Okay Nugget Recall

TOPN	R(V)	R(V)/ N	R(O)	R(O)/ N
TOP1	21.1%	0.211	11.2%	0.112
TOP5	46.7%	0.093	25.6%	0.051
TOP10	54.8%	0.055	34.0%	0.034
TOP20	61.8%	0.031	44.4%	0.022
TOP50	73.1%	0.015	60.1%	0.012
TOP100	82.4%	0.008	68.4%	0.007
TOP200	90.0%	0.005	81.9%	0.004

Table 2. The effect of candidate sentences selection in definitional question answering

Candidate Answer Sentence	Size	R(V)	R(O)	F(β=3)
All sentences without selection	56100K	90.0%	81.9%	0.186
Sentences selected by Basic Ranker	1992K	62.7%	45.0%	0.320

Table3. The comparison of using three Rankers for definitional question answering

Ranking Method	F(β=3)	R(V)	R(O)	Precision
BASIC	0.272	0.388	0.213	0.097
WEB	0.264	0.381	0.199	0.097
RT	0.211	0.297	0.167	0.083
BASIC +WEB	0.311	0.460	0.240	0.105
BASIC +RT	0.305	0.433	0.225	0.108
WEB+RT	0.280	0.412	0.193	0.097
BASIC+WEB+RT (MCR)	0.318	0.467	0.250	0.111

experiments show that the confidence of question answers is determined according to the degree of candidate sentences noise. So it is difficult but crucial to balance well the Vital/Okay nuggets information with the noise information for definitional question answering.

3.3 Effectiveness of Multiple Combined Ranker

For each candidate sentence, three scores are calculated by Basic Ranker, WEB Ranker and Related Terms (RT) Ranker respectively. These scores are then applied to extract the question answers, both respectively and synthetically. In ranking process, weights of the three scores are estimated by our automatic evaluation system. Question with different target type is allocated with different weight. The performance of these ranking procedures, briefly named as BASIC, WEB and RT, have been evaluated and are shown in Table 3. As can been seen from this table, the best single solution is BASIC. This phenomenon is largely due to the fact that, the BASIC method in choosing candidate sentences is not only an important element for answering question, but it is also the foundation of WEB Ranker and RT Ranker. The third Ranker (RT) returns the worst F-measure against other two single method though, it shows competitive performance while working together with the BASIC Ranker and WEB Ranker. We can see from Table 3 that adding Related Terms Ranker to BASIC Ranker and WEB Ranker could improve the system performance up to 12% and 6% respectively, and compared with BASIC+WEB, employing Multiple Combined Ranker (MCR) could enhance system performance by 2%. Generally, the combined solution is much better than separated ones. This could be deduced from the fact that the best solution method BASIC + WEB + RT (MCR), whose F-Measure achieved 0.318, outperformed the best single solution to a great extent (about 17% improvement).

4 Conclusion

Compared with other question answering tasks, definitional question answering has more uncertain factors. There are still many divergences even among experts while answering these questions. Therefore the key of answering these questions is to find reliable knowledge related to the target. So we propose a Multiple Combined Ranker (MCR) approach to rank candidate sentences for definitional question answering. To acquire the reliable and related information, external knowledge from online websites and the related words, phrases and entities were extracted. Using these multiple knowledge, the definitional QA system can rank the candidate answers effectively.

Acknowledgements

This research was supported by the National Natural Science Foundation of China under Grant No. 60435020 and No. 60503070

References

1. Sudo, K., Sekine, S., Grishman, R.: Automatic Pattern Acquisition for Japanese Information Extraction. In: Proc. HLT 2001, San Diego, CA (2001)
2. Cui, H., Kan, M.-Y., Chua, T.-S.: Unsupervised Learning of Soft Patterns for Generating Definitions from Online News. In: Proceedings of WWW (2004)
3. Kouylekov, M., Magnini, B., Negri, M., Tanev, H.: ITC-irst at TREC-2003: the DIOGENE QA system. In: Proceedings of the Twelfth Text REtreival Conference. NIST, GAthersburg, MD, pp. 349–357 (2003)
4. Blarr-Goldensohn, S., McKeown, K.R., Schlaikjer, A.H.: A hybrid approach for QA track definitional questions. In: Proceedings of the Twelfth Text Retrieval Conference. NIST, Gathersburg, MD, pp. 185–192 (2003)
5. Radev, D., Jing, H., Budzikowska, M.: Centroid based summarization of multiple documents. In: ANLP/NAACL 2000 Workshop on Automatic Summarization, Seattle, WA, April 2000, pp. 21–29 (2000)
6. Zhang, Z., Zhou, Y., Huang, X., Wu, L.: Answering Definition Questions Using Web Knowledge Bases. In: The Proceeding of IJCNLP, pp. 498–506 (2005)
7. Echihabi, A., Hermjakob, U., Hovy, E., Marcu, D., Melz, E., Ravichandran, D.: Multiple-Engine Question Answering in TextMap. In: The Twelfth Text REtrieval Conference (TREC 2003) Notebook (2003)
8. Kwok, C., Etzioni, O., Weld, D.S.: Scaling Question Answering to the Web. In: Proceedings of the 10th World Wide Web Conference (WWW 2001), HongKong (2001)

Route Description Using Natural Language Generation Technology

XueYing Zhang

Dept. of Geographical Information System, Nanjing Normal University, 210046, China
zhangsnowy@163.com

Abstract. This paper aims to solve the problems of generating natural language route description in Chinese way-finding systems, on the basis of datasets of geographical information systems and natural language generation technology. The techniques of deriving important information e.g. paths, roads, directions and landmarks from geographical information systems are discussed in detail. Through examples we describe the construction of linguistic knowledge base for route description including the categories of Chinese terms, grammar rules and syntax schemata. The experimental output indicates that there are no more distinguishable from human route description.

Keywords: natural language, route description, electronic map, text planning.

1 Introduction

Natural language generation (NLG) is the subfield of artificial intelligence and computational linguistics. It is concerned with the construction of computer systems that can produce understandable texts in human languages and form some underlying non-linguistic representation of information in order to meet specified communicative goals (McDonald, 1987). The technology has been used in a wide variety of systems and contexts (Reiter, 2000).

Route description is interesting for at least two reasons: first of all, as navigation aides in general they help to solve a real world problem; second, despite their apparent simplicity, especially with regard to surface form, they require the solution of a number of non trivial linguistic and discourse problems, which are intimately rooted in human cognition (Fraczak, 1998). Language can be effective at relating a simple scene of people, objects, and landmarks. Routes are schematized as a point changing direction along a line or a plane, or as a network of nodes and links. Route description can be broken into segments consisting of four elements: start point, reorientation/direction, path/progression and end point (Denis, 1994). Human's route description always misses a start or an end point or path/progression information. However, many communications contain missing information that can be inferred from context or medium (Clark, 1977). Two simple rules of inference allow recovery of most of the missing information: According to "continuity" rule, if a start point is omitted, it is assumed to be the same as the previous end point, or conversely. According to "forward progression" rule, the direction of motion is assumed to be

H. Li et al. (Eds.): AIRS 2008, LNCS 4993, pp. 454–459, 2008.

forward (Tversky, 1998). Gedolf (2002) described an approach to developing a mode of interaction which supports the cognitive involvement of the user in performing the task of following a route description.

There already exists a considerable research in automatic generation of natural language route description. A majority of these approaches are all concerned with establishing a link between GIS knowledge and linguistic realization principles. Pattabhiraman (1990) pointed the importance of salience and relevance in content selection, and reported some aspects on generating descriptions of bus route directions from a given source to an intended destination within a city. Fraczak (1998) developed a system producing variants of subway route directions by mapping the relative importance of information entities onto syntactico-semantic features. A multimodal concerning with the generation of multimodal route description that combine natural language, maps and perspective views is developed for computer assisted vehicle navigation (Maaß, 1993). Maaß (1995) indicated that incremental route description can be classified by a small set of syntactic and semantic structures. Williams (1999) studied the collections of the corpus contains 56 monologue descriptions of routes around a university department building, which was used to improve a natural language generation systems that generates prosodically annotated routes to be spoken by a speech synthesizer.

There are a lot of successful way-finding systems based on NLG technology, which provide navigational services in the context of city public traffic and driving directions, such as MapBlast[1], MapPoint[2] and MapQuest[3]. However, there are only a few such web-based systems in Chinese. With the example of 51map[4], it is the biggest port of electronic maps, in which the start address and target address can be visually identified on the map and natural language descriptions of its route plans can generated automatically.

Although there are some differences of detail in the user interfaces, all these systems are similar in concept and content. The route description approaches commonly-used by Chinese way-finding systems are identified four typical disadvantages:

- The instructions are normally in the simple form of "turn by turn", e.g. left, right, north, west, east and south etc.
- Instructive landmarks and visible features of the environment to identify turning points are omitted, whereas the systems generally describe these points by quantitative distances or times of travel from previous decision points.
- These systems generate route description by one-sentence-per-step mappings using a limited few terms and clause structures, while humans typically produce complex clause structures and gathering related information together into single sentences in order to shorten or ignore some redundant instructive information.
- Only bus route description or driving information is provided. In fact, route description involves knowledge of different granularities. In some cases, the bus route description also includes walking route information.

---

[1] Supported by the Specialized Research Fund of Nanjing Normal University
http://www.mapblast.com

[2] http://www.mapoint.com

[3] http://www.mapquest.com

[4] http://www.51map.com

Our work is concerned with the techniques of Chinese route description for way-finding systems based on the datasets of underlying geographical information systems (GIS), which incorporates NLG technology to produce more natural-sounding route description.

2 Architecture of Route Description

There are many ways of building applied NLG systems. From a pragmatic perspective the most common architecture in applied NLG systems includes three stages.

Text Planning: The most popular technique used for text planning for the development of practical systems in limited domains is schemas. A route description schema provides the semantic of a route description. Basic constituents of a description schema are paths, locations, directions, distance, landmarks, and temporal structure etc., and events and states are used in higher-order structures. A majority of this information can be derived from GIS datasets.

Sentence planning: This stage combines sentence aggregation, lexicalization, and referring expression generation. Once the content of individual sentences has been determined, this still has to be mapped into morphologically and grammatically well-formed words and sentences. Sentence aggregation is the process of grouping messages together into sentences. It can be used to form paragraphs and other higher-order structures as well as sentences.

Linguistic realization: It is the task of generating grammatically correct sentences to communicate messages. From a knowledge perspective, the realizer is the module where knowledge about the grammar of the natural language is encoded. It includes syntactic and morphological knowledge, such as rules about verb group formation, agreement and syntactically required pronominalization.

Route description is always accompanied by intentions. Main intention of the model of route description is to follow a path from a starting point to a destination. By referring to the path finding process, man knows where to go at a decision point. Path-related intentions control the focus of attention. But these intentions are also important for the language generation process.

3 Route Description Information Based on GIS

The GIS datasets represent the world in terms of nodes, arcs and polygons. Nodes (points) typically represent junctions or decision points in a road networks; arcs are the travelable paths between points; and polgygons are used to represent areas such as parks or railway stations. A GIS system typically also provides street names, the lengths of paths, categories of points of interest and so on.

The first step of route description is to get paths mapping the user's query requirements based on a given GIS. The shortest route algorithm is effective to implement this task. In this paper, we select the commonly-used algorithm developed by Dijkstra (Paul, 2006). Based on path plans, other information important for route description can be derived easily.

Road names are one of the most useful reference information from the view of travelers. Road names can be easily extracted on the basis of a definite path.

Distance is usually expressed in terms of its value, i.e. the distance between two points is less than 200m for walking directions or less than 500 m for driving directions. In our system we simplify the task of distance estimation by directly representing distances in terms required by the text planner.

Direction is the most important inductive information in the route description. Absolute directions refer to the cardinal points of the compass marked as north, south, east and west. Relative directions refer to the relative aspects of direction and are locationally and culturally variable. People usually are confused by the identification of absolute directions. Therefore, relative directions are very important, i.e. left and right. However, the identification of left and right directions is dependent on absolute directions, the location relations between the objective point and the moving direction. Computation of direction information requires availability of a co-ordinate system and additional inference modules. The text planner uses additional procedural domain knowledge for retrieving or computing orientation information.

Location information is crucial at transfer points. Additional descriptions of routes in terms of landmarks crossed give the prospective traveler a feel for how long she needs to be on the bus, train, etc, and assure her that she is still on the right back. The additional information such as landmarks, while not essential, may be important for keeping the traveler confidently on track. It anticipates that travelers may become uneasy when there is a relatively long distance without a change of orientation or distinguishing feature or when there is uncertainty about the identity of a landmark (Denis, 1994).

Retrieving landmarks for describing locations of sources and destinations requires examining finer-grainer spatial layout information at the neighborhood of these points. Along with the task of choosing the landmarks comes the task of computing the locative relation between the two objects, and expressing the relation in language as in right by.

4 Linguistic Knowledge Base for Route Description

Route description information derived from GIS must be combined sentences via specific terms. These terms can be divided into five classes according to Chinese language characteristics and human spatial cognition custom.

- Common verbs:步行(walk)、走(walk or go)、行至(get at)…
- Spatial verbs: 穿过(pass)、到达(get)、拐(turn)、转(turn)…
- Connection:然后(then)、那么(then)、再(then)、接着(then)、紧接着(then)…
- Direction:左(left)、右 (right)、东边(east)、南边(south)、西边(west)…
- Preposition:
  沿着(along)、顺着(along)、穿过(across)、往前(forward)、附近(near)、旁边(beside)、靠近(near)…
- Phrases: 从……出发(Start at)、沿途会经过(see along this road)…

The grammar rules for Chinese route description (in short RD) are defined by:

< RD> : : = { < Unit RD > }
< RD > : : = { < Modifier phrase> } < Action> [<Object>]
<Modifier phrase> : : = < Action> < Object> | < Common modifier>
< Object> : : = < Geographical feature> | < Distance> | < Direction>
< Action> : : = < Common verb> | < Spatial verb>
< Geographical feature > : : = < Geometric object > | < Road >
< Distance> : : = < Number> < Metric>
< Direction> : : = < Absolute direction> | < Relative direction>
< Common verb> : : = 行走(walk or go)| 步行(walk)| …
< Spatial verb> : : = 经过(Pass)|到达(Get)|…

A description schema provides the semantic of a route description. Over 60 unit schemata are stored in the linguistic knowledge base, such as follows:

- Phrase + start point
- [Phrase] + [Directional preposition]+Action+[Distance]+Objective point
- [Connection] + [Preposition] + Phrase + Road names
  + Directional preposition + Action.
- [Directional preposition] +Action+ [Phrase] + Progression point.
- Action+ [Directional or other preposition]+ [Phrase]+Landmarks

It should be noted that the elements labeled square brackets are optional, and all the terms can be chosen randomly providing that one term can not used frequently in a route description. According the continuity and forward progression rules described in section 1, a combined schema is defined as the incremental integration of unit schemata. Based on the above mentioned techniques, we develop a prototype of route description system. An example is given as follows:

Departure:金鹰国际购物中心(Jin Ying Shopping Center)

Destination:华联大酒店(Hua Lian Hotel)

Route description: 从金鹰国际购物中心出发,向右到达路口,再向右沿着王府大街行走,路上会看到南京华联集团,南京市现代测绘科技开发公司,步行一会到了石鼓路,沿着石鼓路行走,这就到了最后一个路口了,这时候再向前步行约17米就会在右边看到华联大酒店了。(Start at Jin Ying Shopping Center, turn right at the junction, and walk along Wan Fu Da Jie, see Nanjing Huan Lian Group, Nanjing Modern Survey Technology Company along this road, at Shi Gu LV go forward until the last junction, and then walk about 17m, and finally find Hua Lian Hotel on right hand.)

It is very interesting that the same query may generate a different route description at a different time, because the system can select some terms and schemata randomly in order to generate more natural-sounding descriptions.

5 Conclusion

In this paper we presented an approach that takes as input datasets of underlying GIS, and uses natural language generation techniques in constructing the textual output of route plans. Chinese linguistic knowledge base for route description focuses on the categories of terms, grammar rules and syntax schemata. Further experimentation demonstrates that the presented approach can achieve more satisfactory performance than the existing way-finding systems e.g.51map and some driving navigation systems.

References

1. Clark, H.H., Clark, E.V.: Psychology and language. Harcourt Brace Jovanovich, New York (1977)
2. Denis, M.: La description d'itineraires: Des reperes pour des actions (The description of routes: Landmarks for actions). In: Notes et Documents du LIMSI, Orsay, France, pp. 14–94 (1994)
3. Fraczak, L., Lapalme, G., Zock, M.: Automatic generation of subway directions: salience gradation as a factor for determining message and form. In: Proceedings of the International Workshop on Natural Language Generation, Niagara-on-the-Lake, Canada, pp. 58–67 (1998)
4. Geldofe, S., Dale, R.: Improving route directions on mobile devices. In: Proceedings of the ISCA workshop on Multi-Modal Dialogue in Mobile Environments, Kloster lrsee, Germany (2002)
5. Maaß, W., Wazinski, P.H.: G. VITRA GUIDE: Multimodal route description for computer assisted vehicle navigation. In: Proceedings of the 6th International Conference on Industrial and Engineering Applications of artificial Intelligence and Expert Systems, Edinburgh, Scotland, pp. 144–147 (1993)
6. Maaß, W., Baus, J., Paul, J.: Visual grounding of route descriptions in dynamic environments. In: Srihari, R.K. (ed.) Proceedings of the AAAI Fall Symposium on Computational Models for Integrating Language and Vision, Cambridge, MA (1995)
7. McDonald, D.D.: Natural language generation. In: Stuart, C.S. (ed.) Encyclopedia of Artificial Intelligence, pp. 642–655. John Wiley & Sons, New York (1987)
8. Paul, E.B. (ed.): Dijkstra's algorithm, in Dictionary of Algorithms and Data Structures. US National Institute of Standards and Technology, US (2006)
9. Pattabhiraman, T., Cercone, N.: Selection: Salience, relevance and the coupling between domain-level tasks and text planning. In: Mckeown, K., Moore, J., Nirenburg, S. (eds.) Proceedings of the 5th International Workshop on Natural Language Generation, pp. 79–86. Linden Hall Conference Center (1990)
10. Tversky, B., Lee, P.U.: How space structures language. In: Freksa, C., Habel, C., Wender, K.F. (eds.) Spatial Cognition 1998. LNCS (LNAI), vol. 1404, pp. 157–176. Springer, Heidelberg (1998)
11. Reiter, E., Dale, R.: Building natural language generation systems. J. Spatial Cognition and Computation 1, 227–259 (1999)
12. Williams, S., Watson, C.I.: A profile of the discourse and international structures of route description. In: Proceedings of the 6th European Conference on Speech Communication and Technology, Budapest, Hungary, pp. 1659–1662 (1999)

Some Question to Monte-Carlo Simulation in AIB Algorithm

Sanming Song, Qunsheng Yang, and Yinwei Zhan

Faculty of Computer, Guangdong University of Technology
510006 Guangzhou, P. R. China
SanmSoong@gmail.com, jsjqsy@sina.com, ywzhan@gdut.edu.cn

Abstract. Hierarchical clustering algorithm is efficient in reducing the bytes needed to describe the original information while preserving the original information structure. Information Bottleneck (IB) theory is a hierarchical clustering framework derivative from the information theory. Agglomerative Information Bottleneck (AIB) algorithm is a suboptimal agglomerative clustering procedure designed for optimizing the original computation-exhausted IB algorithm. But the Monte-Carlo simulation formula which is widely adopted for distortion measures in AIB algorithm is problematic. This paper testified that there being a contradiction between the adopted Monte-Carlo formula and IB principle. Extending special distortion measures to common distances, the paper also present several proposals. And Experiments show their efficiency and availability.

1 Introduction

The recently introduced Information Bottleneck (IB) principle, proposed by Tishby, Peresia and Bialek [1], provides a hierarchical clustering framework. But, in searching for the optimal solution, the original approach adopts deterministic annealing strategy, leading to high computation complexity. Later, Slonim suggested a suboptimal "bottom-up" agglomerative clustering procedure, known as AIB [2], and which has been developed into a generalized information clustering method. Some applications can be found in [3-7]. But the Monte-Carlo simulation formula which is adopted for distortion measures in AIB algorithm is problematic. In this paper, we demonstrate that there being a contradiction between the adopted Monte-Carlo formula and IB principle. By extending special distortion measures to common distances, we present several proposals, and experiments show their efficiency and plausibility.

2 IB Principle

IB principle can be depicted as follows [1][2]: Assume a signal $x \in X$ provides information about another signal $y \in Y$. Understanding the signal x requires not only predicting y, but also specifying which relevant features of x play a role in the prediction. The generalized problem was shown to have a natural information theoretic

H. Li et al. (Eds.): AIRS 2008, LNCS 4993, pp. 460–465, 2008.

formulation: Find a compressed representation of the variable X, denoted $\tilde{X}$, such that the mutual information between $\tilde{X}$ and Y, $I(\tilde{X},Y)$, is as high as possible, under a constraint on the mutual information between X and $\tilde{X}$, $I(X,\tilde{X})$.

Denote X, Y and $\tilde{X}$ the object space, the feature space and the compressed representation of X. And $d(x,\tilde{x})$ is the distortion measure between x and $\tilde{x}$. Representing X with no more than R bits, there would be no more than 2^R clusters. It is clear that we can reduce the number of clusters by enlarging the average quantization error. Shannon's rate-distortion theorem states that the minimum log number of clusters needed to keep the average quantization error below D is given by the following rate-distortion function: $R(D) = \min_{\{p(\tilde{x}|x):<d(x,\tilde{x})>\le D\}} I(X,\tilde{X})$.

Using *Lagrangian* multipliers and consider the following distortion function:

$$d(x,\tilde{x}) = D\big(p(y \mid X = x) \,\|\, p(y \mid X = \tilde{x})\big) \tag{1}$$

where $D(f \,\|\, g)$ is the Kullback-Leibler (KL) divergence.

And the loss in the mutual information between X and Y caused by the clustering $\tilde{X}$ is in fact the average of this distortion measure [2]:

$$I(X;Y) - I(\tilde{X};Y) = E\big[D\big(p(y \mid x) \,\|\, p(y \mid \tilde{x})\big)\big] = \langle d(x,\tilde{x})\rangle_{p(x,\tilde{x})}.$$

Optimizing the minimization problem yields the self-consistent equations [1].

3 AIB Algorithm

Since self-consistent equations are transcendental, and its original deterministic annealing strategy leads to high computation complexity. Slonim proposed the suboptimal AIB algorithm, which states that the minimization problem could be approximated by a greedy algorithm based on a bottom-up merging procedure. The algorithm starts with the trivial clustering where each cluster consists of a single point. In order to minimize the overall information loss caused by the clustering, classes are merged in every step, such that the loss in the mutual information caused by merging them is the smallest. That is, merging the clusters C_i, C_j that having the minimum mutual information loss [2].

And, distortion caused by merging two clusters C_1 and C_2 is:

$$d(C_1,C_2) = I(C_{before},Y) - I(C_{after},Y) = \sum_{i=1,2} p(C_i) D\big(p(y \mid C_i) \,\|\, p(y \mid C_1 \cup C_2)\big) \tag{2}$$

where $I(C_{before},Y)$ and $I(C_{after},Y)$ are the mutual information between the classes and the feature space before and after C_1 and C_2 are merged into a single class. Obviously,

$$d(C_1,C_2) \ge 0 \tag{3}$$

As related in [3-7], there is no closed form expression for the KL-divergence between two mixtures of Gauss, and the following Monte-Carlo simulation is widely adopted approximate the KL-divergence,

$$D(f \parallel g) \approx \frac{1}{n}\sum_{t=1}^{n} \log \frac{f(x_t)}{g(x_t)}. \tag{4}$$

4 Dilemma of Monte-Carlo Simulation in AIB

If Monte-Carlo simulation formula (4) really works, substitute it into distortion measure (2). Let $a = p(C_1)$, $f = p(y \mid C_1)$, $b = p(C_2)$, $g = p(y \mid C_2)$, then $f \cup g = p(y \mid C_1 \cup C_2)$. Denote $d(a,b,f,g)$ as $d(C_1,C_2)$, we can abstract a model from (2):

$$d(a,b,f,g) = a \cdot \frac{1}{n}\sum_t \log\left(\frac{f}{f \cup g}\right) + b \cdot \frac{1}{n}\sum_t \log\left(\frac{g}{f \cup g}\right) = \frac{1}{n}\sum_t \log\left(\frac{f^a g^b}{(f \cup g)^{a+b}}\right).$$

Because $f \cup g = \dfrac{af + bg}{a+b}$,

$$d(a,b,f,g) = \frac{1}{n}\sum \log\left(\frac{(a+b)^{a+b} f^a g^b}{(af+bg)^{a+b}}\right). \tag{5}$$

Let $F(a,b,f,g) = \dfrac{(a+b)^{a+b} f^a g^b}{(af+bg)^{a+b}} = \left(1 + b\dfrac{f-g}{af+bg}\right)^a \left(1 - a\dfrac{f-g}{af+bg}\right)^b$,

Let $x = \dfrac{f-g}{af+bg}$, and $-\dfrac{1}{b} \le x \le \dfrac{1}{a}$. Let $f(x) = (1+bx)^a$ and $g(x) = (1-ax)^b$, $F(a,b,f,g)$ could be written as $F(x) = f(x)g(x)$.

Expand $F(x)$ at $x = 0$ with Taylor series,

$$\begin{aligned}
F(x) &= f(x)g(x) \\
&= \left(1 + abx + \frac{ab(a-1)}{2!}x^2 + o(x^2)\right)\left(1 - abx + \frac{ab(b-1)}{2!}x^2 + o(x^2)\right) \\
&= 1 + \left(\frac{ab}{2}(a+b-2) - a^2b^2\right)x^2 + \frac{a^2b^2(b-a)}{2}x^3 + \frac{a^2b^2(a-1)(b-1)}{4}x^4 + o(x^4) \\
&= 1 + a^2b^2\left(\frac{(b-a)x}{2} - 1\right)x^2 + \frac{ab}{2}\left(a+b-1+\frac{ab(a-1)(b-1)}{2}x^2 - 1\right)x^2 + o(x^4)
\end{aligned}$$

When $0 < a,b \le 1$ and $0 < a+b \le 1$ hold, in the field of $x = 0$, $\dfrac{(b-a)x}{2} - 1 \le 0$, $a+b-1 \le 0$,and $\dfrac{ab(a-1)(b-1)}{2}x^2 - 1 \le 0$, so $F(x) \le 1$ holds.

Apply it into equation (5), we obtain $d(C_1,C_2) \le 0$, in contrary to constraint (3)!

5 Several Plausible Proposals

Mutual information loss can be seen as cost, we may consider it to be a distance. With that, we merge the clusters with shortest distance. Such a transformation would bring a beneficiation that we can use those distance measures that has closed-form solution.

Consider two Gauss components $z_\alpha \sim G(p_\alpha;\mu_\alpha,\Sigma_\alpha)$, and $z_\beta \sim G(p_\beta;\mu_\beta,\Sigma_\beta)$. Where p,μ,Σ denotes the a priori probability, mean vector and covariance, and $G(;,)$ stands for Gauss component. When merge them into a new component $\bar{z}$, its parameters could be decided by a method similar to the update step in Hierarchical EM algorithm [8], as suggested in [6].

$$\begin{aligned}
\bar{p} &= p_\alpha + p_\beta \\
\bar{\mu} &= \frac{1}{\bar{p}}\left(p_\alpha\mu_\alpha + p_\beta\mu_\beta\right) \\
\bar{\Sigma} &= \frac{1}{\bar{p}}\left(p_\alpha\Sigma_\alpha + p_\beta\Sigma_\beta\right)+\left(\mu_\alpha-\mu_\beta\right)\left(\mu_\alpha-\mu_\beta\right)^T
\end{aligned} \tag{6}$$

We discuss several comparatively ideal distance measures underneath.

5.1 ALA Distance

ALA (Asymptotic Likelihood Approximation), which was designed to match two Gauss Mixtures, was proposed by Vasconcelos [9]. Consider two mixtures: $P_1(x)=\sum_{j=1}^{C_1} G(x,\mu_{1,j},\Sigma_{1,j})\,p_1(\omega_j)$ and $P_2(x)=\sum_{k=1}^{C_2} G(x,\mu_{2,k},\Sigma_{2,k})\,p_2(\omega_k)$, where C_1 and C_2 represents the number of components. ALA is defined

$$\mathrm{ALA}(P_1 \| P_2)=\sum_j p_1(\omega_j)\left\{\log p_2\left(\omega_{m(j)}\right)+\left[\log G\left(\mu_{1,j},\mu_{2,m(j)},\Sigma_{2,m(j)}\right)-\frac{1}{2}trace\left[\Sigma_{2,m(j)}^{-1}\Sigma_{1,j}\right]\right]\right\} \tag{7}$$

Where $m(j)=k \Leftrightarrow \left\|\mu_{1,j}-\mu_{2,k}\right\|_{\Sigma_{2,k}}^2 -\log p_2(\omega_k) < \left\|\mu_{1,j}-\mu_{2,l}\right\|_{\Sigma_{2,l}}^2 -\log p_2(\omega_l), \forall l \neq k$.

5.2 Divergence

Divergence J_D is defined as the sum of the average discrimination information of class ω_i and ω_j. With Gauss mixtures, and $\omega_i \sim N(\mu_i,\Sigma_i)$, $\omega_j \sim N(\mu_j,\Sigma_j)$ [11]:

$$J_D=\frac{1}{2}tr\left[\left(\Sigma_i^{-1}\Sigma_j+\Sigma_j^{-1}\Sigma_i-2I\right)\right]+\frac{1}{2}\left(\mu_i-\mu_j\right)\left(\Sigma_i^{-1}+\Sigma_j^{-1}\right)\left(\mu_i-\mu_j\right)^T \tag{8}$$

5.3 Bhattacharyya Distance

With Gauss mixtures, Bhattacharyya distance can be written below [11]:

$$J_B = \frac{1}{8}\left\|\mu_i - \mu_j\right\|^2_{\frac{\Sigma_i+\Sigma_j}{2}} + \frac{1}{2}\ln\left|\frac{1}{2}\left(\Sigma_i + \Sigma_j\right)\right| - \frac{1}{4}\ln\left[\left|\Sigma_i\right|\left|\Sigma_j\right|\right] \quad \textbf{(9)}$$

Similarly, in AIB, we use it to select the clusters with the minimum Bhattacharyya and ALA distance or maximum J_D distance, and then merge them into a new cluster.

6 Experiment Evaluation

Now, we apply them to image retrieval. Experiments is conducted with image sets selected from Corel database, including sunset, coast scenes, skiing, mountain, waterfall, elephant, horse, flowers, steamship, and so on.

All images were converted from the RGB to the Luv color space. A Gauss mixture was fitted by EM [10], to the samples extracted from each image. Then, Gauss mixture for all class were computed and stored as feature data respectively. System searches in the best matched class that returned after comparing query image to GMM of each class.As a matter of fact, a match function is also a distance measure function. So those distance measures related above, like ALA, Divergence, and Bhattacharyya, all could be used for clustering and retrieval.

As it can be concluded from above analysis, the essential step is to find the best matched class. We use probability of class match order (PCMO) to compare the performances of those distance measures. Assume class match order counter (CMOC) is an array with length 20. For each picture in the database, with some distance measure, we obtain 20 distance values by matching to each class. After arrangement, if the ith *(i=1,...20)* class is the object class, CMOC[*i*] increase by 1. When the traverse finished, we obtain average PCMO by dividing CMOC with the number of images in database.

Fig.1 shows the probability mass function of PMCO. as it shows, PMCO[1], the probability that the most similar class is the object class lifts from 37.00% to 65.55%! Due to overlap between image classes, the most similar class doesn't necessarily

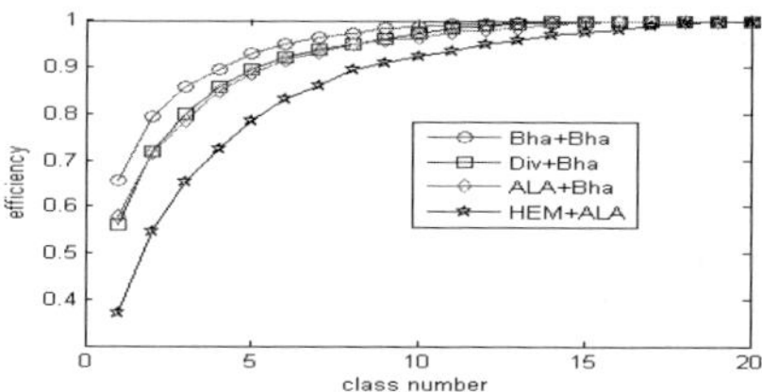

Fig. 1. Probability mass function of PMCO

correspond to the object class. Under that condition, we may search in the most similar classes. The probability that the object class contained in the most similar 4 classes lifts from 72.55%(HEM+ALA) to 89.50%(Bha+Bha, when both image clustering and image retrieval adopts the Bhattacharyya distance)

7 Conclusions

This paper demonstrates the contradiction between the adopted Monte-Carlo formula and IB principle. Extending special distortion measures to common distances, this paper also shows several proposals. And Experiments show their efficiency and plausibility.

Acknowledgments

This work was supported in part by the National Natural Science Foundation of China (Grant No. 60572078) and by Guangdong Natural Science Foundation (Grant No. 05006349).

References

1. Tishby, N., Pereira, F., Bialek, W.: The information bottleneck method. In: Proc.of the 37-th Annual Allerton Conference on Communication, Control and Computing, pp. 368–377 (1999)
2. Slonim, N., Tishby, N.: Agglomerative Information Bottleneck. In: Proceedings of NIPS-12 (1999)
3. Goldberger, J., Greenspan, H., Gordon, S.: Unsupervised Image Clustering Using the Information Bottleneck Method. In: Van Gool, L. (ed.) DAGM 2002. LNCS, vol. 2449, Springer, Heidelberg (2002)
4. Gordon, S., Greenspan, H., Goldberger, J.: Applying the information bottleneck principle to unsupervised clustering of discrete and continuous image representations. In: ICCV (2003)
5. Goldberger, J., Gordon, S., Greenspan, H.: An information theoretic framework for unsupervised image clustering. IEEE Trans. on Image Processing 15, 449–458 (2006)
6. Goldberger, J., Roweis, S.: Hierarchical Clustering of a Mixture Model. In: Neural Information Processing Systems (NIPS) (2004)
7. Goldberger, J., Greenspan, H., Gordon, S.: An efficient similarity measure based on approximations of KL-divergence between two Gaussian mixtures. In: International Conference on Computer Vision (ICCV) (2003)
8. Vasconcelos, N., Lippman, A.: Learning mixture hierarchies. In: Proc. of Neural Information Processing Systems (1998)
9. Vasconcelos, N.: On the Efficient Evaluation of Probabilistic Similarity Functions for Image Retrieval. IEEE Transactions on Information Theory 50(7), 1482–1496 (2004)
10. Dempster, A., Laird, N., Rubin, D.: Maximum likelihood from incomplete data via the EM algorithm. Journal of the Royal Statistical Society, Series B 39(1), 1–38 (1977)
11. Bian, Z.Q., Zhang, X.G.: Pattern Recognition, 2nd edn. Tsinghua University Press, China (2000)

An Opinion Analysis System Using Domain-Specific Lexical Knowledge

Youngho Kim, Yuchul Jung, and Sung-Hyon Myaeng

Information and Communications University,
119, Moonji-ro, Yuseong-gu, Daejeon, 305-714, South Korea
{yhkim,enthusia77,myaeng}@icu.ac.kr

Abstract. In this paper, we describe an opinion analysis system using domain-specific lexical knowledge in Korean economic news. We tested our hypothesis that such domain-specific knowledge helps enhancing the performance of statistically based approaches and obtained a promising result.

1 Introduction

Gathering opinions about a specific subject is important in many areas such as governments to improve their services [1]. Since people often express their thoughts on articles, newspaper is one of the good resources where opinions of various sorts are found. Especially, in the economy domain, opinions about economic events flood into news. Sentiment information (e.g., negative or positive tendency) can be a reflection of people's opinions for a specific subject. Our research is motivated by this and centered around sentiment analysis of news articles on economy. More precisely, we attempt to build a system that determines the polarity of sentiment and its strength.

Previously, researchers have used a statistical learning method and semantically oriented *seed* terms as clues. Pang et al [2] adopted supervised machine learners to predict a document's semantic polarity. Ku et al [3] attempted to develop opinion extraction, summarization and tracking systems. Hatzivassiloglou [6] have attempted to predict semantic orientation of adjectives by analyzing adjective pairs. Turney [7] have bootstrapped from a seed set[1]. Kamps et al [8] have focused on the use of lexical relations defined in WordNet. Esuli [5] proposed semi-supervised learning method started from expanding Turney's seed set. Bradley [4] tried psychological studies which have found measurable associations between words and human emotions.

Following the same line of thought, we take a novel approach that considers domain-specific lexical knowledge to complement generic methods entirely relying on statistical learning. In particular, we built an opinion analysis (OA) system for Korean economic news. Instead of constructing a costly deep knowledge base (KB) in the domain, we collected a set of domain-specific terms that indicate sentiments of an article in the domain. We postulate that domain knowledge is critical because a priori

[1] Seed set contains 7 positive words and 7 negative words as good, nice, excellent, positive, fortunate, correct, superior in positive set and bad, nasty, poor, negative, unfortunate, wrong, inferior in negative set.

H. Li et al. (Eds.): AIRS 2008, LNCS 4993, pp. 466–471, 2008.

knowledge on semantic orientation of domain-specific terms complements statistical learning methods. The KB contains sentiment information of economy terms such as "감자" (a reduction of capitals) and general sentiment clues such as "폭락" (falling).

In order to test our hypothesis that domain-specific lexicon should improve the performance of sentiment analysis, we evaluated our system using Korean economic news. The system consists of a KB that contains information about domain specific terms as well as domain independent terms that express sentiments, agent architecture for crawling news articles and determines their sentiments using the knowledge base, and an information retrieval (IR) system that enables end-users to capture the sentiment tendency of an interesting topic. In order to simulate an operational environment, for which this system was envisioned to begin with, about 170,000 news documents were collected initially. We ran experiments using topic-specific 200 articles of the collection to test our system's effectiveness.

2 Constructing a Knowledge Base

In our system, the KB is a core according to our hypothesis. The lack of machine-readable knowledge in the economy domain drove us to gather sentiment knowledge manually. We employed 5 annotators, who were majoring in economics. They annotated semantic orientation and its strength for a given term, based on its perceived role in determining sentiments in the domain. They were instructed to make judgments in view of the Korean economy. For example, "부채 계정 (liability account)" is considered strongly negative since it has a negative effect on the Korean economy.

Our assumption is that some economy terms have polarity values and degrees of strengths. Several well-known dictionaries[2] in economy domain were the main source. In addition, non-economy terms (context terms) that influence the economy were collected. For example, "북한 핵 실험" (North Korea's nuclear test) in news severely affects the economy as witnessed by North Korean's announcement of their success in underground nuclear test last year, which shook South Korea's stock market.

We also gathered words that frequently carry sentiment information in news because generally the sentiment often depends on an occurrence of *seed* words such as "비난" (denounce). These domain-independent words have some clues as to the polarity of text containing them. Our annotators extracted those terms from a randomly selected 300 sample articles from the collection of news articles collected during a month. All the collected terms were annotated, resulting in 13,564 dictionary terms, 620 context terms, and 176 general words. The annotation process was as follows. First, the annotator selected a term's polarity: positive, negative, or neutral. Next, the annotator decided its strength if the term is polarized. If a term's polarity is not obvious, the annotator was allowed to choose "neutral." We designed the scale of 1 to 5 for polarity values (1 and 2 are weak, 3 is mild, 4 and 5 are strong). Naturally, we found many cases with disagreement among the annotators. For examples, "1월 효과" (January Effect) was judged as two positives, two negatives, and one neural. For the cases with disagreements, we decided a term's semantic orientation by majority

[2] Mae-il Economic terms dictionary, Economic dictionary by the bank of Korea, Dong-a Economic terms dictionary are freely open on the Internet.

(i.e., counting voting results and selecting the most one) and the term's strength by the average of the selected polarity voters' strength values. When two or less annotators viewed a term as either positive or negative, we set it to be a neutral term. Therefore, we gathered 4043 dictionary terms, 531 context terms, and 176 general words which are either a single word or a phrase consisting up to three words. Table 2 shows the final results: the number of positive and negative terms with different strengths.

Table 1. Statistics of all annotation results

	Positive			Negative			Neutral
Annotator	Weak	Mild	Strong	Weak	Mild	Strong	
A	672	186	159	388	276	262	12,417
B	803	403	407	266	284	390	11,807
C	1,112	317	295	368	213	300	11,666
D	1,008	240	255	455	290	316	11,796
E	977	304	244	328	219	182	12,106

Table 2. Gold-standard

	Positive	Negative
Weak	2,739	1,322
Moderate	119	209
Strong	151	210

3 System Architecture

Our system determines a news article's sentiment information for end-users who want to read an opinionated news article on a specific topic, sometime with an interest on a particular sentiment. Due to an enormous number of daily news data, swiftness is an essential in our system. So, firstly, an agent[3] collects news documents from news wires. Next, another agent detects sentiment terms in KB and reacts to determine the sentiment of news in the background of the scoring formula, paragraph segmentation and morphological analysis[4]. After that, the news documents sentiment information are tagged are transferred to IR system [10]. As a result, news articles with sentiment information are viewed by IR system. Also, a document's sentiment value, *v* is determined in Figure 1. For each term (*tc*) in a document (*D*), the formula averages all the detected terms by each term's frequency (*f*(*tc*)) and each term's sentiment value (*s*(*tc*)). We added a bias for the most frequent term (*tm*) since frequent term is an important sentiment factor. (In experiment, β is 0.4, α is 0.6). Also, we promoted one step for sentiment words in the first paragraph as like "weakly positive" to "mildly positive", which reflects the writing style that topics are arranged at the top.

---

[3] Aglet™ agent (http://www.trl.ibm.com/aglets/)

[4] Morphological analyzer developed from ETRI (Korea Electronics and Telecommunications Research Institute) extracts noun portion from raw text and puts the part-of-speech tag on it.

$$v = \alpha \cdot \frac{\sum_{tc \in D} s(tc) \cdot f(tc)}{\sum_{tc \in D} f(tc)} + \beta \cdot tm, \text{where } \alpha + \beta = 1$$

Fig. 1. Document Sentiment Value Determination Function

4 Experimental Results

Our collection covers 21 different newswire sources from July of 2005 to November of 2006. Among them, we tested the documents retrieved within top-200 based on two queries, "정부 부동산 정책" (government's real estate policy) and "정몽준 현대 자동차 회장 구속" (arrest of Jung Mong Jun, President of Hyundai automobile company). Our goal of the experiments was testing two hypotheses: 1) our approach can complement the statistically based approach to OA; 2) our KB containing domain-specific lexical knowledge is useful in determining the polarity and the strength of news article. We assumed an information seeking scenario where a user wanting to get a feeling about the most sensational news in economy enters a query like the above query and reads the retrieved news articles. First, we measured the accuracy of the polarity determination function using three classes: negative, positive, and neutral. Second, we examined the sentiment strength determination function in terms of accuracy and the Mean-Squared Error (MSE) that can capture the problem of the huge distinction between the true strength and the predicted strength. If t_i is the true strength of document i, and p_i is the predicted strength of document i,

$$MSE = \frac{1}{n} \cdot \sum_{i}^{n} (t_i - p_i)$$

where n is the number of documents in the test set. We obtained the results of strengths at the 3-step level (i.e., weak, mild, and strong) and 5-step level (i.e., 1~5).

For the gold-standard, we decided the polarities and their strength of the test articles by majority as used in Section 3. To verify our system's competitiveness over statistical learning methods, we produced results of SVM and Naïve Bayes-based classifiers as the baseline. Those classifiers using uni-gram feature were trained by another 100 retrieved documents with the same queries (i.e., we searched another 100 samples and annotated them in the same manner as in the test set). Pang et al [2] report that SVM has the best performance in English text. In spite of the belief that adjective is more effective contributor than other parts-of-speech in classifying English documents based on their sentiment values [5, 6], noun is a much more critical feature in Korean text (Table 4). We used Joachim's SVM^{light} package [9].

We obtained 74% accuracy which is 9% increase over the better baseline performance and 30% increase over the case where only the general KB was used. Since the terms in KB are only nouns, adjective was not encountered in the cases of KB. We observed, at least in Korean news, that domain specific terms play a more important role in determining the sentiment because the context of words provide additional information in capturing more precise sentiment. For example, general words such as "하락(decrease)" that may be considered as negative in many context have different meanings depending on its context. For example, "분양 가격 하락 (decrease of the

Table 4. Accuracy results of polarity

Features	Naïve Bayes	SVM	KB with domain knowledge	KB with only general knowledge
Adjective	51%	58%	N/A	N/A
Noun	**59%**	**68%**	**74%**	**57%**
Adjective + Noun	58%	68%	N/A	N/A

Table 5. Polarity determination results for different annotators

Methods	A	B	C	D	E
KB with domain knowledge	**68%**	**69%**	**69%**	**74%**	**73%**
KB with only general knowledge	58%	51%	62%	53%	50%

Table 6. Results of determination of polarity strengths

Methods	Accuracy		MSE	
	3 step	5 step	3 step	5 step
KB with domain knowledge	**63%**	**33%**	**1.40**	**2.83**
KB with only general knowledge	51%	21%	2.73	4.37

selling price of an apartment)" is positive, but "부동산 대출금리 하락 (decrease of the interest for real-estate loans)" is negative since it may cause land speculation. To resolve certain variation, we need domain knowledge.

Our approach is more effective across all the annotation results although there were quite a wide range of variations. The results in the table reflect the diversity compared to the gold-standard. However, our approach is shown to outperform the one with general term knowledge across all the cases, varying from 24% to more than 50%.

5 Conclusions

We proposed a knowledge-based OA system that does not rely on heavy natural language processing to figure out the semantics or pragmatics of text, which is not really feasible for the given task at this point. Our approach is practical since the knowledge base contains lexical information, i.e. terms and their polarity information in a specific domain like economy, which can be captured with relative ease. The machine-readable dictionary we developed contains not only domain specific terms, but also domain-independent opinion clue terms. We verified that our approach is effective and promising in an experiment by showing that the proposed system determines the polarity and the strength more accurately than statistically based machine learning methods and the method with only general seed words. While the experiment was done with Korean news articles, the result is valuable since most popular statistical machine learning approaches and the approach with knowledge on general, domain-independent terms only have clear limitations that should be overcome. In the future, a more detailed analysis of the reasons why machine leaning based approaches are inferior would be of great value.

Acknowledgements

This work was supported by KICOS through a grant by Ministry of Science & Technology (K20711000007-07A0100-00710) and 2nd Phase of Brain Korea 21 project sponsored by Ministry of Education and Human Resources Development, Korea.

References

1. Dave, K., Lawrence, S., Pennock, D.M.: Mining the peanut gallery: Opinion extraction and semantic classification of product reviews. In: Proceedings of the 12th International Conference on the World Wide Web, Budapest, Hungary, ACM Press, New York, US (2003)
2. Pang, B., Lee, L., Vaithyanathan, S.: Thumbs up? Sentiment Classification using Machine Learning Techniques. In: Proceedings of the 7th Conference on Empirical Methods in Natural Language Processing, Philadelphia, US, Association for Computational Linguistics, Morristown, US (2002)
3. Ku, L., Liang, Y., Chen, H.: Opinion Extraction, Summarization and Tracking in News and Blog Corpora. In: Proceedings of AAAI Spring Symposium on Computational Approaches to Analyzing Weblogs, AAAI Technical Report SS-06-03, Stanford University, California, US (2006)
4. Bradley, M.M., Lang, P.J.: Affective norms for English words (ANEW): Stimuli, instruction manual and affective ratings. Technical Report C-1, The Center for research in Psychophysiology, University of Florida, Gainesville, Florida, US (1999)
5. Esuli, A., Sebastiani, F.: Determining the Semantic Orientation of Terms through Gloss Classification. In: Proceedings of the 14th ACM International Conference on Information and Knowledge Management, ACM Press, New York, US (2005)
6. Hatzivassiloglou, V., McKeown, K.R.: Predicting the semantic orientation of adjectives. In: Proceedings of the 35th Annual Meeting of the Association for Computational Linguistics, Madrid, Spain, Association for Computational Linguistics (1997)
7. Turney, P.D., Littman, M.L.: Measuring praise and criticism: Inference of semantic orientation from association. ACM Transactions on Information Systems (2003)
8. Kamps, J., Marx, M., Mokken, R.J., Rijke, M.D.: Using WordNet to measure semantic orientation of adjectives. In: Proceedings of the 4th International Conference on Language Resources and Evaluation, Lisbon, Portugal (2004)
9. Joachims, T.: Making large-scale SVM learning practical. In: Scholkopf, B., Burges, C., Smola, A. (eds.) Advances in Kernel Methods –Support Vector Learning, MIT Press, Cambridge (1999)
10. Hatcher, E., Gospodnetic, O.: Lucene IN ACTION. Manning Publications Co. (2004)

A New Algorithm for Reconstruction of Phylogenetic Tree*

ZhiHua Du and Zhen Ji

ShenZhen Unvierstiy, Shenzhen,China
du_zhihua@yahoo.com.cn

Abstract. The abstract should summarize the contents of the paper and should contain at least 70 and at most 150 words. It should be set in 9-point font size and should be inset 1.0 cm from the right and left margins. There should be two blank (10-point) lines before and after the abstract. This document is in the required format. In this paper, we present a new algorithm for reconstructing large phylogenetic tree. This algorithm is based on a family of Disk-Covering Methods (DCMs) which are divide-and-conquer techniques by dividing input dataset into smaller overlapping subset, constructing phylogenetic trees separately using some base methods and merging these subtrees into a single one. Provided the high memory efficiency of RAxML (which the program inherited from fastDNAml) compared to other programs and the good performance on largereal-world data it appears to be best-suited for use as the base method. The experiments clearly show that the proposed algorithm improves over stand-alone RAxML on all datasets, i.e. yields better likelihood values than RAxML in the same amount of time. This results serve as an argument for the choice of the proposed algorithm instead of stand-alone RAxML.

Keywords: Phylogenetic tree,divide-and-conquer, DCM.

1 Introduction

Phylogenetic tree illustrates the evolutionary relationships among a groups of organisms, or among a family of related nucleic acid or protein sequences, e.g., how might have this family been derived during evolution. It plays a fundamental role in many biological problems such as multiple sequence alignment, protein structure and function prediction, and drug design [1].

There are two general categories of methods for calculating phylogenetic trees: distance-based and character-based. Distance-based methods compute a matrix of pairwise distances between sequences in an alignment, and then constructing a tree based entirely on the original distance computations instead of sequences. There exist many methods based on this idea. Such as, Neighbor-Joining [2] and other improved

---

* This work is partially funded by a Research Foundation granted by the Shenzhen University under grant no: 4DZH), the National Natural Science Foundation of China under grant no. 60572100, the Royal Society (U.K.) International Joint Projects 2006/R3 - Cost Share with NSFC (China).

H. Li et al. (Eds.): AIRS 2008, LNCS 4993, pp. 472–477, 2008.

distance methods, WEIGHBOR [3], BIONJ [4], FASTME [5] and a latest approach considering maximum-likelihood estimated triplets of sequences [6]. Disadvantages of distance-based methods include the inevitable loss of evolutionary information when a sequence alignment is converted to pairwise alignment [9].

Character-based methods would examine each column of the alignment separately and look for the tree that best accommodates all of this information, such as maximum parsimony (MP) [7] or maximum likelihood (ML) [8]. MP chooses tree that minimizes number of changes required to explain data. ML, under a model of sequence evolution, finds the tree which gives the highest likelihood of the observed data. Character-based methods are information rich for there is a hypothesis for every column in the alignment. However, The MP method is NP-hard, and ML has unknown complexity [10] and is very hard to solve in practice. Primary sources of phylogenetic tree construction software include the PHYLIP website (http:// evolution. genetics.washington.edu/phylip.html), MrBayes [11] and PAUP [12][13]

Previous studies have shown that large datasets are challenging for MP heuristics implemented in these packages [14][15]. To analyze datasets containing thousands of sequences Disk Covering Methods (DCMs)[14][16][17][18] were introduced. To the best our knowledge, Recursive-Iterative-DCM3 (Rec-I-DCM3), is the best known technique heuristic for solving MP.

AS to ML, one of the recent methods, RAxML [19][20], is among the fastest, most accurate, and most memory-effcient ML heuristics on real biological datasets to the best of our knowledge. Furthermore, the global optimization method (fast Nearest Neighbor Interchange adapted from PHYML [21] is not as efficient on real alignment data as RAxML. Thus, it is not suited to handle large real-data alignments of more than 1,000 sequences.

In this paper, we present a new algorithm for reconstructing large ML phylogenetic tree by integrating algorithmic concepts from Rec-I-DCM3 and RAxML. The experimental results show that the proposed algorithm outperforms the existing methods.

2 Methods

The proposed algorithm is consists of four main steps.

Step 1: Recursively divide the given dataset into smaller, overlapping subproblems, until the subproblems become at most of size Maximum subproblem size (MS) and stores the merging order (subset-guidetree, rurTree) which is required to correctly execute the merging step.

Step 2: Construct phylogenetic tree for subproblems by using the RAxML method.

Step 3: Combine the sub phylogenetic trees into a unique supertree

Step 4: A hill-climbing ML search on the supertree, T', is applied to do a global rearrangement

Some definitions used are shown as following:

Short subtree. Suppose there is a tree T with an edge e in it. Let Q_1, Q_2, Q_3 and Q_4 be the four subtrees around e; q_1, q_2, q_3 and q_4 be the set of leaves closet to e in each of the four subtrees respectively. The distance between them is measured by the hamming distances on the edges. The set of nodes in $q_1 \cup q_2 \cup q_3 \cup q_4$ is the short subtree around e.

Short subtree graph. Short subtree graph is the union of cliques formed on "short subtrees" around each edge in T

Separator. Separator is the short subtree of a special edge, which would produce the most balanced bipartition of the leaves in tree T when removed.

The outline of the proposed algorithm is as following:

1. Problem Initialization
   1.1 Set S={S0, …, Sk-1} of aligned biomolecular sequences. Set k=number of sequences, n=number of iteration, b=base heuristic (TNT), T=starting tree, MS=maximum subproblem size.
   1.2 Initialize a subset guide-tree, rurTree, to record recursive calls as the topology for merge subtrees.
   Initialize, allsubsets, to save a total set of subproblems.
2. For n iterations do
   2.1 /*Construct a recursive DCM3 decomposition using T|S (a guidetree tree on dataset S) as the guide tree, producing a total set of subproblems, allsubsets =A0, A1 ,…, Am-1 (m is the total number of subsets). Produce a subset guide-tree, rurTree, to keep the merge order. The rurTree is expressed in a string format that uses parenthesis to start and end subtree groups, commas to separate group members, and subproblems names to name tree leaves. */

   Recursive_Divide(S, MS, T)
   2.2 Build phylogenetic subtrees of subproblems by using RAxML.
   2.3 Use a postorder tree walk algorithm to search subset-guidetree, rurTree, in order to merge subtrees into a suptertree, T'.
   2.4 Apply hill-climbing heuristic starting from T' until we reach a local optimum. Let T' be the resulting local optimum. Set T=T'.

Function Recursive_Divide(S, MS, T)

Input: Set of k sequences S = S1; S2; ::::; Sk Maximum subset size MS
Starting tree T
Output: Set of subproblems, allsubsets = A1,A2, …,Am
(m is the total number of subsets)
Algorithm: Recursively divide a set of k sequences S into subproblems

(a) Compute edge weighting for each edge by using the hamming distances.
(b) Compose short subtree graph around edges by selecting set of all leaves that are elements in a short quartet around an edge, that is $sub_1,sub_2,\ldots,sub_x$ (where x is the number of subsets).
(c) Find a separator, spt, by selecting an edge that when removed, produces the most balanced bipartition of the leaves as centroid edge, Ec.

The spt is the leaves of the short subtree around Ec. The subsets are then defined to be

$$A_i = spt \cup sub_i, 1 \le i \le x$$

(d) For $A_i (1 \le i \le x)$
If (A_i 'size >MS){
Let T|A_i be the result of restricting tree T to Ai for each i.
/*Recursively compute the subsets for A_i */
Recursive_Divide (A_i, MS, T| A_i)
}
Else{
Add A_i to allsubsets.
Re-build subset-guidetree, rurTree.
/*Produce a subset-guidetree, rurTree, to keep the merge order. The rurTree is expressed in a string format that uses parenthesis to start and end subtree groups, commas to separate group members, and subproblems names to name tree leaves.*/
}

3 Experimental Design

The online version of the volume will be available in LNCS Online. Members of institutes subscribing to the Lecture Notes in Computer Science series have access to all the pdfs of all the online publications. Non-subscribers can only read as far as the abstracts. If they try to go beyond this point, they are automatically asked, whether they would like to order the pdf, and are given instructions as to how to do so.

Methodology

To the best our knowledge, RAxML, is the best known technique heuristic for solving ML. Therefore, we design an experiment to show that the proposed algorithm would improve over stand-alone RAxML on all datasets in order to demonstrate the benefits which arise from using the dividing method.

Datasets

The experiments were done on six large datasets, some of which are obtained from http://www.cs.njit.edu/usman/RecIDCM3.html. The datasets we used are (1) 6281 Eukaryotes ssu rRNA sequences from the European rRNA database, (1661 sites), (2) 6458 firmicutes bacteria 16s rRNA sequences from the RDP (1352 sites), (3) 6722 three-domain rRNA sequences from Robin Gutell (1122 sites) [22], (4) 7769 three-domain + 2 organelle rRNA sequences from Robin Gutell (851 sites), (5) 11361 set of all bacteria ssu rRNA sequences from the European rRNA database (1360 sites)[23], and (6) 13921 proteobacteria 16s rRNA sequences from the RDP (1359 sites) [22].

4 Experimental Results

In our experiments both methods start optimizations on the same starting tree. Due to the relatively long execution time on large alignments we only executed one iteration per dataset. The run time of one the proposed algorithm iteration was then used as inference time limit for RAxML. Table 1 provides the log likelihood values for RAxML and the proposed algorithm after the same amount of execution time. Note that, the apparently small differences in final likelihood values are significant because those are logarithmic values and due to the requirements for high score accuracy in phylogenetics (T.L.Williams et al. 2004).

Table 1. The proposed algorithm versus RAxML log likelihood values after the same amount of inference time

Dataset	Proposed algorithm	RAxML
Dataset1	-99967	-99982
Dataset2	-355071	-355342
Dataset3	-383578	-383988
Dataset4	-1270920	-1271756
Dataset5	-901904	-902458
Dataset6	-541255	-541438

The experiments clearly show that the proposed algorithm improves over stand alone RAxML on all datasets, i.e. yields better likelihood values than RAxML in the same amount of time. This results serve as an argument for the choice of the divide-and-conquer method instead of stand-alone RAxML.

References

1. Smith, T.F., Waterman, M.S.: Identification of Common Molecular Subsequences. J. Mol. Biol. 147, 195–197 (1981)
2. Bull, J.J., Wichman, H.A.: Applied evolution. Annual Review of Ecology and systematics 32, 183–217 (2001)
3. Saitou, N., Nei, M.: The nigehbor-joining method: a new method for reconstructing phylogenetic tree. J Mol Evol 4, 406–425 (1987)
4. Bruno, W.J., Socci, N.D., Halpern, A.L.: Weighted Neighbor Joining: A Likelihood-Based Approach to Distance-Based Phylogeny Reconstruction. Mol Biol Evol 17, 189–197 (2000)
5. Gascuel, O.: BIONJ: an improved version of the NJ algorithm based on a simplemodel of sequence data. Mol Biol Evol 14, 685–695 (1997)
6. Desper, R., Gascuel, O.: Fast and Accurate Phylogeny Reconstruction Algorithms based on the Minimum-Evolution Principle. J Comput Biol 19, 687–705 (2002)
7. Ranwez, V., Gascuel, O.: Improvement of Distance-Based phylogenetic Methods by a Local Maximum Likelihood Approach Using Triplets. Mol Biol Evol 19, 1952–1963 (2002)
8. Camin, J., Sokal, R.: A method for deducing branching sequences in phylogeny. Evolution 19, 311–326 (1965)
9. Felsentein, J.: Evolutionary trees from DNA sequences: a maximum likelihood approach. J Mol Evol 17, 368–376 (1981)
10. Steel, M.A., et al.: Loss of information in genetic distances. Nature 336, 118 (1988)
11. Stelel, M.A.: The maximum likelihood point for a phylogenetic tree is not unique. Systematic Biology 43(4), 560–564 (1994)

12. Huelsenbeck, J.P., Ronquist, F.: MYBAYES: Bayesian inference of phylogenetic trees. Bioinformatics 17, 754–755 (2001)
13. Swofford, D.: PAUP*. Phylogenetic Analysis Using Parsimony (* and other mothods). Version 4.Sinauer Associates (2002)
14. Sjolander, K.: Phylogeneomi inference of protein molecular function: advances and challenges. Bioinformatics 20, 170–179 (2004)
15. Roshan, U.: Algorithm techniques for improving the speed and accuracy of phylogenetic methods. PhD thesis (2004)
16. Roshan, U., Moret, B.M.E., Warnow, T., Williams, T.L.: Rec-i-dcm3: a fast algorithmic technique for reconstructing large phylogenetic trees. In: Proceedings of the IEEE Computational Systems Bioinformatics conference (CSB), Stanford, California, USA (2004)
17. Huson, D., Nettles, S., Warnow, T.: Disk-covering, a fast-converging method for phylogenetic tree reconstruction. Journal of Computational Biology 6, 369–386 (1999)
18. Nakhleh, L., Roshan, U., St. John, K., Sun, J., Warnow, T.: Designing fast converging phylogenetic methods. In: Proc. 9th Int'l Conf. on Intelligent Systems for Molecular Biology (ISMB 2001). Bioinformatics, vol. 17, pp. S190–S198. Oxford U. Press, Oxford (2001a)
19. Warnow, T., Moret, B., St. John, K.: Absolute convergence: True trees from short sequences. In: Proc. 12th Ann. ACM-SIAM Symp. Discrete Algorithms (SODA 2001), pp. 186–195. SIAM Press, Philadelphia (2001)
20. Stamatakis, A., Ludwig, T., Meier, H.: Parallel inference of a 10.000-taxon phylogeny with maximum likelihood. In: Proceedings of 10th International EuroPar Conference, pp. 997–1004 (2004)
21. Stamatakis, A., Ludwig, T., Meier, H.: Raxml-iii:a fast program for maximum likelihood-based inference of large phylogenetic trees. Bioinformatics 21(4), 456–463 (2005)
22. Guindon, S., Gascuel, O.: A simple, fast, and accurate algorithm to estimatelarge phylogenies by maximum likelihood. Syst. Biol. 52(5), 696–704 (2003)
23. Maidak, B., et al.: The RDP (ribosomal database project) continues. Nucleic Acids Research 28, 173–174 (2000)
24. Wuyts, J., Van de Peer, Y., Winkelmans, T., De Watchter, R.: The European database on small subunit ribosomal RNA. Nucleic Acid Research 30, 183–185 (2002)

A Full Distributed Web Crawler Based on Structured Network

Kunpeng Zhu, Zhiming Xu, Xiaolong Wang, and Yuming Zhao

Intelligent Technology and Natural Language Processing Lab, School of Computer Science and Technology, Harbin Institute of Technology, Harbin 150001, China
{kpzhu,xuzm,wangxl,ymzhao}@insun.hit.edu.cn

Abstract. Distributed Web crawlers have recently received more and more attention from researchers. Full decentralized crawler without a centralized managing server seems to be an interesting architectural paradigm for realizing large scale information collecting systems for its scalability, failure resilience and increased autonomy of nodes. This paper provides a novel full distributed Web crawler system which is based on structured network, and a distributed crawling model is developed and applied in it which improves the performance of the system. Some important issues such as assignment of tasks, solution of scalability have been discussed. Finally, an experimental study is used to verify the advantages of system, and the results are comparatively satisfying.

Keywords: Web crawling; full distributed; structured network.

1 Introduction

Due to the exponential growth of the web, an important challenge of web crawler is to efficiently collect massive pages in a limited time frame, one that has received considerable research attention.

Some distributed crawling systems have been worked out to finish Web massive information collecting task. The distributed crawling systems mentioned in [1, 2, 3] use a centralized server to manage the communication and synchronization of crawling nodes. These centralized solutions are known to have problems like link congestion, being a single point of failure, and expensive administration. Some full distributed crawling systems have been proposed in [4, 5, 6], that is, no central coordinator can exist in these systems. In these systems, large numbers of nodes collaborate dynamically in an ad-hoc manner and share information in large-scale distributed environments without centralized coordination. An important issue of the presented crawlers is dynamic load balance. Most systems concern the methods of static load assignment and ignore the unbalance in crawling process. Another issue that has not been well resolved is scalability caused by the arrivals and departures of nodes.

In this paper, our research mainly focuses on how to design a full distributed crawler based on a distributed crawling model. A structured architecture will be proposed and the mechanism to achieve load balance and scalability will be given.

H. Li et al. (Eds.): AIRS 2008, LNCS 4993, pp. 478–483, 2008.

2 Architecture

In our crawler, crawling nodes are organized as a structured ring network to offer the service of collecting Web pages. The ring is composed of several crawling nodes that autonomously coordinate their behaviour in such a way that each of them scans its share of the Web. Such organization has several desirable properties – it is highly resilient to a single point of failure, and incur low overhead at node arrivals and departures. More importantly, they are simple to implement and incur virtually no overhead in topology maintenance. The overview of crawler system can be described by Figure.1.

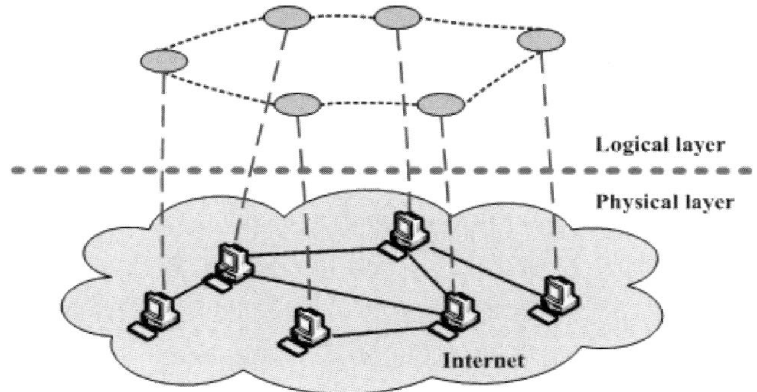

Fig. 1. Overview of crawler system

Each crawling node in system is composed of 2 parts: crawling module and control module. The function of crawling module is to download web pages from Internet according to the URLs queue. The function of control module is to manage the communication and harmony with other crawling nodes.

The inside organization of each crawling node can be described by Figure.2.

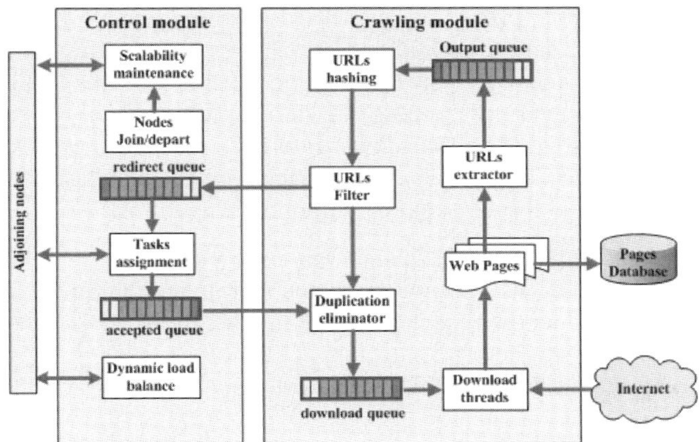

Fig. 2. Architecture of each crawling node

3 Distributed Crawling Model

Our research mainly focuses on the decentralized crawling strategy which has been implemented in control module. The core of the strategy is called distributed crawling model (DCM). As mention above, the model is composed of three parts which will be described as follow.

3.1 Tasks Assignment

The sub-module of tasks assignment is used to divide whole crawling task into different parts, and allocate them to each node in order to achieve a parallel processing. We propose a new method of tasks assignment which is a dynamic consecutive division to the value space of hash function, and we explain why this method makes it possible to decentralize every task and to resolve the above problems.

Let the value space of hash function be a rang from a_0 to a_n. For example, if $H(URL)$ denotes the sum of integer parts of the IP of URL's host, then $a_0 = 0$ and $a_n = 255 \times 4 = 1020$. Let n denote the number of nodes, we can get a division with $n-1$ numbers denoted by $(a_1, a_2, ..., a_{n-1})$ and $a_0 < a_1 < a_2 < ... < a_{n-1} < a_n$, the node i will take charge of the URLs whose $H(URL)$ are located in the range of (a_{i-1}, a_i).

At the beginning, we initialize the value of a_i as follow:

$$a_i = \frac{a_n}{n} \times i \qquad (i = 1, 2, ..., n-1) \tag{1}$$

Obviously, formula (1) is a n equivalent division on the range of (a_0, a_n). We will dynamically change the value of a_i in crawling process to achieve a load balance, the more detail will described in next section.

The crawling nodes are organized as a ring. Each node has two neighbors which called "preceding-node" and "following-node". The hashing value of URLs on preceding-node is smaller than that on following-node. Each node need maintain three URLs queues: local-queue, preceding-queue and following-queue. The URLs in preceding-queue need to be sent to preceding-node, URLs in following-queue need to be sent to following-node and URLs in local-queue need to be sent to local download queue. In order to complete this process, we define two token in our structured network named "forward-token" and "backward-token". The "forward-token" starts off from the first node in network which charges the set of the smallest hashing value of URLs. The node holding "forward-token" will operate as follow:

1. The node sends its following-queue to its following-node.
2. The following-node will accept the queue and divide the queue into two parts, one is added into its own local download queue, the other is added into its following-queue.
3. The node gives the forward-token to its following-node.

The "backward-token" starts off from the last node and the process is the opposite with the "forward-token".

The time of token walking a circle on the ring is called cycle T. In order to avoid frequent communication in the network, we let T equal a longish time, such as one hour. So, the interval of sending token from one node to another is T/n. And a new URL will arrive at the corresponding node within the time T.

3.2 Dynamic Load Balance Management

Load balance means that each node should be responsible for approximately the same number of URLs. But the n equivalent division on the range of (a_0, a_n) can not assure that there are same number URLs located in each part. So we provide a dynamic load balance model to achieve the characteristic of load balance. Our model is based on three principles:

1. A little unbalance is permitted for the communication price of adjusting load balance.
2. At certain moment, the operation of adjusting load balance only occurs between two adjoining nodes.
3. Local balance should comply with the global.

We use a token called DLBT (dynamic load balance token) to perform the function. The operation of adjusting load balance only occurs between the node holding DLBT and its "following-node". The DLBT starts off from the first node and walk on the ring.

3.3 Scalability Maintenance

High scalability means that the more crawling nodes, the higher performance. We should develop the mechanism to maintain the topology of structured ring networks and manage the arrivals and departures of crawling nodes.

The mechanism is rather simple. Each node in the structured network not only keeps the information of its two neighbors, but also saves the information of three closest nodes in up and down direction. If a node is failure, its neighbor will find the next node to rebuild the virtual link. If a node joins in the network, it will request for the connected node and get the information of neighbors to create the link, of course, the redundant link will be removed.

4 Evaluation Methodology and Experiment Results

The goal of this section is to analyze the load balance and scalability features of our crawler. In order to achieve the load balance of the system, we use hash function to dynamically assign URLs to each crawling node. The consequence can be obtained by analyzing colleted Web pages by each node every hour.

In Jan 2007, we utilize our crawler to get experimental data which are about 7771402 Web pages with 21.75GB capacity within ten hours. And the number of parsed URLs is about 58606584. All of our measurements are made on six general Intel PCs with the P4 3.0GHZ Intel processors, 2GB of memory and 400GB hard SATA disk, the bandwidth is 100M. The operating system is Redhat Linux 9.0.

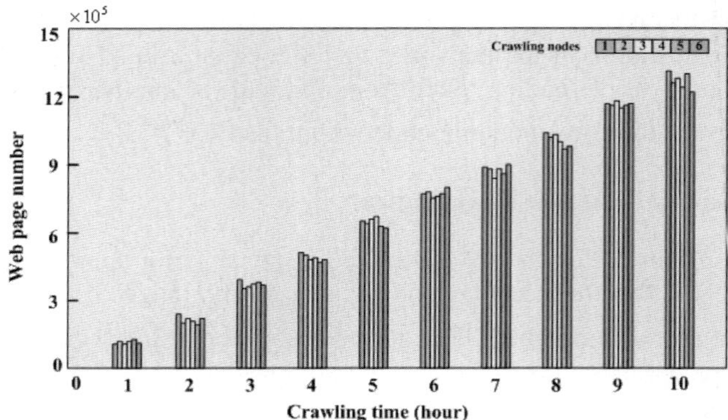

Fig. 3. Evaluation of load balance

Figure.3 shows the performance of load balance of our system. The experimental results show that our dynamic load balance model has a remarkable performance in improving the load balance in distributed crawler systems.

With the more number of crawling nodes, the crawling speed of our system is higher, shown in figure.4. The relation is almost linear. But with the increasing of nodes, the overload of the synchronization and communication among the nodes may decrease the performance.

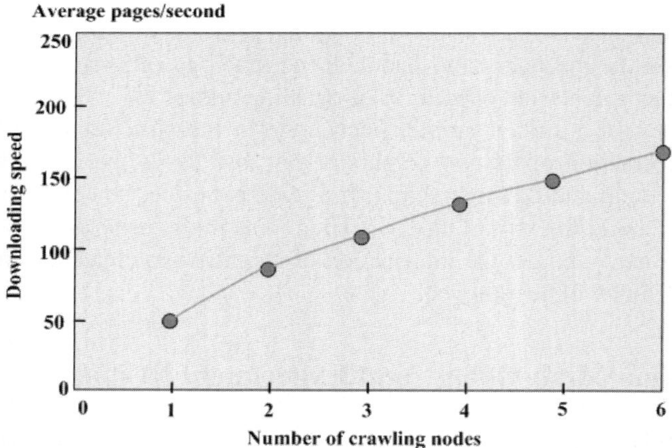

Fig. 4. Evaluation of scalability

5 Conclusions

In this paper we present a full distributed crawler system based on a structured network. A distributed crawling model (DCM) is proposed to achieve the merits of load balance and scalability. And a new method of tasks assignment is presented, which is a dynamic consecutive division on the value space of hash function. Also, a dynamic load balance model is used on the structured ring network. The experiment results show that our methods achieve a well performance to improve the load balance and scalability in distributed crawling environment.

Acknowledgements

This paper is supported by the Key Program of National Natural Science Foundation (No. 60435020).

References

1. Yan, H., Wang, J., Li, X., Guo, L.: Architectural Design and Evaluation of an Efficient Web-crawling System. Journal of System and Software 60(3), 185–193 (2002)
2. Shkapenyuk, V., Suel, T.: Design and Implementation of a High-Performance Distributed Web Crawler. In: Proceedings of the 18th International Conference on Data Engineering (ICDE 2002), pp. 357–368 (2002)
3. Hafri, Y., Djeraba, C.: High performance crawling system. In: Proceedings of the 6th ACM SIGMM international workshop on multimedia information retrieval, pp. 299–306. ACM Press, New York (2004)
4. Boldi, P., Codenotti, B., Santini, M., Vigna, S.: Ubicrawler: A scalable fully distributed Web crawler. Software: Practice & Experience 34(8), 711–726 (2004)
5. Loo, B.T., Cooper, O., Krishnamurthy, S.: Distributed Web Crawling over DHTs. Tech. Rep. UCB//CSD-04-1332, UC Berkeley, Computer Science Division (February 2004)
6. Singh, A., Srivatsa, M., Liu, L., Miller, T.: Apoidea: A Decentralized Peer-to-Peer Architecture for Crawling the World Wide Web. In: The Proceedings of the SIGIR workshop on distributed information retrieval (August 2003)

A Simulated Shallow Dependency Parser Based on Weighted Hierarchical Structure Learning

Zhiming Kang, Chun Chen*, Jiajun Bu, Peng Huang, and Guang Qiu

College of Computer Science, Zhejiang University, Hangzhou, China
{kzm,chenc,bjj,huangp,qiuguang}@zju.edu.cn

Abstract. In the past years much research has been done on data-driven dependency parsing and performance has increased steadily. Dependency grammar has an important inherent characteristic, that is, the nodes closer to root usually make more contribution to audiences than the others. However, that is ignored in previous research in which every node in a dependency structure is considered to play the same role. In this paper a parser based on weighted hierarchical structure learning is proposed to simulate shallow dependency parsing, which has the preference for nodes closer to root during learning. The experimental results show that the accuracies of nodes closer to root are improved at the cost of a little decrease of accuracies of nodes far from root.

1 Introduction

Recently, dependency grammar has gained renewed attention and becomes more prominent. Currently it is dominant that using data-driven approaches to learn parsers automatically from experience, such as probabilistic generative models [3], generative probabilistic parsing models [2] and deterministic discriminative model [7] and so on. Generally speaking, data-driven approaches fall into two categories, i.e. generative models and discriminative models. The latest state-of-the-art dependency parsers are discriminative which are based on classifiers trained to score trees, given a sentence, either via factored whole structure scores [5] or local parsing decision scores [6]. However, seldom work about shallow dependency parsing like shallow phrase-structure parsing has been done. In the paper, a discriminative dependency parser based on weighted hierarchical structure learning is proposed to simulate shallow dependency parsing, aiming at improving dependency parsing for nodes closer to the root node.

The remainder of this paper is organized as follows. Section 2 first makes a brief introduction to dependency grammar, and then describe dependency parsing algorithm in detail. Section 3 gives the details of adopted learning algorithm and some discussion. To demonstrate the usefulness of our algorithm, Section 4 contains the results produced by several dependency parsers. Last section contains some conclusions plus some ideas for future work.

* Corresponding author.

H. Li et al. (Eds.): AIRS 2008, LNCS 4993, pp. 484–489, 2008.

2 Dependency Parsing

2.1 Overview of Dependency Grammar

In Dependency Grammar, individual words in a sentence are considered to be linked together in dependency relations instead of being combined just mechanically. Whenever two words are linked by a dependency relation, we say that one of them is the head and the other is the dependent, and that there is an edge connecting them. In general, the dependent is the modifier or complement; the head plays the larger role in determining the behavior of the pair. The dependent presupposes the presence of the head; the head may require the presence of the dependent. The figure 1 depicts the skeleton of dependency structure of a sentence. The dashed line means the head 'Root' and the relation <'Root', 'had'> both are dummies. Essentially, a dependency link is a directed arc pointing from head to dependent. The dependency structure is a tree with the main verb as its root (head).

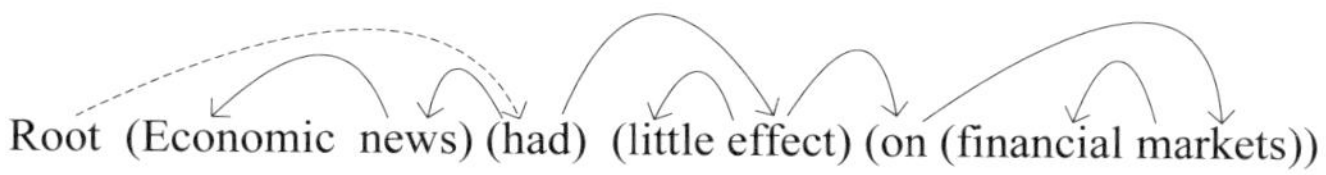

Root (Economic news) (had) (little effect) (on (financial markets))

Fig. 1. An example of annotated image

Similar to shallow phrase-structure parsing, shallow dependency parsing breaks up sentence into 'spans', and then link them with directed arcs. The edges connecting different spans are named 'span-link', and the two nodes linked by 'span-link' are defined as 'span-head' with respect to corresponding span. Different from full parsing, shallow dependency parsing only focuses on 'span-head' and 'span-link', instead of nodes and edges inside spans. Currently there is no standard about what is shallow dependency parsing like shallow phrase-structure parsing, and the following gives a rough guideline: the dummy node 'Root' is the root of a dependency tree; each subtree is treated as a span in shallow parsing. Based on above analysis, it is reasonable to think that 'span-head' and 'span-link' closer to 'Root' are more important than the others in shallow parsing, such as that inside span. Thus it is feasible to improve accuracy of dependency relations closer to 'Root' to simulate shallow dependency parsing, with regular full parsing.

2.2 Parsing Algorithm

The CKY algorithm is a well-known $O(n^3)$ algorithm for PCFG parsing [4]. When applied to dependency parsing, however, the CKY has the time complexity of $O(n^5)$. Eisner proposed an parsing algorithm similar to CKY that has a time complexity of $O(n^3)$ [3]. The idea is to parse the left and right dependents of a word independently and combine them at a later stage. During dependency parsing, there are many spans produced. Among them adjacent spans are possible to be combined into a longer span iteratively. At last one span including all words can be generated as output. This parsing algorithm removes the need for the additional head indices and requires only two additional binary variables that specify the direction of the item and whether an item is complete. For space limitation the parsing algorithm is described here briefly, and for details please refer to [3,5].

3 Learning

As indicated earlier, dependency tree is built bottom-up via combining small spans iteratively. The number of generated spans, however, grows exponentially with the size of sentence length, so the learning task is to, given a sentence, find the best one from numerous candidates. In this paper we adopt a strategy similar to McDonald et al [5], that is to say, every candidate is scored and chooses the one with highest score as final dependency parsing output.

An extension of original binary perceptron for multiclass problem (MPA) is proposed by Collins [1] as follows:

$$w = w + \alpha \cdot (\phi(x_i, y_i) - \phi(x_i, z_i)) \tag{1}$$

where z_i is a prediction for instance x_i and α is a constant positive factor for promotion or demotion, and the definition of ϕ is the same as the previous representations of feature vector. Note that the parameter α is a constant, that means weights of all relations in dependency structure are updated (add or minus) with equal scalar. However, it is not always reasonable. For instance, in figure 2 there are three dependency tree candidates - a correct dependency tree (a), both incorrect dependency trees (b) and (c) (node enclosed by dashed circle has incorrect head) - for sentence "Economic news had little effect on financial markets". The shallow parsing result, "news had effect on markets", can be easily drawn from the right candidate (a) or the wrong candidate (b), except for (c). So from the viewpoint of shallow parsing, (b) is better than (c) in spite of both having one wrong relation. As discussed in subsection 2.1, we only concentrate on 'span-head' and 'span-link' in shallow dependency parsing: the nodes and edges closer to 'Root' transmit more semantic information than others. Based on analysis above, we proposed a simulated shallow dependency parser derived from a full parsing.

To differentiate nodes and edges in dependency tree we replace the scalar factor α with a diagonal matrix $A = (\partial_1, \ldots, \partial_n)$. Assuming that feature vector $\phi(x, y)$ is denoted by $(f_1, \ldots, f_n)$, we obtain $(\alpha f_1, \ldots, \alpha f_n)$ as the result of '$\alpha \cdot \phi(x, y)$', or $(\partial_1 f_1, \ldots, \partial_n f_n)$ as the result of '$A \cdot \phi(x, y)$'. Note that feature f_i is a binary value, i.e. 1 or 0. In what follows we make some assumptions for simplicity. Given a sentence x_i and corresponding dependency tree y_i, let T be the set of candidates. The inner product, $\phi(x_i, y_i) \cdot w^T$, is defined to be the score of candidate y_i of sentence x_i. The error set for instance (x_i, y_i) is defined to be the set of the index of candidates which achieve higher scores than correct dependency tree y_i:

$$E = \{r \neq y_i | r \in T, \ \phi(x_i, r) \cdot w^T > \phi(x_i, y_i) \cdot w^T\} \tag{2}$$

Comparing to original perceptron algorithm and others, the innovation of our algorithm derived from the refinement of update factor, i.e. diagonal matrix A. Given a candidate z drawn from error set E, A_z is defined as follows:

$$A_z = diag(\partial_1, ..., \partial_n), \ \partial_i = 2f_i \cdot (1 + e^{hl(f_i, z)})^{-1} \tag{3}$$

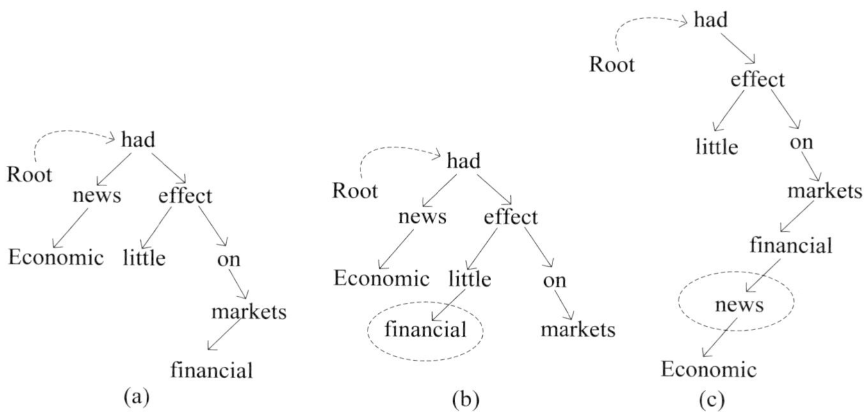

Fig. 2. Dependency tree examples: one correct (a) and two incorrect (b) and (c)

where $hl(f_i, z)$ is a function named 'hierarchical length function' and defined as: (1). The height of a relation $< i, j >$ is the number of passed edges from 'Root' to node j, for example the height of $<$'had', 'news'$>$ in figure 2(a) is 2, while that of $<$'on', 'markets'$>$ is 4; (2). If feature f_i is derived purely from a relation r, then $hl(f_i, z)$ is the height of r; (3). If feature f_i is a mixture of features derived from more than one relations $r_1, r_2, \ldots, r_m$, then $hl(f_i, z)$ is the minimum relation height among $r_1, \ldots, r_m$.

Then we can rewrite formula 1 by substituting 'A' for 'α' and obtain formula 4:

$$w = w + A_{y_i} \cdot \phi(x_i, y_i) - \frac{1}{k} \sum_{i=1}^{k} A_{z_i} \cdot \phi(x_i, z_i), \ z_i \in E \tag{4}$$

where E is the error set of size k and z_i is some candidate belonging to error set E. It is clear that diagonal element ∂_i is zero when corresponding feature f_i is zero, therefore, we can ignore 'zero' features in implementation for simplifying the computation. The definition of A in equation 3 implies that the closer to 'root' nodes and relations are, the more aggressive the update to them are during learning. Consequently, we make a trade-off between relations closer to 'root' and relations far from 'root': improve the learning of the former at the cost of decrease of the latter. We have drawn a conclusion in subsection 2.1 that nodes and edges close to 'root' are more semantic and important than those far from 'root'. Likewise, shallow dependency parsing concentrates on nodes and edges which are close to 'root'. So it is feasible to simulate shallow dependency parser via the proposed approach.

4 Experimental Study

4.1 Data and Task Definition

The original data consisting of 10,000 Chinese instances are provided by Information Retrieval Lab of Harbin Institute of Technology. These instances have been labeled manually with POS, relation and relation type in advance. The total data in the experiments were randomly divided into two groups: one for training with size of 9000 and one for test with size of 1000.

The task in the experiment is to assign labeled dependency edges to Chinese sentences. Each sentence was represented as a sequence of tokens plus POS. For each token, the parser must output its head and corresponding dependency relation. The metrics applying to evaluate parsers are defined as follows (Specially, excluding punctuation from scoring):

- LAS (labeled attachment score). It is the proportion of "scoring" tokens that are assigned both the correct head and the correct dependency relation label
- UAS (unlabeled attachment score). It is the proportion of "scoring" tokens that are assigned the correct head (regardless of the dependency relation label)
- LS (label attachment score). It is the proportion of "scoring" tokens that are assigned the correct dependency relation label (regardless of the head)
- LAS-n (labeled attachment score with height n). The height of a token is the number of passed edges from the dummy node 'Root' of dependency tree to itself. Specially, the height of one token t is equal to the height of one relation whose child is just t. Arabic numeral n indicates a concrete height. For example, in figure 2 (a) the height of "financial" is 5, "news" is 2, and "Economic" is 3. LAS-n is the proportion of "scoring" tokens with height n in dependency tree that are assigned both the correct head and the correct dependency relation label. In this experiment dependency structure are divided into 5 partitions: LAS-1, LAS-2, LAS-3, LAS-4, and LAS-5_+. Specially, LAS-5_+ includes nodes whose height is larger than 5 (including 5). If there are 100 tokens with height 1, and 80 tokens are assigned both the correct head and the correct label, then LVS-1 is 80%.

4.2 Experimental Results and Analysis

In this paper we proposed a variation of multiclass perceptron algorithm (VMPA), using update rule in formula 4, for simulated shallow dependency parsing. To verify the effectiveness of the proposed approach, we carried out the experiment in which two parsers were built from the training dataset of size 9000 using VMPA and original multiclass perceptron algorithm (MPA) described in formula 1, and then applied them to the test dataset of size 1000 respectively. To make a comparison, we evaluated additional two state-of-the-art dependency parsers: MSTParser[1] and MaltParser[2], which were developed by McDonald [5] and Nivre [6] respectively. Note that MaltParser used embedded SVM learner and predefined feature model for Chinese. MSTParser used its default setting in the experiment. All results were reported in table 1. The results of VMPA showed that the accuracies of tokens with lower height, comparing to that of MPA, had some improvement. LAS-1 and LAS-2 increased 5.86% and 2.02% respectively. Of course, the improvement was at the cost of decrease of accuracies of nodes far from 'Root'. At the same time, the results of both VMPA and MPA were a little worst than two state-of-the-art parsers, i.e. MSTParser and MaltParser. We believed that may be due to the difference of learning algorithms.

[1] http://sourceforge.net/projects/mstparser

[2] http://w3.msi.vxu.se/ nivre/research/MaltParser.html

Table 1. Results of three dependency parsers. (MST:MSTParser, Malt:MaltParser)

Algos	LAS	UAS	LS	LAS-1	LAS-2	LAS-3	LAS-4	LAS-5_+
VMPA	77.16%	79.01%	80.23%	76.87%	75.67%	78.69%	79.01	75.83%
MPA	77.73%	80.20%	83.06%	71.01%	73.65%	80.95%	79.93%	76.45%
Malt	81.16%	83.20%	87.06%	75.13%	77.65%	78.15%	84.93%	84.45%
MST	81.51%	83.21%	86.19%	77.93%	84.16%	83.21%	80.63%	78.01%

5 Conclusions and Future Work

In this paper we focus on shallow dependency parsing and propose a discriminative dependency parsing algorithm based on weighted hierarchical structure learning to simulate it. The results demonstrated that accuracies of nodes closer to 'Root' increased at the cost of some decrease in nodes far from 'root'. This improvement, however, is somewhat limited because the learning is based on instance one by one thus could not make overall trade-off over all training instances. Some improvements to the proposed approach may be brought through additional research. First, the definition for shallow dependency parsing in this paper is still rough and simple. Secondly, the trade-off is based on single example one by one instead of the whole examples due to the online framework of the learning algorithm. In future work we may consider applying batch learning algorithm, such as SVM, with trade-off strategy for shallow parsing.

References

1. Collins, M.: Head-driven statistical models for natural language parsing. Computational Linguistics 29(4), 589–637 (2003)
2. Collins, M., Ramshaw, L., Haji, Ccirc, J., Tillmann, C.: A statistical parser for czech. In: Proceedings of the 37th conference on Association for Computational Linguistics, pp. 505–512 (1999)
3. Eisner, J.: Three new probabilistic models for dependency parsing: An exploration. In: Proceedings of the 16th International Conference on Computational Linguistics (COLING 1996), pp. 340–345 (1996)
4. Jurafsky, D., Martin, J.H.: Speech and Language Processing: An Introduction to Natural Language Processing, Computational Linguistics, and Speech Recognition. MIT Press, Cambridge (2000)
5. McDonald, R., Crammer, K., Pereira, F.: Online large-margin training of dependency parsers. In: Proceedings of the 43rd Annual Meeting on Association for Computational Linguistics, pp. 91–98 (2005)
6. Nivre, J., Hall, J., Nilsson, J.: Maltparser: A data-driven parser-generator for dependency parsing. In: Proc. of LREC 2006 (2006)
7. Yamada, H., Matsumoto, Y.: Statistical dependency analysis with support vector machines. In: Proc. IWPT (2003)

One Optimized Choosing Method of K-Means Document Clustering Center

Hongguang Suo, Kunming Nie, Xin Sun, and Yuwei Wang

School of Computer and Communication Engineering, China University of Petroleum, Dongying, China
{suohg,nkm1985,sunxin1000,wyw1101}@163.com

Abstract. A center choice method based on sub-graph division is presented. After constructing the similarity matrix, the disconnected graphs can be established taking the text node as the vertex of the graph and then it will be analyzed. The number of the clustering center and the clustering center can be confirmed automatically on the error allowable range by this method. The noise data can be eliminated effectively in the process of finding clustering center. The experiment results of the two documents show that this method is effective. Compared with the tradition methods, F-Measure is increased by 8%.

Keywords: Document Clustering, K-means, Initial Center, Sub-graph Division.

1 Introduction

K-means has better scalability and higher implementation efficiency to the application of document clustering, and it can achieve good results and is superior to hierarchical clustering[1], but it is difficult to determine the cluster center and it is sensitive to the isolated point in document set. Aiming at the characteristic that the cluster center of the K-means needs to be assigned, there have been some improved methods[2].

Literature [3] made use of genetic algorithm to optimize K which is the number of initial center; Literature [4] mentioned that a new program for choosing the initial clustering center has been raised by the global k-means method, which adds a dynamic clustering center through making use of global searching process. Liu Yuanchao made use of the improvement of the Maximin Principle to decide the clustering number and clustering center[5]. In literature [6] the clustering center was determined by improving the CBC committee algorithm, and the shortcoming of the method is that it has a high time complexity and it needs to use the hierarchy algorithm when selecting a committee, which depresses the accuracy to some extent. Additionally, the algorithm needs to assign the number of the cluster manually, which is always a difficult task.

This paper presents a new initial center choice method, which aims at the characteristic that the K-means algorithm needs to assign the initial clustering center, based on sub-graph division.

2 Initial Center Choosing Method Based On Sub-graph Division

After setting up the similarity matrix of the text, we select the entire document nodes as the vertexes of the graph. For the two documents whose similarity is bigger than the

H. Li et al. (Eds.): AIRS 2008, LNCS 4993, pp. 490–495, 2008.

Fig. 1. Changing Process of Sub-graph (θ=0.25 → θ=0.20)

current similarity threshold, we connect a line between the two corresponding nodes in the graph. Thus a disconnected graph will be formed (if the threshold is too small, the graph may be a connected graph).

In the descending process of the similarity threshold, when the threshold value is proper, some sub-graphs exceeding a dedicated text number in the disconnected graph will come forth. And these sub-graphs perform as: the similarity between the documents of sub-graphs is high; meanwhile, the similarity between the documents included in different sub-graphs is smaller than the current similarity threshold. The purpose of the text clustering is that it makes the similarity of data point in the same cluster maximum and makes the similarity of data point in the different cluster minimum. It is feasible to take these sub-graphs as the candidate initial cluster center.

In the changing process of the threshold, if the mutual similarities of the document of one cluster are lower in the similarity matrix, when the threshold falls to a dedicated value, a sub-graph will be formed possibly between the documents of this cluster, and then we can consider the vector center of the documents in this sub-graph as the candidate clustering centers, which reduces the data noise.

2.1 Preprocess of Clustering

During the document clustering, preprocess work is very important. The usual processing steps include: segmentation, stop word removal, word frequency statistic, feature selection and building vector space model.

In the process of our experiment, we select different number of high frequency words to experiment, and find that the use of 50 high frequency words would have the best experiment result. So, we select the first 50 high frequency words of each clustering document as the feature words of the clustering. Collecting statistic of all the words of the total document and taking them as column, and build the vector space model. The computing of text similarity uses cosine formula.

$$sim(dt, dm) = \frac{dt' \times dm}{\| dt \| \times \| dm \|} \tag{1}$$

2.2 Algorithms

The first stage: Cluster center

The main process of the algorithm is, firstly, we find out all the sub-graphs of the text set and sub-graphs formed in the current threshold from these sub-graphs. We deal with them and then depress the threshold, loop until satisfying the end situation. The input of the algorithm is the storage structure of the adjacent table of N clustering documents and the threshold of similarity is θ, β.

Step 1. Finding out all the sub-graphs of the text set by using of the depth graph traverse method, if the document numbers of all the sub-graphs are smaller than β, turn to **Step 5**;

Step 2. If N sub-graph *sub-graph{*$W_1, W_2, W_3, \ldots, W_n$ *}* are disposed for the first time, we take the N sub-graphs as the candidate initial cluster center of the text set, signature as *old{*$M_1, M_2, M_3, \ldots, M_k$*}* and compute the vector center of each cluster center. Turn to **Step 5**, or else turn to **Step 3**;

Step 3. Collecting statistic of all the sub-graph *sub_graph{* $W_1, W_2, W_3, \ldots, Wn$ *}* produced by traverse, establishing the mapping between it and the old cluster center *old{*$M_1, M_2, M_3, \ldots, M_{k1}$*}*, and then dealing with these sub-graphs in *sub_graph{* $W_1, W_2, W_3, \ldots, W_n$ *}* respectively.

For the sub-graph that has mapping relationship $W_i \rightarrow \{M_i, M_j \ldots\}$ with the old cluster center, if the increased document number of sub-graph W_i, num(W_i)-num(M_i+M_j) is smaller than β, we consider that the cluster center of M_i, M_j doesn't change, and we diverse the newly increased document to the old cluster center having the higher similarity. If the increased document number is bigger than β, we need to judge whether the new element is a new cluster center or not.

For the sub-graph that doesn't have mapping relationship with the old cluster center, taking the sub-graph as the candidate cluster center and turn to **Step4**;

Step4. Firstly, we must judge whether the similarity between each candidate cluster center and other candidate cluster center that doesn't have mapping relationship with the old cluster center is bigger than θ or not, if the similarity is bigger than θ, the two candidate cluster centers can be emerged as one cluster center M_{k+1}, meanwhile, we must judge whether the M_{k+1} is a new cluster center entered to *old{*$M_1, M_2, M_3, \ldots, M_{k1}$*}*. If the maximum of the similarity between M_{k+1} and old cluster center is small than the present threshold of sub-graph division, M_{k+1} needs to be entered to *old {*$M_1, M_2, M_3, \ldots, M_{k+1}$*}*;

Step 5. Reducing the similarity threshold of the sub-graph division by 0.05;

Step 6. Repeating the process **Step 1 ~ Step 5** until the threshold reducing to the dedicated value η, the judged cluster center *old{*$M_1, M_2, M_3, \ldots, M_k$*}* will output and the program is end.

The sub-graph in the algorithm refers to the graph that the number of the interrelated document is over β. In the forth step, if the candidate cluster center W_i and the old cluster center M_i are in the same cluster, after the vector center of W_i being emerged, the similarity between W_i and M_i will be bigger than before, and the similarity is bigger than the similarity between any of the element in W_i and the old cluster center M_i, so, we take the candidate cluster center, which has a lower similarity than the present sub-graph division threshold, as a new cluster center.

The second stage: Clustering

We conduct a k-means clustering by use of the clustering centers, which are found out in the first stage.

Step 1. Take the k cluster center producing in the first stage as the initial cluster center;

Step 2. Assign each document to the most similar cluster according to the average of object in the cluster;

Step 3. Update the average value of the cluster;
Step 4. Repeat **Step 1~Step 3** until the cluster division does not change again;

2.3 Time Complexity Analysis of Algorithms

Prior to the implementation of the algorithm, we need to build the similarity matrix of the document. The time spending is bigger, but most of the document clustering algorithms needs to build the similarity matrix in advance [6]. In the process of building similarity matrix, the storage structure of adjacent table can be built according to the dedicated threshold. The time complexity of the graph's traverse algorithm is $O(n+e)$ in the implementation process, taking d as the feature dimension of text set and taking k as the number of cluster center. The time complexity of the second step is nd and the time complexity of dealing with the newly created sub-graph is k^2d , so the time complexity of the algorithm is $O\left(n+e+nd+k^2d\right)$。 The time complexity of the algorithm can be increased with the increasing of the dimension numbers. Consequently, when we handle the large data set, the time complexity of the algorithm is high.

3 Experiments

We conduct a series of experiments. The first group selects the different articles that have already been categorized from www.sina.com.cn. These articles are classified artificially, so it is convenient to compare the test result. There are 7 classes of documents: studying abroad, real estate, music, automobile, military affairs, college entrance examination, sport. We take out 20 documents for experimentation from each category of these documents. The second group makes use of the classification corpus which is provided by Lee Lurong of FuDan University, selecting 6 classes from these corpus for the experiment: Energy、Electronics、Medical、Communication、Philosophy、Literature.

The evaluation criteria of the experiment result used most commonly F-measure values [7]. Let *P* be average precision. Let *R* be average recall. F-measure is defined to be *2RP/(R+P).*

Table 1. The Experiment Result of the Algorithms

True k values	The Number of Initial Centers Generated on Different Group of Corpus					
	1	2	3	4	5	6
2	2	2	2	2	2	2
3	3	3	3	3	2	3
4	4	4	4	4	4	3
5	5	5	4	5	5	5
6	5	5	6	5	5	6

The threshold of the algorithm is selected according to the experiment experience. When θ is smaller, there will be many density areas and the mutual similarity of these areas will be lower. We conducted several experiments by taking definite standard text set as the training text set. After training, we let θ=0.25, β=N/5 in the process of the experiment (N is the document number of each cluster).

For the different values of k, we use different k cluster's combination to conduct experiment and each k is conducted 6 groups of different experiments. We can see that the experiment results are unanimously with the standard result from table 1. When analyzing the cluster that the division number is lack, it is easy to find that the document theme distributing of the cluster is too loose. When k=6, the military cluster is not found.

We select 2~6 clusters from the 7 clusters to conduct experiment, and three group of experiments are conducted to every k. Corresponding to each group of experiment for each K, the right column data is the cluster result that using the sub-graph division method.

Table 2. Comparision of Two Different Cluster Initial Center Choosing Method (Our Method B)

Experiment Times	Clusters of Experimental Text Set									
	2 Clusters		3 Clusters		4 Clusters		5 Clusters		6 Clusters	
	A	B	A	B	A	B	A	B	A	B
First Group	0.90	1.00	0.88	1.00	0.85	0.92	0.83	0.90	0.74	0.81
Second Group	0.95	1.00	0.83	0.95	0.80	0.86	0.72	0.90	0.75	0.89
Third Group	0.90	0.92	0.87	0.93	0.80	0.82	0.83	0.84	0.71	0.81

From table 2, we can see that the experiment result by the method of automatically determining the cluster center is superior to the method of artificially designating the cluster center. In the first stage of the cluster, the vector center of the cluster are the combination of the documents that are divided into the sub-graph, so each cluster in the test result has a high recall rate. If the number of selected cluster center is equal to the number of the original cluster of document set, the precision of each cluster will be promoted, so the accuracy of the cluster result is higher.

The second experiment (Table 3.) is a comparison between our method in this paper and the original cluster method C. In the original method, the cluster center is generated randomly, and we select the average of the three experiments as the result of this experiment.

Table 3. Comparison of Our Method With Original Clustering Method (Our Method B)

Experiment Times	Clusters of Experimental Text Set							
	2 Clusters		3 Clusters		4 Clusters		5 Clusters	
	B	C	B	C	B	C	B	C
First Group	0.78	0.76	0.64	0.66	0.69	0.54	0.4	0.49
Second Group	0.95	0.73	0.68	0.67	0.59	0.45	0.54	0.44
Third Group	0.69	0.52	0.81	0.59	0.57	0.62	0.59	0.55

From the experiment results of table 3, in most cases, this method can achieve better results. The F-measure value of the automatically determining the cluster center method is promoted by 8% than the original K-means cluster method. We analyze the reasons for several poorer test results and it is mainly because that the selected feature words of experiment are correlated with each other, such as electronics cluster and communication cluster. The other reason is that the length of individual document is too short. When selecting the key words of clustering, the effective words that can distinguish the document are less.

4 Discussion and Conclusions

A method of determining the cluster center is presented, by which the potential cluster center can be discovered using sub-graph division. In the process of searching the cluster center, it removes the noise data successfully. The method, which can improve the results of the cluster remarkably, is proved effective. When our method applying in the short content subjects, the effect of the cluster will decline because of the small amount of information contained in the text. In future, we will continue to research it deeply.

References

1. Zhao, Y., Karypis, G.: Criterion Functions for Document Clustering Experiments and Analysis. J. Department of Comp. Sci & Eng University of Minnesota, 01–40 (2001)
2. Khan, S.S., Ahmad, A.: Cluster center initialization algorithm for K-means clustering. J. Pattern Recognition Letters 25(11), 1293–1302 (2004)
3. Casillas, A., González de Lena, M.T., Martínez, R.: Document clustering into an unknown number of clusters using a Genetic Algorithm. In: A. International Conference on Text Speech and Dialogue TSD, 43–49 (2003)
4. Likas, A., Vlassis, N., Verbeek, J.J.: The global k-means algorithm. Pattern Recognition. J 23, 451–461 (2003)
5. Liu, Y., Liu, X., Liu, B.: An adapted algorithm of choosing initial values for k-means document clustering. J. High Technology Letters 16(1), 11–15 (2006)
6. Zhao, W., Wang, Y., Zhang, X., Li, J.: Variant of K-means algorithm for document clustering: optimization initial centers. J. Computer Applications 25(9), 2037–2040 (2005)
7. Liu, Y., Wang, X., Xu, Z., Guan, Y.: A Survey of Document Clustering. J. Journal of Chinese Information Processing 20(3), 55–62 (2005)

A Model for Evaluating the Quality of User-Created Documents

Linh Hoang, Jung-Tae Lee, Young-In Song, and Hae-Chang Rim

Dept. of Computer and Radio Communications Engineering
Korea University, Seoul, Korea
{linh,jtlee,song,rim}@nlp.korea.ac.kr

Abstract. In this paper, we propose a model for evaluating the quality of general user-created documents. The model is based on supervised classification approach, in which output scores are considered as quality of given document. In order to utilize both textual and non-textual attributes of documents, we incorporated a number of objectively measurable, real-valued features selected upon predefined criteria for quality. Experiments on two datasets of real world documents show that textual features are stable indicators for evaluating documents' quality. Some features are inferred to be effective for general kinds of documents.

1 Introduction

User-created documents are well known types of user-generated contents, which are produced by end-users. For example, user product reviews in shopping sites or answers in community driven Q&A are two common types of user-created documents. This has motivated us to investigate on proposing a quality evaluating model that can be applied to any common types of user-created documents.

Using a supervised classification approach, we first manually labeled experimental documents conforming to three levels of document quality, namely *good*, *fair* and *bad*. A classifier trained from annotated corpus then ranked documents according to their prediction scores. In this work, we concentrate on building a feature combination which does not depend on the type of target documents. Our proposed method empirically worked well, even though documents have been collected from independent sources.

2 Related Work

Recently, [1] studied a task similar to our work, which is specific to user-created answers. Only non-textual features, such as *click-through counts* and *user recommendation counts*, were used for predicting answer's quality. However, it has turned out that the most effective feature is *document length* (which does not refer to non-textual information), whereas the others are less contributed. This conclusion infers that non-textual information considered previously may not always be stable along time; intuitively, data sparseness may often occur for newly created documents because they would be seen less by users.

H. Li et al. (Eds.): AIRS 2008, LNCS 4993, pp. 496–501, 2008.

[3] investigated the task of predicting reviews' helpfulness that considered users' vote as ground-truth evaluation. Firstly, different classes of features are utilized to helpfulness. SVM regression then learned helpfulness function and ranked reviews according to their output scores. In this work, the *length of a review*, *product rating* and *its unigrams* were found to be most useful. However, assessing reviews' helpfulness based on users' rating ground-truth is not always reliable due to several voting biases [4].

Showing three biases of [3]'s approach, [4] presented a framework for detecting low-quality reviews. Instead of using users' vote information, the authors manually annotated a set of ground-truth according to manually predefined specification for reviews' quality. However, many selected features are directly extracted from product's attributes such as the *number of products, product features, brand names*. Such features made this approach domain restricted since they are hardly applicable to other types of user-created documents.

Limitations from prior works have suggested us to employ both textual and non-textual features in the proposed method. To widely exploit this work for almost any types of user-created documents, only general features are chosen regarding intrinsic properties of documents. Our proposed model empirically improved performance in comparison with baseline approach that utilizes only non-textual features.

3 Method

3.1 Features Categories

One of the enhancements in our approach is the combination of objectively measurements selected upon predefined classes. All experimented features are separated into four categories: *authority*, *formality*, *readability* and *subjectivity*.

Features on authority

Among four categories, *authority* is a unique category that relies on non-textual information collected by service providers. Features on *authority* indicate whether document is written by a trustworthy author or not. Some representative examples of features in this category are as follow:

- Number of documents previously written by the same writer (NDOC)
- Number of votes or scores granted by users (NVOT)

Features on formality

This feature category refers to the writing style of target document. A formal document tends to be accessible to the intended audience. Based on this observation, some of consecutive features are considered:

- Number of words in the document (NWRD)
- Number of different words in the document (DWRD)
- Number of sentences in the document (NSNT)
- The fourth root of the number of words in the document (RWRD = $\sqrt[4]{NWRD}$)
- Average length of sentences in the document (SLEN)

Features on readability
Typically, a well-organized document imparts much information to reader. With assumption that format of document contributes to its quality prediction; three described features have been chosen for experiments:

- Lexical density of the document (LXDN = DWRD/NWRD)
- Number of paragraphs in the document (NPRG)
- Average length of paragraphs in the document (PLEN)

Features on subjectivity
Subjectivity refers to opinions of authors in a document. Several following features have been defined based on simple and easy-measurable criteria:

- Ratio of positive sentences (RPST)
- Ratio of negative sentences (RNST)
- Ratio of subjective sentences regardless of positive or negative (RSST)
- Ratio of comparative sentences (RCST)

Basically, most of features in *formality* and *readability* category are similar to the ones used in Project Essay Grade [6]. *Subjectivity* category consists of opinion-based features. Using subjective and comparative languages clues [2,8], we refined a set of opinion words and phrases for each testing corpus. *Subjectivity* features have been extracted by using a simple keyword-based approach. For example, positive sentences are considered as sentences that contain at least one positive opinion word or phrase.

3.2 Quality Evaluation Model

In our proposed model, Maximum Entropy (MaxEnt) is chosen for training a classifier. The main advantage of MaxEnt is that we can easily integrate variety of relevant features since they are expressed in the form of feature functions. For later improving a retrieval system, we intend to build a statistical model of which output scores can be considered as prior information.

The underlying idea of MaxEnt indicates that without external knowledge, one should prefer the most uniform models that also satisfy any given constraints. Once we assume that assessing quality of documents is a random process that observes documents and assign them a quality label y, MaxEnt motivates to find the model p as close to the empirical probability distribution p' of random process as possible. Applying to our classification task, each feature is represented by a feature function $f_i(x, y) = x_{fi}$ where x_{fi} is the value of the ith feature in the document x. MaxEnt then estimate expected value for each feature from training data and take this as constraint of the model distribution.

Firstly, a set of weighting parameters λ for each feature function are estimated by using Limited-Memory Variable method [5]. The model then computes the conditional probability for predicting the quality of document x by the formula:

$$p(y|x) = \frac{1}{Z(x)} exp \left[\sum_i \lambda_i f_i(x, y) \right]$$

where $p(y|x)$ is the output score indicates the quality of document x and $Z(x)$ is a normalization factor to ensure $\sum_y p(y|x) = 1$. We specifically use $p(y{=}good|x)$ as a score output.

4 Experiments

4.1 Experimental Corpus

We experimented our model on two datasets of real world user-created documents. The first one consists of 1000 English reviews on Amazon website (`http://amazon.com`). Twenty products in electronics category were randomly selected for constructing corpus. For each product, we manually accumulated 50 reviews regardless of order. Other relevant information of reviews such as author's rank, users' vote, comments are also saved. Two students were asked for hand-tagging each given document as *good*, *fair* or *bad* (Table 1 shows descriptions of three-level quality on each dataset).

The second dataset includes 2589 Korean Q&A samples collected from Naver's Knowledge Search service (`http://kin.naver.com`). Basically, this corpus has already built and experimented in previous research [7]. The dataset is composed of questions, along with one *best* answer for each question. (Knowledge Search service allows users to select one best answer among all answers corresponding to a question). In this scenario, we used only answers for the experiments. Also, all answers were manually labeled based on three-level quality.

Table 1. Three-level specification for document quality

Level	Document types	
	Review	**Answer**
Good	- Complete, broad, well-organized description of the product - Pros & cons reasonably explained - Objective for most of the time	- Objective with certain basis or subjective but logically explained - Attachment often included ore answer to the question
Fair	- Contains some information about the product - Rather more subjective	- Objective but lack of details - Subjective with no basis but partially logical
Bad	- Contains very little, misleading information or even no description of the product - Many inappropriate words, wrong spellings, or bad readability - Completely subjective	- Abuse languages or spams contained - Libel on someone particular, irrelevant answer to the question - Very speculative or subjective with no basis

Table 2. Effect of feature categories

Features	Reviews	Answers
Authority(baseline)	0.7647	0.9190
+*Formality*	0.9269	0.9705
+*Readability*	0.9269	0.9674
+*Subjectivity*	0.9624	0.9722

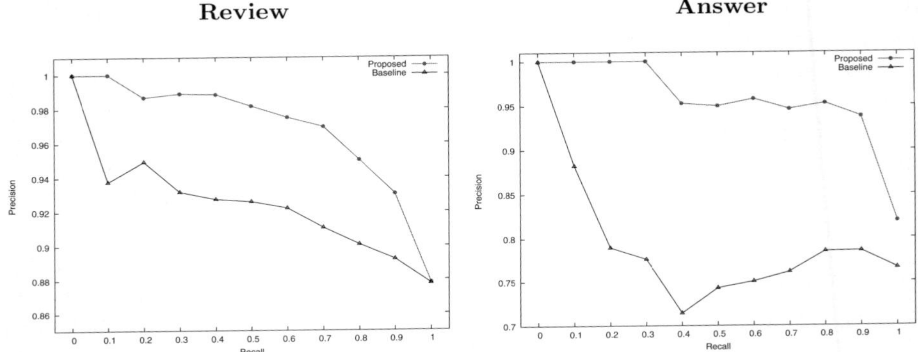

Fig. 1. 11pt recall-precision curves with 2 datasets

4.2 Results

We ranked the documents in descending order of the score output and used the traditional recall and precision metric to evaluate the results. Conforming to the evaluation metric, we consider *good* and *fair* documents as relevant documents while *poor* documents are treated as non-relevant ones. Aimed to measure the effectiveness of textual features in comparison with non-textual features, we take the model that utilizes only the features on *authority* as baseline. The average precision is chosen for measuring the overall performance and the contribution of each feature category.

Table 2 indicates the contribution of each feature category based on average precision score. From the table, *formality* is shown to be the most effective category when incorporated with non-textual features. Features on *readability* make no contribution, and even slightly decreased precision on the answer corpus. *Subjectivity* features conduced a remarkable improvement on review corpus that can be justified because of the imbalance size between two sentiment word sets.

Fig. 1 shows 11pt recall-precision curves for baseline and proposed model as well. On both of experimental datasets, textual features proved to be predictive indicator since our proposed method outperformed the baseline approach that utilizes only non-textual features.

web page, seeks the hyperlinks in the page, and then skims through all the linking pages using recursion or non-recursion algorithm. Recursion is an algorithm that can shift the program logic back to itself. It is simple, but it can not be applied to multi-thread technique. Therefore it can't be adopted in an efficient spider program. Using non-recursion method, spider program puts the hyperlinks it found into the waiting queue instead of transferring to it. When the spider program has finished scanning the current web page, it will link the next URL in the queue according to the algorithm.

A hyperlink would be judged by the commerce-oriented spider if it is related to commerce or not before it is added to the queue. The way to achieve it is as follows:

1. Collect some typical commerce-related documents and transform them to text files as exercise texts originally;

2. Use LSA theory to build up an entry-text matrix of exercise texts. Using LSA model, a text set can be denoted as r*m entry matrix D. "M" means the quantity of texts in the text set , while "r" represents the number of different entries in the text set. That is, each different entry corresponds to a row of the matrix D; and each text file corresponds to a column of the matrix D. $D=[d_{ij}]_{r\times m}$, Here, d_{ij} is the weight of entry I in text j. As is known to all, there are many formulas to calculate weight in the traditional vector representations. Following is a very familiar formula to calculate weight:

$$W(t,\vec{d})=\frac{tf(t,\vec{d})\times\log(N/n_t+0.01)}{\sqrt{\sum_{t\in\vec{d}}\left[tf(t,\vec{d})\times\log(N/n_t+0.01)\right]^2}} \tag{1}$$

In formula (1), $W\left(t,\vec{d}\right)$ means the weight of entry t in the text d, $tf\left(t,\vec{d}\right)$ means the frequency of entry t in the text d, N means the number of exercise texts, n_t means the quantity of texts which include entry t, and the denominator is a normalized factor.

In practice, we noticed that the position where the entry appears is very important. For example, the entries which appear in the headline, beginning or ending parts are often keywords of the file. So we modified the weight calculation formula, it can be described as formula (2):

$$W(t,\bar{d})=\frac{tf(t,\bar{d})\times\log(N/n_t+0.01)}{\sqrt{\sum_{t\in\bar{d}}[tf(t,\bar{d})\times\log(N/n_t+0.01)]^2}}(1+a_t) \tag{2}$$

When entry t appears in the headline, beginning or ending parts, a_t=0.5. It equals to zero if entry t appears in other parts.

After matrix D is calculated, we begin to calculate the K rank approximate matrix D_k of matrix D based on SVD (singular value decomposition). After the singular value decomposition , matrix D can be denoted as: $D=U\Lambda V^T$. Here, U represents the left singular matrix according to the singular value of matrix D and V represents the right. Λ is the Diagonal Matrix that insists of D's singular values in descending order. Then the K rank approximate matrix D_k can be formed with the front "K" columns in U and

V. That is, $D_k = U_k \Lambda_k V_k^T$. The columns in U_k and V_k are orthogonal vectors. D_k is used to represent D approximately, and U_k and V_k represents entry vector and text vector respectively.

3. When the Spider is carrying out searching progress, it draws out the circumferential entries of the hyperlinks and names them X before putting the links into the queue. The characteristic vector for X is: $X = (x_1, x_2, \cdots, x_r)$, x_i can be calculated by the modified weight calculation formula (2). Project X into the space of D_k and come out XX: $XX = X^T U_k \Lambda_k^{-1}$. Now we are to compute the similarity between XX and V_k 's row vectors. Suppose that $XX = (\chi_1, \cdots, \chi_k)$ and a certain row vector of V_k is $V = (\nu_1, \cdots, \nu_k)$, the formula to calculate similarity is:

$$sim(V, X) = \frac{\sum_{i-1}^{k} \nu_i \chi_i}{\sqrt{\sum_{i=1}^{k} \nu_i^2 \sum_{i=1}^{k} \chi_i^2}} \quad (3)$$

Add up all the similarities (with the quantity of m), and compare the sum with a given value. If the sum is higher than the given value, we can consider the hyperlink as commerce-oriented.

3 Procedure Optimization

3.1 Filtration of the Entries in the Text

The experiments showed that the dimension of the entry-text matrix can not be too large. Otherwise, it will take more time to judge whether the hyperlink is commerce-oriented or not. Therefore, we usually decompose the text by using the decomposing program and sign the attribute of an entry while catching it. Delete those that do less contribution to the classification, such as prepositions, and keep nouns, verbs, adjectives and adverbs. Classify the verbs into three groups: connection verbs, state verbs and action verbs. And then delete connection verbs and state verbs. Calculate the weight of the remained entries, array them in descending order and then pick up the top "n" entries to use.

3.2 Using Thread Pool

As experiments showed, multi-thread mechanism can improve the spider's efficiency. By using multi-thread technique, Spider can request or wait for other pages while dealing with a certain page. Usually the Spider program builds a thread for each request. But it costs a lot in building a new thread for each request. When there are many requests, it will cost much more time and system resource to build and destroy threads than to deal with pages. So we use the thread pool to solve this problem. In the thread pool, the number of threads is given in advance. Too many or too few threads will reduce the Spider's efficiency. Tasks are distributed to each thread by the thread pool.

When a thread finishes a task, the thread pool will assign the next task to it. Because the number of thread is certain, additional threads can't be built.

We can use an available thread pool program instead of programming it ourselves. Jeff Heaton provided a thread pool class in his paper "Creating a Thread Pool with Java " which is published in www.informit.com or www.jeffheaton.com. Doug Lea has programmed an excellent open source set of concurrent utility programs: util.concurrent. In the open source set, the PooledExecutor class which is very effective and widely used is the right realization based on working queue. The thread pool used in our experiments is programmed by Jeff Heaton.

Now, the algorithm using thread pool can be described as follows:

(1) Build a thread pool with the given size n;
(2) Check the waiting queue for other URLs. If there are web pages waiting to be dealt with, turn to step (3); If null, the program ends.
(3) Build a working thread MySpiderWorker to deal with the next URL; Check the thread pool for idle thread. If there is an idle thread, use it to carry out MySpiderWorker and then turn to step (2); if null, keep MySpiderWorker waiting until an idle thread appears in the thread pool.

4 Search Strategy

Since the spider is aimed at catching business information, the initial URLs should be connected with commerce. If not, there will be probably no commerce-oriented hyperlinks in the first page of the website, and it will result in an empty queue and the termination of the spider program. So it would be better not to run the spider program over the education network because most of the business information is from commercial network instead of educational website.

5 Experiments and Conclusion

Different computers, different network and different time will leads to different download speeds. For example: The same Spider program running on a computer in the campus network can download only several documents from non-educational networks per second, but it can download even up to 100 documents from the educational network. Following experiments are held in campus network using the same computer. Since the exterior environment is almost the same, the experiment results are comparable.

Table 1. Experiment result 1

Quantity of documents (pcs)	Number of threads	Time (seconds)	Speed (pcs/second)
205	10	19	10.7
205	15	14	14.6
205	20	12	17
205	25	11.5	17.8
205	30	12	17
205	35	13	15.7

1. No judgment on commerce is done to the hyperlinks and the initial URLs are educational websites.

This experiment shows that more threads do not mean higher efficiency. So using thread pool can improve working efficiency.

2. No judgment on commerce is done to the hyperlinks and the initial URLs are non-educational websites.

Table 2. Experiment result 2

Quantity of Document (30-200K) (pcs)	Number of threads	Time(seconds)	Speed (pcs/second)
300	25	298	1

From table 1 and table 2 we can come to a conclusion that the speed of Spider differs widely when it is used to search information from different origin URLs. So in reality, general spiders often run over different networks to get different information from different origin. For example, some spiders are running over the CERNET to deal with ".edu" websites, while other spiders focus on ".com"or ".net" websites.

3. Judgment on commerce is done to the hyperlinks and the initial URLs are a set of business websites.

Table 3. Experiment result 3

Quantity of Document (30-200K) (pcs)	Number of threads	Time(seconds)	Speed (pcs/second)
300	25	310	0.97

From table 2 and table 3 we can conclude that when the initial URLs are from business websites, whether commerce-relating judgment has been carried out or not will have little influence on processing speed.

4. Relevant degree in obtained documents to business affairs

In the experiment, θ was fixed to: 0.3; the length of circumambient text of the URLs is 10-600 words. We obtained 500 documents in the result.

Analyzing the result documents, we can get a general conclusion. There are some irrelevant documents: 2 flashes, 30 picture documents, 4 javascript documents, 1 applications and 10 irrelevant HTML documents. The quantity of relevant documents is 453 and the relevant degree is 90.6%. The pictures are surrounded by business information, which causes lots of the picture documents were downloaded. Most of the 10 irrelevant HTML documents are blank in body. From the source code, we can find that the head and title of the documents is relevant to business while the body has nothing to do with it.

To conclude, using word-filtering and thread pool technology to retrieve business information in the Internet is practicable and effective.

References

1. Xu, B.-w., Zhang, W.-f.: Search engine and Information Fetching Technology. Tsinghua University Press, China (2003)
2. Heaton, J.: Creating a Thread Pool with Java [EB/ OL], http://www.informit.com/articles/article.asp?p=30483&redir=1
3. Heaton, J.: Programming a Spider in Java [EB/OL], http://www.jeffheaton.com/jhmag.shtml
4. Che, D.: Brief Introduction of LUCENE, the whole-length search engine based on JAVA, http://www.chedong.com/tech/lucene.html
5. Ling, Y., Wang, X., Fei, Y.: Intelligent Technology and Information Processing. Science Press (2003)

IR Interface for Contrasting Multiple News Sites

Masaharu Yoshioka

Graduate School of Information Science and Technology, Hokkaido University
N-14 W-9, Kita-ku, Sapporo 060-0814, Japan
National Institute of Informatics
2-1-2 Hitotsubashi, Chiyoda-ku, Tokyo 101-8430, Japan
yoshioka@ist.hokudai.ac.jp

Abstract. In order to utilize news articles from multiple news sites, it is better to understand the characteristics of each news site. In this paper, a concept of contrast set mining is applied for analyzing the characteristic difference between each news site and all others. The News Site Contrast (NSContrast) system is also proposed based on this mining technique. This system is applied to a news article database constructed from multiple news sites from different countries in order to evaluate its effectiveness.

1 Introduction

We have recently been able to access a wide variety of news sites from all over the world through the Internet. Because each country has different opinions and interests, when we use news sites from different countries, we can obtain different points of view on a topic. For example, considering diplomatic issues to do with North Korea, Asian news sites, European sites, and American sites have common interests as well as their own characteristic interests. So, in order to analyze some events by using multiple sites, it is important to clarify the characteristics of each news site.

There are several experimental systems that integrate news articles for a particular event from multiple news sites. For example, PENG [1] and Newsblaster [2] are integrated news aggregation systems from distributed news archives. PENG is good for finding articles that meet users' preferences, but this is not a system that contrasts different news sites. Newsblaster [2] is a system that collects news articles from different resources and generates news article summaries by integrating articles from different resources. This system is good for users to understand particular events using different resources, but does not pay attention to the characteristics of each news resource.

In order to characterize different data sets, the concept of contrast set mining has been proposed by Bay *et. al.* [3]. This framework characterizes each contrast set by finding out conjunctions of attributes and values that have meaningfully different support levels.

By using contrast set mining, Yoshioka *et. al.* proposed a news-sites analysis method that focuses on correlation change between different news sites [4]. This

H. Li et al. (Eds.): AIRS 2008, LNCS 4993, pp. 508–513, 2008.

method demonstrates the possibility of determining the characteristic information of each news site for a given topic, but it is too computationally demanding.

In this paper, the News Site Contrast (NSContrast) system is proposed for accessing news articles from multiple news sites by contrasting each news site's characteristics. This system is applied to a news article database constructed from multiple news sites from different countries in order to evaluate its effectiveness.

2 News-Site Analysis Method Based on Contrast Set Mining

First, news-site analysis method that focuses on correlation change between different news sites [4] is briefly reviewed.

2.1 Contrast Set Mining

Conventional data mining, such as association rule mining based on a support confidence framework [5], tries to find rules that are dominant in the database. These rules assist in understanding the database. However, in many cases, most of the rules are well known and are not so interesting. To solve this problem, the concept of contrast set mining was proposed [3]. This framework compares a global and a conditioned local data set to find characteristic item information that is significantly different from the global characteristic information. Even though this information is not dominant in either the global or local data sets, it can be used to understand the characteristics of the local database.

DC pair mining [6] is an algorithm that is based on the concept of contrast set mining. DC pair mining tries to find the characteristic item pairs in a local database by contrasting correlations between a global and a conditioned local database. The following are definitions used in the DC pair mining problem.

In DC pair mining, the "difference of correlations observed by conditioning a local database" is of particular interest. To quantify this difference, a new measure, $correl(X,Y)$ and $change(X,Y;C)$ are introduced;

$$correl(X,Y) = \frac{P(X \cup Y)}{P(X)P(Y)}$$

$$change(X,Y;C) = \frac{correl_C(X,Y)}{correl(X,Y)}$$

where X and Y represent the item sets and C represents the condition for constructing the local database. $correl(X,Y)$ and $correl_C(X,Y)$ correspond to the correlation between X and Y in the global database and in the C conditioned local database, respectively.

By using this measure, the system can extract item set pairs whose correlations are different from the global database; e.g., an item set pair with higher change means that it is a characteristic correlation in the local database (positive feature), and a pair with lower change means that it is rarely correlated in the local database (negative feature).

2.2 News Sites Analysis System Based on DC Pair Mining

Following is an algorithm for making a news site database for DC pair mining from multiple news sites.

1. Extraction of news articles from news sites
   Since most of the news articles in news sites have additional content, such as indexes or advertisements that are not part of the main content, it is necessary to extract the main content from the news article pages. Webstemmer [1] is used to extract the main content by using layout analysis.
2. Generation of index terms from the articles
   A morphological analysis system is applied to extracted articles to generate the index terms. Noun, adjective, verb, and unknown as categories are used for the unigram index and consecutive nouns are used for the bigram index.
3. Addition of news site information
   The name of the news site and the date when the article was obtained are added to the entry. By using this information as a condition (C), *change* $(X, Y; C)$ can be calculated for each news site.

Based on this database and the DC pair mining algorithm, a news site analysis system was proposed.

However, the original DC pair mining algorithm is quite time consuming because all combinations of item set pairs (X, Y) that satisfy a given minimum support condition are examined to calculate $change(X, Y; C)$ for each news site (C). Therefore, the item set X is restricted to a set of given topic keywords for analyzing the news site.

This news site analysis system demonstrates the possibility of determining the characteristic keywords for each news site for a given keyword set. However, it is still time consuming and it is not possible to find out characteristic keyword sets that have a smaller change.

3 NSContrast: News Site Analysis System

Based on the research results discussed in the previous section, the News Site Contrast (NSContrast) system is proposed for accessing news articles from multiple news sites by using a concept of DC pair mining; i.e., the contrast between a global database of news articles from multiple sites, and a local database of news articles from a single news site.

3.1 System Architecture

NSContrast consists of a news article retrieval system and a DC pair mining system.

In the news article analysis system discussed in previous section, all combinations of item set pairs (X, Y) that satisfy a given minimum support condition

[1] http://www.unixuser.org/%7Eeuske/python/webstemmer/index.html

are examined to calculate $change(X, Y; C)$. However, since it is not so easy to understand complicated item set pair information and it is quite time consuming to calculate them, this algorithm is inappropriate for an interactive system.

To solve this performance problem, NSContrast restricts exploration space. Since NSContrast is a system for accessing news articles, the topic keywords for searching news articles are used as item set pairs X, Y to calculate *change* $(X, Y; C)$. The system checks only items (not item sets) for Y. As a result, NSContrast can generate a list of keywords with $change(X, Y; C)$. By using this list, the user can find out positive and negative characteristic terms about a given topic's keywords for each news site.

The following is the algorithm for extracting characteristic terms.

1. A user inputs topic keywords and the IR system retrieves articles that contain topic keywords.
2. The DC pair mining system selects candidate keywords that exist in the retrieved articles and satisfy minimum support.
3. Correlations between topic keywords and each candidate keyword are calculated for every news site. Since $correl(X, Y)$ used in previous research cannot discriminate the difference between keywords that are totally correlated to the topic keyword (i.e., $correl(X, Y) = \frac{P(X \cup Y)}{P(X)P(Y)} = \frac{1}{P(X)}$ as $P(X \cup Y) = P(Y)$), this $correl(X, Y) = 0$ is not appropriate for this analysis. Therefore, in this paper, the log-likelihood ratio is used to measure correlation. However, when two items in the global database are independent, $correl(X, Y) = 0$ and so $change(X, Y; C) = \infty$. This is not appropriate for this system. In addition, when correlation between X and Y is dominant in condition C and Y is not so frequent in other collection, $correl(X, Y)$ has a higher value. As a result, $change(X, Y; C)$ may have lower value even though Y may be a characteristic terms of C. Therefore, it is better to contrast between conditioned database and rest of the database. The following formula, therefore, is used to calculate change:
$$change(X, Y; C) = \frac{\alpha + correl_C(X, Y)}{\alpha + correl_{\overline{C}}(X, Y)}$$
In this paper, we use $\alpha = 1$ for all experiment.
4. The DC pair mining system calculates the $change(X, Y; C)$ value for each topic keyword set. In this step, C, X, and Y correspond to the news site, topic keywords, and a candidate characteristic keyword, respectively.

Based on this procedure, NSContrast extracts the following keyword lists for given topic keywords:

List with higher correlation in global database. This list is good for understanding common interests.

List with higher change for each news site. This list is good for understanding interests that characteristically exist in a news site.

List with smaller change for each news site. This list is good for understanding topics that are mostly neglected by a news site.

3.2 Analysis Experiment

In order to analyze the effectiveness of the system, an analysis experiment was conducted using a news site database obtained from the Internet. The news article database was populated by English language articles obtained from the following sites from May 1 to November 30, 2006: (Table 1). Since most of the news articles from the news sites have additional nonnews content, such as indexes and advertising, it is necessary to extract the main content from the news article pages using Webstemmer.

Tables 2–4 are tables extracted for analyzing the keyword "North Korea".

Table 1. News Site Information

Site (country)	Abbrev.	Articles	Site (country)	Abbrev.	Articles
Asahi newspaper (Japan)	asahi	3314	Yomiuri newspaper (Japan)	yomiuri	3501
CNN (USA)	cnn	9003	The New York Times (USA)	nyt	11246
Los Angeles Times (USA)	lat	13645	Chosun newspaper (Korea)	chosun	745
Joins newspaper (Korea)	joins	462	People's Daily newspaper (China)	people	2873
Al Jazeera (Qatar)	alja	1499			

Table 2. Terms (Phrases) with Higher Correlation in Each News Site and Total

Total	chosun	cnn	asahi	yomiuri
pyongyang	pyongyang	nuclear	pyongyang	pyongyang
nuclear	nuclear	pyongyang	missile	nuclear
korean	test	nuclear weapon	nuclear	korean

lat	nyt	joins	alja	people
nuclear	korean	nuclear	nuclear	six-party process
pyongyang	nuclear	pyongyang	nuclear test	six-party talk
korean	nuclear test	south	pyongyang	six-party

Table 3. Terms (Phrases) with Highest Change in Each News Site

chosun	cnn	asahi	yomiuri
complex	stiffen	1977	serious threat
wa	two u.s.	dna test	money transfer
department	saturday	director-general	yen

lat	nyt	joins	alja	people
britain	r.	settle	n	department
france	iraq	reporting	n korea	process
russian foreign	american ambassador	trade ministry	acquire	want

Table 2 shows terms (phrases) with highest correlation with keyword "North Korea" in each news site and total. "Pyongyang" and "nuclear" were selected as common topic keywords. From this table, we can see almost all news sites pay attention to the issue of nuclear and missile in this period.

Table 3 shows terms (phrases) with the highest change with keyword "North Korea" in each news site. From this table, we can see the Los Angeles Times

Table 4. Terms (Phrases) with Smaller Change in Each News Site

chosun	cnn	asahi	yomiuri
prime minister	abduction	unification	industrial complex
atomic	abductees	south koreans	north koreas
prime	dae	dismantle	dismantle

lat	nyt	joins	alja	people
abduction	catastrophic failure	japanese	u.s.	nuclear test
yokota	n.	long-range	launch	china
dae	government official	summit	cooperation	security council

pays attention to the relationship between European countries, such as "Britain" and "France".

Table 4 shows terms (phrases) with smaller change with the keyword "North Korea" in each news site. From this table we can see American news sites, such as CNN, Los Angeles Times and The New York Times, pay little attention to "Abduction" in the "North Korea" keyword case. (This is a common topic for Japanese newspapers; e.g., characteristic keyword "1977" in Asahi newspaper is a year that one Japanese woman was abducted.)

4 Conclusion

In this paper, the application of the contrast set mining technique to multiple-news-site analysis is proposed and a news site analysis system NSContrast is demonstrated. This system can find characteristic information of the news site for given topic keywords.

References

1. Baillie, M., Crestani, F., Landoni, M.: Peng: integrated search of distributed news archives. In: SIGIR 2006: Proceedings of the 29th annual international ACM SIGIR conference on Research and development in information retrieval, pp. 607–608. ACM Press, New York (2006)
2. McKeown, K.R., Barzilay, R., Evans, D., Hatzivassiloglou, V., Klavans, J.L., Nenkova, A., Sable, C., Schiffman, B., Sigelman, S.: Tracking and summarizing news on a daily basis with Columbia's Newsblaster. In: Proceedings of the Human Language Technology Conference (2002)
3. Stephen, D., Bay, M.J.P.: Detecting group differences: Mining contrast sets. Data Mining and Knowledge Discovery 5(3), 1213–1246 (2001)
4. Yoshioka, M., Taniguchi, T., Haraguchi, M.: Research on multiple news sites analysis using correlation change. In: Knowledge Media Science. Preparing the Ground: International Workshop, Landsberg Castle, Meiningen Germany, October 2-5, 2006, Revised Selected Papers (to appear, 2007)
5. Agrawal, R., Srikant, R.: Fast algorithms for mining association rules. In: Bocca, J.B., Jarke, M., Zaniolo, C. (eds.) Proc. 20th Int. Conf. Very Large Data Bases, VLDB, pp. 487–499. Morgan Kaufmann, San Francisco (1994)
6. Taniguchi, T., Haraguchi, M.: Discovery of hidden correlations in a local transaction database based on differences of correlations. Data Engineering Applications of Artificial Intelligence 19(4), 419–428 (2006)

Real-World Mood-Based Music Recommendation

Magnus Mortensen[1,2], Cathal Gurrin[1,3], and Dag Johansen[1,4]

[1] Department of Computer Science, University of Tromsø, Breivika, NO-9037 Tromsø, Norway
mortensm@fast.no{gurrin,dag}@cs.uit.no
[2] Senior Engineer at Fast Search & Transfer
[3] Adaptive Information Cluster at Dublin City University, Ireland
cgurrin@computing.dcu.ie
[4] Chief Scientist of Fast Search & Transfer

Abstract. We present a music recommendation system that incorporates both collaborative filtering and mood-based recommendations. The benefits of incorporating mood-based recommendations over both content/genre-based and collaborative filtering-based recommendation are illustrated by means of a real-world user evaluation in which 54 users took part in a one month long evaluation.

Keywords: Collaborative filtering, content filtering, recommendation, music recommendation, mood.

1 Introduction and Background to Music Recommender Systems

In the recent years, we have witnessed the increasing use of personalisation and recommendation systems in order to solve the problem of information seeking and information overload when accessing large archives of content. Whether it is our ever increasing digital music collections, digital photo collections, NEWS stories or even web pages, we are becoming more and more reliant on smart information systems to understand our needs and recommend content according to our interests. In this paper we present a novel music recommendation system that incorporates both collaborative filtering and mood-based recommendations. We illustrate the performance improvements of a music recommender system that incorporates mood-based recommendations and collaborative filtering by evaluating three different approaches for producing recommendations by means of live user experiments over a period of one month.

1.1 Collaborative Recommendation

Typically recommender systems have been extensively used within e-commerce and online communities for recommending items like movies and books. More recently, recommender systems have been deployed in online music players, recommending music to users. Content-based filtering and collaborative filtering are two well-known algorithmic techniques for computing recommendations. A content-based filtering system [1] selects items for recommendation based on the correlation between content and the stored user's preferences. A collaborative filtering system [2] recommends

H. Li et al. (Eds.): AIRS 2008, LNCS 4993, pp. 514–519, 2008.

items not seen by the user, based on a correlation calculated between the user and other users with similar preferences. In addition, hybrid approaches have been developed to avoid the limitations of using either alone [3], such as the tendency of content-filtering systems to recommend only 'more of the same' content to a user.

1.2 Music Recommender Systems

There has been previous work on music recommender systems. For example Ringo [4] is a collaborative filtering recommendation system where the ratings of users similar to a particular user are utilized to suggest music for recommendation. Lee and Lee [5] have developed a music recommender system based on automatically identifying a person's mood (using temporal and context information). This mood-based recommendation is positively evaluated on a closed set of user listening data, retrospectively gathered with recommendations based on user's playback history. Where our research differs from the pre-existing research is in the integration of low-cost, mood-based recommendations with collaborative filtering, based on explicitly gathered mood ratings of both music and users, which we know to be accurate. In addition, our recommender system is positively evaluated using real-world data in a live implementation, with live users interacting with the system and receiving recommendations on an ongoing basis and instantly evaluating them, for a period of just over one month.

2 Mood-Based Recommendation Experiment

For our experiment, the music collection employed was a 6,027 song collection, gathered by the participants in the experiment, and thus representing the range of musical tastes of the participants. This song collection was categorized into five genres, as shown in Table 1, which is used to recommend content to participants. Each song a user played was explicitly rated by that user for mood on a four-point scale (angry, happy, relaxed, sad). In addition users rated each song played on a binary scale, as being positive (liked) or negative (did not like) and these ratings were later used for evaluation and training purposes, as described in subsequent sections.

Table 1. Music Collection, Analysis by Genre

Genre	Rock	R&B	Pop	Jazz	Folk	Other
% of Collection	40%	3.5%	9%	3%	4.5%	40%
Num of Sub-Genres	8	5	3	3	3	6

In total there were 54 users for this experiment, from three locations (in both the US and Norway) and, as stated, this was a live evaluation which lasted for a total of 33 days. This month long evaluation was divided into three phases of experimentation, with the first 19 days (content-based period) being dedicated to gathering a critical mass of user ratings to support subsequent music recommendation using collaborative filtering and collaborative mood filtering. After this first phase, the following two weeks were divided between the non-mood based collaborative filtering (collaborative period) and mood-based (mood-based collaborative period) experiments, as shown in Figure 1.

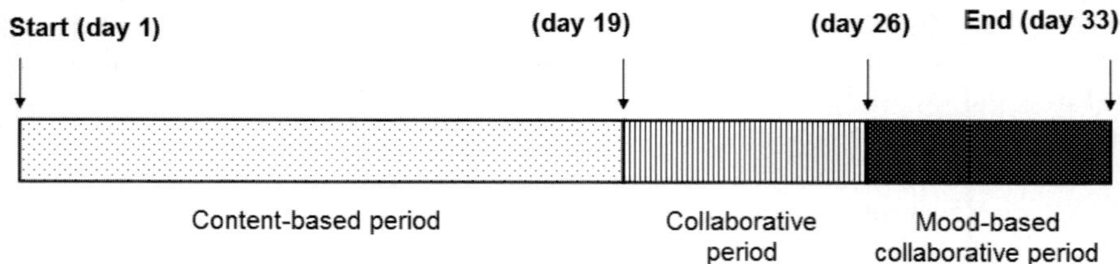

Fig. 1. Experiment Timeline Showing three Periods of the Evaluation

2.1 Genre-Based Content Filtering

During the initial training phase of gathering a critical mass of user ratings, users were recommended music based on their listening history ('more-like-i-heard-already') and genre's that they liked. These recommendation resulted in new (not yet rated) music being recommended from the most important genres for each user. In this way, we maximized the number of ratings stored in the system. The evaluation of recommendation quality was ongoing during this phase, and the results are presented in Figure 1 below. Recall that a typical binary scale (like / not-like) was employed for user judgments and every time a song was played by a user, they rated each song on this binary scale and also annotated each song with one of the four mood descriptors.

The key result of this phase of the experiment was that a critical mass of user ratings (6,159) was now stored in the system, along with the mood annotation of these songs. This provided a source of data for recommendations in the following two phases of recommendation.

2.2 Collaborative Filtering

The content filtering phase (just described) was followed by a week of collaborative filtering recommendations, which recommend music to users on the basis of the similarity between users. User similarity was estimated by calculating the well known Mean Squared Difference measure between any two users, as shown below:

$$msd_{a,u} = \frac{\sum_{i=1}^{m} \left(r_{a,i} - r_{u,i}\right)^2}{m}$$

where m is the number of items that both users have rated, $r_{a,i}$ is the rating by the active user on item i and $r_{u,i}$ is the rating by the user u on item i. We can see that $0 < msd_{a,u} < 1$, and if both users have similar ratings for all items, $msd_{a,u}$ will be 0, or if the ratings differ, $msd_{a,u}$ is 1. Exploiting the ratings of similar users, songs that user a has not rated before, are recommended to user a thereby avoiding recommendation cycles (recommending the same content constantly)[1].

During this (and the following) phase of the experiment, ratings were not stored for recommendation purposes, rather only for later evaluation of recommendation

[1] At the start of each phase of the experiment, the previously-seen-list of songs already played by the user are reset to zero, thereby allowing all songs to be recommended for each new phase, so as to avoid bias against the later phases of the experiment.

performance, so as to avoid any bias in favour of the subsequent collaborative mood filtering phase. In addition, further mood annotation was not performed for this or the following phase.

2.3 Collaborative Mood Filtering

Collaborative Mood Filtering is similar to the collaborative filtering, however with the difference that songs are only recommended to a user that match the user's current mood (explicitly stated by the user in the interface). In effect, mood operates as a filter over the recommendations, thereby focusing the recommendations on the particular mood of the user at any given time. The effect of this is to reduce the number of songs that are available for recommendation and once again users rated songs for evaluation purposes only. Once again, songs that the user had not seen were recommended to the user and the previously-seen-list reset to zero at the start of this phase.

2.4 Results of Evaluation

For the evaluation, instead of using a simple Precision value, we have utilised an accuracy measurement which will give a more intuitive evaluation of the quality of the recommendations, by allowing the number of negatively rated songs to influence the score. The accuracy measurement is shown in the following formula:

$$Acc_n = \frac{\left(S^+_n - S^-_n\right)}{S_n}$$

Where Acc_n is the accuracy score on day n, S^+_n is the number of positively rated songs recommended on day n, S^{-}_n is the number of negatively rated songs recommended on day n, and S_n is the number of songs recommended on day n. The Accuracy measure will always be in the range of -1 to +1, with positive scores illustrating a higher number of positive ratings than negative ratings.

Since we gathered the binary relevance judgement as the user played any given song, the accuracy of the recommendation techniques could be immediately evaluated. The results of the experiment are presented in Figure 2 below, where we plot the accuracy of the recommendations on a day-by-day basis (through the three experiment phases), for all users. Outlier days were removed from the results, i.e. days in which only one user requested recommendations, which were two days during two weekends[2].

As can be seen from Figure 2, the trendline shows that during the content-based phase, that recommendation accuracy increased during the first few days as ratings are built up, but then remains relatively static, suggesting that genre-based user histories for non-collaborative, genre-based recommendations become effective within a week or so.

Upon entering the collaborative period, where recommendations are based on pure collaborative filtering, the accuracy of recommendations drops significantly and shows more variance in performance than mood-based filtering. In the mood-based

[2] Outlier days were in which only one user requested recommendations. This happened on two days, during two weekends.

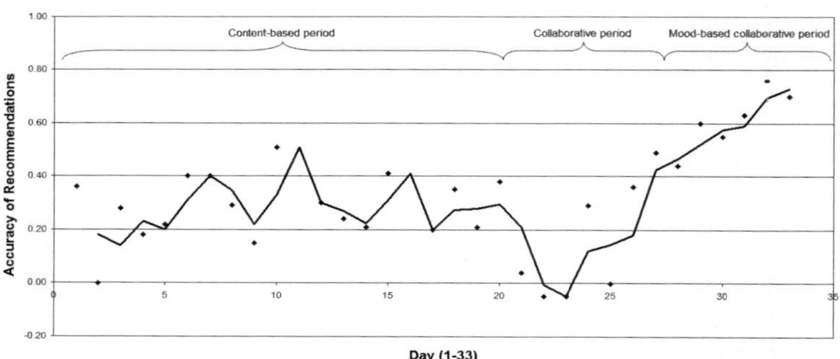

Fig. 2. Accuracy of recommendations, in all three experimentation periods

collaborative filtering phase, the climb in accuracy (seen in the collaborative period) continues steadily, suggesting that mood-based collaborative filtering outperforms conventional collaborative filtering. If we compare the average accuracy of both genre-based content filtering (0.28 for all 19 days, and 0.29 for the last 7 days) and collaborative filtering (0.12) to that of mood-based collaborative filtering (0.6) the benefit of mood-based collaborative filtering is clear.

It is difficult at this point to firmly conclude the reason why the performance of the collaborative period (phase 2) is so low, though we believe that this could be due the requirement of a longer first phase of the experiment to gather even more user ratings than the 6,159 already gathered. What is interesting to note is that the collaborative mood filtering approach does not seem to be affected in the same way by the density of user ratings in the system, which we believe is due to the fact that the mood filtration of recommendations reduces the negative effect of any noisy recommendations from the collaborative filtering technique.

Although this is early work in the area of mood-based recommendations and is based on a coarse five point scale for music genre and a four point scale for evaluating mood, the application of a mood filter clearly shows benefits for recommendation quality on our experiment.

3 Conclusions and Future Work

In this paper we have shown that a low-cost, mood-based collaborative filtering mechanism outperforms both conventional collaborative filtering and the genre-based content filtering system in a real-world music recommendation system. We propose that the mood-based recommendations reduce the number of noisy recommendations that are present when compared to a pure collaborative filtering technique.

In future work, we will maintain a balance between recommending content that the user has not rated and that the user has rated, so as not to artificially deflate system performance. We will also examine scalability issues for real-world collaborative-based recommendations and will examine the effect of automatically estimating user mood. Finally, moving from explicit user feedback of mood and recommendation quality judgments to implicit judgments, we will examine the effect and accuracy of these implicit judgments.

References

1. Schafer, J.B., Konstan, J.A., Riedl, J.: E-Commerce Recommendation Applications. Data Mining and Knowledge Discovery 5, 115–153 (2001)
2. Herlocker, J.L., Konstan, J.A., Borchers, A., Riedl, J.: An algorithmic framework for performing collaborative filtering. In: ACM SIGIR 1999, Berkeley, CA, August 15-19 (1999)
3. Melville, P., Mooney, R.J., Nagarajan, R.: Content-Boosted Collaborative Filtering for Improved Recommendations. In: AAAI 2002, Palo Alto, CA, March 25-27 (2002)
4. Shardanand, U.: Social Information Filtering: Algorithms for Automating Word of Mouth. In: ACM CHI 1995, Denver, CO, May 7-11 (1995)
5. Lee, J.S., Lee, J.C.: Music for my Mood: A Music Recommendation System Based on Context Reasoning. In: Havinga, P., Lijding, M., Meratnia, N., Wegdam, M. (eds.) EuroSSC 2006. LNCS, vol. 4272, Springer, Heidelberg (2006)

News Page Discovery Policy for Instant Crawlers

Yong Wang, Yiqun Liu, Min Zhang, and Shaoping Ma

State Key Lab of Intelligent Tech. & Sys., Tsinghua University
wang-yong05@mails.tsinghua.edu.cn

Abstract. Many news pages which are of high freshness requirements are published on the internet every day. They should be downloaded immediately by instant crawlers. Otherwise, they will become outdated soon. In the past, instant crawlers only downloaded pages from a manually generated news website list. Bandwidth is wasted in downloading non-news pages because news websites do not publish news pages exclusively. In this paper, a novel approach is proposed to discover news pages. This approach includes seed selection and news URL prediction based on user behavior analysis. Empirical studies in a user access log for two months show that our approach outperforms the traditional approach in both precision and recall.

Key words: web log, user behavior analysis, news page discovery.

1 Introduction

Nowadays, there are high freshness requirements for search engines. Many web users prefer reading news from search engines. They type a few key words about a recent event into a search engine, check the returned result list and navigate to pages providing details about the event. If a search engine fails to perform such service, users will be frustrated and turn to other search engines. News pages should be downloaded immediately after they are published. Therefore, many search engines have special crawlers called instant crawlers to download novel news pages. The work flow of an instant crawler is

```
load seed URLs into waiting list                          (1)
while (waiting list is not empty)
{
pick a URL from the waiting list
download the page it points to
write the page to disk
for each URL extracted from the page
if the URL points to a novel news page                    (2)
add the URL to the waiting list
}
```

The performance of an instant crawler is largely determined by two factors: (1) quality of seed URLs; (2) accuracy of prediction about whether a URL points to a news page when its content has not been downloaded yet.

H. Li et al. (Eds.): AIRS 2008, LNCS 4993, pp. 520–525, 2008.

Currently, manually generated rules are provided to solve the problem. An instant crawler administrator writes a news website list for an instant crawler to monitor. The instant crawler takes the homepages of these websites as seed URLs. A newly discovered URL will be added to its waiting list if it is in the monitored websites.

This policy works fine, but there are some problems. Many web sites contain both news pages and non-news ones. For example, *auto.sohu.com* is a website about automobiles. There are news pages reporting car price fluctuation and non-news pages providing car maintenance information. Only news pages in this website should be downloaded by instant crawlers. A web site is too large a granularity to make this discrimination. This problem can be solved with our method.

News pages provide information on recent events. Users are interested in a news page only in a short period after it is published. As more and more users get to know the event, fewer users are likely to read that page. In contrast, non-news pages are not relevant to recent events. Users access them constantly. This feature is used to identify news pages. If a page accumulates a large proportion of click throughs in a short period after publication, it is likely to be a news page.

A policy for instant crawlers to discover news pages is proposed based on user behavior analysis in click through data. In the beginning, news pages are identified based on how their daily click through data evolves. Then web pages which directly link to many news pages are used as seed URLs. Web administrators usually publish news pages under only a few paths, such as /news/. URLs of many news pages in the same folder share the same news URL prefixes. If there are already many news pages sharing the same news URL prefix, it is likely that novel news pages will be stored under that path and their URLs will start with that prefix.

The rest of this paper is organized as follows: Section 2 introduces earlier research in priority arrangement in waiting list of crawlers; Section 3 describes the dataset which will be used later; Section 4 discusses and verifies a few properties of news pages; Section 5 addresses the problems in seed selection and news URL estimation; the approach proposed is applied in the dataset and the result is analyzed in Section 6; Section 7 is the conclusion of this paper.

2 Related Work

Earlier researchers performed intensive studies on evolutionary properties of the web, including the rate of existing page updates and that of novel page appearance [1], [2]. The conclusion is that the web is growing explosively [2] and it is almost impossible to download all novel pages. Web crawlers have to organize a frontier which is consisted of discovered but not downloaded URLs. Priority arrangement in the frontier is important. This problem is studied from several perspectives. Some researchers tried to find a balance between downloading novel pages and refreshing existing pages [3], [4] and [5]. They studied page update intervals and checked existing pages only when necessary. Crawlers downloaded novel pages during the intervals. Focused crawlers only download pages related to a given topic [6], [7], [8] and [9]. They estimate whether a URL is worth downloading mainly based on its anchor text. Other crawlers [10], [11], [12] and [13] predict quality of novel URLs and download candidates of high quality. This work is similar with ours. We also make an order of the frontier, in the perspective of freshness requirements instead of page quality. Pages of high freshness requirement are downloaded with high priority, while others can be downloaded later.

3 News Page Discovery Policy for Instant Crawlers

News hub pages are used as seed URLs to discover novel news pages if they link to many previous news pages. Novel news pages are usually stored in the same location with known news pages. So news pages are identified to find where novel news pages are likely to be stored. A newly discovered URL will be downloaded if its URL starts with one of the news URL prefixes.

3.1 Generate Seed URL List for an Instant Crawler

It is proved in Section 4.1 that ClickThroughConcentration of most news pages is larger than that of most non-news ones. For each web page in the click through log, it is a news page if its ClickThroughConcentration is less than a threshold. Otherwise, it is a non-news page. News pages can be automatically identified with this method.

A seed URL for an instant crawler is of high quality if a large number of news pages can be discovered from it in only one or two hops. It is probable that novel news pages will be linked by pages which already have links to many known news pages. News hub pages which have linked most news pages are included in seed list.

3.2 Estimate Whether a URL Points to a News Page

Some news pages cluster in the same folder and some are dynamically generated from the same program with different parameter values. News URL prefixes can be found from known news pages. Given a website, a URL prefix tree is built according to its folder structure. In this tree, a node stands for a folder. Node A has a child node B if B is the direct subfolder of A. Web pages are leaf nodes. A program is also a non-leaf node. Dynamic pages generated from that program are its leaf nodes. Each non-leaf nodes are labeled by two numbers: the number of news pages and that of non-news

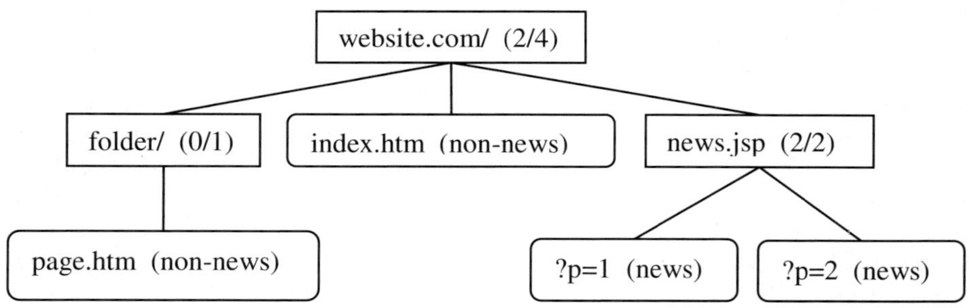

Fig. 1. A URL prefix tree of a sample website

ones directly and indirectly under that node. For example, website.com contains four pages: /index.html, /folder/page.htm, /news.jsp?p=1 and /news.jsp?p=2. Its URL prefix tree is organized as in Fig. 1.

Each non-leaf node is scored with the proportion of the number of news pages to that of all pages under that node. All prefix trees are traversed from the roots. A node is a news node if its score is greater than a threshold. Otherwise, its children nodes are tested. This algorithm is described below.

```
FindNewsNode(TreeNode N){
if (score of N is greater than the threshold){
N is a news node;
return;
}
foreach child in NonLeafChildrenOfN
FindNewsNode(child);
}
```

A news URL prefix consists of all nodes on the path from the root to the news node. Take the tree in Fig. 3 for example, if the node "news.jsp" is a news node, "website.com/news.jsp" is a news URL prefix. It is probable that URLs starting with news URL prefixes point to news pages and is worth downloading.

4 Experiment and Evaluation

Anonymous click through data for consecutive 60 days from November 13th 2006 to January 11th, 2007 is collected by a proxy server. Each record is a structure below:

Request Date and Time	Client IP	Target URL	Referrer URL

A user accesses the Target URL from a hyperlink in the Referrer URL. Referrer URL is null if a user types the address instead of clicking a hyperlink. Daily click through data of all 75,112,357 pages is calculated. Multiple requests to a single page from the same IP in one day are counted as one click through to avoid automatically generated requests by spammers. Pages whose average daily click throughs are less than one are filtered out for lack of reliability, leaving daily click through history of 975,151 pages for later studies.

4.1 Experiment

A page is classified as a news page if its ClickThroughConcentration is greater than a threshold p. Pages from focus.cn which have been annotated manually are used as training set in which there are 2,337 news pages and 703 non-news pages. The best performance is achieved when p=1.91 and the maximized hit (the number of pages correctly classified) is 2,682. This threshold is applied on all pages and 147,927 are labeled news pages and other 827,224 are labeled non-news.

A navigation record from page A to page B indicates that B is linked by A. The number of news pages linked by each page is calculated and the top 1,542 pages which link to the most news pages are included in the seed URL list. The number of seed URLs is the same with that in the baseline used later for comparison.

In URL prefix trees, a node is a news node if the proportion of news pages under that node is larger than a given threshold, where 0.8 is used. 439 nodes are labeled news nodes. Larger threshold can be used if bandwidth is limited and that wasted in downloading non-news pages is unaffordable. If an instant crawler has enough bandwidth and wants to recall more news pages, the threshold can be smaller.

4.2 Evaluation

Sogou Inc. is a search engine company in China. Its instant crawler uses a manually generated website list which contains 1,542 news websites. Homepages of these websites are seed URLs and the instant crawler downloads pages from these websites only. This policy is used as the baseline to be compared with ours.

Table 1. Performance comparison

	Baseline	Our Method
Number of Downloaded News Pages	86,714	101,870
Number of Total Downloaded Pages	177,801	111,934
Precision	48.8%	91.0%
Recall	58.6%	68.9%

46,210 different news pages are directly linked by homepages of news sites in Sogou's list, while 79,292 are directly linked by news hub pages in our seed list. Not all homepages are the best seeds. There are websites which publish both news pages and non-news ones. The index pages of news channels are better candidates for seed URLs. For example, finance.sina.com.cn is a financial website. The web log shows that most of its news pages are linked by finance.sina.com.cn/stock/. This page instead of the homepage should be included in the seed list.

The instant crawler of Sogou Inc. downloads all pages from their site list, while our instant crawler downloads pages whose URLs start with one of the news URL prefixes. The result is shown is Table 2.

There are 147,927 news pages in the dataset. Precision is the proportion of downloaded news pages in all downloaded pages. Recall is the proportion of downloaded news pages in all news pages in the data set. As is shown in Table 2, 86,714 news pages are in the site list, while 101,870 are covered by the URL prefixes. The performance of the efficiency crawler is improved that it downloads more news pages with less burden of non-news ones.

5 Conclusion

In this paper, an effective news page discovery policy is proposed. The current instant crawlers which are assigned to download news pages cannot produce satisfactory result due to news page distribution complexity. In this paper, we propose and verify a few features of news pages. Then these features are used in seed URL selection and news URL prediction. The performance of instant crawlers is improved both in precision and recall because they can discover more news pages with less bandwidth wasted in downloading non-news pages.

References

1. Fetterly, D., Manasse, M., Najork, M., Wiener, J.L.: A Large-scale Study of the Evolution of Web Pages. Software Practice and Experience (2004)
2. Brewington, B., Cybenko, G.: How Dynamic is the Web. In: Proceedings of WWW9 –9th International World Wide Web Conference (IW3C2), pp. 264–296 (2000)
3. Cho, J., Garcia-Molina, H.: Effective Page Refresh Policies for Web Crawlers. ACM Transactions on Database Systems (TODS) (2003)
4. Shkapenyuk, V., Suel, T.: Design and Implementation of a High-performance Distributed Web Crawler. In: Proceedings of the 18th International Conference on Data Engineering, San Jose, Calif. (2002)
5. Barbosa, L., Salgado, A.C., Carvalho, F., Robin, J., Freire, J.: Workshop On Web Information And Data Management. In: Proceedings of the 7th annual ACM international workshop on Web information and data management (2005)
6. Menczer, F., Belew, R.: Adaptive Retrieval Agents: Internalizing Local Context and Scaling up to the Web. Machine Learning 39(23), 203–242 (2000)
7. Pant, G., Menczer, F.: Topical Crawling for Business Intelligence. In: Koch, T., Sølvberg, I.T. (eds.) ECDL 2003. LNCS, vol. 2769, pp. 233–244. Springer, Heidelberg (2003)
8. Stamatakis, K., Karkaletsis, V., Paliouras, G., Horlock, J., et al.: Domain-specific Web Site Identification: the CROSSMARC focused Web crawler. In: Proceedings of the 2nd International Workshop on Web Document Analysis (WDA 2003), Edinburgh, UK (2003)
9. Menczer, F., Pant, G., Srinivasan, P.: Topical Web Crawlers: Evaluating Adaptive Algorithms. ACM Transactions on Internet Technology 4(4), 378–419 (2004)
10. Cho, J., Garcia-Molina, H., Page, L.: Effecient Crawling through URL Ordering. WWW8 / Computer Networks 30(1-7), 161–172 (1998)
11. Eiron, N., McCurley, K.S., Tomlin, J.A.: Ranking the Web Frontier. In: Proc. 13th WWW, pp. 309–318 (2004)
12. Eiron, N., McCurley, K.S.: Locality, Hierarchy, and Bidirectionality in the Web. In: Workshop on Algorithms and Models for the Web Graph, Budapest (2003)
13. Abiteboul, S., Preda, M., Cobena, G.: Adaptive On-line Page Importance Computation. In: Proc. 12th World Wide Web Conference, pp. 280–290 (2003)

An Alignment-Based Approach to Semi-supervised Relation Extraction Including Multiple Arguments

Seokhwan Kim[1], Minwoo Jeong[1], Gary Geunbae Lee[1], Kwangil Ko[2], and Zino Lee[2]

[1] Department of Computer Science and Engineering, Pohang University of Science and Technology, San 31, Hyoja-dong, Nam-gu, Pohang, 790-784, Korea
{megaup,stardust,gblee}@postech.ac.kr
[2] Alticast Corp., 15th Floor, Nara Investment Banking Corp. Bldg. 1328-3, Seocho-Dong, Seocho-Gu, Seoul, 137-858, Korea
{kik,zino}@alticast.com

Abstract. We present an alignment-based approach to semi-supervised relation extraction task including more than two arguments. We concentrate on improving not only the precision of the extracted result, but also on the coverage of the method. Our relation extraction method is based on an alignment-based pattern matching approach which provides more flexibility of the method. In addition, we extract all relationships including two or more arguments at once in order to obtain the integrated result with high quality. We present experimental results which indicate the effectiveness of our method.

1 Introduction

During the past few years, we have been able to obtain a large amount of information about various topics through the Internet. However, the high accessibility of the Internet has caused the trend of information overflow which makes it difficult to obtain valuable information due to excessive amount of information rather than lack of it. In order to improve the efficiency of gathering valuable information, the information extraction task has been actively researched by many researchers, and it has grown into one of the most important topics of natural language processing field.

The area of information extraction is divided into several subtasks by the characteristic and range of target information, and most of them can be generalized by extracting the defined number of relevant arguments from natural language documents. Named entity recognition and binary relation extraction tasks can be considered as special cases of the above-mentioned generalized concept of information extraction, which define the number of extracted arguments as 1 and 2, respectively. Both subtasks are the most widely researched topics in

H. Li et al. (Eds.): AIRS 2008, LNCS 4993, pp. 526–536, 2008.

the information extraction tasks, and several researchers have shown that supervised machine learning based approaches are significantly effective ways to solve these problems. [1] [2]

However, supervised machine learning methods have a cost problem by requiring a considerable amount of training data for achieving good performance. In order to reduce the cost of building required resources with minimal performance loss, recently, semi-supervised machine learning methods have been attempted to solve the problem. Most of existing works for semi-supervised information extraction commonly concentrate on automatically creating context patterns guaranteeing high-precision by integrating statistical characteristics of target documents with grammatical induction methodologies. [3] [4] [5] [6] [7] [8]

In this paper, we will describe our semi-supervised information extraction approach with following two points of views which are little different from other existing works. The first issue is about the coverage of each context pattern. Our approach is based on the bootstrapping methods. In the case of bootstrapping, the high-precision is an absolutely important goal of the method, because even very small number of errors generated in earlier iteration can be enormously harmful to the overall performance due to error accumulation by iterating. Nevertheless, high-precision is not the only prerequisite for achieving high performance of the method. If it is guaranteed that the set of context patterns accumulated by iterating more than considerable times will have the sufficient coverage which is needed to extract all existing information, then reducing errors by improving the precision of context pattern induction is the best way to improve the overall performance. However, this assumption is far from realistic, because expressions indicating even the same information can be entirely different each other and each expression also can be derived into the totally new expressions as time goes by. It is difficult to keep up with the variety of the expressions only depending on the set of precise context patterns, and even if it is possible, it might require huge number of iteration which is limited by current computing power. In actuality, we should consider not only high-precision, but also the way of enhancing coverage of each context pattern for improving the overall performance. In order to encourage coverage of the method, we focused on the task of context pattern matching rather than context pattern induction, and we will present an alignment-based information extraction method as a pattern matching approach in our method.

The other issue is about the number of extracted arguments of the task. Although most of existing works have concentrated on the task of extracting individual named-entities or relationships between just two named-entities, in many cases, we should extract the relationship including more than two arguments. For extracting the n-ary relationships, we applied our alignment-based information extraction method to the task of extracting relationships including not only just two arguments, but also more than two arguments. Moreover, we will present a reinforcement scheme based on the result of bottom-up integration, starting with the result of binary relation extraction.

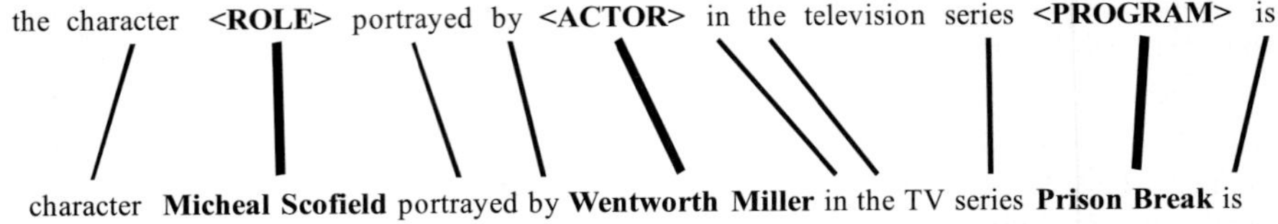

Fig. 1. Sentence alignment for extracting multiple relevant arguments

The remainder of this paper is organized as follows: In the next section, we present a detailed description of the alignment-based information extraction method. In section 3, the overall architecture of our method and detailed descriptions of each subtask are presented. We present the experimental results in section 4, and our conclusions are provided in section 5.

2 Alignment-Based Information Extraction

Kim *et al.* [9] presented an alignment-based named entity recognition method to solve the spoken language understanding problem. We modified the method to extract not individual named entities, but tuples including two or more relevant arguments, and applied this modified method to the task of n-ary relation extraction. As shown in Fig. 1, we align a raw sentence with a context pattern which is a part of sentence containing labels of target arguments. Then, from the result of the best alignment between them, we extract the parts of the raw sentence which are aligned to the argument labels in the context pattern, and incorporate the extracted arguments into a tuple which is a candidate of n-ary relation. In Fig. 1, a tuple (Prison Break, Michael Scofield, Wentworth Miller) is extracted as a candidate of ternary relationship, (PROGRAM, ACTOR, ROLE), which means that an ACTOR acts a role of ROLE on a PROGRAM.

In order to enhance the coverage of each context pattern, we should consider the flexibility of the alignment task. Accordingly, we adapted an alignment scheme based on the Smith-Waterman algorithm [10], which is a widely used biological sequence alignment algorithm providing a systematic way of controling the flexibility of the task. We utilize this algorithm into the sentence alignment task by considering a word or a morpheme as a unit of alignment instead of biological residues.

The alignment algorithm is performed by computing the score for each word pair in the alignment matrix M. Each row in the matrix corresponds to a word in the context pattern, while each column in the matrix corresponds to a word in the raw sentence. Moreover, a point of crossing between a row and a column has the score of aligning the word in the raw sentence with the word in the context pattern.

The first step in the alignment method is to assign the initial value of each position in the matrix M with 0. And then, we find the maximum alignment score by starting on the upper left hand corner in the matrix M and continuing

	the	character	**Michael**	**Scofield**	portrayed	by	**Wentworth**	**Miller**	in	the	TV	series	**Prison**	**Break**	is
character	0	1	1	1	1	1	1	1	1	1	1	1	1	1	1
<ROLE>	1	1	2	2	2	2	2	2	2	2	2	2	2	2	2
portrayed	1	1	2	2	3	3	3	3	3	3	3	3	3	3	3
by	1	1	2	2	3	4	4	4	4	4	4	4	4	4	4
<ACTOR>	1	2	2	3	3	4	5	5	5	5	5	5	5	5	5
in	1	2	2	3	3	4	5	5	6	6	6	6	6	6	6
the	1	2	2	3	3	4	5	5	6	7	7	7	7	7	7
television	1	2	2	3	3	4	5	5	6	7	7	7	7	7	7
series	1	2	2	3	3	4	5	5	6	7	7	8	8	8	8
<PROGRAM>	1	2	3	3	4	4	5	6	6	7	8	8	9	9	9
is	1	2	3	3	4	4	5	6	6	7	8	8	9	9	10

Fig. 2. An example of computed alignment matrix

to find the maximum score $M_{i,j}$ for each position in the matrix according to the lower right direction. The maximum score $M_{i,j}$ is defined as

$$M_{i,j} = \max \begin{pmatrix} M_{i-1,j-1} + \mathrm{sim}_{i-1,j-1} \\ M_{i-1,j} + gp \\ M_{i,j-1} + gp \\ 0 \end{pmatrix}, \tag{1}$$

where $\mathrm{sim}_{i,j}$ is the value of similarity between the i-th word in the context pattern and j-th word in the raw sentence, and gp is the pre-defined penalty for a gap. Fig. 2 shows an example of computed matrix for the alignment which is shown in Fig. 1. In this example, we defined the similarity function as

$$sim_{i,j} = \begin{cases} 1, & \text{if PTN}_i \text{ and RAW}_j \text{ are identical} \\ & \text{or PTN}_i \text{ is an argument label} \\ 0, & \text{otherwise,} \end{cases} \tag{2}$$

where PTN_i is the i-th word in the context pattern, while RAW_j is the j-th word in the raw sentence. And the value of gap penalty, gp, is ignored in this example.

After matrix computation, we trace back the matrix to find the best alignment with the maximum score and extract the relevant arguments from the result of alignment. The traceback task is started at the position with maximum score on the alignment matrix. For each current position $[i, j]$, the next position is determined by the following policies in order:

1. if $M_{i,j} = M_{i,j-1} + gp$, then the next position is $[i, j-1]$.
2. if $M_{i,j} = M_{i-1,j-1} + sim_{i,j}$, then the next position is $[i-1, j-1]$.
3. if $M_{i,j} = M_{i-1,j} + gp$, then the next position is $[i-1, j]$.

The order of applying the policies should be preserved. Although most of sequence alignment methods based on the Smith-Waterman algorithm consider the diagonal advancement corresponding to our second policy as a prior direction, we give a preference to the left direction by applying the policy of the left position first, because we should make it possible to align each argument label in the context pattern with two or more words in the raw sentence in order to extract the arguments which consist of multiple words.

In Fig. 2, the sequence of positions with gray color indicates the best alignment with maximum alignment score. From the result of the alignment, we can extract the words "Michael Scofield", "Wentworth Miller", and "Prison Break" as relevant arguments which have types of ROLE, ACTOR, and PROGRAM respectively.

3 Semi-supervised Relation Extraction Including Multiple Arguments

Most of existing works about relation extraction have concentrated on the task of extracting relationships including just two arguments, regardless of supervised or semi-supervised approach. However, the binary relation extraction might not be sufficient in some circumstances.

Firstly, we consider the case that we should extract a relationship which includes more than two arguments. In the example which is mentioned in the previous section, the binary relationship between ACTOR and ROLE arguments has to be specified by another argument about the corresponding PROGRAM, because an actor can be related to various roles according to the performed programs. In order to extract relationships with multiple arguments, we can consider an approach of integrating several binary relationships into an n-ary relationship. For example, for obtaining relationships including three arguments which are PROGRAM, ACTOR, and ROLE, we should extract the following three binary relationships, (ACTOR,ROLE), (PROGRAM,ACTOR), and (PROGRAM,ROLE) and integrate them together. However, we have a problem that the errors originated from each task of binary relation extraction are accumulated into the integrated result. Added to that, this problem is getting worse as the number of arguments of the target relationship is increased.

The second problem is caused by the tendency of relevant arguments to be contiguously located each other. Although we want to deal with a complete binary relationship which has a specified meaning with just two arguments, other relevant arguments located closely to the target arguments might interfere the task of binary relation extraction. In the case of the (PROGRAM,ROLE) relationship in the previous example, another argument, ACTOR, tends to be closely located to both PROGRAM and ROLE. For example, we consider a context pattern of the (PROGRAM,ROLE) relationship extracted from Fig. 1, 'character ROLE portrayed by *Wentworth Miller* in the TV series PROGRAM is'. In this context pattern, 'Wentworth Miller' is the part of an ACTOR argument, and the coverage of the context pattern is dramatically weakened by this interposed argument.

In order to solve these problems which are caused by depending only on the binary relation extraction, we propose a new semi-supervised relation extraction method including multiple arguments. An overview of our method is shown in Fig 3.

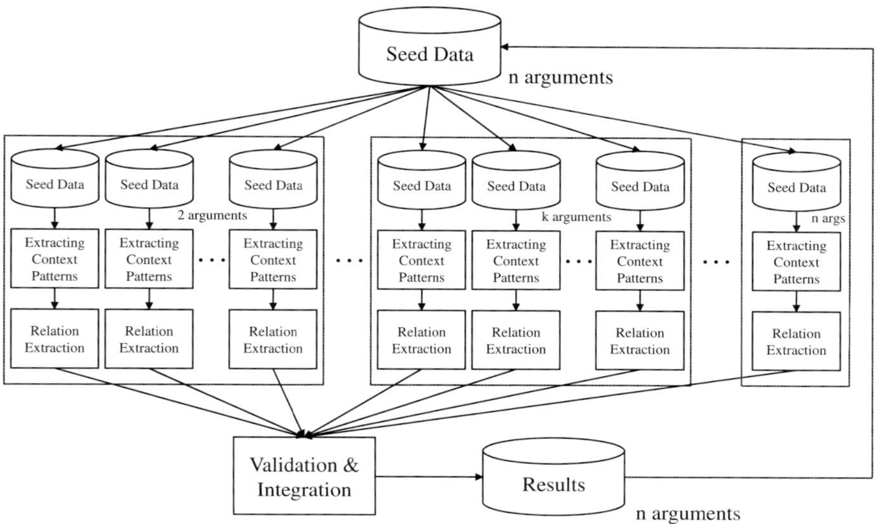

Fig. 3. Overall Architecture of the Method

If the number of arguments in the target relationship is n, we define the variable k which has all values between two and n. For each value of k, we consider all subsets which are organized by combinations of any k relevant arguments among overall n arguments in seed data. For example, if the target relationship is organized by four arguments, CHANNEL, PROGRAM, ACTOR, and ROLE, the value of k can be two, three, or four. The subsets which include two arguments are (CHANNEL,PROGRAM), (PROGRAM,ACTOR), and (ACTOR,ROLE), and the subsets are (CHANNEL,PROGRAM,ACTOR), (CHANNEL,PROGRAM,ROLE), and (PROGRAM,ACTOR,ROLE) for the k value of three. For the k value of four, we use seed data themselves, because k is equivalent to the value of n. For each individual subset, we execute the following series of tasks. Firstly, we separate the sub-seed data including only corresponding arguments from the original seed data. By using these separated data as seeds, we extract context patterns from the source documents. And then, based on the context patterns, we extract new tuples including k relevant arguments from source documents. These series of tasks are executed parallel for all derived subsets. The set of results in each parallel execution is verified and integrated by post-processing methods, and we can obtain the integrated result with n number of arguments which is equal to the number of the arguments in the original seed data. The above-mentioned tasks are performed in iterations of bootstrapping, and the results from the iteration are added to the seed data and affect to the next iteration.

Because our method extract not only binary relationships, but also all intermediate k-ary relationships up to the ultimate n-ary ones, we can reduce the accumulated errors in the integration of extracted tuples with relatively less arguments through the cross-validation process between intermediate k-ary relationships. Also, the problem of interfering by other closely located relevant arguments can be solved by extracting more than two relevant arguments at once.

We present the detailed descriptions about subtasks in our method on the following subsections.

3.1 Context Patterns Extraction

Since, we use an alignment-based approach between raw sentences and context patterns for extracting relevant arguments as stated in section 2, each context pattern should take the form that can be aligned to raw sentences, hence we incorporate context patterns starting from the sentences in source documents.

For each tuple in the seed data, we search for the sentences containing all arguments of the tuple in source documents. Although we can directly utilize the full sentence as a context pattern, we segment out subpart of the sentence which densely contains the arguments for enhancing coverage of the context patterns. The range of subpart is determined by locations of arguments and the value of margin size m. We extract the subpart from the m-th word on the left hand of the leftmost argument to the m-th word on the right hand of the rightmost argument in the sentence. And then, we make a context pattern by replacing the parts of arguments in the sub-sentence with corresponding argument labels.

For example, for a seed tuple (Prison Break, Michael Scofield, Wentworth Miller) of the ternary relationship (PROGRAM, ACTOR, ROLE), we can extract a context pattern, 'character ⟨ROLE⟩ portrayed by ⟨ACTOR⟩ in the TV series ⟨PROGRAM⟩ is' from the raw sentence in Fig. 1 and the margin size m of one.

3.2 Relation Extraction Based on Pairwise Alignment

Each extracted context pattern is aligned pairwisely with the sentences in the source documents for extracting candidate tuples containing relevant arguments. We can compute the alignment score for each alignment based on the alignment matrix M which is introduced in section 2. The alignment score is based on the maximum value in the alignment matrix M, and the position which has the maximum value is the start position of the trace-back task. Since we set the matching reward to 1 and both mismatching and gap penalty to 0, the maximum value on the matrix M means the number of equally aligned words in the best alignment. This value can be normalized by the length of the context pattern, and we define the alignment score as

$$score(PTN, RAW) = \frac{\max(M(PTN, RAW))}{\text{length}(PTN)}, \tag{2}$$

where PTN is a context pattern, RAW is a raw sentence, and $M(PTN, RAW)$ is the matrix which is computed by the task of alignment between them. We regard this score as a measure of reliability of extracted candidates, and we select only candidates with higher score than a threshold value as a result of extraction.

3.3 Alignment-Based Verification

Most of candidate tuples which are extracted parallel for each subset of seed data are still erroneous. The primary factor of the errors is the redundant attachment problem. This problem is caused by aligning not only argument itself but also contiguous words to the argument with an argument label in the context pattern. For example, if a context pattern, 'character ⟨ROLE⟩ portrayed by ⟨ACTOR⟩ in ⟨PROGRAM⟩ is' is aligned with the raw sentence in Fig. 1, we will obtain not 'Prison Break', but 'the TV series Prison Break' as the argument of PROGRAM. The redundant attachment problem is more serious in case that propositional words and particles are frequently omitted, or morpheme-based processing is required, such as in Korean language.

In order to solve the redundant attachment problem, we propose a verification approach which is also based on the alignment method. We consider the alignment score as a measure of similarity between two candidate arguments, and the similarity is defined as

$$similarity(A, B) = \frac{\max(M(A, B)) \times 2}{\text{length}(A) + \text{length}(B)} \tag{3}$$

where both A and B are candidate arguments.

For verifying the candidate tuples, firstly, we organize the clusters of similar tuples based on the tuple similarity measure defined as

$$sim(tuple1, tuple2) = \frac{\sum_{i=1}^{\#args} similarity(tuple1_i, tuple2_i)}{\#\text{ of arguments}} \tag{4}$$

where $tuple1$ and $tuple2$ are candidate tuples being compared each other, and $tuple_i$ is the i-th argument in the tuple. We consider tuples which pairwisely have higher similarity than a threshold value as a cluster.

And then, we perform the task of pairwise alignment of each argument in a cluster of tuples. For each argument, we replace it with the argument which has the maximum summation of similarities. For example, if there are considerable number of (Prison Break, Michael Scofield, Wentworth Miller) and a few (the TV series Prison Break, Michael Scofield, Wentworth Miller) in a cluster of the tuples for the relationship (PROGRAM, ACTOR, ROLE), 'the TV series Prison Break' might be replaced by 'Prison Break' which has the maximum summation of similarities, By this alignment verification, the distribution of candidates is reflected to the final result, and it plays the important role of reflecting the statistical characteristics of data in the overall method.

3.4 Bottom-Up Integration

From the extracted and verified tuples including relatively small number of arguments, we can integrate new tuples with more arguments. For example, we can make a new tuple of (PROGRAM, ACTOR, ROLE) by integrating a tuple of (PROGRAM, ROLE) relationship and a tuple of (ACTOR, ROLE) relationship which have the common argument of ROLE. However, these new integrated

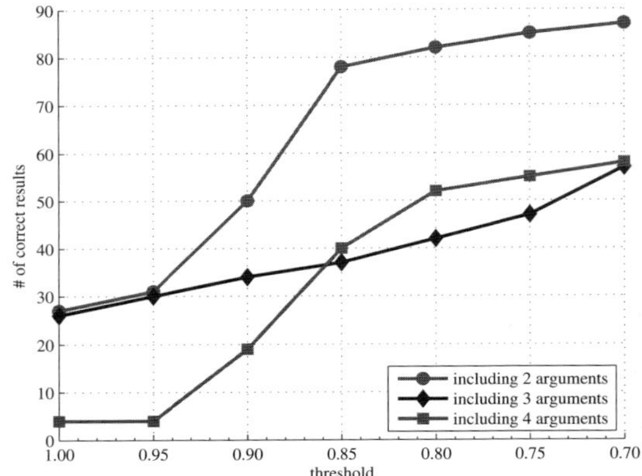

Fig. 4. Comparison of the coverage for various threshold values

tuples can come into collision with existing tuples which are extracted for the consisted arguments at once. In order to resolve the collision, we performed the previously mentioned verification task again to the set of the tuples which contains both existing and integrated tuples.

4 Experiments

We evaluated our method on 930 Korean news documents about TV series which consist of 13,175 sentences. Only a tuple with 4 arguments about the relationship (CHANNEL, PROGRAM, ACTOR, ROLE) is used as a seed information. Each result is collected after the first iteration and evaluated manually.

Firstly, we performed the method for various values of the threshold which affect to the acceptance rate of extracted candidates in the task of relation extraction. We selected all values of the threshold from 1.0 to 0.7 by downing the difference of 0.05. And the results were evaluated by manually counting of correctly extracted argument pairs after the verification task. The tuples with more than two arguments were separated into several pairs of arguments according to the corresponding binary relationships for evaluation. The experimental results are shown in Fig. 4. As the value of threshold decreases, the number of correctly extracted relevant arguments increases regardless of the number of target arguments. It means that the coverage of context patterns can be enhanced by the flexible setting of the threshold in the alignment-based approach.

And then, we evaluated the differences between before and after the verification task for a fixed threshold value, 0.85, which is determined empirically. The compared result shown in Table 1 indicates that the verification task contributes to dramatically improve the precision of the extracted results. From the analysis on the errors, we discovered that only 16.28% of the errors in the verified results are caused by the redundant attachment problem, while 82.0% of the errors occurred by the redundant attachment before verification. We can confirm that the

Table 1. Result of the verification

types of relations	before verification		after verification	
	# of tuples	precision	# of tuples	precision
(ACTOR,ROLE)	249	36.55	79	73.42
(PROGRAM,ROLE)	19	52.63	17	58.82
(PROGRAM,ACTOR)	10	60	10	60
(CHANNEL,PROGRAM)	12	33.33	6	66.67
(PROGRAM,ACTOR,ROLE)	7	42.86	5	60
(CHANNEL,PROGRAM,ROLE)	18	55.56	16	81.25
(CHANNEL,PROGRAM,ACTOR)	8	62.5	8	75
(CHANNEL,PROGRAM,ACTOR,ROLE)	15	60	14	85.71

Table 2. Result of the integration

types of relations	with only binary relations		with all intermediates	
	# of tuples	precision	# of tuples	precision
(PROGRAM,ACTOR,ROLE)	9	77.78	9	88.89
(CHANNEL,PROGRAM,ROLE)	11	81.82	16	87.5
(CHANNEL,PROGRAM,ACTOR)	12	58.33	9	77.78
(CHANNEL,PROGRAM,ACTOR,ROLE)	8	87.5	16	87.5

alignment-based verification elevates the performance of the relation extraction by solving the redundant attachment problem.

As the last experiment, we compared the result of the bottom-up integration using all intermediate sub-tuples with the result of integration which depends on only binary relationships. As shown in Table 2, by using not only binary tuples, but also intermediate sub-tuples with more than two arguments, we can obtain more precise integrated results with wider coverage than the cases of depending on only binary relationships.

5 Conclusions

We have presented an alignment-based approach to semi-supervised relation extraction including multiple arguments. Using the alignment-based pattern matching approach, we improved the coverage of context patterns. And we solved the redundant attachment problem which causes the critical precision loss, by introducing the alignment-based verification method. In the integration phase, we considered not only binary relationships, but also all k-ary intermediate relationships, which produced more improved results than binary relationship-based integration.

On the other hand, there are still more rooms to be improved in our approach such as lack of statistical and linguistic features. Although, the statistical characteristic of the data is used for verification in the current method, it should also be utilized for extracting context patterns in order to obtain more reliable context patterns. In the case of linguistic information, we expect that it can be reflected by defining more systematic policies of the alignment method. Refining the method and applying it to more sophisticated problems such as automatic ontology population are our future works.

Acknowledgements

This research was supported by the Invitation for Proposals on Internet Services Theme funded by Microsoft Research Asia.

References

1. McCallum, A., Li, W.: Early results for named entity recognition with conditional random fields, feature induction and web-enhanced lexicons. In: Proceedings of The Seventh Conference on Natural Language Learning (CoNLL-2003) (2003)
2. Zelenko, D., Aone, C., Richardella, A.: Kernel methods for relation extraction. Journal of Machine Learning Research 3, 1083–1106 (2003)
3. Riloff, E., Jones, R.: Learning Dictionaries for Information Extraction by Multi-level Bootstrapping. In: Proceedings of the Sixteenth National Conference on Artificial Intelligence, pp. 474–479 (1999)
4. Etzioni, O., Cafarella, M., Downey, D., Popescu, A.M., Shaked, T., Soderland, S., Weld, D.S., Yates, A.: Unsupervised named-entity extraction from the web - an experimental study. Artificial Intelligence Journal 165, 91–134 (2005)
5. Lee, S., Lee, G.G.: Exploring phrasal context and error correction heuristics in bootstrapping for geographic named entity annotation. Information Systems 32, 575–592 (2007)
6. Agichtein, E., Gravano, L.: Snowball: extracting relations from large plain-text collections. In: Proceedings of the fifth ACM conference on Digital libraries, pp. 85–94 (2000)
7. Yangarber, R., Grishman, R., Tapanainen, P., Huttunen, S.: Automatic Acquisition of Domain Knowledge for Information Extraction. In: Proceedings of the 18th International Conference on Computational Linguistics (COLING 2000), pp. 940–946 (2000)
8. Sudo, K., Sekine, S., Grishman, R.: An improved extraction pattern representation model for automatic IE pattern acquisition. In: Proceedings of the 41st Annual Meeting of the Association for Computational Linguistics (ACL 2003), pp. 224–231 (2003)
9. Kim, S., Jeong, M., Lee, G.G.: A semi-supervised method for efficient construction of statistical spoken language understanding resources. In: Proceedings of the Interspeech 2007-Eurospeech (2007)
10. Smith, T.F., Waterman, M.S.: Identification of Common Molecular Subsequences. J. Mol. Bwl. 147, 195–197 (1981)

An Examination of a Large Visual Lifelog

Cathal Gurrin, Alan F. Smeaton, Daragh Byrne, Neil O'Hare, Gareth J.F. Jones, and Noel O'Connor

Adaptive Information Cluster, Dublin City University, Dublin 9, Ireland
cgurrin@computing.dcu.ie

Abstract. Lifelogging is the act of recording some aspect of your life in digital format. A basic and common form of lifelogging is the creation and maintenance of blogs, which are typically textual in nature, though often with multimedia elements. In this paper we are concerned with visual lifelogging, a new form of lifelogging based on the passive capture of photos of a person's experiences. We examine the nature of visual lifelogs, and the differences between visual lifelog photos and explicitly captured digital photos. This is done by examining a million lifelog photos encompassing a year of a visual lifelog from the life of one individual.

Keywords: Lifelogging, visual lifelog, photograph, passive capture, SenseCam.

1 Introduction and Background to Visual Lifelogging

Lifelogging is the process of digitally capturing ones life experiences and the most popular form of this is to record a text description of some part of your day in a blog. Most blogging activities are text-only, though increasingly we are seeing bloggers include visual aspects such as deliberately taken digital photos or video clips. These are usually included to illustrate some aspect of the blog such as "*Here is a picture of the place I visited*" or "*Here is a picture of my friend John*". Such inclusion of deliberately posed and deliberately taken photos/videos is distinct from passively taken photos/images which constantly record the wearer's activities, visually. This is somewhat similar to a personal CCTV system, worn by the wearer, for the wearer's own, exclusive use. We call this visual lifelogging. Despite its relative novelty, visual lifelogging is gaining much interest, due to projects such as the Microsoft SenseCam [1] and Reality Mining [2]. Such lifelog data can include text, visual information (video or photos), audio information, biometric data (heart rate, galvanic skin response, blood pressure, etc.), location data, co-presence information and more besides. This information can potentially allow users to revisit the events of their life.

In this paper we examine a collection of photos gathered by the constant wearing of a visual lifelogging device (SenseCam [3]) for a period of more that one year. We explore the composition of these photos and compare how photos from visual lifelogs differ from conventional personal digital photo collections. The motivation for this work stems from fact that much research is ongoing into content analysis of digital photos, while in the absence of large-scale visual lifelogging efforts, there has been little research undertaken for similar analysis of visual lifelogs. Since the SenseCam

H. Li et al. (Eds.): AIRS 2008, LNCS 4993, pp. 537–542, 2008.

captures around 3,000 photos during a typical day, the rapid rate at which enormous archives can be gathered means that the need for understanding the natures of the composition and automatic organization of these photo collections is compelling.

1.1 Visual Search and Retrieval of Photos

There has been much research recently into the organization of personal photo archives [4,5,6]. Some of this research aims to organize personal photos by exploiting the results of visual content analysis of the photos in order to extract some semantic meaning, with the goal of aiding the organization, search and retrieval of the photos. An example is facial analysis of photo content, to enumerate faces or to match faces across an individual's entire photo archive [7]. In addition, photo retrieval research has exploited the context of photo capture, or in some cases both visual content and photo context [5] to aid organisation. Often content analysis of photos requires the development of sets of concept detectors (e.g. face, crowd, building) which are trained on a representative set of photos [8]. In this paper we compare and contrast passively-captured large-scale visual lifelog data with more traditional intentionally-captured personal photo collections. We do this in order to explore the nature of a visual lifelog and to aid our understanding of the contents and composition of such a collection. Consequently we hope this understanding will allow us to leverage and deploy existing knowledge and techniques from the management and content analysis of personal photos in order to aid in automatic organization of visual lifelogs.

1.2 Visual Lifelogging

The device used in this research is a Microsoft SenseCam, which is a small wearable digital camera (worn around the neck) that is designed to take photos passively (without user intervention). Unlike a digital camera, the SenseCam has a fisheye lens, to maximize the field of view, and incorporates multiple sensors including; light sensors, a multi-axis accelerometer, a thermometer and a passive infra red sensor to detect the presence of a person. Used collectively these sensors can trigger the capture of a photo. If capture is not triggered by one of these sensors, by default the SenseCam will take a photo after 30 seconds. This means that a SenseCam will normally capture approximately 3,000 images per day, amounting to one million images per year. The potential benefits of using a SenseCam to generate a personal visual diary have been detailed by Hodges et al. [3] and include the maintenance of personal histories, security benefits, and healthcare benefits, both for healthcare professionals and patients. For this research a SenseCam was worn constantly over one year, amassing over a million images. The device was worn all day, from breakfast until sleep. Sample photos are shown in Figure 1 to illustrate the nature of the visual lifelog data captured in this experiment. Typically in visual lifelogging research, participants use a wearable camera only for short periods of time, to record single activities or significant events. Importantly and uniquely, our study is the first time that an individual has worn a SenseCam for such a prolonged period of time.

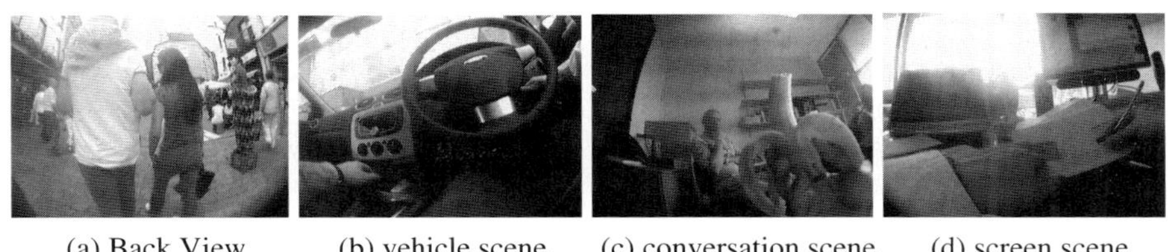

(a) Back View, (b) vehicle scene, (c) conversation scene, (d) screen scene

Fig. 1. Typical Visual Lifelog Photos

2 Analysis of a Large Visual Lifelog

The process of manually analysing all one million photos in the visual lifelog would be too time consuming, therefore one thousand photos were chosen at random for individual examination. The selected photos were examined for the presence of a range of concepts and compared to a conventional digital photo collection of over ten thousand photos, captured using standard digital cameras. The analysis presented here does not focus on particulars of the capture device such as image size or lens quality, rather emphasis has been placed on visual content which has been captured passively compared to that captured intentionally by conventional digital camera.

2.1 General Observations on Visual Lifelog Data

An observation of visual lifelog data captured using a wearable device such as a SenseCam, clearly illustrates the differences between such data and a conventionally gathered photo collection. First, there is the prominence of the wearer's hands and arms (51% of examined images contained visible hands and/or arms, as in Fig 1b, c & d). Second, in most cases, there is often no clear salient object (Fig 1b & d) in the photo (people/objects are often in the periphery), whereas with a conventionally posed photo, there is normally an identifiable salient object typically focused in the center of the image. In addition, the author's office/work environment amounted to 16% of photos (easily identified by the presence of a computer screen, see Fig 1d) while steering wheel photos (see Fig 1b) amounted to 15% of the total. Finally, it is important to note that as the lifelogging device was worn around the neck of the owner, and as a result, the photos noticeably appear to be captured from below eye level, often producing 'headless' shots of individuals and/or an unusual viewpoint. Since the user does not have to initiate capture we often see scenes where the user is actively using both hands and during conversations individuals are often captured mid-gesture adding to the conversational style and reinforcing the 'naturalness' of the photos. This contrasts heavily from conventional photo capture where photos are often staged and individuals and objects 'arranged' prior to capture.

2.2 Photo Quality Analysis

The quality of a photo is important for both end-user viewing and for content analysis algorithms, and for the analysis of visual lifelogs this is no exception. Our assumption is that a conventional digital photo, requiring an explicit user action to capture, will convey rich semantic meaning and will be of higher quality compared with the automatically captured photos from a wearable passive capture device which doesn't possess a flash or adjustable focus. Each of the randomly selected 1,000 photos were manually evaluated on a five-point scale which ranged from very low quality to excellent or 'photo album' quality. The results are summarized in Table 1.

Table 1. Visual quality of lifelog photos

Very Low	Poor	Reasonable	Good	Photo Album
14%	25%	35%	22%	4%

Very low and poor quality photos (40%) are likely to be of little use for people reviewing the happenings of a day and therefore could be automatically removed from the collection prior to organizing the day, into, for example a sequence of distinct events [9] for presentation to the user. In addition, these photos are less likely to be of benefit to content analysis tools and their removal could save processing resources when analysing content. Typical very low and poor quality photos are likely to be dark or blurred[1] photos where the content of the photo is not obvious by casually examining the photo. Of the remaining 60% of photos, the good photos (22%) are ideal candidates for presentation to users or for content analysis (e.g. all photos in Figure 1) in that they are not blurred and visually excellent. The photo album quality photos are visually excellent, but also convey semantic meaning, just like a traditionally posed photo from a photo album and we estimate that almost seventy such photos would be captured in a typical day wearing a Sensecam[2]. Finally, the 35% of photos that are of reasonable quality will visually acceptable, and one could consider that they would also be suitable for content analysis.

2.3 Comparison to Digital Photo Collections

In addition to understanding the quality of the visual lifelog photos, it is important to determine how similar lifelog photos are to personal photos. This is especially important if we plan to deploy existing semantic concept detection techniques to visual lifelogs. Table 2 presents findings from examining the 1,000 random lifelog photos and comparing these to 500 randomly chosen photos from the personal photo collection of the author (2,869 photos) and an additional 500 photos randomly chosen from the photo collections of ten colleagues (10,523 photos). As can be seen, both conventional, photo collections are equivalent (with a small amount of variation), as expected. However, the visual lifelog photos are very different in nature to the conventional photo collections. There is only one similar concept between the lifelog and the conventional photo collections (photos containing people). All other concepts are very different and clearly illustrate the differences between visual lifelogs and conventional photo collections. The typical concept detectors that could be applied to

[1] The SenseCam employs accelerometers sensors to trigger photo capture when the likelihood of blurring is minimised. Without such sensors, more blurred content would be captured.

[2] We have found that the notion of semantic meaning in lifelog photos is a very subjective concept and with inter-annotator reliability being poor.

personal photo collections such as building detectors, cityscape and landscape detectors are less likely to be useful for visual lifelog data. Rather we are presented with a whole new set of concepts that are important for visual lifelogs. Concepts such as computer screens, conversations, vehicle scenes and work scenarios are typical of common concepts found in visual lifelog collections, which if automatically identifiable, can be used to help organise a large visual lifelog.

Table 2. Comparison to Conventional Digital Photo Collections (values in percentages)

	Lifelog (1 million)	**Personal Photos** (2,869)	**Multi-User** (10,523)
People	29.9	*15.0*	*30.5*
Buildings	3.5	*43.3*	*35.0*
Indoor	73.4	*9.4*	*15.2*
Outdoor	5.1	*90.6*	*84.8*
Cityscape	2.1	*26.2*	*22*
Landscape	1.1	*23.2*	*23.8*
Computer Screen/TV	**26.7**	**< 1.0**	**0.0**
Conversation Scenario	**13.3**	**< 1.0**	**< 1.0**
In a Vehicle	**16.2**	**0.0**	**0.0**
Work Scenario	**24.6**	**< 1.0**	**0.0**
Back view of People	**6.8**	**1.0**	**2.0**

Rows in bold are concepts that we know to be very specific to visual lifelogs and as expected, are virtually non-existent in more traditional photo collections due to the fact that people would not often take conventional photos of work scenarios, driving scenarios or rear-views of people. Our findings suggest that the focus of semantic concept detectors for visual lifelog data will need to be tailored for the unique nature of visual life logs. In addition, the vast presence of almost redundant 'junk-content' such as computer screens, steering wheels, gives scope for filtering and summarizing the lifelogs automatically.

3 Conclusions and Future Work

We have illustrated that the photos contained within a large-scale visual lifelog collection are vastly different in visual content to conventional digital photos. It is important to note that the collection examined consists of continuously collected photos for a single user for a one year period, we have no reason to believe that this collection is not typical of what one would expect to find across many people's lifelogs and we anticipate that the major distinctions between traditional and lifelog photo collections highlighted in this work will hold true. We are, however, currently working to collect large scale visual lifelogs from a range of users to confirm these initial findings.

We have found that visual lifelog photos often don't have salient objects, many photos are either low quality or redundant, and the types of scenes/objects captured differ greatly from conventional photo collections. These findings will affect the focus of our future work in developing specific concept detectors appropriate to lifelogs (such as driving, eating, talking or working) and the results of this experiment will help us in the migration of pre-existing concept detection techniques from the domain of conventional photo collections to visual lifelogs. However, simply applying such concept detectors to visual lifelog data will not be sufficient to organize the enormous amounts of data involved. At Dublin City University we are researching new methods to help users organize these vast visual lifelog collections, to summarise and highlight, to filter, search and recommend from a content source that grows by up to 3,000 photos every day for a single user.

Acknowledgements. We acknowledge and thank Microsoft for their support of our lifelogging research. This work was also supported by the Irish Research Council for Science Engineering and Technology and by Science Foundation Ireland under grant 03/IN.3/I361.

References

1. Gemmell, J., Williams, L., Wood, K., Lueder, R., Bell, G.: Passive Capture and Ensuing Issues for a Personal Lifetime Store. In: CARPE 2004, New York, USA (October 2004)
2. Eagle, N., Pentland, A.: Reality Mining: Sensing Complex Social Systems. Personal and Ubiquitous Computing 10(4) (2006)
3. Hodges, S., Williams, L., Berry, E., Izadi, S., Srinivasan, J., Butler, A., Smyth, G., Kapur, N., Wood, K.: SenseCam: A Retrospective Memory Aid. In: UbiComp - 2006, CA (2006)
4. Rodden, K., Wood, K.: How do People Manage their Digital Photographs? In: SIGCHI-2003, April 5-10, 2003, Ft. Lauderdale, Florida, USA (2003)
5. O'Hare, N., Lee, H., Cooray, S., Gurrin, C., Jones, G., Malobabic, J., O'Connor, N., Smeaton, A.F., Uscilowski, B.: MediAssist: Using Content-Based Analysis and Context to Manage Personal Photo Collections. In: Sundaram, H., Naphade, M., Smith, J.R., Rui, Y. (eds.) CIVR 2006. LNCS, vol. 4071, Springer, Heidelberg (2006)
6. Graham, A., Garcia-Molina, H., Paepcke, A., Winograd, T.: Time as essence for photo browsing through personal digital libraries. In: JCDL 2002, Portland, Oregon, USA (2002)
7. Cooray, S., O'Connor, N.: A Hybrid Technique for Face Detection in Color Images. In: AVSS - 2005, Como, Italy, September 15-16 (2005)
8. Naphade, M., Smith, J.R.: Learning Visual Models of Semantic Concepts. In: IEEE International Conference on Image Processing, Barcelona (2003)
9. Doherty, A., Smeaton, A.F., Lee, K., Ellis, D.: Multimodal Segmentation of Lifelog Data. In: RIAO 2007, Pittsburgh, PA, USA, 30 May-1 June (2007)

Automatic Acquisition of Phrase Semantic Rule for Chinese

Xu-Ling Zheng, Chang-Le Zhou, Xiao-Dong Shi, Tang-Qiu Li, and Yi-Dong Chen

College of Information Science and Technology, Xiamen University,
Xiamen 361005, China
{xlzheng,dozero,mandel,tqli,ydchen}@xmu.edu.cn

Abstract. The semantic collocations play important roles in parsing Chinese phrases. They are useful for both semantic disambiguation and structural disambiguation. In this paper, a representation of phrase semantic rules for Chinese is presented to formulate such knowledge, and a corpus-based method was proposed to acquire phrase semantic rules from a Chinese phrase corpus annotated with syntactic and semantic information. The method includes a metarule-guided algorithm for mining cross-level association rules to acquire phrase semantic rules and an optimization algorithm to filter these rules. The experiment results showed the effectiveness of the method. Disambiguation performance of these resulting rules was quiet well.

Keywords: Semantic rules; corpus; association rules; HowNet.

1 Introduction

Phrase parsing is an important component of many natural language processing systems in applications such as machine translation, information retrieval, or text classification. The parsing quality impacts the performance of these systems. However, the existence of ambiguity is an enormous hindrance to phrase parsing. Morphology and grammar knowledge are insufficient to disambiguate, especially when parsing Chinese phrases for Chinese is a typical parataxis language. To improve parsing, semantic knowledge should be introduced.

Currently, there are various types of manually constructed lexical semantic knowledge bases, such as WordNet, FrameNet, and HowNet [1]. In contrast, efforts in constructing semantic collocation rule bases are less. For Chinese, Dong and Dong constructed the HowNet - Chinese Message Structure Base [2], Yu constructed the Phrase Structure Bank (PSB) of Contemporary Chinese [3], Zhan proposed a set of formulized rules on Chinese phrase structures [4]. All of these are done mainly by intuition-based methods. These rule bases are concentrated reflection of human expert's knowledge. They involve too many artificial factors. The coverage and granularity level of these rules are different, the precision of them are difficult to ensure.

This paper reports our study on acquiring phrase semantic rules from a Chinese phrase corpus annotated with syntactic and semantic information automatically.

H. Li et al. (Eds.): AIRS 2008, LNCS 4993, pp. 543–548, 2008.

2 Representation of Phrase Semantic Rules for Chinese

In this paper, phrase semantic rules for Chinese are proposed to formulate the syntactic and semantic knowledge about words having what kind of meaning can compound a phrase in what way and what type the resulting phrase is. According to the properties of the compounding rules of Chinese phrase and the Knowledge Dictionary Mark-up Language (KDML) of HowNet [1], the pattern of phrase semantic rules for Chinese is specified in the following format:

<Rule> ::= (<POS-Serial> <Syntax-Structure-Type> <Semantic-Relation-List>
<DEF-Pattern> {<DEF-Pattern>}$^+$)

where <POS-Serial> is a sequence of POS tags for all components of the phrase, <Syntax-Structure-Type> is a code of syntactic structure type of the phrase, <Semantic-Relation-List> shows the dependency relations between components, and <DEF-Pattern> is a pattern of concept definition (DEF) which describes the requirement of the sememes for each component. If a phrase consist of n components, then the corresponding rule should include n <DEF-Pattern>'s, called a n-gram rule.

To strengthen the description capacity, we introduce a prefix *HYP* and limited variables in KDML. A sememe with the prefix *HYP*, called a generalizer of sememe, denotes it can match any child of the given sememe in the sememe hierarchy of HowNet. It makes rules more recapitulative. And replacing sememes with limited variables makes it easier to describe the semantic relation between components.

3 Acquisition of Phrase Semantic Rules for Chinese

3.1 Problem Description

Regarding examples in the annotated corpus of Chinese phrases as transactions, and the various tags and their generalizers as items, the acquiring phrase semantic rules problem can be transformed into an association rules mining problem [5,6,7].

Each example in our corpus has three kinds of necessary tags: (1) a syntactic structure type tag (denoted β_1); (2) a semantic relation type tag (denoted β_2); and (3) the first sememe tags or its generalizer of each component in the phrase (denoted $\beta_3[j]$, where j is the order number of each component, $1\leq j\leq n$). So anyone of the acquired rules also should have these three kinds of tags. And the constraint can be formulated as the following metarule (denoted P) which is used to guide the mining process.

Let $I=\{i_1, i_2, \ldots, i_m\}$ be the universe of items, $D=\{T_1, T_2, \ldots, T_q\}$ be the universe of transactions. If T is a transaction in D, then $T\in D$ and $T\subset I$. The metarule P is a pattern of n-gram semantic rules of the form

$$A_1 \wedge A_2 \wedge \ldots \wedge A_n \Rightarrow B_1 \wedge B_2$$

where $B_1\in \beta_1$, $B_2\in \beta_2$, $A_j\in \beta_3[j]$, $j=1,2,\ldots,n$.

Then the problem of acquiring phrase semantic rules is a problem of mining cross-level strong association rules guided by the metarule P on the universe of items I and the universe of transactions D. And it is a two-step process: the first step is to find all frequent itemsets which comply with the metarule P; and the second step is to generate strong association rules from these frequent itemsets.

3.2 Finding Frequent Itemsets

Apriori is an influential algorithm for mining frequent itemsets for single-dimensional Boolean association rules. To acquire phrase semantic rules for Chinese, we proposed a variation of Apriori, named P_CLA, which is a metarule-guided algorithm for mining cross-level association rules.

Similar to Apriori algorithm, P_CLA employs an iterative approach known as a level-wise search. In the first iteration of P_CLA, the find_frequent_1-itemsets procedure finds the set of frequent 1-itemsets L_1. Since each itemset in L_1 complies with the metarule P, L_1 also can be denoted by L_1^P. In each subsequent iteration of P_CLA, the set of frequent k-itemsets complying with the metarule P (denoted L_k^P) are used to explore L_{k+1}^P: firstly, the gen_candidates procedure generates a candidate set of frequent (k+1)-itemsets complying with the metarule P (denoted C_{k+1}^P); secondly, the database is scanned to accumulate the support count of each candidate in C_{k+1}^P; finally, all those candidates satisfying minimum support form L_{k+1}^P. The set of frequent itemsets complying with the metarule P in transaction database D is a union of all L_k^P's. Furthermore, to improve the efficiency of P_CLA, we take measures to reduce the number of transactions scanned in future iterations.

The Find_frequent_1-itemsets Procedure

A top-down strategy is employed, where counts are accumulated for the calculation of frequent items at each concept level, starting at the highest level and working towards the lower, more specific concept levels, until no more frequent items can be found.

The Gen_candidates Procedure

The gen_candidates procedure performs two kinds of actions, namely join and prune, to generate C_{k+1}^P.

1. The join step

C_{k+1}^P is generated by joining L_k^P with itself. Let l_1 and l_2 be itemsets in L_k^P (assume items within a transaction or itemset are sorted in lexicographic order), l_1 and l_2 are joinable if their first k-1 items are in common and there are not "ancestor" or "congener" relationships between items. It ensures that not only no duplicates are generated, but also only the itemsets complying with the metarule P can be generated.

2. The prune step

C_{k+1}^P is a superset of L_{k+1}^P, which can be huge. To reduce the size of C_{k+1}^P, the Apriori property [5] is used. For P_CLA, the following two derived properties should be applied: (1) All non-empty subsets of a frequent itemset complying with the metarule P must also be frequent and comply with the metarule P; (2) Using uniform minimum support for all levels, all the ancestors of a frequent itemset must also be frequent, on the contrary, any child of a non-frequent itemset cannot be frequent.

The gen_candidates procedure prunes twice. Applying the first derived property, the early pruning is performed immediately after each candidate is generated in the join step. If any k-item subset of a candidate is not a member of L_k^P, the candidate should be pruned. All the pruned candidates are added to a set (denoted E_t). After the join step is completed, the late pruning is performed. If any ancestor of a candidate in C^P_{k+1} is a member of E_t, the candidate should be pruned. To reduce the duplicate judgment of ancestor relationship between itemsets, only the member of minimal cover of all known non-frequent k-itmesets is remaining in E_t.

Transaction Reduction
In P_CLA algorithm, three transaction reducing measures are taken: (1) After the set of frequent 1-itemsets is found, remove non-frequent items from each transaction in the database D, and eliminate transactions which do not contain any frequent 1-itemsets, we use D_1 to denote the reduced database; (2) Transactions which do not contain any frequent k-itemsets can be remove from D_k while D_k is scanned to determine L^P_{k+1}, the resulting database is D_{k+1}; (3) After L^P_{k+1} is found, if the set-out rate of candidates is greater than the threshold value, D_{k+1} will be reduced again.

3.3 Generating Candidate Rules

Under the constraint of metarule P, there is only one syntactic structure type item (denoted b_1, $b_1 \in \beta_1$) and one semantic relation type item (denoted b_2, $b_2 \in \beta_2$) in any frequent itemset c which is found by P_CLA. If the other items in c are denoted by a_1, $a_2, \ldots, a_k$, then the candidate rule r generates from c is "$a_1 \wedge a_2 \wedge \ldots \wedge a_k \Rightarrow b_1 \wedge b_2$". Through the transformation of rule form, the candidate phrase semantic rules are acquired from these strong candidate association rules.

3.4 Optimizing the Set of Phrase Semantic Rules

In the candidate phrase semantic rule set, some rules are redundant due to "ancestor" relationships between items and "superset" relationships between itemsets. If these rules are directly used in parsing Chinese phrases, it will not only increase the amount of computation, but also produce much interference and misleading.

Let GL be the set of candidate rules which are generated in section 3.3. We define a "cover" relation on GL in order to remove redundant rules [8,9] based on the approximate degree of support count and confidence between rules and the principle of the strongest restrictions.

Let p and q be rules in GL that satisfy both of the following conditions: (1) The confidence of p and q are approximately equal, that is, $|\text{confidence}(p) - \text{confidence}(q)| \leq \varepsilon$ (close to 0);(2) p is an ancestor rule (a rule p is an ancestor of a rule q, if p can be obtained by replacing the items in q by their ancestors in a concept hierarchy) or a subset rule (a rule p is an subset of a rule q, if their right hand side are same and the left hand side of p is a subset of the left hand side of q) of q, or there exists a child rule of p which is a subset rule of q. And if the support count of p and q are approximately equal too, that is, $\text{support}(q)/\text{support}(p) \geq \mu$, then we say q cover p, denoted by $p \prec q$; Otherwise, we say p cover q, denoted by $q \prec p$.

Since the "cover" relation on GL is transitive approximately, $(GL, \prec)$ can be considered as a poset. Illuminated by Hasse diagrams of poset, we propose the following method to draw the cover relation diagram of GL: (1) Each rule in GL is represented by a vertex; (2) Let $p, q \in GL$, if $p \prec q$, then draw an arc from vertex q to vertex p; if there exists $r \in GL$, and $p \prec r \prec q$, then remove the arc from vertex q to vertex p; (3) Remove the self-loop at each vertex. All the rules corresponding to vertexes with zero indegree in the diagram compose the optimal set of GL.

Therefore, the major idea of our semantic rule optimization algorithm is: first, draw the cover relation diagram of GL; then check weather the indegree of each vertex is zero in non-descending order of the length of candidate rules; finally, add rules corresponding to vertexes with zero indegree to the optimal set of GL. The resulting optimal set of GL is the acquired set of phrase semantic rules for Chinese.

4 Experiments and Results

In this research, we use a training corpus of Chinese phrases which consists of 8,516 phrases extracted semi-automatically from the People's Daily Corpus (PDC, 2000) and an electronic version of Reader Magazine (1995-2006). The corpus can be divided into "n+n", "adj+n", "m+n", "adj+adj", "m+adj", "v+n" and "n+v+n" eight subsets. All phrases are annotated with syntactic and semantic information, such as POS, semantic relation and etc. All word tokens are tagged by the HowNet definitions.

Employed the method proposed in this paper to the corpus, we found 3,266 candidate phrase semantic rules. After optimized, only 379 rules remained.

To verify the effectiveness of the proposed method, the following three different rule sets are employed respectively to parsing Chinese phrases: R_1, a rule set consists of rules which are adapted directly from message structures in HowNet - Chinese Message Structure Base; R_2, a rule set consists of rules generated by the proposed algorithm for mining candidate rules; R_3, a rule set acquired by the optimization algorithm. The whole training corpus is used in close test, and a testing corpus of 2,000 phrases (POS tagged only) extracted from newspaper texts is used in open test.

The results are listed in Table 1. Despite the recall performance of R_2 and R_3 are lower than R_1, the precision performance of them are improved remarkably. This is consistent with our assumption that a corpus-based method is more objective and comprehensive than an intuition-based one. By comparing the size of these three rule sets, R_3 is larger than R_1 by 164 rules, but R_3 improve the precision by about 24% relative, and the increase of time consumption is receivable. The recall and precision of R_2 and R_3 are similar, whereas R_2 is 8.6 times larger than R_3, and it is even worse that the close test of R_2 spends considerable time and space. These experiments show that the acquisition algorithm and the optimization algorithm for acquiring phrase semantic rules are effective and feasible.

Table 1. Results of different rule-sets

Rule set	Rule number	Close test		Open test	
		Recall (%)	Precision (%)	Recall (%)	Precision (%)
R_1	215	80.4	69.5	78.3	62.9
R_2	3266	74.9	84.1	73.7	79.2
R_3	379	73.2	86.1	71.4	77.5

5 Conclusions

The semantic collocations play important roles in parsing Chinese phrases, which are useful for both semantic and syntactic disambiguation. In this paper, a representation of phrase semantic rules for Chinese is presented, and a corpus-based method was proposed to acquire phrase semantic rules from a Chinese phrase corpus annotated with syntactic and semantic information. The method is shown to be very effective in both avoiding subjectivity in constructing the rule set and improving the precision performance of the rule set.

As future work, we plan to investigate a better way to reduce the workload of semantic tagging. Another interesting research direction is to apply phrase semantic rules to machine translation, information retrieval and other NLP applications.

Acknowledgments. This work was supported by the National Nature Science Foundation of China (60373080).

References

1. Dong, Z.D., Dong, Q.: HowNet, http://www.keenage.com/zhiwang/e_zhiwang.html
2. Dong, Z.D., Dong, Q.: Introduction to HowNet - Chinese Message Structure Base, http://www.keenage.com/html/aboutMessage.html
3. Yu, S.W.: The Specification of the Phrase Structure Bank of Contemporary Chinese. Journal of Chinese Language and Computing 13, 215–226 (2003)
4. Zhan, W.D.: A Study of Constructing Rules of Phrases in Contemporary Chinese for Chinese Information Processing. Tinghua University Press, Beijing (2000)
5. Han, J.W., Kamber, M.: Data Mining: Concepts and Techniques. Morgan Kaufmann Publishers, San Francisco (2001)
6. Han, J.W., Fu, Y.: Discovery of Multiple-Level Association Rules from Large Databases. In: Proceedings of 21st International Conference on Very Large Data Bases, pp. 420–431. Morgan Kaufmann Publishers, Zurich (1995)
7. Ouyang, W.M., Cai, Q.S.: Meta-pattern Guided Discovery of Multiple-Level Association Rule in Large Databases. Journal of Software 12, 920–927 (1997)
8. Aggarwal, C.C., Sun, Z., Yu, P.S.: Online Generation of Profile Association Rules. In: Proceedings of the 14th International Conference on Knowledge Discovery and Data Mining, pp. 129–133. AAAI Press, Orlando (1998)
9. Aggarwal, C.C., Yu, P.S.: A New Approach to Online Generation of Association Rules. IEEE Transactions on Knowledge and Data Engineering 13, 527–540 (2001)

Maximum Entropy Modeling with Feature Selection for Text Categorization

Jihong Cai and Fei Song

Department of Computing and Information Science
University of Guelph, Guelph, Ontario, Canada N1G 2W1
{jcai,fsong}@uoguelph.ca

Abstract. Maximum entropy provides a reasonable way of estimating probability distributions and has been widely used for a number of language processing tasks. In this paper, we explore the use of different feature selection methods for text categorization using maximum entropy modeling. We also propose a new feature selection method based on the difference between the relative document frequencies of a feature for both relevant and irrelevant classes. Our experiments on the Reuters RCV1 data set show that our own feature selection performs better than the other feature selection methods and maximum entropy modeling is a competitive method for text categorization.

Keywords: Text Categorization, Feature Selection, Maximum Entropy Modeling.

1 Introduction

Text categorization is the process of assigning predefined categories to textual documents. One of its many useful applications is for web content filtering, since only when we know the categories of the web pages can we decide whether they are offensive/inappropriate in order to block them from user access. Text categorization has been conducted with different machine learning techniques, including Naïve Bayes [4], k-Nearest Neighbors [3], Linear Least Squares Fit [8], Support Vector Machines [1], and Maximum Entropy Modeling [5].

This paper explores the use of different feature selection methods for text categorization in the context of maximum entropy modeling. In addition to some commonly-used feature selection methods, we propose a new feature selection method, called Count Difference, which is based on the difference between the relative document frequencies of a feature for both relevant and irrelevant classes. As Nigam et al. [5] points out, maximum entropy modeling may be sensitive to poor feature selection, and we want to see how these feature selection methods can affect the performance of a text classifier based on maximum entropy modeling and how is maximum entropy modeling is compared with other popular text categorization methods.

2 Feature Selection Methods

Most text categorization systems use word types and their frequency counts for document representation. We refer to these word types as features. Feature selection

H. Li et al. (Eds.): AIRS 2008, LNCS 4993, pp. 549–554, 2008.

not only reduces computational cost in space and time, but also improves performance by carefully selecting good features for classification [9].

2.1 Existing Feature Selection Methods

Document frequency stands for the number of documents in which a feature occurs in a document collection. This method favors features whose document frequencies fall into a mid-range, since low-frequency features do not contribute much to the distinction of most documents and high-frequency features are so common that they reduce the distinction between documents.

χ^2 ranking [3] favors features that are strongly dependent on relevant or irrelevant classes. One problem with this method is that it may give a high score to a rare feature. For example, a feature may only appear in 5 documents in a collection of 100,000 documents, but if all these 5 documents belong to the relevant class, the feature may still get a high score, which is counter-intuitive.

Likelihood ratio attempts to address the issue of assigning high scores to rare features in χ^2 ranking [3]. For a large sample size, it tends to behave similarly to χ^2 ranking, but it also works well for a small sample size.

Mutual Information only measures the dependency between a feature and its relevant class, and as a result, it tends to favor rare terms if they are mostly used for relevant documents [9].

Information Gain is a measure based on entropy, which has been successfully applied to the construction of an optimal decision tree [4]. Features that reduce the entropy the most are favored for this method.

Orthogonal Centroid chooses features by an objective function based on transformations on centroids. To overcome the time and space complexity of the original orthogonal centroid algorithm, an optimal orthogonal centroid algorithm was proposed to provide a simple solution for feature selection [7].

Term Discrimination tries to measure the ability of a feature for distinguishing one document from the others in a collection [6]. A very popular feature often has a negative discrimination value, since it tends to reduce the differences between documents, while a rare feature usually has a close-to-zero value, since it is not significant enough to affect the space density.

2.2 Count Difference

A feature whose document frequency for one class is higher than that for the other class is desirable since it helps distinguishing between the two classes. However, if the feature is rare in the training documents, its use will be limited since it only affects a small number of documents. This leads us to propose a new feature selection method called Count Difference (CD), which tries to reflect the above two factors in ranking features.

Given a feature, we can partition the set of training documents into four regions in the following contingency table.

Table 1. Feature-Class Contingency Table

	Relevant	Irrelevant
Feature Used	a	b
Feature Not Used	c	d

We first introduce the notation of relative document frequency, which is the ratio of the document frequency of a feature for one class over the average document frequency for the same class:

$$relativeDF(t, y) = a_t / \overline{a} \quad \text{and} \quad relativeDF(t, \tilde{y}) = b_t / \overline{b}$$

Here, $\overline{a}$ and $\overline{b}$ denote the average document frequencies for the relevant and irrelevant classes, which are computed as follows:

$$\overline{a} = \frac{1}{M}\sum_{t=1}^{M} a_t \quad \text{and} \quad \overline{b} = \frac{1}{M}\sum_{t=1}^{M} b_t$$

where M is the number of original features before the selection process.

With the relative document frequencies, we can then define the count difference score of a feature as the difference between its two relative document frequencies:

$$CD(t) = (a_t / \overline{a} - b_t / \overline{b})^2$$

Intuitively, the relative document frequency measures the importance of a feature against the average feature for one class. If a feature is rare, its relative document frequency will be low, whereas if a feature is popular, its relative document frequency will be high. The count difference tends to favor features whose relative document frequencies for one class are higher than those for the other class. If a feature is popular for both classes, its count difference score will be reduced.

3 Maximum Entropy Modeling

Maximum entropy provides a reasonable way of estimating probability distributions from training data. The key principle is that we should agree with everything that is known, but carefully avoid assuming anything that is unknown. In particular, when nothing is known about certain features, the distribution for them should be as uniform as possible (thus the maximum entropy).

3.1 Feature Functions

Following Nigam et al. [5], we represent features as feature functions. For text categorization, we can define a feature function for each word-class combination:

$$f_{w,c}(d, y) = \begin{cases} 0 & \text{if } y \neq c \\ N(d, w) / N(d) & \text{otherwise} \end{cases}$$

Here, N(d, c) is the number of times word w occurring in document d, and N(d) is the number of words in document d.

3.2 Log-Linear Models

A maximum entropy model generally takes the following parametric form:

$$p(x,c) = \frac{1}{Z}\prod_{i=1}^{K}\alpha_i^{f_i(x,c)}$$

where K is the number of feature functions, α_i is a weight for feature function f_i, and Z is normalization constant. The model is also called the log-linear model, since by taking the logarithm on both sides, we get a linear combination for the feature functions.

The log-linear model above allows overlapping/dependent features. Although we use individual words as features for text categorization, we could easily extend the set with word pairs, longer phrases, and even non-text features (such as links in web pages). Even for individual words, we do not require them to be independent as the case for Naïve Bayes method.

In addition, the log-linear model provides further differentiations among features by carefully assigning weights α_i to different feature functions. In particular, by setting $\alpha_i = 1$, we essentially eliminate that feature in the combination process.

4 Experimental Results

Reuters RCV1 data set is a benchmark collection for evaluating text categorization systems [2]. It has over 800,000 news articles collected over one year's time. We focus on the topic scheme, which is a hierarchy of 103 categories, with the top four categories being CCAT (Corporate/Industrial), ECAT (Economics), GCAT (Government/Social), and MCAT (Markets). We use $F_{1.0}$ (harmonic mean of recall and precision) to measure the classification performance. For multiple categories, we use both macro-average (per-class average) and micro-average (per-document average).

4.1 Classification Performance

There are several factors that can affect the classification results. For a hierarchical scheme, there is a choice of flat classification (testing all categories independently) and hierarchical classification (only the documents belonging to a parent category are tested further for its sub-categories). We choose the hierarchical classification, since it is naturally suited for a hierarchical scheme.

We use the Improved Iterative Scaling algorithm [5], which iteratively updates the weighting parameters until they are converged or a certain number of iterations are reached. To avoid over fitting to the training data, we can terminate the process after a certain number of iterations or monitor the performance with a validation data set and stop the training when the performance starts to decrease. For simplicity, we terminate the training after 50 iterations.

Table 2 shows the classification results based on the above setting, where the values are the micro-averages of $F_{1.0}$ measures. First, we see that the classification performance goes up for all selection methods as we increase the number of features but only to a certain degree. After that, the performance starts to decrease but very slowly. This indicates the need for feature selection, since it not only reduces the computational cost but also improves the performance.

Table 2. Effects of Feature Selection on Text Categorization

	DF	χ^2	CD	LR	MI	IG	OC	TD
100	.601	.267	.676	.292	.061	.448	.628	.599
500	.741	.553	.769	.703	.067	.736	.764	.728
1k	.774	.717	.795	.745	.094	.757	.791	.748
1.5k	.789	.755	.793	.752	.112	.762	.789	.748
2k	.790	.760	.791	.754	.135	.764	.788	.746
4k	.786	.763	.780	.756	.223	.763	.778	.740
8k	.776	.758	.768	.749	.514	.757	.766	.728

Secondly, the rates of performance increase are different for different selection methods. We can roughly divide the selection methods into several groups. The best group contains Document Frequency, Count Difference, and Optimal Orthogonal Centroid. They start off with relatively high performance values even with just 100 features, and reach the highest performance values between 1000 and 2000 features. Our own feature selection method, Count Difference, gets the best performance of 0.795 with 1000 features. The group with the highest overlaps of featuers, including 춘춘. Ranking, Likelihood Ratio, and Information Gain, start with low performance values and improve slowly to reach their peaks. Note that Mutual Information (used in Nigam et al. [5]) has the worst performance since it tends to favor rare features.

Finally, we see that the performance between different selection methods become close as we reach 1000 features and above (except for Mutual Information). This leads us to conclude that maximum entropy modeling is not too sensitive to feature selection as long as most important features are included in the pool of features, since the weights for log-linear combination provide further differentiations between the features used for text categorization.

4.2 Comparison with Other Methods

Reuters RCV1 data set provides benchmark results for several text categorization methods, including Support Vector Machines, K-Nearest Neighbors, and Rocchio-style classifiers.

Table 3. Comparison with Other Methods

	SVMs	K-NN	Rocchio	MaxEnt
Macro-avg	.619	.560	.504	.472
Micro-avg	.816	.765	.693	.794

Table 3 summarizes the classification results for the entire topic scheme in both macro- and micro-averages. We see that MaxEnt (maximum entropy modeling) is a competitive method for text categorization: its performance is better than that for K-Nearest Neighbors and Rocchio-style classifiers in terms of micro-averages, although the performance is still not as good as that for Support Vector Machines. In terms of macro-averages, MaxEnt is the lowest among the four methods. This leads us to conclude that MaxEnt tends to perform better when a category is adequately covered by training documents. Otherwise, the performance will be affected considerably. This is perhaps not surprising since the feature functions are essentially defined in terms of maximum likelihood estimate (N(d,w)/N(d)). Future work remains to smooth this probability with data from parent categories.

5 Conclusions

We compared eight different feature selection methods for text categorization using maximum entropy modeling. We showed that feature selection is an effective way of reducing the computational cost while at the same time improving the classification performance. We demonstrated that our own feature selection method, Count Difference, is promising for text categorization, not only achieving the best performance but also working reasonably well for very aggressive feature cutoffs. We further illustrated that maximum entropy modeling is a competitive method for text categorization and has potential for further improvements.

References

1. Joachims, T.: Text categorization with Support Vector Machines: Learning with many relevant features. In: Tenth European Conference on Machine Learning, pp. 137–142 (1998)
2. Lewis, D.D., Yang, Y., Rose, T.G., Li, F.: RCV1: A new benchmark collection for text categorization research. Journal of Machine Learning Research 5, 361–397 (2004)
3. Manning, C., Schütze, H.: Foundations of Statistical Natural Language Processing. The MIT Press, Cambridge (1999)
4. Mitchell, T.: Machine Learning. The McGraw-Hill Companies, Inc., New York (1997)
5. Nigam, K., Lafferty, J., McCallum, A.: Using maximum entropy for text classification. In: IJCAI 1999 Workshop on Machine Learning for Information Filtering, pp. 61–67 (1999)
6. Salton, G.: Automatic Text Processing: The Transformation, Analysis, and Retrieval of Information by Computer. Addison-Wesley, Reading (1989)
7. Yan, J., Liu, N., Zhang, B., Yan, S., Chen, Z., Cheng, Q., Fan, W., Ma, W.-Y.: OCFS: Optimal Orthogonal Centroid Feature Selection for text categorization. In: 28th Annual International ACM SIGIR Conference on Research and Development in Information Retrieval, pp. 122–129 (2005)
8. Yang, Y., Chute, C.G.: An example-based mapping method for text categorization and retrieval. ACM Transactions on Information Systems 12(3), 252–277 (1994)
9. Yang, Y., Pedersen, J.O.: A comparative study of feature selection in text categorization. In: Fisher, J.D.H. (ed.) The Fourteenth International Conference on Machine Learning, pp. 412–420 (1997)

Active Learning for Online Spam Filtering

Wuying Liu and Ting Wang

School of Computer, National University of Defense Technology
No.137, Yanwachi Street, Changsha, Hunan 410073, P.R. China
{wyliu,tingwang}@nudt.edu.cn

Abstract. Spam filtering is defined as a task trying to label emails with spam or ham in an online situation. The online feature requires the spam filter has a strong timely generalization and has a high processing speed. Machine learning can be employed to fulfill the two requirements. In this paper, we propose a SVMEL (SVM Ensemble Learning) method to combine five simple filters for higher accuracy and an active learning method to choose training emails for less training time. The experiments results show the filter applying active learning method can reduce requirements of labeled training emails and reach steady-state performance more quickly.

Keywords: Spam Filtering, Machine Learning, Active Learning, Ensemble Learning, SVM.

1 Introduction

Spam filtering can be considered as a two-class classification problem in an online situation. The online feature manifests in two aspects. One is the behavior of emails changes continuously along the time which requires the spam filter has a strong timely generalization. Machine learning can help the filter to refine constantly according to user's response. The other is the online application must have a high processing speed which requires the spam filter to reduce training time. So, choosing appropriate training emails is the most essential task.

In many machine learning tasks, gathering training data is time-consuming and costly; thus, finding ways to minimize the number of training data is beneficial. To reach steady-state performance more quickly, active learning can be employed. The active learner chooses some training emails from a large unlabeled emails set, asks user for their labels and trains the filter with those labeled ones.

There are three main approaches in the spam filtering -- the IP protocol based one, the SMTP protocol based one and the content based one. The previous research shows that each filtering method has its own advantage but can only process certain class problem well. We add active learning to a SVMEL spam filter for a higher learning speed and less requirements of labeled training emails.

2 Ensemble Learning

Applying ensemble learning we can divide a complex spam filtering problem into several simple aspects. To divide and conquer, the advantage of ensemble learning

H. Li et al. (Eds.): AIRS 2008, LNCS 4993, pp. 555–560, 2008.

can reduce complexity efficiently in analyzing and computing each aspect. Our previous research shows the SVMEL filter has the highest performance among several ensemble learning filters [1].

Features of spam filtering include linguistic ones of email text and behavior ones of filtering problem. In this paper, we use the linguistic ones (*body* full word, *subject* full word, *body* keyword) and behavior ones (sending and receiving feature, *subject Re:* feature) to build five simple filters. In order to gain better performance than the simple filter could do, the SVMEL filter makes final decision by combining multi-result of several simple ones. Using kernel function mapping SVM is fit for the non-linear numerical classification problem [2]. Also each simple filter can generate a numerical spam confidence score (SCS) which is a real number between 0 and 1.

The SVM ensemble method is showed below. Let the object space of the SVM classifier is {spam, ham}, each training vector is (C_1, C_2, …, C_n) which C_i denotes the SCS output by the *i*th simple filter. In the learning process a SVM model can be gotten from the training vectors while in the filtering process the SVM classifier can classify the email with the SVM model, and output a SCS.

It is time-consuming of training a SVM model. Our experiments also show that adding a few training emails can not improve SVM performance obviously after 4,000 training emails added. In order to reduce the training time we propose a memory window which denotes the number of distinct training emails. In our experiments the memory window is 4,000, that is to say we only remember last 4,000 distinct training emails and forget earlier ones. This improvement reduces the training time and makes the SVM model has a strong timely generalization with less performance drop.

3 Active Learning

For a higher learning speed and less requirements of labeled training emails we add active learning to above SVMEL spam filter. Active learning is a machine learning method which can choose training samples dynamically [3]. Using current knowledge active learner does not receive training samples passively but actively select the samples which can train a more optimal model.

3.1 Filtering Architecture

Figure 1 shows a spam filtering architecture with active learning, in which each pair simple filter and learner is an independent adapted learning system. The ensemble filter combines the SCS generated by those simple ones to a new SCS, with that labels the email with spam or ham by comparing the new SCS against the threshold.

According to the new ensemble SCS of an email active learner makes a decision. If the learner thinks the email is a more informative sample and can improve training performance then it actively queries user for a response and sends the response to each simple learners and ensemble one. After learning those learners online refine their knowledge.

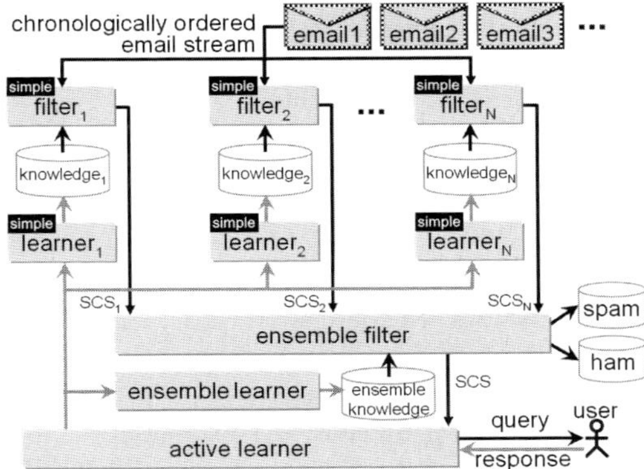

Fig. 1. Spam filtering architecture with active learning

3.2 Choosing Training Email

It is a key point for active learning how to choose more informative training email. There are already some methods such as uncertainty-based sampling (UBS) method [4] [5], committee method [6], version space and margin-based method [7], statistical method [8] and so on.

UBS method selects those samples to train which are easily wrong filtered. The reason is the more uncertain sample can train the more improvement. This method must have a pre-filter to estimate the likelihood of wrong filtering. This likelihood is a key factor which indicates how much possibility the email will be wrong filtered.

Spam filter is refined in an online situation whose performance is bad early in the filter's deployment. But the pre-filter may not be more tolerant of errors that are made early. In our experiments, we can build an early better pre-filter by training a model from other previous emails. Or we can use SVMEL method to train a better model with some early user's responses and with that we apply this model in UBS method. Then the current version of filter can be used as a pre-filter to estimate the likelihood of wrong filtering. Using UBS method we choose those emails whose ensemble SCS is about 0.5.

3.3 Cache Improvement

For bulk feature of spam we can improve SCS more by comparing new email content with memorized labeled email one. But we can not memorize all labeled emails because content comparison one by one is time-consuming. We find a principle that the same spam often appears in some period of time. So we can use a cache to memorize the last ones whose size is the previous memory window.

For a higher content comparison speed this paper compromises to model email content with SCS vector like section 3.2 introduced. Each item in cache includes a SCS vector and a user response label. If output SCS vector generated by simple filters is equal to one vector in cache then the SCS is 1 or 0 according to cached spam label or ham one.

The improved performance of cache is dependent with the quantity of repeated emails. If there are a lot of repeated emails then the cache improvement will be obvious. Contrarily there will be less improvement.

4 Experiments and Discussion

Firstly we implement five above mentioned simple filters and then build a SVMEL spam filter as a baseline system. Secondly we add active learning and cache improvement to the baseline. At last we run some experiments to compare those filters in TREC07p corpus and TREC07 active learning task [9] [10]. The TREC07p corpus contains 75,419 messages: 25,220 hams and 50,199 spams.

4.1 Experiments

We do 2 groups experiments to evaluate the proposed methods. Table 1 shows the parameters details of those experiments. The *allowance* indicates the allowed query numbers and the *consume* indicates the actually used query numbers. The *active* is *No* means the run uses baseline system. The *Del* and *Par* runs are baseline system run in first 10,000 feedbacks and random 30,388 feedbacks.

Table 1. Parameters details of experiments

group	test run	active	cache	allowance/consume
	Del	No	No	10000/10000
1	Act10000	Yes	No	10000/10000
	ActCac10000	Yes	Yes	10000/10000
	Par	No	No	30388/30388
2	Act30388	Yes	No	30388/30388
	ActCac30388	Yes	Yes	30388/30388

Group 1 and 2 test filtering performance of there filtering methods in 10,000 allowances and 30,388 allowances. After that we do some performance comparison tests in 10,000 and full allowances between our best system (active learning and cache SVMEL filter) and Bogo-0.93.4 system[1] which had high performance in TREC06 spam track.

4.2 Results

Figure 2(a) and 2(b) show the active filter is better than baseline system in performance and the cache technique can improve more from the active filter. The UBS method is fit for spam filtering since it can choose informative emails and cache improvement is useful since bulk feature of spam. Figure 2(a) and 2(b) also show before 30,000 messages the cache improvement is obvious and after 30,000 messages the cache improvement is unobvious because less repeated emails after 30,000 messages.

[1] http://bogofilter.sourceforge.net/

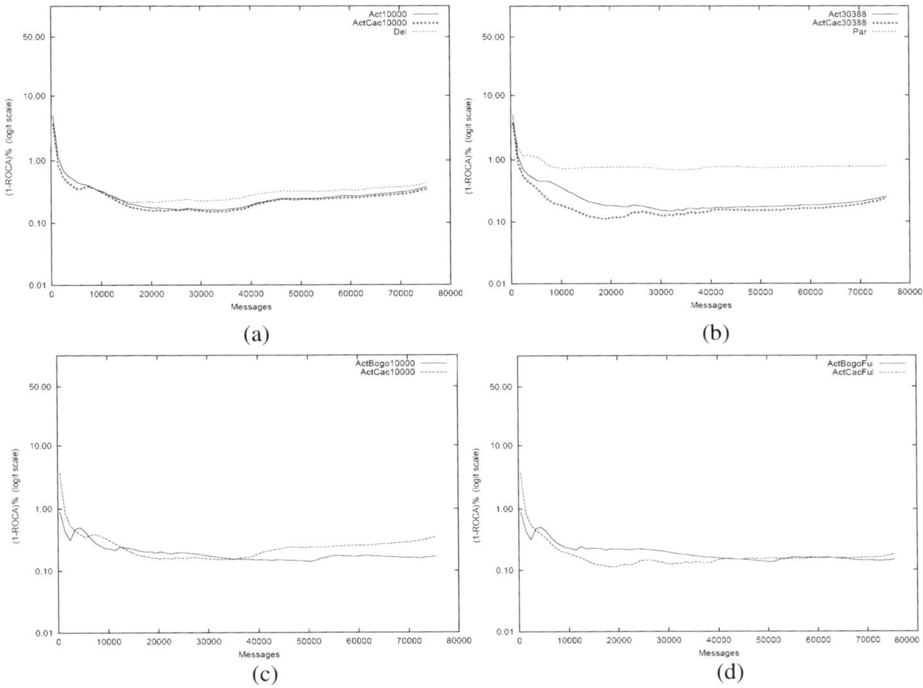

Fig. 2. Curves of experiments results

Figure 2(c) and 2(d) show the whole performance of our best filter is better than that of Bogo-0.93.4. Moreover the learning speed of *ActCac10000* and *ActCacFul* is faster than Bogo-0.93.4. Especially in [0, 45000] zone, our *ActCacFul* run reaches steady-state performance more quickly. In addition, it takes our filter half of the wall clock time that Bogo-0.93.4 costs to filter.

Our experiments results show active learning is useful to decrease computation complexity, save training time, resist useless samples and increase filtering accuracy. Active learning strategy allows the learner to dynamically select more informative samples from the candidate training set to compose the training set and remove noise samples. One side the size of the training set is smaller then before when they reach the same performance, so the training time can be decreased. On the other side that even can increase filtering accuracies. Furthermore the cache improvement is also a useful one.

5 Conclusions

This paper studied the online feature of the spam filtering, analyzed various features of email and built a baseline spam filter applying SVM ensemble learning. In order to achieve a strong timely generalization and a high filtering speed, we apply active learning for spam filtering, which included choosing training email actively and cache improvement technology. Our experiment result shows the filter applying active learning can reduce requirements of labeled training emails and reach steady-state performance more quickly. Furthermore, the cache improvement can reach higher performance.

In the future, the research will focus on the improvement of active learning. We hope to apply hierarchical active learning, not only in ensemble filter level but also in each simple filter one. Moreover, the emails behavior generated from the body text semantic mining by natural language processing is also very interesting.

Acknowledgements

This research is supported by the National Natural Science Foundation of China (60403050), Program for New Century Excellent Talents in University (NCET-06-0926) and the National Grand Fundamental Research Program of China under Grant (2005CB321802). Many thanks to Prof. Gordon V. Cormack for his evaluation system.

References

1. Liu, W., Wang, T.: An Ensemble Learning Method of Multi-filter for Spam Filtering. In: The Third National Conference on Information Retrieval and Content Security (NCIRCS 2007), Suzhou, China (2007)
2. Cristianini, N., Shawe-Taylor, J.: An Introduction to Support Vector Machines and Other Kernel-based Learning Methods, pp. 26–29. Cambridge Univ. Press, Cambridge (2000)
3. Engelbrecht, A.P., et al.: Incremental Learning Using Sensitivity Analysis. In: IJCNN 1999, International. Joint Conference on Neural Networks, pp. 1350–1355 (1999)
4. David, D., Lewis, W.A.: A Sequential Algorithm for Training Text Classifiers. In: Proceedings of the 17the Annual ACM-SIGIR Conference, London U.K, pp. 3–12 (1994)
5. David, D., Lewis, C.J.: Heterogeneous Uncertainty Sampling for Supervised Learning. In: Proceedings of the Eleventh International Conference on Machine Learning, New Brunswick, NJ, pp. 48–156 (1994)
6. Seung, H.S., Opper, M., Somepolinsky, H.: Query by Committee. In: Proc.5th Annu. Workshop on Computation Learning Theory, pp. 287–294 (1992)
7. Tong, S., Koller, D.: Support Vector Machine Active Learning with Applications to Text Classification. Journal Machine Learning Research, 999–1006 (2001)
8. Cohn, D.A., Ghahramani, Z., Jordan, M.I.: Active Learning with Statistical Models. Artificial Intelligence Research, 129–145 (1996)
9. Cormack, G., Lynam, T.: TREC 2005 Spam Track Overview. University of Waterloo, Waterloo, Ontario, Canada (2005)
10. Cormack, G.: TREC 2006 Spam Track Overview. University of Waterloo, Waterloo, Ontario, Canada (2006)

Syntactic Parsing with Hierarchical Modeling

Junhui Li, Guodong Zhou, Qiaoming Zhu, and Peide Qian

Jiangsu Provincial Key Lab of Computer Information Processing Technology
School of Computer Science & Technology, Soochow University, China 215006
{lijunhui,gdzhou,qmzhu,pdqian}@suda.edu.cn

Abstract. This paper proposes a hierarchical model to parse both English and Chinese sentences. This is done by iteratively constructing simple constituents first, so that complex ones could be detected reliably with richer contextual information in the following processes. Evaluation on the Penn WSJ Treebank and the Penn Chinese Treebank using maximum entropy models shows that our method can achieve a good performance with more flexibility for future improvement.

Keywords: syntactic parsing, hierarchical modeling, POSTagging.

1 Introduction

Syntactic parser takes a sentence as input and returns a syntactic parse tree that reflects structural information about the sentence. However, with ambiguity as the central problem, even a relatively short sentence can map to a considerable number of grammatical parse trees. Therefore, given a sentence, there are two critical issues in syntactic parsing: how to represent and score a parse tree.

In the literature, several approaches have been proposed in parsing by representing a parse tree as a sequence of decisions with different motivations. Among them, (lexicalized) PCFG-based parsers usually represent a parse tree as a sequence of explicit context-free productions (grammatical rules) and multiply their probabilities as its score (Charniak 1997; Collins 1999). Alternatively, some other parsers represent a parse tree as a sequence of implicit structural decisions instead of explicit grammatical rules. (Magerman et al. 1995) maps a parse tree into a unique sequence of actions and applies decision trees to predict next action according to existing actions. (Ratnaparkhi 1999) further applies maximum entropy models to better predict next action according to existing actions.

In this paper, we explore the above two issues with a hierarchical parsing strategy by constructing a parse tree level by level. This can be done as follows: given a forest of trees, we recursively recognize simple constituents first and then form a new forest with a less number of trees until there is only one tree in the newly produced forest.

2 Hierarchical Parsing

Similar to (Ratnaparkhi 1999), our parser is divided into three consequent modules: POS tagging, chunking and structural parsing. One major reason is that

H. Li et al. (Eds.): AIRS 2008, LNCS 4993, pp. 561–566, 2008.

previous modules can decrease the search space significantly by providing n-best results only. Another reason is that POS tagging and chunking have been well solved in the literature and we can concentrate on structural parsing by incorporating the start-of-the-art POS taggers and chunkers. In the following, we will concentrate on structural parsing only.

Let's first look into more details at structural parsing in (Ratnaparkhi 1999). It introduces two procedures (*BUILD* and *CHECK*) for structural parsing, where *BUILD* decides whether a tree starts a new constituent or joins the incomplete constituent immediately to its left and *CHECK* finds the most recently proposed constituent and decides if it is complete, and alternates between them. In order to achieve the correct parse tree in Fig.1, the first two decisions on *NP(IBM)* must be *B-S* and *NO*. However, as the other children of *S* have not constructed yet at that moment, there lacks reliable contextual information on the right of *NP(IBM)* to make correct decision. One solution to this problem is to delay the *B-S* decision on *NP(IBM)* until its right brother *VP(bought Lotus for $200 million)* has already constructed.

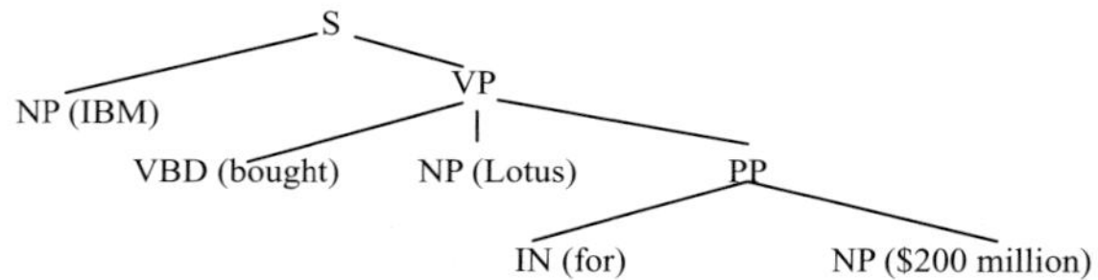

Fig. 1. The parse tree for *IBM bought Lotus for $200 million*

Motivated by above observation, this paper proposes a hierarchical parsing strategy by constructing a parse tree level by level. The idea behind the hierarchical parsing strategy is to parse easy constituents first and then leave those complex ones until more information is ready.

Table 1. BIESO tags used in our hierarchical parsing strategy

Tag	Description	Tag	Description
B-X	start a new constituent X	I-X	joint the previous one
E-X	end the previous one	S-X	form a new constituent X alone
O	hold the same		

Table 1 shows various tags in the hierarchical parsing strategy. In each pass, starting from left, the parser assigns each tree in a forest with a tag. Consequent trees with tags *B-X, I-X, .., E-X* from left to right would be merged into a new constituent *X*. Especially, *S-X* indicates to form a constituent *X* alone. The newly formed forest usually has less number of trees and the process will repeat until there is only one tree in the new forest. Moreover, maximum entropy models are used for predicting probability distribution and Table 2 shows the contextual information employed in our model.

Table 2. Templates for making predicates & Predicates used for prediction

Template	Description
cons(n)	Combination of the headword, constituent (or POS) label and action annotation of the n-th tree. Action annotation omitted if $n \geq 0$
cons(n*)	Combination of the headword's POS, constituent (or POS) label and action annotation of the n-th tree. Action annotation omitted if $n \geq 0$
cons(n**)	Combination of the constituent (or POS) label, and action annotation of the n-th tree. Action annotation omitted if $n \geq 0$
Type	**Templates used**
1-gram	cons(n), cons(n*), cons(n**), $-2 \leq n \leq 3$
2-gram	cons(m, n), cons(m*, n), cons(m, n*), cons(m*, n*), cons(m**, n), cons(m**, n*), cons(m*, n**), cons(m, n**), cons(m**, n**), (m,n)=(-1, 0) or (0, 1)
3-gram	cons(0, m, n), cons(0, m*, n*), cons(0, m*, n), cons(0, m, n*), cons(0*, m*, n*), (m, n)= (1, 2), (-2, -1) or (-1, 1), and cons(1, 2, 3), cons(1*, 2*, 3*), cons(1**, 2**, 3**), cons(2*, 3*, 4*), cons(2**, 3**, 4**)
4-gram	cons(0, 1, 2, 3), cons(0, 1*, 2*, 3*), cons(0*, 1*, 2*, 3*), cons(1*, 2*, 3*, 4*), cons(1**, 2**, 3**, 4**)
5-gram	cons(0*, 1*, 2*, 3*, 4*), cons(0**, 1**, 2**, 3**, 4**)

The decoding algorithm attempts to find the best parse tree T^* with highest score. The breadth-first search (BFS) algorithm introduced in (Ratnaparkhi 1999) with a compuation complexity of $O(n)$ is revised to seek possible sequences of tags for a forest. In addition, heaps are used to store intermediate forests in the evolvement. The BFS-based hierarchical parsing algorithm has a computational complexity of $O(n^2N^2M)$, where n is the number of words, N is the size of a heap and M is the number of actions.

3 Experiments and Results

In order to test the performance of this hierarchical model proposed in this paper, we conduct experiments both on Penn WSJ Treebank (PTB) and Penn Chinese Treebank (CTB).

3.1 Parsing Penn WSJ Treebank

In this section, all the evaluations are done on English WSJ Penn Treebank. Here, Sections 02-21 are used as the training data for POS tagging and chunking while Section 02-05 are used as the training data for structural parsing. Meanwhile, Section 23 (2,416 sentences) is held-out as the test data. All the experiments are evaluated using measures of LR(Labeled recall), LP(Labeled precision) and F1. And POS tags are not included in the evaluation.

Table 3 compares the effect of different window sizes. It shows that, while the window size of 5 is normally used in the literature, extending the window size to 7 (from -2 to 4) can largely improve the performance.

Table 3. Performance of hierarchical parsing on Section23. (Note: Evaluations below collapse the distinction between labels ADVP and PRT, and ignore all punctuation)

windows size	#events	#predicates	LR	LP	F1
5	471,137	229,432	82.01	83.21	82.61
6	520,566	302,410	84.48	85.79	85.13
7	559,472	377,332	85.21	86.59	85.89

One advantage of hierarchical parsing is its flexibility in parsing a fragment with higher priority. That's to say, it is practicable to parse easy (or special) parts of a sentence in advance, and then the remaining of the sentence. The problem is how to determine those parts with high priority, such as appositive and relative clauses. Here, we define some simple rules (such as finding *(LRB, RRB)* pairs or *"–"* symbols in a sentence) to figure out the fragments with high priority. As a result, 163 sentences with appositive structure are found with the above rules. The experiment shows that it can improve the F1 by 1.53 (from 77.42 to 78.59) on those sentences, which results in performance improvement from 85.89 to 86.02 in F1 on the whole Section 23.

3.2 Parsing Penn Chinese Treebank

The Chinese Penn Treebank (5.1) consists of 890 data files, including about 18K sentences with 825K words. We put files 301-325 into the development sets, 271-300 into the test set and reserve the other files for training. All the following experiments are based on gold standard segmentation but untagged. The evaluation results are listed in Table 4. The accuracy of automatic POS is 94.19% and POS tags are not included in the evaluation.

Table 4. Evaluation results (<=40 words) by hierarchical parsing. Gold Standard POS means using gold standard POS tags; Automatic POS using the best one automatic POS result; Automatic POS* using multiple automatic POS results with $\lambda = 0.20$.

	LR	LP	F1
Gold Standard POS	88.28	89.79	89.03
Automatic POS	81.02	82.61	81.81
Automatic POS*	82.19	83.96	83.07

Impact of Automatic POS. As shown in Table 4, the performance gap posed by automatic POS is up to 7.22 in F1, which is much wider than that of English parsing performance. The second column in Table 5 shows top 5 POS tagging errors on the test set. Mistaggings between verbs (VV) and common nouns (NN) occur frequently and make up 28% of all POS tagging errors.

In order to verify the effect of those POS tagging errors on the whole performance, for each error, we obtain the F1 on the test set and the corresponding decline rate (the last two columns in Table 5) by supposing other POS tags are

Table 5. The top 5 POS tagging errors on the test set and their influence. Based on Gold Standard POS, the F1 on the test set (348 sentences) is 86.38.

Num.	mistagging errors	#errors(rate%)	F1	decline rate(%)
1	VV→NN	70 (15.05)	85.02	1.57
2	NN→VV	60 (12.90)	84.78	1.85
3	DEC→DEG	40 (8.60)	84.77	1.86
4	JJ→NN	38 (8.17)	85.79	0.67
5	DEG→DEC	26 (5.59)	85.55	0.96

all correct. In particular, both POS tagging errors between verbs and nouns, such as *VV→NN* and *NN→VV*, and *de5* tagging errors (*DEC→DEG*, *DEG→DEC*) significantly deteriorate the performance. This is not surprising because: 1) All nouns are immediately merged into NP, and all verbs into VP; 2) *de5* has different structural preferences if tagged as *DEC* or *DEG*.

In order to lower the side effect caused by POS tagging errors, the top K POS results are served as the input of chunking model. Here K is defined as following, where $\lambda\,(0 \leq \lambda \leq 1)$ is the factor for deciding the number of automatic POS tagged results. The third row in Table 6 shows the performance when $\lambda = 0.20$.

$$K = \min\left(20, \left|\left\{result_i \middle| P\left(result_i\right) \geq \lambda * P\left(result_0\right)\right\}\right|\right) \qquad (1)$$

Table 6. Results on CTB parsing for sentences of at most 40 words

Parsers	LP/LR/F1	Parsers	LP/LR/F1
Bilke & Chiang 2000	77.2/76.2/76.7	Ours	80.0/76.5/78.2
Levy & Manning 2000	78.4/79.2/78.8	Xiong et al.	80.1/78.7/79.4
Chiang & Bilke 2000	81.1/78.8/79.9		

Compare with Other CTB Parsers. (Bikel & Chang 2000) implemented two parsers: one based on the modified BBN model and the other based on TIG. (Chiang & Bikel 2002) used the EM algorithm on the same TIG-parser to detect head constituents by mining latent information. (Levy & Manning 2003) employed a factored model and improved the performance by error analysis. Likewise, (Xiong et al. 2005) integrated the head-driven model with several re-annotations into a model with external semantic knowledge from two Chinese electronic semantic dictionaries. Table 6 compares above systems. For fair comparisons, we also train our three models (POStagging, chunking and parsing) and test the performance with the same training/test sets as theirs. Table 6 shows that our system only performs slightly worse than the best reported system. This may be due to our low performance in chunking. With further analysis on the parsing results, our chunking model only achieves 80.82 in F1 on basic constituents, which make up 40.9% of all constituents. Therefore, there is still much room for performance improvement by employing a better chunking model.

4 Conclusions

This paper represents an attempt at applying hierarchical parsing with machine learning techniques. In the parsing process, we always try to detect constituents from simple to complex. Evaluation on the Penn WSJ Treebank shows that our method can achieve a good performance with more flexibility for future improvement. Moreover, our experiments on Penn Chinese Treebank suggest that there is still much room for performance improvement by employing a better chunking model.

Acknowledgements

This research is supported by Project 60673041 under the National Natural Science Foundation of China and Project 2006AA01Z147 under the "863" National High-Tech Research and Development of China.

References

1. Bikel, D.M., Chiang, D.: Two statistical parsing models applied to the Chinese Treebank. In: Proceedings of 2nd Chinese Language Processing Workshop (2000)
2. Charniak, E.: Statistical parsing with a context-free grammar and word statistics. In: Proceedings of AAAI 1997 (1997)
3. Chiang, D., Bikel, D.M.: Recovering latent information in treebanks. In: Proceedings of COLING 2002, pp. 183–189 (2002)
4. Collins, M.: 1999. Head-driven statistical model for natural language parsing [D]. Ph. D. Thesis, the University of Pennsylvania (1999)
5. Levy, R., Manning, C.: Is it harder to parse Chinese, or the Chinese Treebank? In: Dignum, F.P.M. (ed.) ACL 2003. LNCS (LNAI), vol. 2922, Springer, Heidelberg (2004)
6. Magerman, D.M.: Statistical decision-tree models for parsing. In: Proceedings of the 33rd Annual Meeting of the Association for Computational Linguistics (1995)
7. Ratnaparkhi, A.: Learning to parse natural language with maximum entropy models. Machine Learning 341(2/3), 151–176 (1999)
8. Xiong, D., Li, S.L., Liu, Q., et al.: Parsing the Penn Chinese treebank with semantic knowledge. In: Proceedings of the 2nd IJCNLP, pp. 70–81 (2005)

Extracting Hyponymy Relation between Chinese Terms*

Yongwei Hu and Zhifang Sui

Institute of Computational Linguistics (ICL), Peking University, Beijing, China
yongweihu@pku.edu.cn, szf@pku.edu.cn

Abstract. This paper studies the problem of the automatic acquisition of the hyponymy (is-a) relation in sentences and develops a new method for it. In this paper, we treat the task of identifying hyponymy relation as two separate problems and solve them based on the following three techniques: term type's commonality, sequential patterns, property nouns and domain verbs.

Keywords: hyponymy, relation extraction, pattern-based, sequential pattern, property noun, domain verb.

1 Introduction

Detecting terms and hyponymy relation among them in text data has many applications. The previous work on identifying hyponymy relation commonly used pattern-based methods and had several problems. We can classify terms that have hyponymy relation into two types: hyponym and hypernym. This paper develops a novel approach for acquiring hyponymy relation by modeling commonality of hyponyms and that of hypernyms separately. This method is different from traditional approaches in that terms having hyponymy relation don't need to occur syntactically near one another. Intuitively, this method could extract more relation instances from corpora. In order to solve the second problem, we introduce the Sequential Patterns (SP), which is another pattern representation method and is well-known in Data Mining field.

The rest of this paper is structured as follows. Section 2 discusses related works. Section 3 defines the problem. Section 4 records our preliminary experiments we ran. Finally, section 5 makes a conclusion of our work.

2 Related Work

Research on recognizing relations can be classified into three categories. The first category uses statistical techniques, such as (Miller et al., 2000), (Zhao and Grisman, 2005), and (Zhou et al., 2006). Statistical approaches perform well on large corpora, but for their good performance a large number of features have to be explored and

* This paper is supported by 863 High Technology Project of China (No.2006AA01Z144), NSFC Project 60503071 and Beijing Natural Science Foundation 4052019.

H. Li et al. (Eds.): AIRS 2008, LNCS 4993, pp. 567–572, 2008.

many training examples must be labeled, which is expensive and time-consuming for some domain.

The second category makes use of hand-crafted or automatically extracted rules. This type of approaches is pioneered by Hearst (1992). Manually selecting as seed instances a list of term pairs for which the target relation is known to hold, Hearst sketched an algorithm to learn patterns that indicate the relation of interest, and then use these pat-terns to extract more instances. These methods extract patterns from sentences containing both terms of seed instances, which limit the number of relation instances we can get because that not all relation instances would occur syntactically near one another.

Another related work is about Sequential Patterns (SP). SPs have been used in many fields to solve quite different problems, such as, (Sun et al. 2007), (Jindal and Liu, 2006). The work in (Sun et al. 2007) focuses on the problem of detecting erroneous/correct sentences.

3 Proposed Technique

This section first defines the problem in a formal way and then presents our solution.

3.1 Problem Statement

Let T be a set of terms in a domain D. Given a corpus, we could treat all terms in it as T. We say term t1 in T is a hyponym of term t2 if people accept sentences constructed from the frame *A/An t1 is a (kind of) t2.* Here, t2 is said to be a hypernym of t1. Let T_{hypo} be the set of all hyponyms in T and T_{hyper} the set of all hypernyms in T. A hyponymy relation, r, is in the form of <t1, t2>, where term t1 is a hyponym of term t2. Let R_T be a set of relations among terms in T and $T_{hypo} \times T_{hyper}$ represent the set of all term pairs composed of terms in T_{hypo} and T_{hyper}, and, obviously, $R_T \subseteq T_{hypo} \times T_{hyper}$.

We treat the task of identifying hyponymy relation as two separate problems. The first problem is defined as follows. Note that terms are already labeled in corpora and given to us as input.

Problem 1(Term Type Recognition). Suppose T is the set of terms in corpora D; recognize the set of hyponyms T_{hypo} and the set of hypernyms T_{hyper} in D. Problem 1 is solved in next subsection. After identifying terms' type, the next problem at hand is that of determining whether a term pair has the hyponymy relation.

Problem 2(Relation Identification). Given two sets T_{hypo} and T_{hyper}, identify legal term pairs. A term pair (t1, t2) is legal if it satisfies the following constraints: t1 $\in T_{hypo}$, t2 $\in T_{hyper}$ and t1 is a hyponym of t2.

3.2 Term Type Recognition

To solve this problem, we first present the following assumption.

Hypothesis 1. *If two terms in T hold the same term type (either hyponym or hypernym), their occurrences in text data tend to have similar context.*

For many domains, this assumption is intuitively true. Based on the assumption, for a given corpus T, ideally, we could recognize all terms that are hyponyms and all those that are hypernyms, and get the two sets, namely, T_{hypo} and T_{hyper}. The strategy we adopt for this recognition problem is similar in spirit to the pattern-based techniques used in earlier relation extraction works. The difference lies in that patterns here are composed of distant words in sentences and that we want to extract patterns indicating term types (i.e. hyponym and hypernym) rather than hyponymy relation.

In order to extract patterns from sentences, we introduce the idea of Sequential Patterns (SP) from Dining Mining. The definitions of sequence and sequential pattern and the algorithms for extracting such patterns are introduced in (Sun et al. 2007).

3.3 Relation Identification

Terms in a specified domain are usually associated with meaningful phrases which could be used to show their semantic features and are usually domain-specific. For example, the noun phrase容量(*volume*) describes a property of the term 随机存储器 (*RAM*) in sentence "随机存储器的容量是大多数任务的关键参数(*RAM volume is a critical parameter for the majority of tasks*）." In sentence, "这种驱动使用SCSI子系统存取USB存储器(*This driver uses the SCSI subsystem to access to the USB storage device*）", the verb phrase 存取(*access*) indicate the action we can take on the term USB存储器(*USB storage device*), a property of the term USB存储器(*USB storage device*).

In terms of Part of Speech (POS), we classify phrases that could show terms' properties into two categories: property noun and domain verb. Phrase 容量(*volume*) is one example of property noun. As other examples, phrase 速度(*speed*) describing term 处理器(*CPU*), phrase 大小(*size*) describing term 笔记本电脑(*notebook*). Phrase 存取(*access*) is one example of domain verb and 关闭(*turn off*) associated with 计算机(*computer*) is another example.

Note that two terms having hyponymy relation are often described by similar property nouns and domain verbs. Take relation r = <笔记本电脑*(notebook),* 计算机*(computer)*> as an example. Term 计算机(*computer*) can be described with property noun大小(*size*), so can term笔记本电脑*(notebook)*, and they both can be described with domain verb关闭(*turn off*). Therefore, if we found property nouns and domain verbs connected with every term in term set T, it would be easy to solve the second problem, by just selecting all those term pairs described by similar property nouns and domain verbs.

Property nouns and domain verbs in a specific domain D1 could be specified manually. In this paper, we get all the verbs and nouns relatively specific to corpus T1 in D1 and use them as the domain verbs and property nouns. We treat all extracted phrases as property nouns and domain verbs in T1. This is because property nouns and domain verbs are domain-specific and corpus T2 is used to filter out all those phrases. After dividing terms into hyponym and hypernym and extracting phrases which show properties of terms, we construct for each term a feature vector which consists of all the phrases we extracted. If a term includes a phrase, the corresponding feature is set at 1. Term pair <t1, t2> having hyponymy relation must satisfy some constraints. For example, term t1 and t2 cannot be the same; the similarity between t1 and t2 must be bigger than a threshold *min_sim*. We sort the identified relation instances according to the similarities of their terms at last.

4 Experiments

The following subsections describe the experiments we ran in computer domain and the experimental results.

4.1 Experimental Setup

In order to evaluate our algorithm, we first collected sentences from the book 计算机科学技术百科全书(Encyclopedia of Computer Science and Technology), which are mostly technical essays in computer domain, and tagged all terms in these sentences. Among the collected sentences, 3623 sentences contain terms and 740 terms are labeled. There are about 1282 hyponymy relation instances. In order to extract property nouns and domain verbs in target domain, we collected 1000 sentences from the Chinese broadcast news training data for ACE 2004, which are mainly daily news and definitely a different domain.

4.2 Experimental Results

Term Type Recognition. The experiment needs some relation instances as seeds to bootstrap. The seeds we selected are: <笔记本计算机(notebook), 计算机(computer)> , <磁带存储器(tape), 存储器(storage)>, <键盘(keyboard), 输入设备(input device)>, <环网(ring network), 局域网 (LAN)>. We adopted the frequent sequence mining algorithm in (Pei et al., 2001) for learning patterns. In order to ensure that our discovered pattern p is not too general, this mining algorithm needs us to specify an argument, min_sup, denoting the minimum number of terms whose context contains the pattern p. In our experiment, min_sup is empirically set to 5 for hyponym and 7 for hypernym. At last, we get two sets T_{hypo} and T_{hyper}. T_{hypo} contains 452 terms and T_{hyper} contains 523 terms. Note that the number of terms in T_{hypo} T_{hyper} is larger that 740, the total number of terms in the corpus. This is because some terms are actually both hyponym and hypernym. In addition, there are also terms that are not contained in any set, such as term临界区. This is mainly due to the data sparseness problem in the corpus and few sentences contain these terms. The performance of the step is showed in Table 1.

Table 1. Result of Term Type Recognition

Type	P	R	F
hyponym	70.82	92.14	80.08
hypernym	62.34	85.78	72.21

Table 2. Performance of Relation Identification Effected by *k*

k	P	R	F
300	87.67	20.51	33.25
400	83.75	26.13	39.83
500	77.60	30.27	43.55
600	74.83	35.02	47.72
800	76.38	41.42	51.01
1000	64.10	50.00	56.18
1200	57.00	53.35	55.12

Table 3. Performance of Relation Identification Effected by *min_sim*

min_sim	#instances	P	R	F
0.9	121	88.43	8.35	15.25
0.8	543	57.83	24.49	34.41
0.7	1028	61.67	49.45	54.89
0.5	8231	8.65	55.54	14.97
0.3	21384	3.85	64.20	7.26

Property Nouns and Domain Verbs. This step is relatively simple. For the parameter, freq, we empirically set at 10. Some examples of the extracted property nouns: 类型(type), 价格(price), 性能(performance), 体积(size), 速度(speed), 复杂性(complexity), 效率(efficiency). Some examples of the discovered domain verbs: 计算(calculate), 运算(operate), 加(add), 转换(transform), 命中(hit), 执行(execute), 检索(search), 存储(store), 储存(store), 保存(save), 存放(put), 输入(input), 输出(output), 传送(send), 传输(transfer),共享(share), ,分布(distribute),通信(communicate). Due to space limitation, we do not show all the phrases we extracted.

Relation Identification. The experimental results are presented in Table 2, Table 3. We calculated the precision, recall, and F-score. There are two different ways to affect the number of relation instances our algorithm extract, by setting parameter k, the amount of relations our algorithm outputs, or setting another parameter min_sim, which determines when two terms should be identified as a hyponymy relation. Table 2 reports the performance of the first method. And the performance of the second method is presented in Table 3. As can be seen from Table 3, the highest precision is achieved when min_sim is set at 0.9 and with large threshold, the performance deterioration is significant. At the same time, this proves our assumption that terms having hyponymy relation are usually described by similar property nouns and domain verbs. As shown in Table 2, our technique got the best performance, e.g. 56.18%, when we set k at 1000. When k is relatively small, we can achieve high precision. This is because we sorted all the extracted instances according to their terms' similarities and then the k-top instances have the largest similarities.

Comparing with Other Methods In this paper, we compare our technique with (Hearst, 1992). As discussed in Section 2, Hearst (1992) pioneered the pattern-based relation extraction method, and proposed a relation extraction framework which is used by nearly all pattern-based like methods. The best result achieved by this approach is: precision: 42.24% recall: 39.78%, f-measure: 40.97%. It is obvious that our method outperforms Hearst(1992) in terms of precision, recall and f-measure. After comparing the relation instances they found, we realize that many instances got by our method don't necessarily contain terms that occur in the same sentence. That is to say, even though two terms appear far enough in the corpus, our technique could still determine whether they have the hyponymy relation. As stated in Section 2, in all earlier pattern-based like methods we know of, terms having the target relation must occur syntactically near one another. Therefore, these methods could not find term instances far away in the corpus as well.

5 Conclusions

This paper proposed a new method to identify hyponymy relation. Empirical evaluating in Computer domain demonstrated the effectiveness of our techniques. This method is actually based on two assumptions. One is that the same term type has similar context. The other is that two terms having the hyponymy relation will be described by similar property nouns and domain verbs in the corpus. Our method could find relation instances on a global level, which is its improvement over other pattern-based methods.

References

1. Sun, G., Cong, G., Liu, X., Lin, C.-Y., Zhou, M.: Mining sequential patterns and tree patterns to detect erroneous sentences. In: AAAI (2007)
2. Hearst, A.: Automatic Acquisition of Hyponyms from Large Text Corpora. In: Proc. Of COLING 1992, pp. 23–28 (1992)
3. Miller, S., Fox, H., Ramshaw, L., Weischedel, R.: A novel use of statistical parsing to extract information from text. In: Proc of 6th Applied Natural Language Processing Conference, Seattle, USA, 29 April- 4 May (2000)
4. Zhao, S.B., Grisman, R.: Extracting relations with integrated information using kernel methods. In: Proc. Of ACL 2005, pp. 419–426 (2005)
5. GuoDong., Z., Jian, S., Min, Z.: Modeling Commonality among Related Classes in Relation Extraction. In: Proc. Of ACL 2006, pp. 121–128 (2006)

A No-Word-Segmentation Hierarchical Clustering Approach to Chinese Web Search Results

Hui Zhang, Liping Zhao, Rui Liu, and Deqing Wang

State Key Lab.Of Software Development Environment, Beihang University, 100083
{hzhang,zhaolp,liurui,wangdeq}@nlsde.buaa.edu.cn

Abstract. In this paper, we present a No-Word-Segmentation Hierarchical Clustering Approach (NWSHCA) to Chinese Web search results. The approach uses a new similarity measure between two documents based on a variation of the Edit Distance, and then it generates preliminary clusters using a partitioning clustering method. Next it ranks all common substring in a cluster using a cluster-discriminative metric with the top K as cluster description labels. Finally it uses HAC to cluster the top K cluster labels to form a navigational tree. NWSHCA can generate overlapping clusters contrast to most clustering algorithms. Experimental results show that the approach is feasible and effective.

Keywords: hierarchical clustering, Chinese Web search results, no-word segmentation, Edit Distance.

1 Introduction and Related Work

Nowadays people frequently use Web search engines such as Google to look things up on the World Wide Web. However, many Web page snippets of the long list returned by search engines are irrelevant to users' query. To Cluster Web search results is an effective method to help people to find the information from several groups with relevant topics at a glance.

As a post-retrieval document clustering algorithm, there are several key requirements, such as coherent clusters, efficiently browsable and speed [1]. Hierarchical Agglomerative Clustering algorithms are probably the most commonly used, but they are quadratic in the number of documents and therefore too slow for online requirements [2]. Linear time clustering algorithms are the better candidates, for example K-means, Single-Pass [3]. However, there are some shortcomings with these partitioning algorithms. For example, they can not provide overlapping clusters. In addition, most of these traditional algorithms are based on Vector Space Model (VSM) [4] and need to segment Chinese texts to many words as vectors, so the clustering results are influenced by the performance and precision of Chinese word-segmentation algorithms.

2 NWSHCA Description

NWSHCA is composed of five logic steps: document preprocessing, similarity computing, partitioning clustering, cluster label generating and hierarchical agglomerative clustering.

H. Li et al. (Eds.): AIRS 2008, LNCS 4993, pp. 573–577, 2008.

2.1 Step 1: Document Preprocessing

Web pages' titles and snippets in search results are parsed and split into sentences according to punctuations (period, comma, semicolon, question mark etc.), and all non-Chinese characters are deleted.

2.2 Step 2: Similarity Computing

In NWSHCA, we use a new similarity measure between two sentences based on a variation of the Edit Distance [6]. Edit Distance is usually used for fast approximate string matching between sentences, and the smaller distance, the greater similarity between sentences. But edit distance is not similarity. First, edit distance could not represent the similarity between sentences. Second, the operations to calculate edit distance are not flexible and the minimum operating unit is character, which is not applicable to Chinese texts.

It is obvious that for two sentences the more co-occurrence strings (common substrings) they have, the more similar they are. We use maximum matching algorithm KMP (Knuth-Morris-Pras) [7] to search the common substrings between two document snippets. When a maximum matching substring is acquired by using KMP, the algorithm deletes this substring from the document and continues to find the next maximum matching substring. This process will iterate till the length of matching string is 2. At last, sequences of the maximum matching substrings $(S_1, S_2, S_3, \ldots S_{n-1}, S_n)$ between sentence A and sentence B are generated, where S_i is i-th common substring between A and B.

The length of common substring is another factor that affects the similarity between two sentences. The longer common substring the two sentences have, the more similar they are. The formula (1) defines the similarity between two sentences:

$$Sim_{(A,B)} = (\mathrm{Imp}_{(A,B)} + \mathrm{Imp}_{(B,A)})/2 = \left(\sum_{i=1}^{n}(times_{(i,A)} \times length_{(Si)}^{\alpha})/length_{(A)}^{\alpha} + \sum_{i=1}^{n}(times_{(i,B)} \times length_{(Si)}^{\alpha})/length_{(B)}^{\alpha}\right)/2 \quad (1)$$

where $Sim_{(A,B)}$ is the similarity of sentence A and B, $times_{(i,A)}$ is the frequency of the i-th substring in sentence A, $length_{(Si)}$ is the length of the i-th substring, $length_{(A)}$ is the length of sentence A and Parameter α is an adjustment factor. In the paper, $\alpha \in [1,2]$ and it affects the number of initial clusters.

2.3 Step 3: Partitioning Clustering

According to the above calculating method of similarity, the similarity matrix of document snippets is obtained. NWSHCA uses partitioning clustering method based on the similarity matrix to produce the initial clusters. Firstly we select two most similar documents into a cluster. The cluster is represented by a centroid which is the average similarity among the documents within the cluster. Next expecting document is selected and compared to the centroid, and if the similarity between them exceeds an empirically chosen threshold, the document belongs to the cluster. If there are no such documents, other two nearest documents are grouped to another cluster and the process iterates till no document left.

2.4 Step 4: Cluster Label Generating

After partitioning clustering, several clusters are generated and each cluster has several documents in it. We compute the weight of each common substring of these documents within each cluster by TF*IDF, and then we rank the weight and choose top K common substrings as cluster labels. In this paper, we define:

$$w_{ij} = TF * IDF(S_i, d_j) = TF(S_i, d_j) \times IDF(S_i) \tag{2}$$

$$IDF(S_i) = \left(\log \frac{m}{DF(S_i)} \right) \tag{3}$$

Where S_i is the i-th common substring in the cluster; w_{ij} is the weight of S_i ; d_j is the j-th document that contains S_i in the cluster; $TF(S_i, d_j)$ is the frequency that S_i occurs in the d_j; m is the total number of documents in the cluster; $DF(S_i)$ is the number of documents that only includes S_i in the cluster.

2.5 Step 5: Hierarchical Agglomerative Clustering

After the above process, some initial clusters with their own cluster label are generated. We adopt hierarchical agglomerative clustering (HAC), which allows one document belonging to different clusters. However, HAC is typically slow when applied to large number of Web documents. To improve the performance of clustering, we only input the generated cluster labels instead of full texts of documents to HAC. The experiments in section 3 proved that the improvement is effective.

The algorithm is described as follows:

a. Select two unprocessed cluster description labels which have largest similarity in similarity matrices, merge these two clusters together with most common substrings in their labels, and take these common substrings as the new label of the new merged super cluster. Then examine all other unprocessed clusters. If their labels are included in the super cluster's label phrases, merge these clusters into the former super cluster. Then set the similarity between every two of these cluster as 0 and mark them as processed state. Turn to b.
b. If there still exist unprocessed clusters then turn to a, or turn to c.
c. Take the new generated super clusters with the new labels as the initial object for next iteration, and turn to a. The whole algorithm ends until total number of the generated super clusters remains unchanged.

3 Evaluation of NWSHCA

3.1 Speed

The purpose of NWSHCA is to improve the speed of clustering, especially in similarity computing and hierarchical clustering. In the experiments we also find the HAC is quickly finished after a few iterations, mostly six times. For 100 queries and first 100

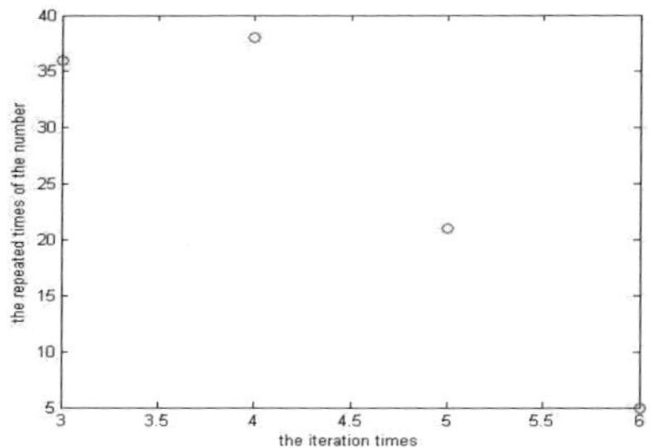

Fig. 1. Iteration times

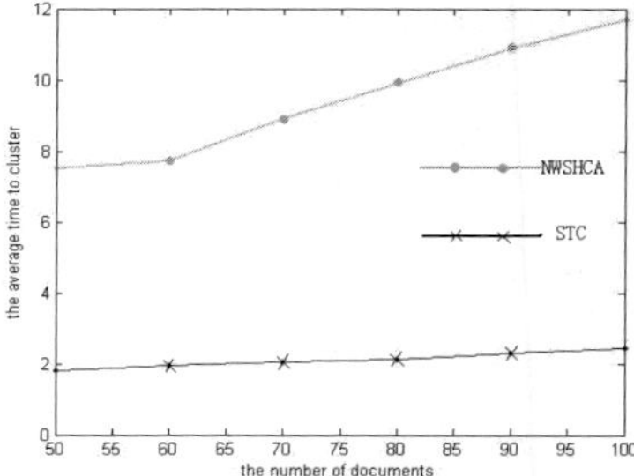

Fig. 2. Average time

returned documents of each query, the distribution of iteration times is shown in Figure 1. In addition, we separately record the clustering average time for STC (Suffix Tree Clustering) and NWSHCA shown in Figure 3. The Figure 3 shows that NWSHCA has higher performance than STC, as STC needs segmenting words for Chinese document.

Table 1. Results comparison with Vivisimo

keyword	苹果	六方	上海	清华	四川	科技	航天	聚类
The number of vivisimo cluster	7	17	10	10	7	5	15	13
The number of NWSHCA cluster	7	16	10	9	7	5	13	13
keyword	搜索	北航	计算机	川大	中国	研究所	科学院	パンダ
The number of vivisimo cluster	11	19	4	15	5	12	7	15
The number of NWSHCA cluster	10	19	4	11	5	12	6	13

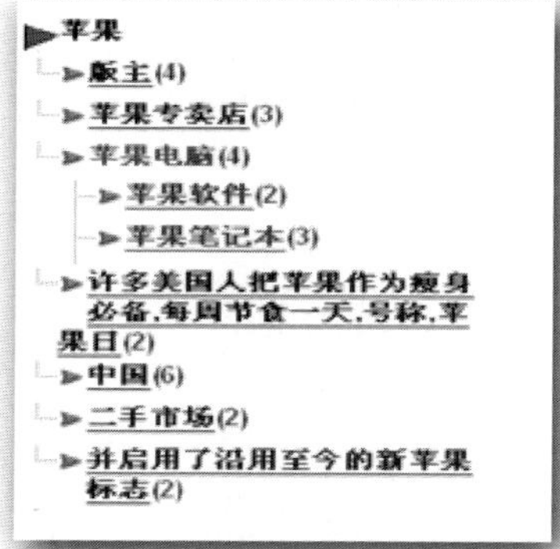

Fig. 3. Display the results of NWSHCA

3.2 Efficiently Browsable

We use Baidu [8] as Web search engine in NWSHCA and the result is displayed as a navigational tree shown in Figure 3. We also compare cluster number formed by NWSHCA with those by Vivisimo [9], using the same 16 Chinese keywords to search. The comparison results are shown in Table1. It is easy to see that the cluster number generated by NWSHCA is very close to Vivisimo on Chinese results.

4 Conclusions

NWSHCA can cluster the Chinese Web search results efficiently without word segmentation. Experiments show that NWSHCA is faster than the linear time clustering algorithm STC on Chinese texts. In addition, final cluster description labels are displayed as a navigational tree for browsing with ease. However, NWSHCA is independent to Chinese dictionary and the common substrings are not semantic, so sometimes NWSHCA may generate inaccurate cluster labels, for example "日上午". Extensive work is needed to improve the performance of NWSHCA and try to solve these problems.

Acknowledgement

The project is supported by National Infrastructure for Science and Technology Foundation of China under Grant No.2005DKA63901.

References

1. Zamir, O., Etzioni, O.: Grouper: A dynamic clustering interface to Web search results. Computer Networks 31(11-16), 1361–1374 (1999)
2. Voorhees, E.M.: Implementing agglomerative hierarchical clustering algorithms for use in document retrieval. Information Processing and Management 22, 465–476 (1986)
3. Hill, D.R.: A vector clustering technique. In: Samuelson (ed.) Mechanized Information Storage, Retrieval and Dissemination, North-Holland, Amsterdam (1968)
4. Raghavan, V.V., Wong, S.K.M.: A critical analysis of vector space model for information retrieval. Journal of the American Society for Information Science (1986)
5. Zamir, O., Etzioni, O.: Grouper: A dynamic clustering interface to Web search results. Computer Networks 31(11-16), 1361–1374 (1999)
6. Ristad, Yianilos, E.S.: Learning string-edit distance. Pattern Analysis and Machine Intelligence, IEEE Transactions (1998)
7. Navarro, G., Fredriksson, K.: Average complexity of exact and approximate multiple string matching[J]. Theoretical Computer Science 321(2-3), 283–290 (2004)
8. http://www.vivisimo.com
9. http://www.baidu.com

Similar Sentence Retrieval for Machine Translation Based on Word-Aligned Bilingual Corpus

Wen-Han Chao[1] and Zhou-Jun Li[2]

[1] School of Computer Science, National University of Defense Technology, China, 410073
cwh2k@163.com

[2] School of Computer Science and Engineering, Beihang University, China, 100083
lizj@buaa.edu.cn

Abstract. In this paper, we present a novel method to retrieve the similar examples from the corpus when given an input which should be translated, in which we use the word alignment between the bilingual sentence pair to measure the similarity of the input and the example.

1 Introduction

Measures of the sentence similarity are very important in many NLP applications, such as machine translation (MT), especially the example-based machine translation (EBMT) [1], information retrieval (IR) and text summarization etc.

Recently, some researchers present a hybrid corpus-based machine translation system[2], which is a statistical machine translation (SMT), while using an example-based decoder, in which the similar translation examples will avoid translating from the scratch, and the similar examples retrieved will affect the results greatly.

When retrieving the similar examples, most of the metrics measure the similarity of the two monolingual sentences. However, in a hybrid machine translation system, the examples in the training corpus provide rich information, such as word alignment.

In this paper, we provide a novel approach, which use the bilingual information of the examples to retrieve the similar examples for an input sentence.

2 The Word-Aligned Bilingual Corpus for MT

There are two types of corpus-based MT systems: Statistical Machine Translation (SMT) and EBMT. SMT obtains the translation models during training, and does not need the training corpus when decoding; while EBMT retrieves the similar examples in the corpus when translating.

For both of them use the same corpus, we can generate a hybrid MT, which is a SMT system, while using an example-based decoder. Generally, the hybrid MT system uses a word-aligned bilingual corpus, which is a collection of the word-aligned sentence pair, represented as (*C*, *A*, *E*), where *C* is the source language sentence, and *E* is the target language sentence and *A* is the word alignment between *C* and *E*.

H. Li et al. (Eds.): AIRS 2008, LNCS 4993, pp. 578–585, 2008.

A word alignment is defined as follows: given a sentence pair (*C*, *E*), we define a link $l = (i..j, s..t)$, where $i..j$ represents a sequence of words in sentence *C*, where i and j are position indexes, i.e. $i, i+1, ..., j$; and $s..t$ represents a sequence of words in sentence *E*. So each link represents the alignment between the consecutive words in both of the sentences. A word alignment between a sentence pair is a set of links $A = \{l_1, l_2, ..., l_n\}$. In the paper, the example refers to a triple (*C*, *A*, *E*).

We can merge two links $l_1 = (i_1..j_1, s_1..t_1)$ and $l_2 = (i_2..j_2, s_2..t_2)$ to form a larger link, if the two links are adjacent in both of the sentences, i.e. $i_1..j_1$ is adjacent to $i_2..j_2$ where $i_2 = j_1 + 1$ or $i_1 = j_2 + 1$, or $i_1..j_1$ (or $i_2..j_2$) is ε , so do the $s_1..t_1$ to $s_2..t_2$. If the region $(i..j, s..t)$ can be formed by merging two adjacent links gradually, we call the region is **independent**.

Figure 1(a) shows a word alignment example; the number below the word is the position index. In the example, the region (1..3, 3..5) is independent.

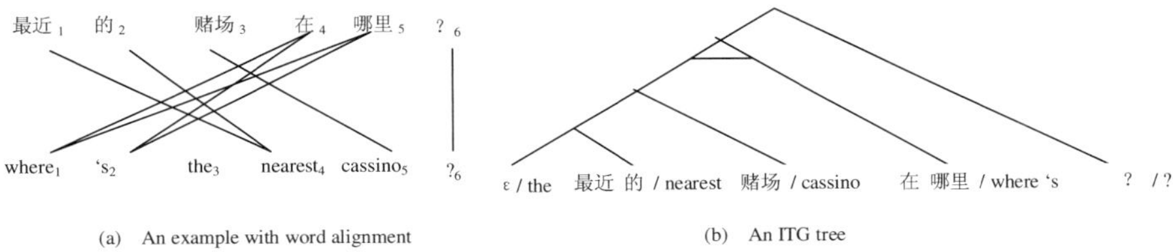

(a) An example with word alignment (b) An ITG tree

Fig. 1. An ITG tree which is derived from the ITG. A line between the branches means an inverted orientation, otherwise a straight one.

In our hybrid MT system, the word alignment in the corpus must satisfy the Inversion Transduction Grammar (ITG) constraint[3], so that each word alignment will form a binary branching tree, which provides a weak but effective way to constrain the re-ordering of the words. Figure 1b illustrates an ITG tree formed from the alignment.

Using the word-aligned bilingual corpus, we can extract the phrase pairs, called as **blocks**. Each block is formed by combining one or more links, and must be independent. Then, we build the corresponding translation models using the relative frequencies.

3 The Fast Retrieval of Similar Sentences

We divide the retrieval of similar examples into two phases:

- Fast Retrieval Phase: retrieving the similar examples from the corpus quickly, and take them as candidates. The complexity should be not too high.
- Refining Phase: refining the candidates to find the most similar examples.

In this section, we will describe how to retrieve the similar examples quickly through the following three similarity metrics.

3.1 The Matched Words Metric

Given an input sentence $I = w_1 w_2 ... w_n$ and an example (*C*, *A*, *E*), we calculate the number of the matched source words between the input sentence and the source sentence C in the example firstly.

$$Sim_{word}(Input, Example) = \frac{2 * Match_{word}}{Len(I) + Len(C, A, E)} \tag{1}$$

where $Match_{word}$ is the number of the matched words, $Len(I)$ is the number of words in I , and $Len(C, A, E)$ is the number of the words in the in C .

3.2 The Matched Blocks Metric

Given an input sentence $I = w_1 w_2 ... w_n$, we search the blocks for each word $w_i (i \in \{1,2,...n\})$. We use B^i_{k-gram} to represent the blocks, in which for each block (c, e) , the source phrase c use the word w_i as the first word, and the length of c is k , i.e. the $c = w_{i..(i+k-1)}$. In order to retrieve the B^i_{k-gram} , we first get all the source phrases with w_i as the first word, and then find the blocks for each source phrase c in the translation model. Considering the complexity, we set *M* as the maximum length of the source phrase (here *M*=3 or 5), i.e., $k \le M$.

For each c , there may exists more than one blocks with c as the source phrase, so we will sort them by the probability and keep the best N (here set N=5) blocks. Now we represent the input sentence as:

$$\sigma(I) = \{block \mid block \in B^i_{k-gram}, 1 \le i \le n, 1 \le k \le M\} \tag{2}$$

For example, in an input sentence "我 来自 中国', the $B^1_{1-gram} = \{(我, i), (我, me), (我, my), (我, Mine)\}$. Note, some B^i_{k-gram} may be empty, e.g. $B^2_{2-gram} = \phi$, since no blocks with "来自 中国' as the source phrase.

In the same way, we represent the example (C, A, E) as:

$$\varphi(C, A, E) = \{block \mid block \in B^i_{k-gram}, block \in A^*, 1 \le k \le M\} \tag{3}$$

where A^* represents the blocks which are links in the alignment A or can be formed by merging adjacent links independently. In order to accelerate the retrieval of similar examples, we generate the block set for each example during the training process and store them in the corpus.

Now, we can use the number of the matched blocks to measure the similarity of the input and the example:

$$Sim_{block}(Input, Example) = \frac{2 * Match_{block}}{B_{gram}^{Input} + B_{gram}^{Example}} \tag{4}$$

where $Match_{block}$ is the number of the matched blocks and B_{gram}^{Input} is the number of B_{k-gram}^{i} ($B_{k-gram}^{i} \neq \phi$) in $\sigma(I)$, and $B_{gram}^{Example}$ is the number of the blocks in $\varphi(C, A, E)$.

Since each block is attached a probability, we can compute the similarity in the following way:

$$Sim_{prob}(Input, Example) = \frac{2 * \sum_{b} \Pr ob(b)}{B_{gram}^{Input} + B_{gram}^{Example}} \tag{5}$$

So the final similarity metric for fast retrieval of the candidates is:

$$Sim_{fast}(Input, Example) = \alpha Sim_{block} + \beta Sim_{word} + \gamma Sim_{prob} \tag{6}$$

where $\alpha + \beta + \gamma = 1$. Zhao et al.[4] provides a method to tune the weights, but here we use mean values, i.e. $\alpha = \beta = \gamma = 1/3$. During the fast retrieval phase, we first filter out the examples using the Sim_{word}, then calculate the Sim_{fast} for each example left, and retrieve the best *N* examples.

4 Refine the Candidates

After retrieving the candidate similar examples, we refine the candidates using the swallow structure of the sentences, to find the best *M* similar examples.

4.1 The Alignment Structure Metric

Given the input sentence I and an example (C, A, E), we first search the matched blocks, at this moment the order of the source phrases in the blocks must correspond with the order of the words in the input.

As Figure 2 shows, the matching divides the input and the example respectively into several regions, where some regions are matched and some un-matched. And we take each region as a whole and align them between the input and the example according to the order of the matched regions. For example, the region (1..3,3..5) in (C, A, E) is un-matched, which aligns to the region (1..1) in I. In this way, we can use a similar edit distance method to measure the similarity. We count the number of the Deletion/Insertion/Substitution operations, which take the region as the object.

We set the penalty for each deletion and insertion operation as 1, while considering the un-matched region in the example may be independent or not, we set the penalty for substitution as 0.5 if the region is independent, otherwise as 1. E.g., the distance is 0.5 for substituting the region (1..3,3..5) to (1..1).

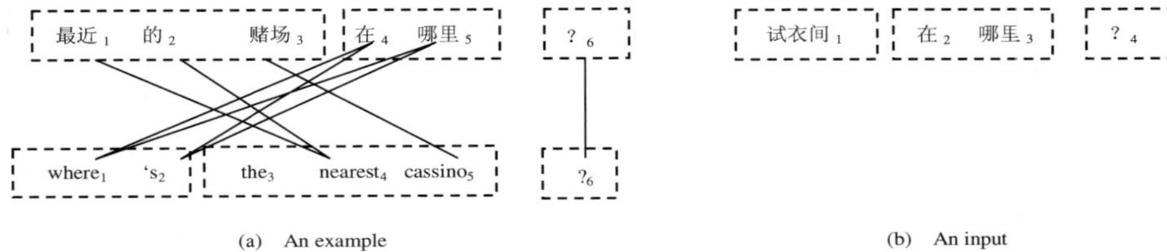

(a) An example (b) An input

Fig. 2. An input and an example. After matching, there are three regions in both sides, which are included in the line box, where the region (4..5,1..2) in the example matches the region (2..3) in the input, so do (6..6,6..6) to (4..4). And the region (1..3,3..5) in the example should be substituted to (1..1) in the input.

We get the metric for measuring the structure similarity of the I and (C, A, E):

$$Sim_{align}(Input, Example) = 1 - \frac{D + I + S}{R_{input} + R_{exmaple}} \tag{7}$$

where D, I, S are the deletion, insertion and substitution distances, respectively. And the R_{input} and $R_{exmaple}$ are the region numbers in the input and example.

4.2 The Semantic Metric

We calculate the semantic distance using the "Tong_Yi_Ci_Ci_Lin"[6], which is a Chinese semantic lexicon. In the lexicon, the semantics of the words are divided into three levels, called large, middle and small level, represented as the uppercase letter, lowercase letter, and integer number respectively, so that it can represent a semantic using a unique code. Each code may include one or more words, and each word may have one or more codes, for it may have multiple semantics. For example, "我' and "咱' have the same code "Aa02", the code for "你' is "Aa03".

Now, we can compute the semantic distance between two words using the codes they are marked: If the large level is different, the distance is 3; the middle is 2; and the small is 1; if the codes are equal, the distance is 0. Since each word has more than one code, we will choose the nearest distance.

Using the semantic distance of words, we can measure the semantic distance of two regions in the input and the example.

$$Dist_{reg\_sem}(reg_1, reg_2) = \frac{\sum\limits_{w_i \in reg_1} \min\limits_{w_2 \in reg_2} [Dist_{word\_sem}(w_1, w_2)]/3}{Len(reg_1) + Len(reg_2)} \tag{8}$$

Given the same match as section 4.1, the substitution distance is the sum of semantic distance of the substitution regions:

$$S_{sem}(Input, Example) = \sum_{regions} Dist_{reg\_sem}(region_{input}, region_{example}) \tag{9}$$

Then, we obtain the semantic metric:

$$Sim_{sem}(Input, Example) = 1 - \frac{D + I + S_{sem}}{R_{input} + R_{exmaple}} \tag{10}$$

In the end, we obtain the similarity metric, which considers all of the above metrics:

$$Sim_{final}(Input, Example) = \alpha' Sim_{fast} + \beta' Sim_{align} + \gamma' Sim_{sem} \tag{11}$$

where $\alpha'+\beta'+\gamma'=1$. Here we also use the mean values as the weights.

5 Experiments

We carried out experiments on a Chinese-English bilingual corpus, which is the training corpus in the open Chinese-English translation task of IWSLT2007, consisting of the sentence-aligned spoken language text for traveling. The training corpus consists of 39,953 sentence pairs, and the test set 489 Chinese sentences. We lowercased and stemmed the words in the English sentences for preprocessing.

During the training, we obtained firstly the word alignments, which satisfy the ITG constraint, for the sentence pairs in the corpus; then built the translation models, and retrieved and stored the blocks in each examples, i.e. $\varphi(C, A, E)$.

5.1 Evaluation of the Retrieval of Similar Examples

We evaluate our retrieval method in manual way: obtain 200 input sentences from the test set randomly, take them as inputs to retrieve the similar examples (10 examples for each input), and verify manually whether the retrieved examples are similar to the inputs.

We define a correct similar example as the following way: if the retrieved example can be turned to the input sentence by modifying x independent regions, where each region consists of at most y blocks (here set $y = 2$).

To evaluate the two retrieval phases, we also get the best 10 examples for the fast retrieval phrase in the candidates, and compare with the final 10 examples. Table 1 shows the results. It shows the correct number and accuracies for both phases by changed the x. With x becomes larger, the accuracies are higher. When $x = 1$, the accuracies are both 45%, because some of the input sentences are very short, and both of the two phases will find the exactly similar examples. But when x becomes larger, the refined retrieval will get more accurate examples.

5.2 Evaluation of the Translation

Our goal is to improve the translation quality, so we combined the retrieval module into our hybrid MT system, and compared the translation results with other SMT systems. The results list in Table 2.

The first column lists the MT systems: the first two systems are two SMT systems, where Moses[7] is a state-of-the-art SMT system, and SMT-CKY is our SMT system with a CKY style decoder; the last two systems are hybrid MT systems with fast retrieval module and refined retrieval module. The second column lists the Bleu scores[8] for these systems. The results show that our hybrid MT system achieves an improvement when using the new retrieval module. We conclude our method to retrieve the similar examples is effective.

Table 1. The results for retrieval of similar sentences

	Test Number	$x=1$		$x=2$		$x=3$	
		Corr.	Accu.	Corr.	Accu.	Corr.	Accu.
Fast Retrieval	200	90	45%	141	70.5%	155%	77.5%
Refined Retrieval	200	90	45%	153	76.5%	174	87%

Table 2. The results of translation

System	Bleu(%)
Moses	22.61
SMT-CKY	28.33
Hybrid MT with fast retrieval	30.03
Hybrid MT with refined retrieval	33.05

6 Conclusions

We present a method to retrieve the similar examples for the input sentence in the hybrid MT systems. The approach makes good use of the word alignment information contained in the corpus. Since the method considers the internal translation process contained implicitly in each example, it helps to improve the translation quality.

In the future, we will improve the refining metric further, especially the semantic metric, and consider how to tune the weights. Also, we should consider how to use the retrieved examples more effectively.

Acknowledgements

This work is supported by the National Science Foundation of China under Grants No. 60573057, 60473057 and 90604007.

References

1. Mandreoli, F., Martoglia, R., Tiberio, P.: Searching Similar (Sub)Sentences for Example-Based Machine Translation. In: Atti del Decimo Convegno Nazionale su Sistemi Evoluti per Basi di Datt (SEBD 2002), Isola d'Elba, Italy (2002)
2. Watanabe, T., Sumita, E.: Example-based Decoding for Statistical Machine Translation. Machine Translation Summit IX, 410–417 (2003)

3. Wu, D.: Stochastic Inversion Transduction Grammars and Bilingual Parsing of Parallel Corpora. Computational Linguistics 23(3), 374 (1997)
4. Zhao, Y.Y., Qin, B., Liu, T., Su, Z.-Z.: Sentence Similarity Computing Based on Multi-Features Fusion. In: The Proceedings of JSCL 2005, August 2005, pp. 168–174 (2005)
5. Li, B., Liu, T., Qin, B., Li, S.: Chinese Sentence Similarity Computing Based on Semantic Dependency Parsing (2003)
6. Mei, J.J., Zhu, Y.M., Gao, Y.Q., Yin, H.X.: Tong Yi Ci Ci Lin, 2nd edn., Shanghai (1996)
7. Moses, http://www.statmt.org/moses/
8. Papineni, K., Roukos, S., Ward, T., Zhu, W.-J.: BLEU: a Method for Automatic Evaluation of Machine Translation. In: Proceedings of the 40th Annual Meeting of the Association for Computational Linguistics (ACL), Philadelphia, July 2002, pp. 311–318 (2002)

An Axiomatic Approach to Exploit Term Dependencies in Language Model

Fan Ding[1,2] and Bin Wang[2]

[1] Graduate University, Chinese Academy of Sciences, Beijing, 100080, China
[2] Institute of Computing Technology, Chinese Academy of Sciences, Beijing, 100080, China
{dingfan,wangbin}@ict.ac.cn

Abstract. One of important problems of dependence retrieval model is the challenge to integrate both single words and dependencies in one weighting schema. Although there are many retrieval models that exploit term dependencies in language modeling framework, seldom of them simultaneously study the problem on different query types, e.g., short queries and verbose queries. In this paper, we derive an axiomatic dependence model by defining several basic desirable constraints that a retrieval model should meet. The experiment results show that our model significantly and robustly improves retrieval accuracy over the baseline (unigram model) in verbose queries and achieves better performance than some state-of-art dependence models.

Keywords: Term Dependency, Language Model, Axiomatic Approach, Retrieval Heuristics.

1 Introduction

When incorporating term dependencies into the retrieval model, one of the important problems of dependence retrieval model is the over-scored problem. It states that without a theoretically motivated integration model, documents containing dependencies may be over-scored if they are weighted in the same way as single words.

The scoring model is the core of retrieval model. Research had been conducted to study the performance of different scoring models. For example, [4] has studied some heuristics that a scoring model should meet in generative models. [7] has studied different feature selection methods in discriminant models. In the Language Modeling framework, there has been some research to resolve the over-scored problem by introducing some ways to incorporate the dependence model with the independent unigram model [1,2,5,6]. For example, the Markov Random Field model [2] defines sequential dependence and full dependence to incorporate term dependencies into a unigram model in keyword-based short queries. The Dependency Language Model [5] use a generative model P(L|D) as a normalized factor to balance the unigram model and dependence model in sentence-based verbose queries. The scoring functions in these works are theoretically sound. However, for different query types, the over-scored problem may behave differently. There are few works to investigate the dependence retrieval model in both short queries and verbose queries.

H. Li et al. (Eds.): AIRS 2008, LNCS 4993, pp. 586–591, 2008.

In this paper, we attempt to solve the over-scored problem using an axiomatic approach through defining several heuristic constraints. The experiment results on three TREC collections show that our model achieves robust improvement both in short queries and verbose queries. In the rest of the paper, Section 2 reviews some previous relevant work; Section 3 presents the definition of several retrieval heuristics and our axiomatic dependence model; a series of experiments on TREC collections is presented in Section 4; some conclusions are summarized in Section 5.

2 Related Works

In the dependence retrieval model, the final relevance score of a query and a document consists of both the independence score and dependence score, such as Bahadur Lazarsfeld expansion [8] in classical probabilistic IR models. With respect to language model, the situation is same.

In the simplest bi-gram model [3], the probability of bi-gram (q_{i-1},q_i) in document D is smoothed by its unigram:

$$P_{smoothed}(q_i \mid q_{i-1}, D) = \lambda \times P(q_i \mid D) + (1-\lambda) \times P(q_i \mid q_{i-1}, D)$$
$$where,\ P(q_i \mid q_{i-1}, D) \equiv \frac{P(q_{i-1}q_i \mid D)}{P(q_{i-1} \mid D)} \tag{1}$$

If $P(q_i|q_{i-1},D)$ is smoothed as Equation (1), the relevance score of query $Q=\{q_1q_2\ldots q_m\}$ and document D is:

$$\begin{aligned}
&\log P(Q \mid D) = \log P(q_1 \mid D) + \sum_{i=2\ldots m} \log P_{smoothed}(q_i \mid q_{i-1}, D) \\
&= \log P(q_1 \mid D) + \sum_{i=2\ldots m} \log(\lambda \times P(q_i \mid D) + (1-\lambda) \times P(q_i \mid q_{i-1}, D)) \\
&\propto \sum_{i=1\ldots m} \log P(q_i \mid D) + \sum_{i=2\ldots m} \log(1 + \frac{1-\lambda}{\lambda} \times \frac{P(q_{i-1}q_i \mid D)}{P(q_i \mid D) \times P(q_{i-1} \mid D)}) \\
&= \sum_{i=1\ldots m} \log P(q_i \mid D) + \sum_{i=2\ldots m} MI_{smoothed}(q_{i-1}, q_i \mid D) \\
&usually,\ MI(q_{i-1}, q_i \mid D) \equiv \log \frac{P(q_{i-1}q_i \mid D)}{P(q_i \mid D) \times P(q_{i-1} \mid D)}
\end{aligned} \tag{2}$$

In Equation (2), the first score term is the independence unigram score and the second score term is the smoothed dependence score. Usually λ is set to 0.9, i.e., the dependence score is given less weight than the independence score.

Dependence Model [5], which can be regarded as the generalization of the bi-gram model, gives the relevance score of a document as:

$$\log P(Q \mid D) = \sum_{i=1\ldots m} \log P(q_i \mid D) + \log P(L \mid D) + \sum_{(i,j)\in L} MI(q_i, q_j \mid L, D) \tag{3}$$

In Equation (3), L is the set of term dependencies in query Q. The score function consists of three parts: a unigram score $\log P(q_i|D)$, a smoothing factor $\log P(L|D)$, and a dependence score $MI(q_i,q_j|L,D)$. It is the parsing score P(L|D) that serves as a normalization factor (or penalty) to balance the impact of single terms and term dependencies. To estimate P(L|D), the authors use many smoothing strategies.

MRF [2] combines the score of full independence T, sequential dependence O and full dependence U in an interpolated way using coefficient (λ_T,λ_O,λ_U). They are set to (0.8,0.1,0.1), which means that the dependence model is given less weight than the independence model.

Though the above models are derived from different theories, smoothing is an important part when incorporating term dependencies. And it also our belief that smoothing strategy will play a important role when dealing with over-scored problem. The scoring functions in the above models have different forms, which are dedicated to a specific type of queries, i.e., short queries or verbose queries. In the next section, we attempt to define some scoring functions by several retrieval heuristics. Some smoothing strategies will be used in the definition of these scoring functions.

3 Axiomatic Dependence Model

In smoothed unigram language model [9], the RSV formula has the general form of:

$$RSV_{UG}(Q,D) = \sum_{w \in Q \cap D} p(w,Q) \log \frac{p_{DML}(w \mid D)}{\alpha_D \, p(w \mid C)} + |Q| \log \alpha_D \tag{4}$$

α_D is the smoothed coefficient of document D. There are two parts in Equation(4): $P_{DML}(w|D)$, which is discounted maximum likelihood estimation of word w in D, and P(w|C), which is the weight of word w in collection C. When it comes to the dependence model, we define the RSV formula as:

$$RSV_{DEP}(Q,D) = \sum_{(w_i,w_j) \in L \cap D} p(w_i,w_j,Q) \log \frac{p(w_i,w_j \mid D)}{p(w_i,w_j \mid C)}) \tag{5}$$

Here we only consider the term dependency as a term pair. L is the set of term pairs in Q. $P(w_i,w_j|D)$ and $P(w_i,w_j|C)$ are the probabilities of term pair (w_i,w_j) in document D and Collection C. We follow the spirit of the axiomatic approaches in [4] and define several constraints to derive the two parts of $P(w_i,w_j|D)$ and $P(w_i,w_j|C)$.

3.1 Constraints on Weight of Term Dependency in Collection

For a query of three terms (q_1,q_2,q_3), we define term dependencies as term pairs (q_1,q_2) and (q_2,q_3).

- C1: If the document frequency of the three terms satisfies: $DF(q_1)=DF(q_2)=DF(q_3)$, $DF(q_1,q_2)>DF(q_2,q_3)$, then $P(q_1,q_2|C)>P(q_2,q_3|C)$
- C2: If $DF(q_1,q_2)=DF(q_2,q_3)$ and $DF(q_1)>DF(q_2)>DF(q_3)$, then $P(q_1,q_2|C)>P(q_2,q_3|C)$

$DF(q_i)$ is the document frequency of term q_i, $DF(q_i, q_j)$ is the document frequency of term q_i and q_j. From the above two constraints, we define $P(w_i,w_j|C)$ in a interpolated way with λ as the weight:

$$P(w_i, w_j \mid C) = \lambda \times \frac{DF(w_i, w_j)}{DocCount(C)} + (1-\lambda) \times \frac{\min(DF(w_i), DF(w_j))}{DocCount(C)} \tag{6}$$

3.2 Constraints on Weight of Term Dependency in Document

- C3: Let $C_D(q_1,q_2,R)=C_D(q_2,q_3,R)$, and $C_D(q_1)= C_D(q_2)= C_D(q_3)$. If term pair (q_1,q_2) is a common collocation while (q_2,q_3) is not, then $P(q_2,q_3|D)>P(q_1,q_2|D)$.

$C_D(q_i,q_j,R)$ denotes the number of times that q_i and q_j have a relation in Document D and $C_D(q_i)$ is the number of times q_i appears in D. We define the non-trivial degree of a term pair(w_i,w_j), $NTD(w_i,w_j)$, as $\log[(1+DF(w_i,w_j))/(1+DF(w_i,w_j,R))]$, where $DF(w_i,w_j,R)$ is the count of documents in which w_i and w_j have a relation.

- C4: If $NTD(q_1,q_2)=NTD(q_2,q_3)$, $C_D(q_1,q_2,R)>C_D(q_2,q_3,R)$, $C_D(q_1)> C_D(q_2) >C_D(q_3)$, then $P(q_1,q_2|D)>P(q_2,q_3|D)$.

  From the constraints C3 and C4, we define the $P(w_i,w_j|D)$ as

$$P(w_i, w_j \mid D) = \log\left(\frac{1+DF(w_i, w_j)}{1+DF(w_i, w_j, R)}\right) \times \frac{C_D(w_i, w_j, R)}{doclength(D)} + \frac{\min(C_D(w_i), C_D(w_j))}{doclength(D)} \tag{7}$$

Finally, we smooth the $P(w_i,w_j|D)$ using a discounted method, and get the final RSV formula of the dependence model as:

$$RSV_{DEP}(Q, D) = \sum_{(w_i, w_j) \in L \cap D} p(w_i, w_j, Q) \log\left(1 + \alpha \times \frac{p(w_i, w_j \mid D)}{p(w_i, w_j \mid C)}\right) \tag{8}$$

In equation (8), there are two parameters: α and λ. The relevance score of (Q,D) is the sum of $RSV_{UG}(Q,D)$ and $RSV_{DEP}(Q,D)$. $P(w_i,w_j|Q)$ is set to $1/|L|$ in the following experiments; $|L|$ stands for the number of term pairs in L.

4 Experiments and Results

We evaluated our axiomatic dependence model (ADM) using three TREC collections. Table1 shows some statistics of these collections. In the documents' indexing phase, terms are stemmed using WordNet and stop words are not removed. There are several retrieval models in the experiments. UG is the implementation of unigram language model using Dirichlet Prior smoothed KL-divergence (KLD) model [10]. BG is a smoothed bigram language model which sets the weight of the empirical bigram probability to 1% and weight of unigram probability to 99% (Differently from [3], we find 99% achieves better performance than 90%). CULM is our implementation of Concept Unigram Language Model described in [6]. In CULM, the estimation of bigram is same as BG. ADM is implemented by incorporating above UG model and our axiomatic dependence model. All these models were implemented in Lemur.

Table 1. Statistics of TREC collections

Collection	Description	Size (MB)	#doc.	Query
WSJ	Wall Street Journal (1990,1991,1992), in Disk 2	248	74,520	51-200
AP	Associated Press (1988,1989), in Disk 1&2	489	164,597	51-200
TREC7-8	Disk 4&5 (no CR)	3,120	528,155	351-450

Table 2. Results on Short Queries

Models		WSJ	AP	TREC7&8
UG	AvgPr.	0.2498	0.2336	0.1982
	RPr.	0.2611	0.2702	0.2511
BG	AvgPr.	0.2567	0.2235	0.2021
	RPr.	0.2736	0.2626	0.2471
ADM	AvgPr.	0.2565	0.2346	0.2022*
(0.4,0.1)	RPr.	0.2655	0.2718	0.2528
	Optimal	(0.4,0)	(0.3,0.3)	(0.4,0.1)

Table 3. Results on Verbose Queries

Models		WSJ	AP	TREC7&8
UG	AvgPr.	0.211	0.2159	0.1893
	RPr.	0.2315	0.258	0.2387
CULM	AvgPr.	0.2286*	0.2193	0.1918
	RPr.	0.2389	0.2597	0.243
ADM	AvgPr.	0.2272*	0.2364*	0.2052*
(0.2,0.2)	RPr.	0.2417	0.2768*	0.2509*
	Optimal	(0.2,0.2)	(0.2,0.6)	(0.2,0.2)

We evaluated our ADM model on both short queries (title field of topic) and verbose queries (description field). On short queries, the set of term pairs L is treated as adjacent words in the queries (queries that have more than two words are used). Thus, ADM and BG share the same query term dependencies. On verbose queries, Minipar [11] is used to parse the queries and derive the concepts/phrases. The identified phrases of more than two words are decomposed to sequential relations between adjacent words. ADM and CULM share the same query term dependencies.

In the equation(7) of ADM model, we define $C_D(w_i,w_j,R)$ as count of term pair (w_i,w_j) appears with the window of size N in document D. We tried the window size N of 5, 10, 20 and 40 and find the optimal N is 10. This size is close to sentence length and it is used in the following experiments. The main evaluation metric in this study is the non-interpolated average precision (AvgPr.) and R-Precision(RPr.). We train the parameters (α,λ) by directly maximizing AvgPr on three respective collections. The settings of (α, λ) used in the experiments are listed in the first column. The last rows list the optimal settings of (α, λ). An asterisk denotes where the improvement over UG is statistically significant according to paired-samples t-test (significant level is 0.05). We find the optimal settings of (α,λ) are very close to each other. Especially, WSJ and TREC7&8 collections get the same optimal setting of (α,λ) for verbose queries experiments. This shows the robustness of our ADM model.

We can see from Table 3 that the improvement of ADM over UG is statistical significant on verbose queries. While its improvement on short queries is not as significant, it is better than BG. This shows the effectiveness of our ADM model.

5 Conclusions

In order to deal with the over-scored problem in dependence retrieval model, we have derived a new score function using the axiomatic retrieval framework. Firstly, we formally define four heuristic constraints that any reasonable retrieval function should satisfy. Secondly, some smoothing strategies are used to derive the score function by these constraints. The experiments on both short queries and verbose queries show some promising improvements over some state-of-art dependence retrieval model.

In our future works, other reasonable constraints could also be considered. And more comparison can be conducted to verify the effectiveness of our ADM model.

Acknowledgments. This work is supported by a China 973 Hi-tech project (No. 2004CB318109), a project of Natural Foundation of Sciences of China (No. 60603094) and a China 863 Hi-tech project (No. 2006AA010105).

References

1. Alvarez, C., Langlais, P., Nie, J.-Y.: Word Pairs in Language Modeling for Information Retrieval. In: Proceedings of RIAO 2004, pp. 686–705 (2004)
2. Metzler, D., Croft, W.B.: A Markov random field model for term dependencies. In: Proceedings of SIGIR 2005, pp. 472–479 (2005)
3. Song, F., Croft, W.B.: A general language model for information retrieval. In: Proceedings of SIGIR 1999, pp. 279–280 (1999)
4. Fang, H., Zhai, C.: An Exploration of Axiomatic Approaches to Information Retrieval. In: Proceedings of SIGIR 2005, pp. 480–487 (2005)
5. Gao, J., Nie, J.-Y., Wu, G., Cao, G.: Dependence Language Model for Information Retrieval. In: Proceedings of SIGIR 2004, pp. 170–177 (2004)
6. Srikanth, M., Srihari, R.: Exploiting Syntactic Structure of Queries in a Language Modeling Approach to IR. In: Proceedings of CIKM 2003, pp. 476–483 (2003)
7. Geng, X., Liu, T.-Y., Qin, T., Li, H.: Learning to Rank II: Feature selection for ranking. In: Proceedings of SIGIR 2007, pp. 407–414 (2007)
8. Losee, R.M.: Term dependence: Truncating the Bahadur Lazarsfeld expansion. Information Processing and Management 30(2), 293–303 (1994)
9. Zhai, C., Lafferty, J.: A study of smoothing methods for language models applied to ad hoc information retrieval. In: Proceedings of SIGIR 2001, pp. 334–342 (2001)
10. Zhai, C., Lafferty, J.: Model-based feedback in the KL-divergence retrieval model. In: Proceedings of CIKM 2001, pp. 403–410 (2001)

A Survey of Chinese Text Similarity Computation

Xiuhong Wang[1], Shiguang Ju[1], and Shengli Wu[2]

[1] Jiangsu University, Zhenjiang, China
[2] University of Ulster, Northern Ireland, UK
Lib510@ujs.edu.cn, Jushig@ujs.edu.cn, S.wu1@ulster.ac.uk

Abstract. There is not a natural delimiter between words in Chinese texts. Moreover, Chinese is a semotactic language with complicated structures focusing on semantics. Its differences from Western languages bring more difficulties in Chinese word segmentation and more challenges in Chinese natural language understanding. How to compute the Chinese text similarity with high precision, recall and low cost is a very important but challenging task. Many researchers have studied it for long time. In this paper, we examine existing Chinese text similarity measures, including measures based on statistics and semantics. Our work provides insights into the advantages and disadvantages of each method, including tradeoffs between effectiveness and efficiency. New directions of the future work are discussed.

Keywords: Chinese text similarity, Chinese information processing, Similarity algorithm.

1 Introduction

Text similarity computation plays important roles in the fields of duplicate detection, document classification, automatic question answering, filtering, and so on. It is important to compute similarity as effectively and efficiently as possible, and some efforts have been made to compare the quality of various similarity measures in some contexts [1].

Chinese is more difficult to understand by computers than Western language, such as English. English is a merplotactic language focusing on syntax, while Chinese is a semotactic language focusing on semantics. Since Chinese texts do not have a natural delimiter between words, Chinese word segmentation and feature vector spaces in high-level dimensions are expensive. Moreover, difficulties of Chinese text similarity computation also lie in extraction of keywords, processing of synonymies, polysemies and combination among concepts, distinguishing commendatory and derogatory remarks, etc. For example, sentence 1 “坚决反对青少年过度上网’ and sentence 2 “每天12个小时在网上冲浪是一种享受！” are very different in expression, but much related in content. All these characteristics of Chinese language cause problems in efficiency and effectiveness, which are essential for evaluating a similarity algorithm of Chinese text similarity computation.

In this paper, we provide a survey of Chinese text similarity computation, and the advantages and disadvantages of both measures based on statistics and semantics understanding are evaluated.

H. Li et al. (Eds.): AIRS 2008, LNCS 4993, pp. 592 – 597, 2008.

2 Chinese Text Similarity Measures Based on Statistics

2.1 Measures Based on VSM

The Vector Space Model (VSM) [2] [3] is an algebraic model widely used not only for Western languages but also for Chinese document similarity computation. It represents each natural language document in a formal manner as a vector with one real-valued component, which is computed using the TF-IDF (Term Frequency-Inverted Document Frequency) weighting scheme for each term (keywords) in a multi-dimensional linear space. The set of documents in a collection then turns into a vector space, with one axis for each term. The set of terms is a predefined collection of terms, occurring in the document corpus. To compensate for the effect of different document/query length, the standard way of quantifying the similarity between a query/document vector and a document vector is to compute the cosine similarity coefficient between them, which reflects the degree of similarity in the corresponding terms and term weights.

The IF-IDF algorithm based on VSM has some disadvantages. On the one hand, it can achieve good effectiveness only if enough number of words are included. On the other hand, only the statistic of words in text is considered and all the structural and semantic information is lost. In addition, VSM assumes that all the terms are independent (orthogonal) to each other, which is incorrect regarding natural language that causes problems with synonyms or strong related terms. Meanwhile, messages have to pass through a stopword-list, stemming and thesaurus-algorithms before they are forwarded to the VSM.

Generalized Vector Space Model (GVSM) [3] [4], an improved VSM, assigns a document vector to each one without the assumption of orthogonal terms. For the GVSM, term-angles are based on the computation of co-occurrence of terms. Jorg Becker [5] explored a topic-based vector space model (TVSM), which does not assume independence between terms and is flexible regarding the specification of term-similarities. Stemming and thesaurus can be fully integrated into the model. Cheng [6] established a component frequency model (CFM) based on components, in which Chinese texts were expressed in the aspect of component granularities. The text attribute vector space model was established according to the statistic of component frequency. This measure has higher precision and recall than algorithms based on keywords. The most advantage of CFM is that it avoids difficulties in Chinese word segmentation.

2.2 Measures Based on Attribute Theory and on Hamming Distance

Pan [7] analyzed the relationship between textual attributes and the attribute barycenter coordinate model, and established a text attribute barycenter coordinate model, in three-dimension Cartesian coordinates, and based on attribute theory. A text vector and a query vector in the attribute coordinate were represented. After deciding the criterion and computing the distance between the vectors, the similarity between the texts and the queries was elicited. This measure could express more semantic information than VSM and improve precision and recall. However, difficulties in automatic Chinese word segmentation cannot be avoided.

Hamming distance, named after Richard Hamming, is the number of positions in two strings of equal length, for which the corresponding elements are different. In other words, it measures the number of substitutions required to change one into the other. For example, the Hamming distance between 1011101 and 1001001 is 2. Zhang [8] presented a Chinese text similarity algorithm based on the theory of Hamming distance. The advantage of this algorithm is that it greatly simplified the similarity computation, not like many modern genomics algorithms taking gaps into consideration. We cannot simply see how many characters are different because some differences are more significant than others.

3 Measures Based on Semantic Understanding

With more effort, inter-document similarity can be measured in more sophisticated ways, in which semantics are considered.

3.1 Words Similarity Measures

Ontology, which usually uses a thesaurus, is one of approaches to compute the distance between words. Early in 1995, Agirre E. [9] gave a. a proposal for word sense disambiguation using conceptual distance based on WordNet. Wang [10] computed the Chinese word similarity on "Cilin". Liu and Li [11] presented a word semantic similarity algorithm based on HowNet. Suppose there are two Chinese words W_1 and W_2, if W_1 has n concepts: $S_{11}, S_{12}, \ldots, S_{1n}$, W_2 has m concepts as: $S_{21}, S_{22}, \ldots, S_{2m}$, then

$$Sim(W_1, W_2) = \max_{i=1..n, j=1..m} Sim(S_{1i}, S_{2j}) . \tag{1}$$

Primitive is a minimal meaning unit of describing a concept. HowNet describes each concept by using a series of primitives. The similarity of two primitives P_1 and P_2,

$$Sim(P_1, P_2) = \frac{\alpha}{d + \alpha} . \tag{2}$$

Here d is the distance between P_1 and P_2, and α is a smoothing parameter, which is the distance when the similarity equals to 0.5.

Xia [12] proposed a method based on HowNet, geared to semantic and could be expanded. The proposed method defined a similarity computation formula among HowNet's sememes according to information theory, finding a way out of the difficulty that out-of-vocabulary (OOV) words cannot participate in semantic computation by implementing concept segmentation and automatic semantic production to OOV words, and realized the similarity computation on the semantic level among arbitrary words. Although semantic lexicon based on structure relations between conceptions is simple and effective, building a Chinese semantic lexicon is a large-scale system project.

Individual Chinese characters, bi-grams, n-grams (n>2), and words are the most widely used indexing units. Kwok [13] concluded that single character indexing is good but not sufficiently competitive, while bi-gram indexing works surprisingly well and it is as precise as short-word indexing.

Zhao [14] proposed a measure based on Chinese Character Association Measurement (CCAM) matrix. To reduce the higher complexity, using the association of Chinese characters for text similarity analysis was probed with feature words. CCAM is a better solution than bi-gram and keywords indexing. Since word segmentation is not needed, the algorithm is useful in massive Chinese data corpus.

3.2 Sentence Similarity Measures

Che [15] improved the original edit-distance approach by computing Chinese sentence semantic similarity with more information in structure. Both HowNet and "Cilin" thesauruses were used as the semantic resource to compute the semantic similarity between two words. Jing [16] proposed a model of semantic-based text formalization - context framework model (CFM). In this model, a text was represented in three demensions: domain, situation and background. Based on the context framework, the algorithm dealt with the domain of the text and the semantic role of the object, computed synonymies, polysemies, combinations of concepts, and distinguishing commendatory and derogatory remarks. The algorithm can improve the efficiency of text filtering practically.

3.3 Paragraphs and Documents Similarity Measures

Jin [17] proposed a set of textural similarity algorithm to study paragraphs similarity. HowNet was used to compute words similarity. Only substantives such as noun, verb, adjective, numeral, measure word and pronoun, were extracted, it could avoid the complicated Chinese syntax analysis. Suppose there are two paragraph texts: t_1 and t_2. Each paragraph consists of a group of sentences, $t_1 = \{s_{11}, s_{12}, \ldots, s_{1m}\}, t_2 = \{s_{21}, s_{22}, \ldots s_{2n}\}$. The similarity between paragraphs t_1 and t_2 is:

$$sim(t_1, t_2) = \frac{1}{k}\sum_{i=1}^{k} simS_{\max_i} . \tag{3}$$

Afterward, Jin [18] proposed a set of document-structure-based algorithms to detect plagiarism of Chinese academic articles, with the help of document-structure analysis, fingerprinting and word-frequency techniques.

Compared with measures based on statistics, measures based on semantic understanding can lead to more accurate judgment. But the processing needed for them is usually much more expensive.

4 Conclusive Remarks

For Chinese information retrieval, Chinese natural language processing, and many other tasks, a measure for text similarity is a key element of it.

In this paper we have reviewed a set of different measures which can be used to evaluate inter-document similarity. All those measures can be divided into two categories: statistics-based and semantics-based. Compared with each other, they have pros and cons. For the statistics-based on measures, they need training on massive corpus to achieve reasonably good performance since contextual information is not kept, while this is not needed for the semantics-based measures. On the other hand, using semantics-based measures is more expensive on computational cost than statistics-based measures, though the former can often lead to more accurate judgment than the latter.

Up to now, how to find a Chinese text similarity measure which can be performed effectively and efficiently remains to be a challenging issue. In our opinion, language models which can support certain semantic structures is likely to be a good research direction. In order to achieve this, some measures such as information-theorectic measures [19, 20] and corresponging algorithms are desirable.

Acknowledgments. This work was financially supported by the Natural Science Foundation of Jiangsu Province (grant No. BK2006073).

References

1. McGill, M., Koll, M., Norrreault, T.: An Evaluation of Factors Affecting Document Ranking by Information Retrieval Systems. Technical Report, Syracuse University School of Information Studies (1979)
2. Lesk, M.E.: Computer Evaluation of Indexing and Text Processing. Journal of the ACM 1, 8–36 (1968)
3. Beaza-Yates, R., Ribeiro-Neto, B.: Modern Information Retrieval. ACM Press, New York (1999)
4. Wong, S.: On Modeling of Information Retrieval Concepts in Vector Spaces. ACM Transactions on Database Systems 2, 299–321 (1987)
5. Becker, J., Kuropka, D.: Topic-based Vector Space Model. Business Information Systems. In: Proceedings of BIS 2003, Colorado Springs, USA (2003)
6. Cheng, Y., Wu, S.: Text Similarity Computing Based on Components. Computer Engineering and Design 18, 3444–3446 (2006)
7. Pan, Q., Wang, J., Shi, Z.: Text Similarity Computing Based on Attribute Theory. Chinese Journal of Computers 6, 653–655 (1999)
8. Zhang, H., Wang, G., Zhong, Y.: Text Similarity Computing Based on Hamming Distance. Computer Engineering and Applications 19, 21–22 (2001)
9. Agirre, E., Rigau, G.: A Proposal for Word Sense Disambiguation Using Conceptual Distance. In: International Conference on Recent Advances in Natural Language Processing, Velingrad, pp. 258–264 (1995)
10. Wang, B.: Study on Chinese-English Bi-language Corpus Automatic Ordering. Institute of Computing Technology, Chinese Academy of Science (1999)
11. Liu, Q., Li, S.: Words Semantic Similarity Computation Based on HowNet. In: Proceedings of the 3rd Symposium on Chinese Words Semantics, vol. 5 (2002)
12. Xia, T.: Study on Chinese Words Semantic Similarity Computation. Computer Engineering 6, 191–194 (2003)
13. Kwok, K.L.: Comparing Representations in Chinese Information Retrieval. In: Proceedings of the ACM SIGER 1997 Conference, pp. 34–41 (1997)
14. Zhao, Y., Li, Q.: Chinese Character Association Measurement Method and Its Application on Chinese Text Similarity Analysis. Computer Applications 6, 1396–1397, 1400 (2006)
15. Che, W.: Chinese Sentences Similarity Computation Oriented the Searching in Bilingual Sentence Pairs. In: The 7th National JSCH, pp. 81–88. Tsinghua University press, Beijing (2003)

16. Jin, Y.: Text Similarity Computing Based on Context Framework Model. Computer Engineering and Applications 16, 36–39 (2004)
17. Jin, B., Shi, Y., Teng, H.: Similarity Algorithm of Text Based on Semantic Understanding. Journal of Dalian University of Technology 2, 291–297 (2005)
18. Jin, B., Shi, Y., Teng, H.: Document-structure-based Copy Detection Algorithm. Journal of Dalian University of Technology 1, 125–130 (2007)
19. Javed, A., Aslam, M.F.: An Information-theoretic Measure for Document Similarity. ACM SIGIR 3, 449–450 (2003)
20. Lin, D.: An Information-theoretic Definition of Similarity. In: Proc. 15th International Conf. on Machine Learning (1998)

Study of Kernel-Based Methods for Chinese Relation Extraction

Ruihong Huang[1,2], Le Sun[1], and Yuanyong Feng[1,2]

[1] Institute of Software, Chinese Academy of Sciences, South Fourth Street, Zhong Guan Cun, Hai Dian. 4, 100190, Beijing, China
[2] Graduate University of Chinese Academy of Sciences, Yu Quan Street, Shi Jin Shan. 19, 100049, Beijing, China
{Ruihonghuang.china,lesunle}@gmail.com, yuanyong02@iscas.ac.cn

Abstract. In this paper, we mainly explore the effectiveness of two kernel-based methods, the convolution tree kernel and the shortest path dependency kernel, in which parsing information is directly applied to Chinese relation extraction on ACE 2007 corpus. Specifically, we explore the effect of different parse tree spans involved in convolution kernel for relation extraction. Besides, we experiment with composite kernels by combining the convolution kernel with feature-based kernels to study the complementary effects between tree kernel and flat kernels. For the shortest path dependency kernel, we improve it by replacing the strict same length requirement with finding the longest common subsequences between two shortest dependency paths. Experiments show kernel-based methods are effective for Chinese relation extraction.

1 Introduction

The aim of relation extraction as a subtask of information extraction is to find various predefined semantic relations between pairs of entities in text. The research of relation extraction has been advanced by the Message Understanding Conferences (MUC) [1] and the NIST Automatic Content Extraction (ACE) program (ACE, 2000-2007) [2] in Phase 2. As a subtask of Information Extraction, relation extraction can be utilized in many applications such as Question Answering and information retrieval.

To our knowledge, no work has been done to examine the performance of the tree kernel or the shortest path dependency kernel on Chinese corpus. Since more errors exist in Chinese syntax analysis compared with English, whether these kernel methods are applicable for Chinese relation extraction is uncertain.

2 Related Work

Since relation extraction task was first introduced in MUC6, many methods, such as feature-based [3, 4], tree kernel-based [11, 12, 13, 14] and composite kernel-based [5, 15, 16], have been proposed in literature.

Feature-based methods for relation extraction employ explicitly various linguistic features, from lexical features, syntactic information to dependency trees and

H. Li et al. (Eds.): AIRS 2008, LNCS 4993, pp. 598–604, 2008.

semantic knowledge in Max Entropy model [3] or SVM model [4]. The feature-based methods have achieved the state-of-art performances. However, the feature selection is heuristic, so it needs much manual efforts, besides, it is difficult to improve since the features in nearly all linguistic levels have been examined [5, 6], furthermore, feature-based methods lack the ability to explore the structural syntactic or dependency information which should be quite important for relation extraction in our first sight.

In contrast, kernel-based methods [7, 8] has the potential for further performance improvements as it gives us an elegant way to explore structural features implicitly by computing the similarity between two objects via a kernel function, thus give us a good chance to explore the parsing or dependency information of the sentence where the entity pair occur. In recent years, different kernel types have been proposed for English relation extraction, from the hierarchical tree kernels [11] and [12] defined on shallow parse tree or dependency trees, shortest path dependency kernel [13] to the current convolution parse tree kernel[14,16]. Moreover, composite kernel which generally is a combination of a tree kernel and a feature-based kernel has advanced the performance further [15, 16]. Up to now, the kernel-based methods have achieved and recently exceeded the best performance of the feature-based methods for relation extraction.

For Chinese entity relation extraction, various features and different classification algorithms, for example, SVM [17] and bootstrapping [18], have been proposed in feature-based framework, and the reported results are usually alluring just on certain types of relations or on non-standard dataset. Besides, [19] proposed a novel improved Edit kernel for Chinese relation extraction, in which, the author improved the edit distance algorithms considering the Chinese word property and applied it to the Chinese relation extraction. However, as to structural kernels, which have been explored extensively recently for English, few work has been done up to now.

When we reflect over the two hierarchical kernels, the shortest path dependency kernel and the convolution parse tree kernel, we can see undoubtedly the convolution parse tree kernel has achieved better performance than the other kernel types and the shortest path dependency kernel is the most efficient since it is so fast and still achieves general performance. Therefore, in this paper, we explore different feature spaces involved in convolution kernel and improve the original shortest path dependency kernel with an aim to loosen its strict constraints of same lengths on shortest dependency paths.

3 Tree Spans in Convolution Kernel for Relation Extraction

A convolution kernel [7] aims to capture structural information in terms of substructures. As a special convolution kernel, the convolution tree kernel proposed in [9] counts the number of common sub-trees as the structural similarity between two parse trees. Furthermore, [10] has proved that the convolution tree kernel can be computed in $O(|N1|*|N2|)$ where N1 and N2 equal to the number of nodes in two trees.

Same as most of previous work on kernel-based relation extraction, we first employ the parse tree segment within the entity pair (called Path-enclosed tree in [14]) in convolution kernel for entity relation extraction. Then, considering the limited parse tree spans within the nested entity pair, we extend the feature spaces for the nested relation instances to incorporate a verb factor in two strategies.

The first strategy (figure 1) is to extend the cases by including the highest level of verb (the main predicate) of the sub-sentence where the entity pair occurs when there are not any verb between the entity pair. The second strategy (figure 2) is to include the nearest verb to the entity pair (similar to the dynamic span expansion method proposed in [16]). The two strategies are out of our wonder that whether the main predicate which is powerful to determine the semantics of the whole sub-sentence or the nearest verb which is more relevant to the entity pair semantic relation will contribute more to the final entity relation identification.

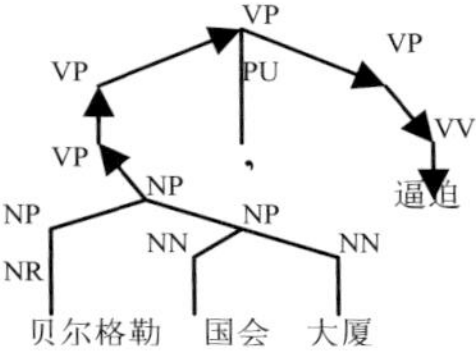

Fig. 1. strategy one

Fig. 2. strategy two

4 Shortest Path Kernel Based on Longest Common Subsequence

The shortest path kernel proposed in [13] requires two shortest dependency paths to have the same length which may contribute to the low recall. To loosen the constraint, we improve the kernel by summing up the common word classes on the longest common subsequences of two shortest dependency paths other than the original shortest dependency paths.

In implementation (figure 3), the general part-of-speech features (the italic) of the nodes is used to match the longest common subsequences while the part-of-speech features, the word features and the entity type features of the nodes in the resulted longest common subsequences are utilized to compute the similarity. Same as in [13], no normalization is done in the improved shortest path dependency kernel.

Besides, in our experiments, the dependency information is generated by the same CFG (Context Free Grammar) parser employed to generate the parse trees, so the dependency links form a tree naturally. Thus, the shortest dependency path is

民国 →	驻 ←	捐赠 ←	代表 ←	透过 ←	大使馆
NOUN	***VERB***	*NOUN*	*NOUN*	*VERB*	***NOUN***
NN	**VV**	NN	NN	VV	**NN**
GPE					**ORG**

民国 →	驻	← 大使馆
NOUN	***VERB***	***NOUN***
NN	**VV**	**NN**
GPE		**ORG**

Fig. 3. two relations of type "PART-WHOLE" with the similarity equal to 4*3*4=48

somewhat different from the path used in [13], the meanings slightly differ accordingly. There is no unified direction distribution in the original shortest dependency path. In meaning, the shortest dependency paths of Bunescu and Mooney mainly describe the predicate-argument interactions, that is, the dependency relation of the two entities as the arguments to the predicates. In contrast, we always have a parent, pointed by dependency nodes on both sides, in our shortest dependency path. In meaning, our shortest dependency paths convey an ordered dependency series connecting the entity pair. Accordingly, the improved dependency path kernel will operate on the two longest common subsequences belonging to the two sides from the two ends to the central parent.

5 Experiments

5.1 Experimental Setting

Data: we use the Chinese portion of ACE (Automatic Content Extraction) 2007 corpora from LDC to conduct our experiments. In the ACE 2007 data, the training set include 689 documents and 6900 relation instances while the testing set include 160 documents and 1977 relation instances. Specifically, there are 2030 non-nested relation instances in the training set and 620 in the testing set. The ACE 2007 data defines 7 major entity types: Facility, GPE, Location, Organization, Person, Vehicle and Weapon. In this paper, we assume that the entities and their types are already known. Besides, all pairs of entities occurring in the same sentence are treated as potential relation instances. The data imbalance and sparseness are potential problems in ACE corpus, for example, the "Employment" subtype has 1265 positive instances while the "Artifact" subtype has only 6 instances in the training data, besides, in both the training part and the testing part, the negative samples are 10 times more than the positive samples.

Data processing: We select the Stanford syntactic parser to generate the sentence parse tree and dependency list. For we don't find any appropriate POS tagger preserving the Penn standard required by the above parser, we use the POS tagger internal to the parser. Besides, we utilize the PKU Chinese word segmenter. During parsing, we segment sentences which are too long to be parsed at one time. For shortest path dependency kernel, we construct the sentence dependency tree utilizing the output dependency list and extract the shortest path from it.

Classifier: we select the LibSVM [20] as our classifier and insert into the convolution tree kernel [21] and our shortest dependency path kernel based on the longest common subsequences. Besides, we adopte one vs. one strategy for the multi-class classification. The parameters are selected using 5-fold cross-validation.

Kernel normalization: the convolution tree kernel, the linear entity kernel and its expansion occurring in all experiments are all normalized while the two shortest path dependency kernels are not. The normalizing method is:

$$\hat{K}(T_1,T_2) = K(T_1,T_2) / \sqrt{K(T_1,T_1) \bullet K(T_2,T_2)}$$

Evaluation Methods: we adopt Recall (R), Precision (P) and F-measure (F) standards.

5.2 Experimental Results

(1) Table 1 compares the performances of three different parse tree spans involved in the pure convolution tree kernel. We threshold the output to get best performances. We can see the Path-enclosed parse tree achieves best performance, although much lower than 74.12/54.90/63.07 (P/R/F) attained in our feature-based experiments. Especially, strategy one gets the lowest F-measures which generally will involve larger part of the parse tree in the convolution tree kernel. Thus, on the one hand, we may infer that more noises have been introduced into the kernel computation which may counteract the benefits of involving somewhat more complete syntactic structures in the kernels. On the other hand, we are confirmed that more efforts are needed to find appropriate and effective feature spaces.

(2) With the aim to examine the complementary effect of the tree features and the flat features, we also experimented with two composite kernels which are combinations of the above convolution kernel with the linear entity kernel (Comp-linear) and its polynomial expansion (Comp-poly) as stated in [15]: $K_c(R_1,R_2)=\alpha \cdot K_T(R_1,R_2)+(1-\alpha)K_{Ent}$.

In table 2, the F-measures show both composite kernels (with α set to 0.5 in Comp-linear and 0.7 in Comp-poly) advance the performances to a extent not as evident as that on English dataset [15]. Especially, unlike [15, 16], the polynomial entity kernel embodying bi-gram entity information doesn't improve any performance compared with the linear entity kernel, ruins the evaluations reversely.

Table 1. Three different feature spaces involved in the pure convolution tree kernel

Feature spaces	Avg. P(%)	Avg. R(%)	Avg. F
Path-enclosed tree	40. 05	33. 04	35.03
Strategy one	22. 99	26. 68	22.06
Strategy two	29. 31	40. 19	32.66

Table 2. Performance comparison of different kernel setups on the path-enclosed parse trees

Path-enclosed	Avg. P(%)	Avg. R(%)	Avg. F
Pure Conv	40.05	33.04	35.03
Comp-linear	41.67	34.75	36.73
Comp-poly	41.59	34.75	36.72

(3) Our original motivation to study kernel-based methods is to improve the extraction performance of the non-nested relations, they have longer distances and more complex syntactic structures between entity pairs, thus are more difficult to identify using flat features in feature-based methods. The initial results of kernel method on non-nested relations are presented in table 3. Besides, we give our previous experimental results using feature-based methods. The comparison shows that structural

Table 3. Performance comparison on non-nested relations of ACE 2007 data

Non-nested	Avg. P(%)	Avg. R(%)	Avg. F
Conv Kernel	40.09	33.22	34.13
Feature-based	54.81	19.15	29.57

kernel's performance utilizing only the parse tree information has exceeded the feature-based methods on non-nested relation instances a bit.

Table 4. Classification performance between the original shortest path dependency kernel and the improved version based on the longest common subsequences over real relations

Shortest-path dep-kernel	Avg. P(%)	Avg. R(%)	Avg. F
Original	4.52	16.67	7.12
Improved	17.89	18.62	15.20

(4) The experimental results on the shortest dependency path kernel are somewhat disappointing. The improved kernel shows poor performance on the relation detection while the original shortest dependency path kernel has even not any relation detection ability since the trained model treats all the potential relations as positive or negative instances under different parameters. So, in table 4, we only show their multi-classifying performances on all the 1977 positive instances. We can see although our improved kernel has better performance, in general, the classification performances are both too low. When we examined the dependency path representations of relation instances, we find that a large part of Chinese relations lack clear predict-argument dependencies which is presumed in [13] and we may reason that it's really difficult to extract Chinese relations using this type of kernel.

6 Conclusion

In this paper, we explore the effectiveness of kernel-based methods for Chinese relation extraction. The experimental results show that although the current tree kernels haven't achieved as good performance as the feature-based methods, it has given reasonable measures, especially it has overrun feature-based methods on non-nested relations which is difficult to deal with by just using flat features.

References

1. MUC (1987-1998), http://www.itl.nist.gov/iaui/894.02/related_projects/muc/
2. ACE (2002-2007), urlhttp://projects.ldc.upenn.edu/ace/
3. Nanda, K.: Combining lexical, syntactic and emantic features with Maximum Entropy models for extracting relations. In: ACL 2004 (poster)
4. Zhou, G.D., Su, J., Zhang, J., Zhang, M.: Exploring Various Knowledge in Relation Extraction. In: ACL 2005 (2005)

5. Shubin, Z., Ralph, G.: Extracting Relations with Integrated Information Using Kernel Methods. In: ACL 2005 (2005)
6. Jiang, J., Zhai, C.: A systematic exploration of the feature space for relation extraction. NAACL-HLT (2007)
7. Haussler, D.: Convolution Kernels on Discrete Structures. Technical Report UCS-CRL-99-10, University of California, Santa Cruz (1999)
8. Schölkopf, B., Smola, A.J.: Learning with Kernels: SVM, Regularization, Optimization and Beyond, pp. 407–423. MIT Press, Cambridge (2001)
9. Collins, M., Duffy, N.: Convolution Kernels for Natural Language. In: NIPS (2001)
10. Collins, M., Duffy, N.: New ranking algorithms for parsing and tagging: Kernels over discrete structures, and the voted perceptron. In: ACL 2002(2002)
11. Zelenko, D., Aone, C., Richardella, A.: Kernel Methods for Relation Extraction. Journal of Machine Learning Research 2, 1083–1106 (2003)
12. Culotta, A., Sorensen, J.: Dependency Tree Kernel for Relation Extraction. In: ACL 2004 (2004)
13. Bunescu, R.C., Mooney, R.J.: A Shortest Path Dependency Kernel for Relation Extraction. In: EMNLP 2005 (2005)
14. Zhang, M., Zhang, J., Su, J.: Exploring syntactic features for relation extraction using a convolution tree kernel. In: Proceedings of HLT/NAACL (2006a)
15. Zhang, M., Zhang, J., Su, J., Zhou, G.D.: A Composite Kernel to Extract Relations between Entities with both Flat and Structured Features. In: COLINGACL (2006b)
16. Zhou, G.D., Zhang, M., Ji, D.H., Zhu, Q.M.: Tree Kernel-based Relation Extraction with Context-Sensitive Structured Parse Tree Information. In: ACL 2007 (2007)
17. Che, W.X.: Automatic Entity Relation Extraction. Journal of Chinese Information Processing 19(2) (2004)
18. Zhang, S.X.: Study about automatic entity relation extraction. Journal of Harbin Engineering University (July 2006)
19. Che, W.X.: Improved-Edit-Distance Kernel for Chinese Relation Extraction. In: Dale, R., Wong, K.-F., Su, J., Kwong, O.Y. (eds.) IJCNLP 2005. LNCS (LNAI), vol. 3651, Springer, Heidelberg (2005)
20. LibSVM. http://www.csie.ntu.edu.tw/~cjlin/libsvm/
21. Moschitti. Convolution Tree kernel, http://ai-nlp.info.uniroma2.it/moschitti/TK1.2-software/Tree-Kernel.htm

Enhancing Biomedical Named Entity Classification Using Terabyte Unlabeled Data

Yanpeng Li, Hongfei Lin, and Zhihao Yang

Department of Computer Science and Engineering,
Dalian University of Technology, Dalian, China, 116024
lyp_8218@163.com, {hflin,yangzh}@dlut.edu.cn

Abstract. This paper presents a semi-supervised learning method to enhance biomedical named entity classification using features generated from labeled and terabyte unlabeled data, called Feature Coupling Degree (FCD) features. Highly discriminative context words are obtained from labeled free text using Chi-square method and queries formed by combining the named entity and context words are retrieved by search engine. Then the retrieved web page counts are converted into binary features by discretization. We investigate the effect of this type of feature in a biomedical corpus generated from several online resources. Support Vector Machine (SVM) is used as classifier and the performances of different features with various kernels and discretization methods are compared. The results show that the method enhances the classification performance especially for Out-of-Vocabulary (OOV) terms and relative small size of training data. In addition, only using FCD features with polynomial kernels, the performance is competitive to classical features.

Keywords: semi-supervised learning, biomedical named entity, classification, discretization, SVM, polynomial kernel.

1 Introduction

Named entity classification is an important component of text mining, which can be distinguished from named entity recognition. The former is to classify named entities that may not be in free text into right classes and the latter needs to recognize names in free text. Named entity classification can be a powerful auxiliary method for named entity recognition, for instance, building a domain specific dictionary or ontology automatically or entity disambiguation and can also be used to remove noises introduced by query expansion in Information Retrieval (IR) tasks. Named entity classification/recognition in biomedical domain is difficult mainly due to the vast vocabulary of biomedical terms (e.g., gene or protein names) and many new entities are being described all the time. The state-of-the-art performance reported by BioCreative 2004 [1] does not exceed 85%.

For the two tasks, recently most researchers focus on supervised machine learning approach [2][3][4] due to its flexibility and robustness. Classical features used in these systems are similar, which can be categorized into three types:

H. Li et al. (Eds.): AIRS 2008, LNCS 4993, pp. 605–612, 2008.

Surface feature: words, context words, and regular expressions, etc.
Linguistic feature: POS tags, semantic tags, and syntactic tags, etc.
Dictionary feature: external dictionaries, stop words, and common word lists, etc.

Predicting the category of Out-of-Vocabulary (OOV) terms is an essential issue in named entity classification/recognition. All the three types of feature are effective to identify unknown terms, but there are limitations in each of them. For surface feature, surface lexical information must be an indicator to entity class and big size of training data is needed when the vocabulary of entity is large. Context word feature can be useful in some case, but it is not robust since it will not work unless the context is indicative. The linguistic feature is not a strong indicator to entity class and is often used in combination with other features to make prediction according to linguistic role such as POS tags, semantic tags, and syntactic tags, etc. However the current best performance of many of these techniques itself is low and most can only work in free texts. Dictionary feature is widely used in many domains such as [5][6], but it has two drawbacks. First, it is difficult to enumerate all the names in a lexicon for named entities with large vocabulary. Second, noise will be introduced if definitions of the same entity class are inconsistent in the lexicon and training corpus or just due to errors during building lexicon (many can be found in biomedical domain since some lexicons are generated automatically).

In this paper, we introduce a new type of feature, called Feature Coupling Degree (FCD) features, which utilize the co-occurrence information of strong indicative features for instances and classes in labeled and unlabeled data. We also propose a method based on FCD features for biomedical named entity classification. Features are generated from discretized co-occurrence counts of a candidate named entity phrase and typical gene/protein related contexts on Web pages. We investigate its effect with different kernels and discretization methods comparing with classical features.

This method has used both labeled and unlabeled data (Web), so it can be assumed as a semi-supervised learning approach. Semi-supervised machine learning is to improve the performance of supervised learning by incorporating unlabeled data. Many semi-supervised algorithms are proposed in recent years [7][8][9][10]. Most of these methods focus on how to generate "virtual examples" or estimate a better structure of prediction functions utilizing unlabeled data. While we focus on using unlabeled data for feature generation, which is similar to the idea proposed in [11]. Our method also reflects a general framework to learn a new feature representation from unlabeled data.

2 Methods

2.1 Feature Coupling Degree

In supervised machine learning, an instance x is usually represented by a vector of features $(x1, \ldots, xn) \in X \subset Rn$ and one label $y \in \{y1, \ldots, ym\}$. Assume that they are independently generated according to some unknown probability distribution D. The goal is to choose a function $f(x)$ based on training examples where the label y is known so that its prediction error with respect to D is as small as possible. While most feature generation methods are done manually. Intuitively when selecting a feature, two factors should be considered, i.e. its discrimination abilities for instances and for classes. If either of the abilities is too weak, the feature is bound to be irrelevant. Considering these two factors we categorize features into two types: Instance Discriminative Features (IDFs) and Class Discriminative Features (CDFs). Given an instance **x**, an IDF x_i is defined as the feature with relative high ability to discriminate **x** from other instances. Usually the conditional probability $P(\mathbf{x}|x_i)$ is relative high. For a class y, a CDF x_c is the feature with relative high ability to discriminate y from other classes. Usually the conditional probability $P(y|x_c)$ is relative high.

For x_i and x_c, Feature Coupling Degree (FCD) is defined as the value to measure the joint probability of the pair of IDF and CDF.

$$FCD(x_i, x_c) = P(x_i, x_c)\ . \quad (1)$$

It is easy to understand that high FCD values tend to be strong indictors for classification, so they can be used as features. However, usually many of them are not available in limited amount of training data especially when the distribution of the feature space is sparse (e.g., named entity classification). Fortunately, large scale of unlabeled data can be obtained without much effort. Especially the World Wide Web (WWW) provides terabyte unlabeled data for machine learning in NLP (Natural Language Processing). In this context much more FCDs that do not exist in labeled data can be estimated from unlabeled data, which provides the opportunity to make full use of this type of feature.

For the biomedical entity classification task, we choose the normalized term itself as IDF and strong indicative context words as CDFs, which are obtained by computing Chi-square values of terms in training corpus. FCD value is approximately estimated by the Web page count where IDF and CDF co-occur in a phrase. Also a "feature cluster" approach is employed by merging all pairs with the same CDF but different IDF into one dimension of feature. For example, given two instances "NF-kappaB", "PRNP gene", and a CDF "expression of", the feature functions will be $f_{expression\text{-}of}$. The two feature values can be computed as follow:

$$f_{expression\text{-}of}(\text{NF-kappaB}) = FCD(\text{nf kappab, expression of}) = Count(\text{"expression of nf kappab"})\ .$$

$$f_{expression\text{-}of}(\text{PRNP gene}) = FCD(\text{prnp gene, expression of}) = Count(\text{"expression of prnp gene"})\ .$$

In this way, feature space is mapped into a much smaller dimension which is determined by the number of CDFs. In addition it has the ability to identify unknown terms. For instance, if "NF-kappaB" is not in the training corpus, the feature $f_{\text{expression-of}}$ can also be predictive. Finally the Web page counts are converted into binary features by discretization method.

2.2 Feature Generation

Assume that a sentence from training data is denoted as S = {..., t_{-2}, t_{-1}, e_1, e_2, ..., e_n, t_{+1}, t_{+2}, ...} where each element is a token separated by white space, { e_1, e_2, ..., e_n} refers to the entity, t_{+n} is the next nth token and t_{-n} is the previous nth token. The algorithm is described as follow:

1. Context terms are categorized into three types: left contexts, right contexts and surrounding contexts (Table 1)
2. For each type of context, calculate the Chi-square values of context terms. A context term that appears at the context position defined above is treated as positive, otherwise as negative.
3. Select the top context terms with high Chi-square scores as CDFs. Common terms like “the” and “of the” are removed manually. Query is formed using the phrase of named entity plus CDF. Also the entity itself is a query. All the queries are strict match.
4. Get the returned Web page count of the query from Google.
5. Features are obtained by discretizing the Web page counts.

Table 1. Types of indicative contexts used in named entity classification

Context type	Context terms
Left	t_{-1}, (t_{-2}, t_{-1}), (t_{-3}, t_{-2}, t_{-1})
Right	t_{+1}, (t_{+1}, t_{+2}), (t_{+1}, t_{+2}, t_{+3})
Surrounding	(t_{-1}, t_{+1}), (t_{-2}, t_{-1}, t_{+1}), (t_{-1}, t_{+1}, t_{+2})

2.3 Discretization Methods

The range of page count is extremely large and it is impossible to use them directly as features, so discretization techniques must be used to convert them to several binary features. In our experiment we investigate five methods: non-zero value method, simple binary method, equal frequency, CAIM (Class-Attribute Interdependence Maximization) algorithm [12] and manual threshold:

Non-zero value method: only non-zero counts are used as binary features.

Simple binary method: non-zero and zero counts are selected as two binary features respectively.

Equal frequency: zero counts are used as feature, and non-zero counts are discretized equally according to width or frequency.

CAIM: zero counts are used as feature, and non-zero counts are discretized by maximizing the CAIM value iteratively. Detail of the algorithm can be found in [12].

Manual threshold: zero counts are used as feature, and non-zero counts are divided into two intervals manually.

2.4 Classical Feature

We attempt to use stat-of-the-art classical features as the baseline. We make use of most effective features reported in two well-known challenges BioCreative 2004 and JNLPBA 2004. The features include words, normalized terms, n-grams, regular expressions, and stop words list, etc. In addition, some “entity-level” features are added (e.g., length of name and characters in a sliding window), since the entity classification is a little different from entity recognition and these features are not easy to incorporate in most NER systems.

3 Results

3.1 Experimental Design

Training and test corpus in our experiment is derived from the data collection in BioCreative2004 Task 1A [1], where system is required to recognize gene/protein names from MEDLINE abstracts. Our named entity classification experiment was to classify a list of terms not considering the context. The instances were obtained by the following steps:

First, build a lexicon of gene/protein names from four biomedical databases: LocusLink, EntrezGene, BioThesaurus, and ABGene lexicon. We got a lexicon of around 8 million terms. Note that this lexicon is very "noisy", since there are many common English words and other type of named entities.

Table 2. Number of instance in training and test corpus

Training/Test	Positive	Negative
Training	15607	6302
Test	4081	4172

Then, we use maximum match scheme to search the text in training and test data set in BioCreative2004. For both training and test corpus, if the term is in the external dictionary and free text but not in the gold standard, it is treated as negative instance. For training corpus, all terms in gold standard are selected as positive instance. For test corpus, if the term is both in the gold standard and the external dictionary, it is selected as positive instance. Table 2 shows the information in training and test data.

In the procedure of feature generation, we obtained 93 left context terms, 98 right context terms and 63 surrounding context terms from training data using method proposed in Section 2. Classifier is SVM-light [13]. We tested our approach in different size of training data comparing with classical features. Also we show its ability to recognize unknown terms and the performances on different kernels and discretization methods are investigated.

3.2 Performance of FCD Feature

Table 3 shows the comparison of classical features, FCD features and a linear combination of the two features by SVM. Here OOV terms are defined as terms in which any token is not in the training data. Surprisingly, we find that only the FCD features with 6-degree polynomial kernel performs better than the classical features especially for OOV term identification (nearly 4 percent absolute improvement). This has proved our analysis in Section 1. In the experiment, we find that for classical features the feature space is highly sparse mainly due to surface features, and SVM does not work well using non-linear kernel. However, FCD features are able to work well in a high-order polynomial kernel. Also the number of feature and VC-dimension estimated by SVM-light is significantly reduced. The idea behind is to "compress" features space in terabyte of unlabeled data. In addition, a linear combination of the two types of feature with SVM gives an improvement (84.47% to 86.12%, and 80.15% to 84.74%), which indicates that the FCD feature does enhance the classification performance.

Table 3. Comparison of classical features and FCD features

Feature	#of feature	Kernel	VC-dim	F-score	F-score(OOV)
Classical feature	211570	Linear	198582	84.47%	80.15%
Classical feature	211570	Polynomial (6th order)	9339702	67.23%	77.18%
FCD feature	768	Linear	2353	82.06%	82.28%
FCD feature	768	Polynomial (6th order)	2805	84.73%	84.18%
Combination	212338	Linear	317576	**86.12%**	**84.74%**

Table 4. Performance comparison of discretization methods on FCD features

Discretization method	Precision	Recall	F-score
Non-zero value	77.08%	89.83%	82.97%
Simple binary	78.65%	89.61%	83.77%
Equal frequency	78.75%	90.88%	84.38%
CAIM	78.79%	**90.91%**	84.41%
Manual threshold	**81.20%**	88.58%	**84.73%**

Table 4 shows the comparison of different discretization methods using polynomial kernel with degree 6. It can be seen that the performance of most methods differ not much. Non-zero value method is relatively inferior since it discards zero counts that could be indicative sometimes. Manual threshold method is the best of all. In this method, the threshold was assigned 250 for left or right context FCDs, and 50 for surrounding context FCDs. For equal frequency method, we found that the performance was the best when the number of intervals is 2. For CAIM method, all the non-zero counts were automatically divided into 2 intervals. For all the discretization methods, the non-zero Web page count of the name itself was discretized into 8 equal frequency intervals and then converted to binary features.

From the learning curves (Fig. 1), we can see that the advantage of FCD features is even more apparent when the size of training data is small. Using around 1000 named

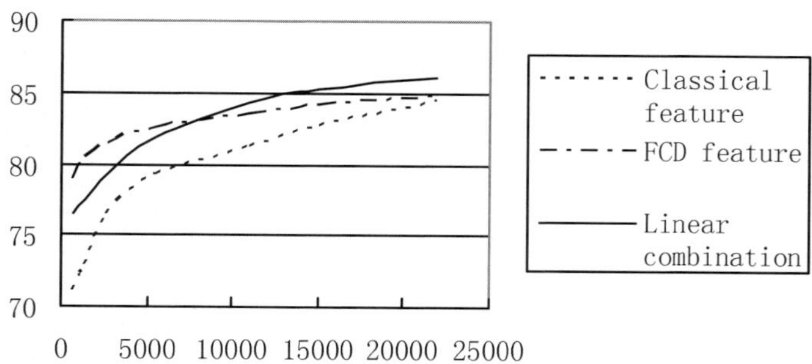

Fig. 1. Learning curves of three types of feature

entities for training, the F-score of FCD features is over 80%, which is about 7 percent absolute higher than the performance of classical features. It indicates that when vocabulary of entity is large and labeled data is little, utilizing unlabeled data is a powerful way to enhance the classification performance. In addition, it can be seen that combining all the features in a linear kernel does not always provide best result depending on the scale of training data. This can be explained by the change of feature distribution with the size augment of training data and in some case the improvement produced by non-linear kernel is bigger than combination with the classical features.

4 Conclusions and Future Work

We investigate the characteristic of a novel type of feature, called Feature Coupling Degree (FCD) features, and its application in biomedical named entity classification. The experimental results show that it is effective in non-linear kernel method and has better generalization ability to recognize OOV terms than classical features. The method can be viewed as a general framework for semi-supervised learning, which seeks to generate new feature representations from unlabeled data. We think it has three advantages comparing with other semi-supervised learning methods: first, efficient search engine technique make it able to utilize terabyte unlabeled data such as Web. Second, using the statistical figure as features is most robust to deal with noise in unlabeled data. Third, the feature space can be dense and dimension can be low, so the performance can be enhanced by kernel mapping.

Furthermore, there is much room for this method to be further improved. For example, the method to generate IDFs and CDFs can be further researched. The type of these features can be expanded. For FCD features, it is important to find a better method for feature cluster, discretization, or combination with original features. We believe that the idea proposed in this paper can be expanded to other problems. Also it is interesting to make a theoretical analysis, since it has provided a more automatic and general way for feature generation.

Acknowledgments. This work is supported by grant from the Natural Science Foundation of China (No.60373095 and 60673039) and the National High Tech Research and Development Plan of China (2006AA01Z151).

References

1. Lynette, H., Alexander, Y., Christian, B., Alfonso, V.: Overview of BioCreAtIvE: critical assessment of information extraction for biology. BMC Bioinformatics 6(suppl. 1), 1 (2005)
2. Finkel, J., Dingare, S., Manning, C.: Exploring the boundaries: gene and protein identification in biomedical text. BMC Bioinformatics 6(suppl. 1), 5 (2005)
3. McDonald, R., Pereira, F.: Identifying gene and. protein mentions in text using conditional random fields. BMC Bioinformatics 6(suppl. 1), 6 (2005)
4. Guodong, Z., Jie, Z., Jian, S., et al.: Recognizing names in biomedical texts: a machine learning approach. Bioinformatics 20(7), 1178–1190 (2004)

5. Cohen, W.W., Sarawagi, S.: Semi-Markov Conditional Random Fields for Information Extraction. In: Eighteenth Annual Conference on Neural Information Processing Systems (NIPS) (2004)
6. Tomohiro, M., Sevrani, F., Masaki, M., Kouichi, D., Hirohumi, D.: Gene/protein name recognition based on support vector machine using dictionary as features. BMC Bioinformatics 6(suppl. 1), 8 (2005)
7. Vapnik, V.: Statistical learning theory. Wiley-Interscience, Chichester (1998)
8. Blum, A., Mitchell, T.: Combining labeled and unlabeled data with co-training. In: 11th Annual Conference on Computational Learning Theory (COLT), pp. 92–100 (1998)
9. Zhu, X., Ghahramani, Z., Lafferty, J.: Semi-supervised learning using Gaussian fields and harmonic functions. In: 20th International Conference on Machine Learning (ICML) (2003)
10. Ando, R., Zhang, T.: A framework for learning predictive structures from multiple tasks and unlabeled data. Journal of Machine Learning Research 6, 1817–1853 (2005)
11. Rajat, R., Alexis, B., Honglak, L., Benjamin, P., Andrew, Y.N.: Self-taught learning: Transfer learning from unlabeled data. In: 24th International Conference on Machine Learning (ICML) (2007)
12. Lukasz, K., Krzysztof, C.: CAIM Discretization Algorithm. IEEE Transactions on Knowledge and Data Engeering 16(2), 145–153 (2004)
13. Joachims, T.: Making large-Scale SVM Learning Practical. In: Schölkopf, B., Burges, C., Smola, A. (eds.) Advances in Kernel Methods - Support Vector Learning, MIT-Press, Cambridge (1999)

An Effective Relevance Prediction Algorithm Based on Hierarchical Taxonomy for Focused Crawling

Zhumin Chen, Jun Ma, Xiaohui Han, and Dongmei Zhang

School of Computer Science & Technology, Shandong University, Jinan, 250061, China
chenzhumin@mail.sdu.edu.cn

Abstract. How to give a formal description for a user's interested topic and predict the relevance of unvisited pages to the given topic effectively is a key issue in the design of focused crawlers. However, almost all previous known focused crawlers do the Relevance Predication based on the Flat Information (RPFI) of topic only, i.e. regardless of the context between keywords or topics. In this paper, we first introduce an algorithm to map the topic described in a keyword set or a document written in natural language text to those described in hierarchical topic taxonomy. Then, we propose a novel approach to do the Relevance Predication based on the Hierarchical Context Information (RPHCI) of the taxonomy. Experiments show that the focused crawler based on RPHCI can obtain significantly higher efficiency than those based on RPFI.

Keywords: Focused Crawling, Relevance Prediction, Hierarchical Topic Taxonomy, Topic Description.

1 Introduction

Focused crawling is a potential solution to scaling problems for these general-purpose crawlers caused by the limited resources and rapid growth of the World Wide Web. Focused crawlers traverse a subset of the Web to only gather documents relevant to a given topic [1-9]. They are activated in response to particular information needs described as topics. Thus, how to describe the topics clearly and predict the relevance based on the topics is the key of focused crawling. Current topics are often denoted by keywords [3], documents in natural language [4] and hierarchical taxonomy [5-8].

To the best of our knowledge, all these relevance computing methods in [1-9] are RPFI in spit of topic representation models. In this paper, we first introduce a method to map the topics described in keywords or documents in natural language to those represented in hierarchical topic taxonomy, which contains hierarchical context information. Furthermore, we propose a novel approach to predict the relevance of the unvisited page to a given topic based on the topic and its hierarchical context, i.e. RPHCI. We also present a simple but effective method to weight the given topic and its contextual topics in terms of their relative hierarchies in the taxonomy. Finally, experiments show the focused crawler based on RPHCI can obtain significantly higher efficiency than that based on RPFI.

H. Li et al. (Eds.): AIRS 2008, LNCS 4993, pp. 613–619, 2008.

2 Topics Description and Mapping

We use ODP to describe topics to improve the relevance prediction. Every internal node in the ODP is a *topic node* (N_l) corresponding to a topic where l is the name of the node. Once N_l is identified, *topic path* (P_l) is the path from the root of ODP to N_l. *Topic path depth* (D_l) is the number of nodes in the P_l. Then, $P_l = \{ l_1, l_2, ..., l_{D_l} \}$. *Topic subtree* is the subtree rooted at N_l and including its all descendants. All nodes description, anchor text and description of these example pages are regarded as *topic description* (Des_l) used as the document in natural language to describe the topic N_l.

Fig. 1 illustrates an example. Every internal node (marked as bold line ellipse) may be a topic. There are many example Web pages (marked as dot line rectangle). For a topic "NBA" corresponding to N_{NBA}, D_{NBA} = 3, P_{NBA} = {Sports, Basketball, NBA} where Sports and Basketball are regarded as the hierarchical context of topic "NBA".

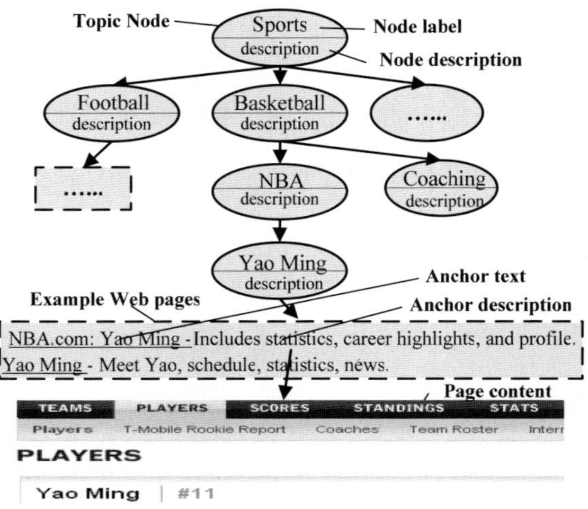

Fig. 1. Topics description example in ODP

We give an algorithm to map the topics described in keywords or documents in natural language text (denoted by T_{KW} and T_{NLT}) to those described in ODP (denoted by T_{ODP}). T_{NLT} is first translated to keywords by TF (Term Frequency). Then, the mapping process of T_{NLT} is the same with that of T_{KW}. Let $T_{KW} = \{kw_1, kw_2, \ldots, kw_J\}$ where kw_j is the jth keyword. *Matched Topic Paths* (*MTP*) are all those topic paths containing at least one keyword of T_{KW}. Let $MTP = \{\{P_1,wt_1\},\{P_2,wt_2\},\ldots,\{P_K,wt_K\}\}$ where P_k is the kth topic path and wt_k is the corresponding weight. Let MaxD denotes the maximal depth of all topic paths in *MTP*. The mapping algorithm is as follows:

Algorithm 1. Topic Mapping

Input: T_{KW}, T_{ODP}
Output: *MTP*
1 for every keyword kw_j in T_{KW} do map kw_j to T_{ODP};

2 for every matched topic path P_k do

3 { $wt_k = \alpha * \sum_{i=1}^{\text{MaxD}} (i/\text{MaxD}) * b_i$ where $b_i = \begin{cases} 1 & \exists\, kw_j,\ l_i \text{ matche } kw_j \\ 0 & \forall\, kw_j,\ l_i \text{ not matche } kw_j \end{cases}$ and

$\alpha = 1/\sum_{i=1}^{\text{MaxD}} (i/\text{MaxD})$;

4 $\{P_k, wt_k\} \rightarrow MTP$;}

5 return MTP;

3 Relevance Prediction and Focused Crawling

In this section, we propose a novel relevance predicting approach, i.e. RPHCI, which makes full use of the hierarchical context information of T_{ODP}. Let T represent the given topic. Let url_p be the parent URL and point to page p_p which has been fetched. c_p is the textual content of p_p. url_c is an unvisited URL on p_p and points to the page p_c that has not been crawled. a_c represents the anchor text of url_c and ac_c extracted between the predefined boundaries is the textual context of a_c. Then, the triple $\{c_p, \{a_c, ac_c\}, url_c\}$ is used to predict the relevance of p_c to the T.

Let $v_t = \{\{kw_1,wt_1\},\{kw_2,wt_2\},\ldots,\{kw_Q,wt_Q\}\}$ represent the vector of T in which kw_q and wt_q denote a term's name and weight respectively. In this paper, T is described by three models: (1) Keywords, i.e. T_{KW}, $v_t = v_t^{KW} = \{\{kw_1,1\}, \{kw_2,1\}, \ldots \{kw_Q,1\}\}$. (2) Documents in natural language text, i.e. T_{NLT}, $v_t = v_t^{NLT} = \{\{kw_1,wt_1\},\{kw_2,wt_2\}, \ldots \{kw_Q,wt_Q\}\}$ where wt is calculated by TF. (3) ODP, i.e. T_{ODP}, $v_t = v_t^{ODP} = P_l = \{\{l_1,1/D_l\},\{l_2,2/D_l\},\ldots,\{l_{D_l},D_l/D_l\}\}$. It is obvious that v_t^{kw} and v_t^{NLT} only contain flat information, while v_t^{ODP} includes some hierarchical topic context information. A simple but effective approach is used to weight T and its contextual topics in terms of their relative hierarchies in the ODP. T's weight is $D_l/D_l = 1$. The weights will decrease when the distance between the contextual topics and T increases. For example, $v_t^{ODP} = \{\{\text{Sports}, 1/3\}, \{\text{Basketball}, 2/3\}, \{\text{NBA}, 3/3\}\}$ for the topic "NBA".

3.1 Relevance Prediction

We use three kinds of available information, namely the triple $\{c_p, \{a_c, ac_c\}, url_c\}$, to predict the relevance of p_c to a given topic T.

The page content relevance prediction denoted by R_{PC} is that using c_p to evaluate the relevance of p_c to T where v_c is the vector of c_p represented by TF.

$$R_{PC} = (v_c \bullet v_t)/(v_c \times v_t) \tag{1}$$

The anchor text relevance prediction is that using a_c and ac_c to estimate the relevance of p^c to T, represented in R_a and R_{ac} respectively. Therefore, the combining relevance R_{AT} of R_a and R_{ac} is as follows in which v_a and v_{ac} are the vectors of a_c and ac_c represented by TF respectively.

$$R_{AT}=\begin{cases}R_a & if\ R_a>0\\ R_{ac} & if\ R_a=0\end{cases}\ \text{in which}\ R_a=\frac{v_a\bullet v_t}{|v_a|\times|v_t|}\ \text{and}\ R_{ac}=\frac{v_{ac}\bullet v_t}{|v_{ac}|\times|v_t|} \tag{2}$$

URLs of pages are not randomly created but associate semantic meanings with the pages content. The tokens in a known URL may be used to predict the relevance. Suppose that a candidate url_c, it is parsed into tokens stored in the list *TL* (Token List) in term of the "." and "/". Note that it is not case sensitive in the match process. Finally, the relevance score R_{url} is as follows where β is used to normalize the R_{url}.

$$R_{url}=\sum_{q=1}^{Q} wt_q * t_q * \beta \ \text{in which}\ t_q=\begin{cases}1 & \text{if}\ kw_q\in TL\\ 0 & \text{if}\ kw_q\notin TL\end{cases}\ \text{and}\ \beta=1/\sum_{q=1}^{Q} wt_q \tag{3}$$

We get the aggregate Relevance Score (*RS*) by summing the weighted individual relevance score.

$$RS = wt_1 * R_{PC} + wt_2 * R_{AT} + wt_3 * R_{url} \tag{4}$$

Here wt_1, wt_2, and wt_3 are weights used to normalize the different relevance scores. In our particular implementation, we chose to use weights such that each individual relevance score is almost equally balanced.

3.2 Focused Crawling Algorithms

This section presents two focused crawling algorithms based on RPHCI and RPFI respectively.

Algorithm 2. Focused Crawling based on RPHCI

Input: url_{seed}, depth (*D*) and the number (*N*) of the pages to be crawled, a topic *T* (T_{KW}, T_{NLT} or T_{ODP})
Output: *PS* (crawled Pages Set)
if Topic is T_{NLT} then translate it to T_{KW};
if topic is T_{KW} then call algorithm1 to get *MTP*; else {{the query topic path, 1}} in T_{ODP} → *MTP*;
url_{seed}.depth = *D*; url_{seed}→*UL*; //*UL* is URLs priority List
while *UL* is not empty and *PS*.size < *N*
  remove the first element of *UL* to url_p; crawl p_p; p_p →*PS*;
  for each uncrawled outgoing hyperlink url^c in p_p do
    for every Topic Path P_k in *MTP* do compute R^k_{PC}, R^k_{AT}, R^k_{url} and RS^k in terms of function (1), (2), (3) and (4) respectively;
    $RS=\sum_{k=1}^{K}(wt_k * RS^k)$;
    if $RS > \delta$, then set url_c.depth = *D*; else url_c.depth = url_p.depth - 1;
    if url_c exists in *UL* then *RS* = Max{the existing *RS*, the new *RS*} and reorder *UL* if necessary; url_c.depth = Max{the existing depth in *UL*, the new depth};
    else if url_c.depth > 0 then insert url_c at its right location in *UL*;
endwhile;
return *PS*;

We get algorithm3, i.e. focused crawling based on RPFI, by replacing step 2 of algorithm2 with "{{T_{KW}, 1}} → *MTP*". The algorithm3 will be as the baseline crawler in the following experiments. The differences between algorithm2 and algorithm3 are as follows: the input topic T for algorithm2 can be represented in three ways, T_{KW}, T_{NLT} and T_{ODP}. If the topic is T_{KW} or T_{NLT}, it will be firstly mapped to the ODP by calling the topic mapping algorithm1. While the input topic for algorithm3 can be T_{KW} or T_{NLT}.

4 Experiments

We implemented algorithm2 and algorithm3 and conducted experiments on real data set on different topics and websites. The experiments show that the algorithm2 based on RPHCI outperforms the algorithm3 based on RPFI significantly.

Two measures were used to evaluate the performance. *Precision* (*harvest*) *rate* measures the query result at page level. A page is a "relevant page" if its R_{PC} is greater than a certain threshold. Therefore, $precision\_rate = n_1 / N$ where n_1 is the number of relevant pages retrieved. The sum of information evaluates the result regarding all collected pages as a whole. Thus, $sum\_of\_info = \sum_{(\text{every page in } PS)} R_{PC}$.

We collected 20 topics described in T_{KW}, T_{NLT} and T_{ODP} respectively. At run time, $D = 3$, $N = 5000$, $Q = 5$, $\delta = 0.1$, and $wt_1 = wt_2 = wt_3 = 1/3$. Table 1 shows the experimental results. It is obvious that the algorithm2 outperforms the algorithm3 significantly.

Table 1. Average *precision_rate* and *sum_of_info* for 20 topics

	Algorithm3		Algorithm2		Improvement Ratio
precision_rate	T_{KW}	12.68%	T_{KW} / T_{NLT} /	18.72%	1.48
	T_{NLT}	11.46%			1.63
sum_of_info	T_{KW}	58.46	T_{ODP}	83.09	1.59
	T_{NLT}	49.79			1.67

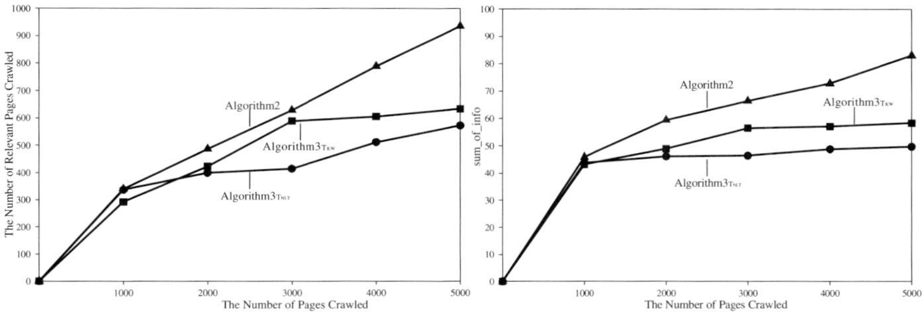

Fig. 2. The number of relevant pages and sum of information crawled for 20 topics

Fig. 2 illustrates dynamic performance during different stages of the crawling process. We see that the performance of algorithm2 outperforms that of algorithm3 during the completely crawling process.

We also evaluate the weighting method. Let $v_t^{ODP}{}' = P_l = \{\{l_1,1\},\{l_2,1\},...,\{l_{D_l},1\}\}$ without weight and $v_t^{ODP} = P_l = \{\{l_1,1/D_l\},\{l_2,2/D_l\}, ..., \{l_{D_l},D_l/D_l\}\}$ with weight. We use $v_t^{ODP}{}'$ and v_t^{ODP} as the input topic of algorithm2 represented as alrorithm2$_{nw}$ and algorithm2$_w$ respectively. As shown in Fig. 3, algorithm2$_w$ outperforms alrorithm2$_{nw}$.

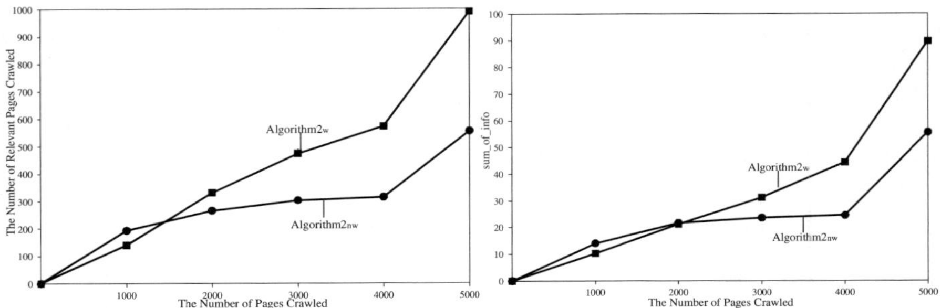

Fig. 3. The number of relevant pages and sum of information of algorithm2$_w$ and alrorithm2$_{nw}$

5 Conclusions and Future Work

In this paper, we studied the problem of topic description by mapping the topic described in keywords or documents in natural language text to that in hierarchical topic taxonomy. We proposed an improved focused crawling algorithm which predicts the relevance more accurately by using of the hierarchical context. Finally, we empirically showed that the new focused crawler based on hierarchical context outperforms that based on flat information. For the future research, it would be interesting to evaluate the impact of different topic specificities to the crawling performance.

References

1. Novak, B.: A Survey of Focused Web Crawling Algorithms. In: Proceedings of SIKDD 2004 at Muticonference IS, Slovennia, pp. 55–58. ACM, New York (2004)
2. Chau, M., Chen, H.: Comparison of Three Vertical Search Spiders. IEEE Computer 36(5), 56–62 (2003)
3. Hersovici, M., Jacovi, M., Maarek, Y., et al.: The Shark-Search Algorithm - An Application: Tailored Web Site Mapping. In: Proceedings of the Seventh International World Wide Web Conference, Brisbane, Australia (1998)
4. TREC, http://trec.nist.gov/pubs/trec12/papers/WEB.OVERVIEW.pdf
5. Chen, Z., Ma, J., et al.: An Improved Shark-Search Algorithm based on Link Analysis. Journal of Computational Information Systems 3(4) (2007)

6. Srinivasan, P., Menczer, F., Pant, G.: A General Evaluation Framework for Topical Crawlers. Information Retrieval 8(3), 417–447 (2005)
7. Menczer, F., Pant, G., Srinivasan, P.: Topical Web Crawlers: Evaluating Adaptive Algorithms. ACM TOIT 4(4), 378–419 (2004)
8. Sengor, I., Altingovde, Ulusoy, O.: Exploiting Interclass Rules for Focused Crawling. IEEE Intelligent Systems 19(6), 66–73 (2004)
9. Ehrig, M., Maedche, A.: Ontology-focused crawling of Web documents. In: Proc. of the 2003 ACM symposium on Applied computing, Melbourne, Florida (2003)
10. Open Directory Project, `http://dmoz.org/`

Short Query Refinement with Query Derivation

Bin Sun, Pei Liu, and Ying Zheng

Institute of Computational Linguistics, Peking University, Beijing 100871, China
{bswen,zhengying,liupei}@pku.edu.cn

Abstract. In this paper we present a method for short query refinement. The method includes a query retrieval model that constructs multiple derived queries for the user's query, where derived queries denote a set of queries that are closely related to the query submitted by the user. Derived queries can be efficiently obtained using the indexing and retrieval of a small-unit index, which has index terms that are commonly used words and word senses. Each of the derived queries can be associated with a rank value according to its similarity to the user's search query. Derived queries are useful for improving the current query refinement method and for constructing the final results.

Keywords: Short query problem, derived query, word sense.

1 Introduction

One of the major challenges for present-day IR systems and search engines is the short query problem, namely, the users have become more and more accustomed to submitting short queries that consist of very few keywords, which are typically of broad use and meanings, and thus often include many possible purposes. Since the current mainstream technology of IR systems and search engines is keyword-based document indexing and retrieval [1], the returned results in response to short queries comprising ambiguous keywords are often heterogeneous in topics, genres and quality, which makes great difficulties for the users to efficiently find interested information. For example, the query "virus" (or "notebook", "mp3", etc.) is a highly ambiguous search query, with which different users may express very different meanings: the search for biologic viruses, or a computer virus (software); and for each of the possible meanings, there may be various kinds of usage, e.g., in the computer virus case, the user's search topic may be one of multiple possibilities: virus prevention, download of virus cleaning software, virus library updating, elements of computer viruses, etc.

Some search engines and enterprise retrieval systems have employed a method to address the short query problem, which is called *query refinement*, or equivalently, *related queries*, or *suggested searches* (or even *experts suggestions* [2]). The idea is that, when a submitted query is very ambiguous, the system will indicate that the user may try a few more refined, narrowed and less ambiguous queries for better results. Such method can be very helpful in many scenarios, especially when the user has only some vague idea to search. Although such query refinement method may handle the short query problem successfully to some extent, the method has not been popular in mainstream applications.

H. Li et al. (Eds.): AIRS 2008, LNCS 4993, pp. 620–625, 2008.

In our opinion, this may be due to several limitations and shortcomings associated with such kind of query refinement. First, the refined, related, similar or suggested queries are usually queries that were submitted by other users in the search log, instead of mining from document contents, and hence may be still heterogeneous in topics. Understanding, properly selecting and retrying these refined queries would provide significant additional difficulties and burden to the user. Second, such refined queries are not utilized to generate or improve the final results presented to the user, and hence do not lead to perceived improvement in the new search results list. And thirdly, the refined queries per se are obtained by keyword matching against the original short query, and thus are unrelated or not grouped in meaning, or still very ambiguous. The user would have to face many more longer but unrelated queries. With these and other reasons, current query refinement tends to be insufficient for handling the short query problem.

In this paper, we present a new method for short query refinement. The method includes a query retrieval model that constructs multiple derived queries for the user's query, where *derived queries* denote a set of queries that are closely related to the query submitted by the user. Each of the derived queries may represent a more specific meaning or a more concrete form of usage of the user's query. Derived queries are obtained from many sources (including document contents) and can be efficiently obtained using the indexing and retrieval of a small-unit index. Each of the derived queries can be associated with a rank value according to its similarity to the user's search query, its frequency of search, the number and ranks of the documents in its corresponding search results, etc. Describe that derived queries can be employed to improve or supplement the current query refinement method, or to provide a new mechanism for constructing the final results to return to the user.

2 Query Derivation

In this paper, we use derived queries to denote a set of queries that are closely related to a search query submitted by the user. Each of the derived queries represents a more specific meaning, or a more concrete form of usage, or a derived or auxiliary semantic, or a collocation with other associated words of the user's query. For a given query, to obtain its derived queries effectively, a query set consisting of a large number of candidate queries, called a *candidate query set*, can be pre-constructed, wherein each of the queries may be used as a derived query of some search query. Such query sets can be constructed by extracting candidate queries from multiple sources, using semantic dictionaries, collocation libraries, phrase rules, document contents and corpus statistics, in addition to mining the user search query logs. The resulting candidate query set may consist of millions of candidate queries to sufficiently cover most of the closely related forms of each of its elements.

With such a candidate query set, the process to obtain the derived queries of a given query becomes the process to find out its closely related candidate queries from the candidate query set, corresponding to various synonyms, semantic equivalents, ambiguous forms and collocations. There are multiple methods that can be used to implement such a lookup process. Since the number of query strings in the candidate query set may be very large, for the reason of efficiency, a special candidate query retrieval method using small index units can be employed, which comprises the following steps:

- Indexing each of the queries in the candidate query set as a small document with an index lexicon including only small index units;
- Building an inverted index for the whole candidate query set;
- Processing the user's search query as a small document the same way as the candidate queries, and via the inverted index of the candidate query set, selecting the candidate queries with a certain similarity with the search query.

The retrieval method actually used for selecting similar candidate queries can be one of any retrieval models well known in the field, such as the Vector Space Model (VSM) [3], a probabilistic retrieval model, or a language model. The particular point in this method is the use of small (fine-grained) index units for the indexing of the candidate query set. In order to perform the retrieval of synonymy and equivalent use, the document vectors corresponding to candidate queries can be transformed into a set of semantic index units, which comprises the semantic classification tags of the entries of the above fine-grained index lexicon. With such transformation, the index terms are changed to the semantic classification tags, and the inverted index may be accordingly built with these tags. Such retrieval method is a semantic-based retrieval model.

The semantic classification system used for indexing the candidate query set can be adapted from the lexical sense set of the WordNet project [4], where a semantic classification tag is denoted by a *synset* (synonym set). WordNet identifies a large number of semantic classes for commonly used words and denotes them with well-formed numerical tags, and further organizes these semantic classes with multiple semantic relations. Currently there are multilingual versions of the WordNet database [5], which can be used to support multilingual query derivation.

The sense space of the candidate query set is constructed with the *synset_ids* being its dimensions. For example, if some query **Q** containing the word "bank", then **Q** will have non-zero components in the above 17 dimensions corresponding to the *synset_ids* of the word "bank". The concrete value of a component is determined by the term weighting method of the model. In this paper, the conventional VSM weighting scheme of "term frequency - inverse document frequency" ($tf \cdot idf$ weighting) is adopted to determine the component values on the sense dimensions, with the index term being the sense tags *synset_ids*, and thus the term frequency *tf* being the sense frequency *sf*. The similarity of any two candidate queries $\mathbf{Q}_i$ and $\mathbf{Q}_j$, denoted by $\mathrm{sim}(\mathbf{Q}_i, \mathbf{Q}_j)$, is measured by the cosine of the angle between their vectors on the sense space,

$$\mathrm{sim}(\mathbf{Q}_i, \mathbf{Q}_j) = \cos(\mathbf{Q}_i, \mathbf{Q}_j) .$$

As in conventional case, such similarity may be further adjusted by other factors like term proximity, Boolean relations, etc. In addition, other term weighting schemes and similarity measures may be adopted the same way.

With such settings, the process of selecting multiple derived queries with a user's search query **Q** would comprise the following steps:

- Decomposing the query **Q** into small index units using the special index lexicon;
- Looking up the inverted index of the candidate query set using **Q**'s small index units to obtain a set of relevant candidate queries;

- Computing the similarities of these relevant candidate queries with **Q** using the above formula, and selecting some candidate queries $\mathbf{Q}_1$, $\mathbf{Q}_1$, …, $\mathbf{Q}_n$ to be the derived queries of **Q**, which have the largest similarity values (or have a similarity value that is larger than a given threshold).

3 Computing Derived Query Rank

After obtaining the derived queries $\mathbf{Q}_1$, $\mathbf{Q}_2$, …, $\mathbf{Q}_n$ of a user's search query **Q**, as elaborated above, the search results of any $\mathbf{Q}_i$ of these derived queries can be individually constructed according to the conventional document retrieval processing, and then a search result list of $\mathbf{Q}_i$ is generated by sorting the results by their estimated similarities with the query $\mathbf{Q}_i$. The number of derived queries, however, may be very large, usually around the scale of thousands, namely $n \sim 10^3$. It would take an exceedingly long time of processing if all the search results of these derived queries are individually constructed. On the other hand, the number of queries that can be simultaneously processed by the retrieval system is limited. Thus it is usually unfeasible to construct the search results for all the selected top n derived queries $\mathbf{Q}_{1,\ldots,n}$. To make the query derivation method practical, each of the derived queries $\mathbf{Q}_{1,\ldots,n}$ can be associated with a rank, and at each time of user interaction, only a few derived queries with higher ranks are selected to actually generate a search result list for each query. In the following, the ranks of the derived queries $\mathbf{Q}_1$, $\mathbf{Q}_2$, …, $\mathbf{Q}_n$ of a search query **Q** are denoted by QueryRank($\mathbf{Q}_i|\mathbf{Q}$), i=1, 2, …, n. QueryRank($\mathbf{Q}_i|\mathbf{Q}$) represents the priority degree that the system presents the derived query $\mathbf{Q}_i$ together with its search results when the user's search query is **Q**.

The rank QueryRank($\mathbf{Q}_i|\mathbf{Q}$) can be simply defined to be the similarity of the queries $\mathbf{Q}_i$ and **Q**, $\quad \text{QueryRank}_1(\mathbf{Q}_i|\mathbf{Q}) = \text{sim}(\mathbf{Q}_i , \mathbf{Q})$.

In a more comprehensive implementation, QueryRank($\mathbf{Q}_i|\mathbf{Q}$) would be determined with an additional factor $f_{\text{History}}(\mathbf{Q}_i)$, which is the frequency of query $\mathbf{Q}_i$ in the historical search log of a search engine:

$$\text{QueryRank}_2(\mathbf{Q}_i|\mathbf{Q}) = a \cdot \text{sim}(\mathbf{Q}_i , \mathbf{Q}) + b \cdot v\,(f_{\text{History}}(\mathbf{Q}_i)) ,$$

where a and b are two adjustable parameters, representing the importance of the similarity and the search frequency respectively. In our system, $a = b = 0.5$, and the function $v\,(f)$ takes a linear form as follows:

$$v\,(f_{\text{History}}(\mathbf{Q}_i)) = f_{\text{History}}(\mathbf{Q}_i) \cdot u(\mathbf{Q}_i) ,$$

$$u(\mathrm{Q}_i) = \frac{1 + \log\left(tf(\mathrm{Q}_i)\right)}{1 + \log\left(\frac{1}{n}\sum_{j=1}^{n} tf(\mathrm{Q}_j)\right)} \times \log\frac{N}{1 + df(\mathrm{Q}_i)}$$

where tf ($\mathbf{Q}_i$) and df ($\mathbf{Q}_i$) are the term frequency (total times of occurrence) and document frequency (number of documents containing $\mathbf{Q}_i$) in current document collection of the query $\mathbf{Q}_i$, and N is the total number of documents in the collection.

After obtaining the derived queries $\mathbf{Q}_1$, $\mathbf{Q}_2$, …, $\mathbf{Q}_n$ of a user's search query **Q**, the derived queries $\mathbf{Q}_{1,\ldots,n}$ are then ranked and sorted by the above QueryRank_1 or

QueryRank$_2$. In the first time of user interaction, the first group of the top $m < n$ derived queries $\mathbf{Q'}_1$, $\mathbf{Q'}_2$, …, $\mathbf{Q'}_m$ with higher ranks are selected to search the inverted index of the document collection, and a search result list is generated for each of them. In the second time of user interaction, when user chooses to look up more following derived queries, the next group of m derived queries $\mathbf{Q'}_{m+1}$, $\mathbf{Q'}_{m+2}$, …, $\mathbf{Q'}_{2m}$ $(2m \leq n)$ are selected and processed accordingly. So on and so forth. In our system, the range of m is among 5 ~ 15, which is the number of derived queries that are selected to actually generate search results at each time.

4 Applications

The above mechanism of query derivation can be used in many applications. The first application would be a useful method to improve or supplement query refinement currently equipped in some IR systems. Another important application is to provide a new mechanism for constructing the final results to return to the user. In this section, we present a brief description of these two applications.

Using the query derivation method as presented above, we design a Chinese short query refinement prototype system. For an input query, the system searches its derived queries with word senses as index terms. The sense set mostly consists of the top 2 level synsets of a Chinese version of WordNet [8]. For an input query $\mathbf{Q}$ = "病毒(virus)", its derived query list is returned as follows:

$\mathbf{Q}_i$, i = 1, 2, …	QueryRank($\mathbf{Q}_i$|$\mathbf{Q}$)
• 计算机病毒 (computer virus)	10.8%
• 病毒程序 (virus program)	8.1%
• 网络病毒 (network virus)	4.2%
• 病毒疾病 (virus disease)	2.2%
• 病毒扫描 (virus scan)	1.4%
• 病毒升级 (virus update)	1.1%
• 病毒防护 (virus shield)	0.8%
• 病毒性肝炎 (virus hepatitis)	0.6%
• 艾滋病毒 (AIDS virus)	0.5%
• ……	

where the rank of each derived query $\mathbf{Q}_i$ is determined with QueryRank$_1$($\mathbf{Q}_i$|$\mathbf{Q}$) above, which is the similarity of $\mathbf{Q}_i$ and the original query $\mathbf{Q}$. The rank values are normalized to be a percentage. As opposed to conventional query refinement, the candidate queries used by the system are constructed from many sources (including dictionaries and corpus statistics), and are indexed by word senses. The derived queries can be grouped (classified or clustered) with respect to word senses at different levels. Thus the user may use the system for better and comprehensive understanding of short queries.

Derived queries can also be used to generate, organize or present the search results that are to return to the user. We may use the ranked derived queries to provide a mechanism for grouping the search results, which presents top-ranked derived queries together with their search results to the user, such that derived queries with higher ranks and top-ranked documents of each derived query are preferentially presented. The final result list may be ordered with the above ClassRank($\mathbf{Q}'_j|\mathbf{Q}$) values determined for each derived query's results. This will allow the user to browse the results in classes and hopefully in a more efficient manner. (The details of such application are beyond the scope of this paper and will be described elsewhere.)

5 Conclusion

In this paper, we present a new method for disambiguating short queries. It uses a query retrieval model that constructs multiple derived queries for the user's query, which may represent more specific meanings or more concrete forms of usage of the original query. Derived queries can be efficiently obtained using the indexing and retrieval of a small-unit index, which has index terms that are commonly used words and word senses. Each of the derived queries can be associated with a rank value according to its similarity to the user's search query. We believe that derived queries are useful for improving the current query refinement method and for constructing the final results. In our future work, we will apply this query derivation method to a large-scale Web search engine to present better search results.

The work described here was supported by National Natural Science Foundation of China (No. 60435020 and 60475020).

References

1. Baeza-Yates, R., Ribeiro-Neto, B.: Modern Information Retrieval. Addison-Wesley, Reading (2000)
2. E.g., see the Ask.com Search Engine, http://www.ask.com
3. Salton, G.: The SMART retrieval system – Experiments in automatic document processing. Prentice Hall Inc., Englewood Cliffs (1971)
4. WordNet - Princeton University Cognitive Science Laboratory, http://wordnet.princeton.edu
5. The Global Wordnet Association, http://www.globalwordnet.org
6. Heinz, S., Zobel, J.: Efficient Single-Pass Index Construction for Text Databases. Journal of the American Society for Information Science and Technology 54(8), 713–729 (2003)
7. Greengrass, E.: Information Retrieval: A Survey, http://www.csee.umbc.edu/cadip/readings/IR.report.120600.book.pdf
8. CCD, The Chinese Conception Dictionary. Institute of Computational Linguistics, Peking University, http://icl.pku.edu.cn/

Applying Completely-Arbitrary Passage for Pseudo-Relevance Feedback in Language Modeling Approach

Seung-Hoon Na[1], In-Su Kang[2], Ye-Ha Lee[1], and Jong-Hyeok Lee[1]

[1] Pohang University of Science and Technology (POSTECH), AITrc, Republic of Korea
{nsh1979,sion,jhlee}@postech.ac.kr
[2] Korea Institute of Science and Technology Information (KISTI), Republic of Korea
dbaisk@kisti.re.kr

Abstract. Different from the traditional document-level feedback, passage-level feedback restricts the context of selecting relevant terms to a passage in a document, rather than to the entire document. It can thus avoid the selection of non-relevant terms from non-relevant parts in a document. The most recent work of passage-level feedback has been investigated from the viewpoint of the fixed-window type of passage. However, the fixed-window type of passage has limitation in optimizing the passage-level feedback, since it includes a query-independent portion. To minimize the query-independence of the passage, this paper proposes a new type of passage, called *completely-arbitrary passage*. Based on this, we devise a novel two-stage passage feedback – which consists of passage-retrieval and passage-extension as sub-steps, unlike previous single-stage passage feedback relying only on passage retrieval. Experimental results show that the proposed two-stage passage-level feedback much significantly improves the document-level feedback than the single-stage passage feedback that uses the fixed-window type of passage.

Keywords: pseudo-relevance feedback, passage-level feedback, completely-arbitrary passage, language modeling approach.

1 Introduction

Traditional pseudo-relevance feedback is document-level feedback which assumes that the entire content of a feedback document are relevant to the query [1-2]. However, this assumption is unreasonable, since normal documents are topically diverse so that they may contain non-relevant parts as well as relevant parts to a given query, even though they are relevant documents. Hence, document-level feedback cannot prevent the feedback process from selecting non-relevant expansion terms from non-relevant parts, undermining the retrieval performance. To handle this problem, passage-level feedback has been investigated by restricting the context for selecting expansion terms to a query-relevant passage, in order to minimize the risk of selecting non-relevant terms, and to increase the possibility of selecting relevant terms [3,4].

Previous works on passage-level feedback have explored window-type passages [5], in which a passage is defined as a window of W contiguous words where W is pre-fixed regardless of documents [3,4]. However, this fixed-length window passage

H. Li et al. (Eds.): AIRS 2008, LNCS 4993, pp. 626–631, 2008.

can be dangerous since lengths of relevant parts can be various in documents. Thus, we need to introduce the concept of variable-length passages, in which the length of a passage can be flexibly determined according to the portion of relevant parts in a document.

To this end, we propose a new type of passage, called a *completely-arbitrary passage*, in which all possible word sequences of arbitrary-length are included to the set of candidate passages. The concept of a completely-arbitrary passage is different from the arbitrary passage proposed by Kaszkiel [6]. Kaszkiel's arbitrary passage (either fixed-length or variable-length) starts at an arbitrary position within a document, but still has the restriction of fixing the length of passage. Based on this new type of passage, we propose a two-stage passage-level feedback approach which consists of 1) passage retrieval and 2) passage extension. The first step finds the best relevant passage to a given query. Due to the definition of our passage, a passage of extremely small length can be selected as the best passage. Such a short passage may not contain a sufficient context to select expansion terms. Thus, we put the second step, the passage extension, in which the boundary of the best relevant passage is extended in forward and backward directions until the best relevant passage contains a sufficient amount of useful expansion terms. We call the extended passage the *maximally relevant passage* (MRP). However, this second step raises a serious issue, the determination of increment of length for constructing MRP, indicating "how many additional contexts should we extend from the best passage, to optimize the selection of expansion terms?" We convert this issue to an optimization problem, by pre-defining a criterion and determining the maximally relevant passage as one that maximizes the criterion.

Experimental results show that the proposed two-stage feedback using the completely-arbitrary passage significantly improves the document-level feedback, as well as slightly increasing the performance of the passage-level feedback using window-passage. Considering that Liu' work [4], which first applied the passage-level feedback in the language modeling approaches, was not successful, our work is the first successful report that passage-level feedback is better than document-level feedback in language modeling approaches.

2 Two-Stage Passage-Level Feedback

2.1 Passage Retrieval: Finding the Best Passage

The first stage, the passage retrieval, is a process to find the best passage among all possible set of passages which is assumed to be the most relevant to a given query. Suppose that Q, D, and SP(D) are a query, a document, and the set of all passages of document D, respectively. Let Score(Q,P) be the similarity score between passage P (or document) and query Q, Then, the passage retrieval is summarized as follows:

$$P_{best} = \underset{P \in \mathrm{SP}(D)}{\arg\max}\, Score(Q,P) \quad (1)$$

In the above Eq. (1), P_{best} indicates the best passage. According to the definition of passages, there are several options for setting SP(D) – a semantic passage, a window passage, etc. We refer to the window passage as the *fixed-length arbitrary passage* in

Kaszkiel's work [5], which is notated by $SP_{WIN(W)}(D)$ where W is the length of window. *Completely-arbitrary passage* (notated by $SP_{COMPLETE}(D)$), which we propose as a new type of passage, is denoted as an arbitrary subsequence of adjacent words in document D [6]. The ranking of completely-arbitrary passages can be efficiently implemented through *cover-set ranking* by significantly reducing the number of completely-arbitrary passages to be checked in a document [6].

2.2 Passage Extension: Extending the Best Passage

After the first stage, we find the best passage which is the most relevant snippet to a query. In the second stage, our goal is to extend the best passage into a maximally relevant passage (MRP). The proposed extension method is a center-oriented extension, which assumes that the best passage is located at a center-position of a final MRP and then enlarges it towards forward and backward directions by the same length. In other words, let us suppose that l is the length of the best passage, and L is the increment of the length (the *expansion length*). Our expansion strategy is given as follows:

Center-oriented expansion: Put the best passage at center-position of final MRP. And, extend it by $L/2$ towards both forward and backward directions. When L is an odd number, the best passage is extended by $(L-1)/2$ and $(L+1)/2$ in forward and backward directions, respectively.

The above expansion strategy has some trivial exceptional cases. When the best passage is near the boundaries of start or end position of a document, we cannot put the best passage as center of MRP. For all exceptional cases, we adopt the following principle – 1) The expansion length should be L, and 2) the best passage should be at the center of MRP, as possible as we can.

2.3 Automatic Determination of *L* for Each Document

For center-oriented expansion, we need to determine L (the expansion length). Fixing L for all feedback documents is unreasonable, since their relevant portions would be different in terms of the length. Thus, we pursue a procedure to automatically determine L dependently of each document. To this end, we first define a criterion for evaluating the degree of appropriateness of a *candidate relevant passage* (CRP) for MRP, and then convert the determination of L to an optimization problem which finds the best one to maximize the criterion.

As for such criterion, we employ Sim(P,F), which means a similarity between a candidate relevant passage P and the set of feedback documents F. Then, our optimization problem for determining L is formulated as follows:

$$P_{ext} = \arg\max_{P \in E(P_{best})} Sim(P, F)$$

where P_{best} is the best passage selected from the first passage retrieval stage, and P_{ext} is MRP. E(P_{best}) is a set of CRP by centering P_{best}. To define E(P_{best}), let e(P_{best}, L) be the expanded passage obtained after applying our center-oriented expansion strategy with the expansion length parameter L. Then, E(P_{best}) is formulated using parameter ΔL as follows:

$$E(P_{best}) = \{e(P_{best}, n\Delta L) | W_{\min} \leq nL,\ 1 \leq n\} \quad (2)$$

E(P_{best}) can be incrementally constructed by first extending the best passage by ΔL to make the first expanded passage, which is again extended by ΔL to make the next expanded passage. Note that we additionally introduce W_{min} which indicates the minimum length of MRP. This parameter is necessary for preventing an extremely short passage from having unreasonably high similarity value. By using W_{min}, many non-query terms are sufficiently contained for all CRPs so that the fair evaluation of Sim(P,F) can be made. As a result, MRP of reasonable-length can be selected only by evaluating Sim(P,F). A default value for W_{min} is 100.

3 Language Modeling Setting for Two-Stage Passage-Level Relevance Feedback

The proposed framework has two model-dependent parts - Score(Q,P) and Sim(P,F). Score(Q,P) is defined as query-likelihood from passage language models [7]. For estimating passage language model, we select Jelinek-Mercer smoothing method with smoothing parameter λ. Let us suppose that θ_Q, θ_P, and θ_C are query language model for a given query, passage language model for passage P, and collection language model, respectively. Then, Score(Q,P) is defined as follows:

$$Score(Q,P) = \sum_w P(w \mid \theta_Q) \log\left(\frac{(1-\lambda)P(w \mid \hat{\theta}_P)}{\lambda P(w \mid \theta_C)} + 1.0 \right) \quad (3)$$

where $\hat{\theta}_P$ is MLE (Maximum Likelihood Estimation) of passage model for P.

For Score(Q,P), suppose that θ_F is feedback language models from feedback documents. We propose the following *log-likelihood ratio* $\mathrm{Sim}_{\mathrm{Ratio}}(P,F)$ between generation probabilities of passage P from θ_F and θ_C:

$$Sim_{Ratio}(P,F) = \sum_w P(w \mid \hat{\theta}_P) \log\left(\frac{P(w \mid \theta_F)}{P(w \mid \theta_C)} \right) \quad (4)$$

We further smooth the feedback model P(w|θ_F) with collection language model θ_C by using Jelinerk-Mercer smoothing: $(1-\lambda_F)P(w \mid \hat{\theta}_P) + \lambda_F P(w \mid \theta_C)$. Note that smoothing parameter λ_F is close to the smoothing parameter of document model, not smoothing parameter λ of passage model. A default value for λ_F is 0.25.

4 Experimentation

For evaluation, we used four TREC test collections – TREC4-AP, TREC7, TREC8 and WT2G. TREC4-AP is the sub-collection of Associated Press in disk 1 and disk 2 for TREC4 (the number of documents is 158,240). Other test collections are the same as the standard set for ad-hoc retrieval track at corresponding year. For queries, we used description field for TREC4-AP, and title field for other test collections. The standard method was applied to extract index terms.

For all experiments, the baseline run used Dirichlet-prior for the document model due to its superiority over Jelinek-Mercer smoothing, using smoothing parameter μ as follows. We use acronym DM for this baseline run. To determine the smoothing parameter μ, we performed about 15 different runs on μ between 100 and 30,000, and selected the best performed one for each test collection. For passage language models, Jelinek-Mercer smoothing is applied. For smoothing passage language model, λ is fixed to 0.9 for TREC4-AP, and 0.1 for other test collections. In a similarity metric of Section 3 ($sim_{Ratio}(P,F)$), the smoothing parameter λ_F of feedback language model is set to 0.25. We used MAP (Mean Average Precision) for evaluating all experiments. For feedback model, we adopted Zhai's model-based feedback (i.e. the generation method using two-component mixture model [2]). For the best interpolation parameter α, we performed feedback runs using different interpolation parameter α between 0.05 and 0.9, and selected the best performed one. For most runs, the best α was less than 0.4.

Table 1 shows results of the baseline (DM), document-level feedback (Doc), two passage-level feedbacks using window passage (W as parameter) and completely-arbitrary passage (ΔL as parameter), across the different numbers of feedback documents (R). To see whether or not a passage-level feedback improves Doc, we performed Wilcoxon signed rank test to examine whether the improvement was statistically significant or not, at 95% and 99% confidence level. We attached ↑ and to the figure (performance number) of each cell in the table when the test passes at 95% and

Table 1. Performances of passage-level feedback using *Window Passage* (W = 150) and *Completely-Arbitrary Passage* across unit of expansion length (ΔL) and the number of feedback documents (R). Doc indicates the performance of the document-level feedback.

Collection	R	DM	Doc	W=200	ΔL=10	ΔL=25	ΔL=50	ΔL=75
TREC4-AP	R = 5	0.2560	0.2989	0.3036	0.3073	0.3063	**0.3157**	0.3100
	R = 10		0.2848	0.2828	0.2908↑	**0.2915**↑	0.2907↑	0.2933
	R = 20		0.2910	0.2981↑	0.3015¶	**0.3051**	0.3015	0.3018¶
	R = 30		0.2874	0.2978	0.2991	**0.3025**	0.3015	0.3015¶
	R = 50		0.2775	0.2889	**0.2926¶**	0.2902¶	0.2922¶	0.2900¶
TREC7	R = 5	0.1786	0.2110	**0.2131**	0.2103	0.2096	0.2088	0.2139
	R = 10		0.2152	0.2227	0.2251	**0.2251**	0.2243	0.2234
	R = 20		0.2032	0.2228↑	0.2291↑	0.2293↑	**0.2298**↑	0.2278↑
	R = 30		0.1995	0.2181↑	0.2231¶	0.2221↑	**0.2230¶**	0.2213¶
	R = 50		0.1883	0.2075¶	**0.2144¶**	0.2141¶	0.2140¶	0.2131↑
TREC8	R = 5	0.2480	0.2805	0.2807	0.2752	0.2780	0.2757	0.2767
	R = 10		0.2834	0.2876	**0.2885**	0.2874	0.2871	0.2862
	R = 20		0.2817	0.2924¶	0.2933¶	0.2938¶	**0.2939¶**	0.2932¶
	R = 30		0.2764	0.2847¶	**0.2868¶**	0.2867¶	0.2866¶	0.2854¶
	R = 50		0.2739	**0.2798¶**	0.2774↑	0.2776↑	**0.2783**↑	0.2772↑
WT2G	R = 5	0.3101	0.3398	0.3508	0.3472	0.3473	0.3482	**0.3546**↑
	R = 10		0.3488	0.3604↑	0.3603↑	0.3622¶	0.3612↑	**0.3627¶**
	R = 20		0.3354	0.3548↑	0.3578¶	**0.3585¶**	0.3573¶	0.3574¶
	R = 30		0.3269	0.3448	0.3473¶	**0.3517¶**	0.3517¶	0.3473↑
	R = 50		0.3205	0.3210	0.3291↑	0.3318¶	**0.3334¶**	0.3313

99% confidence level, respectively. Both of passage-level feedback improve document-level feedback. However, the window-passage does not frequently show a statistically significant improvement over the Doc for some test collections such as TREC4-AP and WT2G. At the best *W* (*W=150*), among the total of 20 runs, only 9 runs show a significant improvement at 95% confidence level, and only 4 runs at 99% confidence level. On the other hand, the complete-arbitrary passage does show statistically significant improvement over Doc, for all test collections, including TREC4-AP and WT2G which are not statistically significant in case of window passage. At the best ΔL ($\Delta L=75$), among total 20 runs, 13 runs shows a significant improvement at 95% confidence level, and 8 runs at 99% confidence level.

Summing up, the proposed two passage-level feedbacks clearly show significant improvement over the baseline (DM) at a high confidence level. Compared with document-level feedback (Doc), the proposed completely-arbitrary passage is much more effective than the window passage, by having many runs with significant improvements for all test collections. Especially, the proposed completely-arbitrary passage is almost close to a parameter-less approach due to much-less-sensitivity of ΔL.

Acknowledgements

This work was supported by the Korea Science and Engineering Foundation (KOSEF) through the Advanced Information Technology Research Center (AITrc), also in part by the BK 21 Project and MIC & IITA through IT Leading R&D Support Project in 2007.

References

1. Rocchio, J.J.J.: Relevance feedback in information retrieval. In: SMART Retrieval System: Experiments in Automatic Document Processin, ch. 14, pp. 313–323 (1971)
2. Zhai, C., Lafferty, J.: Model-based feedback in the language modeling approach to information retrieval. In: CIKM 2001, pp. 403–410 (2001)
3. Allan, J.: Relevance feedback with too much data. In: SIGIR 1995, pp. 337–343 (1995)
4. Liu, X., Croft, W.B.: Passage retrieval based on language models. In: CIKM 2002, pp. 375–382 (2002)
5. Kaszkiel, M., Zobel, J.: Effective ranking with arbitrary passages. JASIST 52(4), 344–364 (2001)
6. Na, S.H., Kang, I.S., Lee, Y.H., Lee, J.H.: Completely-arbitrary passage retrieval in language modeling approach. In: AIRS 2008 (to be appear)
7. Zhai, C., Lafferty, J.: A study of smoothing methods for language models applied to ad hoc information retrieval. In: SIGIR 2001, pp. 334–342 (2001)

Automatic Generation of Semantic Patterns for User-Interactive Question Answering

Tianyong Hao[1], Wanpeng Song[1,2], Dawei Hu[1,2], and Wenyin Liu[1]

[1] Department of Computer Science, City University of Hong Kong, Hong Kong, China
[2] Department of Computer Science and Technology,
University of Science & Technology of China, Hefei, China
{tianyong,wanpesong2,50008964}@student.cityu.edu.hk,
csliuwy@cityu.edu.hk

Abstract. An automatic method for generation of semantic patterns from free-text questions is proposed in this paper. An evaluation method is also proposed to estimate the suitability of the generated patterns and implemented in our user-interactive question answering (QA) system. Experiments with 5500 questions show that 63.9% generated patterns are satisfactory in the average.

Keywords: Semantic pattern, Question answering, Tagger ontology.

1 Introduction

Automatic question answering (QA) targets at providing more precise answers to users' questions than search engines. Although more preferable, they cannot outperform currently well-known search engines. Hence, more and more user-interactive QA systems, including Google Answers [1], Microsoft QnA [2] and Yahoo Answers [3], have been launched, serving as interactive platforms for users to help each other with human-provided answers. However, most of these QA systems request users to input entire questions, even though many similar questions have been asked many times. Moreover, these free text questions are difficult to be analyzed and understood by computers. Hence, we propose to use question patterns to improve the performance of these QA systems.

A structural question pattern is a generalization of a group of questions which have similar structures. It has been demonstrated that patterns can facilitate machine understanding [4]. However, there also exist some shortcomings in this structural pattern, such as lacking of semantic information and flexibility. Semantic pattern which is based on structural pattern can reduce the ambiguity of the questions and enhance semantic representation of the question. Meanwhile, this pattern allows QA systems to locate answers based on its semantic type and filter out irrelevant answers by matching the semantic parts of question patterns.

Although semantic patterns are very useful in QA systems, manually building high-quality semantic patterns is a difficult, time-consuming task. Hence, automatically and efficiently constructing these patterns is more desirable and becomes an important topic in the QA area. Several research efforts have focused on learning extraction rules from training examples provided by users [5]. However,

H. Li et al. (Eds.): AIRS 2008, LNCS 4993, pp. 632–637, 2008.

many of them abandon the semantic information and only keep the structure information when generating patterns, which make the patterns hardly understood by machines.

In this paper, we propose an automatic generation method for the semantic pattern defined by Hao et al. [6]. It can generate high-quality semantic patterns from a set of free-text questions. This method uses structural processing and name entity recognition to obtain main structure of a question. WordNet is also used to acquire upper concepts for certain terms and map with our tagger ontology to obtain semantic labels, which can be used to annotate or constrain the semantics of the terms in generated pattern. An entropy-based model is used to evaluate and select suitable parts for generalization such that generated patterns are controlled at suitable granularity level. In our experiments, we implemented the proposed method in our user-interactive QA system -BuyAns [6]. 5500 questions available at [7] are chosen to generate patterns and evaluate their qualities. Experiment results show that 63.9% generated patterns are satisfactory in the average according to our evaluation method.

The rest of this paper is organized as follows: Section 2 presents the proposed automatic generation method in detail, including structure processing, Entropy based selecting for generalization, and semantic mapping and tagging. Section 3 shows our experiment results. Section 4 summarizes this paper and discusses future work.

2 The Automatic Generation Method for Semantic Patterns

We propose a method for automatic generation of semantic patterns based on the previous definition. It can analyze a user's question to form its main structure. WordNet and a tagger ontology are then used to tag suitable nouns and verbs to obtain corresponding semantic labels. The main structure combined with labeled semantic information forms a new semantic question pattern. The whole procedure is shown in Fig. 1. It consists of three main modules: (1) Structure processing and name entity recognition, (2) Entropy based selection of suitable nouns/verbs for generalization, (3) Semantic mapping and tagging.

2.1 Structure Processing and Name Entity Recognition

Given a free-text question, this step is to analyze the sentence by structure processing and name entity recognition (NER) to obtain the main structure of the question.

The NER step is to identify certain atomic elements of information in text, including person names, company/organization names, locations, dates & times, percentages and monetary amounts. We mainly focus on recognizing people names and location names in this paper. Since NER affects the result of part of speech and the main structure obtained, we design a kind of tagger for name entity recognition, which can identify the common entity names based on our entity dictionaries.

The main structure can be seen as a simplified representation of the original question. It includes all the important parts of the question, such as question type, nouns and verbs, which are most useful in pattern generation. In addition, with the main structure we can ignore some useless information for pattern generation such as the stop words and some meaningless words.

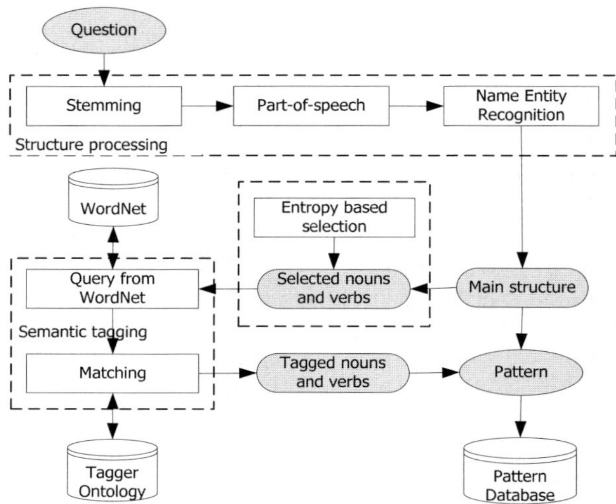

Fig. 1. Flowchart of automatic semantic pattern generation

This procedure mainly includes the following four steps:

Step 1: *Acquire the question type by a basic keyword searching.*
Step 2: *Process question to obtain of the basic structure with nouns and verbs*
Step 3: *Label words using name entity recognizer.*
Step 4: *Obtain the main structure of this question.*

For example, given a question "who is the mayor of Beijing?", we first acquire the question type "who" according to our question type list and label it as "<Type:Who>". Structural processing then remove "the" as a stop word and find noun "mayor". Using NER, we obtain name entity of "Beijing" as a location. Finally we obtain the main structure as "<Type:Who> is [Noun=mayor] of [NE(location)= Beijing]?".

2.2 Selection of Nouns/Verbs for Generalization

All nouns and verbs in the main structure are candidates for generalization to form new patterns. However, some of them are more suitable to be generalized such that the generated patterns can cover more questions and are easily understood by users. In this section, we develop a method based on the entropy model to evaluate the sensitivity/suitability of a certain term in the question to be generalized.

Before we describe our evaluation strategy, we would like to give some definitions first. Let Q refer to a question, a term T_i (noun or verb) is the *i-th* nouns or verbs in Q, the upper concept of T_i is SC_j, the probability of T_i occurs in all terms which have the same upper concept SC_j is $p(T_i)$. The entropy value of upper concept SC_j can be calculated by Equation (1).

$$H(SC_j) = -\sum_{(T_i \in SC_j)} p(T_i) \log p(T_i) \tag{1}$$

The higher entropy value $H(SC_j)$ is, the bigger the variety of words which can be used to replace T_i at its position, which is also more appropriate for generalization and tagging. When it exceeds a threshold δ, we can select it as a candidate for semantic tagging. For better usage for common users in our system, we use γ to limit the number of selected nouns/verbs in the same question or the number of blanks in the pattern to be generalized. That is, if the number of selected nouns/verbs is more than γ, we only select top γ terms whose upper concepts have best entropy values.

From the example "Who is mayor of Beijing?", we have noun "mayor" and "Beijing". The upper concept of "Beijing" is "location\city", which have instances such as "Washington", "Boston" in our training base. The possibilities of these two instances are "0.00154" and "0.00308", respectively. The entropy of "location\city" is 0.12757, which is greater than the threshold of entropy value we set (0.05). Hence we select it as a good noun for generalization in the main structure.

2.3 Semantic Mapping and Tagging

After the main structure is obtained, the selected nouns/verbs are sent to the semantic tagging module to be assigned with semantic labels, which include two key steps: 1) Query from WordNet [9], and 2) Matching from the tagger ontology. The purpose of using WordNet is to label the nouns and verbs by their upper concepts. In the main structure, each noun and verb will be searched in WordNet to get all of their upper concepts. The upper concepts obtained in the first step are mapped to the labels in the tagger ontology in the second step. The most important role of the tagger ontology is to tag nouns and verbs in the main structure with suitable semantic keywords. These tagged keywords contain semantic description and relationship of the hierarchical concepts in the ontology and can therefore be understood by machine easily. It is also useful for manual questioning or answering. Users can understand the content of the blanks in the patterns by their semantic labels as tips of the filled content.

The tagger ontology contains upper concept hierarchy of WordNet and a list of question target. For better understanding and using by users, it only includes two-level concepts. Matching from tagger ontology is to build the relation between concepts in WordNet and labels in the tagger ontology.

For the example mentioned above, we select the noun "mayor" and "Beijing" for generalization. "Mayor" is queried in the WordNet to obtain upper concepts such as "person" and "authority". These concepts are then matched in our tagger ontology to get the semantic label [Human\Title]. 'Beijing' can be matched with '[Location\City]'. We then can obtain a semantic pattern as "<Type:Who> is [Human\Title] of [Location\City]".

3 Experiments and Evaluation

The quality of the generated patterns mainly relies on the quality of the semantic labels. Our evaluation method is to get the user's satisfaction of each pattern and get average satisfaction of all patterns by asking users to determine whether each part of the generated patterns is suitable or not. Since determining whether each part is suitable need much experience based on our pattern definition, we check it manually.

Given a question set $QS = \{Q_1, Q_2, \ldots Q_n\}$, we can generate a corresponding pattern set $PS = \{ P_1, P_2, \ldots P_n\}$. For each pattern P_i $(1 \leq i \leq Number\_of\_questions(N))$, we use *RT*, *RN* and *RV* to represent whether its question type, nouns and verbs is right in P_i. For example, if the question type part is correct, the *RT* of P_i is 1, otherwise it is 0. Suppose there are *m* nouns and *n* verbs in Q_i, the satisfaction of the noun part in P_i is $\frac{1}{m}\sum_{j=1}^{m} RN(j)$ and the verb part is $\frac{1}{n}\sum_{k=1}^{n} RV(k)$.

The satisfaction of a pattern means whether a generated pattern is satisfactory to our pattern definition and user's demand, which is represented as *S(P)*. It can be calculated by the following formula:

$$S(P_i) = \alpha \times RT + \frac{\beta}{m}\sum_{j=1}^{m} RN(j) + \frac{(1-\alpha-\beta)}{n}\sum_{k=1}^{n} RV(k) \quad (2)$$

Where α and$\beta$are two weights for evaluation and they can be fixed by a serial of experiments. Finally, the average satisfaction of the whole pattern set *AS(PS)* can be easily obtained from the satisfaction of all patterns in it.

We implemented and applied our proposed method into our QA system-BuyAns [7]. When a user posts a question, system can analyzes its structure and matchs patterns from our pattern database. If no pattern matched, system will generate a few patterns automatically. The user only needs to select and use one relevant pattern to ask the question. The user interface is also implemented for this purpose.

In experiments, we select 500 questions from our 5500 training questions randomly [8]. Our system generates patterns for these questions automatically. We select 10 generated patterns randomly as a group for five times to check the result manually. The average satisfaction can be calculated using the evaluation method in Section 5.2 automatically.

α is set to 0.2 and β is set to 0.6 in our experiments empirically. The results for these five groups are shown in Table 1. Finally, the average satisfaction for all generated patterns in the five groups is 63.9%.

Table 1. Experiment results for five groups

PS	$S(P_i)$ $(1 \leq i \leq 10)$										*AS(PS)*
Group 1	0.8	0.6	0.4	0.6	0.6	0.6	0.73	0.6	0.6	0.6	63.6%
Group 2	0.7	0.7	0.59	1	1	0.7	0.32	0.55	1	0.7	72.6%
Group 3	0.7	1	0.7	0.59	0.4	0.7	0.7	0.796	0.4	0.4	63.86%
Group 4	0.7	0.4	0.4	0.4	0.4	0.7	0.59	0.7	0.59	0.7	55.8%
Group 5	0.4	0.4	1	1	0.7	0.7	0.4	0.4	0.7	0.7	64%

4 Conclusion and Future Work

Pattern generation for QA systems is a tedious, difficult, time-consuming task. Hence, we propose an automatic generation method for the semantic patterns defined by Hao et al. [6]. The generation procedure first process structural and recognize name entitis for free-text questions. The upper concepts of the obtained nouns/verbs in the questions are then mapped to the semantic labels in the tagger ontology. An entropy based evaluation method is finally used to evaluate which nouns and verbs are more suitable to be generalized to form a new question pattern. We implement the proposed method and evaluate it with experiments using 5500 questions. The results show that our method return patterns with 63.9% satisfaction in the average. We think it can be applied for practical usage and can save a lot of human efforts of manual generation of patterns.

Though this method can be applied in user-interactive QA system easily and can work well, it is not very satisfactory in recognition of name entities and hence the quality of the acquired main structures is not very good. In future work, we will improve the method of structural processing to obtain more accurate structures and update parameters of our evaluation method according to experiments to make the generated pattern more satisfactory.

Acknowledgement

The work in this paper was fully supported by the China Semantic Grid Research Plan (National Grand Fundamental Research 973 Program, Project No. 2003CB317002).

References

1. Google Answers (2007), http://answers.google.com/answers/
2. Microsoft QnA (2007), http://qna.live.com/
3. Yahoo Answers (2007), http://answers.yahoo.com/
4. Deepak, R., Eduard, H.: Learning Surface Text Patterns for a Question Answering System. In: Proceedings of the 40th ACL Conference, Philadelphia (2002)
5. Cowie, J., Ludovik, E., Molina-Salgado, H., Nirenburg, S., Sheremetyeva, S.: Automatic Question Answering. In: Proceedings of the Rubin Institute for Advanced Orthopedics Conference, Paris (2000)
6. Hao, T.Y., Hu, D.W., Liu, W.Y., Zeng, Q.T.: Semantic Patterns for User-Interactive Question Answering. Journal of Concurrency and Computation: Practice and Experience 20(1) (2007)
7. BuyAns (2007), http://www.buyans.com
8. 5500 Question list (2007), http://l2r.cs.uiuc.edu/~cogcomp/Data/QA/QC/train_5500.label
9. WordNet (2007), http://wordnet.princeton.edu/

A Comparison of Textual Data Mining Methods for Sex Identification in Chat Conversations

Cemal Köse, Özcan Özyurt, and Cevat İkibaş

Department of Computer Engineering, Faculty of Engineering,
Karadeniz Technical University, 61080 Trabzon, Turkey
{ckose,oozyurt,cikibas}@ktu.edu.tr

Abstract. Mining textual data in chat mediums is becoming more important because these mediums contain a vast amount of information, which is potentially relevant to a society's current interests, habits, social behaviors, crime tendency and other tendencies. Here, sex identification is taken as a base study in information mining in chat mediums. In order to do this, a simple discrimination function and semantic analysis method are proposed for sex identification in Turkish chat mediums. Then, the proposed sex identification method is compared with the Support Vector Machine (SVM) and Naive Bayes (NB) methods. Finally, results show that the proposed system has achieved accuracy over 90% in sex identification.

Keywords: Mining Chat Conversations, Sex Identification, Information Extraction, Text Mining, Machine Learning.

1 Introduction

Chat mediums are becoming an important part of human life in societies and provide quite useful information about people such as their current interests, habits, social behaviors and tendencies [1], [2], [3], [4]. Users may spend a large portion of their time to find out information in chat mediums. A system can be developed to help users find the interested information in the mediums [1], [4], [5]. Here, one of our final goals is to develop a system that automatically determines persons with criminal tendencies in chat mediums. Thus, in this study, sex identification is taken as a preliminary study in chat mediums. In order to do this, conversations are acquired from a specially designed chat medium and real chat mediums and then statistical and semantic information are obtained from the conversations [3], [4], [5], [6]. These are used to determine weighting coefficients of the proposed discrimination function and to evaluate the semantic analyzer for identification.

In literature, most of the related studies focus on topic categorizations rather than sex-identification [6], [7], [8], [9]. In this paper, a comparative study about sex identification techniques is presented to automatically monitor chat conversations. The rest of this paper is organized as follows. Some notable characteristics of Turkish chat language are given in Section 2. Proposed discrimination function for sex identification is presented in Section 3. A detailed description of methods used in the system is

H. Li et al. (Eds.): AIRS 2008, LNCS 4993, pp. 638–643, 2008.

given in the same section. The implementation and results are discussed in Section 4. The conclusion and future work are given in Section 5.

2 A Summary of Turkish Chat Language

Many world languages have a common noticeable characteristic so that they could expose the sexual identity of the chatter in a conversation. Some of these languages have three grammatical genders (masculine, feminine, and neuter) but Turkish language has only one (neuter). Hence, unlike Turkish, some words may reveal the sex of chatters in English. For example, "Ben onun kardeşiyim" in Turkish may be translated into English as "I am his brother/sister". Therefore, the identification of chatters' sex from a conversation in Turkish may be more difficult than in English.

In real-time and informal environment of IM (Instant Messages) systems, Turkish chat messages are very different from conventional Turkish. Therefore, chat language includes acronyms, short forms, polysemes, synonyms and misspelled terms.

3 A Simple Method for Sex Identification Systems

A simple identification system is proposed for binary subject. This system consists of discrimination and semantic analyzer units. Here, the proposed function includes some important parameters such as words, word groups, and weighting coefficients. A semantic analyzer is also employed to enhance accuracy of the system. This simple identification system is explained under the following titles.

3.1 Classifying Words for Sex Identification

In a chat medium, many word groups may be defined to identify chatters` sex in a dialogue. In this study, eighth word groups are defined to cover as many sex related concepts and subjects as possible in a chat medium [4], [5]. These groups are abbreviations and signs, slang and jargon words, politeness and delicacy words, interjections and shouting, sex and age related words, question words, particle and conjunction words, and other words group. Word groups are built by considering the conceptual relations and usage frequency of words used by female or male chatters.

Due to the nature of chats, contents of conversations vary completely. Hence, the feature vectors and words in the conversations also changes dynamically. Therefore, new words are added to each word group when a new male or female dominant word is accounted in a conversation.

3.2 A Discrimination Function for Sex Identification

A simple discrimination function is designed to identify sex of a person in a chat medium. This function considers each word in conversations separately and collectively. Therefore, statistical information related to each chosen word is collected from the Specially Designed Chat Medium (SDCM) and Real Internet Medium (RIM) [3], [4], [5]. By using the statistical information, a weighting coefficient is determined for each word in each word group. Practically, a normalized weighting coefficient (in 0.0-1.0 interval) of each word is determined. For each conceptually related word group, a sexual identity value is calculated by equation (1).

$$g_i = (\alpha_{i1} w_1 + \alpha_{i2} w_2 +,...,+ \alpha_{ik} w_k)/(\alpha_{i1} + \alpha_{i2} +,..+ \alpha_{ik}). \quad (1)$$

where, g_i varies from 0.0 to 1.0 and determines the chatters' sexual identities as female or male for *ith* word group, α_{ij} is the weighting coefficient of *jth* word in *ith* word group and varies related to the number of words in the interested text, w_j represent the existing *jth* words in the interested text (if a word exists in the text, then w_j=1.0 else w_j= 0.0), and k is the number of word in *ith* word group. Here, if a word is female dominant, α varies from 0 to 0.5, but if the word is male dominant, α varies from 0.5 to 1.0.

As explained before, words are also classified in several groups according to their conceptual relations. A weighting coefficient is also determined for each word group. Then, the proposed discrimination function is formed for sex identification as Equation (2).

$$\gamma = (\lambda_{g1} * g_1 + \lambda_{g2} * g_2 +,...,+ \lambda_{gn} * g_n)/(\lambda_{g1} + \lambda_{g2} +,... + \lambda_{gn}). \quad (2)$$

where, γ varies from 0 to 1 and determines the chatters sexual identity as female or male, and λ_{gi} is the weighting coefficient for *ith* female or male word group. These weighting coefficients of each group are determined according to dominant sexual identity of the group. Then, the sex of the chatters may be identified as female when γ is determined between 0.0 and 0.5. On the other hand, chatters may be identified as male when γ is determined between 0.5 and 1.0. Here, the accuracy of the results increases so that it shows female or male gender when γ approaches to 0.0 and 1.0 respectfully.

3.3 A Simple Semantic Analysis Method to Enhance the Sex Identification

For further improvement of the accuracy of the identification, a semantic analyzer is added to the system. Semantically, some sentence structures in a conversation such as

Table 1. A typical conversation for semantic analysis

	Chatter	Sentence	English
(1)	Baskan	sen naber e	What is the news, and so
(2)	TaRaNTuLa	ben murat değilim abisiyim	I am not Murat, His older brother
(3)	TaRaNTuLa	slm	Hi
(4)	Baskan	ömer abi ?	Ömer?
(5)	TaRaNTuLa	zekeriya	Zekeriya
(5)	Baskan	ooo aslan zekeriya abi	Dear Lion Zekeriya
(6)	Baskan	ben ahmet	I am Ahmet

questions, answers and addressed sentences may reveal the sex of the chatter [3], [4]. For example, "What is your name? (ismin neydi?) → Ahmet", "Name? (ismin?)→ Ali", "Who are you? (sen kimsin?) → Davut", "U, (U) → Buket" and etc. In some conversations, many addressed sentences may also be used such as "Hi Ahmet (Merhaba Ahmet)", "How are you Davut (Nasılsın? Davut)" and etc.

In the semantic analysis, idiomatic expressions, phrases, expressions and the relations between subjects and suffixes are especially analyzed for the identification. For example, "At this rate you won't be able to get married *(O zaman sen evde kal-ır-sın bu gidişle)*". In the sentence, the suffix "-sın" expresses that other chatter is female because the sentence, "you won't be able to get married", is used for female persons in Turkish. Here, the personal suffixes used in the semantic analysis are –m and -(y)Im (im, ım, um, üm, yim, yım, yum, yüm), and -n and -sIn (sin, sın, sun, sün) for singular first person and singular second person respectively. As another example, "Yakışıklıyım (I am handsome)" can morphologically be analyzed as yakışıklı-(y)ım. Here, "handsome" determines the dominant sex as male and the suffix "-(y)ım" determines the singular first person. Then, the chatter can be identified as male.

In the application, the semantic analyzer analyses each sentence in the conversation and generates outcomes as male, possible male, no results (neuter), possible female and female [3]. Finally, the analyzer takes average of all decisions to generate the single semantic decision about sex of the chatter. Hence, semantic relations may contribute to the final decision and strengthen the accuracy of the identification system. Equation (3) combines statistical and semantic identification outputs and produce the final identification output.

$$\lambda = (\lambda_{sta} * \gamma_{sta} + \lambda_{sem} * \gamma_{sem}) / (\lambda_{sta} + \lambda_{sem}). \tag{3}$$

where λ is the final result that identifies the sex of the chatter, λ_{sta} and λ_{sem} are statistical and semantic weighting coefficients, and γ_{sta} and γ_{sem} are statistical and semantic identifications respectively.

4 Results

In this paper, we have presented performances of different similarity measurement methods. About two hundreds conversations have been collected from SDCM. Forty-nine of the conversations including ninety-eight chatters (forty-four female and fifty-four male) are chosen as the training set for testing. Tests results for the same data sets and the same feature vectors are also obtained on WEKA's SVM and NB implementations [9], [10]. The test results are given in Table 2. These approaches including the proposed method show similar performances without the semantic analysis. On the other hand, experimental results show that our system with the semantic analysis method performs better than the other systems in sex identification. The decision accuracy of our system reaches to 88.8%.

More than one hundred conversations are collected from RIM: mIRC (mIRC is a shareware Internet Relay Chat) and forty-two of them are chosen randomly. The test results are given in Table 2. The decision results of the identification systems are listed in the table. These results show that general accuracy of our system reaches to 92.9%.

Table 2. The general result of sex identification for the specially designed medium

	NB		SVM		Our Method	
	SDCM	RIM	SDCM	RIM	SDCM	RIM
Number of chatters	98	42	98	42	98	42
Number of correct decision	80	35	83	38	87	39
Number of wrong decisions	18	7	15	4	11	3
Percentage of correct decision	81.7%	83.3%	84.7%	90.5%	88.8%	92.9%
Percentage of wrong decisions	18.3%	16.7%	15.3%	9.5%	11.2%	7.1%

About 1.27 MB of text data were obtained from more than two hundred conversations. Duration of conversations varies from few minutes to few hours. The identification system ran on PC with P4-3.2 GHz CPU and 512 MB RAM. For sex identification, these conversations are processed in 4.34, 8.75 and 10.93 minutes on the proposed, NB and SVM systems respectively. These results prove that our system is quite promising for large-scale mining applications.

5 Conclusions and Future Work

Mining chat conversations are becoming more important and provide quite useful information about people in a society. Hence, a simple discrimination function with semantic analysis method is defined for sex identification in conversations. Our simple and computationally less expensive sex identification system with semantic analysis method provides better performance comparing to the other methods. The results show that the proposed identification function is quite useful for binary classification such as sex identification. This identification system with the discrimination function achieves accuracy about 90% in the sex identification.

Although some satisfactory results are obtained, the system still needs to be improved and tested on larger data sets. Our experiments and results show that the proposed methods for sex identification may also be applied to other concepts and subjects such as topic detection.

In this application, misleading questions and answers are not taken into account. Misleading questions and answers may affect the identification results negatively. In the future implementation of the system, the problem will be considered to minimize or eliminate these misleading sentences.

References

1. Khan, F.M., Fisher, T.A., Shuler, L.A., Tianhao, W., Pottenger, W.M.: Mining Chat-room Conversations for Social and Semantic Interactions. Lehigh University Technical Report LU-CSE-02-011 (2002)
2. Elnahrawy, E.: Log-Based Chat Room Monitoring Using Text Categorization: A Comparative Study. In: The International Conference on Information and Knowledge Sharing, US Virgin Islands (2002)

3. Kose, C., Özyurt, O., Amanmyradov, G.: Mining Chat Conversations for Sex Identification. In: Zhou, Z.-H., Li, H., Yang, Q. (eds.) PAKDD 2007. LNCS (LNAI), vol. 4426, pp. 106–117. Springer, Heidelberg (2007)
4. Kose, C., Ozyurt, O.: A Target Oriented Agent to Collect Specific Information in a Chat Medium. In: Levi, A., Savaş, E., Yenigün, H., Balcısoy, S., Saygın, Y. (eds.) ISCIS 2006. LNCS, vol. 4263, pp. 697–706. Springer, Heidelberg (2006)
5. Kose, C., Nabiyev, V., Özyurt, O.: A statistical approach for sex identification in chat mediums. In: The international scientific conference on Problems of Cybernetic and Informatics (PCI) (October 2006)
6. Bengel, J., Gauch, S., Mittur, E., Vijayaraghavan, R.: Chattrack: chat room topic detection using classification. In: Chen, H., Moore, R., Zeng, D.D., Leavitt, J. (eds.) ISI 2004. LNCS, vol. 3073, pp. 266–277. Springer, Heidelberg (2004)
7. Haichao, D., Siu, C.H., Yulan, H.: Structural analysis of chat messages for topic detection. Online Information Review 30(5), 496–516 (2006)
8. Koppel, M., Argamon, S., Shimoni, A.R.: Automatically Categorizing Written Texts by Author Gender. Oxford Journals, Humanities, Literary and Lingustic Computing 17, 401–412 (2003)
9. Joachims, T.: Text categorization with support vector machines: learning with many relevant features. In: Nédellec, C., Rouveirol, C. (eds.) ECML 1998. LNCS, vol. 1398, pp. 137–142. Springer, Heidelberg (1998)
10. http://www.cs.waikato.ac.nz/ml/weka/

Similarity Computation between Fuzzy Set and Crisp Set with Similarity Measure Based on Distance

Sang H. Lee[1], Hyunjeong Park[2], and Wook Je Park[1]

[1] School of Mechatronics, Changwon National University
#9 sarim-dong, Changwon, Gyeongnam 641-773, Korea
{leehyuk,parkwj}@changwon.ac.kr
[2] Dept. of Mathematics Education, Ewha Womans University
#11-1 daehyun-dong, seodaemun-ku, Seoul 120-750, Korea
hyunjp@ewhain.net

Abstract. The computation procedure of similarity between fuzzy set and crisp set is derived. The proposed similarity measure is constructed through distance measure. And our results are compared with those of previous similarity which is based on fuzzy number.

Keywords: Fuzzy entropy, similarity measure, distance measure.

1 Introduction

Computation of similarity between two or more data is very interesting for the fields of decision making, pattern classification, or *etc*.. Until now the research of designing similarity measure has been made by numerous researchers [1-6]. Most studies are emphasized on designing similarity measure based on membership function, and those studies are also mainly carried out for the fuzzy membership functions. As the previous similarity results it is vague to obtain degree of similarity between fuzzy set and crisp set or crisp set and crisp set. In this paper we try to obtain degree of similarity between fuzzy set and crisp set. Hence we first derived similarity measure via well known-Hamming distance. We introduce the similarity measure which is derived previously from fuzzy number, and the computation results are discussed. Two similarity measure obtaining methods have their own strong points, fuzzy number methods is simple and easy to compute similarity if membership function is trapezoidal or triangular. Whereas similarity with distance measure needs more time and consideration, however it can be applied to the general membership function. At this point, it is interesting to study and compare two similarity measures between fuzzy set and crisp set.

In the next chapter, fuzzy number, center of gravity, and the similarity measure are introduced. In Chapter 3, similarity measures with distance measure and fuzzy number are derived and proved. Also two similarity measures are compared and discussed in Chapter 4. In the example, we consider the degree of similarity between fuzzy membership function and singleton. By the comparison, we obtain similarity measure that has proper meaning. Conclusions are followed in Chapter 5. Notations of Liu's are used in this paper [7].

H. Li et al. (Eds.): AIRS 2008, LNCS 4993, pp. 644–649, 2008.

2 Similarity Measure Preliminaries

In this chapter, we introduce some preliminary results for the degree of similarity. Fuzzy number, center of gravity, and axiomatic definitions of similarity measure are included. A generalized fuzzy number $\tilde{A}$ is defined as $\tilde{A}=(a,b,c,d,w)$ where $0<w\leq 1$ and a , b , c and d are real numbers [1,2]. Trapezoidal membership function $\mu_{\tilde{A}}$ of fuzzy number $\tilde{A}$ satisfies the following conditions [4]:

1) $\mu_{\tilde{A}}$ is a continuous mapping from real number R to the closed interval $[0\sim 1]$
2) $\mu_{\tilde{A}}(x)=0$, where $-\infty<x\leq a$
3) $\mu_{\tilde{A}}(x)$ is strictly increasing on $[a,b]$
4) $\mu_{\tilde{A}}(x)=w$, where $b\leq x\leq c$
5) $\mu_{\tilde{A}}(x)$ is strictly decreasing on $[c,d]$
6) $\mu_{\tilde{A}}(x)=0$, where $d\leq x<\infty$.

If $b=c$ is satisfied, then it is natural to satisfy triangular type. Four fuzzy number operations are also found in literature [4]. Traditional center of gravity (COG) is defined by

$$x_{\tilde{A}}^{*}=\frac{\int x\mu_{\tilde{A}}(x)dx}{\int \mu_{\tilde{A}}(x)dx}$$

where $\mu_{\tilde{A}}$ is the membership function of the fuzzy number $\tilde{A}$, $\mu_{\tilde{A}}(x)$ indicates the membership value of the element x in $\tilde{A}$, and generally, $\mu_{\tilde{A}}(x)\in[0,1]$. Chen and Chen presented a new method to calculate COG point of a generalized fuzzy number [4]. They derived the new COG calculation method based on the concept of the medium curve. These COG points play an important role in the calculation of similarity measure with fuzzy number. Liu suggested axiomatic definition of similarity measure as follows [7]. By this definition, we study the meaning of similarity measure.

Definition 2.1. [7] A real function s: $F^{2}\rightarrow R^{+}$ is called a similarity measure, if s has the following properties:

(S1) $s(A,B)=s(B,A),\ \forall A,B\in F(X)$

(S2) $s(D,D^{C})=0,\ \forall D\in P(X)$

(S3) $s(C,C)=\max_{A,B\in P} s(A,B),\ \forall C\in F(X)$

(S4) $\forall A,B,C\in F(X)$, or , if A,B,C, then $s(A,B)\geq s(A,C)$ and $s(B,C)\geq s(A,C)$.

where $R^{+}=[0,\infty)$, X is the universal set, $F(X)$ is the class of all fuzzy sets of X, $P(X)$ is the class of all crisp sets of X , and D^{C} is the complement of D .

3 Similarity Measure by Fuzzy Number and Distance Measure

In this chapter we introduce the degree of similarities which were contained in the previous literatures [1-4]. Which are all based on the fuzzy number. And the similarity measure construction with the distance measure is also included.

3.1 Similarity Measure Via Fuzzy Number

We introduce the conventional fuzzy measure that is based on the fuzzy number. Chen introduced the degree of similarity for trapezoidal or triangular fuzzy membership function of $\tilde{A}$ and $\tilde{B}$ as [1]

$$S(\tilde{A},\tilde{B}) = 1 - \frac{\sum_{i=1}^{n} | a_i - b_i |}{4} \tag{1}$$

where $S(\tilde{A},\tilde{B}) \in [0,1]$. If $\tilde{A}$ and $\tilde{B}$ are trapezoidal or triangular fuzzy numbers, then the n can be 4 or 3, respectively. Hsieh *et. al.* also proposed similarity measure for the trapezoidal and triangular fuzzy membership function as follows [2]:

$$S(\tilde{A},\tilde{B}) = \frac{1}{1 + d(\tilde{A},\tilde{B})} \tag{2}$$

where $d(\tilde{A},\tilde{B}) = | P(\tilde{A},) - P(\tilde{B}) |$, and for triangular fuzzy number the graded mean integration of $\tilde{A}$ and $\tilde{B}$ are $P(\tilde{A}) = (a_1 + 4a_2 + a_3)/6$ and $P(\tilde{B}) = (b_1 + 4b_2 + b_3)/6$, whereas for trapezoidal fuzzy number, $P(\tilde{A}) = (a_1 + 2a_2 + 2a_3 + a_4)/6$ and $P(\tilde{B}) = (b_1 + 2b_2 + 2b_3 + b_4)/6$ are satisfied. Lee derived the trapezoidal similarity measure using fuzzy number operation and norm definition. That is

$$S(\tilde{A},\tilde{B}) = 1 - \frac{\| \tilde{A} - \tilde{B} \|_{l_p}}{\| U \|} \times 4^{-1/p}, \tag{3}$$

where $\| \tilde{A} - \tilde{B} \|_{l_p} = (\sum_i (|a_i - b_i|))^{-1/p}$, $\| U \| = \max(U) - \min(U)$, and p is the natural number greater or equal 1, finally U is the universe of discourse. Chen and Chen propose similarity measure to overcome the drawbacks of existing similarity:

$$S(\tilde{A},\tilde{B}) = [1 - \frac{\sum_{i=1}^{n} | a_i - b_i |}{4}] \times (1 - | x^*_{\tilde{A}} - x^*_{\tilde{B}} |)^{B(S_{\tilde{A}}, S_{\tilde{B}})} \times \frac{\min(y^*_{\tilde{A}}, y^*_{\tilde{B}})}{\max(y^*_{\tilde{A}}, y^*_{\tilde{B}})} \tag{4}$$

where $(x^*_{\tilde{A}}, y^*_{\tilde{A}})$ and $(x^*_{\tilde{B}}, y^*_{\tilde{B}})$ are the COG of fuzzy number $\tilde{A}$ and $\tilde{B}$, $S_{\tilde{A}}$ and $S_{\tilde{B}}$ are expressed by $S_{\tilde{A}} = a_4 - a_1$ and $S_{\tilde{B}} = b_4 - b_1$ if they are trapezoidal. $B(S_{\tilde{A}}, S_{\tilde{B}})$ is denoted by 1 if $S_{\tilde{A}} + S_{\tilde{B}} > 0$, and 0 if $S_{\tilde{A}} + S_{\tilde{B}} = 0$.

3.2 Similarity Measure with Distance Function

Hamming distance is commonly used as distance measure between fuzzy sets A and B,

$$d(A,B)=\frac{1}{n}\sum_{i=1}^{n}|\mu_A(x_i)-\mu_B(x_i)|$$

where $X=\{x_1,x_2,\cdots,x_n\}$, $|k|$ is the absolute value of k. With Definition 2.1, we propose the following theorem as the similarity measure.

Theorem 3.1. For any set $A,B\in F(X)$, if d satisfies Hamming distance measure and $d(A,B)=d(A^C,B^C)$, then

$$s(A,B)=1-d((A\cap B^c),[0])-d((A\cup B^C),[1]) \tag{5}$$

is the similarity measure between set A and set B.

Theorem 3.2. For any set $A,B\in F(X)$ if d satisfies Hamming distance measure, then

$$s(A,B)=2-d((A\cap B),[1])-d((A\cup B),[0]) \tag{6}$$

is also a similarity measure.

Proofs of Theorem 3.1 and 3.2 are found in [10].

From Theorem 3.1 and 3.2 we can compute the degree of similarity between fuzzy sets. Then how can we compute the degree of similarity between fuzzy set and crisp set ? In Fig. 1 there are three membership function pairs. All three pairs have the same degree of similarity ? Naturally it must have the different degree of similarity.

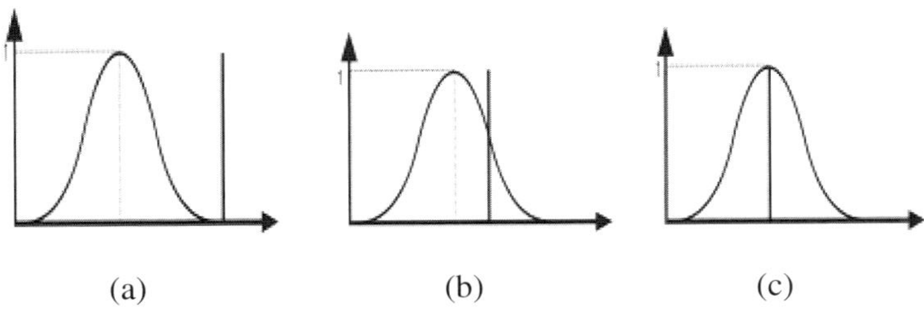

Fig. 1. Similarity between fuzzy set and crisp set

Now we replace fuzzy set B (5) and (6) to crisp set A_{near}, membership function of A is defined 1 when $\mu_A(x)\geq 1/2$, and 0 when $\mu_A(x)<1/2$. Now it is possible to represent the degree of similarity between fuzzy set and crisp set. We replace fuzzy set B into crisp set A_{near}.

$$s(A,A_{near})=1-d((A\cap A^c_{near}),[0])-d((A\cup A^c_{near}),[1]) \tag{7}$$

Similarly,

$$s(A, A_{near}) = 2 - d((A \cap A_{near}),[1]) - d((A \cup A_{near}),[0]) \tag{8}$$

is also the similarity measure of fuzzy set A and crisp set A_{near}. Proofs of (7) and (8) are found in [6].

4 Computation of Similarity Measures

In this chapter, we compare our similarity measure (7) with (4) by Chen and Chen [4]. As mentioned before, similarity measures based on fuzzy number have to depend on the membership function type. They consider only trapezoidal or triangular. Whereas proposed similarity measure can be applied to fuzzy membership function generally. In [4], Chen and Chen illustrate twelve membership function pairs [4]. Among them, 8-th pair similarity degrees are illustrated in Table 1. 4 previous results are all obtained through fuzzy number.

Table 1. Comparison with the result of Chen and Chen

Methods	Lee[3]	Hsieh and Chen[2]	Chen [1]	Chen and Chen[4]
8-th pair	0.5	0.909	0.9	0.54

Chen and Chen compute the degree of similarity as follows

$$S(\tilde{A}, \tilde{B}) = [1 - \frac{0.2 + 0.1 + 0}{3}] \times (1 - (0.1)^1 \times \frac{\min(1/3, 0.5)}{\max(1/3, 0.5)} = 0.54$$

This result is obtained through fuzzy number, so computation is easy to obtain, however the result is strictly limited for the trapezoidal or triangular membership functions. Whereas with similarity measure we also compute the similarity measure, and results are generally applied for the arbitrary shape of membership functions. Our computation conditions, Universe of discourse : 0.1~1.0, Data points : 100, Sample distance : 0.01. For fuzzy set A, domain can be from 0.1 to 0.3 among universe of discourse, whereas crisp set B can only be 0.3. We apply similarity measure (7) instead (8). Because there are no interchanging points between A and B. We have obtained the similarity measure of 8-th pair of Chen and Chen as 0.476 [4]. Our proposed similarity measures are possible to compute the degree of similarity for the membership function pairs like Fig. 1.

5 Conclusions

We have introduced the fuzzy number and the similarity measure that is derived from fuzzy number. We also proposed a similarity measure based on the distance measure. The usefulness of proposed similarity measure is proved. By the comparison with previous example, we can see that proposed similarity measure can be applied to the general types of fuzzy membership functions.

Acknowledgments. This work was supported by 2nd BK21 Program, which is funded by KRF(Korea Research Foundation).

References

1. Chen, S.M.: New methods for subjective mental workload assessment and fuzzy risk analysis. Cybern. Syst.: Int. J. 27(5), 449–472 (1996)
2. Hsieh, C.H., Chen, S.H.: Similarity of generalized fuzzy numbers with graded mean integration representation. In: Proc. 8th Int. Fuzzy Systems Association World Congr., vol. 2, pp. 551–555 (1999)
3. Lee, H.: An optimal aggregation method for fuzzy opinions of group decision. In: Proc. 1999 IEEE Int. Conf. Systems, Man, Cybernetics, vol. 3, pp. 314–319 (1999)
4. Chen, S.J., Chen, S.M.: Fuzzy risk analysis based on similarity measures of generalized fuzzy numbers. IEEE Trans. on Fuzzy Systems 11(1), 45–56 (2003)
5. Lee, S.H., Cheon, S.P., Kim, J.: Measure of certainty with fuzzy entropy function. In: Huang, D.-S., Li, K., Irwin, G.W. (eds.) ICIC 2006. LNCS (LNAI), vol. 4114, pp. 134–139. Springer, Heidelberg (2006)
6. Lee, S.H., Kim, J.M., Choi, Y.K.: Similarity measure construction using fuzzy entropy and distance measure. In: Huang, D.-S., Li, K., Irwin, G.W. (eds.) ICIC 2006. LNCS (LNAI), vol. 4114, pp. 952–958. Springer, Heidelberg (2006)
7. Liu, X.: Entropy, distance measure and similarity measure of fuzzy sets and their relations. Fuzzy Sets and Systems 52, 305–318 (1992)
8. Fan, J.L., Xie, W.X.: Distance measure and induced fuzzy entropy. Fuzzy Set and Systems 104, 305–314 (1999)
9. Fan, J.L., Ma, Y.L., Xie, W.X.: On some properties of distance measures. Fuzzy Set and Systems 117, 355–361 (2001)
10. Park, D.H., Lee, S.H., Song, E.H., Ahn, D.K.: Similarity computation of fuzzy membership function pairs with similarity measure. In: Huang, D.-S., Heutte, L., Loog, M. (eds.) ICIC 2007. LNCS (LNAI), vol. 4682, pp. 485–492. Springer, Heidelberg (2007)

Experimental Study of Chinese Free-Text IE Algorithm Based on W_{CA}-Selection Using Hidden Markov Model

Qian Liu, Hui Jiao, and Hui-bo Jia

Optical Memory National Engineering Research Center,
State Key Laboratory of Precision Measurement Technology and Instruments,
Tsinghua University, Beijing100084, China
{liuqian00,jiaoh04}@mails.tsinghua.edu.cn,
jiahb@mail.tsinghua.edu.cn

Abstract. This paper proposes the extraction task of the Chinese Sci-tech journal text and presents a W_{CA}-Selection Chinese free-text HMM IE algorithm. The HMM IE algorithm takes the Chinese Sci-tech journal abstract text as the extraction text. According to the features of W_{CA}, an idea of W_{CA} selection model re-optimization is proposed. And a W_{CA} selection optimization strategy is concreted. Then the experimental verification is conducted with a satisfied result. The experiment results show that the designed extraction algorithm and W_{CA} selection optimization strategy have good performance in the the Chinese Sci-tech journal abstract text.

1 Introduction

The goal of Information Extraction (IE) is to extract desired information from natural language texts. Unlike information retrieval, IE is interested in extracting the actual relevant facts and representing them in some useful form. In the past decade, IE technology has become an important branch of Nature Language Processing (NLP) technology. IE systems have been developed for writing styles ranging from structured text, semi-structured text and free text [1]. The desired information can easily be correctly extracted form the structured text. But for the free text this process is more complex and difficult. The output of IE system would be represented as hierarchical attribute-value structures called Templates.

These days, the Chinese IE system is generally limited to extract structured text and semi-structured text only. The words in these texts are ungrammatical and often following a predefined format or style. The research of IE system for Chinese free text is still in the stage of beginning.

There are two basic approaches [2] to design the modules of an IE system, which can be called the *Knowledge Engineering approach*, and the *Machine Learning approach*. The Knowledge Engineering approach [3] can be hand crafted. But the development of this approach is very laborious and time-consuming. The Machine learning approach [4][5] is relied on the annotated corpus providing examples on which learning algorithms can operate. Statistical machine learning techniques, while well proven in fields such as speech recognition, have become increasingly popular in the last several years. Hidden Markov model (HMM) [6] is a powerful statistical

H. Li et al. (Eds.): AIRS 2008, LNCS 4993, pp. 650–655, 2008.

machine learning technique that is just beginning to gain use in information extraction tasks. HMM offers the advantages of having strong statistical foundations that are well-studied to natural language tasks, handling new data robustly, and being computationally efficient to develop and evaluate.

The CAJ network(www.chinajournal.net.cn) is the biggest Chinese Sci-Tech journal full text data in the world at present. It embodies almost 8000 kinds of important full text of Sci-Tech journal now. The Chinese Sci-Tech documents are stored separately in the data according to the topic of document, abstract, author information, full text and so on.

To each of Sci_Tech journal, the abstract is the summarized introduction and includes lots of important information. And the words in these abstract texts are grammatical and syntactical. It has different description ways for the same fact information. It is the standard Chinese free text. So it has the significance of both representational and actual for the research of IE algorithm for the abstract of Chinese Sci_Tech journal.

So we take the Chinese Sci-tech journal abstract text as the object and design the W_{CA}-Selection Chinese free-text HMM IE algorithm. Then we'll introduce the extraction task.

2 Confirmation of Extraction Task

Before the introduction to extraction algorithm, firstly, we introduce the extraction objective of the Chinese Sci-Tech journal abstract text. We use the abstract text of Sci-Tech journal in the same field as the extraction object and extract the information which can describe and represent the characteristics of this field that is called as characteristic information. The characteristic information can be artificially designated according to the characteristic of some fields. It also can be obtained by using Semi-surprised statistical method. Then, we create the output template of information extraction of its field by using characteristic information. We use *slot_i* to denote the *i*th characteristic information of the output template. We use *info_i* to denote the relevant information that is supposed to be extracted and matches to *slot_i*. The area where the relevant information is supposed to be extracted is called as extraction area. We use *E-A* to denote the extraction area.

In the following, we give an introduction to the designed structure of extraction model.

3 The Structure of Extraction Model

HMM is a double embedded stochastic process with an underlying stochastic process that is not observable (it is hidden), but can only be observer through another set of stochastic processes that produce the sequence of observations. An HMM can be specified by a five-tuple $\mu(S,V,\Pi,A,B)$ [7], where S and V are the set of states and the output observation symbols, and Π,A,B are the probabilities for the initial state, state transitions, and symbol emissions, respectively.

In the designed model of HMM extraction algorithm, the limited set of observation value is the same type adjacent character string, called as Same_Str. We conduct the same type merge to the adjacent character string in the text. The types of the character string are differentiated based on the following types: Chinese word, number string, English character string, all-shape punctuation string and half-shape punctuation string.

In the designed extraction algorithm model, the limited set of state is denoted as below:

$$S = \{s_q | s_q = s_{\_i} \text{ or } s_{\_U} \text{ or } s_{\_WTi} \text{ or } s_{\_WTi} \text{ or } s_{\_WTi\_BL} \text{ or } s_{\_WTi\_BR} \cdot \}$$

Where the explanation for each of state is in table 1:

Table 1. The Explanation of State Logo and *O*

Stage Logo	the attribute of *O* (Observation)
s_i	*O* belongs to *info_i*
$s_{\_U}$	irrelevant *O*
s_w_{Pi}	the Positioning- Word w_{Pi}
$s\_w_{Pi\text{-L}}$	the *O* located between w_{Pi} and *E-A*, when w_{Pi} is located to the left of *E-A*
$s\_w_{Pi\text{-R}}$	the *O* located between w_{Pi} and *E-A*, when w_{Pi} is located to the right of *E-A*

After the introduction to designed model structure of extraction algorithm, we take a look to the training process of algorithm.

4 The Training and Extraction Processes of Extraction Algorithm

The training process of designed IE algorithm is mainly divided into 3 steps: statistical process of W_{CA}, the confirmation process of model state parameters, the training process of model.

Step 1: the statistical process of $W_{CA.}$

Firstly, we give the definition of W_{CA}, W_{CA} (Co-Appearing-Word) is the same type adjacent character string which appears together with the extraction fact, *info_i*, in the same sentence. According to the relative position between W_{CA} and *info_i*, we divide W_{CA} into two types: W_T (Triggering-Word) and W_P (Positioning- Word). W_T is the W_{CA} which appears inside of E-A and W_P is the W_{CA} which appears outside of E-A. So, we conduct the statistics to the each sentence of the abstract text of Chinese Sci-Tech document in the same field which includes the set of *info_i* so that we can get the W_{Ti} set $W_{Ti}(w_{T1},w_{T2},\ldots,w_{Tr})$ and W_{Pi} set $W_{Pi}(w_{P1},w_{P2},\ldots,w_{Pk})$ which match to character information, *slot_i*.

Step 2: the training process of model state parameters

In this training process, the extraction model is divided into two states only: state s_i and state $s_{\_U}$. Then, we use ML algorithm to train the extraction model to get the initial extraction model, called as μ_0. Then, we can make sure how many states needed to be increased to differentiate the w_{Pi} which is corresponded to character information by using the extraction results of initial model.

Step 3: the training process of extraction model

We obtain the final state parameter from the extraction model which is trained in step 2. Then we use the each state in table 1 to label the training text and use ML algorithm to train the train text, and obtain the final extraction model.

In the extraction processes, we select the Veterbi algorithm as the decoding algorithm.

5 Experiment Verification and Result Discussion

We use the high power solid laser as the subject to search the Chinese journal full text database and obtain 213 different records. We take the abstract part of Sci-Tech document in 110 records whose content is laser experiment as the object text, taking 81 records as training text and taking 31 records as testing text. We can get information extraction template which has 5 slots in this field by using the method of semi-supervised word frequency statistic. It is shown in table 2.

Table 2. The Output Template of the IE system

laser crystal	Slot_1
input power	Slot_2
output power	Slot_3
output wavelength	Slot_4
System Efficiency	Slot_5

Firstly, in our designed Word-based Chinese Document Experimental System [8], the training text and testing text are transformed into the text of CDM format. Then, for the 5 slots of output template, we can obtain the 10 sets of triggering word W_T and positioning word W_P from training text by using the method of semi-supervised word frequency statistic separately.

Then, we go to the training stage of extraction model state value. We use the initial extraction model from the training to do the initial test for the training text and find out that the states of slot2 and slot3 are easily mixed up in the error analysis. So, we need to add the state label which is used to differentiate slot2 and slot3 into the process of training. We choose the largest differentiation degree w_{Pi} in the W_P set of slot2 and slot3, and put the corresponding state of s_w_{P2} and s_w_{P3} into the training process. So, there are totally 12 states in the training process of extraction model.

Next, we go to the process of W_{CA} selection model re-optimization. We conduct the re-optimization to the parameter of extraction model in training text by using the designed W_{CA} selection strategy. During this process, the changing of F value with the W_{CA} selection is shown in figure 1:

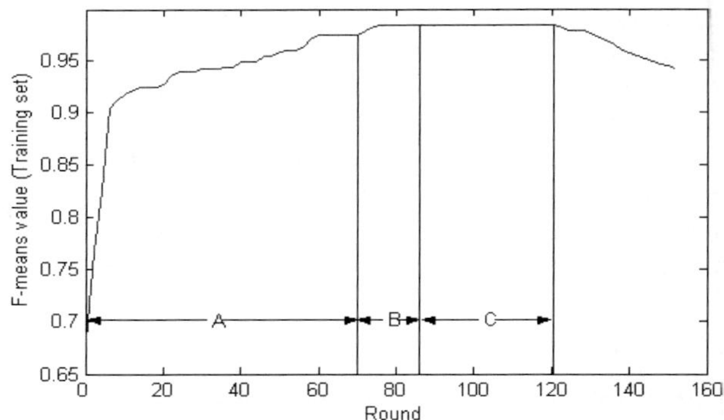

Fig. 1. The changing of F value in the optimization process

From the figure, we can see that the whole optimization process is divided into 3 processes which are shown as A, B, C. The process A is the first stage of increasing word strategy. We set a threshold and choose the word frequency of W_{CA} whose value is higher than the threshold to put them into the training process according to the order of word frequency in 10 sets of W_{CA}, and in every training round we choose a W_{CA} which makes F value increase fastest as the observation value to be kept in the set of observation value of extraction model. After F value stop increasing with 5 W_{CA} increasing, the process A is end and go to process B, there are 70 W_{CA} to be selected at this time.

The process B is the second stage of increasing word strategy. In this stage, we put the rest of W_{CA} in the 10 sets of W_{CA} as the observation value into the training process and choose a W_{CA} which makes F value increase fastest as the observation value every training round to be kept in the set of observation value of extraction model. After F value stop increasing with 5 W_{CA} increasing, the process B is end and go to process C, there are 86 W_{CA} to be selected at this time.

Process C is the stage of decreasing word strategy. During this process, we delete W_{CA} from the set of observation value one by one according to the opposite order in the stage of increasing word under the situation of keeping F value invariable. After decreasing some W_{CA} to cause F value decrease, the process C is ended and the whole optimization process is ended. There are 36 W_{CA} decreased in stage C and there are 42 W_{CA} kept for the extraction model after finishing the whole optimization process.

Then, we use the extraction model after optimizing to do the extraction for the testing text and get the values of R, P, F-means for R=96.38%, P=88.67%, F-means=92.39% respectively.

From the analysis above, we can see that both process A and B are the processes to optimize the extraction model by increasing W_{CA} in every training round. With the increasing of W_{CA} more extraction area can be found in the process of decoding. So, for the abstract text with less redundancy information, F value will increase with the W_{CA} increasing. At same time, for the process A and B, the word frequency selected in the process A is higher than that of process B so that the speed of increasing for F value in process A is faster than that in process B. By the end of process B, we already obtained the most optimized F value in the training text, but the parameter of extraction model is not the most optimized at this moment. We need to do the optimization to the dimension of observation layer of extraction model under the situation of keeping F value invariable by using decreasing word strategy. In the process C, for A,B,C three W_{CA}, because the information provided by A and C already made some extraction area to be found in the process of decoding, the decreasing of B will not cause F value to decrease.

6 Conclusion and Future Work

We take the Chinese Sci-tech journal abstract text as the object and design the W_{CA}-Selection Chinese free-text HMM IE algorithm And then, we proposed the idea of W_{CA} selection model re-optimization and concrete selection optimization strategy according to the features of W_{CA}.

Then, we conduct the experimental verification to the abstract texts which belong to the same subject in CAJ network and get the satisfied result: Recall ratio is 96.38%, Precision ratio is 88.67%, and F-means value is 92.39%. The experiment results show that the designed extraction algorithm and W_{CA} selection optimization strategy have good performance in the Chinese free text with less redundancy information.

Next, we will do some research separately to 2 kinds of W_{CA} for the different effect during the process of information extraction and the different influence to the extraction result so that we can further complete the designed W_{CA} selection strategy.

References

1. Eikvik, L.: Information Extraction from World Wide Web A survey. Norweigan Computing Center, Norway (1999)
2. Appelt, D.D.: Introduction to Information Extraction. J. AI Magazine 12, 161–172 (1999)
3. Aone, C., Halverson, L., Hampton, T., Ramos-Santacruz, M.: SRA: Description of the IE2 system used for MUC-7. In: Proceedings of MUC-7 (1998)
4. Miller, S., Crystal, M., Fox, H., et al.: Algorithms that learn to extract information - BBN: Description of the SIFT system as used for MUC-7. In: Proceedings of MUC-7 (1998)
5. Zelenko, D., Aone, C., Richardella, A.: Kernel methods for relation extraction. J. Machine Learning Research 3, 1083–1106 (2003)
6. Seymore, K., Mccallum, A., Rosenfeld, R.: Learning Hidden Markov Model Structure for Information Extraction. In: AAAI 1999 Workshop on Machine Learning for Information Extraction (1999)
7. Christopher, D., Manning, H.: Foundations of Statistical Natural Language Processing. The MIT Press, London, England (2001)
8. Liu, Q., Jiao, H., Jia, H.-b.: Study of Word-based Chinese Document Experimental System and Chinese Free-text Information Extraction Experiment Based on it. In: ICNC 2007, vol. 5, pp. 120–123. IEEE Press, Haikou (2007)

Finding and Using the Content Texts of HTML Pages

Jun MA, Zhumin Chen, Li Lian, and Lianxia Li

The Colledge of Computer Science and Technology, Shandong University, Jinan, China
majun@sdu.edu.cn

Abstract: A novel algorithm to find the content text in an HTML page is proposed based on a number of features of textual blocks in the page. Experiments show the new algorithm is better than known ones in terms of the ratios of the correctly removed noise blocks and the correctly found content blocks respectively. The application of the algorithm in hidden web classification is demonstrated as well.

Keywords: page clearning, page segmentation, content extraction.

1 Introduction

The text blocks whose contents are not related to the topic of a Web page is called *noise text* otherwise *content text*. To find the *content text* is an important issue in Information Retrieval (IR) and has many applications [1,2,4,5].

The known methods for finding *content text* are page templates [17], production rules [12], DOM trees [16] and VIPS (Vision Based Page Segmentation) [1,2]. In this paper an algorithm based on the statistical analysis of the features of blocks gotten by VIPS style segmentation is proposed. Experiments were carried out to compare performance of the new algorithm with some known ones in terms of the ratios of correctly removed noise blocks and correctly found content blocks. An application of the algorithm in hidden web classification is demonstrated as well.

2 Finding Content Block in an HTML Page

A segmentation program based on the idea of VIPS was employed to find the textual blocks and give the digital presentation of their features, which include the tags, punctuations, fonts of letters in a block, positions and area counted by the pixel numbers or word numbers. About one thousand pages chosen from CWT200G [3] were used to compute the probability that a text block is a *content text*. These pages came from different web sites and distributed in more than 10 domains. Based on our observation six features can be considered as the features for probability computation.

1) The co-appearance of some special pairs of tags in a block, e.g. <P>
.
2) The fonts of letters used in a block.
3) The punctuation set used in a block.

H. Li et al. (Eds.): AIRS 2008, LNCS 4993, pp. 656–662, 2008.

4) The intersection area of a text block and that of a *content text* block in a given template.
5) Area of a block counted by its pixel number.
6) Block length counted by the number of words in a block.

Let P_i be the conditional probability calculated based on feature i, $1 \le i \le 6$, and each P_i is calculated by Bayes' rule respectively. P_1 is the conditional probability that a block is a *content text* when it has the tags in a special established tag set. P_3 is calculated in a similar way. P_2 is computed according to its average letter size, which may be 8,10,12,14, or in the intervals of (0,8) or (16,+∞); P_4 is calculated based on the intersection of area of a block and that of the *content text* block in a template given in [16]; P_5 is computed based on which intervals the area counted by pixel number of a block falls into, where the intervals are (0,10], (10,20], (20,30], (30,+∞). P_6 is calculated in a similar way. Because we assume the events that a block has feature i and j, $i \neq j$, are mutually disjoint, then the final probability is the sum of the 6 probabilities.

We implemented an algorithm to find the content text of a page by choosing the textual block with the maximum probability. We named it MFM (Multi-Feature Model) and compared the outputs of the algorithm with the followings.

1) SM: extract all texts in a HTML page as the *content texts*;
2) IM[10]: choose the text block with the maximum information entropy;
3) PM[5]: extract the texts according to the text position in a HTML page;
4) DTM [16] : recognizes the *content texts* based on the DOM trees.

NR and OR are used to evaluate the performance, which are defined below.

$$NR = \frac{\textit{the area of the deleted noise blocks}}{\textit{the area of the total noise blocks}} \tag{1}$$

NR is the ratio of correctly removed noise blocks and OR is the ratio of correctly obtained *content text*s.

$$OR = \frac{\textit{the area of the correctly obtained content text blocks}}{\textit{the area of the all content text blocks}} \tag{2}$$

NR and OR were computed in a semi-automatic way. Table. 1 gives the comparison of the *content text* extraction by the four algorithms.

Table 1. The comparison on NR and OR

Algorithm	NR	OR
IM	86.3%	93.74%
PM	93.5%	91.25%
DTM	84.1%	94.62%
MFM	92.8%	97.87%

Furthermore we studied the recall and precision of IR on the "clean-up" pages by above algorithms. We re-indexed about cleaned 10,000 pages by above 4 algorithms respectively and compared the IR performance based on following measurements.

1) MPOS : the Mean of POSition of the first relevant result for different given queries;
2) MRR (the Mean of the Reciprocal Ranking)
3) P@5 and P@10 : Precision at the top 5 and top 10 results ：
4) MAP: Mean Average Precision

About 50 hot topics are selected as queries for information retrieval. The MRR of each individual query is the reciprocal of the rank at which the first correct response was returned, or 0 if none of the first 15 responses contained a correct answer. The score for a sequence of queries is the mean of the individual query's reciprocal ranks. The MAP for multi-queries is average of the means of precision of different searching topics.

Table 2. The Comparison on multi-measurements

Method	MPOS	MRR	P@5	P@10	MAP
SM	6.78	0.4732	0.245	0.201	0.175
IM	4.21	0.5389	0.371	0.229	0.190
PM	5.78	0.5096	0.286	0.188	0.168
DTM	5.23	0.5109	0.343	0.223	0.182
MFM	4.00	0.5952	0.411	0.257	0.217

Clearly MFM is the best in terms of MRR, P@5, P@10 and MAP. Let us take algorithm SM as the base algorithm, and compute the Gain of other algorithms relative to the performance of SM. Figure 1 shows the relative Gains of IM,PM, DTM and MFM to SM based on formula 3.

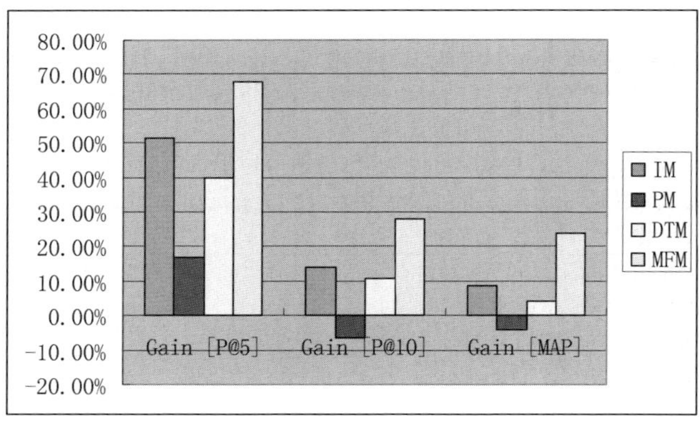

Fig. 1. The Gain of four algorithms to SM

$$Gain(MAP)_{IM} = \frac{MAP(IM) - MAP(SM)}{MAP(SM)} \times 100\% \quad (3)$$

MFM is the best among all others on P@5, P@10 and MAP.

3 An Application in Hidden Web Classification

Hidden Web usually means the databases avalable through the forms embedded in HTML pages. Hidden Web classification is the main way in organizing Hidden Web. Now most employed methods are pre-query [7] and post-query[12]. The former techniques issue probe queries and classifies the databases based on the retrieved outputs of the database. The latter techniques classify the databases based on the visible feature of forms, e.g. the available attribute labels and values in these database forms. However, most approaches did not use the text in HTML pages. Clearly the texts usually provide the context of the content of the database whose entry form in the same page. In our approach let P be a HTML pages with a database form. We obtain FC(form feature content) and PC(Page content text) by MFM algorithm. Further let FCV and PCV be the vector presentation of FC and PC respectively, the similarity between two pages P_1,P_2 is calculated by

$$sim(P_1, P_2) = \frac{k_1 \cos(PCV_1, PCV_2) + k_2 \cos(FCV_1, FCV_2)}{k_1 + k_2} \quad (4)$$

where PCV_i and FCV_i are the vectors of the content texts and attribute labels of P_i, i=1, 2, respectively.

We use the k-NN algorithm for the classification on the test data of UIUC [15]. We name the classes in UIUC in Table. 3 for conciseness

Table 3. The Classes in UIUC data set

Class	airfares	automobiles	Books	CarRentals	Hotels	Jobs	Movies	Music
Name	C_1	C_2	C_3	C_4	C_5	C_6	C_7	C_8

For each class C_i we choose about 10 to 30 sample pages carefully as the initial elements in C_i. $1 \le i \le 8$. The k-NN classification algorithm is given below.

Algorithm CAFC (Context-Aware Form Classifying)

Input: The pages needed to be classified, parameter k_1, k_2 and non-negative integer k, where k is the number of the nearest neighbors of the algorithm, it can be determined based on the number of classes and the number of the samples in the initial $C_1, C_2, \ldots, C_m$. C_i has the example pages for class i as well as their PCV and FCV vectors, $1 \le i \le M$.

Output: m classes stored in $C_1, C_2, \ldots, C_m$.

Step 0 //initialization//. $S = C_1 \cup C_2 \cup \ldots \cup C_M$ 。

Step 1: input page P, get the content texts as well as the attribute labels of P according to the methods discussed in section 3 and 4, then calculate the *tf* and *idf* and establish the vector PCV and FCV for P.
Step 2: ∀P' ∈ S, calculate *sim*(P,P') by formula 4;
Step 3: Choose the first nearest neighbor pages based on the values of function *sim* from S and store then in set A.
for $i = 1$ to m

$$\{\text{compute } score(C_i \mid p) = \sum_{p' \in A} sim(p, p') I(p', C_i)$$

where function $I(p', C_i) = 1$ if $p' \in C_i$; $I(p', C_i) = 0$ otherwise;}
Step 4: put p into class C_j if score $(C_j \mid p) \geq$ score $(C_i \mid p)$, $1 \leq i,j \leq m$;
Step 5: output $C_1, C_2, \ldots, C_m$ and stop if all pages have been dealt with otherwise go to step 1。

Table. 4 gives the recall, precision and F1 value for classification results based on FCV only.

Table 4. The classification based on FCV only

Class	C_1	C_2	C_3	C_4	C_5	C_6	C_7	C_7
Recall	93.33%	86.49%	76.92%	33.33%	62.50%	72.22%	66.67%	77.78%
Precision	73.68%	88.89%	90.91%	100.00%	83.33%	38.24%	74.07%	84.00%
F1	0.82	0.88	0.83	0.50	0.71	0.50	0.70	0.81

Now let consider how to determine the parameters of k_1 and k_2 in formula 4 when we use PCV to enhance the classification. In principle k_1, k_2 can be any positive number, for conciseness we only consider the case $k + k_2 = 1$, and assume they are in set {0.1,0.2,…,0.9}. Experiments were carried out to find the optimal combination of the pair k_1 and k_2 . They showed when $k_1 = 0.4$ and $k_2 = 0.6$ the values of F1 are almost the best for most classes. Therefore we only show the experiment results on $k_1 = 0.4$ and $k_2 = 0.6$ in the following discussion. Table. 5 shows the classification based on both PCV and FCV.

Table 5. The classification based on both PCV and FCV

Class	C1	C2	C3	C4	C5	C6	C7	C8
recall	100.00%	87.50%	80.77%	88.89%	68.75%	90.00%	93.55%	28.57%
precision	88.24%	83.33%	91.30%	88.89%	100.00%	45.00%	82.86%	100.00%
F1	0.94	0.85	0.86	0.89	0.81	0.60	0.88	0.44

Table 6. The classification based on all text and FCV

Class	C1	C2	C3	C4	C5	C6	C7	C8
Recall	100.00%	85.00%	100.00%	44.44%	81.25%	90.00%	38.71%	96.43%
Precise	75.00%	97.14%	60.47%	66.67%	76.47%	75.00%	100.00%	96.43%
F1	0.86	0.91	0.75	0.53	0.79	0.82	0.56	0.96

Clearly the content texts really enhance the classification of hidden web in terms of recall, precision as well as the values of F1.

Table. 6 shows the database classification based on all texts in an HTML pages except the HTML tags. If we compare the data in Table. 5 with those in Table. 6, clearly the performance of the algorithm CAFC with *content texts* are better than those with all texts for classes C1, C3, C4,C5,C7 in terms of F1, while are worse in terms of F1 for class C2,C6,C8. It is because the HTML pages in class C2,C6, and C8 have too less textual information. Therefore we believe that the performance of Deep Web classification with content text are better than those with all texts in Web pages in general.

References

1. Cai, D., Yu, S., Ma, J.W.W.: VIPS: a Vision-based Page Segmentation Algorithm, MSR-TR_2003-79
2. Cai, D., Yu, S., Wen, J.-R., Ma, W.-Y.: Extracting Content Structure for Web Pages Based on Visual Representation. In: APWeb 2003, pp. 406–417 (2003)
3. CWT200G: http://www.cwirf.org/SharedRes/DataSet/cwt.html
4. Debnath, S., Mitra, P., Giles, C.L.: Identifying Content Blocks from Web Documents. In: Hacid, M.-S., Murray, N.V., Raś, Z.W., Tsumoto, S. (eds.) ISMIS 2005. LNCS (LNAI), vol. 3488, p. 2005. Springer, Heidelberg (2005)
5. Feng, H., Liu, B., Liu, Y.: A framework for extracting the content and analysis for the Web pages with the position coordinates tree. Tsinghua Science and technology 45(S1), 1767–1771 (2005)
6. Gravano, L., Ipeirotis, P.G., Sahami, M.: QProber: A system for automatic classification of hidden-Web databases. ACM TOIS 21(1), 1–41 (2003)
7. He, B., Tao, T., Chang, K.C.-C.: Organizing structured web sources by query schemas: a clustering approach. In: CIKM, pp. 22–31 (2004)
8. Liu, W., Meng, X., Meng, W.: Vision-based Web Data Records Extraction. In: Proceedings of the 9th SIGMOD International Workshop on Web and Databases (SIGMOD-WebDB 2006), Chicago, Illinois, June 30 (2006)
9. Liu, B., Zhao, K., Yi, L.: Eliminating Noisy Information in Web Pages for Data Mining. In: Proc. Ninth ACM SIGKDD Int'l Conf. Knowledge Discovery and Data Mining, pp. 296–305 (2003)
10. Simon, K., Lausen, G.: Augmenting Automatic Information Extraction with Visual Perceptions. 2005 ACM 1595931406/05/0010 (2005)
11. Ou, J., Dong, S., Cai, B.: A method to extract the topic information from the HTML pages with design model. Tsinghua Science and technology 45(S1), 1743–1747 (2005)

12. Ru, Y., Horowitz, E.: Indexing the invisibleWeb: a survey. Online Information Review 29(3), 249–265 (2005)
13. Song, R., Liu, H., Wen, J., Ma, W.: Learning important models for web page blocks based on layout and content analysis. SIGKDD Explorations 6(2), 14–23 (2004)
14. Song, Y., Ma, S., Chen, G., li, J.: A Parse method for HTML pages to enhance the quality of Chinese Search Engine. J. of Chinese Information Process, 1003–1077 (2003) 04-0019-08
15. The UIUC Web Integration repository, http://metaqerier.cs.uiuc.edu/repository
16. Wang, J., Loehovsky, F.: Data-rich section extraction from HTML pages. In: Proc. 3rd Int. Conf. On Web Info. Syst. Eng., Singapore, pp. 1–10. IEEE Computer Society Press, Los Alamitos (2002)
17. Yi, L., Liu, B.: Web Page Cleaning for Web Mining through Feature Weighting. In: The Proceedings of Eighteenth International Joint Conference on Artificial Intelligence (IJCAI 2003), Acapulco, Mexico (August 2003)

A Transformation-Based Error-Driven Learning Approach for Chinese Temporal Information Extraction*

Chunxia Zhang[1], Cungen Cao[2], Zhendong Niu[1], and Qing Yang[1]

[1] School of Computer Science and Technology, Beijing Institute of Technology, Beijing 100081, China
[2] Institute of Computing Technology, Chinese Academy of Sciences, Beijing 100080, China
cxzhang@bit.edu.cn, cgcao@ict.ac.cn, zniu@bit.edu.cn, qingyang2005@bit.edu.cn

Abstract. Temporal information processing plays an important role in many application areas such as information retrieval, question answering, machine translation, and text summarization. This paper proposes a transformation-based error-driven learning approach to extracting temporal expressions from Chinese unstructured texts. The temporal expression annotator used in the approach is developed based on a Chinese time ontology, which includes concepts of temporal expressions and their taxonomical relations. Experiments in three domains show that our algorithm obtained promising results.

Keywords: temporal information extraction, Chinese temporal expressions, transformation-based error-driven learning.

1 Introduction

Temporal information processing is a crucial issue in many natural language processing applications such as information retrieval, information extraction, question answering, and text summarization. Recently, there has been considerable progress in the area of temporal information extraction and temporal reasoning in English [1-7]. Unfortunately, few efforts have been made in this area in Chinese. This impels us to investigate temporal information processing in Chinese. This paper focuses on extracting temporal expressions (TE) from Chinese texts. Here, TE is defined as chunks of text, which refer to time points or intervals.

Unlike English, there are no orthographic boundaries between words in Chinese. Furthermore, Chinese is devoid of morphological alterations and change tags of case, number, tense, and part-of-speech. Hence, the difficulty of the TE identification problem is to identify TE in Chinese texts that have no any inflections at all.

---

* The first, third and fourth authors are supported by the Natural Science Foundation (grant no.60705022), the Program for New Century Excellent Talents in Universities of China, and Beijing Institute of Technology Basic Research Foundation (grant no.411002). The second author is supported by the Natural Science Foundation (grant no.60273019, 60496326, 60573063, and 60573064) and the National 973 Program (grant no. 2003CB317008 and G1999032701).

H. Li et al. (Eds.): AIRS 2008, LNCS 4993, pp. 663–669, 2008.

Currently, most works [8-11] concentrate on extracting TE of the Gregorian calendar from news text. However, Chinese society uses both the Gregorian calendar and the traditional Chinese calendar. In addition, Chinese people have their own ways of representing temporal entities such as the Stem-Branch timing system and the Emperor's Title and Reign Title timing system. There exist two kinds of methods to recognize TE: rule-based methods and machine learning based methods. The former relies on hand-written rules and dictionary [8-12], while the latter depends on annotated corpus [13]. The rule-based approach is simple and effective, but it requires intensive knowledge resources and an exhaustive linguistic study over temporal information. Wu et al. [11] present a temporal parser based on temporal grammar rules and constraint rules built manually to extract Chinese TE.

The purpose of this paper is to identify temporal expressions from Chinese unstructured text, which are based on the Gregorian calendar and the traditional Chinese calendar. We propose a transformation-based error-driven learning method to recognize TE. A TE annotator used in this approach is developed in terms of a Chinese time ontology, which consists of concepts of temporal expressions and their taxonomical relations. Experiments in three domains of news, history, and archaeology show that our algorithm obtained promising results.

The rest of the paper is organized as follows. Section 2 presents a Chinese time ontology. Section 3 explains how to identify TE from Chinese texts. Experiments of the algorithm are given in Section 4. Section 5 concludes the paper.

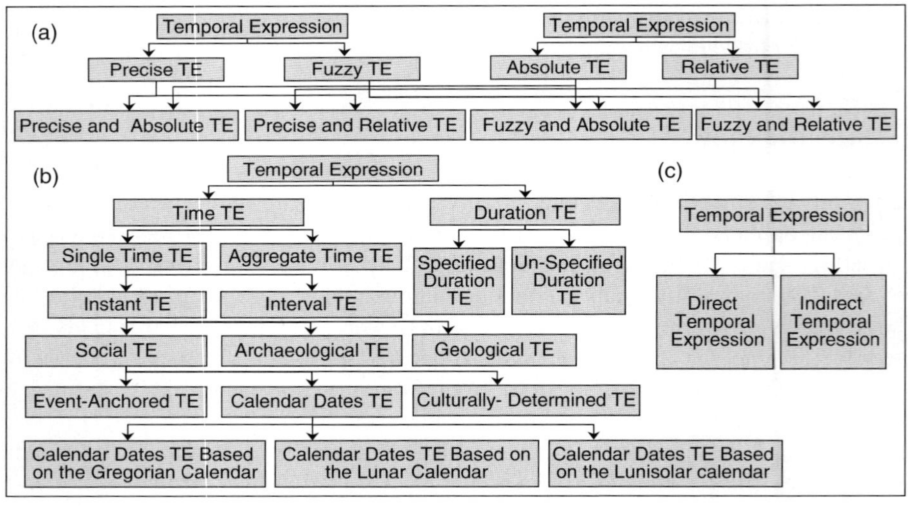

Fig.1. Taxonomy Structure of Temporal Expressions

2 A Chinese Time Ontology

Temporal concepts are time and durations, and time is an instant or interval on the time line. A duration is a distance between two different times. It can be anchored by the start time and the end time. If a duration is referred to as a length, it can not be anchored on the time line. We build a Chinese time ontology, which gives the taxonomy hierarchy of TE including four classifications of TE, as shown in Fig.1. In Fig.1.(a), TE is divided into precise/definite TE and fuzzy/indefinite TE, based on the uniqueness of the starting and ending times of temporal entity. TE is also separated into absolute/explicit/non-anaphoric TE and relative/implicit/anaphoric/indexical TE according to the existence of time references. For example, 'Jan.1.2007' is a precise and absolute TE, 'spring 2007' is a fuzzy and absolute TE. In Fig.1.(b) and Fig.1.(c), TE is classified into time TE and duration TE in terms of times and spans, and direct and indirect TE grounded on the existence of lexical triggers within TE.

3 A Transformation-Based Error-Driven Learning Method of Temporal Expressions Extraction

The transformation-based error-driven learning is a corpus-based and symbolic machine learning technique, which has been applied to many areas of natural language processing such as shallow parsing [14].

Our Chinese TE extraction algorithm first identifies TE in the training corpus using the TE annotator. During the first iteration, transformation rules, generated from incorrectly annotated TE, are used to update the EDL, which is an executable declarative language to write programs for identifying TE. The procedure repeats until no transformation rules can be built from inaccurate TE in the training corpus. Therefore, new components and composition methods of TE can be continually learned to amend the EDL. The TE annotator is to compile EDL and to recognize TE through executing EDL. The process of the error-driven learning method to identify TE is illustrated with Fig.2.

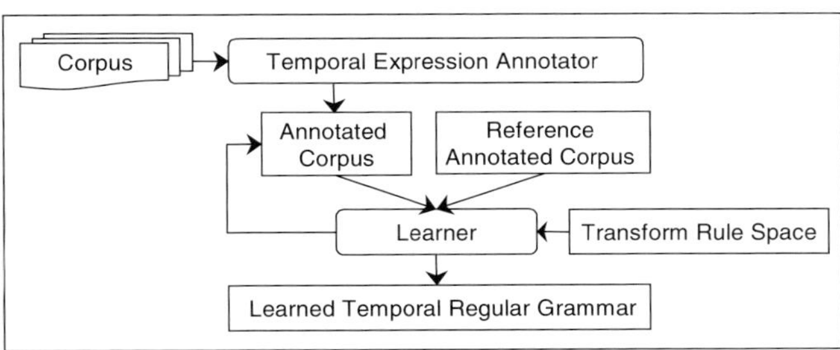

Fig. 2. The Process of the Error-Driven Learning Algorithm of TE Identification

Algorithm 3.1.1. An error-driven learning algorithm of TE identification

Input: training free text corpus Ct and test free text corpus Cp;
Output: Texts annotated with TE.
Step1: Input Ct into the TE annotator, which identifies TE based on the EDL;
Step2: Manually annotate Ct with TE as the reference corpus;
Step3: A learner compares the automatically annotated corpus with the reference corpus. And the learner applies transformation rule templates to the

annotated corpus to make it better resemble the reference corpus, where a template specifies a triggering environment and a series of actions;

Step4: Find transformation rules whose application results in the best score according to the objective function, and execute actions of these rules to update the EDL;

Step5: Step3 and step4 are iterated until no transformation rule can be found whose application results in an improvement to the annotated corpus;

Step6: Use the TE annotator to identify TE in Cp based on the learned EDL.

The EDL program is composed of agents including contexts and operations. Contexts comprise a basic one and a constraint one, in which an agent can be activated. The content of a basic context is a regular grammar system, which is built based on the taxonomy of TE, and gives constitutive methods of every category of TE. The content of a constraint context consists of predicates used to restrict TE and its contextual features. An agent performs its operations if its contexts are met.

```
defagent agent Instant TE
{
    Basic Context: <Syntax_Instant_TE>
    Constraint Context: NotContain(<Syntax_Instant_TE>, “。|.  |?|!|,  |,|.|?|!|;  |;”)
    Operation1: FrontInsert (<Syntax_Instant_TE>,"<Instant_TE>")
    Operation2: BackInsert (<Syntax_Instant_TE>,"</Instant_TE>")
   ......
}
defclass Syntax <Syntax_Instant_TE>
{
  <Syntax_Instant_TE>::=<Geological_TE>|<Archaeological_TE>|<Social_TE>
  <Social_TE>::=<Calendar_Dates_TE>|<Event-Anchored_TE>|<Culturally_Determined_TE>
  <Calendar_Dates_TE>::=<Gregorian_Calendar_Dates_TE>|<lunar_Calendar_Dates_TE>|<lunisolar_Calendar_Dates_TE>
  <Gregorian_Calendar_Dates_TE>::=<Gregorian_Calendar_Dates_TE_Type1>|<Gregorian_Calendar_Dates_TE_Type2>
  <Gregorian_Calendar_Dates_TE_Type1>::=[<A.D.Tag>][<Pre_Quantifier_Modifier>]{<Integer>
                        [<Middle_Quantifier_Modifier>]<Temporal_Unit>}*[<Post_Quantifier_Modifier>]
  <Gregorian_Calendar_Dates_TE_Type2>::=<Integer><Temporal_Unit>[<Temporal_Unit>]<Span_Modifier>
  <A.D._Tag>::=公元|公元前|前
  <Pre_Quantifier_Modifier>::=大约|约|大体|大概|大抵|大致|大致上|将近|近于|近乎
  <Middle_Quantifier_Modifier>::=多|余|弱|把|强|来
  <Post_Quantifier_Modifier>::=左右|上下|前后|内外|里外|开外|冒尖儿|出头|有零|有余|挂零|光景|模样|前|后|左|右|里|中|外|内|
                        间|头里|当中|中间|之上|之下|之前|之后|之里|之外|之内|之中|之间|以上|以下|以前|以后|以里|以外|以内
  <Temporal_Unit>::=世纪|时代|年代|年|岁|季|季度|月|月份|礼拜|星期|周|日|小时|点|刻|分钟|分|秒钟|秒
  <Span_Modifier>::=早期|初期|中期|晚期|末期|前期|后期|初|中|底|末|终|上叶|中叶|下叶|上半叶|中半叶|下半叶|前叶|后叶|
                        前半叶|后半叶
  ......
}
```

Fig. 3. An Agent of Instant TE

For example, ‘agent Instant TE’ in Fig.3 takes ‘*<Syntax_Instant_TE>*’ as its content value of the basic context, which is defined in ‘*defcalss Syntax <Syntax_Instant_TE >*’. Here, ‘|’ expresses logical ‘or’, an optional item <I> is enclosed in ‘[<I>]’, and NotContain(x,y) denotes that string y is not a sub-string of string x. The set of all word terminals on the right side of a production rule form a semantic class whose elements have the same semantic meaning or function, and whose appellation is the symbol on the left side of this rule. A transformation rule template consists of a triggering environment and actions. A set of rule templates determines a space of possible transformation rules. The transformation rule templates are given below, which mean that if one of five conditions is satisfied, then one of six series of actions is executed.

Triggering Environment:

(a) IF TempExp(X, Ct) $\wedge$ NonTempExp(X, Cr)
(b) IF NonTempExp(X, Ct) $\wedge$ TempExp(X, Cr)
(c) IF TempExp(X, Ct) $\wedge$ TempExp(Y, Cr) $\wedge$ Contain(Y, X)
(d) IF TempExp(X, Ct) $\wedge$ TempExp(Y, Cr) $\wedge$ Contain(X, Y)
(e) IF TempExp(X, Ct) $\wedge$ TempExp(Y, Cr) $\wedge$ $\exists$Z(Contain(X, Z) $\wedge$ Contain(Y, Z) $\wedge$ NonEqual(X, Z) $\wedge$ NonEqual(Y, Z))

Obligatory Actions: SegmentingWord(X) $\wedge$ SegmentingWord(Y)
Optional Action:

(a) AddSemanticCalssElement(X, Y, E)
(b) AddSemanticClassRule(X, Y, R)
(c) AddRule(X, Y, R)
(d) AddRule(X, Y, R) $\wedge$ AddSemanticCalssElement(X, Y, E)
(e) AddRule(X, Y, R) $\wedge$ AddSemanticClassRule(X, Y, R)
(f) AddConstraintPredicate(X, Y, P)

Here, (a) X and Y are strings, Ct is the automatically annotated corpus, and Cr is the reference corpus. (b) TemExp(X, Ct) and NonTemExp(X, Cr) denote X in Ct is or is not identified as a TE. Contain(X,Y) expresses Y is a sub-string of X. Equal(X, Z) and NonEqual(X, Z) indicates that X is or is not equal to Z. (c) The action of SegmentingWord(X) is to do word segmentation of X. Actions of AddRule(X, Y, R), AddSemanticClasseRule(X, Y, R) and AddSemanticCalssElement(X, Y, E) are to add a production rule R, a production rule R with all elements of a semantic class, an element E of a semantic class, generated by X and Y, to the EDL, respectively. Action AddConstraintPredicate(X, Y, P) adds the constraint predicate P to the EDL.

Evaluation of a candidate transformation rule aims to quantitate positive and negative changes that are caused by applying the transformation rule to the training corpus. The objective function O(R) of a candidate rule R is C(R)-E(R), where C(R) and E(R) denote the number of identified TE that cause positive or negative changes.

4 Experiments

Precision (P), recall(R) and F-measure (F) are used to evaluate experimental results obtained by our TE identification algorithm. Here, recall=N_2/N, precision=N_2/N_1, F=2RP/(R+P), where N is the total number of TE in the Corpus Cp, N_1 is the total number of identified TE, and N_2 is the total number of correctly identified TE.

Three domains of news, history, and archaeology, and four corpora are used to test the performance of our Chinese TE extraction algorithm. These corpora are web pages from the website of Chinese Government News, Chinese History, and China Huaxia, and texts about relics, sites, cultures and the ancients of Archaeology Volume of Chinese Encyclopedia. The corpus collection is about 1.5 million characters, consisting of about 720 articles and 500 web pages. 70 articles and 50 web pages are randomly selected as the training corpus. The initial temporal grammar consists of about 200 production rules and about 160 transformation rules are built to update the

EDL. Four human judges were asked to evaluate experiment results, and the averages of four precisions and recalls are the final precision and recall. The score of all corpora are 86.8% precision, 88.3% recall and 87.5% F-measure. Only the work in [11] identified Chinese TE, while other works [8-10,12,13] recognize TE from English, German, French, and Spanish texts, respectively. The result reported in [11] reaches 78.5% precision, 84.5% recall, and 81.38% F-measure. The reasons that our transformation-based learning method attains high performance are as follows: (a) the grammar system of Chinese TE is constructed based on the taxonomy hierarchy of TE. (b) The EDL can be continually updated by adding new production rules, terminals and constraint predicates driven by the error-driven learning approach. The main causes of incorrect TE are ambiguities of word segmentation and word senses.

5 Conclusion

Time is a crucial dimension in information retrieval and extraction, and it is an indispensable element to achieve an activity and an event. In this paper, we present a transformation-based error-driven learning technique to identify temporal expressions in Chinese unstructured texts, and have shown its effectiveness through experiment results. In this learning approach, the temporal grammar and transformation rule templates are independent of languages. Hence our method can be applied to extracting temporal expressions in any language. In future, we would like to add the disambiguation of word senses to improve the performance of TE identification.

References

1. ACL 2001 (2001), http://ucrel.lancs.ac.uk/acl/W/W01/
2. LREC2002 (2002), http://www.lrec-conf.org/lrec2002
3. TREN2004 (2005), http://timex2.mitre.org/tern.html
4. TIME (2007), http://Time.dico.unimi.it
5. Mani, I.: Recent Developments in Temporal Information Extraction. In: Proceedings of the Conference on Recent Advances In Natural Language Processing (2004)
6. Mani, I., et al.: Introduction to the Special Issue on Temporal Information Processing. ACM Transactions on Asian Language Information Processing 1, 1–10 (2004)
7. Wong, K., Xia., Y.: An Overview of Temporal Information Extraction. International Journal of Computer Processing of Oriental Language 2, 137–152 (2005)
8. Mani, I., Wilson, G.: Robust Temporal Processing of News. In: Proceedings of the Annual Meeting of the Association for Computational Linguistics, Hong Kong, China (2000)
9. Saquete, E., et al.: Recognizing and Tagging Temporal Expressions in Spanish. In: Workshop on Annotation Standards for Temporal Information in Natural Language, LREC (2002)
10. Schilder, F., Habel, C.: From Temporal Expressions to Temporal Information: Semantic Tagging of News Messages. In: Proceedings of the ACL 2001 Workshop on Temporal and Spatial Information Processing, Toulouse, pp. 65–72 (2001)
11. Wu, M., et al.: CTEMP: A Chinese Temporal Parser for Extracting and Normalizing Temporal Information. In: Proceedings of the Second International Joint Conference on Natural Language Processing, pp. 694–706 (2005)

12. Vazov, N.: A System for Extraction for Temporal Expression from French Texts based on Syntactic and Semantic Constraints. In: Proceedings of the ACL Workshop on Temporal and Spatial Information Processing, Toulouse, pp. 65–72 (2001)
13. Ahn, D., Adafre, S.F., Rijke, M.: Towards Task-Based Temporal Extraction and Recognition. In: Procedings of the Workshop on Annotating, Extracting and Reasoning about Time and Events (2005)
14. Lager, T.: A Logic Programming Approach to Word Engineering. In: Proceedings of the International Conference on Artificial and Computational Intelligence for Decision, Control and Automation in Engineering and Industrial Applications, Monastir (2000)

An Entropy-Based Hierarchical Search Result Clustering Method by Utilizing Augmented Information

Kao Hung-Yu[1], Hsiao Hsin-Wei[1], Lin Chih-Lu[1], Shih Chia-Chun[2], and Tsai Tse-Ming[2]

[1] Department of Computer Science and Information Engineering, National Cheng Kung University, Tainan, Taiwan, ROC
{hykao,p7694103,p7895106}@mail.ncku.edu.tw
[2] Innovative Digitech-Enabled Applications & Services Institute (IDEAS), Institute for Information Industry, Taiwan
{chiachun,eric}@iii.org.tw

Abstract. Because of the improvement of the technology of search engines, and the massively increase of the number of web pages, the results returned by the search engines are always mixed and disordered. Especially for the queries with multiple topics, the mixed and disorderly situation of the search results would be more obvious. The search engines can return information of several hundred to thousand of the pages' titles, snippets and URLs. Almost all of the technologies about search result clustering must attain further information from the contents of the returned lists. However, long execution time is not permitted for a real-time clustering system.

In this paper we propose some methods with better efficiency to improve the previous technologies. We utilize and augment information that search engines returned and use entropy calculation to attain the term distribution in snippets. We also propose several new methods to attain better clustered search results and reduce execution time. Our experiments indicate that these proposed methods obtain the better clustered results.

Keywords: Search Engine, Clustering, Snippet, Entropy, Augmented Information.

1 Introduction

Search engines like Google, Yahoo and MSN can provide the search results from wide Internet based on the typical user queries. However, those search engines always display a long and flat returned list and the list has been ranked by their inner ranking methods. Although the returned list has been ranked, generally it was not applicable to users to browse. For example, a query like "jaguar", search engines do not know which topic of the query the user wants to know. Hence the search engines seek and obtain the more popular web sites which they supposed about the query word "jaguar" without considering the multiple meanings of query word. "Jaguar" has a few meanings such as "a big cats", "MAC OS", "car brand"…etc. The clustering on search results is then useful for those queries which are short, non-specific or imprecise and it could ease the disorder of search result.

H. Li et al. (Eds.): AIRS 2008, LNCS 4993, pp. 670–675, 2008.

The method for clustering of search results always utilize short web snippets as the information and then extracting the representative words from the snippets to be candidate category names. In this paper, we enhanced the method proposed by Krishna Kummamuru[4] by adding some other augmented information to extracting the candidate concepts. We also integrate the entropy calculation with Kummamuru's clustering method to refine the terms selection steps. We consider that it is not enough to utilize the web snippets only to attain the final clustered result. We also make the structure of hierarchical clustering result by utilizing the collected candidate concepts. The hierarchical structure would represent the relation between one concept and another concept. We modify the hierarchical tree construction step and simplify the relation calculation to string comparison.

2 Related Work

For search result categorization, VIVISIMO (http://www.vivisimo.com) is one of the examples of clustered search engines on Internet. In [1], we know that the AOL portal adopted VIVISIMO on top of the search results provided by Google. Some of the other well-known clustered search engines like Clusty (http://clusty.com) and Kartoo (http://www.kartoo.com/) display their search results by applying relevant pictures or the graphic user interface. By offering these relevant figures, users could attain what information the query relates to more easily. Search results returned by Clusty accompany some pictures related to the query users submit.

Zamir and Etzioni presented an on-line search result clustered system called Grouper [6]. Grouper is an interface to group search results into clusters dynamically. Microsoft [5] collects the titles and the snippets information returned from one or more search engines. This method extracts salient phrases from the titles and the snippets to calculate the appropriate cluster name instead of consider all contents of the web pages. Ferragina and Gulli proposed a personalized search engine based on web-snippet hierarchical clustering [1]. They make the clustering search result focusing on user rather on query words. CAARD [3]utilize the relation between pair of concepts to determine that which concept is one of the sub-topics of the other clusters. DSP [2] apply a greedy method to build the concepts hierarchy and it helps confirm that all salient concepts would be included in the hierarchy. Krishna Kummamuru proposed an innovative hierarchical clustering method called "DisCover" [4]. The word "DisCover" means the short writing of "Distinctiveness" and "Coverage". Clearly this method uses documents' distinctiveness and documents' coverage of each concept to calculate the ranking of all concepts extracted from web documents.

3 Proposed Method for Clustering

We use one of the features called Cluster Entropy that proposed from [5]. In function g_c and g_d of DisCover, it only applies the conception of subtraction to determine the difference between a concept and a concept set. However, the intersection of a concept and a concept set should be considered too. Hence we add a cluster entropy feature into original DisCover algorithm to adjust the clustered result. The formula for integration of the method is shown as follows:

$$g'(S_{\alpha,k-1},c_j) \equiv w_1 \times g_c(S_{\alpha,k-1},c_j) \times (1+Entropy_{Doc}) + w_2 \times g_d(S_{\alpha,k-1},c_j) \times (1+Entropy_{UnderConcepts}) \quad (1)$$

$$Entropy_{Doc} = -\sum_{w \in S_{\alpha,k-1}} \frac{|d(c_j) \cap d(w)|}{\sum_{w \in S_{\alpha,k-1}} |d(c_j) \cap d(w)|} \log \frac{|d(c_j) \cap d(w)|}{\sum_{w \in S_{\alpha,k-1}} |d(c_j) \cap d(w)|} \quad (2)$$

$$Entropy_{UnderConcept} = -\sum_{w \in S_{\alpha,k-1}} \frac{|t(c_j) \cap t(w)|}{\sum_{w \in S_{\alpha,k-1}} |t(c_j) \cap t(w)|} \log \frac{|t(c_j) \cap t(w)|}{\sum_{w \in S_{\alpha,k-1}} |t(c_j) \cap t(w)|} \quad (3)$$

The value of $Entropy_{Doc}$ indicates the value of documents covered by c_j distribute in documents covered by $S_{\alpha,k-1}$. The value of $Entropy_{UnderConcepts}$ denotes the entropy value of the under concepts of a concept c_j distribute in the under concepts of a concept set $S_{\alpha,k-1}$. It indicates the value of concepts under a concept c_j distribute in concepts under a concept set $S_{\alpha,k-1}$. The value of $Entropy_{UnderConcepts}$ indicates the value of under concepts of c_j distribute in under concepts of $S_{\alpha,k-1}$.

3.1 Concept Hierarchy Modification

The method to construct the concept hierarchy in the CAARD [3] is to calculate each pair of concepts and clusters to distribute each rest concept that is not top level cluster into the most appreciate category. In our new hierarchy method, we do not utilize the traditional method to calculate the relation for each pair of concepts by using vector model. For example, we give $V_k = (V_{k1}, V_{k2}, \ldots V_{kN})$ as a binary vector of a cluster *k*, $V_c = (V_{c1}, V_{c2}, \ldots V_{cN})$ as a binary vector of a concept *c*. Each mapping elements in the same position of V_k and V_c would determine the relation value of cluster *k* and concept *c*. In traditional computation of relation about two vectors, it would get the value 0 ($0\times1 = 1\times0 = 0$). However we consider that the meaning of corresponding pair (0,0) and (0,1), (1,0) are different.

4 System Description and Experiments

We implement an on-line system that could cluster the top 100 search results returned by Google into several categories for our enhanced clustering methods. After users send their queries, our system applies the proposed clustering method to generate a concept hierarchy as shown in Fig. 1. User can choose the concepts they interest in to obtain the document results related to the concept.

In our experiments, for each query we collect the top 100 search results returned by Google. We choose five Chinese ambiguous queries and five English ambiguous queries as shown in Table 1 to be our testing queries. We also choose five Chinese queries and five English queries with more specific meaning to be the other test queries. We collect the top 100 Google search results of these queries as our testing data set.

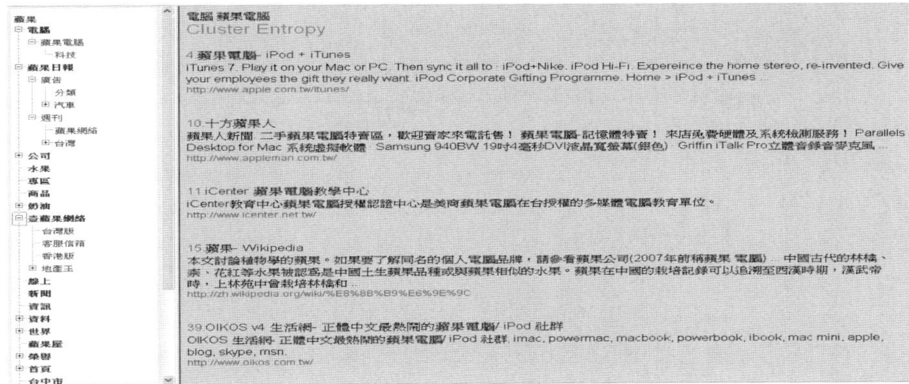

Fig. 1. Our system interface after querying the Chinese term "蘋果"

Table 1. The queries in experiments

Chinese Queries (Ambiguous)	蘋果, 天堂, 爪哇, 飛碟, 地中海
English Queries (Ambiguous)	Jaguar, Latex, Java, Panther, Lotus
Chinese Queries (Specific)	蘋果電腦, 天堂線上遊戲, 爪哇咖啡, 飛碟電台, 地中海型貧血
English Queries (Specific)	Jaguar car, Latex editor, Java Program, Panther OS, Lotus IBM

In the experiment we examine for the coverage of each methods. We select the average coverage ratio of the top1 to top10 clusters of each method. Fig. 2 and Fig. 3 indicate the average coverage improvement extent for Chinese ambiguous and specific queries. In these two figures we find that almost all of our methods would enhance the coverage for a little degree. However the M1 is always with lower coverage ratio than DisCover method. M1 is the method that applying the heavier weighting to title terms. In M1 method it would attain more title concepts to be the cluster concepts. The title concept generally would not cover too many related documents. Therefore the coverage of M1 is almost lower than DisCover method. Fig. 4 and Fig. 5 represent the coverage improvement for English ambiguous and specific queries. For English queries we would also enhance the coverage ratio.

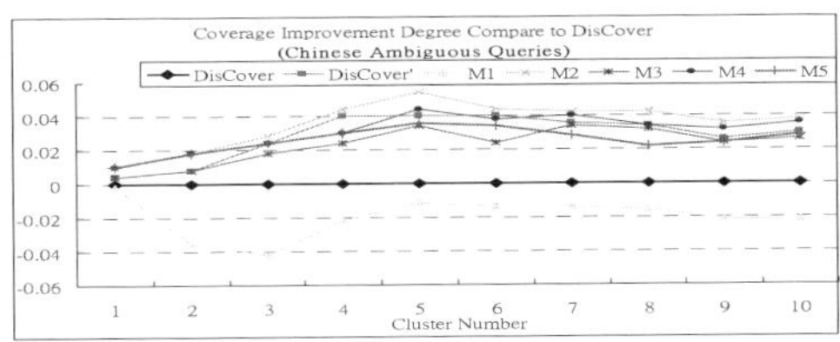

Fig. 2. Coverage improvement (Chinese ambiguous queries)

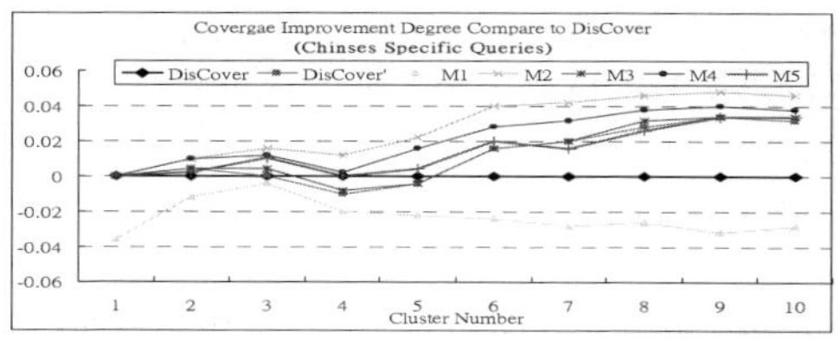

Fig. 3. Coverage improvement (Chinese specific queries)

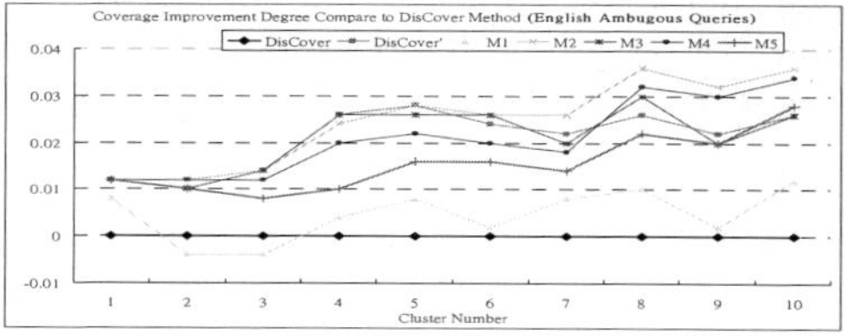

Fig. 4. Coverage improvement (English ambiguous queries)

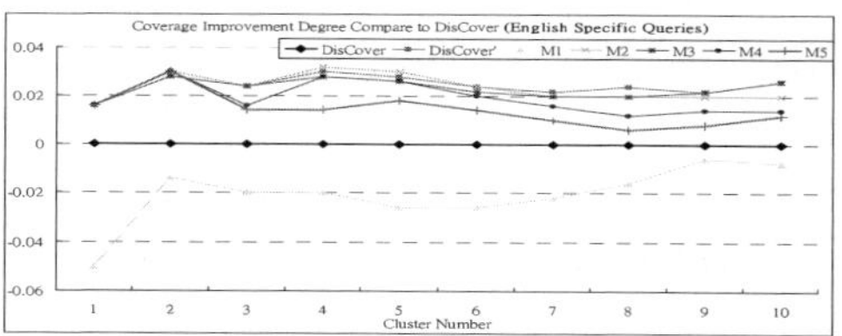

Fig. 5. Coverage improvement (English specific queries)

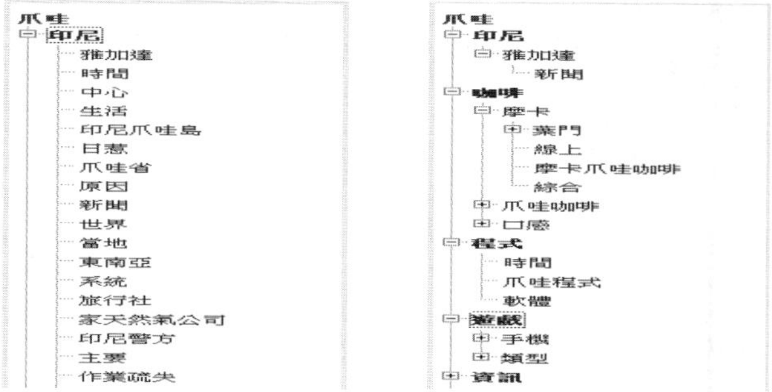

Fig. 6. Comparison for the concept hierarchy of CAARD and our method

We propose a new manner to evaluate our new concepts hierarchy. We attempt to calculate the concepts number in level 2, the coverage ratio, and the average coverage for each concept in level 2. Fig. 6 is an example of the concept hierarchy. The left part of Fig. 6 is the hierarchical result of CAARD [3] while the right part is the hierarchical result of our proposed hierarchical method. We could confirm that our hierarchical method could make the salient concepts more structural. From the hierarchical result we could attain the relation between each pair of the concept. The concepts which are in the deeper level of the hierarchy might be the specific meaning concept for the query.

5 Conclusion and Future Work

In this paper, we proposed several new methods to enhance the final clustered results. We apply information theory to get the messy degree between a concept and a concept set and integrate the entropy theory with the intersection and subtraction of the information of a concept and a concept set. From our experiment, these new methods could not only attain good clustered results but also save a lot execution time. We also proposed a new method to construct the concepts hierarchy.

Acknowledgments. The authors are supported in part by the Project No. 96C076 of IDEAS, Institute for Information Industry, Taiwan, Republic of China.

References

1. Ferragina, P., Gulli, A.: A Personalized Search Engine Based on Web-Snippet Hierarchical Clustering. In: WWW 2005, pp. 801–810 (2005)
2. Lawrie, D.J., Croft, W.B.: Generating Hierarchical Summaries for Web Searches. In: SIGIR 2003, pp. 457–458 (2003)
3. Kummamuru, K., Krishnapuram, R.: A Clustering Algorithm for Asymmetrically Related Data with Application to Text Mining. In: CIKM 2001, pp. 571–573 (2001)
4. Kummamuru, K., Lotlikar, R., Roy, S., Singal, K., Krishnapuram, R.: A Hierarchical Monothetic Document Clustering Algorithm for Summarization and Browsing Search Results. In: WWW 2004, pp. 658–665 (2004)
5. Zeng, H.-J., He, Q.-C., Chen, Z., Ma, W.-Y., Ma, J.: Learning to Cluster Web Search Results. In: SIGIR 2004, pp. 210–217 (2004)
6. Zamir, O., Etzioni, O.: Grouper: A Dynamic Clustering Interface to Web Search Results. In: WWW 1999, pp. 1361–1374 (1999)

Pattern Mining for Information Extraction Using Lexical, Syntactic and Semantic Information: Preliminary Results

Christopher S.G. Khoo, Jin-Cheon Na, and Wei Wang

Division of Information Studies, Wee Kim Wee School of Communication & Information, Nanyang Technological University, Singapore 637718
{assgkhoo,tjcna,w060001}@ntu.edu.sg

Abstract. A method is being developed to mine a text corpus for candidate linguistic patterns for information extraction. The candidate patterns can be used to improve the quality of extraction patterns constructed by a pseudo-supervised learning method—an automated method in which the system is provided with a high quality seed pattern or clue, which is used to generate a training set automatically. The study is carried out in the context of developing a system to extract disease-treatment information from medical abstracts retrieved from the Medline database. In an earlier study, the Apriori algorithm had been used to mine a sample of sentences containing a disease concept and a drug concept, to identify frequently occurring word patterns to see if these patterns could be used to identify treatment relations in text. Word patterns and statistical association measures alone were found to be insufficient for generating good extraction patterns, and need to be combined with syntactic and semantic constraints. In this study, we explore the use of syntactic, semantic and lexical constraints to improve the quality of extraction patterns.

Keywords: Information Extraction, Pattern Mining, Apriori Algorithm.

1 Introduction

Information extraction systems use automated methods to extract from natural-language text facts or pieces of information related to a particular topic or event. The facts are used to fill pre-defined templates or to populate a database for various purposes. Information extraction is usually performed using pattern matching—searching for certain linguistic patterns in the text that indicate the presence of the desired information. These extraction patterns can be constructed automatically or semi-automatically by the system by analyzing sample relevant text and the associated answer key (the training corpus) that is usually constructed by human analysts. An information extraction system requires an extensive training corpus or review of the extraction patterns by a human expert to achieve good accuracy.

The challenge is to develop user-friendly personalizable information extraction systems that can be trained by end-users to give reasonable accuracy with a small training set. Since the training set is small, the system needs to use other sources of information to supplement the small amount of information provided by the user

H. Li et al. (Eds.): AIRS 2008, LNCS 4993, pp. 676–681, 2008.

in the construction of extraction patterns. The text corpus itself represents the most conveniently available source of supplementary information to exploit to improve the extraction patterns.

In this study, we attempt to develop a method to mine candidate linguistic patterns from the text corpus for information extraction. The candidate patterns mined from the corpus can be used in two ways in the development of extraction patterns: (Not clear about the following two approaches)

1. Machine-assisted pattern construction: given a sentence containing a target piece of information to extract, the system can present the most promising candidate patterns (this term is not clear) for the user to select and customize to form an extraction pattern.
2. Pseudo-supervised learning method: an automated method in which the system is provided with a high quality seed pattern or clue, which is used to automatically generate a training set. Since the training set generated by the seed pattern is not as good as a manually constructed training set, the patterns learnt will include a higher proportion of erroneous patterns. Limiting the patterns learnt to those in the set of candidate patterns will serve to filter out the more promising patterns.

Though we are interested in both these uses of pattern mining, this report focuses on the second application of mining patterns that can be used in pseudo-supervised learning. We retrieved a sample of medical abstracts from the Medline database on the topic of colon cancer therapy and attempted to develop patterns for extracting treatment relations from the abstracts. Instead of manually constructing a training set, we assumed that any sentence that contains a treatment concept (e.g. drug) and a disease concept expresses a treatment relation between the treatment and disease. In other words, we used a semantic pattern to retrieve all sentences containing a treatment and a disease, and assumed that the treatment and disease are to be extracted. These sentences then represent the training set, and extraction patterns are constructed to represent the linguistic context of *treatment* and *disease*. The extraction patterns can later be applied to other sentences to extract new treatments, new diseases and new relations. Figure 1 shows the overall process for mining information extraction patterns; detailed explanations follow in later sections.

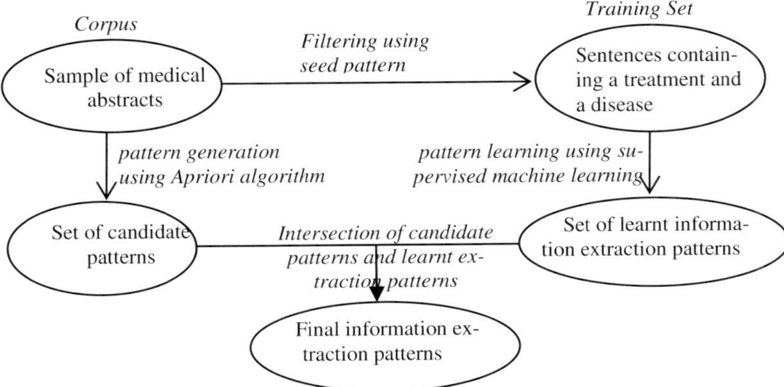

Fig. 1. The overall process of generating information extraction patterns

2 Previous Studies

In an earlier study [1], we had found that mining patterns to extract drug and disease in a sentence is a non-trivial task. We used the Apriori algorithm [2] to mine sample sentences containing a disease concept and a drug concept, to identify frequently occurring word patterns to see if these patterns could be used to identify treatment relations in sentences. Various measures were used to rank the rules, such as Rule Confidence, Normalized Chi Square, Confidence Difference and Confidence Ratio. The results were not convincing as the rules contained few terms that signified a treatment relation.

Word patterns and statistical association measures alone were not good enough to construct extraction patterns. Statistical association measures need to be combined with syntactic and semantic constraints. To obtain some insights into what kind of syntactic and semantic constraints might be helpful, we manually constructed extraction patterns for identifying sentences containing drug-disease relations based on 100 abstracts. We found that the patterns could be grouped into the following domain-specific semantic categories:

- Administration of treatment, e.g. *exposure to, use of, using, clinically used, administered,* and *receiving treatment with.*
- Treatment dosage, e.g. *low-dose, dose of,* and *dosage schedule.*
- Mortality and survival, e.g. *mortality, death rate, survival benefit,* and *extends the survival.*
- Therapy, e.g. *chemotherapy, treatment, regimen, adjuvant, drug,* and *pro-drug.*
- Clinical trial, e.g. *tested on, feasibility trial,* and *clinical trial.*
- Effect, e.g. *outcome, responsive, influence, results, sensitivity,* and *effective*. Words referring to an effect can be subdivided into 11 subtypes, including agent of effect (e.g. *anti-cancer agent*), target of effect (e.g. *targeting*), effect action (e.g. *anti-tumor activity*), effect against something (e.g. *anti-cancer, anti-tumor,* and *antagonist*), etc.

From this, we compiled a dictionary of words belonging to these semantic categories.

We continue to investigate what kind of syntactic and semantic constraints can be imposed on linguistic patterns mined from a text corpus to generate good quality extraction patterns. In particular, we wanted to find out to what extent adding the constraints from the domain-specific semantic categories would improve the extraction patterns.

3 Method for Generating the Extraction Patterns

1570 abstracts were downloaded from the MEDLINE database [3] via the PubMed interface using "colon cancer/therapy" as query. These articles were then parsed using the MMTx (MetaMap Transfer) program developed by the National Library of Medicine [4] to produce an output text file. MMTx is a part-of-speech and semantic tagger which takes biomedical text as input and identifies Unified Medical Language System (UMLS) concepts in the text by mapping relevant phrases to the UMLS Metathesaurus

[5]. The output was further processed to tag the tokens (either words or phrase chunks) in the sentence with part-of-speech and semantic tags (if available) and stored the information in a relational database.

The Apriori Algorithm was then used to generate all possible 2, 3, 4 and 5-token patterns that occur at least 5 times in the corpus. A token can be represented by either of the following attributes:

- Lexical token (L), i.e. word or phrase chunk,
- Part-of-speech (P), or
- Semantic concept (C).

Example candidate patterns generated are shown in Table 1. Note that there is an implied wildcard (representing 0 to 3 tokens) between the tokens in each pattern—i.e. the patterns are sequential patterns but not adjacent patterns.

Table 1. Example candidate patterns

Pattern Type	Example candidate patterns
CLC	[Neoplastic Process] treats [Therapeutic or Preventive Procedure]
CPLC	[Neoplastic Process] *noun* underwent [Therapeutic or Preventive Procedure]
CLCPC	[Patient or Disabled Group] underwent [Therapeutic or Preventive Procedure] *prep* [Neoplastic Process]

Note: Terms in square brackets represent UMLS Metathesaurus concept. Terms in italics represent part-of-speech tag.

Without any lexical, syntactic or semantic constraints, a large number of candidate patterns were generated from the 1570 abstracts. For any particular sentence in the training set, there were on average more than 1000 candidate patterns that match parts of the sentence and have to reviewed by a human analyst to select an extraction pattern. Furthermore, the very few useful patterns were often buried deep in the set of candidate patterns.

Based on an informal error analysis, we introduced the following constraints to improve the quality of the candidate patterns:

- There must be at least 1 lexical item that is not a stopword in the candidate pattern
- The pattern must not contain a preposition as the first or last token

We then filtered out the patterns containing a treatment concept and a disease concept. These are candidate patterns for identifying disease-treatment relations in sentences. The treatment and disease concepts were identified using the UMLS semantic types annotated by the MMTx program.

On examining the candidate patterns, we found some useless patterns that contain the treatment concept and disease concept in adjacent positions (with no tokens between them). These were eliminated since the tokens between the treatment and disease concepts seem to be the most useful for identifying disease-treatment relations.

Next, we separated the candidate patterns into 2 subsets:

- Subset 1 contains a word token that matches an entry in the domain-specific semantic dictionary described in the last section. These are words associated with the semantic categories of *treatment administration*, *dosage*, *mortality and survival*, *therapy*, *clinical trial* and *effect.*
- Subset 2 does not contain a word in the dictionary.

Subset 1 contains 62 candidate patterns, some of which are listed in Table 2.

Table 2. Some candidate patterns

Token1	Token2	Token3	Token4	Token5	Type
[Therapeutic or Preventive Procedure]	Patients	[Neoplastic Process]			CLC
[Neoplastic Process]	noun	underwent	[Therapeutic or Preventive Procedure]		CPLC
[Therapeutic or Preventive Procedure]	in	[Neoplastic Process]	cells		CLCL
[Neoplastic Process]	prep	[Therapeutic or Preventive Procedure]	prep	Apoptosis	CPCPL

The candidate patterns were converted to final extraction patterns. Converting a candidate pattern to an extraction pattern involves indicating where the extraction slots are in the pattern—the placeholders for the information of interest to extract. To extract a disease-treatment relation, two slots—a disease slot and a treatment slot need to be created, and this can easily be accomplished by converting the disease concept token and treatment concept token to slot tokens. Three types of extraction patterns can be constructed:

- Type 1: patterns with a disease slot only
- Type 2: patterns with a treatment slot only
- Type 3: patterns with a disease and a treatment slot.

We have investigated only the last two types of patterns.

4 Extraction Results

114 extraction patterns (Type 2 and Type 3) were derived from the 61 candidate patterns. The patterns were applied back to the corpus, i.e. the 1570 medical abstracts, to extract *disease* and *treatment* from each matched sentence through pattern matching. We computed the estimated precision measure for each pattern based on the first 20 extractions by the pattern.

The average precision for the two-slot (*disease+treatment*) patterns (Type 3) were:

- 47% for sentences containing both *treatment* and *disease* concepts
- 47% for sentences containing only *treatment* concepts
- 54% for sentences containing only *disease* concepts.

The average precision for single-slot (*treatment* only) patterns (Type 2) were:

- 58% for sentences containing both *treatment* and *disease* concepts

Wrong extractions were mainly due to six causes:

1. Erroneous Parsing. The MMTx parser and an auxiliary annotation module failed to tag noun phrases with correct part of speech class and hence incurred missing hits.
2. Inadequate chunking. The current preprocessing is unable to recognize major phrasal units such as noun phrases and hence causes a lot of inaccurate extractions.
3. Complex entity names. Names of most drugs or therapies, especially their abbreviations and acronyms, often could not be tagged as noun or adjective and hence ended up with "unknown" as their part-of-speech tag.
4. Complex syntactic structures. Coordination, relative clauses, prepositional phrases, etc. reduce the extraction accuracy of a pattern.
5. Coreference problem. Pronouns and definite references are common in medical articles probably because of the complex names of many medical terms. Another type of reference is an is-a reference, e.g. "TS-1 is expected to be an effective agent for the treatment of colon cancer with peritoneal dissemination." The extracted treatment was "agent" instead of "TS-1".
6. Semantic uncertainty. This refers to sentences expressing relations like "A has/causes C which cures B". "C" can be extracted instead of "A". However, this is not necessarily true when "C" is just a chemical or biological function or reaction, but not a therapy or drug. Similarly, relations like "A inhibits/causes C of B" would not always be extracted correctly.

Solutions for these six issues such as adding a second layer parsing and a phrase identifier, designing domain-specific patterns to identify entities and so on will be investigated and implemented in the future work. We are currently analyzing each extraction pattern and the text extracted by the pattern to see what further constraints can be added to improve their accuracy. We have also noticed that out of the 176 entries in the domain-specific dictionary, only 22 matched with patterns mined from the corpus. We are investigating the usefulness of the other 154 entries to see in what context they appear in the text.

References

1. Lee, C.H., Khoo, C., Na, J.-C.: Automatic Identification of Treatment Relations for Medical Ontology Learning: an Exploratory Study. In: McIlwaine, I.C. (ed.) Knowledge Organization and the Global Information Society: Proceedings of the Eighth International ISKO Conference, pp. 245–250. Ergon Verlag, Wurzburg, Germany (2004)
2. Agrawal, R., Imielinski, T., Swami, A.: Mining Association Rules between Sets of Items in Large Databases. In: 1993 ACM SIGMOD International Conference on Management of Data, pp. 207–216. ACM Press, New York (1993)
3. National Library of Medicine. MEDLINE Fact Sheet (Retrieved November 14 2007), http://www.nlm.nih.gov/pubs/factsheets/medline.html
4. National Library of Medicine. MetaMap Transfer (MMTx): Documentation (Retrieved November 14 2007), http://mmtx.nlm.nih.gov/docs.shtml
5. National Library of Medicine. Unified Medical Language System Fact Sheet (Retrieved November 14 2007), http://www.nlm.nih.gov/pubs/factsheets/umls.html

Author Index

Printing: Mercedes-Druck, Berlin
Binding: Stein+Lehmann, Berlin

Lecture Notes in Computer Science

Sublibrary 3: Information Systems and Application, incl. Internet/Web and HCI

For information about Vols. 1–4601
please contact your bookseller or Springer

Vol. 5021: S. Bechhofer, M. Hauswirth, J. Hoffmann, M. Koubarakis (Eds.), The Semantic Web: Research and Applications. XIX, 897 pages. 2008.

Vol. 5017: T. Nanya, F. Maruyama, A. Pataricza, M. Malek (Eds.), Service Availability. XII, 225 pages. 2008.

Vol. 5013: J. Indulska, D.J. Patterson, T. Rodden, M. Ott (Eds.), Pervasive Computing. XIV, 315 pages. 2008.

Vol. 5006: R. Kowalczyk, M. Huhns, M. Klusch, Z. Maamar, Q.B. Vo (Eds.), Service-Oriented Computing: Agents, Semantics, and Engineering. X, 154 pages. 2008.

Vol. 4997: B. Monien, U.-P. Schroeder (Eds.), Algorithmic Game Theory. XI, 363 pages. 2008.

Vol. 4993: H. Li, T. Liu, W.-Y. Ma, T. Sakai, K.-F. Wong, G. Zhou (Eds.), Information Retrieval Technology. XIII, 685 pages. 2008.

Vol. 4976: Y. Zhang, G. Yu, E. Bertino, G. Xu (Eds.), Progress in WWW Research and Development. XVIII, 699 pages. 2008.

Vol. 4956: C. Macdonald, I. Ounis, V. Plachouras, I. Ruthven, R.W. White (Eds.), Advances in Information Retrieval. XXI, 719 pages. 2008.

Vol. 4952: C. Floerkemeier, M. Langheinrich, E. Fleisch, F. Mattern, S.E. Sarma (Eds.), The Internet of Things. XIII, 378 pages. 2008.

Vol. 4947: J.R. Haritsa, R. Kotagiri, V. Pudi (Eds.), Database Systems for Advanced Applications. XXII, 713 pages. 2008.

Vol. 4936: W. Aiello, A. Broder, J. Janssen, E.. Milios (Eds.), Algorithms and Models for the Web-Graph. X, 167 pages. 2008.

Vol. 4932: S. Hartmann, G. Kern-Isberner (Eds.), Foundations of Information and Knowledge Systems. XII, 397 pages. 2008.

Vol. 4928: A.H.M. ter Hofstede, B. Benatallah, H.-Y. Paik (Eds.), Business Process Management Workshops. XIII, 518 pages. 2008.

Vol. 4903: S. Satoh, F. Nack, M. Etoh (Eds.), Advances in Multimedia Modeling. XIX, 510 pages. 2008.

Vol. 4900: S. Spaccapietra (Ed.), Journal on Data Semantics X. XIII, 265 pages. 2008.

Vol. 4892: A. Popescu-Belis, S. Renals, H. Bourlard (Eds.), Machine Learning for Multimodal Interaction. XI, 308 pages. 2008.

Vol. 4882: T. Janowski, H. Mohanty (Eds.), Distributed Computing and Internet Technology. XIII, 346 pages. 2007.

Vol. 4881: H. Yin, P. Tino, E. Corchado, W. Byrne, X. Yao (Eds.), Intelligent Data Engineering and Automated Learning - IDEAL 2007. XX, 1174 pages. 2007.

Vol. 4877: C. Thanos, F. Borri, L. Candela (Eds.), Digital Libraries: Research and Development. XII, 350 pages. 2007.

Vol. 4872: D. Mery, L. Rueda (Eds.), Advances in Image and Video Technology. XXI, 961 pages. 2007.

Vol. 4871: M. Cavazza, S. Donikian (Eds.), Virtual Storytelling. XIII, 219 pages. 2007.

Vol. 4858: X. Deng, F.C. Graham (Eds.), Internet and Network Economics. XVI, 598 pages. 2007.

Vol. 4857: J.M. Ware, G.E. Taylor (Eds.), Web and Wireless Geographical Information Systems. XI, 293 pages. 2007.

Vol. 4853: F. Fonseca, M.A. Rodríguez, S. Levashkin (Eds.), GeoSpatial Semantics. X, 289 pages. 2007.

Vol. 4836: H. Ichikawa, W.-D. Cho, I. Satoh, H.Y. Youn (Eds.), Ubiquitous Computing Systems. XIII, 307 pages. 2007.

Vol. 4832: M. Weske, M.-S. Hacid, C. Godart (Eds.), Web Information Systems Engineering – WISE 2007 Workshops. XV, 518 pages. 2007.

Vol. 4831: B. Benatallah, F. Casati, D. Georgakopoulos, C. Bartolini, W. Sadiq, C. Godart (Eds.), Web Information Systems Engineering – WISE 2007. XVI, 675 pages. 2007.

Vol. 4825: K. Aberer, K.-S. Choi, N. Noy, D. Allemang, K.-I. Lee, L. Nixon, J. Golbeck, P. Mika, D. Maynard, R. Mizoguchi, G. Schreiber, P. Cudré-Mauroux (Eds.), The Semantic Web. XXVII, 973 pages. 2007.

Vol. 4823: H. Leung, F. Li, R. Lau, Q. Li (Eds.), Advances in Web Based Learning – ICWL 2007. XIV, 654 pages. 2008.

Vol. 4822: D.H.-L. Goh, T.H. Cao, I.T. Sølvberg, E. Rasmussen (Eds.), Asian Digital Libraries. XVII, 519 pages. 2007.

Vol. 4820: T.G. Wyeld, S. Kenderdine, M. Docherty (Eds.), Virtual Systems and Multimedia. XII, 215 pages. 2008.

Vol. 4816: B. Falcidieno, M. Spagnuolo, Y. Avrithis, I. Kompatsiaris, P. Buitelaar (Eds.), Semantic Multimedia. XII, 306 pages. 2007.

Vol. 4813: I. Oakley, S.A. Brewster (Eds.), Haptic and Audio Interaction Design. XIV, 145 pages. 2007.

Vol. 4810: H.H.-S. Ip, O.C. Au, H. Leung, M.-T. Sun, W.-Y. Ma, S.-M. Hu (Eds.), Advances in Multimedia Information Processing – PCM 2007. XXI, 834 pages. 2007.

Vol. 4809: M.K. Denko, C.-s. Shih, K.-C. Li, S.-L. Tsao, Q.-A. Zeng, S.H. Park, Y.-B. Ko, S.-H. Hung, J.-H. Park (Eds.), Emerging Directions in Embedded and Ubiquitous Computing. XXXV, 823 pages. 2007.

Vol. 4808: T.-W. Kuo, E. Sha, M. Guo, L.T. Yang, Z. Shao (Eds.), Embedded and Ubiquitous Computing. XXI, 769 pages. 2007.

Vol. 4806: R. Meersman, Z. Tari, P. Herrero (Eds.), On the Move to Meaningful Internet Systems 2007: OTM 2007 Workshops, Part II. XXXIV, 611 pages. 2007.

Vol. 4805: R. Meersman, Z. Tari, P. Herrero (Eds.), On the Move to Meaningful Internet Systems 2007: OTM 2007 Workshops, Part I. XXXIV, 757 pages. 2007.

Vol. 4804: R. Meersman, Z. Tari (Eds.), On the Move to Meaningful Internet Systems 2007: CoopIS, DOA, ODBASE, GADA, and IS, Part II. XXIX, 683 pages. 2007.

Vol. 4803: R. Meersman, Z. Tari (Eds.), On the Move to Meaningful Internet Systems 2007: CoopIS, DOA, ODBASE, GADA, and IS, Part I. XXIX, 1173 pages. 2007.

Vol. 4802: J.-L. Hainaut, E.A. Rundensteiner, M. Kirchberg, M. Bertolotto, M. Brochhausen, Y.-P.P. Chen, S.S.-S. Cherfi, M. Doerr, H. Han, S. Hartmann, J. Parsons, G. Poels, C. Rolland, J. Trujillo, E. Yu, E. Zimányie (Eds.), Advances in Conceptual Modeling – Foundations and Applications. XIX, 420 pages. 2007.

Vol. 4801: C. Parent, K.-D. Schewe, V.C. Storey, B. Thalheim (Eds.), Conceptual Modeling - ER 2007. XVI, 616 pages. 2007.

Vol. 4797: M. Arenas, M.I. Schwartzbach (Eds.), Database Programming Languages. VIII, 261 pages. 2007.

Vol. 4796: M. Lew, N. Sebe, T.S. Huang, E.M. Bakker (Eds.), Human–Computer Interaction. X, 157 pages. 2007.

Vol. 4794: B. Schiele, A.K. Dey, H. Gellersen, B. de Ruyter, M. Tscheligi, R. Wichert, E. Aarts, A. Buchmann (Eds.), Ambient Intelligence. XV, 375 pages. 2007.

Vol. 4777: S. Bhalla (Ed.), Databases in Networked Information Systems. X, 329 pages. 2007.

Vol. 4761: R. Obermaisser, Y. Nah, P. Puschner, F.J. Rammig (Eds.), Software Technologies for Embedded and Ubiquitous Systems. XIV, 563 pages. 2007.

Vol. 4747: S. Džeroski, J. Struyf (Eds.), Knowledge Discovery in Inductive Databases. X, 301 pages. 2007.

Vol. 4744: Y. de Kort, W. IJsselsteijn, C. Midden, B. Eggen, B.J. Fogg (Eds.), Persuasive Technology. XIV, 316 pages. 2007.

Vol. 4740: L. Ma, M. Rauterberg, R. Nakatsu (Eds.), Entertainment Computing – ICEC 2007. XXX, 480 pages. 2007.

Vol. 4730: C. Peters, P. Clough, F.C. Gey, J. Karlgren, B. Magnini, D.W. Oard, M. de Rijke, M. Stempfhuber (Eds.), Evaluation of Multilingual and Multi-modal Information Retrieval. XXIV, 998 pages. 2007.

Vol. 4723: M. R. Berthold, J. Shawe-Taylor, N. Lavrač (Eds.), Advances in Intelligent Data Analysis VII. XIV, 380 pages. 2007.

Vol. 4721: W. Jonker, M. Petković (Eds.), Secure Data Management. X, 213 pages. 2007.

Vol. 4718: J. Hightower, B. Schiele, T. Strang (Eds.), Location- and Context-Awareness. X, 297 pages. 2007.

Vol. 4717: J. Krumm, G.D. Abowd, A. Seneviratne, T. Strang (Eds.), UbiComp 2007: Ubiquitous Computing. XIX, 520 pages. 2007.

Vol. 4715: J.M. Haake, S.F. Ochoa, A. Cechich (Eds.), Groupware: Design, Implementation, and Use. XIII, 355 pages. 2007.

Vol. 4714: G. Alonso, P. Dadam, M. Rosemann (Eds.), Business Process Management. XIII, 418 pages. 2007.

Vol. 4704: D. Barbosa, A. Bonifati, Z. Bellahsène, E. Hunt, R. Unland (Eds.), Database and XML Technologies. X, 141 pages. 2007.

Vol. 4690: Y. Ioannidis, B. Novikov, B. Rachev (Eds.), Advances in Databases and Information Systems. XIII, 377 pages. 2007.

Vol. 4675: L. Kovács, N. Fuhr, C. Meghini (Eds.), Research and Advanced Technology for Digital Libraries. XVII, 585 pages. 2007.

Vol. 4674: Y. Luo (Ed.), Cooperative Design, Visualization, and Engineering. XIII, 431 pages. 2007.

Vol. 4663: C. Baranauskas, P. Palanque, J. Abascal, S.D.J. Barbosa (Eds.), Human-Computer Interaction – INTERACT 2007, Part II. XXXIII, 735 pages. 2007.

Vol. 4662: C. Baranauskas, P. Palanque, J. Abascal, S.D.J. Barbosa (Eds.), Human-Computer Interaction – INTERACT 2007, Part I. XXXIII, 637 pages. 2007.

Vol. 4658: T. Enokido, L. Barolli, M. Takizawa (Eds.), Network-Based Information Systems. XIII, 544 pages. 2007.

Vol. 4656: M.A. Wimmer, J. Scholl, Å. Grönlund (Eds.), Electronic Government. XIV, 450 pages. 2007.

Vol. 4655: G. Psaila, R. Wagner (Eds.), E-Commerce and Web Technologies. VII, 229 pages. 2007.

Vol. 4654: I.-Y. Song, J. Eder, T.M. Nguyen (Eds.), Data Warehousing and Knowledge Discovery. XVI, 482 pages. 2007.

Vol. 4653: R. Wagner, N. Revell, G. Pernul (Eds.), Database and Expert Systems Applications. XXII, 907 pages. 2007.

Vol. 4636: G. Antoniou, U. Aßmann, C. Baroglio, S. Decker, N. Henze, P.-L. Patranjan, R. Tolksdorf (Eds.), Reasoning Web. IX, 345 pages. 2007.

Vol. 4611: J. Indulska, J. Ma, L.T. Yang, T. Ungerer, J. Cao (Eds.), Ubiquitous Intelligence and Computing. XXIII, 1257 pages. 2007.

Vol. 4607: L. Baresi, P. Fraternali, G.-J. Houben (Eds.), Web Engineering. XVI, 576 pages. 2007.

Vol. 4606: A. Pras, M. van Sinderen (Eds.), Dependable and Adaptable Networks and Services. XIV, 149 pages. 2007.

Vol. 4605: D. Papadias, D. Zhang, G. Kollios (Eds.), Advances in Spatial and Temporal Databases. X, 479 pages. 2007.

Vol. 4602: S. Barker, G.-J. Ahn (Eds.), Data and Applications Security XXI. X, 291 pages. 2007.